Methods in Enzymology

Volume 351
GUIDE TO YEAST GENETICS AND MOLECULAR
AND CELL BIOLOGY
Part C

METHODS IN ENZYMOLOGY

EDITORS-IN-CHIEF

John N. Abelson   Melvin I. Simon

DIVISION OF BIOLOGY
CALIFORNIA INSTITUTE OF TECHNOLOGY
PASADENA, CALIFORNIA

FOUNDING EDITORS

Sidney P. Colowick and Nathan O. Kaplan

Methods in Enzymology

Volume 351

Guide to Yeast Genetics and Molecular and Cell Biology

Part C

EDITED BY

Christine Guthrie

DEPARTMENT OF BIOCHEMISTRY AND BIOPHYSICS
UNIVERSITY OF CALIFORNIA
SAN FRANCISCO, CALIFORNIA

Gerald R. Fink

WHITEHEAD INSTITUTE FOR BIOMEDICAL RESEARCH
MASSACHUSETTS INSTITUTE OF TECHNOLOGY
CAMBRIDGE, MASSACHUSETTS

ACADEMIC PRESS

An imprint of Elsevier Science

Amsterdam Boston London New York Oxford Paris
San Diego San Francisco Singapore Sydney Tokyo

Front cover photograph: Mex67-GFP localization in *Saccharomyces cerevisiae.*
Photograph courtesy of Anne de Bruyn Kops.

This book is printed on acid-free paper. ∞

Copyright © 2002, Elsevier Science (USA).

All Rights Reserved.
No part of this publication may be reproduced or transmitted in any form or by any means, electronic or mechanical, including photocopy, recording, or any information storage and retrieval system, without permission in writing from the Publisher.

The appearance of the code at the bottom of the first page of a chapter in this book indicates the Publisher's consent that copies of the chapter may be made for personal or internal use of specific clients. This consent is given on the condition, however, that the copier pay the stated per copy fee through the Copyright Clearance Center, Inc. (222 Rosewood Drive, Danvers, Massachusetts 01923), for copying beyond that permitted by Sections 107 or 108 of the U.S. Copyright Law. This consent does not extend to other kinds of copying, such as copying for general distribution, for advertising or promotional purposes, for creating new collective works, or for resale. Copy fees for pre-2002 chapters are as shown on the title pages. If no fee code appears on the title page, the copy fee is the same as for current chapters.
0076-6879/2002 $35.00

Explicit permission from Academic Press is not required to reproduce a maximum of two figures or tables from an Academic Press chapter in another scientific or research publication provided that the material has not been credited to another source and that full credit to the Academic Press chapter is given.

Academic Press
An imprint of Elsevier Science.
525 B Street, Suite 1900, San Diego, California 92101-4495, USA
http://www.academicpress.com

Academic Press
84 Theobalds Road, London WC1X 8RR, UK
http://www.academicpress.com

International Standard Book Number: 0-12-182254-0 (case)
International Standard Book Number: 0-12-310672-9 (comb)

PRINTED IN THE UNITED STATES OF AMERICA
02 03 04 05 06 07 SB 9 8 7 6 5 4 3 2 1

Table of Contents

CONTRIBUTORS TO VOLUME 351 ix

PREFACE xv

VOLUMES IN SERIES xvii

Section I. Cytology

1. Digital Time-Lapse Microscopy of Yeast Cell Growth — STEPHEN J. KRON — 3

2. Quantitative Microscopy of Green Fluorescent Protein-Labeled Yeast — DANIEL R. RINES, XIANGWEI HE, AND PETER K. SORGER — 16

3. Fluorescence Resonance Energy Transfer Using Color Variants of Green Fluorescent Protein — DALE W. HAILEY, TRISHA N. DAVIS, AND ERIC G. D. MULLER — 34

4. Immunoelectron Microscopy of Aldehyde-Fixed Yeast Cells — JON MULHOLLAND AND DAVID BOTSTEIN — 50

5. Electron Tomography of Yeast Cells — EILEEN T. O'TOOLE, MARK WINEY, J. RICHARD MCINTOSH, AND DAVID N. MASTRONARDE — 81

6. Cryomethods for Thin Section Electron Microscopy — KENT MCDONALD AND THOMAS MÜLLER-REICHERT — 96

Section II. Biochemistry

7. Vacuolar Proteases and Proteolytic Artifacts in *Saccharomyces cerevisiae* — ELIZABETH W. JONES — 127

8. Analysis of the Size and Shape of Protein Complexes from Yeast — SCOTT C. SCHUYLER AND DAVID PELLMAN — 150

9. Purification of Glutathione *S*-Transferase Fusion Proteins from Yeast — AVITAL A. RODAL, MARA DUNCAN, AND DAVID DRUBIN — 168

10. Protein- and Immunoaffinity Purification of Multiprotein Complexes — DOUGLAS R. KELLOGG AND DANESH MOAZED — 172

v

11. *In Vitro* DNA Replication Assays in Yeast Extracts PHILIPPE PASERO AND SUSAN M. GASSER 184

12. Yeast Pre-mRNA Splicing: Methods, Mechanisms, and Machinery SCOTT W. STEVENS AND JOHN ABELSON 200

13. Analysis and Reconstitution of Translation Initiation *in Vitro* KATSURA ASANO, LON PHAN, THANUJA KRISHNAMOORTHY, GRAHAM D. PAVITT, EDITH GOMEZ, ERNEST M. HANNIG, JOSEPH NIKA, THOMAS F. DONAHUE, HAN-KUEI HUANG, AND ALAN G. HINNEBUSCH 221

14. Assaying Protein Ubiquitination in *Saccharomyces cerevisiae* JEFFREY D. LANEY AND MARK HOCHSTRASSER 248

15. Vesicle Budding from Endoplasmic Reticulum YUVAL SHIMONI AND RANDY SCHEKMAN 258

16. Mapping Phosphorylation Sites in Proteins by Mass Spectrometry WENYING SHOU, RATI VERMA, ROLAND S. ANNAN, MICHAEL J. HUDDLESTON, SUSAN L. CHEN, STEVE A. CARR, AND RAYMOND J. DESHAIES 279

17. Identification of Yeast Proteins by Mass Spectrometry ALEXANDRE V. PODTELEJNIKOV AND MATTHIAS MANN 296

Section III. Cell Fractionation

18. Subcellular Fractionation of Secretory Organelles CHRIS A. KAISER, ESTHER J. CHEN, AND SASCHA LOSKO 325

19. Plasma Membrane Biogenesis AMY CHANG 339

20. Separation of Golgi and Endosomal Compartments GYÖRGY SIPOS AND ROBERT S. FULLER 351

21. Visualization and Purification of Yeast Peroxisomes RALF ERDMANN AND STEPHEN J. GOULD 365

22. Studying the Behavior of Mitochondria JODI NUNNARI, EDITH D. WONG, SHELLY MEEUSEN, AND JENNIFER A. WAGNER 381

23. Isolation of Nuclear Envelope from *Saccharomyces cerevisiae*	JULIA KIPPER, CATERINA STRAMBIO-DE-CASTILLIA, ADISETYANTARI SUPRAPTO, AND MICHAEL P. ROUT	394
24. Studying Yeast Vacuoles	ELIZABETH CONIBEAR AND TOM H. STEVENS	408
25. Purification of Yeast Actin and Actin-Associated Proteins	BRUCE L. GOODE	433
26. Identifying Functional Interactions with Molecular Chaperones	JILL L. JOHNSON AND ELIZABETH A. CRAIG	442

Section IV. Cell Biology

27. Synchronization Procedures	ANGELIKA AMON	457
28. Separation of Mother and Daughter Cells	PETER U. PARK, MITCH MCVEY, AND LEONARD GUARENTE	468
29. Assays of Cell and Nuclear Fusion	ALISON E. GAMMIE AND MARK D. ROSE	477
30. Analysis of Prion Factors in Yeast	YURY O. CHERNOFF, SUSAN M. UPTAIN, AND SUSAN L. LINDQUIST	499
31. Assaying Replication Fork Direction and Migration Rates	ANJA J. VAN BRABANT AND M. K. RAGHURAMAN	539
32. Analysis of RNA Export	CHARLES N. COLE, CATHERINE V. HEATH, CHRISTINE A. HODGE, CHRISTOPHER M. HAMMELL, AND DAVID C. AMBERG	568
33. Nuclear Protein Transport	MARC DAMELIN, PAMELA A. SILVER, AND ANITA H. CORBETT	587
34. How to Monitor Nuclear Shuttling	ELAINE A. ELION	607
35. Flow Cytometry/Cell Sorting for Isolating Membrane Trafficking Mutants in Yeast	THOMAS VIDA AND BEVERLY WENDLAND	623
36. Protein Synthesis Assayed by Electroporation of mRNA in *Saccharomyces cerevisiae*	ANJANETTE M. SEARFOSS, DANIEL C. MASISON, AND REED B. WICKNER	631
37. Monitoring Protein Degradation	DANIEL KORNITZER	639

38. Analyzing mRNA Decay in *Saccharomyces cerevisiae* — MICHELLE A. STEIGER AND ROY PARKER — 648

39. Use of Green Fluorescent Protein in Living Yeast Cells — KELLY TATCHELL AND LUCY C. ROBINSON — 661

AUTHOR INDEX 685

SUBJECT INDEX 715

Contributors to Volume 351

Article numbers are in parentheses following the names of contributors.
Affiliations listed are current.

JOHN ABELSON (12), *Division of Biology, California Institute of Technology, Pasadena, California 91125*

DAVID C. AMBERG (32), *Department of Biochemistry and Molecular Biology, State of New York Health Sciences Center, Syracuse, New York 13210*

ANGELIKA AMON (27), *Center for Cancer Research, Howard Hughes Medical Institute, Massachusetts Institute of Technology, Cambridge, Massachusetts 02139*

ROLAND S. ANNAN (16), *Proteomics and Biological Mass Spectrometry Laboratory, GlaxoSmithKline, King of Prussia, Pennsylvania 19406*

KATSURA ASANO (13), *Department of Biology, Kansas State University, Manhattan, Kansas 66506*

DAVID BOTSTEIN (4), *Department of Genetics, Stanford University School of Medicine, Stanford, California 94305*

STEVE A. CARR (16), *Millennium Pharmaceuticals, Cambridge, Massachusetts 02139*

AMY CHANG (19), *Department of Anatomy and Structural Biology, Albert Einstein College of Medicine, Bronx, New York 10461*

ESTHER J. CHEN (18), *Department of Biology, Massachusetts Institute of Technology, Cambridge, Massachusetts 02139*

SUSAN L. CHEN (16), *Proteomics and Biological Mass Spectrometry Laboratory, GlaxoSmithKline, King of Prussia, Pennsylvania 19406*

YURY O. CHERNOFF (30), *School of Biology and Institute for Bioengineering and Bioscience, Georgia Institute of Technology, Atlanta, Georgia 30332*

CHARLES N. COLE (32), *Departments of Biochemistry and Genetics, Dartmouth Medical School, Hanover, New Hampshire 03755*

ELIZABETH CONIBEAR (24), *Institute of Molecular Biology, University of Oregon, Eugene, Oregon 97403*

ANITA H. CORBETT (33), *Department of Biochemistry, Emory University School of Medicine, Atlanta, Georgia 30322*

ELIZABETH A. CRAIG (26), *Department of Biomolecular Chemistry, University of Wisconsin-Madison, Madison, Wisconsin 53706*

MARC DAMELIN (33), *Department of Biological Chemistry and Molecular Pharmacology, Harvard Medical School, Boston, Massachusetts 02115*

TRISHA N. DAVIS (3), *Department of Biochemistry, University of Washington, Seattle, Washington 98195*

RAYMOND J. DESHAIES (16), *Division of Biology, Howard Hughes Medical Institute, California Institute of Technology, Pasadena, California 91125*

THOMAS F. DONAHUE (13), *Department of Biology, Indiana University, Bloomington, Indiana 47405*

DAVID DRUBIN (9), *Department of Molecular and Cell Biology, University of California, Berkeley, California 94720*

ix

MARA DUNCAN (9), Department of Molecular and Cell Biology, University of California, Berkeley, California 94720

ELAINE A. ELION (34), Department of Biological Chemistry and Molecular Pharmacology, Harvard Medical School, Boston, Massachusetts 02115

RALF ERDMANN (21), Institute for Biochemistry-Chemistry, Freie Universität Berlin, D-14195 Berlin, Germany

ROBERT S. FULLER (20), Department of Biological Chemistry, University of Michigan, Ann Arbor, Michigan 48109

ALISON E. GAMMIE (29), Department of Molecular Biology, Princeton University, Princeton, New Jersey 08544

SUSAN M. GASSER (11), Department of Molecular Biology, University of Geneva, CH-1211 Geneva 4, Switzerland

EDITH GOMEZ (13), School of Life Sciences, University of Dundee, Dundee DD1 5EH, United Kingdom

BRUCE L. GOODE (25), Department of Biology, Rosenstiel Basic Medical Sciences Research Center, Brandeis University, Waltham, Massachusetts 02454

STEPHEN J. GOULD (21), Department of Biological Chemistry, Johns Hopkins University School of Medicine, Baltimore, Maryland 21205

LEONARD GUARENTE (28), Department of Biology, Massachusetts Institute of Technology, Cambridge, Massachusetts 02139

DALE W. HAILEY (3), Department of Cell Biology and Molecular Genetics, University of Maryland, College Park, Maryland 20742

CHRISTOPHER M. HAMMELL (32), Department of Genetics, Dartmouth Medical School, Hanover, New Hampshire 03755

ERNEST M. HANNIG (13), Department of Molecular and Cell Biology, University of Texas at Dallas, Richardson, Texas 75083

XIANGWEI HE (2), Department of Biology, Massachusetts Institute of Technology, Cambridge, Massachusetts 02139

CATHERINE V. HEATH (32), Department of Biochemistry, Dartmouth Medical School, Hanover, New Hampshire 03755

ALAN G. HINNEBUSCH (13), Laboratory of Gene Regulation and Development, National Institute of Child Health and Human Development, National Institutes of Health, Bethesda, Maryland 20892

MARK HOCHSTRASSER (14), Department of Molecular Biophysics and Biochemistry, Yale University, New Haven, Connecticut 06520

CHRISTINE A. HODGE (32), Department of Biochemistry, Dartmouth Medical School, Hanover, New Hampshire 03755

HAN-KUEI HUANG (13), Molecular Biology and Virology Laboratory, The Salk Institute for Biological Studies, La Jolla, California 92037

MICHAEL J. HUDDLESTON (16), Proteomics and Biological Mass Spectrometry Laboratory, GlaxoSmithKline, King of Prussia, Pennsylvania 19406

JILL L. JOHNSON (26), Department of Biomolecular Chemistry, University of Wisconsin-Madison, Madison, Wisconsin 53706

ELIZABETH W. JONES (7), Department of Biological Sciences, Carnegie Mellon University, Pittsburgh, Pennsylvania 15213

CHRIS A. KAISER (18), Department of Biology, Massachusetts Institute of Technology, Cambridge, Massachusetts 02139

DOUGLAS R. KELLOGG (10), Department of Biology, Sinsheimer Laboratories, University of California, Santa Cruz, California 95064

JULIA KIPPER (23), *Laboratory of Cellular and Structural Biology, Rockefeller University, New York, New York 10021*

DANIEL KORNITZER (37), *Department of Molecular Microbiology, Rappaport Faculty of Medicine, Technion-Israel Institute of Technology, 31096 Haifa, Israel*

THANUJA KRISHNAMOORTHY (13), *Laboratory of Gene Regulation and Development, National Institute of Child Health and Human Development, National Institutes of Health, Bethesda, Maryland 20892*

STEPHEN J. KRON (1), *Department of Molecular Genetics and Cell Biology, The University of Chicago, Chicago, Illinois 60637*

JEFFREY D. LANEY (14), *Department of Molecular Biophysics and Biochemistry, Yale University, New Haven, Connecticut 06520*

SUSAN L. LINDQUIST (30), *Whitehead Institute and Department of Biology, Massachusetts Institute of Technology, Cambridge, Massachusetts 02142*

SASCHA LOSKO (18), *Biomax Informatics AG, D-82152 Martinsried, Germany*

MATTHIAS MANN (17), *Protein Interaction Laboratory, Department of Biochemistry and Molecular Biology, University of Southern Denmark, DK-5230 Odense M, Denmark*

DANIEL C. MASISON (36), *Laboratory of Biochemistry and Genetics, National Institute of Diabetes, Digestive and Kidney Diseases, National Institutes of Health, Bethesda, Maryland 20892*

DAVID N. MASTRONARDE (5), *Boulder Laboratory for 3-D Fine Structure, Department of Molecular, Cellular, and Developmental Biology, University of Colorado, Boulder, Colorado 80309*

KENT MCDONALD (6), *Electron Microscope Laboratory, University of California, Berkeley, California 94720*

J. RICHARD MCINTOSH (5), *Boulder Laboratory for 3-D Fine Structure, Department of Molecular, Cellular, and Developmental Biology, University of Colorado, Boulder, Colorado 80309*

MITCH MCVEY (28), *Department of Biology, University of North Carolina, Chapel Hill, North Carolina 27599*

SHELLY MEEUSEN (22), *Section of Molecular and Cellular Biology, University of California, Davis, California 95616*

DANESH MOAZED (10), *Department of Cell Biology, Harvard Medical School, Boston, Massachusetts 02115*

JON MULHOLLAND (4), *Cell Sciences Imaging Facility, Stanford University School of Medicine, Stanford, California 94305*

ERIC G. D. MULLER (3), *Department of Biochemistry, University of Washington, Seattle, Washington 98195*

THOMAS MÜLLER-REICHERT (6), *Electron Microscope Facility, Max Planck Institute of Molecular Cell Biology and Genetics, D-01307 Dresden, Germany*

JOSEPH NIKA (13), *Department of Molecular and Cell Biology, University of Texas at Dallas, Richardson, Texas 75083*

JODI NUNNARI (22), *Section of Molecular and Cellular Biology, University of California, Davis, California 95616*

EILEEN T. O'TOOLE (5), *Boulder Laboratory for 3-D Fine Structure, Department of Molecular, Cellular, and Developmental Biology, University of Colorado, Boulder, Colorado 80309*

PETER U. PARK (28), *Department of Biology, Massachusetts Institute of Technology, Cambridge, Massachusetts 02139*

ROY PARKER (38), *Department of Molecular and Cellular Biology, Howard Hughes Medical Institute, University of Arizona, Tucson, Arizona 85721*

PHILIPPE PASERO (11), *Institute of Molecular Genetics, National Center for Scientific Research, Mixed Research Unit 5535, F-34293 Montpellier Cedex 05, France*

GRAHAM D. PAVITT (13), *Department of Biomolecular Sciences, University of Manchester Institute of Science and Technology, Manchester M60 1QD, United Kingdom*

DAVID PELLMAN (8), *Dana-Farber Cancer Institute, Harvard Medical School, Boston, Massachusetts 02115*

LON PHAN (13), *National Center for Biotechnology Information, National Institutes of Health, Bethesda, Maryland 20894*

ALEXANDRE V. PODTELEJNIKOV (17), *MDS Proteomics A/S, DK-5230 Odense M, Denmark*

M. K. RAGHURAMAN (31), *Department of Genome Sciences, University of Washington, Seattle, Washington 98195*

DANIEL R. RINES (2), *Department of Biology, Massachusetts Institute of Technology, Cambridge, Massachusetts 02139*

LUCY C. ROBINSON (39), *Department of Biochemistry and Molecular Biology, Louisiana State University Health Sciences Center, Shreveport, Louisiana 71130*

AVITAL A. RODAL (9), *Department of Molecular and Cell Biology, University of California, Berkeley, California 94720*

MARK D. ROSE (29), *Department of Molecular Biology, Princeton University, Princeton, New Jersey 08544*

MICHAEL P. ROUT (23), *Laboratory of Cellular and Structural Biology, Rockefeller University, New York, New York 10021*

RANDY SCHEKMAN (15), *Department of Molecular and Cell Biology, Howard Hughes Medical Institute, University of California, Berkeley, California 94720*

SCOTT C. SCHUYLER (8), *Dana-Farber Cancer Institute, Harvard Medical School, Boston, Massachusetts 02115*

ANJANETTE M. SEARFOSS (36), *National Center for Biotechnology Information, National Library of Medicine, National Institutes of Health, Bethesda, Maryland 20892*

YUVAL SHIMONI (15), *Genentech, Inc., South San Francisco, California 94080*

WENYING SHOU (16), *Rockefeller University, New York, New York 10021*

PAMELA A. SILVER (33), *Department of Biological Chemistry and Molecular Pharmacology, Harvard Medical School, Boston, Massachusetts 02115*

GYÖRGY SIPOS (20), *Department of Biological Chemistry, University of Michigan, Ann Arbor, Michigan 48109*

PETER K. SORGER (2), *Department of Biology, Massachusetts Institute of Technology, Cambridge, Massachusetts 02139*

MICHELLE A. STEIGER (38), *Department of Molecular and Cellular Biology, Howard Hughes Medical Institute, University of Arizona, Tucson, Arizona 85721*

SCOTT W. STEVENS (12), *Division of Biology, California Institute of Technology, Pasadena, California 91125*

TOM H. STEVENS (24), *Institute of Molecular Biology, University of Oregon, Eugene, Oregon 97403*

CATERINA STRAMBIO-DE-CASTILLIA (23), *Laboratory of Cellular and Structural Biology, Rockefeller University, New York, New York 10021*

ADISETYANTARI SUPRAPTO (23), *Laboratory of Cellular and Structural Biology, Rockefeller University, New York, New York 10021*

KELLY TATCHELL (39), *Department of Biochemistry and Molecular Biology, Louisiana State University Health Sciences Center, Shreveport, Louisiana 71130*

SUSAN M. UPTAIN (30), *Department of Molecular Genetics and Cell Biology, Howard Hughes Medical Institute, University of Chicago, Chicago, Illinois 60637*

ANJA J. VAN BRABANT (31), *NaPro BioTherapeutics, Genomics Division, Delaware Biotechnology Institute, Newark, Delaware 19711*

RATI VERMA (16), *Division of Biology, Howard Hughes Medical Institute, California Institute of Technology, Pasadena, California 91125*

THOMAS VIDA (35), *Department of Microbiology and Molecular Genetics, University of Texas Health Science Center, Houston, Texas 77030*

JENNIFER A. WAGNER (22), *Section of Molecular and Cellular Biology, University of California, Davis, California 95616*

BEVERLY WENDLAND (35), *Department of Biology, Johns Hopkins University, Baltimore, Maryland 21218*

REED B. WICKNER (36), *Laboratory of Biochemistry and Genetics, National Institute of Diabetes, Digestive and Kidney Diseases, National Institutes of Health, Bethesda, Maryland 20892*

MARK WINEY (5), *Boulder Laboratory for 3-D Fine Structure, Department of Molecular, Cellular, and Developmental Biology, University of Colorado, Boulder, Colorado 80309*

EDITH D. WONG (22), *Section of Molecular and Cellular Biology, University of California, Davis, California 95616*

Preface

Volumes 350 and 351 of *Methods in Enzymology,* "Guide to Yeast Genetics and Molecular and Cell Biology," Parts B and C, reflect the enormous burst of information on *Saccharomyces cerevisiae* since publication of Part A, Volume 194. The ten years between these publications witnessed the emergence of *Saccharomyces cerevisiae* as the most technically advanced experimental organism, extending its versatility as a system to drug discovery, cancer research, and aging. As the first eukaryotic genome to be completely sequenced (April 1996), yeast provided the inaugural view of the basic functions common to all nucleated cells. The availability of the complete yeast genome sequence ($\sim$13 mbp) coupled with facile databases that are easily accessible on the internet quickly fueled the discovery of new techniques such as two hybrid analysis, transcriptional and protein arrays, and sophisticated microscopic techniques, all of which have completely changed the landscape of today's biology. Because of these neoteric advances, Volumes 350 and 351 contain chapters on proteomics and genomics that provide convenient links to reliable sites on the internet. These information-based tools extend the power intrinsic to the traditional yeast genetic system. This vibrancy is evident in the creation of a library containing a null allele for each of the 6100 yeast genes predicted to encode a protein of 100 amino acids or more. This library permits a comprehensive screen of all such genes in the genome for any loss of function phenotype without the biases of random mutant hunts.

These remarkable advances like a searchlight in a cave also reveal many unexplored areas. Some are technical; there is still no reliable method for obtaining pure yeast nuclei. And of the 6000 genes in the genome there are $\sim$35% whose function is not known. Other unexplored areas are of a more theoretical nature. Though we know much about the information coding capacity of yeast genomic DNA there are many molecules whose information content is unknown. The work on yeast prions shows that some proteins contain heritable information not coded in the DNA. How widespread is this phenomenon? Do the lipids, polysaccharides, and RNA molecules passed mitotically from mother to daughter cell or through meiosis to the progeny also exist in alternative states that influence phenotype? It is our hope that the techniques recounted in these volumes will help answer these many questions.

CHRISTINE GUTHRIE
GERALD R. FINK

METHODS IN ENZYMOLOGY

VOLUME I. Preparation and Assay of Enzymes
Edited by SIDNEY P. COLOWICK AND NATHAN O. KAPLAN

VOLUME II. Preparation and Assay of Enzymes
Edited by SIDNEY P. COLOWICK AND NATHAN O. KAPLAN

VOLUME III. Preparation and Assay of Substrates
Edited by SIDNEY P. COLOWICK AND NATHAN O. KAPLAN

VOLUME IV. Special Techniques for the Enzymologist
Edited by SIDNEY P. COLOWICK AND NATHAN O. KAPLAN

VOLUME V. Preparation and Assay of Enzymes
Edited by SIDNEY P. COLOWICK AND NATHAN O. KAPLAN

VOLUME VI. Preparation and Assay of Enzymes (*Continued*)
Preparation and Assay of Substrates
Special Techniques
Edited by SIDNEY P. COLOWICK AND NATHAN O. KAPLAN

VOLUME VII. Cumulative Subject Index
Edited by SIDNEY P. COLOWICK AND NATHAN O. KAPLAN

VOLUME VIII. Complex Carbohydrates
Edited by ELIZABETH F. NEUFELD AND VICTOR GINSBURG

VOLUME IX. Carbohydrate Metabolism
Edited by WILLIS A. WOOD

VOLUME X. Oxidation and Phosphorylation
Edited by RONALD W. ESTABROOK AND MAYNARD E. PULLMAN

VOLUME XI. Enzyme Structure
Edited by C. H. W. HIRS

VOLUME XII. Nucleic Acids (Parts A and B)
Edited by LAWRENCE GROSSMAN AND KIVIE MOLDAVE

VOLUME XIII. Citric Acid Cycle
Edited by J. M. LOWENSTEIN

VOLUME XIV. Lipids
Edited by J. M. LOWENSTEIN

VOLUME XV. Steroids and Terpenoids
Edited by RAYMOND B. CLAYTON

VOLUME XVI. Fast Reactions
Edited by KENNETH KUSTIN

VOLUME XVII. Metabolism of Amino Acids and Amines (Parts A and B)
Edited by HERBERT TABOR AND CELIA WHITE TABOR

VOLUME XVIII. Vitamins and Coenzymes (Parts A, B, and C)
Edited by DONALD B. MCCORMICK AND LEMUEL D. WRIGHT

VOLUME XIX. Proteolytic Enzymes
Edited by GERTRUDE E. PERLMANN AND LASZLO LORAND

VOLUME XX. Nucleic Acids and Protein Synthesis (Part C)
Edited by KIVIE MOLDAVE AND LAWRENCE GROSSMAN

VOLUME XXI. Nucleic Acids (Part D)
Edited by LAWRENCE GROSSMAN AND KIVIE MOLDAVE

VOLUME XXII. Enzyme Purification and Related Techniques
Edited by WILLIAM B. JAKOBY

VOLUME XXIII. Photosynthesis (Part A)
Edited by ANTHONY SAN PIETRO

VOLUME XXIV. Photosynthesis and Nitrogen Fixation (Part B)
Edited by ANTHONY SAN PIETRO

VOLUME XXV. Enzyme Structure (Part B)
Edited by C. H. W. HIRS AND SERGE N. TIMASHEFF

VOLUME XXVI. Enzyme Structure (Part C)
Edited by C. H. W. HIRS AND SERGE N. TIMASHEFF

VOLUME XXVII. Enzyme Structure (Part D)
Edited by C. H. W. HIRS AND SERGE N. TIMASHEFF

VOLUME XXVIII. Complex Carbohydrates (Part B)
Edited by VICTOR GINSBURG

VOLUME XXIX. Nucleic Acids and Protein Synthesis (Part E)
Edited by LAWRENCE GROSSMAN AND KIVIE MOLDAVE

VOLUME XXX. Nucleic Acids and Protein Synthesis (Part F)
Edited by KIVIE MOLDAVE AND LAWRENCE GROSSMAN

VOLUME XXXI. Biomembranes (Part A)
Edited by SIDNEY FLEISCHER AND LESTER PACKER

VOLUME XXXII. Biomembranes (Part B)
Edited by SIDNEY FLEISCHER AND LESTER PACKER

VOLUME XXXIII. Cumulative Subject Index Volumes I-XXX
Edited by MARTHA G. DENNIS AND EDWARD A. DENNIS

VOLUME XXXIV. Affinity Techniques (Enzyme Purification: Part B)
Edited by WILLIAM B. JAKOBY AND MEIR WILCHEK

VOLUME XXXV. Lipids (Part B)
Edited by JOHN M. LOWENSTEIN

VOLUME XXXVI. Hormone Action (Part A: Steroid Hormones)
Edited by BERT W. O'MALLEY AND JOEL G. HARDMAN

VOLUME XXXVII. Hormone Action (Part B: Peptide Hormones)
Edited by BERT W. O'MALLEY AND JOEL G. HARDMAN

VOLUME XXXVIII. Hormone Action (Part C: Cyclic Nucleotides)
Edited by JOEL G. HARDMAN AND BERT W. O'MALLEY

VOLUME XXXIX. Hormone Action (Part D: Isolated Cells, Tissues, and Organ Systems)
Edited by JOEL G. HARDMAN AND BERT W. O'MALLEY

VOLUME XL. Hormone Action (Part E: Nuclear Structure and Function)
Edited by BERT W. O'MALLEY AND JOEL G. HARDMAN

VOLUME XLI. Carbohydrate Metabolism (Part B)
Edited by W. A. WOOD

VOLUME XLII. Carbohydrate Metabolism (Part C)
Edited by W. A. WOOD

VOLUME XLIII. Antibiotics
Edited by JOHN H. HASH

VOLUME XLIV. Immobilized Enzymes
Edited by KLAUS MOSBACH

VOLUME XLV. Proteolytic Enzymes (Part B)
Edited by LASZLO LORAND

VOLUME XLVI. Affinity Labeling
Edited by WILLIAM B. JAKOBY AND MEIR WILCHEK

VOLUME XLVII. Enzyme Structure (Part E)
Edited by C. H. W. HIRS AND SERGE N. TIMASHEFF

VOLUME XLVIII. Enzyme Structure (Part F)
Edited by C. H. W. HIRS AND SERGE N. TIMASHEFF

VOLUME XLIX. Enzyme Structure (Part G)
Edited by C. H. W. HIRS AND SERGE N. TIMASHEFF

VOLUME L. Complex Carbohydrates (Part C)
Edited by VICTOR GINSBURG

VOLUME LI. Purine and Pyrimidine Nucleotide Metabolism
Edited by PATRICIA A. HOFFEE AND MARY ELLEN JONES

VOLUME LII. Biomembranes (Part C: Biological Oxidations)
Edited by SIDNEY FLEISCHER AND LESTER PACKER

VOLUME LIII. Biomembranes (Part D: Biological Oxidations)
Edited by SIDNEY FLEISCHER AND LESTER PACKER

VOLUME LIV. Biomembranes (Part E: Biological Oxidations)
Edited by SIDNEY FLEISCHER AND LESTER PACKER

VOLUME LV. Biomembranes (Part F: Bioenergetics)
Edited by SIDNEY FLEISCHER AND LESTER PACKER

VOLUME LVI. Biomembranes (Part G: Bioenergetics)
Edited by SIDNEY FLEISCHER AND LESTER PACKER

VOLUME LVII. Bioluminescence and Chemiluminescence
Edited by MARLENE A. DELUCA

VOLUME LVIII. Cell Culture
Edited by WILLIAM B. JAKOBY AND IRA PASTAN

VOLUME LIX. Nucleic Acids and Protein Synthesis (Part G)
Edited by KIVIE MOLDAVE AND LAWRENCE GROSSMAN

VOLUME LX. Nucleic Acids and Protein Synthesis (Part H)
Edited by KIVIE MOLDAVE AND LAWRENCE GROSSMAN

VOLUME 61. Enzyme Structure (Part H)
Edited by C. H. W. HIRS AND SERGE N. TIMASHEFF

VOLUME 62. Vitamins and Coenzymes (Part D)
Edited by DONALD B. MCCORMICK AND LEMUEL D. WRIGHT

VOLUME 63. Enzyme Kinetics and Mechanism (Part A: Initial Rate and Inhibitor Methods)
Edited by DANIEL L. PURICH

VOLUME 64. Enzyme Kinetics and Mechanism (Part B: Isotopic Probes and Complex Enzyme Systems)
Edited by DANIEL L. PURICH

VOLUME 65. Nucleic Acids (Part I)
Edited by LAWRENCE GROSSMAN AND KIVIE MOLDAVE

VOLUME 66. Vitamins and Coenzymes (Part E)
Edited by DONALD B. MCCORMICK AND LEMUEL D. WRIGHT

VOLUME 67. Vitamins and Coenzymes (Part F)
Edited by DONALD B. MCCORMICK AND LEMUEL D. WRIGHT

VOLUME 68. Recombinant DNA
Edited by RAY WU

VOLUME 69. Photosynthesis and Nitrogen Fixation (Part C)
Edited by ANTHONY SAN PIETRO

VOLUME 70. Immunochemical Techniques (Part A)
Edited by HELEN VAN VUNAKIS AND JOHN J. LANGONE

VOLUME 71. Lipids (Part C)
Edited by JOHN M. LOWENSTEIN

VOLUME 72. Lipids (Part D)
Edited by JOHN M. LOWENSTEIN

VOLUME 73. Immunochemical Techniques (Part B)
Edited by JOHN J. LANGONE AND HELEN VAN VUNAKIS

VOLUME 74. Immunochemical Techniques (Part C)
Edited by JOHN J. LANGONE AND HELEN VAN VUNAKIS

VOLUME 75. Cumulative Subject Index Volumes XXXI, XXXII, XXXIV–LX
Edited by EDWARD A. DENNIS AND MARTHA G. DENNIS

VOLUME 76. Hemoglobins
Edited by ERALDO ANTONINI, LUIGI ROSSI-BERNARDI, AND EMILIA CHIANCONE

VOLUME 77. Detoxication and Drug Metabolism
Edited by WILLIAM B. JAKOBY

VOLUME 78. Interferons (Part A)
Edited by SIDNEY PESTKA

VOLUME 79. Interferons (Part B)
Edited by SIDNEY PESTKA

VOLUME 80. Proteolytic Enzymes (Part C)
Edited by LASZLO LORAND

VOLUME 81. Biomembranes (Part H: Visual Pigments and Purple Membranes, I)
Edited by LESTER PACKER

VOLUME 82. Structural and Contractile Proteins (Part A: Extracellular Matrix)
Edited by LEON W. CUNNINGHAM AND DIXIE W. FREDERIKSEN

VOLUME 83. Complex Carbohydrates (Part D)
Edited by VICTOR GINSBURG

VOLUME 84. Immunochemical Techniques (Part D: Selected Immunoassays)
Edited by JOHN J. LANGONE AND HELEN VAN VUNAKIS

VOLUME 85. Structural and Contractile Proteins (Part B: The Contractile Apparatus and the Cytoskeleton)
Edited by DIXIE W. FREDERIKSEN AND LEON W. CUNNINGHAM

VOLUME 86. Prostaglandins and Arachidonate Metabolites
Edited by WILLIAM E. M. LANDS AND WILLIAM L. SMITH

VOLUME 87. Enzyme Kinetics and Mechanism (Part C: Intermediates, Stereochemistry, and Rate Studies)
Edited by DANIEL L. PURICH

VOLUME 88. Biomembranes (Part I: Visual Pigments and Purple Membranes, II)
Edited by LESTER PACKER

VOLUME 89. Carbohydrate Metabolism (Part D)
Edited by WILLIS A. WOOD

VOLUME 90. Carbohydrate Metabolism (Part E)
Edited by WILLIS A. WOOD

VOLUME 91. Enzyme Structure (Part I)
Edited by C. H. W. HIRS AND SERGE N. TIMASHEFF

VOLUME 92. Immunochemical Techniques (Part E: Monoclonal Antibodies and General Immunoassay Methods)
Edited by JOHN J. LANGONE AND HELEN VAN VUNAKIS

VOLUME 93. Immunochemical Techniques (Part F: Conventional Antibodies, Fc Receptors, and Cytotoxicity)
Edited by JOHN J. LANGONE AND HELEN VAN VUNAKIS

VOLUME 94. Polyamines
Edited by HERBERT TABOR AND CELIA WHITE TABOR

VOLUME 95. Cumulative Subject Index Volumes 61–74, 76–80
Edited by EDWARD A. DENNIS AND MARTHA G. DENNIS

VOLUME 96. Biomembranes [Part J: Membrane Biogenesis: Assembly and Targeting (General Methods; Eukaryotes)]
Edited by SIDNEY FLEISCHER AND BECCA FLEISCHER

VOLUME 97. Biomembranes [Part K: Membrane Biogenesis: Assembly and Targeting (Prokaryotes, Mitochondria, and Chloroplasts)]
Edited by SIDNEY FLEISCHER AND BECCA FLEISCHER

VOLUME 98. Biomembranes (Part L: Membrane Biogenesis: Processing and Recycling)
Edited by SIDNEY FLEISCHER AND BECCA FLEISCHER

VOLUME 99. Hormone Action (Part F: Protein Kinases)
Edited by JACKIE D. CORBIN AND JOEL G. HARDMAN

VOLUME 100. Recombinant DNA (Part B)
Edited by RAY WU, LAWRENCE GROSSMAN, AND KIVIE MOLDAVE

VOLUME 101. Recombinant DNA (Part C)
Edited by RAY WU, LAWRENCE GROSSMAN, AND KIVIE MOLDAVE

VOLUME 102. Hormone Action (Part G: Calmodulin and Calcium-Binding Proteins)
Edited by ANTHONY R. MEANS AND BERT W. O'MALLEY

VOLUME 103. Hormone Action (Part H: Neuroendocrine Peptides)
Edited by P. MICHAEL CONN

VOLUME 104. Enzyme Purification and Related Techniques (Part C)
Edited by WILLIAM B. JAKOBY

VOLUME 105. Oxygen Radicals in Biological Systems
Edited by LESTER PACKER

VOLUME 106. Posttranslational Modifications (Part A)
Edited by FINN WOLD AND KIVIE MOLDAVE

VOLUME 107. Posttranslational Modifications (Part B)
Edited by FINN WOLD AND KIVIE MOLDAVE

VOLUME 108. Immunochemical Techniques (Part G: Separation and Characterization of Lymphoid Cells)
Edited by GIOVANNI DI SABATO, JOHN J. LANGONE, AND HELEN VAN VUNAKIS

VOLUME 109. Hormone Action (Part I: Peptide Hormones)
Edited by LUTZ BIRNBAUMER AND BERT W. O'MALLEY

VOLUME 110. Steroids and Isoprenoids (Part A)
Edited by JOHN H. LAW AND HANS C. RILLING

VOLUME 111. Steroids and Isoprenoids (Part B)
Edited by JOHN H. LAW AND HANS C. RILLING

VOLUME 112. Drug and Enzyme Targeting (Part A)
Edited by KENNETH J. WIDDER AND RALPH GREEN

VOLUME 113. Glutamate, Glutamine, Glutathione, and Related Compounds
Edited by ALTON MEISTER

VOLUME 114. Diffraction Methods for Biological Macromolecules (Part A)
Edited by HAROLD W. WYCKOFF, C. H. W. HIRS, AND SERGE N. TIMASHEFF

VOLUME 115. Diffraction Methods for Biological Macromolecules (Part B)
Edited by HAROLD W. WYCKOFF, C. H. W. HIRS, AND SERGE N. TIMASHEFF

VOLUME 116. Immunochemical Techniques (Part H: Effectors and Mediators of Lymphoid Cell Functions)
Edited by GIOVANNI DI SABATO, JOHN J. LANGONE, AND HELEN VAN VUNAKIS

VOLUME 117. Enzyme Structure (Part J)
Edited by C. H. W. HIRS AND SERGE N. TIMASHEFF

VOLUME 118. Plant Molecular Biology
Edited by ARTHUR WEISSBACH AND HERBERT WEISSBACH

VOLUME 119. Interferons (Part C)
Edited by SIDNEY PESTKA

VOLUME 120. Cumulative Subject Index Volumes 81–94, 96–101

VOLUME 121. Immunochemical Techniques (Part I: Hybridoma Technology and Monoclonal Antibodies)
Edited by JOHN J. LANGONE AND HELEN VAN VUNAKIS

VOLUME 122. Vitamins and Coenzymes (Part G)
Edited by FRANK CHYTIL AND DONALD B. MCCORMICK

VOLUME 123. Vitamins and Coenzymes (Part H)
Edited by FRANK CHYTIL AND DONALD B. MCCORMICK

VOLUME 124. Hormone Action (Part J: Neuroendocrine Peptides)
Edited by P. MICHAEL CONN

VOLUME 125. Biomembranes (Part M: Transport in Bacteria, Mitochondria, and Chloroplasts: General Approaches and Transport Systems)
Edited by SIDNEY FLEISCHER AND BECCA FLEISCHER

VOLUME 126. Biomembranes (Part N: Transport in Bacteria, Mitochondria, and Chloroplasts: Protonmotive Force)
Edited by SIDNEY FLEISCHER AND BECCA FLEISCHER

VOLUME 127. Biomembranes (Part O: Protons and Water: Structure and Translocation)
Edited by LESTER PACKER

VOLUME 128. Plasma Lipoproteins (Part A: Preparation, Structure, and Molecular Biology)
Edited by JERE P. SEGREST AND JOHN J. ALBERS

VOLUME 129. Plasma Lipoproteins (Part B: Characterization, Cell Biology, and Metabolism)
Edited by JOHN J. ALBERS AND JERE P. SEGREST

VOLUME 130. Enzyme Structure (Part K)
Edited by C. H. W. HIRS AND SERGE N. TIMASHEFF

VOLUME 131. Enzyme Structure (Part L)
Edited by C. H. W. HIRS AND SERGE N. TIMASHEFF

VOLUME 132. Immunochemical Techniques (Part J: Phagocytosis and Cell-Mediated Cytotoxicity)
Edited by GIOVANNI DI SABATO AND JOHANNES EVERSE

VOLUME 133. Bioluminescence and Chemiluminescence (Part B)
Edited by MARLENE DELUCA AND WILLIAM D. MCELROY

VOLUME 134. Structural and Contractile Proteins (Part C: The Contractile Apparatus and the Cytoskeleton)
Edited by RICHARD B. VALLEE

VOLUME 135. Immobilized Enzymes and Cells (Part B)
Edited by KLAUS MOSBACH

VOLUME 136. Immobilized Enzymes and Cells (Part C)
Edited by KLAUS MOSBACH

VOLUME 137. Immobilized Enzymes and Cells (Part D)
Edited by KLAUS MOSBACH

VOLUME 138. Complex Carbohydrates (Part E)
Edited by VICTOR GINSBURG

VOLUME 139. Cellular Regulators (Part A: Calcium- and Calmodulin-Binding Proteins)
Edited by ANTHONY R. MEANS AND P. MICHAEL CONN

VOLUME 140. Cumulative Subject Index Volumes 102–119, 121–134

VOLUME 141. Cellular Regulators (Part B: Calcium and Lipids)
Edited by P. MICHAEL CONN AND ANTHONY R. MEANS

VOLUME 142. Metabolism of Aromatic Amino Acids and Amines
Edited by SEYMOUR KAUFMAN

VOLUME 143. Sulfur and Sulfur Amino Acids
Edited by WILLIAM B. JAKOBY AND OWEN GRIFFITH

VOLUME 144. Structural and Contractile Proteins (Part D: Extracellular Matrix)
Edited by LEON W. CUNNINGHAM

VOLUME 145. Structural and Contractile Proteins (Part E: Extracellular Matrix)
Edited by LEON W. CUNNINGHAM

VOLUME 146. Peptide Growth Factors (Part A)
Edited by DAVID BARNES AND DAVID A. SIRBASKU

VOLUME 147. Peptide Growth Factors (Part B)
Edited by DAVID BARNES AND DAVID A. SIRBASKU

VOLUME 148. Plant Cell Membranes
Edited by LESTER PACKER AND ROLAND DOUCE

VOLUME 149. Drug and Enzyme Targeting (Part B)
Edited by RALPH GREEN AND KENNETH J. WIDDER

VOLUME 150. Immunochemical Techniques (Part K: *In Vitro* Models of B and T Cell Functions and Lymphoid Cell Receptors)
Edited by GIOVANNI DI SABATO

VOLUME 151. Molecular Genetics of Mammalian Cells
Edited by MICHAEL M. GOTTESMAN

VOLUME 152. Guide to Molecular Cloning Techniques
Edited by SHELBY L. BERGER AND ALAN R. KIMMEL

VOLUME 153. Recombinant DNA (Part D)
Edited by RAY WU AND LAWRENCE GROSSMAN

VOLUME 154. Recombinant DNA (Part E)
Edited by RAY WU AND LAWRENCE GROSSMAN

VOLUME 155. Recombinant DNA (Part F)
Edited by RAY WU

VOLUME 156. Biomembranes (Part P: ATP-Driven Pumps and Related Transport: The Na, K-Pump)
Edited by SIDNEY FLEISCHER AND BECCA FLEISCHER

VOLUME 157. Biomembranes (Part Q: ATP-Driven Pumps and Related Transport: Calcium, Proton, and Potassium Pumps)
Edited by SIDNEY FLEISCHER AND BECCA FLEISCHER

VOLUME 158. Metalloproteins (Part A)
Edited by JAMES F. RIORDAN AND BERT L. VALLEE

VOLUME 159. Initiation and Termination of Cyclic Nucleotide Action
Edited by JACKIE D. CORBIN AND ROGER A. JOHNSON

VOLUME 160. Biomass (Part A: Cellulose and Hemicellulose)
Edited by WILLIS A. WOOD AND SCOTT T. KELLOGG

VOLUME 161. Biomass (Part B: Lignin, Pectin, and Chitin)
Edited by WILLIS A. WOOD AND SCOTT T. KELLOGG

VOLUME 162. Immunochemical Techniques (Part L: Chemotaxis and Inflammation)
Edited by GIOVANNI DI SABATO

VOLUME 163. Immunochemical Techniques (Part M: Chemotaxis and Inflammation)
Edited by GIOVANNI DI SABATO

VOLUME 164. Ribosomes
Edited by HARRY F. NOLLER, JR., AND KIVIE MOLDAVE

VOLUME 165. Microbial Toxins: Tools for Enzymology
Edited by SIDNEY HARSHMAN

VOLUME 166. Branched-Chain Amino Acids
Edited by ROBERT HARRIS AND JOHN R. SOKATCH

VOLUME 167. Cyanobacteria
Edited by LESTER PACKER AND ALEXANDER N. GLAZER

VOLUME 168. Hormone Action (Part K: Neuroendocrine Peptides)
Edited by P. MICHAEL CONN

VOLUME 169. Platelets: Receptors, Adhesion, Secretion (Part A)
Edited by JACEK HAWIGER

VOLUME 170. Nucleosomes
Edited by PAUL M. WASSARMAN AND ROGER D. KORNBERG

VOLUME 171. Biomembranes (Part R: Transport Theory: Cells and Model Membranes)
Edited by SIDNEY FLEISCHER AND BECCA FLEISCHER

VOLUME 172. Biomembranes (Part S: Transport: Membrane Isolation and Characterization)
Edited by SIDNEY FLEISCHER AND BECCA FLEISCHER

VOLUME 173. Biomembranes [Part T: Cellular and Subcellular Transport: Eukaryotic (Nonepithelial) Cells]
Edited by SIDNEY FLEISCHER AND BECCA FLEISCHER

VOLUME 174. Biomembranes [Part U: Cellular and Subcellular Transport: Eukaryotic (Nonepithelial) Cells]
Edited by SIDNEY FLEISCHER AND BECCA FLEISCHER

VOLUME 175. Cumulative Subject Index Volumes 135–139, 141–167

VOLUME 176. Nuclear Magnetic Resonance (Part A: Spectral Techniques and Dynamics)
Edited by NORMAN J. OPPENHEIMER AND THOMAS L. JAMES

VOLUME 177. Nuclear Magnetic Resonance (Part B: Structure and Mechanism)
Edited by NORMAN J. OPPENHEIMER AND THOMAS L. JAMES

VOLUME 178. Antibodies, Antigens, and Molecular Mimicry
Edited by JOHN J. LANGONE

VOLUME 179. Complex Carbohydrates (Part F)
Edited by VICTOR GINSBURG

VOLUME 180. RNA Processing (Part A: General Methods)
Edited by JAMES E. DAHLBERG AND JOHN N. ABELSON

VOLUME 181. RNA Processing (Part B: Specific Methods)
Edited by JAMES E. DAHLBERG AND JOHN N. ABELSON

VOLUME 182. Guide to Protein Purification
Edited by MURRAY P. DEUTSCHER

VOLUME 183. Molecular Evolution: Computer Analysis of Protein and Nucleic Acid Sequences
Edited by RUSSELL F. DOOLITTLE

VOLUME 184. Avidin-Biotin Technology
Edited by MEIR WILCHEK AND EDWARD A. BAYER

VOLUME 185. Gene Expression Technology
Edited by DAVID V. GOEDDEL

VOLUME 186. Oxygen Radicals in Biological Systems (Part B: Oxygen Radicals and Antioxidants)
Edited by LESTER PACKER AND ALEXANDER N. GLAZER

VOLUME 187. Arachidonate Related Lipid Mediators
Edited by ROBERT C. MURPHY AND FRANK A. FITZPATRICK

VOLUME 188. Hydrocarbons and Methylotrophy
Edited by MARY E. LIDSTROM

VOLUME 189. Retinoids (Part A: Molecular and Metabolic Aspects)
Edited by LESTER PACKER

VOLUME 190. Retinoids (Part B: Cell Differentiation and Clinical Applications)
Edited by LESTER PACKER

VOLUME 191. Biomembranes (Part V: Cellular and Subcellular Transport: Epithelial Cells)
Edited by SIDNEY FLEISCHER AND BECCA FLEISCHER

VOLUME 192. Biomembranes (Part W: Cellular and Subcellular Transport: Epithelial Cells)
Edited by SIDNEY FLEISCHER AND BECCA FLEISCHER

VOLUME 193. Mass Spectrometry
Edited by JAMES A. MCCLOSKEY

VOLUME 194. Guide to Yeast Genetics and Molecular Biology
Edited by CHRISTINE GUTHRIE AND GERALD R. FINK

VOLUME 195. Adenylyl Cyclase, G Proteins, and Guanylyl Cyclase
Edited by ROGER A. JOHNSON AND JACKIE D. CORBIN

VOLUME 196. Molecular Motors and the Cytoskeleton
Edited by RICHARD B. VALLEE

VOLUME 197. Phospholipases
Edited by EDWARD A. DENNIS

VOLUME 198. Peptide Growth Factors (Part C)
Edited by DAVID BARNES, J. P. MATHER, AND GORDON H. SATO

VOLUME 199. Cumulative Subject Index Volumes 168–174, 176–194

VOLUME 200. Protein Phosphorylation (Part A: Protein Kinases: Assays, Purification, Antibodies, Functional Analysis, Cloning, and Expression)
Edited by TONY HUNTER AND BARTHOLOMEW M. SEFTON

VOLUME 201. Protein Phosphorylation (Part B: Analysis of Protein Phosphorylation, Protein Kinase Inhibitors, and Protein Phosphatases)
Edited by TONY HUNTER AND BARTHOLOMEW M. SEFTON

VOLUME 202. Molecular Design and Modeling: Concepts and Applications (Part A: Proteins, Peptides, and Enzymes)
Edited by JOHN J. LANGONE

VOLUME 203. Molecular Design and Modeling: Concepts and Applications (Part B: Antibodies and Antigens, Nucleic Acids, Polysaccharides, and Drugs)
Edited by JOHN J. LANGONE

VOLUME 204. Bacterial Genetic Systems
Edited by JEFFREY H. MILLER

VOLUME 205. Metallobiochemistry (Part B: Metallothionein and Related Molecules)
Edited by JAMES F. RIORDAN AND BERT L. VALLEE

VOLUME 206. Cytochrome P450
Edited by MICHAEL R. WATERMAN AND ERIC F. JOHNSON

VOLUME 207. Ion Channels
Edited by BERNARDO RUDY AND LINDA E. IVERSON

VOLUME 208. Protein–DNA Interactions
Edited by ROBERT T. SAUER

VOLUME 209. Phospholipid Biosynthesis
Edited by EDWARD A. DENNIS AND DENNIS E. VANCE

VOLUME 210. Numerical Computer Methods
Edited by LUDWIG BRAND AND MICHAEL L. JOHNSON

VOLUME 211. DNA Structures (Part A: Synthesis and Physical Analysis of DNA)
Edited by DAVID M. J. LILLEY AND JAMES E. DAHLBERG

VOLUME 212. DNA Structures (Part B: Chemical and Electrophoretic Analysis of DNA)
Edited by DAVID M. J. LILLEY AND JAMES E. DAHLBERG

VOLUME 213. Carotenoids (Part A: Chemistry, Separation, Quantitation, and Antioxidation)
Edited by LESTER PACKER

VOLUME 214. Carotenoids (Part B: Metabolism, Genetics, and Biosynthesis)
Edited by LESTER PACKER

VOLUME 215. Platelets: Receptors, Adhesion, Secretion (Part B)
Edited by JACEK J. HAWIGER

VOLUME 216. Recombinant DNA (Part G)
Edited by RAY WU

VOLUME 217. Recombinant DNA (Part H)
Edited by RAY WU

VOLUME 218. Recombinant DNA (Part I)
Edited by RAY WU

VOLUME 219. Reconstitution of Intracellular Transport
Edited by JAMES E. ROTHMAN

VOLUME 220. Membrane Fusion Techniques (Part A)
Edited by NEJAT DÜZGUÜNES

VOLUME 221. Membrane Fusion Techniques (Part B)
Edited by NEJAT DÜZGÜNES

VOLUME 222. Proteolytic Enzymes in Coagulation, Fibrinolysis, and Complement Activation (Part A: Mammalian Blood Coagulation Factors and Inhibitors)
Edited by LASZLO LORAND AND KENNETH G. MANN

VOLUME 223. Proteolytic Enzymes in Coagulation, Fibrinolysis, and Complement Activation (Part B: Complement Activation, Fibrinolysis, and Nonmammalian Blood Coagulation Factors)
Edited by LASZLO LORAND AND KENNETH G. MANN

VOLUME 224. Molecular Evolution: Producing the Biochemical Data
Edited by ELIZABETH ANNE ZIMMER, THOMAS J. WHITE, REBECCA L. CANN, AND ALLAN C. WILSON

VOLUME 225. Guide to Techniques in Mouse Development
Edited by PAUL M. WASSARMAN AND MELVIN L. DEPAMPHILIS

VOLUME 226. Metallobiochemistry (Part C: Spectroscopic and Physical Methods for Probing Metal Ion Environments in Metalloenzymes and Metalloproteins)
Edited by JAMES F. RIORDAN AND BERT L. VALLEE

VOLUME 227. Metallobiochemistry (Part D: Physical and Spectroscopic Methods for Probing Metal Ion Environments in Metalloproteins)
Edited by JAMES F. RIORDAN AND BERT L. VALLEE

VOLUME 228. Aqueous Two-Phase Systems
Edited by HARRY WALTER AND GÖTE JOHANSSON

VOLUME 229. Cumulative Subject Index Volumes 195–198, 200–227

VOLUME 230. Guide to Techniques in Glycobiology
Edited by WILLIAM J. LENNARZ AND GERALD W. HART

VOLUME 231. Hemoglobins (Part B: Biochemical and Analytical Methods)
Edited by JOHANNES EVERSE, KIM D. VANDEGRIFF, AND ROBERT M. WINSLOW

VOLUME 232. Hemoglobins (Part C: Biophysical Methods)
Edited by JOHANNES EVERSE, KIM D. VANDEGRIFF, AND ROBERT M. WINSLOW

VOLUME 233. Oxygen Radicals in Biological Systems (Part C)
Edited by LESTER PACKER

VOLUME 234. Oxygen Radicals in Biological Systems (Part D)
Edited by LESTER PACKER

VOLUME 235. Bacterial Pathogenesis (Part A: Identification and Regulation of Virulence Factors)
Edited by VIRGINIA L. CLARK AND PATRIK M. BAVOIL

VOLUME 236. Bacterial Pathogenesis (Part B: Integration of Pathogenic Bacteria with Host Cells)
Edited by VIRGINIA L. CLARK AND PATRIK M. BAVOIL

VOLUME 237. Heterotrimeric G Proteins
Edited by RAVI IYENGAR

VOLUME 238. Heterotrimeric G-Protein Effectors
Edited by RAVI IYENGAR

VOLUME 239. Nuclear Magnetic Resonance (Part C)
Edited by THOMAS L. JAMES AND NORMAN J. OPPENHEIMER

VOLUME 240. Numerical Computer Methods (Part B)
Edited by MICHAEL L. JOHNSON AND LUDWIG BRAND

VOLUME 241. Retroviral Proteases
Edited by LAWRENCE C. KUO AND JULES A. SHAFER

VOLUME 242. Neoglycoconjugates (Part A)
Edited by Y. C. LEE AND REIKO T. LEE

VOLUME 243. Inorganic Microbial Sulfur Metabolism
Edited by HARRY D. PECK, JR., AND JEAN LEGALL

VOLUME 244. Proteolytic Enzymes: Serine and Cysteine Peptidases
Edited by ALAN J. BARRETT

VOLUME 245. Extracellular Matrix Components
Edited by E. RUOSLAHTI AND E. ENGVALL

VOLUME 246. Biochemical Spectroscopy
Edited by KENNETH SAUER

VOLUME 247. Neoglycoconjugates (Part B: Biomedical Applications)
Edited by Y. C. LEE AND REIKO T. LEE

VOLUME 248. Proteolytic Enzymes: Aspartic and Metallo Peptidases
Edited by ALAN J. BARRETT

VOLUME 249. Enzyme Kinetics and Mechanism (Part D: Developments in Enzyme Dynamics)
Edited by DANIEL L. PURICH

VOLUME 250. Lipid Modifications of Proteins
Edited by PATRICK J. CASEY AND JANICE E. BUSS

VOLUME 251. Biothiols (Part A: Monothiols and Dithiols, Protein Thiols, and Thiyl Radicals)
Edited by LESTER PACKER

VOLUME 252. Biothiols (Part B: Glutathione and Thioredoxin; Thiols in Signal Transduction and Gene Regulation)
Edited by LESTER PACKER

VOLUME 253. Adhesion of Microbial Pathogens
Edited by RON J. DOYLE AND ITZHAK OFEK

VOLUME 254. Oncogene Techniques
Edited by PETER K. VOGT AND INDER M. VERMA

VOLUME 255. Small GTPases and Their Regulators (Part A: Ras Family)
Edited by W. E. BALCH, CHANNING J. DER, AND ALAN HALL

VOLUME 256. Small GTPases and Their Regulators (Part B: Rho Family)
Edited by W. E. BALCH, CHANNING J. DER, AND ALAN HALL

VOLUME 257. Small GTPases and Their Regulators (Part C: Proteins Involved in Transport)
Edited by W. E. BALCH, CHANNING J. DER, AND ALAN HALL

VOLUME 258. Redox-Active Amino Acids in Biology
Edited by JUDITH P. KLINMAN

VOLUME 259. Energetics of Biological Macromolecules
Edited by MICHAEL L. JOHNSON AND GARY K. ACKERS

VOLUME 260. Mitochondrial Biogenesis and Genetics (Part A)
Edited by GIUSEPPE M. ATTARDI AND ANNE CHOMYN

VOLUME 261. Nuclear Magnetic Resonance and Nucleic Acids
Edited by THOMAS L. JAMES

VOLUME 262. DNA Replication
Edited by JUDITH L. CAMPBELL

VOLUME 263. Plasma Lipoproteins (Part C: Quantitation)
Edited by WILLIAM A. BRADLEY, SANDRA H. GIANTURCO, AND JERE P. SEGREST

VOLUME 264. Mitochondrial Biogenesis and Genetics (Part B)
Edited by GIUSEPPE M. ATTARDI AND ANNE CHOMYN

VOLUME 265. Cumulative Subject Index Volumes 228, 230–262

VOLUME 266. Computer Methods for Macromolecular Sequence Analysis
Edited by RUSSELL F. DOOLITTLE

VOLUME 267. Combinatorial Chemistry
Edited by JOHN N. ABELSON

VOLUME 268. Nitric Oxide (Part A: Sources and Detection of NO; NO Synthase)
Edited by LESTER PACKER

VOLUME 269. Nitric Oxide (Part B: Physiological and Pathological Processes)
Edited by LESTER PACKER

VOLUME 270. High Resolution Separation and Analysis of Biological Macromolecules (Part A: Fundamentals)
Edited by BARRY L. KARGER AND WILLIAM S. HANCOCK

VOLUME 271. High Resolution Separation and Analysis of Biological Macromolecules (Part B: Applications)
Edited by BARRY L. KARGER AND WILLIAM S. HANCOCK

VOLUME 272. Cytochrome P450 (Part B)
Edited by ERIC F. JOHNSON AND MICHAEL R. WATERMAN

VOLUME 273. RNA Polymerase and Associated Factors (Part A)
Edited by SANKAR ADHYA

VOLUME 274. RNA Polymerase and Associated Factors (Part B)
Edited by SANKAR ADHYA

VOLUME 275. Viral Polymerases and Related Proteins
Edited by LAWRENCE C. KUO, DAVID B. OLSEN, AND STEVEN S. CARROLL

VOLUME 276. Macromolecular Crystallography (Part A)
Edited by CHARLES W. CARTER, JR., AND ROBERT M. SWEET

VOLUME 277. Macromolecular Crystallography (Part B)
Edited by CHARLES W. CARTER, JR., AND ROBERT M. SWEET

VOLUME 278. Fluorescence Spectroscopy
Edited by LUDWIG BRAND AND MICHAEL L. JOHNSON

VOLUME 279. Vitamins and Coenzymes (Part I)
Edited by DONALD B. MCCORMICK, JOHN W. SUTTIE, AND CONRAD WAGNER

VOLUME 280. Vitamins and Coenzymes (Part J)
Edited by DONALD B. MCCORMICK, JOHN W. SUTTIE, AND CONRAD WAGNER

VOLUME 281. Vitamins and Coenzymes (Part K)
Edited by DONALD B. MCCORMICK, JOHN W. SUTTIE, AND CONRAD WAGNER

VOLUME 282. Vitamins and Coenzymes (Part L)
Edited by DONALD B. MCCORMICK, JOHN W. SUTTIE, AND CONRAD WAGNER

VOLUME 283. Cell Cycle Control
Edited by WILLIAM G. DUNPHY

VOLUME 284. Lipases (Part A: Biotechnology)
Edited by BYRON RUBIN AND EDWARD A. DENNIS

VOLUME 285. Cumulative Subject Index Volumes 263, 264, 266–284, 286–289

VOLUME 286. Lipases (Part B: Enzyme Characterization and Utilization)
Edited by BYRON RUBIN AND EDWARD A. DENNIS

VOLUME 287. Chemokines
Edited by RICHARD HORUK

VOLUME 288. Chemokine Receptors
Edited by RICHARD HORUK

VOLUME 289. Solid Phase Peptide Synthesis
Edited by GREGG B. FIELDS

VOLUME 290. Molecular Chaperones
Edited by GEORGE H. LORIMER AND THOMAS BALDWIN

VOLUME 291. Caged Compounds
Edited by GERARD MARRIOTT

VOLUME 292. ABC Transporters: Biochemical, Cellular, and Molecular Aspects
Edited by SURESH V. AMBUDKAR AND MICHAEL M. GOTTESMAN

VOLUME 293. Ion Channels (Part B)
Edited by P. MICHAEL CONN

VOLUME 294. Ion Channels (Part C)
Edited by P. MICHAEL CONN

VOLUME 295. Energetics of Biological Macromolecules (Part B)
Edited by GARY K. ACKERS AND MICHAEL L. JOHNSON

VOLUME 296. Neurotransmitter Transporters
Edited by SUSAN G. AMARA

VOLUME 297. Photosynthesis: Molecular Biology of Energy Capture
Edited by LEE MCINTOSH

VOLUME 298. Molecular Motors and the Cytoskeleton (Part B)
Edited by RICHARD B. VALLEE

VOLUME 299. Oxidants and Antioxidants (Part A)
Edited by LESTER PACKER

VOLUME 300. Oxidants and Antioxidants (Part B)
Edited by LESTER PACKER

VOLUME 301. Nitric Oxide: Biological and Antioxidant Activities (Part C)
Edited by LESTER PACKER

VOLUME 302. Green Fluorescent Protein
Edited by P. MICHAEL CONN

VOLUME 303. cDNA Preparation and Display
Edited by SHERMAN M. WEISSMAN

VOLUME 304. Chromatin
Edited by PAUL M. WASSARMAN AND ALAN P. WOLFFE

VOLUME 305. Bioluminescence and Chemiluminescence (Part C)
Edited by THOMAS O. BALDWIN AND MIRIAM M. ZIEGLER

VOLUME 306. Expression of Recombinant Genes in Eukaryotic Systems
Edited by JOSEPH C. GLORIOSO AND MARTIN C. SCHMIDT

VOLUME 307. Confocal Microscopy
Edited by P. MICHAEL CONN

VOLUME 308. Enzyme Kinetics and Mechanism (Part E: Energetics of Enzyme Catalysis)
Edited by DANIEL L. PURICH AND VERN L. SCHRAMM

VOLUME 309. Amyloid, Prions, and Other Protein Aggregates
Edited by RONALD WETZEL

VOLUME 310. Biofilms
Edited by RON J. DOYLE

VOLUME 311. Sphingolipid Metabolism and Cell Signaling (Part A)
Edited by ALFRED H. MERRILL, JR., AND YUSUF A. HANNUN

VOLUME 312. Sphingolipid Metabolism and Cell Signaling (Part B)
Edited by ALFRED H. MERRILL, JR., AND YUSUF A. HANNUN

VOLUME 313. Antisense Technology (Part A: General Methods, Methods of Delivery, and RNA Studies)
Edited by M. IAN PHILLIPS

VOLUME 314. Antisense Technology (Part B: Applications)
Edited by M. IAN PHILLIPS

VOLUME 315. Vertebrate Phototransduction and the Visual Cycle (Part A)
Edited by KRZYSZTOF PALCZEWSKI

VOLUME 316. Vertebrate Phototransduction and the Visual Cycle (Part B)
Edited by KRZYSZTOF PALCZEWSKI

VOLUME 317. RNA–Ligand Interactions (Part A: Structural Biology Methods)
Edited by DANIEL W. CELANDER AND JOHN N. ABELSON

VOLUME 318. RNA–Ligand Interactions (Part B: Molecular Biology Methods)
Edited by DANIEL W. CELANDER AND JOHN N. ABELSON

VOLUME 319. Singlet Oxygen, UV-A, and Ozone
Edited by LESTER PACKER AND HELMUT SIES

VOLUME 320. Cumulative Subject Index Volumes 290–319

VOLUME 321. Numerical Computer Methods (Part C)
Edited by MICHAEL L. JOHNSON AND LUDWIG BRAND

VOLUME 322. Apoptosis
Edited by JOHN C. REED

VOLUME 323. Energetics of Biological Macromolecules (Part C)
Edited by MICHAEL L. JOHNSON AND GARY K. ACKERS

VOLUME 324. Branched-Chain Amino Acids (Part B)
Edited by ROBERT A. HARRIS AND JOHN R. SOKATCH

VOLUME 325. Regulators and Effectors of Small GTPases (Part D: Rho Family)
Edited by W. E. BALCH, CHANNING J. DER, AND ALAN HALL

VOLUME 326. Applications of Chimeric Genes and Hybrid Proteins (Part A: Gene Expression and Protein Purification)
Edited by JEREMY THORNER, SCOTT D. EMR, AND JOHN N. ABELSON

VOLUME 327. Applications of Chimeric Genes and Hybrid Proteins (Part B: Cell Biology and Physiology)
Edited by JEREMY THORNER, SCOTT D. EMR, AND JOHN N. ABELSON

VOLUME 328. Applications of Chimeric Genes and Hybrid Proteins (Part C: Protein-Protein Interactions and Genomics)
Edited by JEREMY THORNER, SCOTT D. EMR, AND JOHN N. ABELSON

VOLUME 329. Regulators and Effectors of Small GTPases (Part E: GTPases Involved in Vesicular Traffic)
Edited by W. E. BALCH, CHANNING J. DER, AND ALAN HALL

VOLUME 330. Hyperthermophilic Enzymes (Part A)
Edited by MICHAEL W. W. ADAMS AND ROBERT M. KELLY

VOLUME 331. Hyperthermophilic Enzymes (Part B)
Edited by MICHAEL W. W. ADAMS AND ROBERT M. KELLY

VOLUME 332. Regulators and Effectors of Small GTPases (Part F: Ras Family I)
Edited by W. E. BALCH, CHANNING J. DER, AND ALAN HALL

VOLUME 333. Regulators and Effectors of Small GTPases (Part G: Ras Family II)
Edited by W. E. BALCH, CHANNING J. DER, AND ALAN HALL

VOLUME 334. Hyperthermophilic Enzymes (Part C)
Edited by MICHAEL W. W. ADAMS AND ROBERT M. KELLY

VOLUME 335. Flavonoids and Other Polyphenols
Edited by LESTER PACKER

VOLUME 336. Microbial Growth in Biofilms (Part A: Developmental and Molecular Biological Aspects)
Edited by RON J. DOYLE

VOLUME 337. Microbial Growth in Biofilms (Part B: Special Environments and Physicochemical Aspects)
Edited by RON J. DOYLE

VOLUME 338. Nuclear Magnetic Resonance of Biological Macromolecules (Part A)
Edited by THOMAS L. JAMES, VOLKER DÖTSCH, AND ULI SCHMITZ

VOLUME 339. Nuclear Magnetic Resonance of Biological Macromolecules (Part B)
Edited by THOMAS L. JAMES, VOLKER DÖTSCH, AND ULI SCHMITZ

VOLUME 340. Drug–Nucleic Acid Interactions
Edited by JONATHAN B. CHAIRES AND MICHAEL J. WARING

VOLUME 341. Ribonucleases (Part A)
Edited by ALLEN W. NICHOLSON

VOLUME 342. Ribonucleases (Part B)
Edited by ALLEN W. NICHOLSON

VOLUME 343. G Protein Pathways (Part A: Receptors)
Edited by RAVI IYENGAR AND JOHN D. HILDEBRANDT

VOLUME 344. G Protein Pathways (Part B: G Proteins and Their Regulators)
Edited by RAVI IYENGAR AND JOHN D. HILDEBRANDT

VOLUME 345. G Protein Pathways (Part C: Effector Mechanisms)
Edited by RAVI IYENGAR AND JOHN D. HILDEBRANDT

VOLUME 346. Gene Therapy Methods
Edited by M. IAN PHILLIPS

VOLUME 347. Protein Sensors and Reactive Oxygen Species (Part A: Selenoproteins and Thioredoxin)
Edited by HELMUT SIES AND LESTER PACKER

VOLUME 348. Protein Sensors and Reactive Oxygen Species (Part B: Thiol Enzymes and Proteins)
Edited by HELMUT SIES AND LESTER PACKER

VOLUME 349. Superoxide Dismutase
Edited by LESTER PACKER

VOLUME 350. Guide to Yeast Genetics and Molecular and Cell Biology (Part B)
Edited by CHRISTINE GUTHRIE AND GERALD R. FINK

VOLUME 351. Guide to Yeast Genetics and Molecular and Cell Biology (Part C)
Edited by CHRISTINE GUTHRIE AND GERALD R. FINK

VOLUME 352. Redox Cell Biology and Genetics (Part A) (in preparation)
Edited by CHANDAN K. SEN AND LESTER PACKER

VOLUME 353. Redox Cell Biology and Genetics (Part B) (in preparation)
Edited by CHANDAN K. SEN AND LESTER PACKER

VOLUME 354. Enzyme Kinetics and Mechanism (Part F: Detection and Characterization of Enzyme Reaction Intermediates) (in preparation)
Edited by DANIEL L. PURICH

VOLUME 355. Cumulative Subject Index Volumes 321–354 (in preparation)

VOLUME 356. Laser Capture Microscopy and Microdissection (in preparation)
Edited by P. MICHAEL CONN

VOLUME 357. Cytochrome P450 (Part C) (in preparation)
Edited by ERIC F. JOHNSON AND MICHAEL R. WATERMAN

Section I

Cytology

[1] Digital Time-Lapse Microscopy of Yeast Cell Growth

By STEPHEN J. KRON

Introduction

The vast number of yeast experiments are performed with macroscopic colonies, which may contain 10^8 or more cells. The behavior of a colony when replica plated to nonpermissive conditions such as an auxotrophic tester plate can be remarkably decisive—growth or no growth. Here, one imagines all the cells behaving in a relatively coordinated fashion. However, observations such as papillation, uneven growth of cells when struck out, "slow" growth, and inhomogeneous colony phenotypes inevitably reflect the underlying variations in growth and survival of individual cells that often are not determined genetically but stochastically. For many experiments, the "tyranny of the petri plate" prevents full appreciation of the biological complexity of cell-autonomous regulatory pathways. Although often less decisive, examining individual cells and observing their responses to changing environmental conditions can be illuminating. Solely examining large ensembles of cells may lead to inaccurate models for the behavior of individual members of the population. From the first reports of time-lapse observation of yeast growth until the present, a significant cell-to-cell variability in cell cycle parameters and other responses among genetically identical cells cultured under identical conditions has been observed. Interestingly, some of these differences appear to be inherited by daughter cells,[1] suggesting a relatively unexplored aspect of epigenetics. Experiments in which large populations of yeast cells are sampled at various times during steady-state growth or after a shift in conditions, fixed, and then imaged are of proven value. However, only by imaging individual cells can the statistical variations in rates of cellular transitions be fully appreciated—these features are washed out in population-based experiments. Observing individual cells is far more powerful than is often appreciated.

The small size of yeast cells, their amotility, their cell walls, and their relatively bland cytoplasm make them intrinsically less "interesting" to watch than an amoeba or a ciliate. However, because budding is a marker of cell cycle entry and septation is a clear marker of cytokinesis, cell cycle studies are relatively straightforward in this organism. Further, bud size and shape offer valuable cues to cell cycle position. Beyond analysis of vegetative growth rates, several other types of experiments clearly benefit from observing individual cells. Establishing bud-site selection patterns is a natural application of this approach.[2,3] Developmental responses to signals

---

[1] A. E. Wheals and P. G. Lord, *Cell. Prolif.* **25,** 217 (1992).
[2] S. J. Kron, C. A. Styles, and G. R. Fink, *Mol. Biol. Cell* **5,** 1003 (1994).
[3] J. Chant and J. R. Pringle, *J. Cell Biol.* **129,** 751 (1995).

such as pheromone arrest and mating or to nutrient stress such as filamentous differentiation,[2] sporulation, and entry to stationary phase are each readily documented by time-lapse imaging. In turn, checkpoint responses to genomic damage,[4] spindle dysfunction, or noxious environmental stimuli are also well studied by time-lapse imaging. Looking within the cell to follow organelle inheritance by bright-field imaging or vital fluorescent staining[5–8] and to track abundance and/or localization of proteins through the cell cycle via fluorescent markers such as green fluorescent protein (GFP) fusion proteins[9] offers another class of experiments that apply time-lapse microscopy. As each of these events occurs over many minutes to several hours, being able to record images at regular intervals and then reconstruct a time-lapse movie can offer powerful insight into cell autonomous behavior.

Excellent discussions of principles of video microscopy and its applications in live-cell microscopy,[10–12] history and practice of time-lapse microscopy,[13] current approaches to four-dimensional microscopy,[14] and high-resolution microscopy of yeast[15,16] have been presented previously. The reader may want to have access to the Internet when reading this chapter to find and bookmark sites for manufacturers of products mentioned here. In turn, excellent commercial and academic sites offer detailed discussions of criteria for choosing charge-coupled device (CCD) cameras and other microscopy equipment and provide ready access to freeware and shareware of great value to microscopists.

Nearly all investigators have an upright research microscope available to them with a trinocular phototube. Many of these systems are configured to record images with a conventional photographic camera or typical video camera. Video cameras may be adequate imagers for most time-lapse recording, but this chapter assumes that the reader will perform digital recording using a megapixel CCD or other solid-state digital imager connected directly to a microcomputer as their photographic recording device.

Time-Lapse Observation of Living Yeast Cells

Successful time-lapse imaging of living yeast cells can require significant compromises that affect the balance between performance of the microscope and

[4] V. W. Burns, *Radiat. Res.* **4**, 394 (1956).
[5] L. S. Weisman, R. Bacallao, and W. Wickner, *J. Cell Biol.* **105**, 1539 (1987).
[6] H. D. Jones, M. Schliwa, and D. G. Drubin, *Cell Motil. Cytoskel.* **25**, 129 (1993).
[7] A. J. Koning, P. Y. Lum, J. M. Williams, and R. Wright, *Cell Motil. Cytoskel.* **25**, 111 (1993).
[8] M. P. Yaffe, *Methods Enzymol.* **260**, 447 (1995).
[9] S. L. Shaw, E. Yeh, K. Bloomand, and E. D. Salmon, *Curr. Biol.* **7**, 701 (1997).
[10] Y. L. Wang and D. L. Taylor, *Methods Cell Biol.* **29** (1989).
[11] S. Inoue and K. R. Spring, "Video Microscopy." Plenum, New York, 1997.
[12] G. Sluder and D. E. Wolf (eds.), *Methods Cell Biol.* **56** (1998).
[13] S. Paddock, *BioTechniques* **30**, 283 (2001).
[14] A. T. Hammond and B. S. Glick, *Traffic* **1**, 935 (2000).
[15] E. D. Salmon, E. Yeh, S. Shaw, B. Skibbens, and K. Bloom, *Methods Enzymol.* **298**, 317 (1998).
[16] S. D. Kohlwein, *Microsc. Res. Tech.* **51**, 511 (2000).

integrity of the biological preparation. An experiment can become impractical, uninformative, or even misleading unless appropriate allowances are made that optimize both imaging and cell viability. The following sections examine some of these considerations.

Rates of yeast growth are very sensitive to incubation temperature. Growth at room temperature for many strains is painfully slow. Some method to allow the preparation to be warmed may be necessary to promote sufficiently rapid growth to make filming practical. We have found that forced-air heaters are simple and effective. One device, the Air-Therm heater (World Precision Instruments), can be used to heat an enclosure around the stage (Fig. 1a) or used as an open system with the airstream pointed at the sample and objective. The temperature is maintained by a feedback circuit that maintains the set temperature via a sensor placed on or near the sample.

Like most cells, yeast are sensitive to the intense light that a condenser focuses onto the subject. This is not simply due to infrared irradiation. Heat filters do not allow yeast to better tolerate very bright illumination. Further, the absolute light level may be more important than the integrated dose. In general, attenuating the transmitted illumination as much as possible is advisable for long-term observation. The variable exposure time and linear response of CCDs allows them to take useful images at illumination levels too low for comfortable visual imaging. Where bright illumination is required, use of shutters to gate the light to intervals when images will be collected may be required even for short experiments.

Nutrients for Long-Term Growth Experiments

Many practical experiments do not require yeast cells to be imaged for more than a few minutes or a few hours. These situations allow far more latitude in experimental setup than for long-term imaging. To collect images at low numerical aperture (N.A.), a few microliters of cells in media can be applied to a slide and a coverslip placed over the drop. However, to record images at high N.A., particularly in fluorescence imaging, a significant difference in resolution can be detected between cells close to the coverslip and close to the slide surface. Using very small volumes (1–3 μl) allows cells to be effectively trapped and slightly flattened between coverslip and slide. However, cells may rapidly deplete media and/or be stressed by this treatment. Overall, it is generally preferable to attach the cells to the coverslip. To secure cells to the coverslip (or slide), the glass can be treated with 1 mg/ml concanavalin A (Con A) in water and then rinsed off. Cells allowed to settle onto the treated surface will adhere strongly. The unattached cells can be washed away with excess media and the blotted coverslip should then be laid onto a droplet of media on the slide. Rich media may eventually dislodge the cells from the surface, whereas cells applied in synthetic media will adhere more securely. Unfortunately, if any appreciable volume is left under the coverslip, it can float and shift during focusing and scanning. Adding a gelling agent

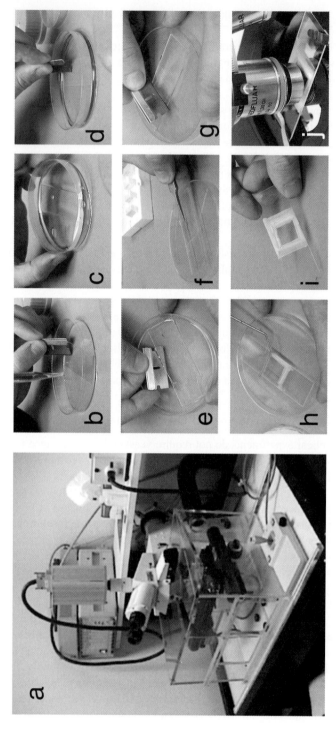

FIG. 1. Microscope system and preparation for digital time-lapse imaging of yeast growth. (a) Microscope with incubator stage, motorized X, Y, and Z controls, and CCD digital camera. (b–j) Protocol for immobilizing yeast cells on an agar pad in a microscope slide and coverslip chamber [S. J. Kron, C. A. Styles, and G. R. Fink, *Mol. Biol. Cell* **5**, 1003 (1994)] for long-term observation of cell growth (see text).

(low-melt agar, alginate, gelatin, etc.) to the media can be inconvenient but helps stabilize the preparation even for short-term observation.

The refractile nature of the cell wall can interfere with imaging the plasma membrane, nuclear membrane, or other organelles and subcellular structures. For high-resolution imaging of intracellular detail, refractive index matching can be used to make the cell wall "disappear." Here, use of a gelling agent can have added benefit. A common approach is to use 20–25% (w/v) gelatin (e.g., Difco, Detroit, MI) in YPD media to match the refractive index of the cell wall. Once melted by stirring into liquid media heated to ~55°, gelatin remains liquid at 37° but gels rapidly at room temperature and remains gelled at 30°. Liquid YPD/gelatin can be applied to cells immobilized with Con A on a coverslip. The excess liquid can be dabbed off and the coverslip placed down onto a warmed slide and then allowed to set at room temperature. Alternatively, cells can be mixed directly into the warm gelatin and then the liquid applied to a warm slide and immediately overlaid with a coverslip. Once gelled, these preparations may be stable for up to several hours. The slide can also be sealed with nail polish to provide mechanical stability and to slow drying.

Long-term imaging of yeast cells at high magnification under the microscope to assess budding, cell shape, cell cycle parameters, and so on presents some special problems. In general, it is easier to work with a solid agar media or other gelled media when performing long-term experiments with yeast cells. Low magnification imaging can be performed readily on a thin agar petri plate, with or without a coverslip overlying the growing colony. However, imaging cells at high resolution requires a standard working distance condenser and objective. Further, any thickness of solid media that scatters light, such as agar, degrades the image, so a thinner layer of agar is preferable.

A convenient method for applying a smooth, even layer of agar is to place a single slide into an empty plastic petri plate (Fig. 1b). Then, 8–10 ml of melted 1% agar media is pipetted onto the slide (Figs. 1c and 1d). The agar is allowed to cool. Then cells are spread onto the agar surface in liquid or gently with a toothpick or micromanipulated onto the center of the embedded slide. Then, the slide is cut from the petri plate (Fig. 1e) and a coverslip is placed over the immobilized cells (Fig. 1f). The sheet of agar is trimmed to the borders of the coverslip (Fig. 1g) and the edges are sealed with melted VALAP (1 : 1 : 1 w/w/w, Vaseline : lanolin : paraffin, Figs. 1h and 1i). Such a preparation can be placed under the microscope within minutes of inoculating the cells onto the agar (Fig. 1j). During incubation, as cells grow in number, microcolonies of noninvasive yeast strains generally spread over the agar surface in a monolayer (Fig. 2), allowing quantitation of cell number. If the density of cells is very low, the amount of nutrient sealed in the chamber is sufficient for many hours of continuous growth. Although yeast are moderately aerobic under normal growth conditions and produce copious amounts of carbon dioxide as a fermentation product, as long as the burden of cells is low enough, the slow exchange of gas through the VALAP seal is likely sufficient to prevent

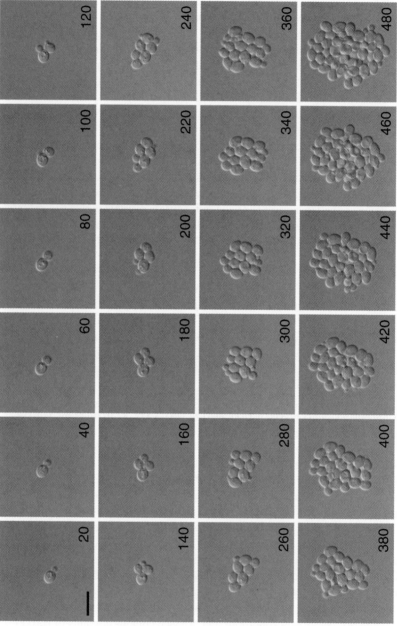

FIG. 2. Time-lapse series of vegetative growth of yeast strain W303-1a at 30° on YPD agar, recorded with a digital autofocus system [LeSage *et al.*, unpublished results (2001)] controlling the microscope shown in Fig. 1 using a 20× objective. Each image is the same small section of the field. Times indicated are minutes after immobilization of cells in a chamber as shown in Fig. 1. During the first 60 min, the cells drifted in the Z axis approximately 15 μm before remaining relatively stable, drifting at no more than 1 μm

dramatic changes in yeast growth until many hours into a typical experiment. Our best results have consistently come from using micromanipulation to place small numbers of cells in closely spaced arrays onto the thin layer of agar lying over the slide before finishing preparation of the mount. Cells can be spread onto the adjoining agar and selected based on their genotype, bud size, etc. We have found that large numbers of pheromone-arrested cells can be manipulated into rows, sealed in, and imaged in the microscope well before a visible bud emerges.

High-Resolution Transmitted Light Microscopy of Living Yeast

A discussion of microscopy optics and instrumentation particularly relevant to imaging living yeast cells at high resolution is presented next. Complete coverage of microscopy practice and principles is found in the book by Inoue and Spring,[11] in the volumes edited by Taylor and Wang,[10] and by Sluder and Wolf[12] and online at the Molecular Expressions web site (http://micro.magnet.fsu.edu). This site also has a comprehensive list of links to microscopy manufacturers, research groups, and other sources.

To a great degree, the quality of an image depends on the quality of the optics. This does not extend to the sturdiness of the microscope stand or other components that do not produce, focus, reflect, or filter light. Instead of a heavy-weight microscope, stability and resistance to vibration are best achieved with a vibration isolation table. A simple upright stand that accepts the top-quality objectives and a turret condenser and allows a range of choices regarding transmitted illumination source, specimen holder, camera mounting, and so on is perfectly adequate.

The stand should not have an integral transmitted light illuminator, a fixed diffuser, or other simplifications of the transmitted light path. An aplanatic turret condenser with N.A. ~0.9 that can be adjusted for Kohler illumination[17] is generally adequate for high-resolution imaging of yeast cells. In Kohler illumination, a single optical axis runs through the center of the illuminator, condenser, objective, camera-coupling optics, and camera so that pairs of apertures are focused in conjugate planes that link the illuminator, the subject, and the imager. In a standard microscope, when the field diaphragm is focused and centered so that it just vignettes the imaging plane, the condenser aperture diaphragm will then be focused in the same conjugate plane as the back aperture of the objective. Then, for full resolution, the condenser aperture should be adjusted to just vignette the objective back aperture, visualized by removing an eyepiece. Nonetheless, resolution can be lost if light does not fill the condenser aperture evenly. One solution is to use an inexpensive halogen illuminator coupled to a ~1-cm fiber optic bundle. This will fill the condenser aperture with much more even illumination than a bulb filament in a standard housing.

[17] H. E. Keller, *Methods Cell Biol.* **56**, 135 (1998).

Selection of objective is a major determinant of image quality. The single most important specification of an objective is numerical aperture, a number proportional to angle of capture. Like a low f stop in a photographic lens, accepting a wide cone of light leads to less contrast and brighter images with increased precision in all three dimensions, observed as a tighter depth of field and improved resolution. The optical artifacts caused by the large refractive index difference between yeast cell wall and cytoplasm and the curvature of the cell wall can be countered with the resolution and depth of field advantages of high N.A. objectives. In practice, a dry objective with nominal magnification of 20× should have an N.A. of 0.7 or greater. Oil or water immersion objectives that allow an N.A. of 1.2 or greater at 40× offer significant advantages. For 60× or greater, typically oil immersion objectives with N.A. of 1.3 or higher are recommended. Immersion objectives designed for fluorescence imaging (e.g., Fluar objectives from Zeiss) generally offer a practical balance of high N.A., an adequately flat field, minimal aberrations, and image brightness. However, some high N.A. immersion objectives designed for bright field have properties that may degrade performance. Plan apochromatic objectives are corrected for flat field and to eliminate chromatic aberrations across the full aperture, but at a cost to image brightness and contrast. Similarly, if the cell being imaged cannot be positioned close to the coverslip, a highly corrected 1.4 N.A. oil-immersion objective may perform poorly in comparison to a 1.2 N.A. water-immersion objective.

Imaging Methods for Yeast Live-Cell Microscopy

The high refractive index of the cell wall has important effects on optical properties that define the success of different imaging modes. Just as yeast colonies appear white rather than gelatinous, the light-scattering ability of the individual yeast cell causes dramatic effects in the microscope. Different imaging methods can be used to exploit or compensate for this large refractive index difference. The Molecular Expressions website is highly recommended for its clear descriptions and diagrams of different imaging methods.

Most yeast biologists are familiar with phase contrast as a method for visualizing live yeast cells. The principle of phase microscopy, grossly simplified, is that a ring of light admitted through the condenser aperture is mostly blocked by a dark ring in the objective aperture. By passing through the condenser, the ring of light is evenly spread in the focus plane and appears as an even, gray background. Light diverted by refractive index inhomogeneity escapes the blocking ring and recombines to form the characteristic halo around yeast cells. For proper phase imaging, the microscope is adjusted for Kohler illumination and then matching rings in the condenser and the objective are brought into alignment. Unfortunately, the phase rings on most objectives are positioned at small radii, leading to high contrast but lost resolution. Thus, most intracellular detail is obscured and subtle events such as emergence of buds and formation of the septum can be missed.

Nomarski differential interference contrast (DIC) is a high-resolution refractive index imaging modality that works well with yeast cells. Nomarski imaging takes advantage of the rotation of polarization created by passing a split beam of polarized light through two paths of differing refractive index and then recombining the beam to detect the offset in phase from the beams traveling at different speeds. Typically, a polarizer is positioned in the light path between the illuminator and the condenser. A fixed Wollaston prism is placed in the condenser aperture and a second, adjustable Wollaston prism is placed close to the objective back aperture. A second polarizer (the analyzer) is positioned in the light path to the ocular. When appropriately adjusted, a null signal of the two paths provides the characteristic gray background while a refractive index difference appears as a black-to-white "shadow" along that edge. As the full condenser and objective apertures remain illuminated, Nomarski allows use of nearly the full N.A. of the objective. However, the shadowed appearance reveals a limitation of the technique as it derives from the prisms only performing interference in one direction so that resolution is maximal in "X" but absent in "Y". Again, the condenser must be set for Kohler illumination. With both prisms out of the light path, the polarizer and analyzer are set for full extinction—closest to black. The prisms are then replaced and the upper prism is adjusted until the optimal balance of resolution and contrast is achieved.

Digital Imaging of Yeast Cell Growth

The growing popularity of CCD-based cameras continues to propel CCD chip prices rapidly downward and push pixel numbers and performance upward. In a recent trend, many specifications for consumer CCD cameras exceed those of the available research equipment. However, this chapter is written assuming use of the currently available 1 to 5 megapixel research and professional CCD cameras in which the pixel size may still be large enough to affect any appreciable zoom-in views of subjects.

Unfortunately, standard microscopes are poorly matched to current CCD cameras, which generally have physically small detectors by comparison to film or tube cameras and may have a nonstandard eye point. As a result, many principles of light microscopy that may be appropriate for film or visual imaging do not apply to digital imaging. For example, concern about "empty" magnification is misplaced until the size and pixel numbers of CCDs approaches those of film. Most research microscopes provide diffraction-limited images when properly adjusted. To reconstruct the full resolution of an image with a typical 1 megapixel array, it would be appropriate to project no more than a 50-μm field onto it, requiring a system magnification of about 500. Nonetheless, the intrinsic resolution and optical sectioning capability of the microscope will still be evident as sharp edges, even in an undersampled image.

In turn, there is a tendency to adjust microscopes to higher contrast than is required to obtain an optimal digital image. Contrast gained by shutting down the condenser aperture is "empty" as it comes at the loss of N.A. and therefore resolution. In general, given a high enough N.A. and a digital imager, even insignificant differences in refractive index within a biological sample can be converted into sufficient contrast to image the subject at the diffraction limit.[11]

Cameras and Computers for Time-Lapse Imaging

Many considerations go into purchasing a digital imager for time-lapse recording. A relatively insensitive color CCD camera that can easily be adapted for use on multiple microscopes (photographing colony phenotypes) and can be readily fit with photographic or video lenses for macrophotography (plate phenotypes) or general photography may be extremely valuable in the yeast laboratory. Alternatively, if the camera will be used for sensitive fluorescence imaging of GFP fusion dynamics or immunolocalization, a cooled CCD may be justified.

In short, all CCDs share a simple technology of converting light to stored electrons in each well or pixel. At the end of an exposure, pixels must be read off one at a time, creating a bottleneck. Strategies such as progressive scan interline transfer allow low-noise, high-resolution, rapid readoff of large CCD arrays. Interline transfer arrays are compatible with electronic shuttering, permitting frame rates of >10 fps and therefore "continuous" imaging. Given the small delay between the camera and the digital display, this allows for easy focusing. While the geometry of interline transfer chips is intrinsically less sensitive than frame transfer, this is often compensated for by microlenses that concentrate incident light onto the active part of each pixel.

Most scientific cameras are designed for fluorescence imaging and their sensitivity and signal-to-noise ratio far exceed what is necessary for bright-field imaging. These devices are generally monochrome cameras and are unavailable with the largest CCD arrays. Often the optics used to mount the camera are a weakness of the system. Poor coupling optics can dramatically degrade the image, introduce vignetting or distortions, or so reduce the field of view that little correspondence exists between eyepieces and camera image. Several companies offer coupling optics to match CCDs to common research microscopes. Diagnostic Instruments offers a wide selection of couplers and a selection guide online.

A practical compromise among price, wide application, and sensitivity is to purchase a scientific color CCD camera. Zeiss (Axiocam), Kodak (MDS 290), Diagnostic Instruments (Spot), and many others offer relatively sensitive color CCDs that are coupled to microscopes easily via C-mount (video) adapters. Compromising lower sensitivity for significantly higher resolution and lower price, SLR "professional" CCD cameras (Nikon, Pentax, Fuji, Olympus, Cannon, etc.) are readily mated to microscopes via a lens-mount adapter available from the microscope manufacturer. All these cameras have LCD viewfinders, are easily controlled, download rapidly, and have excellent specifications. Another cost-effective compromise is to purchase a consumer (fixed lens) megapixel camera and adapt it for use on a microscope. Olympus offers a simple adapter for mounting their C-series cameras to a standard C-mount adapter.

Time-Lapse Digital Image Acquisition and Processing

A recent trend in scientific and professional CCD still cameras has been to take advantage of the data transfer speeds of fast digital interfaces that obviate the requirement for frame grabber boards. Unlike consumer cameras, CCD cameras marketed to the research community (a notable exception is the Kodak MDS 290) do not typically come bundled with a Photoshop (Adobe) plug-in or other image acquisition and processing software package. Although the free program NIH Image can be adapted to perform this job, it may be necessary to purchase a package from a commercial vendor to view, record, and analyze images. Microscopy-oriented packages for use with scientific CCD cameras include QED (QED Imaging), IPLab Spectrum (Scanalytics), Metamorph (Universal Imaging), Openlab (Improvision), Image-Pro (Media Cybernetics), SimplePCI (Compix), and others. QED is a particularly user-friendly program that is compatible with a wide range of cameras and can work as a stand alone or as a plug in for Photoshop or other image-processing programs.

Recording time-lapse sequences is possible, if not straightforward, with many scientific image acquisition software packages. The Kodak MDS 290 is supplied with time-lapse software. QED, IPLab, Metamorph, SimplePCI, and other packages offer this option. Several cost-effective alternatives exist such as macros for NIH Image that are also likely to perform well. Freeware, shareware, and commercial programs that control consumer CCD camera functions from a computer are readily available.

In general, collecting images at 1- to 3-min intervals is sufficient to capture most features of yeast growth. Recording images with a >2 megapixel color CCD and no data compression may entail storing up to 1 gigabyte per hour of recording and can quickly overwhelm hard-disk storage capacity. Digital image compression, affected by the careful choice of file format for saving data during image acquisition (lossy vs lossless image compression via JPEG, TIFF, PICT, GIF, etc.), can help dramatically. However, if high-resolution images are needed, a simple, cost-effective, and high-capacity image archiving system such as a DVD burner is obviously required.

Raw image data can be very unwieldy as a source from which to extract the timing of events during a time-lapse experiment, particularly because image-processing programs such as Photoshop do not allow a large stack of high-resolution raw images to be animated so that they can be flipped through sequentially. In general, it is straightforward to identify software (e.g., shareware,[18] consumer movie-making software, NIH Image, or commercial scientific image-processing packages) that can process stacks of time-lapse still images to create a movie in QuickTime (Apple) or another animation format. Typically, one might distill each hour of yeast growth, comprising 60 images, into 4 sec of video. Given the limitations of current microcomputers, much of the resolution captured in raw data may have to be sacrificed via digital compression in order to make a movie that plays smoothly at 15 frames per second. Such movies can be readily posted on websites or embedded into PowerPoint (Microsoft) presentations. Helpful insights can be gained from examining low-resolution time-lapse movies at slow and full-speed playback and in reverse and then ticking through frame by frame. Inevitably though, for precise measurements, examining high-resolution raw rather than compressed video files data may be critical.

Focus Control during Time-Lapse Experiments

With living yeast cells, the goal is typically to collect a well-focused optical section through the middle of the cell. At high N.A., even small amounts of drift may appear as motion of small features within a cell, making control over focus very significant. With all their moving parts, microscopes are intrinsically unstable systems and magnification amplifies every motion. Mechanical and thermal drift of the microscope and drying or other changes in the specimen can cause the desired plane to rapidly drift out of focus unpredictably. Thermal drift is particularly marked at the beginning of an experiment and when a temperature shift is being performed. Unfortunately, the magnitude of the drift is often quite large by comparison to the depth of field of a microscope objective, let alone the thickness of a yeast cell. Thus, continuous attention to the experiment may be necessary to maintain focus.

Even for an experienced microscopist, stable time-lapse recording at high magnification is nearly impossible when using the fine focus knob to adjust the image through the eyepieces. A fast digital camera with continuous LCD and/or computer display, a reference image to match the current image against, and a motorized focus control to allow rapid and smooth scanning can each make a difference. Several manufacturers (Ludl Electronic Products, Applied Scientific Instrumentation, Prior Scientific, Conix Research, etc.) offer computer-controlled motorized focus drives built with stepper motors or DC servo motors, designed to attach

[18] C. F. Thomas and J. G. White, *Methods Mol. Biol.* **135,** 263 (2000).

to the fine focus knob of Zeiss, Olympus, Leica, Nikon, and other microscopes or, in the case of Autoscan Systems, to replace the stage of such microscopes. A piezoelectric focus drive (e.g., Physik Instrumente PIFOC objective positioner) can be attached between the objective and the nosepiece, allowing very rapid and precise motion (10 nm resolution and reproducibility), ideal for Z-series collection. If purchasing a new microscope body is possible, the harmonic drive focus mechanism available on Zeiss Axioskop microscopes offers high resolution and low hysteresis (backlash) by comparison to a motor connected to the conventional gear-driven stage of the microscope.

Maintaining precise focus during a long time-lapse experiment even with access to a focus motor is quite difficult. Computerized autofocus is clearly the best option for experiments of an hour or more. There is extensive literature on determining focus in machine vision. In general, rapid focusing algorithms control the imaging lens to scan in and out in search of a plane in the Z axis in which a criterion such as image contrast (measured by detecting high spatial frequency content of an image) is maximized. As yet, few commercial CCD image acquisition packages offer an autofocus time-lapse recording option with a simple user interface. SimplePCI (Compix) offers an automated image capture module that may be quite satisfactory for yeast imaging. Alternatively, relatively simple macros will implement simple time-lapse autofocus based on image contrast in any of the commercial image acquisition software packages or NIH Image. These may be freely available from other users and/or the authors of the software. This appears the most cost-effective option. Our approach has been to write a time-lapse autofocus software package in C++ with which we perform long-term observations that can maintain near-ideal focus on colonies of growing yeast cells for as long as the preparation remains stable.[19] In addition to recording yeast growth kinetics (e.g., Fig. 2), we have used this system to examine different image-processing methods for focus determination and to compare different algorithms for focus tracking. Although such a system offers the ultimate in freedom to optimize and customize the system for different experiments, we cannot recommend this option because of the significant time and effort involved in writing and debugging.

Acknowledgments

Allan LeSage is acknowledged for developing several novel time-lapse methods discussed here and for valuable comments on the manuscript and help in preparing the figures. The author is indebted to Shinya Inoue for training and guidance in the art of microscopy. Research in the author's laboratory on live cell microscopy has been supported by a Young Investigator grant from the Arnold and Mabel Beckman Foundation, a Scholar Award from the James S. McDonnell Foundation, and NSF CAREER Grant MCB-9875976.

---

[19] A. LeSage and S. Kron, unpublished results (2001).

[2] Quantitative Microscopy of Green Fluorescent Protein-Labeled Yeast

By DANIEL R. RINES, XIANGWEI HE, and PETER K. SORGER

Introduction

With the development of methods to tag proteins using green fluorescent protein (GFP),[1] fluorescence microscopy has become increasingly important for characterizing protein function in yeast. This chapter describes the use of three-dimensional (3D) deconvolution microscopy to perform fixed and live-cell analysis of cells carrying GFP-tagged proteins. Despite our focus on high-performance imaging, the methods described are applicable to a wide range of experiments using conventional wide-field microscopes.

The goal of fluorescence microscopy is to determine the position and time-dependent distributions of one or more molecules within the cell. Such an analysis is often referred to as five dimensional, having three spatial, one time, and one wavelength dimension. A microscope image is only two dimensional (2D), however, and 3D images must be reconstructed from a stack of 2D image planes, each acquired at different focal planes. A major problem that arises in reconstructing 3D representations from image stacks is that each 2D image contains not only in-focus light, but also out-of-focus light from objects that lie above and below the focal plane.[2] This occurs because light becomes smeared, particularly along the vertical axis (parallel to the path of illumination), when it passes through microscope lenses. Two methods to address the problem of out-of-focus light are in widespread use. In confocal microscopy, scanning point source illumination (typically a laser) combined with a pinhole at the photodetector is used to ensure the collection of in-focus light. In contrast, deconvolution microscopy applies a computational correction to a stack of 2D images to correct for the smearing imposed by the microscope optics. The 2D stack is composed of wide-field images in which the entire focal plane is illuminated and the emitted light is captured on a camera. Comparing the performances of confocal and deconvolution microscopy is complicated by the fact that the instruments are quite different. Confocal microscopes use lasers to scan the sample, illuminating each point successively, and photomultiplier tubes (PMTs) to digitize the emitted light pixel by pixel. Deconvolution microscopes use wide-field illumination from mercury or xenon burners (lamps) and charge-coupled device (CCD) cameras to record images. Iterative deconvolution algorithms generally

[1] M. Chalfie, Y. Tu, G. Euskirchen, W. W. Ward, and D. C. Prasher, *Science* **263,** 802 (1994).
[2] J. R. Swedlow, J. W. Sedat, and D. A. Agard, *in* "Deconvolution of Images and Spectra" (P. A. Jansson, ed.), p. 284. Academic Press, San Diego, 1997.

perform better than confocal microscopy when the sample is thin, the magnification high [objectives over 20× and 0.85 numerical aperature (N.A.)], and the amount of light scattering small.[3] Deconvolution microscopy has higher spatial resolution, greater sensitivity, and causes less photodamage to samples than confocal microscopy. Yeast cells represent a near-optimal sample for deconvolution microscopy, and this chapter focuses on wide-field methods.

Deconvolution in Fluorescence Microscopy

The extent to which microscope optics smear the light emitted by an object is described by the point spread function (PSF). This blurring is intrinsic to optical systems, occurs in both confocal and wide-field microscopes, and is greatest along the Z axis (perpendicular to the plane of the slide). The PSF can be determined by examining the distribution of blurred light emitted from a point source object, a small bead, for example, when the objective is focused through the object (Fig. 1A). Almost all of the smearing in a modern microscope occurs in the objective lens, and a PSF must be experimentally determined for each objective. Blurring can then be computationally corrected by deconvolving the stack of 2D images with the PSF (see Ref. 2). The result is an accurate 3D representation of the object. In applying deconvolution methods, it is important to distinguish between constrained iterative deconvolution,[2,4,5] in which the quality of data is fundamentally improved, and nearest-neighbors deconvolution, which is little more than a filtering method. For high-quality work, constrained iterative deconvolution must be performed, usually using commercial packages from Applied Precision (SoftWoRx; www.api.com), Intelligent Imaging Innovations (SlideBook; www.intelligent-imaging.com), or AutoQuant (AutoDeblur; www.autoquant.com) to name a few. It is also important to note that careful sample preparation and acquisition are essential for good results. Deconvolution is designed to correct computationally for a well-understood smearing effect imposed by microscope optics, not to make poor-quality data look good.

Three characteristics distinguish good- and bad-quality images. The first is optical distortion. With modern apochromat objectives, the field of view is typically very flat and the biggest concern is spherical aberration, a problem that can be controlled by optimizing the immersion oil. In multicolor imaging, one must also worry about shifts in the position of objects from one channel to the next.[6,7] If this chromatic aberration is large, something is probably wrong with the objective.

[3] P. C. Goodwin, *Scanning* **18**, 144 (1996).
[4] P. A. Jansson, R. H. Hunt, and E. K. Plyler, *J. Opt. Soc. Am.* **60**, 596 (1970).
[5] R. Gold, Report No. ANL-6984, Argonne National Laboratory, Chicago, 1964.
[6] M. Kozubek and P. Matula, *J. Microsc.* **200**(Pt. 3), 206 (2000).
[7] K. Dunn and F. R. Maxfield, *Methods Cell Biol.* **56**, 217 (1998).

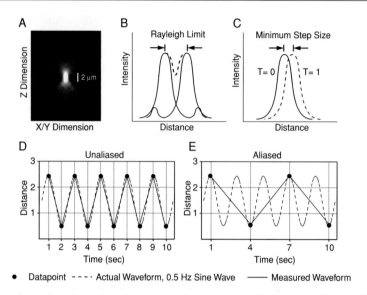

FIG. 1. Image formation and resolution in an optical microscope. (A) Point spread function (PSF). Cross-sectional view of a subresolution bead. A stack of images is obtained by incrementally moving the focus through the fluorescent bead (along Z) while collecting 2D (X–Y) images at each step. The collection of 2D images is rotated 90° or so and a single cross-sectional plane through the center of the Z stack is shown. The conical pattern of light above and below the bead shows the smearing effect imposed by the objective lens and represents the PSF (B) The Rayleigh limit. Cross section along the X or Y axis of the emitted light from two point source objects (this distribution is approximated by a bessel function). Rayleigh criteria determine the minimum distance between the two objects at which they can be distinguished. More precisely, it is the point at which the first trough in the intensity distribution of one object is coincident with the intensity peak of the second object. For two equal intensity objects, this occurs at about 42% of their maximal intensity (S. Bradbury and B. Bracegirdle, "Introduction to Light Microscopy," Vol. 42, p. 66. Springer-Verlag, New York, 1998.) (C) Minimum step size. Even if a bead is displaced in a time series by less than the Rayleigh limit, its motion can still be detected. Under optimal conditions the minimum step size that can be measured is on the order of 10 nm. (D and E) Nyquist criteria in a time series. Plots representing the sinusoidal movement of an object relative to a fixed point over time with a frequency of 0.5 Hz. In (D), data are collected every second and the frequency is correctly determined (even if the waveform is not known). In contrast, when samples are taken every 3 sec (E) the frequency is incorrectly determined. The possibility of aliasing makes it important to collect data points at a minimum of twice the highest frequency of the motion.

Small shifts are observed even with very good objectives, however, and can be dealt with computationally. A second characteristic of good images is accuracy in intensity. Modern scientific-grade CCD cameras typically have a dynamic range of 10^3 to 10^4 (12 bit depth), low thermal noise, and the capacity for quantitative recording from biological structures.[8] The key to recording intensities accurately is to correct for out-of-focus light (by deconvolution, for example) and to select recording

---

[8] Y. Hiraoka, J. W. Sedat, and D. A. Agard, *Science* **238,** 36 (1987).

conditions that maximize the signal-to-noise (S/N) ratio. A third characteristic of a good image or image series (e.g., a movie) is high spatial and temporal resolution. The most commonly used definition of spatial resolution is the Rayleigh limit, the minimum separation for two point sources such that their intensity distributions overlap approximately at their half-maximal values (Fig. 1B). Using a DeltaVision microscope and the methods described in this chapter, we can achieve a Rayleigh resolution of 225 nm in $X-Y$ and 400 nm in Z. It should be noted, however, that objects as close as 150 nm can be discriminated and that changes in position as small as 5–10 nm can be resolved when objects are tracked from one time point to the next (Fig. 1C).[9,10] In time-lapse studies, temporal resolution is also an issue, and a good rule of thumb is that periodic motion must be sampled at a minimum of twice its highest frequency (the Nyquist limit; Figs. 1D and 1E).

Preparing Biological Samples

Strain Construction

A first step in obtaining good images is to use a bright label. A large number of GFP variants have been isolated and new ones are reported all the time (reviewed in Ref. 11). Currently, the brightest GFP molecule is enhanced GFP (EGFP) and its variants, with an excitation peak at 488 nm and an emission peak at 508 nm.[12] Cyan (CFP), yellow (YFP), and blue (BFP) derivatives can also be used in yeast. The CFP–YFP combination has the noteworthy property of forming a fluorescence resonance energy transfer (FRET) pair.[13] In FRET, the light emitted by one fluorophore has the potential to excite a nearby second fluorophore so that the physical proximity of the two fluorophores can be studied. Typically, we fuse GFP to the extreme C termini of proteins by one-step integration into the genome, thereby replacing the endogenous gene with a tagged version. In at least 20 of the 25 essential genes we have examined, functional C-terminal fusions could be generated.[14]

Growth Conditions

Autofluorescence is a significant problem with yeast, particularly with strains that are ade^-. An intermediate in adenine biosynthesis, phosphoribosylamino-imidazole[15] accumulates to high levels in ade^- cells[16] and is highly fluorescent in

---

[9] N. Bobroff, *Rev. Sci. Instrum.* **57**, 1152 (1986).
[10] G. Danuser, P. Tran, and E. Salmon, *J. Microsc.* **198**, 34 (2000).
[11] R. Y. Tsien, *Annu. Rev. Biochem.* **67**, 509 (1998).
[12] R. Heim and R. Y. Tsien, *Curr. Biol.* **6**, 178 (1996).
[13] A. Miyawaki, J. Llopis, R. Heim, J. M. McCaffery, J. A. Adams, M. Ikura, and R. Y. Tsien, *Nature* **388**, 882 (1997).
[14] X. He, D. R. Rines, C. W. Espelin, and P. K. Sorger, *Cell* **10**, 195 (2001).
[15] A. Stotz and P. Linder, *Gene* **95**, 91 (1990).
[16] J. Ishiguro, *Curr. Genet.* **15**, 71 (1989).

GFP channels. Autofluorescence can be minimized in both ade^- and ADE^+ strains by growing them in SD media supplemented with essential amino acids, 20 μg/ml adenine, and a carbon source. Cultures must be maintained below 5×10^6 cells/ml for 4–10 generations and the medium should be refreshed prior to imaging.

Mounting Cells for Microscopy

Two methods are available for mounting cells, depending on the length of time that they are to be kept growing. For short duration observations of live cells (1 hr or less) and fixed samples, cells are suspended in a small amount of medium, the medium is applied directly to a microscope slide, and a cover glass is pressed in place and sealed with petroleum jelly. Mounting substrates [such as poly(L-lysine)] are not necessary because surface tension is enough to exert a slight positive pressure on the sample and hold the cells in place. For long duration observations, cells are maintained on a pad of 1.2% agarose formed in a slide with a shallow depression. The use of an agarose pad is especially important for time-lapse experiments that run for several generations. However, the pad-free method is simpler, suitable for fixed cells, and also optically superior.

Preparing Slides with Agar Pads for Live-Cell Microscopy

1. Prepare a solution of 1.2% (w/v) agarose in SD medium supplemented with a complete mixture of essential amino acids (Bio 101; www.bio101.com), 20 μg/ml of additional adenine, and a carbon source. Make sure that the agarose is completely melted.

2. Add approximately 200 μl of melted agarose to a slide fabricated with a shallow 18 mm hemispherical depression (VWR, www.vwr.com) and prewarmed to 60°. Quickly cover the agarose-filled depression with a regular microscopy slide (the top slide) by placing the top slide at one end of the depression slide and moving the top slide over the agarose while pressing down with your thumbs. It is essential that no air pockets are trapped in the agarose during this process and that the agarose forms a smooth and very flat bed above the well. Hold the top slide tightly over the depression slide for 2 min while the agarose hardens. Leave the top slide in place until the cell culture is ready for mounting. With a little practice, this method will become routine. We have found it to be more reliable than the competing method of cutting out small agarose blocks and transplanting them into slides.

3. Prepare a 22 × 22-mm, No. 1.5 (0.16–0.19 mm), cover glass (VWR) for mounting by first cleaning any dust particles from the cover glass using a precision wipe (Kimwipe) and hand blower (e.g., Bergeon Blower 3B-750 from www.watchmakertools.com). Cleaning is important because dust and other particles prevent a tight seal between the cover glass and the agarose pad. Never use Dustoff or similar compressed gas products because they usually contain small, highly fluorescent particles.

4. Apply a very fine band of pure petroleum jelly to the extreme edge of the cover glass. This should be done to all four sides by adding a small amount of jelly with a pipette tip to one corner and running a finger lightly along the edge to distribute the jelly evenly. The jelly prevents rapid evaporation of the media and creates a better seal when the cover glass is applied to the depression slide. Be careful not to get any jelly on the microscope objectives.

Mounting Cells on Agar Pads

1. Transfer 1–2 ml of log-phase cell culture to an Eppendorf tube and spin at the highest speed (14,000 rpm, ~20,000g) for 1 min to pellet the cells.
2. Remove the supernatant from the Eppendorf tube and resuspend the cells in 0.05–0.10 ml of fresh media. The volume used to suspend the cells depends on the size of the cell pellet in the previous step. For 1×10^7 cells, use 0.75 ml. After removing the top slide from the depression slide and exposing the agarose bed (this is done by sliding the top slide toward one end of the depression slide to release the seal between the two slides), transfer 2.2 μl of cell culture to the agarose and gently place the cover glass over the cell culture. To create a tight seal, apply a very slight amount of pressure at the extreme edges of the cover glass. Care should be taken because pressing too hard can damage or crush the cells between the cover glass and microscope slide.
3. If necessary, spot a little nail polish at the four corners of the cover glass (this is only required for extended time courses of greater than 4–5 hr, as small bubbles are generated by the yeast and cause the cover glass to pull away from the agarose bed).

Mounting Live Cells without Agar Pads

For short duration observations, cells are mounted directly between a slide and cover glass. The use of a depression slide is not required.

1. Transfer 1–2 ml of log-phase cell culture to an Eppendorf tube and spin at the highest speed (14,000 rpm, ~20,000g) for 1 min to pellet the cells.
2. Remove the supernatant from the Eppendorf tube and resuspend the cells in 0.05–0.10 ml of fresh media.
3. Mount washed cells by spotting 2.2 μl on a dust-free slide and then pressing the cover glass firmly in place. Surface tension holds the cover glass in place fairly well, but it helps to seal it in place using nail polish. The sample is good for about an hour until carbon dioxide production causes the coverslip to bow outward.

Fixing Cells with Paraformaldehyde and Mounting

Proteins tagged with GFP and its variants can also be localized in cells fixed with paraformaldehyde. Imaging fixed cells is advantageous when performing colocalization with dim signals and does not require the use of a depression slide. Additionally, because GFP is less sensitive to photobleaching than other chemical fluorophores, the use of antifade or glycerol-based mounting media is usually not required.

1. To 0.875 ml of fresh culture, add 0.125 ml of EM grade 16% aqueous paraformaldehyde (Electron Microscopy Sciences, www.emsdiasum.com/ems) for a final concentration of 2% (w/v) and mix by inversion for 10 min at 25°.
2. Pellet cells in an Eppendorf centrifuge for 2 min at the highest speed (20,000g).
3. Remove the supernatent and resuspend the cell pellet in 1.0 ml of 0.1 M KPO_4, pH 6.6. Wash the cell pellet for 10 min at 25° to remove excess formaldehyde before repelleting cells again and resuspending in 0.05–0.10 ml of KPO_4 buffer (volume varies with cell density).
4. To image cellular DNA, 4,6-diamidino-2-phenylindole (DAPI)[17] is added to a final concentration of 3 μg/ml into the KPO_4 buffer and incubated for 30 sec. Excess DAPI is then washed out by performing three additional rounds of centrifugation and resuspension in KPO_4 buffer.
5. Mount 2.2 μl of washed cells on a dust-free slide as described earlier.

Optimizing Microscope Optics

Obtaining high-quality images involves optimizing the optics of the microscope, illumination conditions, and camera settings. We do not discuss the selection of objectives and cameras, referring the reader to the chapter by Kron elsewhere in this volume,[17a] and instead concentrate on user-adjustable settings and conditions.

Selecting Filters

In an epifluorescence microscope, the excitation and emission wavelengths are selected by three optical elements: an excitation filter, a beam splitter, and an emission filter (see Ref. 18, www.microscopy.fsu.edu, for a detailed discussion of microscope optics and architecture). Broad-spectrum light from the mercury or xenon burner (lamp) passes through the excitation filter and is reflected by the beam splitter into the objective so that it illuminates the sample via the objective. Light emitted by the sample is collected by the objective and passes through the beam splitter and then the emission filter before being recorded by the camera (Fig. 2A). This optical path relies on a special property of beam splitters: they reflect light

---

[17] R. E. Palmer, M. Koval, and D. Koshland, *J. Cell. Biol.* **109,** 3355 (1989).
[17a] S. J. Kron, *Methods Enzymol.* **351,** [1], 2002 (this volume).
[18] M. W. Davidson and M. Abramowitz, "Optical Microscopy," Vol. 2001. The Florida State University & Olympus America Inc., 2001.

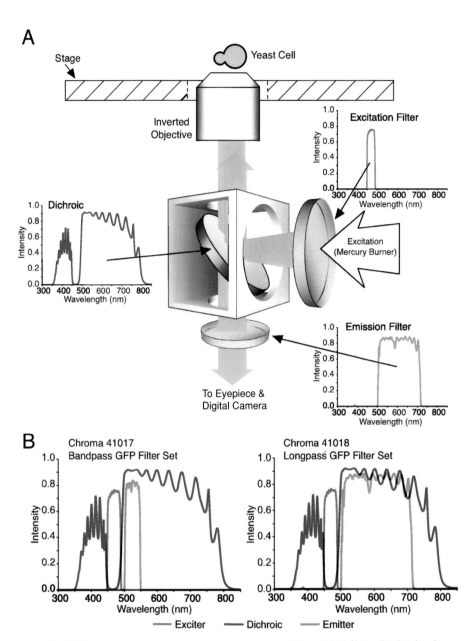

FIG. 2. Filtering elements in an epifluorescent microscope. (A) A typical filter cube showing the positions and spectral properties of the excitation filter, beam splitter, and emission filter. Light from the mercury or xenon burner (lamp) enters the microscope from right. Only those photons between 450 and 500 nm pass through the excitation filter before being reflected by the angled beam splitter up through the objective lens and to the yeast cell. Excitation causes the GFP molecule in the yeast to emit photons at 508 nm, which are then collected by the objective and focused down through the beam slitter and onto the emission filter. Only photons above 508 nm pass through the emission filter and reach the eyepiece and CCD camera. (B) Graphs representing Chroma's band-pass and long-pass filter sets. The 41018 filter set is also illustrated in A.

shorter than a characteristic wavelength but are transparent to longer wavelengths. Dichroic beam splitters (or dichroic mirrors) discriminate between wavelengths at one transition whereas polychroic beam splitters by more than one. Long-pass and short-pass filters are described by a number denoting the wavelength in nanometers at which the filter cuts on or off (more precisely, the wavelength for 50% transmission). Band-pass filters are described by the wavelength of peak emission (the center wavelength) and the bandwidth (the full width at half-maximum transmission) (e.g., 520/20). These features are most easily understood with reference to a plot of filter performance (Fig. 2B).

Filter manufacturers such as Omega Optical (www.omegafilters.com) and Chroma (www.chroma.com) combine these elements into sets (sometimes called filter cubes) suitable for various applications (Table I). With each set the goal is to achieve the maximum signal strength, particularly in the emission channel, while minimizing interference from autofluorescence and, in multispectral images,

TABLE I
SELECTED FILTER SETS FOR EPIFLUORESCENCE MICROSCOPY WITH GFP MOLECULES[a]

Fluorophores	Excitation	Beam splitter	Emission	Manufacturer/ model number[d]	Notes[e]
EGFP (bandpass)	HQ470/40x	Q495LP	HQ525/50m	Chroma/41017	Wide emission range for highest intensity
EGFP (longpass)	HQ470/40x	Q495LP	HQ500LP	Chroma/41018	Higher selectivity in cases of autofluorescence problems
EGFP and YFP	S460/20x S523/20x	86001bs[b] 86001bs[b]	S500/22m S568/50m	Chroma/86001	For EGFP For EYFP
YFP and CFP	S500/20x S436/10x	86002bs[c] 86002bs[c]	S535/30m S470/30m	Chroma/86002	For EYFP For ECFP
BFP and EGFP	S380/30x S485/40m	86003bs 86003bs	S445/40m S535/50m	Chroma/86003	For BFP For EGFP
BFP and GFP	XF1048	XF2041	XF3054	Omega/XF50	Designed for simultaneous imaging of BFP and GFP
CFP and YFP	XF1078	XF2065	XF3099	Omega/XF135	Multiband set allowing simultaneous visualization of CFP and YFP

[a] Prefix and suffix letters indicate special features of the filters. HQ denotes a high-efficiency filter, LP is long pass, S is single band, x is excitation, m is emission, bs is beam splitter, and XF is standard fluorescence.
[b] JP3—multiband (polychroic) beam splitter for EGFP and EYFP.
[c] JP4—multiband (polychroic) beam splitter for ECFP and EYFP.
[d] Chroma provides an excellent downloadable guide to these filter sets.
[e] Information on fluorescent proteins can be found in G. Patterson, R. N. Day, and D. Piston, *J. Cell Sci.* **114,** 837 (2001).

from other fluorophores. However, a trade-off is made between sensitivity and selectivity with every set. Because the optimal choice is dictated by the application, companies such as Chroma sell many different EGFP filter sets, each with a different combination of elements. By examining the characteristics of each element in these sets, it is possible to make a good guess about which will work best for a particular application, although some empirical experimentation is often required as well.

For single-color recording from cells carrying EGFP, we use Chroma filter set 41018 whose 500LP long-pass emission filter provides good sensitivity (Fig. 2B). When autofluorescence is a problem, particularly with ade^- strains, use of the 41017 set with a band-pass emission filter is helpful (Table I). With appropriate filters, dual-color recording of CFP and EGFP can be accomplished with relatively little bleed through and with good signal strength. If you are examining proteins present in a cell at different amounts, the less abundant one should be tagged with EGFP, as it is considerably brighter.[19,20] An alternative approach to dual-color imaging is to use CFP and YFP with the Chroma 86002 filter set (which contains two emission and excitation filters). The signal from YFP is considerably less intense than EGFP, but bleed through into the CFP channel is negligible. It should also be noted that CFP and YFP can be used as a FRET pair with the Chroma 86002 set.[13] Chroma provides an excellent downloadable guide to these filter sets and additional information on fluorescent proteins can be found in Patterson et al.[21]

As discussed later, phototoxicity is often a problem in live-cell analysis, and any unwanted light that leaks through the excitation filter (particularly in the UV range) should be blocked. We typically add a HQ500LP long-pass emission filter and an infrared blocking filter in series with the excitation filter. This can be accomplished by inserting filters in the cube, the excitation filter wheel, or the auxiliary positions in the epifluorescence module, depending on the arrangement of your microscope. We have also noticed that, over time, filters develop small imperfections and pinhole defects. For critical live-cell work, filters should be replaced after about 6 months of heavy use.

Kohler vs Critical Illumination

Normally, microscopes are aligned so that the light source is focused at the condenser aperture to produce an even and unfocused source of light in the specimen plane (Kohler illumination; see Ref. 17a). However, with small objects, such as yeast, in which the entire field does not need to be illuminated, it can be advantageous to focus the light source on the image plane (critical illumination). By using critical illumination, the partial confocal effect of the microscope

[19] R. Heim, D. C. Prasher, and R. Y. Tsien, *Proc. Natl. Acad. Sci. U.S.A.* **91,** 12501 (1994).
[20] R. Heim, A. B. Cubitt, and R. Y. Tsien, *Nature* **373,** 663 (1995).
[21] G. Patterson, R. N. Day, and D. Piston, *J. Cell Sci.* **114,** 837 (2001).

is maximized and the resolution in Z increased.[2] Critical illumination produces unacceptably uneven light with standard collector lenses, but works well if a fiber optic scrambler is used to direct the light into the microscope from the burner, as in a DeltaVision microscope.

Selecting Cover Glass

The cover glass is a component of the optical train of a microscope, and objectives are typically designed to work with a particular thickness of cover glass, usually number 1.5 (0.16–0.19 mm thick). For the most precise work, it is a good idea to measure the thickness of the cover glass and to use only those within ±0.02 mm of nominal dimension. A Mititoyo micrometer suitable for this purpose can be purchased inexpensively from www.mscdirect.com.

Temperature Control

For many live-cell microscope experiments, particularly those involving temperature-sensitive alleles, it is necessary to maintain cells at elevated temperatures. In our experience, it is necessary to heat both the agarose depression slide with a stage heater and the objective with an objective heater (Fig. 3). A commercial stage heater with resistive heating and a temperature controller can be purchased from Instec (www.instec.com). Alternatively, a simple stage heater can be fabricated by drilling channels in a brass block and then circulating warm water from a heated bath through the block. Because the objective and slide are coupled thermally via the immersion oil, the objective acts as a heat sink and causes unwanted local cooling. To avoid this, a heating element (available from Bioptechs Inc., www.bioptechs.com) is mounted to the objective and maintains the objective at the same temperature as the stage heater. The temperature of the sample is monitored using a subminiature RTD probe mounted on the slide adjacent to the sample (available from Omega, www.omega.com).

Minimizing Spherical Aberration through Oil Matching

Spherical aberration causes light from the sample to be focused at different positions depending on where the light passes through the objective (Fig. 4A). Thus, spherical aberration diminishes the quality of the image and causes a substantial loss of data. Spherical aberration can be minimized, however, through the choice of appropriate immersion oils. (These can be obtained in convenient kits or individual bottles with refractive values between 1.500 and 1.534 in increments of 0.002 from Applied Precision, Inc., www.api.com.) By changing the refractive index of the immersion oil, it is possible to correct for differences in the mounting media, the thickness of the cover glass, and, most importantly, the

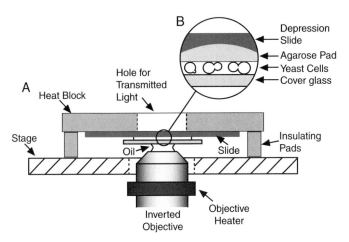

FIG. 3. Schematic of the temperature control apparatus. (A) Elevated temperatures are maintained by a heat block and objective heater. Yeast cells are first mounted on an agarose pad formed in a "depression slide" that has a small hemispherical depression. (B) The yeast cells are held tightly between the agarose and cover glass without damage. Although we have depicted an inverted microscope configuration, this system can work equally well with an upright microscope. The slide is held on the underside of a stage heater by metal clips. A small amount of immersion oil is added to the cover glass and objective. Both the slide and the heat block are then placed above the objective lens. Check that the objective is not adjusted too high because the weight of the heat block on the objective can damage the yeast cells or even crack the cover glass. The objective heater is adjusted to the same temperature as the stage heater and acts to prevent the objective from cooling the sample.

temperature. The choice of mounting oil is based on an analysis of the 3D PSF under different conditions. Image a point source, such as a fluorescent bead [0.1-μm beads in the TetraSpeck fluorescent sampler kit, Molecular Probes, Eugene, OR, www.molecularprobes.com, and mounted to poly(L-lysine)-coated slides] or a small bright object in a cell (we routinely use GFP-tagged spindle pole bodies) by acquiring a stack of 40 Z sections spaced 0.25 μm apart. To examine the PSF, rotate the image stack 90° so that the slices are viewed edge on and then move through the Z–X planes until the bead is in view (Fig. 4B). Compare the cone of light above and below the brightest point. A symmetrical pair of cones indicates optimal conditions, and asymmetric cones indicate the presence of a spherical aberration.[2,22] Adjust the immersion oil until the PSF is symmetric. Once optimal conditions have been determined, they should remain constant for a given objective, mounting medium, and temperature, but be prepared to make adjustments when a new batch of coverslips is used.

[22] Y. Hiraoka, J. W. Sedat, and D. A. Agard, *Biophys. J.* **57,** 325 (1990).

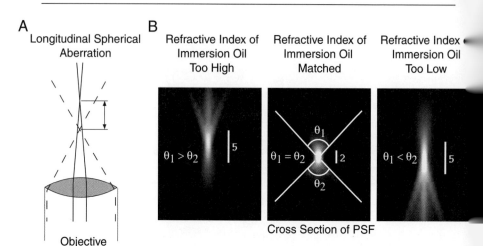

FIG. 4. The effects of spherical aberration on point spread function (PSF). (A) Spherical aberration occurs when the rays of light collected from the outer edges of the lens (dashed lines) and those collected from the central portions (solid lines) do not focus to the same longitudinal position. The amount of aberration is affected by refraction at the interfaces between the immersion oil and the cover glass and between the cover glass and the mounting media. It is important to match these indices. (B) Cross sections of PSFs (from subresolution beads) obtained with refractive oils having index values that are too high, matched, and too low. The angle or cone of light above and below the fluorescent bead is represented by θ_1 and θ_2, respectively. When the values of θ_1 and θ_2 are equal, the amount of spherical aberration is minimized. When the refractive index of the immersion oil is incorrectly matched to the conditions, the values of θ_1 and θ_2 are not equal, and the fluorescent signal is greatly lengthened (images kindly provided by Paul Goodwin of Applied Precision Inc.).

Acquiring Images

In live-cell microscopy, photobleaching and phototoxicity impose a trade-off between the exposure time of each image and the total number of images that can be acquired. Photobleaching, which is also a problem with fixed cell samples, is readily apparent as a reduction in signal strength.[23–25] Phototoxicity can be judged by mounting cells on agarose pads, exposing them to various amounts of light, and then following their growth over several generations. If the goal of the experiment is accurate observation over a limited time period, we use longer exposures and brighter illumination (the amount of illumination can be varied using neutral density filters or an adjustable intensity burner). However, if the goal

---

[23] A. B. Cubitt, R. Heim, S. R. Adams, A. E. Boyd, L. A. Gross, and R. Y. Tsien, *Trends Biochem. Sci.* **20,** 448 (1995).
[24] R. Swaminathan, C. P. Hoang, and A. S. Verkman, *Biophys. J.* **72,** 1900 (1997).
[25] G. H. Patterson, S. M. Knobel, W. D. Sharif, S. R. Kain, and D. W. Piston, *Biophys. J.* **73,** 2782 (1997).

is long-duration observation or high-temporal resolution, we reduce exposure time to the shortest period in which an interpretable image can be acquired.

Camera Settings

The speed, size, and resolution of the image must also be weighed when programming the camera. The fastest high-sensitivity CCD cameras are capable of 4–6 full frames per second but 1–2 per second is more typical. However, the frame rate can be increased by reducing the image size. Epifluorescence microscopes are usually equipped with a "megapixel" camera whose CCD contains a photosensitive array of about 1024×1024 pixels. Even with a 100× objective, a yeast cell fills only a small part of the image (about 128×128 pixels). The camera can be programmed to record data only from a subregion of the CCD, in some cases as small as 64×64 pixels, increasing the speed severalfold (to as much as 10–12 frames per second, although fixed delays in the camera hardware prevent a linear increase in speed as the frame size is reduced) and reducing the file size dramatically. This latter consideration is important, as a full-frame 3D movie can require as much as a gigabyte of hard disk storage.

A second way to increase frame rate and decrease file size is to use on-chip binning. During the acquisition period, each pixel in a CCD converts incident photons into photoelectrons and stores them in a "potential" well. During the subsequent readout period, the photoelectrons are shifted through the array until they reach a digitizer [an analog to digital converter (ADC)] where they are converted into a digital signal. Binning refers to a process whereby the CCD chip adds together the photoelectrons that have accumulated in several adjacent pixels prior to digitizing them. A binned image can be read out more quickly than an unbinned image because ADC speed is usually rate limiting in data acquisition and the binned image has a higher signal-to-noise ratio (see Roper Scientific at www.roperscientific.com for more detailed information). These advantages do, however, come at the cost of lower resolution.

The lower limit for a useful exposure is determined by the S/N ratio of the image. This noise has several sources but can be summarized as arising either from the CCD itself—read noise—or from random fluctuations in the number of photons that are counted—shot noise (dark current is negligible in cooled CCDs). Read noise is essentially constant with variations in signal intensity, whereas shot noise varies with the square root of the signal strength. Thus, very low intensity signals are read noise limited, and stronger signals are shot noise limited. Even with a perfect CCD, the S/N ratio obtained by counting 100 photoelectrons is about threefold better than with 10 photoelectrons. If CCD noise is $10e^-$ (a realistic value), then the stronger signal will have an S/N ratio at least eightfold higher. The goal is usually to work at an intensity range in which shot noise rather than read noise predominates. The image S/N ratio is then bound by physical and not electronic limitations.

Viewing and Printing the Image

In general, GFP-tagged yeast cells are not very bright and exposures are often in the lower region of the camera's dynamic range. Twelve-bit CCD cameras have 16 times the dynamic range of cathode ray tube (CRT) computer screens, and low-intensity images therefore appear to be very faint when viewed on-screen, even though they are well within the acceptable range for the CCD. Thus, it is very important that the digital image be adjusted correctly for viewing. If the signal on the CCD ranges from 8 to 128 digital units (DU), this will appear on the monitor as less than 3% of maximum intensity (e.g., virtually black). The range in the digital image, 8 to 128 in this case, must be adjusted to fit the monitor's gray-scale range of 0–256. Usually, this is accomplished via a histogram tool in the image acquisition software. The result is an on-screen image with a broad dynamic range. We mention this point because it is our observation that many inexperienced microscopists overexpose their images, usually on the basis of its initial appearance on a monitor. Obviously, the noise floor limits the extent with which low-intensity images can be enhanced.

A frequently encountered difficulty with multispectral images is producing a satisfactory printed copy for publication. In converting an image from an on-screen display to print, one must contend with the problem of gamut conversion. CRT screens are based on a red–green–blue (RGB) color space, whereas printed images conform to a cyan–magenta–yellow–black (CMYK) color space. The gamut conversion maps the RGB color values to CMYK color values, but not all RGB colors can be mapped onto CMYK color space. These colors are replaced by approximations, most of which appear very muddy. This is a particular problem with the bright primary colors typical of fluorescent images. Images also look quite different on Macintosh, PC, and Unix systems. This arises because the default brightness, or gamma, of monitors differs with platform but can be fixed by manually adjusting the gamma.

Gamut conversion from RGB to CMYK is a complex topic with many solutions (see "Color Management in Photoshop," www.adobe.com, for more information), but a simple approach is to export each wavelength in the microscope image as a separate gray-scale TIFF file. The individual TIFF files from one image are imported into a desktop publishing or image manipulation program, such as Adobe Photoshop (www.adobe.com), as individual layers, allowing them to be manipulated and colorized independently. The key to this process is to choose CMYK colors for colorization and not RGB colors. We find that bright aqua, dark green, and orange-red are usually the most satisfactory choices. Done correctly, gamut conversion generates an image that can be printed correctly. However, if the monitor is not calibrated to the color printer, the image that appears on screen will still differ from the image that is printed. This calibration is accomplished through a color management system (CMS, see www.color.org).

Example of Imaging GFP-Labeled Yeast

Our analysis of kinetochore function and chromosome dynamics provides an example of applying the methods described earlier.[14,26-30] Kinetochores are DNA–protein complexes that assemble on centromeric DNA and mediate the attachment of chromosomes to the microtubules of the mitotic spindle.[31] We study the recruitment of proteins to kinetochores by tagging them with GFP (Figs. 5A–5C).[32] To study the effects of inactivating kinetochore proteins on chromosome movement, mutant alleles are introduced into cells carrying the tetracycline repressor fused to GFP (TetR-GFP) and a tetracycline operator (TetO) array integrated at one location in the genome.[26,27] These cells also contain a GFP-tagged spindle pole protein so that both the chromosome and the spindle poles are marked by bright fluorescent dots (Figs. 5D and 5E).

To characterize a kinetochore protein, we first localize it in paraformaldehyde-fixed cells. One-step integration is used to create C-terminal fusions between the kinetochore protein and EGFP and between the spindle pole body (SPB) protein Spc42p and CFP.[32] Kinetochores can then be localized relative to the spindle axis. Imaging is performed on an Applied Precision DeltaVision microscope with a Nikon TE300 base, a 1.4 N.A. 100× objective, a Princeton Instrument's MicroMAX camera (Roper RTE/CCD-1300Y), and some of the filters described in Table I. Z slices are spaced by 0.15 μm and are acquired without binning from a 256 × 256 pixel region of the camera (Fig. 5A). Typical exposures are 1–2 sec. The raw image is then deconvolved using eight iterations. Under these conditions, kinetochore and spindle staining can be distinguished quite clearly (Figs. 5B and 5C).

Chromosome dynamics are routinely examined in live cells in which both chromosomes and SPBs are labeled with GFP. We acquire twelve to eighteen 50-msec 128 × 128 unbinned Z slices, separated by 0.20 μm. The extremely short exposures prevent cellular damage due to photobleaching, making it possible to collect as many as 100 3D data sets with as little as 3 sec between each set. Deconvolution is then performed (Fig. 5E), allowing the centroids of the spots to be determined accurately.[14,28] Unfortunately, live-cell analysis at two wavelengths is problematic with most wide-field microscopes. Switching the filter wheel from the first to the second excitation wavelength requires approximately 1 sec. Under optimum

---

[26] C. Michaelis, R. Ciosk, and K. Nasmyth, *Cell* **91**, 35 (1997).
[27] A. F. Straight, W. F. Marshall, J. W. Sedat, and A. W. Murray, *Science* **277**, 574 (1997).
[28] X. He, S. Asthana, and P. K. Sorger, *Cell* **101**, 763 (2000).
[29] G. Goshima and M. Yanagida, *Cell* **100**, 619 (2000).
[30] C. G. Pearson, P. S. Maddox, E. D. Salmon, and K. Bloom, *J. Cell Biol.* **152**, 1255 (2001).
[31] A. A. Hyman and P. K. Sorger, *Annu. Rev. Cell Dev. Biol.* **11**, 471 (1995).
[32] A. D. Donaldson and J. V. Kilmartin, *J. Cell Biol.* **132**, 887 (1996).

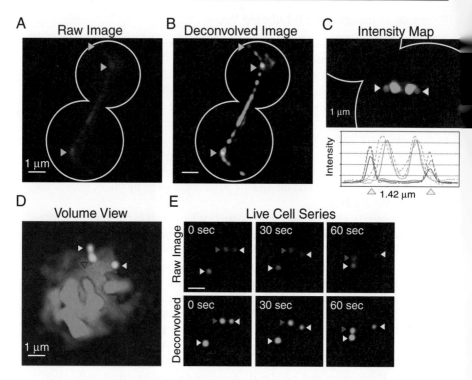

FIG. 5. Live and fixed cell examples of imaging kinetochore proteins. (A and B) Localization of the microtubule-binding protein, Stu2p (in red), and α-tubulin, Tub1p (in green). Both proteins were fused to variants of GFP. (A) Maximum intensity projection of an unprocessed image stack while (B) shows the same image after constrained iterative deconvolution (SoftWoRx; www.api.com). The original image was collected as a stack of twenty, 0.15-μm sections with 2.0-sec exposures for GFP and 2.5-sec exposures for CFP. Light blue arrows indicate the position of the kinetochores and cortical attachment site. (C) Image showing the localization of a kinetochore protein, Ndc80p (in green), and a spindle pole body protein, Spc42p (in red). The cell was fixed using paraformaldehyde and processed as described earlier. The graph shows the distributions of Ndc80p-GFP and Spc42p-CFP signal intensities along the spindle axis. (D) Image of a cell fixed in paraformaldehyde in which the green dots represent the position of the spindle pole bodies (yellow arrows) and a chromosome tagged (red arrow) with the TetO/TetR-GFP system. DNA is shown in blue. (E) Live-cell imaging of a yeast strain containing GFP tags similar to those shown in D. The top image series shows a maximum intensity projection of raw data whereas the bottom shows the same data set after iterative deconvolution. The original time series was collected as a stack of 16 sections, 0.25 μm thick; each section was exposed for 50 msec, and time points were taken every 30 sec.

conditions, if one switches wavelengths at each Z slice, overall acquisition time for a 15-slice image stack increases to 15 sec as compared to approximately 2–3 sec for one color. In our experiments, fluorescent structures move substantially in this time frame and colocalization of multicolor organelles is highly unreliable.[28] One way to reduce the acquisition time is to acquire the EGFP image stack first and the CFP stack subsequently. In this case, however, displacement is observed from one channel to the next and colocalization is very difficult. No easy solution to this problem exists for wide-field microscopy (in contrast, a confocal microscope acquires all channels simultaneously using multiple PMTs). Two recent developments are very fast solid-state wavelength switchers and methods to record two wavelengths simultaneously (see Optical Insights at www.optical-insights.com). At the moment, most of our live-cell analysis is performed with cells in which several structures are labeled with EGFP variants and we exploit differences in their time-dependent motions to discriminate among them.

Future Developments in High-Performance Microscopy

Over the next few years we can expect high-quality imaging software to improve and iterative deconvolution to become routine on inexpensive workstations. Cameras can increase about 30% in sensitivity before they hit a theoretical limit, but there is potential for increasing speed up to 10-fold and for significantly reducing read noise. One exciting development would be the commercial release of wide-field microscopes specialized for multiwavelength live-cell microscopy. These microscopes would be designed with digital imaging in mind and could collect images at two or three wavelengths simultaneously. Finally, we believe that the development of new image analysis software will be essential to unlock the full potential of microscope images (see: www.openmicroscopy.com). This software will be modular and highly configurable by the user. Using such software, it will be possible to analyze and track very dim objects with a precision and speed far greater than is possible using current manual methods.

Web Resources

> Chroma Filters, www.chroma.com: An excellent resource for information on microscope filter sets and GFP imaging.
> Molecular Expressions Website, www.microscopy.fsu.edu: A must-see site with a wide range of information on optical microscopy. Includes a large number of animations on basic microscope concepts.
> Nikon Microscopy U, www.microscopyu.com: A commercial site with several good animations.
> Omega Filters, www.omegafilters.com: One of the two major filter manufacturers. Contains information on filter spectra.
> Roper Scientific, www.roperscientific.com: Contains a good library of articles on cameras and their use.
> Zeiss, www.zeiss.com: The microscope section contains some helpful guides on microscope alignment and use.

Acknowledgments

The authors thank Dominic Hoepfner of the Phillipsen Laboratory for developing and sharing the agarose pad mounting technique; Paul Goodwin of Applied Precision for providing the PSF images; Jason Swedlow, Paul Goodwin, and Carl Brown for advice on deconvolution methods using the DeltaVision microscope; Gaudenz Danuser and Dominik Thomann for discussions on superresolution theory; and the members of the Sorger laboratory for critical review of this manuscript. This work was funded by GM51464 from the NIH and by the Searle Scholars Fund.

[3] Fluorescence Resonance Energy Transfer Using Color Variants of Green Fluorescent Protein

By DALE W. HAILEY, TRISHA N. DAVIS, and ERIC G. D. MULLER

Introduction

The recent discovery that the human genome encodes only five times the number of genes found in the yeast genome suggests that the fundamental cellular biology of a human cell is not that much more complex than the cell biology of a yeast cell. As a step toward a greater understanding of yeast cell biology, the Yeast Resource Center (YRC) at the University of Washington (http://depts.washington.edu/~yeastrc/) integrates four technologies to study the components, the function, the structure, and the dynamic modifications of protein complexes in yeast. Those technologies are mass spectrometry, two-hybrid analyses, optical sectioning microscopy, and protein structure prediction. The goal is to identify not only the constituents of protein complexes, but also to determine how protein complexes interact to form networks of interconnected pathways.

Protein Localization in Live Cells

The function of a protein is intimately associated with its location in the cell. Fluorescence microscopy of live cells expressing proteins fused to the green fluorescent protein (GFP) allows the visualization of proteins and protein complexes in their natural context.[1] In yeast, the replacement of a protein with a GFP–protein fusion is easy to accomplish.[2,3] If the target protein is essential for viability, the function of the GFP–protein fusion can be tested for its ability to support growth. We have tagged 17 essential genes and found that only one fusion protein could not support growth and two conferred slow growth. In addition, driving gene expression from the native promoter ensures that the GFP–protein fusion does not flood the

[1] R. Y. Tsien, *Annu. Rev. Biochem.* **67,** 509 (1998).
[2] B. Prein, K. Natter, and S. D. Kohlwein, *FEBS Lett.* **485,** 29 (2000).
[3] A. Wach, A. Brachat, C. Alberti-Segui, C. Rebischung, and P. Philippsen, *Yeast* **13,** 1065 (1997).

cell with more protein than the cell can properly assimilate. Given these simple precautions, localization of the GFP fusion will likely reflect the natural distribution of the protein in the cell and yield important information on protein function.

In principle, several proteins could be tagged and resolved simultaneously using the different color variants of GFP and dsRed.[1,4] However, we have found that proteins tagged with either dsRed or the blue fluorescent protein yield weak or undetectable fluorescent signals in yeast (unpublished results, 2000; see also Baird et al.[5]). Thus, in practice, only the two color variants cyan fluorescent protein (CFP) and yellow fluorescent protein (YFP)[1] are available for simultaneous expression and colocalization studies in yeast. A full description of the plasmids and methods for generating CFP and YFP fusions is described later. The reader is referred to several papers that provide excellent overviews on fluorescent microscopy in yeast using GFP.[6–8]

Fluorescence microscopy with modern hardware and software achieves spatial resolution close to the theoretical limits of optical resolution, 200–300 nm. Thus proteins within this distance of each other will colocalize. For example, the yeast spindle pole body, a structure that is embedded in the nuclear envelope and nucleates microtubules, is about 200 nm across.[9] The fluorescence intensity from any two proteins of the spindle pole body tagged with CFP and YFP coincide; the distance between them cannot be measured directly.

Localization to defined subcellular structures at a resolution of 200–300 nm is often sufficient to assign a broad function to many proteins. For example, mapping proteins to the spindle pole body implies a shared role in microtubule organization, whereas a bud tip localization implies a role in polarized growth. However, some localizations, such as the cytoplasm, are not very informative. In addition, to probe deeper into the mechanism of action of a protein complex requires a deeper understanding of molecular structure that only comes at higher spatial resolution.

Fluorescence Resonance Energy Transfer

Fluorescence resonance energy transfer (FRET) is the exchange of energy between two fluorescent molecules; one acting as a donor and the other acting as an acceptor. FRET occurs only if the acceptor is within a radial distance of 10 nm (100 Å) from the donor. According to theory proposed by Förster and verified experimentally in model systems (reviewed in Stryer[10]), the efficiency of

---

[4] A. F. Fradkov, Y. Chen, L. Ding, E. V. Barsova, M. V. Matz, and S. A. Lukyanov, *FEBS Lett.* **479**, 127 (2000).
[5] G. S. Baird, D. A. Zacharias, and R. Y. Tsien, *Proc. Natl. Acad. Sci. U.S.A.* **97**, 11984 (2000).
[6] S. L. Shaw, E. Yeh, K. Bloom, and E. D. Salmon, *Curr. Biol.* **7**, 701 (1997).
[7] J. L. Carminati and T. Stearns, *Methods Cell Biol.* **58**, 87 (1999).
[8] S. D. Kohlwein, *Microsc. Res. Tech.* **51**, 511 (2000).
[9] E. T. O'Toole, M. Winey, and J. R. McIntosh, *Mol. Biol. Cell.* **10**, 2017 (1999).
[10] L. Stryer, *Annu. Rev. Biochem.* **47**, 819 (1978).

energy transfer depends on the inverse sixth power of the distance between the donor and the acceptor. Thus the detection of FRET is a very sensitive measure of the proximity of two fluorophores. When applied to fluorescence microscopy, a FRET signal narrows the predicted distance between two fluorophores from within 200 nm to as little as 10 nm. In effect, FRET increases the spatial resolution of fluorescence microscopy by 20-fold.

FRET transfer efficiency depends on several factors in addition to distance. An orientation factor, K^2, expresses the dependence on the angular relationship between donor and acceptor transition moments. K^2 can have a value from 0 to 4 and could, in theory, have a large impact on FRET efficiency. However, in biological systems, K^2 is usually considered to be 2/3 because of the rotational freedom of the fluorophores.[1,10,11]

In addition to these geometric variables, FRET efficiency also depends on several spectroscopic variables. These include the spectral overlap of the emission spectrum of the donor with the excitation spectrum of the acceptor, the rate constant for fluorescence emission of the donor, the quantum yield of fluorescence of the donor in the absence of acceptor, and, to a minor extent, the refractive index of the medium.

For the CFP donor/YFP acceptor FRET pair, the distance at which the energy transfer efficiency is 50% was calculated to be 50 Å.[11] This value assumes a K^2 of 2/3. It should be noted that the shortest possible distance between CFP and YFP fluorophores is 30 Å, as the fluorophores are buried 15 Å beneath the surface of the proteins. Thus the efficiency of transfer can never be 100%. To enable FRET to occur at efficiencies greater than 25%, proteins tagged with CFP and YFP must not only be in contact, but the protein–protein interaction must bring CFP and YFP together to within 30 Å of each other, again assuming a K^2 of 2/3.

Practical Approach to FRET Microscopy in Yeast

This chapter is meant as a step-by-step introduction to the use of FRET to identify and localize interacting proteins in yeast using CFP and YFP protein fusions. Several reviews on FRET microscopy using GFP variants are excellent sources for further information.[11–14]

Equipment

High-performance microscopes are required for detecting FRET in yeast. The small size of yeast and the faint signals from the fluorescent proteins place exacting demands on the optics and image acquisition. We use a Zeiss Axiovert S-100 TV

[11] R. Heim, *Methods Enzymol.* **278,** 408 (1999).
[12] A. Periasamy and R. N. Day, *Methods Cell Biol.* **58,** 293 (1999).
[13] B. A. Pollok and R. Heim, *Trends Cell Biol.* **9,** 57 (1999).
[14] A. Miyawaki and R. Y. Tsien, *Methods Enzymol.* **327,** 472 (2000).

incorporated into the DeltaVision Restoration Microscopy System. Our primary objective is the 100× oil immersion Plan-Apochromat with a numerical aperture of 1.4. To check adjustments, calibrations, and sensitivity on our microscope, we employ InSpeck Green and FocalCheck fluorescent microbeads from Molecular Probes (Eugene, OR) as external standards.

Images are acquired on a cooled charge-coupled device (CCD) camera (Roper Scientific, Tucson, AZ; Quantix camera with a Kodak 1401 chip). Again the dim signals argue for a cooled CCD camera with low electronic noise, high quantum efficiency in the green region of the spectrum, and fast readout speeds. The binning features in a camera are also important. Binning combines charges from adjacent pixels in a CCD during readout. We routinely increase the signal-to-noise ratio by 2 × 2 binning, which also permits shorter exposure times and reduces fluorophore bleaching.

Our optical filter sets are from Omega Optical. For CFP, we use 440AF21 for excitation and 480AF30 for emission. For YFP we use 500AF25 for excitation and 545AF35 for emission. To observe FRET, the sample is excited with the 440AF21 excitation filter and emission is examined with the 545AF35 emission filter. We use a dual-pass dichroic mirror, 436-510DBDR. The dual-pass dichroic mirror is required if YFP fluorescence is measured during the FRET experiment, as suggested later. Optical filters are placed in filter wheels and rotated into position through the software that operates the microscope.

Several software packages are available for image analysis and manipulation. We use softWoRx from Applied Precision (Issaquah, WA), but MetaMorph (Universal Imaging Corp., Downingtown, PA), ISee (Inovision Corp., Raleigh, NC), and NIH Image are also commonly used for image processing.

Technological advances steadily improve the capabilities of fluorescence microscopy. Therefore, this description of equipment is meant only as a guide and not as a shopping list. In particular, improvements in the sensitivity and speed of cameras and the spectral properties of optical filters often call for upgrades of equipment.

FRET Ratio

Several methods to quantify FRET have been published (reviewed in Miyawaki and Tsien[14]). We have found that methods that rely on extensive mathematical corrections[15,16] can compound experimental error in ways that are difficult to rigorously control and derive. Numbers of nearly equal value are subtracted and divided, making the methods very sensitive to small errors in the correction factors.

We prefer a more conventional and straightforward measure of FRET, a FRET ratio, as an index to the extent of energy transfer.[14] The FRET ratio is simply the

---

[15] G. W. Gordon, G. Berry, X. H. Liang, B. Levine, and B. Herman, *Biophys. J.* **74,** 2702 (1998).
[16] M. Damelin and P. A. Silver, *Mol. Cell* **5,** 133 (2000).

ratio of the emission from the acceptor fluorophore divided by the emission from the donor fluorophore while the sample is excited at the excitation wavelengths of the donor. For the CFP and YFP pair, the FRET ratio is obtained by dividing the fluorescence intensities at 545 nm (YFP emission) by the fluorescence intensities at 480 nm (CFP emission) while the sample is excited at 440 nm (CFP excitation) using the excitation and emission filters described earlier. We will refer to the 545-nm emission upon 440 excitation as the FRET channel.

The FRET ratio takes advantage of two effects of FRET: increased acceptor emission and decreased donor emission. First, the fluorescence emission from YFP will increase as YFP is excited by the transfer of excitation energy from CFP. Therefore, FRET between CFP and YFP will increase the emission intensity detected at 545 nm. A FRET interaction also diminishes the signal from CFP. Without a FRET partner, an excited CFP molecule returns to its ground state, emitting photons with wavelengths that peak around 480 nm. With a FRET partner, the excited CFP molecule transfers a fraction of that energy to the YFP acceptor, depending on the efficiency of transfer. Thus, emission at 480 nm decreases. These two effects cause the numerator of the FRET ratio to increase and the denominator to decrease, driving the ratio up.

As described later, the FRET ratio is influenced by several factors in addition to energy transfer. Some CFP emission tails into the FRET channel. YFP is excited directly, although inefficiently, at the excitation wavelengths of CFP. Finally, changes in the relative amounts of CFP and YFP will alter the ratio. Confidence in the FRET ratio as a measure of FRET requires a careful assessment of the contributions of the spectral overlap and YFP/CFP stoichiometry.

Strain Construction

The first step is the construction of appropriate strains. For FRET analysis, the levels of expression of CFP and YFP fusions are best controlled by endogenous promoters. This avoids mislocalization of the fusion, fluctuations in protein levels that often accompany episomal expression, and the potential to drive artificial protein–protein interactions by mass action under conditions of overexpression. Several polymerase chain reaction (PCR) integration systems use homologous recombination to target a fluorescent protein open reading frame (ORF) to the 5' or 3' end of a gene and leave gene expression under the control of the native promoter. We have mutated plasmids designed by Wach et al.[3] and Prein et al.[2] so that C-terminal or N-terminal fusions can be made to CFP or YFP. CFP is F64L/S65T/Y66W/N146I/M153T/V163A/N164H relative to GFP[17] and YFP is S65G/V68L/Q69K/S72A/T203Y.[17] Plasmids used to generate the fusions are available from the YRC. The YRC has made over 50 functional fusions using these

---

[17] A. Miyawaki, O. Griesbeck, R. Heim, and R. Y. Tsien, *Proc. Natl. Acad. Sci. U.S.A.* **96**, 2135 (1999).

systems. We are evaluating other YFP mutants,[11] including the new citrine,[18] for improved spectral characteristics over the current YFP derivative.

Our protocol for making C-terminal fusions is available at our web site: http://depts.washington.edu/~yeastrc/. Noteworthy is our experience that 40 bp of homology is sufficient for recombination and insertion of the PCR product into the genome. Cost savings are also achieved by polyacrylamide gel electrophoresis (PAGE) purifying only the forward primer containing the homology to the 3' end of the ORF. The reverse primer, with homology to the 3' UTR, is just desalted. Our primers are synthesized by Integrated DNA Technologies, Inc. (Coralville, IA).

To make N-terminal fusions, a PCR cassette is targeted by homology to the 5' end of a gene.[2] Both primers are PAGE purified. Integration of this cassette at the 5' end inserts the selectable marker, G418 resistance, and the GFP variant between the target gene and its promoter. Because integration temporarily disrupts expression, a diploid strain is transformed and G418 selection isolates hemizygous cells carrying the cassette. Flanking the G418 resistance marker are a pair of loxP sites. A plasmid with creA recombinase under control of the galactose (*GAL1*) promoter is then transformed into these cells. Induction of creA on YP gal plates excises the G418 resistance marker to produce a CFP or YFP ORF fused to the 5' end of the gene of interest with the sequence of one 32-bp lox site added just upstream of the fluorescent protein ORF. The strain is then sporulated to yield a haploid strain in which the fusion protein has now replaced the wild-type protein.

FRET is exceptionally sensitive to the distance between CFP and YFP. In the absence of detailed knowledge of the structure of the proteins, one cannot predict *a priori* which fusions will better position the fluorophores for FRET. Neither can one predict which fusion will be tolerated by the protein and remain functional. Optimal fusion must be determined empirically for each protein.

Interpreting the FRET ratio requires several careful controls. For a positive control, a protein of interest fused to a tandem CFP–YFP should give a strong FRET signal. We have designed a plasmid, pDH18, that contains a tandem YFP and CFP ORF separated by a glycine–alanine linker. This plasmid is based on the Wach *et al.*[3] plasmids and has the *Schizoseccharomyces pombe his5* gene as a selectable marker. We used pDH18 as a PCR template to construct a positive control strain, DHY138, that expresses the tandem YFP–CFP fused to the histone H4 protein Hhf2p. DHY138 is a heterozygous diploid strain with the genotype of HHF2::YFP–CFP/HHF2. Fluorescent protein fusions to Hhf2p are bright and easy to quantify.

Two negative control strains are needed to characterize the spectral overlap between excitation and emission spectra of CFP and YFP. One strain contains a CFP fusion to the donor protein and the other strain contains a YFP fusion to

[18] A. A. Heikal, S. T. Hess, G. S. Baird, R. Y. Tsien, and W. W. Webb, *Proc. Natl. Acad. Sci. U.S.A.* **97**, 11996 (2000).

the acceptor protein. The strains are isogenic except for the genes encoding the recombinant fusions. These strains serve two purposes. First, because the fusions are expressed independently, the intensity values in the FRET channel represent a negative control in which the signal is derived solely from the spectral overlap and the limitations of the optical filter sets. A second purpose evaluates the relative signal intensities from the CFP and YFP fusions when the fluorophores are excited at their optimum wavelengths. From these values, one assesses the fluctuation in intensity values from cell to cell for each fusion and the comparative amount of each fusion. If the intensity values fluctuate greatly or if the YFP signal is greater than ~5-fold brighter than the CFP signal, the evaluation of FRET will be severely compromised. For our general purpose negative controls, we use two heterozygous diploid strains, DHY57 and DHY199, containing HHF2-YFP and HHF2-CFP, respectively. Both have a wild-type copy of HHF2.

Finally, a set of experimental strains is designed. Each strain carries a donor CFP fusion to one protein and an acceptor YFP fusion to a protein suspected to closely interact. We tested for contact between the C-terminal end of Hhf2p and the C-terminal end of an adjacent Hhf2p molecule in the histone complex by constructing a diploid strain containing Hhf2p-YFP and Hhf2p-CFP in the strain DHY137 (HHF2-CFP/HHF2-YFP).

Sample Preparation

Imaging of live cells demands that all manipulations are gentle and do not disrupt cell physiology. We grow cells overnight on YPD plates[19] and, in cases where cells carry *ade2* and *ADE3*, we add 0.2 ml of 5 mg/ml adenine to the plates before streaking the strain. The cells have uniformly low autofluorescence and are growing exponentially as determined by the budding index. The cells are carefully scraped from the plate and suspended in 300 μl of SD complete medium.[19] In this manner, a high density suspension of cells in log phase is obtained without concentrating the cells by centrifugation.

To make slides, SD complete medium containing 1.2% agarose (SeaKem LE agarose, BioWhittaker Molecular Applications, Rockland, ME) is melted and poured into the 0.5-mm depression in concavity slides (PGC Scientifics Corp., Frederick, MD). The depression holds roughly 250 μl. A second flat slide is placed over the filled depression and the agar is cooled. Once cooled, the top slide is slid off, leaving a depression filled with a bed of solid SD complete agar at an appropriate temperature. The concavity slides are cleaned and reused. For more routine work, a thin-film 60-nm-thick agarose medium is created on a microscope slide by sandwiching the agarose between two slides on which Scotch tape is

---

[19] F. Sherman, G. R. Fink, and J. B. Hicks, "Methods in Yeast Genetics." Cold Spring Harbor Laboratory Press, Cold Spring Harbor, NY, 1986.

applied across both ends of one side. Other techniques include replacing the agarose with 25% (w/v) gelatin and including Oxyrase (Oxyrase, Inc., Mansfield, OH) in the medium to deplete oxygen and reduce bleaching.[6]

An aliquot (10 μl) of cells is placed onto the agarose bed, and a coverslip is pressed down over the cells. For long observation times, the coverslip is sealed at the edges with nail polish to prevent drying of the agarose. An alternative to nail polish is valap (lanolin : petroleum jelly : paraffin, 1 : 1 : 1 w/v). The cells lie in one focal plane at the coverslip surface and are embedded in the agar. The agarose pad supports the cells nutritionally.

As an alternative to live cell imaging, cells can be fixed with formaldehyde or paraformaldehyde. Fixation results in decreased signals from the fluorescent proteins, but effective antifade agents can be added to prevent photobleaching. We add formaldehyde to a cell suspension (prepared as described earlier) to a final concentration of 3.7%. This solution is placed in a 30° rotator for 15–20 min. Formaldehyde-fixed cells are adhered to polylysine-coated coverslips, and glycerol with an antifade agent such as Citifluor (Ted Pella Inc., Redding, CA) is used as a mounting medium.

Optimization of Imaging Parameters

We strongly recommend starting with a positive control and optimizing the imaging technique to maximize the FRET ratio. Limiting exposure time is crucial for detecting FRET signals. The YFP acceptor is very sensitive to bleaching by exposure to the 440-nm light used for CFP excitation, with a half-life of approximately 600 msec in our experiments (Fig. 1). Bleaching of CFP is not as rapid, but still occurs with a half-life of about 3 sec (Fig. 1). Clearly limiting exposure of the sample to 440-nm light will assist in FRET detection. Even the strong FRET ratio from the Hhf2p–YFP–CFP tandem fusion is barely detectable after 5 sec of total exposure (1 sec at 500 nm and 4 sec at 440 nm, Fig. 2). We do the following to protect YFP from bleaching:

1. Use transmitted light for focusing. Light intensity from a transmitted light source (typically a halogen bulb) is negligible compared to the light intensity generated by a mercury lamp. We have not observed effects on FRET signals from exposure to transmitted light.
2. Take very short YFP exposures if additional focusing is necessary. We bin our camera either 2 × 2 or 3 × 3 and take 100-msec exposures if focusing the fluorescent signal is necessary. The FRET signal is bleached slowly by YFP excitation (ex) light (500 nm) compared to the 440-nm exposure. Therefore, the YFP filter set ($ex_{500}|em_{545}$) is used for additional fine focusing. The optimal focal plane is found by raising and lowering the stage in small increments (<400 nm) and locating the position where the pixel values from the YFP emission (em) signal are maximized.

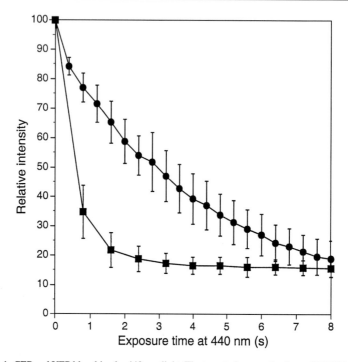

FIG. 1. CFP and YFP bleaching by 440-nm light. The two strains examined were DHY57 (HHF2–YFP/HHF2) and DHY199 (HHF2–CFP/HHF2). The strains were grown and imaged as described in the text. Each curve summarizes data collected from four different cells. Data were sampled by summing total intensity inside a 5 × 5 pixel square. Background was determined from a 5 × 5 pixel square inside each cell but away from the histone signal. ●, CFP bleaching at 440 nm, repeated 400-msec CFP exposures were captured. ■, YFP bleaching at 440 nm. To mimic the conditions during a FRET experiment, we repeatedly exposed the sample to a 200-msec YFP exposure followed by an 800-msec CFP exposure. Bleaching of the YFP during YFP exposures is negligible. Error bars show the standard deviation.

3. Take short exposures during data acquisition. Exposure times are minimized when acquiring data. Many CCD cameras generate 12-bit files with 4096 gray levels. We do not extend the exposure time to make use of the full range because the FRET signal is destroyed within 5 sec of exposure.
4. Take the FRET image before the CFP image. The order of data acquisition affects the strength of a measured FRET signal. The YFP image ($ex_{500}|em_{545}$) is taken first, the FRET image ($ex_{440}|em_{545}$) is taken second, and the image of CFP emission ($ex_{440}|em_{480}$) third to avoid bleaching the YFP. Typically, only minor bleaching of the CFP occurs while the FRET image is taken (Fig. 1).

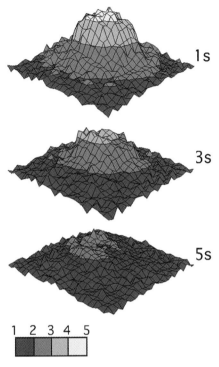

FIG. 2. Rapid bleaching of the FRET signal during sequential image capture. Strain DHY138 (HHF2::YFP–CFP/HHF2) was grown and imaged as described in the text. Six sequential time points were taken. At each time point, a 200-msec YFP channel, a 400-msec FRET channel, and a 400-msec CFP channel exposure were captured. Each FRET channel image was divided by the CFP channel image to generate six sequential FRET ratio images. Pixel values in the ratio images are the FRET ratios at each position in the image. A 25 × 25 pixel square centered on the histone signal was sampled. Data from the first, third, and fifth ratio images are shown. The numerical range of the FRET ratios in the sampled region is indicated by the gray scale. Elapsed total exposure time is indicated next to the plots.

Using all of these techniques to maximize the FRET signal, the Hhf2p–YFP–CFP tandem positive control strain generates a mean FRET ratio of 3.55 with a standard deviation of 0.20 (see "Data Analysis"). This ratio will vary depending on the particular optics and camera used for detection.

Spectral Overlap

Some fluorescence from CFP reaches into the FRET channel, as the broad emission spectrum of CFP extends beyond 545 nm.[14] Thus even without an acceptor YFP present, the FRET ratio has a value that reflects the amount of fluorescence from CFP that spills over into the FRET channel. In addition, the excitation spectrum of YFP extends below 440 nm into the excitation wavelengths of CFP.[14] Fluorescence from YFP that comes from direct excitation of YFP also appears in the FRET channel. Negative control strains identify the contribution of the spectral overlap to the signal in the FRET channel.

Overlap from CFP

The CFP alone strain is imaged to determine the FRET ratio$_{CFP}$ (Fig. 3). The signal intensity in the FRET channel (ex$_{440}$|em$_{545}$) is divided by the signal intensity in the CFP channel (ex$_{440}$|em$_{480}$). For HHF2–CFP/HHF2 diploid strain DHY199, we observe a mean ratio of 1.22 with a standard deviation of 0.072. This ratio will differ from system to system, as it depends in part on the optical filter sets and the quantum efficiency of the CCD camera. Note that CFP does not contribute any fluorescence to the YFP channel (Fig. 3).

Overlap from YFP

The amount of fluorescence in the FRET channel for a given amount of YFP (the YFP overlap factor) is characterized by imaging the YFP–protein fusion strain that does not express the CFP fusion (Fig. 3). Two images are compared: ex$_{500}$|em$_{545}$ (YFP channel) and ex$_{440}$|em$_{545}$ (FRET channel). The YFP overlap factor is the amount of fluorescence in the FRET channel divided by the amount in the YFP channel. Using HHF2–YFP/HHF2 diploid strain DHY57, the YFP overlap factor is 0.26 with a standard deviation of 0.020. This factor is a function of the relative FRET and YFP exposure times. Therefore, the exposure times must remain constant throughout the experiments for this factor to retain significance. Also note that YFP does not contribute any fluorescence to the CFP channel (Fig. 3).

The Experiment

1. Grow experimental and control strains overnight as described under "Sample Preparation."
2. Mount cells on slides as described under "Sample Preparation."
3. Focus with transmitted light as described under "Optimizing Imaging Parameters."
4. Take three images of each sample in the following order: (a) YFP filter set (ex$_{500}$|em$_{545}$), (b) FRET filter set (ex$_{440}$|em$_{545}$), and (c) CFP filter set (ex$_{440}$|em$_{480}$). The parameters for image capture will depend on the strength of the signal above noise and the pattern of intracellular localization. For Hhf2p fusions, the exposure times were 200 msec for the YFP channel, 400 msec for the FRET channel, and 400 msec for the CFP channel. The camera was binned 2 × 2. Images obtained for the positive control strain and the experimental strain are shown in Fig. 3.

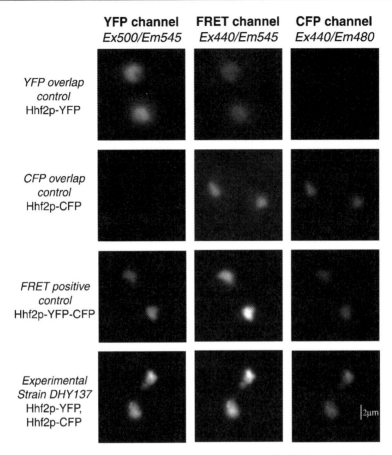

FIG. 3. Images of negative and positive control strains compared with the experimental strain. Strains DHY57 (YFP overlap control), DHY199 (CFP overlap control), and DHY138 (positive control) were compared to DHY137 (HHF2–CFP/HHF2–YFP) (brief strain descriptions in the text). Samples were prepared for microscopy as described in the text. Three images of each strain were captured in this order: a 200-msec YFP channel exposure, a 400-msec FRET channel exposure, and a 400-msec CFP channel exposure. Spectral overlap from both CFP and YFP is observed in the FRET channel. Overlap from CFP into the YFP channel and overlap from YFP into the CFP channel is not observed. DHY57 images are scaled from 100 to 2500. DHY199 images are scaled from 100 to 900. Images from DHY137 and DHY138 are scaled 100–2000. Both YFP and CFP fusion proteins are functional, and the fluorescent protein fusions to Hhf2p do not obviously affect the localization of Hhf2p.

Data Analysis

Quantification of Signal Intensities

Analysis of the image depends on the pattern of intracellular localization. For histones, total intensity is summed from a region in the center of the histone signal. We sampled a 700 × 700-nm square (a 5 × 5 pixel square, 2 × 2 camera

TABLE I
FRET RATIOS FOR POSITIVE CONTROL STRAIN AND EXPERIMENTAL STRAIN[a]

Cell	YFP channel	FRET channel	CFP channel	YFP/CFP	FRET ratio
Positive control: Hhf2p–YFP–CFP/Hhf2p					
1	20781	40326	11615	1.79	3.47
2	15174	32506	9791	1.55	3.32
3	12723	27948	7230	1.76	3.87
4	13965	26629	7303	1.91	3.65
5	16752	33622	9531	1.76	3.53
6	12172	23268	6842	1.78	3.40
7	23523	42753	11183	2.10	3.82
8	16247	29272	8821	1.84	3.32
Mean				1.81	3.55
Standard deviation				0.15	0.20
Experimental: Hhf2p–YFP/Hhf2p–CFP					
1	33649	42030	23355	1.44	1.80
2	28993	38653	22821	1.27	1.69
3	30814	34492	17648	1.75	1.95
4	29355	37999	21875	1.34	1.74
5	22720	27367	15853	1.43	1.73
6	22661	29201	17293	1.31	1.69
7	19662	27871	15990	1.23	1.74
8	18493	23409	14043	1.32	1.67
Mean				1.39	1.75
Standard deviation				0.16	0.084

[a] Images were captured as described in Fig. 3.

binning). An example of the data acquired is presented in Table I. Topographical representations of FRET ratios (Figs. 2 and 4) show a 25 × 25 pixel square centered on the histone signal.

Data Presentation

As a first step in the analysis, the easiest method to display the FRET ratio is to divide the intensity values in the FRET channel by the intensity values in the CFP channel across the image. This manipulation is performed by the image analysis software to generate a three-dimensional topographical display (Fig. 4). The view of the display is rotated to emphasize the value of the FRET ratio at different positions in the image. Different perspectives visually portray the FRET ratios across the image. In Fig. 4, the positive control (HHF2–YFP–CFP/HHF2) shows FRET ratios rising from the plane, while the experimental sample (HHF2–YFP/HHF2–CFP) remains relatively flat. The image reveals significant energy transfer in the positive control, but none or very little from the experimental sample. The background noise was not subtracted because background subtracted values for the CFP channel approach zero in regions void of CFP and cause the FRET ratio around the edges of the image to fluctuate greatly.

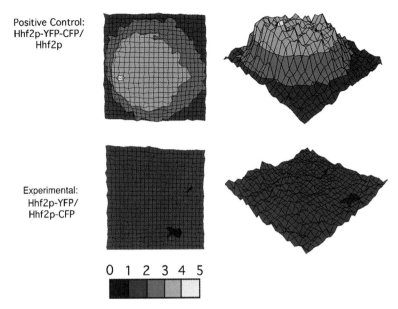

FIG. 4. Graphic displays of FRET ratios in sampled regions. Images of DHY137 and DHY138 were captured as described in Fig. 3. Ratio images were generated as described in Fig. 2. A 25 × 25 pixel square centered on the histone signals in the FRET ratio images was sampled. The left-hand side shows data in the same orientation as the raw images. Squares representing pixels are shaded gray to indicate the FRET ratio at each position. The images on the right-hand side are rotations of the left panels.

A quantitative analysis requires extracting intensity values from a number of images and calculating a mean FRET ratio. Table I shows data from eight cells of the positive control expressing the tandem fusion of Hhf2p–YFP–CFP. For these values, a background value, determined for each cell, has been subtracted. The technique used to determine background values for an image may vary depending on the nature and distribution of the fluorescent signal. For Hhf2p fusions, intensity values from an area of equal size to the sampled histone signal, i.e., a 25 pixel square, were summed. The area was inside the cell but away from the histone signal. Data for YFP, FRET, and CFP channels were analyzed to yield a mean FRET ratio of 3.55 with a standard deviation of 0.20. The mean YFP/CFP ratio was 1.81 with a standard deviation of 0.15.

Table I also shows corresponding data for the experimental strain expressing Hhf2p–YFP and Hhf2p–CFP. Absolute values for YFP and CFP channels are higher than the values from the positive control, as there is no wild-type Hhf2p present to diminish the signal in the experimental strain. In the experimental strain, the calculated mean FRET ratio was 1.75 with a standard deviation of 0.084. The mean YFP/CFP ratio was 1.39 with a standard deviation of 0.16.

Data Interpretation

The final step analyzes data to determine whether the FRET ratio obtained from the experimental strain reflects energy transfer between CFP and YFP. In the absence of energy transfer, a mixture of CFP and YFP proteins will give a signal in the FRET channel. The intensity of that signal depends in part on the amount of CFP and YFP present. Because that could vary for each donor and acceptor pair, it must be taken into account for each experiment. The following formula calculates the ratio expected for a given donor and acceptor if no energy transfer is occurring:

$$\text{FRET ratio}_{\text{baseline}} = \text{FRET ratio}_{\text{CFP}} + \text{YFP overlap factor} \times \frac{\text{YFP channel}}{\text{CFP channel}}$$

The FRET ratio$_{\text{CFP}}$ accounts for the contribution from the CFP fusion protein and is determined by imaging the strain carrying the CFP fusion alone as described earlier. The YFP overlap factor accounts for the contribution from the YFP fusion protein and is determined by imaging the strain carrying the YFP fusion protein alone (see "Spectral Overlap"). The FRET ratio$_{\text{CFP}}$ was 1.22 and the YFP overlap factor was 0.26 for our microscope system and controls. The ratio of the YFP channel/CFP channel is determined from the strains containing both YFP and CFP fusion proteins.

Energy transfer is clearly indicated between CFP and YFP in the positive control. Given the YFP/CFP ratio of 1.81 (Table I), the calculated FRET ratio$_{\text{baseline}}$ is 1.67 with a standard deviation of 0.011. For Hhf2p–YFP–CFP-labeled histones, the observed FRET ratio of 3.55 with a standard deviation of 0.20 is significantly higher (Table I). Under ideal *in vitro* conditions, purified CFP–YFP pairs give a three- to four-fold increase in the FRET ratio when the individual proteins are linked by a 25 amino acid linker.[11] The twofold increase in the observed FRET ratio of the Hhf2p–YFP–CFP over the calculated FRET ratio$_{\text{baseline}}$ is lower but still easily detectable. The short Gly–Ala linker between YFP and CFP in the Hhf2p fusion may impose conformational constraints that limit FRET. However, a YFP–CFP tandem Hhf2p fusion with a 4× (Gly–Ala) linker between the YFP and CFP did not significantly change the FRET ratio.

Based on the crystal structure of the nucleosome from *Xenopus*,[20] energy transfer is predicted between Hhf2p–CFP and Hhf2p–YFP in the experimental strain. Histones 1 through 4 are assembled into an octamer that forms the core of the nucleosome. Each octamer contains a pair of each of the four histones. The C-terminal ends of the two histone H4 proteins are 30 Å apart in the solved structure. This distance would allow energy transfer between CFP and YFP.

The mean observed FRET ratio for the Hhf2p–YFP/Hhf2p–CFP complex was 1.75 with a standard deviation of 0.084 (Table I). The calculated FRET ratio$_{\text{baseline}}$ was 1.58 with a standard deviation of 0.11. The 10% increase in the observed FRET ratio suggests energy transfer between CFP and YFP. The increase is less

[20] K. Luger, A. W. Mader, R. K. Richmond, D. F. Sargent, and T. J. Richmond, *Nature* **389**, 251 (1997).

than might be predicted from the proximity of the two histone H4 molecules in the crystal structure, but as described later, several factors could contribute to a lower than expected value. For comparison, FRET ratio increases of 4, 10, and 250% were used to predict protein–protein interactions in other published work.[16,21,22]

Several factors can decrease the FRET signal for proteins that interact. First, although the proteins may be close, orientation of the CFP and YFP may not be optimal for FRET. Second, when examining the pairing of a protein with itself, random mixing will dilute the signal by half because a tagged protein cannot FRET with a protein tagged similarly. In the case of Hhf2p, the experimental strain is a diploid in which one copy of HHF2 is tagged with CFP and one copy is tagged with YFP. Thus, statistically the maximum theoretical FRET signal will be reduced by half simply by the random process of selection of histones into the nucleosome. Finally, the FRET ratio can be depressed if untagged proteins can substitute for the tagged proteins and dilute the amount of tagged proteins in the complex. In the case of Hhf2p, the FRET signal will be reduced by the presence of Hhf1p. Hhf1p is a histone H4 protein that is identical to Hhf2p. *HHF1* expression is five- to seven-fold less than *HHF2*,[23] so the reduction in FRET should be minor. These points indicate that a lack of FRET between proteins does not argue against interaction; however, the observation of FRET between two tagged proteins demonstrates their close proximity.

Conclusion

Live cell FRET detection in yeast is in its infancy, and there are significant complications with the CFP and YFP pair. Spectral overlap between excitation and emission spectra complicates analysis. The extreme sensitivity of the current YFP to bleaching constrains image acquisition. Detection of a FRET signal is limited by background cellular autofluorescence and the relatively weak fluorescent signal intensities. Dynamic intracellular conditions, such as changes in pH or protein concentrations, can complicate the interpretation of experimental data. However, with careful controls, FRET is a powerful indication of protein–protein interaction. In instances where interactions give robust FRET signals, FRET is a valuable tool used to study the dynamic spatial and temporal behavior of protein–protein interactions in living cells. Future improvements in the spectral properties of CFP and YFP will increase the general applicability of FRET to study a broad range of protein–protein interactions in yeast.

Acknowledgment

This work was supported by a grant from the National Center for Research Resources of the National Institutes of Health, PHS P41 RR11823.

[21] R. N. Day, *Mol. Endocrinol.* **12**, 1410 (1998).
[22] J. Llopis, S. Westin, M. Ricote, Z. Wang, C. Y. Cho, R. Kurokawa, T. M. Mullen, D. W. Rose, M. G. Rosenfeld, R. Y. Tsien, C. K. Glass, and J. Wang, *Proc. Natl. Acad. Sci. U.S.A.* **97**, 4363 (2000).
[23] S. L. Cross and M. M. Smith, *Mol. Cell Biol.* **8**, 945 (1988).

[4] Immunoelectron Microscopy of Aldehyde-Fixed Yeast Cells

By JON MULHOLLAND and DAVID BOTSTEIN

Introduction

Immunolocalization of antigens in cells requires simultaneous preservation of structure and antigenicity. The difficulties of immunolocalization methods are centered around the methods used to preserve structure, which as a rule tend to degrade the ability of antigens (usually proteins) to be recognized by their cognate antibodies. The value of immunolocalization for cell biology was early recognized in the case of yeast at the level of the light microscope: immunolocalizations and, even more important, colocalizations using reagents labeled differentially with different fluorophores were developed and came into general use in the early 1980s.[1–4] However, application of immunolocalization in electron microscopy of yeast lagged behind, largely because of the difficulty of adequately preserving both structure and antigenicity with the sample preparation methods then in general use.

The advent of useful immunoelectron microscopy (immuno-EM) for yeast was the realization by van Tuinen and Riezman[5] that sodium metaperiodate ($NaIO_4$) can be used to facilitate the infiltration of resin into intact yeast cells (i.e., cells with their cell walls in place). Robin Wright (working in Jasper Rine's laboratory) used this method with a variety of new acrylic resins just coming into use and found that one of them, L.R. White, gives particularly good results with metaperiodate-treated yeast cells.[6] We adopted Wright's technique and, over the last decade, have optimized and extended it so that we can now routinely preserve structures and still see most antigens, with a frequency of successful localization comparable to that reported for other cell types, including mammalian cells.[7–11] The critical element

[1] M. N. Hall, L. Hereford, and I. Herskowitz, *Cell* **36**, 1057 (1984).
[2] A. E. Adams and J. Pringle, *J. Cell Biol.* **98**, 934 (1984).
[3] J. Kilmartin and A. E. Adams, *J. Cell Biol.* **98**, 922 (1984).
[4] P. Novick and D. Botstein, *Cell* **40**, 405 (1985).
[5] E. van Tuinen and H. Riezman, *J. Histochem. Cytochem.* **35**, 327 (1987).
[6] R. Wright and J. Rine, in "Methods in Cell Biology" (A. Tartakoff, ed.), Vol. 31, p. 473. Academic Press, New York, 1989.
[7] D. Preuss, J. Mulholland, C. A. Kaiser, P. Orlean, C. Albright, M. D. Rose, P. W. Robbins, and D. Botstein, *Yeast* **7**, 891 (1991).
[8] D. Preuss, J. Mulholland, A. Franzusoff, N. Segev, and D. Botstein, *Mol. Biol. Cell* **3**, 789 (1992).
[9] J. Mulholland, D. Preuss, A. Moon, A. Wong, D. Drubin, and D. Botstein, *J. Cell Biol.* **125**, 381 (1994).
[10] J. Mulholland, A. Wesp, H. Riezman, and D. Botstein, *Mol. Biol. Cell* **8**, 1481 (1997).
[11] J. Mulholland, J. Konopka, B. Singer-Kruger, M. Zerial, and D. Botstein, *Mol. Biol. Cell* **10**, 799 (1999).

that we found necessary to preserve the larger structures and organelles was the maintenance of a suitable osmotic environment during fixation, which we accomplish with the addition of slightly submolar concentrations of sorbitol.

As suggested earlier, the ability to colocalize different antigens on the same specimens is a very important technology for modern cell biology. For the case of immuno-EM, we have reduced to practice two general approaches that can accomplish such colocalizations. One of these exploits the possibility of using antibodies derived from different animal species (e.g., rabbit and guinea pig) as primary recognition reagents; these can then be visualized using secondary antibodies, differentially labeled with different sizes of gold particles, that specifically recognize the antibodies of the two animal species. The other method exploits the ability, in serial sections, of capturing the opposing faces of adjacent sections from the microtome and labeling them differentially, followed by analysis of separate and merged images. This allows the use of primary antibodies raised in a single organism for colocalization at the level of ultrastructure.

In what follows, we specify the current methods we use in detail; we discuss only those issues that relate directly to yeast. It should be mentioned that we do not cover cryopreservation methods, which are given in the chapter by McDonald and Müller-Reichert[12] elsewhere in this volume. We see these methods as complementary and distinguished in considerable part by the reality that the chemical methods use equipment and materials generally found in electron microscopy laboratories, whereas the cryopreservation methods require special, often expensive, equipment as well as expertise. Thus for many routine studies, chemical methods are both adequate and accessible. The literature on chemical methods for immuno-EM in general is extensive and has been reviewed excellently and thoroughly by Griffiths.[13]

Overview and Rationale for Chemical Fixation and Processing of Yeast Cells for Immuno-EM

Culture Density and Media

For most immuno-EM studies of yeast, the culture density (OD_{600}) should be equal to or less than 0.5 at the time of fixation. There are several reasons for this recommendation. First, at this density the morphology and molecular content of organelles and structures will be those of actively growing cells; in rich media (e.g., YEPD) at the relatively low OD_{600} of 0.8, yeast cells are slowing their exponential growth rate and already have significantly altered their gene expression patterns.[14]

---

[12] K. McDonald and T. Müller-Reichert, *Methods Enzymol.* **351**, [6], 2002 (this volume).
[13] G. Griffiths, "Fine Structure Immunocytochemistry." Springer-Verlag, Berlin, 1993.
[14] J. DeRisi, *Science* **278**, 680 (1997).

Second, the yeast cell wall becomes less porous as the culture density increases,[15] making cells more refractory to processing for EM. This is particularly important because our method, when applied to exponentially growing cells, does not require removal of the cell wall. To ensure that cells are really growing exponentially, it is recommended that cultures have been maintained at a low density (<0.5) for three or more generations.

Aldehyde Fixation

To provide adequate fixation with minimal loss of antigenicity, immuno-EM methods generally employ a mixture of formaldehyde and glutaraldehyde. Both aldehydes are uncharged and can diffuse rapidly into cells and both react with the amino groups of proteins. However, the monoaldehyde formaldehyde produces many fewer cross-links than the dialdehyde glutaraldehyde. The reaction of aldehydes with amines also produces protons and thus a drop in intracellular pH is expected during fixation. Neither aldehyde reacts significantly with carbohydrates or lipids, and any retention of these molecules in the fixed cell is through cross-linking of the associated or surrounding protein components. The rationale behind using both aldehydes is based on the reality that the extensive cross-linking that would occur if adequate concentrations of glutaraldehyde were used would severely limit the immunoreactivity of many antigens. For this reason, glutaraldehyde is used in relatively low concentrations in the presence of ca. 10-fold higher concentrations of formaldehyde that serve to complete fixation with a minimal loss of antigenicity. We recommend starting with a glutaraldehyde concentration of 0.4% together with 4% formaldehyde. If this regime fails to provide an expected level of signal, decreasing the glutaraldehyde concentration to 0.2 or 0.1% should be tried. This tactic is limited by the consequence that lower concentrations of glutaraldehyde may result in displacement or complete loss of antigen as well as insufficient preservation of morphology.

The chemistry of aldehyde fixation is extremely complex and therefore the use of aldehydes in EM protocols has, for the most part, been determined empirically. However, there are a few facts about the chemistry of formaldehyde relative to fixation that are worth pointing out. At low concentrations (less than 2%), formaldehyde exists predominately as monomeric methylene glycol. At higher concentrations (10% or more), formaldehyde exists predominately as methylene glycol polymers. Large methylene glycol polymers form polymethylene cross bridges more quickly and more stably than monomeric methylene glycol,[16] thus making it the preferred reagent for fixation. When diluted from high concentrations (10–37%), formaldehyde initially retains a high proportion of these large

---

[15] J. De Nobel and J. Barnett, *Yeast* **7**, 313 (1991).
[16] T. Johnson, *J. Electron Microsc. Tech.* **2**, 129 (1985).

polymers. This argues for preparing formaldehyde for fixation from concentrated stocks. Concentrated formaldehyde can be purchased as a 37% methanol-stabilized stock (100% formalin) or at lower, methanol-free, concentrations of 10 or 16%. Good results can be obtained with commercial stocks of 10 or 16%, but methanol containing formaldehyde is not recommended for histochemical studies. We make fresh formaldehyde stocks (30 to 35%) from paraformaldehyde (the solid, polymerized form of formaldehyde). We have found that freshly prepared formaldehyde gives better and more consistent results than those obtained from commercially prepared formaldehyde.

Osmolarity

The yeast plasma membrane, like those of other eukaryotic cells, appears to remain semipermeable during fixation with the low concentrations of aldehydes typically used in immuno-EM methods.[13,17,18] Therefore, changes in the osmolarity of the external medium during fixation can cause movement of water in either direction across the membrane, resulting in loss of both morphology and antigens. These changes can be observed not only in the electron microscope, but also in changes in the appearance of the sensitive structures (e.g., the vacuole and the actin cytoskeleton[19]) in immunofluorescence light microscopy. We investigated the effects of fixative osmolarity on the preservation of yeast morphology and antigenicity in immuno-EM by fixing wild-type yeast cells under different osmotic conditions starting at 0 M sorbitol and increasing in 0.25 M increments to 1.5 M sorbitol; fixation was accomplished with 4% formaldehyde and 0.4% glutaraldehyde in 0.04 M potassium phosphate buffer, pH 6.7. Using affinity-purified antibodies directed against carboxypeptidase Y (CPY) and antibodies directed against actin, we found that aldehyde fixation with osmotic support greatly improves antigenic and morphological preservation in *Saccharomyces cerevisiae* (Fig. 1). We can summarize the importance of Fig. 1 and many subsequent applications of this technique as follows.

1. Aldehyde concentrations (ca. 4% formaldehyde and 0.4% glutaraldehyde in phosphate buffer) and fixation times (i.e., 30–60 min) used for immuno-EM leave the yeast plasma membrane semipermeable and osmotically active.
2. Fixation in hypotonic conditions (i.e., no added sorbitol) causes rupture of the vacuole in a high percentage of cells (ca. 60%). Hypotonic fixation also disrupts the cortical actin patch and its attachment to the plasma membrane. We

[17] Q. Bone, *J. Cell Biol.* **49,** 571 (1971).
[18] A. M. Glauert, "Practical Methods in Electron Microscopy," Vol. 3. North-Holland, New York, 1975.
[19] J. Pringle, R. Preston, A. Adams, T. Stearns, D. Drubin, B. Haarer, and E. Jones, *in* "Methods in Cell Biology" (A. Tartakoff, ed.), Vol. 31, p. 357. Academic Press, New York, 1989.

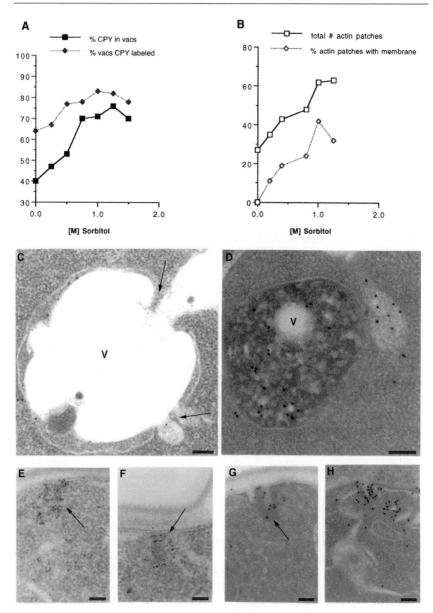

FIG. 1. Fixation in the presence of sorbitol improves the preservation of yeast ultrastructure. (A) Quantitative analysis of anti-CPY (carboxypeptidase Y) localization demonstrates an increase in the localization of CPY (as a percentage of the total number of CPY localizations) to vacuolar structures that correlates with fixation in high osmolarity. A total of 39 cells per each of the seven

attribute these artifacts to osmotically induced swelling of the cell and its organelles (especially the vacuole).

3. Cells fixed in buffered fixative containing 0.2 M sorbitol show a degree of ultrastructural preservation equivalent to cells fixed directly in growth medium, using buffered fixative containing no sorbitol. We find that still higher concentrations of osmotic support produce an even better preservation of the organelle structure.

4. Buffered fixative (see later) containing 1 M sorbitol appears to approximate the internal osmolarity (isotonic) of yeast cells growing in YPD (2% glucose).[20-22]

[20] J. Conway and W. Armstrong, *Biochem. J.* **81**, 631 (1961).
[21] W. N. Arnold and J. S. Lacy, *J. Bacteriol.* **131**, 564 (1977).
[22] W. N. Arnold, *in* "Yeast Cell Envelope: Biochemistry, Biophysics, and Ultrastructure" (W. N. Arnold, ed.), Vol. 1, Chapter 3. CRC Press, Boca Raton, FL, 1981.

---

different fixative osmolarities were observed and quantified for CPY localization. Of the cells fixed in buffered fixative (0 M sorbitol), only 40% of the CPY localizations were to vacuole structures. Of the cells fixed in high osmolarity (0.8 to 1.25 M sorbitol), 70% of the CPY localizations were to vacuolar structures. The percentage of vacuole structures that had CPY labeling also increased with increasing amounts of sorbitol. The level of CPY localization to the vacuole, as well as the number of vacuole structures that label, appears to plateau at approximately 1 M sorbitol, perhaps representing saturation of the available CPY antigen. The total number of anti-CPY localizations per cell section, on average, did not change relative to the fixative osmolarity. (B) Cells fixed in the presence of added sorbitol have an increased number of antiactin localizations and an increased number of cortical actin patches compared to cells fixed without or little added sorbitol. While actin-localizing patches were observed in cells fixed in buffered fixative (0 M sorbitol), no plasma membrane-associated patches (% actin patches with membrane) were observed. Plasma membrane-associated patches were evident only after osmotic support was added to the fixative. Quantification of antiactin localization was done on 15 different cells, at each of the seven different fixative osmolarities. All cells counted were at the same approximate point in the cell cycle as judged by relative bud and neck size. (C and D) Good morphological preservation of vacuolar structures is indicated by vacuoles that are surrounded by an intact membrane and are filled with electron-dense polyphosphate. (C) The cell shown was fixed in buffered fixative without sorbitol; note that the vacuole membrane is ruptured and autolysis of the surrounding cytoplasm is evident (arrows). Also, note the low level of CPY localization and that some of the CPY is located in the cytoplasm. (D) In contrast, the cell shown was fixed with the same protocol as (C) except that 0.75 M sorbitol was added to the fixative. Note that the vacuole membrane of (D) is intact and that the vacuole contains dense polyphosphate and is well labeled with anti-CPY antibodies; also note the well-labeled and preserved late endosome to the right of the vacuole. (E–H) Good preservation of the actin cytoskeleton is indicated by actin patches that are electron dense and surround a finger-like invagination of the plasma membrane [J. Mulholland, J. Konopka, B. Singer-Kruger, M. Zerial, and D. Botstein, *Mol. Biol. Cell* **10**, 799 (1999)]. (E and F) The cortex of cells that were fixed in buffered fixative without sorbitol. Note that antiactin localizations (arrows) are in dispersed patches at the cell cortex; these actin patches do not contain any plasma membrane. (G and H) In contrast, the cells shown were fixed in the presence of 0.75 M sorbitol. Note that the cortical actin patches of these cells are electron dense and contain invaginations of the plasma membrane; note also that actin localization is not as dispersed as it is in (E) and (F). V, vacuole. Bars: 0.1 μm.

5. The best overall ultrastructural preservation of yeast organelles is achieved when using a slightly hypotonic fixative. The exact concentration of sorbitol used will vary with growth conditions, strain background, and mutants. We currently start at 0.75 M sorbitol in 0.08 M potassium phosphate buffer, 4% (w/v) formaldehyde, and 0.4% (w/v) glutaraldehyde for cells grown in YPD (2% sugar). Cells grown in minimal media tend to require less osmotic support than those grown in rich media, and we recommend starting with 0.5 M sorbitol and the same concentration of phosphate buffer. Cells grown in high concentrations of sugar will require more osmotic support during fixation (e.g., cells grown in 5% sugar generally require 1 M sorbitol). Osmotically hypersensitive mutants may require less osmotic support (e.g., actin mutants have better ultrastructure using 0.25 to 0.5 M sorbitol).

Buffers

As pointed out earlier, fixation with aldehydes causes a drop in intercellular pH and therefore the buffering capacity of the fixative vehicle should be important. We have used mainly phosphate buffer (0.08 M, pH 6.8) for immuno-EM, although comparable concentrations of cacodylate (0.1 M, pH 6.8) and PIPES (0.1 M, pH 6.8) have also given good results. To help minimize the extraction of cellular components and to stabilize membranes and polypeptides, many EM methods recommend the addition of millimolar amounts of Mg^{2+} and Ca^{2+} to buffers.[23] However, these cations tend to precipitate out of phosphate buffer and therefore we avoid adding them. In fact, the absence of these cations and a concentration of phosphate buffer of 0.08 M appear to give better overall ultrastructural preservation than the more commonly recommended 0.04 M phosphate buffer with diverse cations. For an extensive discussion about buffers and their use in EM methods, see Johnson[24] and Griffiths.[13]

Fixation Time and Temperature

At the aldehyde concentrations used for immuno-EM of yeast, fixation should be adequate within 30 min at 25°.[25] We generally fix for 60 min. However, low molecular weight antigens may not be retained with short fixation times, especially at lower concentrations of glutaraldehyde, and the retention of some antigens may require longer fixation times (several hours). On the other hand, extended (i.e., overnight) incubation of fixed cells in phosphate buffer could result in an extraction of cellular proteins (particularly low molecular weight antigens) and

---

[23] M. A. Hayat, "Principles and Techniques of Electron Microscopy: Biological Applications," 3rd Ed., p. 17. CRC Press, Boca Raton, FL, 1989.
[24] T. Johnson, *J. Electron Microsc. Tech.* **2**, 129 (1985).
[25] F. Flitney, *J. R. Microsc. Soc.* **85**, 353 (1966).

phospholipids.[26] It is therefore recommended that fixation not exceed 2 hr, with 1 hr being the standard, and that cells be processed into resin as soon as possible. Overnight incubation in phosphate buffer and fixative is not recommended.

There has been no systematic analysis of the effect of fixation temperature on yeast ultrastructure. Theoretical considerations can be used to argue that fixation should be done on ice to slow or block any deleterious physiological reactions that might not be immediately inhibited by fixation. While this may be important in samples where penetration of the fixative is slow, such as in tissue, it should not be a consideration when fixing yeast because infiltration of the fixative is rapid. We generally fix cells at room temperature. Cultures grown or shifted to elevated temperatures can also be harvested and fixed, as described later, at room temperature without any noticeable effects on ultrastructure.

However, in the special case of rapidly reversible, temperature-sensitive mutants, it may be important to fix at the same temperature as the culture or, even better, employ cryofixation methods.[12] For example, when studying a reversible, temperature-sensitive mutant blocked in membrane fusion, it could be imagined that shifting the cells to a permissive temperature during fixation might release the block, allowing membrane fusion to occur prior to sufficiently fixing the cells. Similarly, the mutant protein could be associated with a particular complex and organelle at nonpermissive temperature and the shift to room temperature may reverse this association. This might allow the protein to be relocated during fixation and subsequent processing steps.

Metaperiodate Treatment, Dehydration, Resin Infiltration, and Polymerization

As described earlier, one of the early innovations in the immuno-EM of yeast was the use of sodium metaperiodate to facilitate the infiltration of resin.[5] Sodium metaperiodate attacks the glycol groups of the mannan fibrils of the cell wall, cleaving 1,2-diols, thereby making the wall more permeable. This "loosening" of the cell wall is essential for the adequate infiltration of resin.[6] However, the action of sodium metaperiodate generates free aldehyde groups that can be a source of nonspecific antibody background. Treatment with 50 mM ammonium chloride is used to block these free aldehydes.[5]

There is some concern that treatment with periodate might result in the oxidation and denaturation of some proteins and eliminate sensitive epitopes. These concerns are based on early studies that demonstrated oxidation of amino acids by sodium metaperiodate.[27] Those studies, however, were done at concentrations of metaperiodate sixfold higher than those used for immunohistochemistry with incubation times on the order of hours, not minutes. We suggest using metaperiodate for the minimum time required to provide good infiltration of resin, typically

[26] R. Salema and I. Brandao, *J. Submicrosc. Cytol.* **5,** 79 (1973).
[27] J. Clamp and L. Hough, *Biochem. J.* **94,** 17 (1965).

10 min. If there is reason to suspect sodium metaperiode for the loss of antigen, other methods of making the wall permeable should be considered[28] and of course traditional enzymatic removal of the wall remains possible.[29]

After cells are fixed they must be dehydrated; all cellular water is removed by first replacing it with ethanol and then resin. Once polymerized, the resin provides support for the ultrastructure of the cell so that it can withstand sectioning, introduction into the high vacuum of the electron microscope, and interaction with the electron beam. How water is removed from the cell is very important to the ultimate results obtained, and the regime of dehydration should be tailored to the experimental aims of the researcher. For example, EM methods originally designed for structural studies of the yeast microtubule cytoskeleton initiate dehydration with 95% ethanol, and then proceed to dehydration at 100% ethanol. This rapid dehydration moves water quickly out of the cell, thereby maximizing the extraction of ribosome-dense cytoplasm and thus improving the resolution of microtubules. For immuno-EM, retention of antigens is a main objective and slower, more gradual removal of cellular water is required to preserve them in place. Therefore, dehydration is done on ice and is started at 50% ethanol, moving stepwise through higher concentrations until 100% ethanol is reached.

Once completely dehydrated, the cells are infiltrated with resin. LR White is a low viscosity acrylic resin that can be polymerized with heat or at low temperatures with the use of a chemical accelerant or ultraviolet light. Because polymerization of LR White is inhibited by oxygen, LR White polymerization is done in dry gelatin capsules. Standard plastic embedding capsules (Beem) are not used because they tend to adsorb gases. It should be noted that LR White resin shrinks both during polymerization and in the EM. Infiltration of the resin is done in steps of increasing concentrations, and proper infiltration is a slow process. As described later, infiltration starts with two parts of 100% ethanol and once again proceeds stepwise to 100% resin. Cell wall mutants or strains found to be difficult to infiltrate require longer infiltration times and perhaps more gradual steps before 100% resin is achieved. Unlike the original protocol of Wright and Rine,[6] we do not recommend the application of vacuum during infiltration because it appears somehow to disrupt the ultrastructure of the vacuole.

LR White is polymerized at 47° for 48 to 72 hr. A stable polymerization temperature ($\pm 2°$) is required for proper polymerization.[6] A standard heat block heater that can maintain this temperature, equipped with a custom-drilled aluminum block that fits the gelatin capsules, is recommended for this step. The capsules in the block are covered with foil during polymerization. We have made several aluminum heat blocks and have learned that the diameter of the holes should just allow the nondried

---

[28] O. Rossanese, J. Soderholm, B. Bevis, I. Sears, J. O'Connor, E. Williamson, and B. Glick, *J. Cell Biol.* **145,** 69 (1999).

[29] B. Byers and L. Goetsch, *Methods Enzymol.* **194,** 602 (1991).

gelatin capsule to fit. Drilling the holes to this specification will assure that the dried gelatin capsule will fit properly and can be removed easily. It is important to fully dry the capsules before use (see protocol later) lest they become permanently stuck in the heat block during polymerization.

Common Artifacts and Problems

There are at least five common artifacts that are observed when using chemically fixed yeast cells and LR White resin for immuno-EM studies. The first is plasmolysis, which is the separation of the plasma membrane from the cell wall (Fig. 2A). Plasmolysis generally occurs when cells are transferred to hypertonic media. In immuno-EM, plasmolysis will occur in fixative that contains too much sorbitol. If a large percentage of cells exhibit plasmolysis, the first step is to redo the EM procedure, reducing the amount of sorbitol added to the fixative.

However, it is generally the case that some cells (usually less than 10%) fixed and embedded in LR White resin will have areas where the plasma membrane has separated from the cell wall. This separation is caused by shrinkage of the LR White resin during polymerization as well as when the section is first exposed to the electron beam. If the section is shrinking severely and cells are developing holes between the wall and plasma membrane, it may be useful to first condition the sections. To condition the section, the beam is defocused so that the amount of energy hitting the section is decreased and spread over the whole section. After a few seconds, the beam is focused more and the area of view is moved around so that the section is allowed to shrink as evenly as possible. Once stabilized, higher magnifications and a more focused beam can be used.

However, if there are many holes or it is observed that many of the cells are separating from their cell walls and falling out of the section, then the resin has not been infiltrated properly. To remedy this problem, sections can be picked up on Formvar-coated grids and, after immunolocalization, they can be further stabilized by coating with carbon.[23] Alternatively (and better overall), the experiment can be repeated with fresh resin and longer infiltration times. Also note that extended incubation (>30 sec) in lead citrate stain (pH 12, see later) may cause cells to fall out of the section. Therefore, if cells are falling out of the sections, it may be helpful to check the sections for stability in the EM prior to lead staining.

A second artifact is blebbing and vesiculation of membranes. Vesiculation and blebbing of membranes, particularly the plasma membrane, can occur when little or no glutaraldehyde is used with the standard 4% formaldehyde. Weakly fixed cells are especially sensitive to membrane rearrangements if the fixative is hypertonic, and plasma membrane blebs are often observed at areas of the cell surface that have suffered plasmolysis (Fig. 2A).

The third common artifact is lumenal distention of the nuclear membrane and endoplasmic reticulum (ER) (Fig. 2B). We believe that this artifact is caused by a

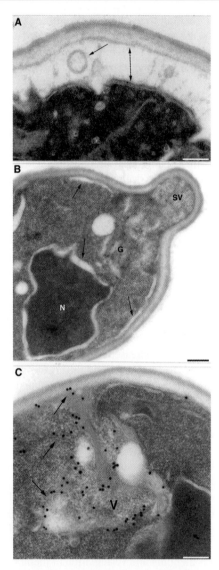

FIG. 2. Artifacts of aldehyde fixation and LR White resin embedding (see text for details). (A) Cell shown exhibits an extreme example of plasmolysis; double-ended arrow indicates the separation between the plasma membrane and the cell wall. Also note the blebbing and vesiculation of plasma membrane (single-ended arrow). (B) Cell shown exhibits artifactual distention of ER and nuclear membranes (arrows). Also note that the Golgi membranes (G) are poorly preserved, whereas the secretory vesicles (SV) of the bud are well preserved. (C) Cell shown illustrates the ultrastructural effects of vacuole rupture. Note that the fusion of the vacuole with the cell surface and autolysis of the cytoplasm results in the localization of CPY (anti-CPY, 15 nm gold) to the cell surface and cytoplasm (arrows). N, nucleus; V, vacuole. Bars: 0.1 μm.

combination of weak fixation and LR White shrinkage. When there is little protein content within the ER lumen, the extent of cross-linking to and across the ER lumen is minimal and is perhaps the most weakly fixed area of the cell. During the initial exposure to the electron beam, the resin shrinks and pulls the weakly fixed ER and nuclear membranes apart. This artifact is frequently eliminated by postfixing the sections in 8% glutaraldehyde after immunolocalization reactions but before electron-dense staining (see later).

The fourth artifact is lysis of the vacuole (Fig. 1C). Fixation in either hypotonic or hypertonic fixative can result in lysis of the vacuole. In an extremely hypertonic fixative, shrinkage of the cell may cause the plasma membrane to contact and fuse with the vacuole membrane, causing rupture and mixing of vacuolar and cell surface proteins (Fig. 2C).

A fifth artifact, or limitation, of our method is the loss of microtubule fine structure while preserving the immunoreactivity of tubulin (Figs. 4C and 4D). Fixation appears to be sufficient to keep the tubulin subunits in place but no individual tubules can be resolved; they appear to have disassembled *in situ*. We have yet to find a fixation condition that preserves both microtubule fine structure and immunoreactivity. This limitation may be overcome by using freeze substitution techniques in combination with light (0.01%) OSO_4 staining (see Ref. 12).

Standard Protocol for Aldehyde Fixation and LR White Resin Embedding

Fixation with Osmotic Support

1. Grow culture to early exponential phase ($OD_{600} < 0.5$) in YEPD or SD medium (2% sugar). We typically process 50–100 ml per sample; this will produce two to three blocks of material for sectioning. However, as little as 5 ml of sample can be processed with slight modification of this protocol (see later).

2. Quickly harvest cells using a disposable 0.45-μm filter unit (Corning) with a vacuum of 20 to 15 in Hg ("house vacuum"). Swirl while filtering to a final volume of 5 ml; important—do not filter dry. Disconnect vacuum and quickly add 25 ml of fixative (see later) directly to the cells. Swirl to cover cells, and use a 10-ml disposable pipette to resuspend the cells by pipetting up and down. The less time between harvesting the sample and adding the fixative the better; it should be possible to take a sample through filtration and addition of fixative in about 1 min.

3. Transfer the cell suspension from the filter unit to 50-ml conical tubes (Corning) and incubate for 60 min at room temperature with occasional mixing.

Sodium Metaperiodate Treatment

4. Collect cells in a low-speed centrifuge (e.g., clinical IEC on setting 6 for 2 min or in a Sorval at 3000 rpm, SS34 rotor for 5 min) at room temperature. Use minimal speed to avoid packing the cells too densely. Pour off supernatant.

5. Resuspend and wash cells by centrifugation (same speed and time) using different buffers as follows:

1× (5 ml) 0.50 M sorbitol in 0.08 M potassium phosphate buffer (pH 6.7)
1× (5 ml) 0.25 M sorbitol in 0.08 M potassium phosphate buffer (pH 6.7)
1× (5 ml) 0.08 M potassium phosphate buffer (no sorbitol) (pH 6.7)

The cell pellet can be resuspended using very gentle vortexing.

6. Transfer cells to disposable glass test tubes (13× 100 mm) with a Pasteur pipette. Centrifuge and resuspend cells in 5 ml of 1% (w/v) NaIO$_4$ (in H$_2$O, Fluka, Ronkonkoma, NY). Make solution just prior to incubation. Incubate for 10 min at room temperature.

Ammonium Chloride (NH$_4$Cl) Treatment

7. Centrifuge and wash cells at room temperature using double-distilled water (or water of equivalent or higher purity obtained by other means). Cells will now clump together due to modification of their cell walls. These clumps can be resuspended easily by mixing with a wooden applicator (VWR, West Chester, PA). Cell clumps should be <1 mm.

8. Centrifuge and resuspend cells in 5 ml of 50 mM NH$_4$Cl (in H$_2$O, Sigma, St. Louis, MO). Incubate for 15 min at room temperature.

Dehydration

9. Centrifuge and resuspend cells serially in an ice-cold ethanol/H$_2$O series (2–5 ml each) as follows: 50% ethanol, 70% ethanol, 80% ethanol, 85% ethanol, 90% ethanol, 95% ethanol, 100% ethanol, 100% ethanol (this is the second such resuspension), and 100% ethanol (for this final resuspension, use a fresh bottle of ethanol at room temperature).

Dehydration should be done for 5 min/step on ice with ice-cold ethanol (except the final 100% step as noted earlier). Centrifugation should be very brief (ca. 2 min) and can be done at room temperature; we use an old-fashioned table-top IEC clinical centrifuge. Do not compact cells too densely. Pellet should be a little loose. Use wooden applicators to resuspend cells in ethanol. In the higher concentrations of ethanol, cells will become less clumped so care must be taken not to pour off cells when recovering the pellets. To avoid damage to cells due to the rapid evaporation of ethanol, cells at the 90% ethanol step and beyond should have the ethanol removed and immediately covered with the next ethanol concentration (or resin).

Infiltration

10. Centrifuge cells and pour off ethanol. Immediately cover and resuspend in 2 ml of a 2 : 1 (v/v) mixture of ethanol : LR White resin (medium hard, Polysciences

Inc., Warrington, PA). Place, covered with Parafilm, on a rotator for 1 hr at room temperature. We use a multipurpose rotator (VWR) with a drum diameter of 6 inches rotating at approximately 20 rpm. The cell/resin mixture can be stored at $-20°$ overnight if needed, although it is preferable to bring the cells through the next step.

11. Pellet cells and pour off ethanol/resin and resuspend in 2 ml of 1:1 ethanol:LR White resin. Place, covered with Parafilm, on a roller overnight at room temperature.

12. Pellet cells and pour off ethanol/resin and resuspend in 1 ml of 1:2 ethanol:LR White resin. Cover with Parafilm and place on a roller for 1–2 hr at room temperature.

13. Pellet cells and remove resin with a Pasteur pipette; resuspend in 1 ml 100% LR White resin. Place on a roller for 1 hr at room temperature. After 1 hr, the cell/resin mixture can be stored at $-20°$ for at least 1 month with a little noticeable change in morphology. However, storage may result in increased extraction of the cytoplasm and may result in movement or extraction of the antigen.

14. Pellet cells and aspirate resin with a Pasteur pipette and resuspend in approximately 0.5 ml 100% LR White resin. If the cells are stored at $-20°$, bring them to room temperature prior to uncovering and proceeding with step 14, otherwise condensation will wet the sample.

Embedding and Polymerization

15. Transfer cells and resin using a Pasteur pipette to gelatin capsules (size 00, Ted Pella, Redding, CA); capsules should be about half full. A 100-ml culture at an OD_{600} of 0.5 should produce enough material for at least three capsules. It is very important to dry gelatin capsules in an oven (60–70°) for 24 hr prior to using; dry capsules will shatter when crushed between fingers. Labels should be written with pencil (No. 2 works best) because some inks appear to be soluble in LR White resin.

16. Top off the capsules with fresh 100% LR White resin and add labels and gelatin caps. Allow cells to settle to the bottom of the capsules for approximately 15 to 30 min at room temperature. Place capsules in a heat block set at 47°. Cover the heat block (several layers of aluminum foil will work) to keep the temperature stable. Polymerize for 2–3 days at 47°.

Fixation of Low Numbers of Cells

As little as 5 ml of a cell culture ($OD_{600} < 0.5$) can be fixed and processed for immuno-EM. However, the main problem with processing small numbers of cells is that when they are transferred into gelatin capsules, the cells spread out over the whole surface of the capsule tip. This forms a very thin layer of cells that can be trimmed away easily or sectioned through. To avoid this problem, we employ two methods that concentrate the cells at the tip of the capsule.

In both methods, cells are first fixed by taking an aliquot of the culture and adding it, with quick mixing, to a equal volume of 2× fixative and buffer. Because medium is being carried over, the amount of osmotic support needed during fixation can be reduced to a final concentration of about 0.5 M sorbitol for rich media and 0.25 M for minimal media. Fixation is then as described earlier; there is no need to resuspend the cells in fresh fixative.

For the first method, after fixation and the final wash, the cell pellet is resuspended in approximately 0.1 ml of 2% low melt agarose (in 0.08 M potassium phosphate buffer, pH 6.7) that has been held at 37°. The pellet is resuspended quickly and then placed on ice to solidify the agarose. The agarose cell pellet is then broken up into pieces no larger than 1 × 1 mm and processed as described earlier. Several of these agarose pieces can be place in a gelatin capsule and polymerized; the rest can be stored in 100% resin at −20°.

For the second method, after being fixed in medium as described earlier, cells can be processed normally until the final resin step. At the final resin step, cells are resuspended in approximately 0.2 ml of 100% resin and transferred to dry, size 4 gelatin capsules (Polysciences, Inc.). This capsule is then placed in the larger standard size 00 capsule and is filled with resin, labeled, and polymerized as usual. This technique produces a thin column of cells at the tip and center of the larger capsule (we call this a "corn dog"). The polymerized block is then trimmed down to the cells at the center. This method can also be used in combination with the agarose method and will make locating agarose-embedded cells a little easier.

Reagents

We use double-distilled water or water purified through ion-exchange systems (e.g., Milli-Q). Although it may not matter for many of the steps, we believe that the immuno-staining protocols, at least, require very pure water.

Buffered fixative (prepare in a hood):
  0.08 M potassium phosphate buffer (pH 6.7),
  0.75 M sorbitol (Sigma),
  4% Formaldehyde (make fresh, see later), and
  0.4% Glutaraldehyde (8% EM grade, snap vial; Polysciences). The final pH should be between pH 6.5 and 7. Check it.

These are the final concentrations for the fixative (i.e., 5 ml of sample plus 25 ml of fixative). Therefore, calculate amounts based on the 30-ml/sample but bring the volume of fixative to 25 ml/sample; the 5 ml of sample will bring the final volume to 30 ml. Fixative is made within 2 hr of processing cultures and is divided into 25-ml aliquots in 50-ml polypropylene conical centrifuge tube (Corning).

1 M potassium phosphate buffer (pH 6.7):
75.77 g dibasic potassium phosphate (K$_2$HPO$_4$) [0.435 M],
76.89 g monobasic potassium phosphate (KH$_2$PO$_4$) [0.565 M], and
800 ml doubly distilled H$_2$O or Milli-Q water;
Mix. Check pH (should be pH 6.7). Bring volume to 1 liter.
Filter sterilize.
Formaldehyde stock (weigh and prepare in a hood):
Weigh 6–8 g paraformaldehyde (Polysciences, Inc.) in a 50-ml conical tube (Corning, Inc.).
Add 20 ml double-distilled (or equivalent purity) water; shake vigorously.
Add 0.5–1 ml 5N NaOH (freshly made, less than 1 week old) and shake vigorously.
Heat to 65–70° in water bath (approximately 15 min).
Shake. Solution should be clear; if not, return to water bath or add a little more NaOH.
Measure volume and calculate (w/v) concentration.
Add required amount of formaldehyde stock to fixative.
Repeat in separate tubes if more formaldehyde is needed.

Overview and Rationale for Postembedding, Immunogold Localization on Yeast Cell Sections

Antibodies and Gold Conjugates

Immuno-EM localization experiments are done using either polyclonal [usually immunoglobulin G (IgGs)] or monoclonal antibodies (usually IgMs). It is essential that the antibody be well characterized before immuno-EM is undertaken. At a minimum, Western blot analysis should be done to confirm that the antibody reacts with a single (preferably) protein band of the expected molecular weight. Immunofluorescence light microscope characterization of the antibody should be the next step. Light microscopy provides the easiest and fastest method for characterizing antigen–antibody interactions in the fixed cell. Optimal fixation conditions and antibody dilution can be determined efficiently at the light level and, once optimized, immunofluorescent light microscopy can be used to provide an overview of the concentration and distribution of the antigen. This overview is virtually essential if one is to interpret successfully immunolocalization results at the level of ultrastructure. If an antigen cannot be localized at the light level, there is usually no point in attempting localization in the electron microscope.

It is strongly recommended that affinity-purified antibodies be used in immuno-EM experiments. In our experience, unfractionated rabbit immunosera produce so much nonspecific background that they are essentially useless for immuno-EM. However, we have had some success using guinea pig immunosera; nevertheless, we would not depend on such a reagent for primary characterization of an antigen. Instead, we have used guinea pig sera in double-label experiments (the

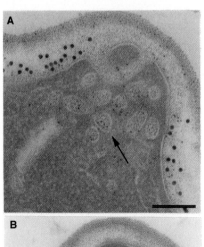
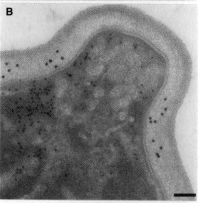
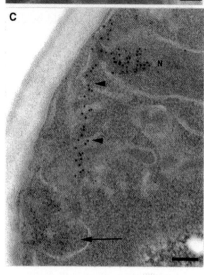

other antibody is typically an affinity-purified rabbit antibody) where the antigen has already been characterized biochemically and immunohistochemically using an affinity-purified antibody (typically also rabbit).

For immuno-EM, antibodies are usually used at concentrations 10 times greater than those used for immunofluorescent microscopy; we titrate every preparation for both purposes. IgG antibodies in a typical purification can be used at a dilution between 1:1 and 1:50, and monclonal IgM antibodies are used at a dilution between 1:1 and 1:10 (supernatant) or between 1:1 and 1:1000 (ascites fluid). These are general rules; we have had success with monoclonal culture supernatant at 1:100 (see Figs. 3C and 5C for example) and affinity-purified IgG antibody at 1:250. It is essential to find the optimum concentration for each preparation. Antibodies directed against the carbohydrate moieties of glycoproteins (e.g., anti-1,6-mannose) are often so effective that they can be used at high dilution (1:2000). This may reflect a relative lack of degradation of the antigenic structures by fixation or, more simply, the relative accessibility of the epitopes themselves.

It should be noted that we have had limited success employing monoclonal antibodies for immuno-EM. It may well be that the reason for this is the uniqueness of the epitopes recognized; in contrast, polyclonal antibodies generally recognize several epitopes on target proteins. If monclonal antibodies are to be used, it is helpful to prepare samples fixed at different concentrations of glutaraldehyde (e.g., 0.4% and 0.1%) to determine whether the single epitope is sensitive to glutaraldehyde fixation.

Colloidal gold has become the most popular particulate marker for immuno-EM, mainly because gold particles of defined size are produced easily and can be adsorbed to a number of affinity molecules such as immunoglobulins and protein A. The dense gold particles are recognized easily and resemble no natural cellular structures. The production of colloidal gold and its adsorption to proteins for immuno-EM has been well described by Roth[30] and reviewed in Jan et al.[31]

[30] J. Roth, Histochem. J. **14**, 791 (1982).
[31] L. Jan, M. Leunissen, and J. De May, in "Immunogold Labeling in Cell Biology" (A. Verkleij and J. Leunissen, eds.), Chapter 1, p. 3. CRC Press, Boca Raton, FL, 1981.

FIG. 3. Examples of double localizations using lectins and antibodies (see text for details). (A) A high magnification image of a small bud demonstrating double localization of two lectins. Con A (Con A–biotin detected with antibiotin–5-nm gold secondary) labels the secretory vesicles (arrow) and cell surface; WGA (WGA–15-nm gold conjugate) labels the chitin ring at the neck of the small bud. Note the lack of WGA labeling of the cell wall of the small bud; also note the difficulty in visualizing 5 nm gold due to the granularity of the lead citrate staining. (B) Image of a small bud demonstrating the antibody localization of actin (antiactin, detected with a 10-nm gold secondary) together with the localization of chitin (WGA–15 nm gold). (C) Double localization of two antibodies shows the localization of nuclear and cytoplasmic microtubule spindles (arrow heads) and the actin cytoskeleton (cortical actin patch, arrow). Tubulin was localized using a mouse monoclonal antitubulin supernatant applied at a 1:100 dilution and detected with an antimouse IgG and IgM–10-nm gold secondary applied at a 1:50 dilution. Actin was localized using affinity-purified (rabbit) antiactin applied at a 1:30 dilution and detected with an antirabbit IgG–5-nm gold secondary applied at a 1:50 dilution. Bar: 0.2 μm.

We use gold particles of different sizes (from 5 to 15 nm in diameter) adsorbed to a secondary antibody [most commonly goat IgG or IgM directed against the immunoglobulin(s) of the organism (rabbit, guinea pig, mouse, or rat) the primary antibody was raised in]. Alternatively, gold particles can be adsorbed to protein A, which binds specifically and tightly enough to IgG molecules of most mammalian species. We always perform such indirect immunolocalization reactions sequentially, first incubating the section with the primary antibody, followed by extensive washing to remove antibodies bound nonspecifically, and only then incubating with the secondary affinity molecule to which the gold particles are bound. High-quality secondary gold conjugates of various sizes are available commercially (see later) and are generally used at a dilution of 1 : 50–1 : 100.

Lectins

Lectins are proteins that bind a diverse group of carbohydrate molecules. They are especially reactive with the sugar groups of glycoproteins and lipids. Very high levels of labeling can be achieved with lectins (as well as antibodies) because the reactive carbohydrate molecules are abundant and are not cross-linked by aldehyde fixation. However, unlike labeling with affinity-purified antibodies, lectins can exhibit a range of avidity for different sugar linkages under diverse conditions. This potential lack of specificity must be kept in mind when interpreting the localization of lectins.

Concanavalin A (Con A) and wheat germ agglutinin (WGA) are the most commonly used lectins for affinity localization in yeast.[19] Con A has a high-avidity interaction with α-mannose linkages (less for α-glucose and α-N-acetyl-D-glucosamine) of glycosylated yeast proteins and can be used to label the secretory pathway, from the Golgi to the plasma membrane and cell wall. WGA has an affinity for β1,4 linkages between N-acetylglucosamine residues and therefore is useful for labeling chitin-rich components of the yeast cell wall (e.g., bud neck and bud scars). WGA labeling is particularly useful for discriminating between mother and daughter cells; newly formed daughter buds and cells have little to no detectable chitin (see Figs. 3A and 3B).

Both Con A and WGA can be purchased directly conjugated to gold of various sizes. We have tried Con A and WGA gold conjugates manufactured by BBInternational Inc. (Ted Pella, Inc.) and have had good labeling with the WGA–gold conjugates (Figs. 3A and 3B). In our hands, Con A–gold conjugates from BBInternational and from Sigma failed to produce an adequate signal. We have been successful in using a Con A–biotin conjugate detected with antibiotin–gold antibody, both from BBInternational (Fig. 3A). It is also possible to purchase purified lectin and colloidal gold sols with which lectin–gold conjugates can be made easily.[30,31]

Blocking

As with any immunohistochemical procedure, nonspecific antibody interactions must be blocked to reduce the background and optimize the signal-to-noise ratio of the antibody. Nonspecific interactions are generally caused by molecules in the section that are "sticky" and tend to adsorb the antibody. This stickiness can be caused by a variety of weak interactions, such as charge and hydrophobicity, or by reactive groups, such as unblocked free aldehydes generated by periodate treatment. Reduction of nonspecific background is achieved in two ways: (1) by beginning the process with reagents that block the sticky molecules so that nonspecific interactions with the immunoglobulins cannot occur and (2) by washing sections extensively after antibody localizations to remove any weak interactions that manage to occur despite the blocking.

Blocking is best done with a mixture of molecules, typically proteins, that do not react with the antibodies used in the immunolocalization. Ovalbumin, gelatin, and bovine serum albumin (BSA) are often used, as is "normal" serum; our experience favors mixtures of the pure proteins. Salt and weak detergents are used to inhibit further nonspecific interactions. Caution is warranted, as blocking agents, particularly salts and detergents, can efficiently block the desired, specific antigen–antibody interactions if used at a high enough concentration. It should also be remembered that some "background" localization can be specific. For example, unwanted "spurious" localization of the antibody could be caused by the presence of an authentic epitope in a protein other than the one of interest. Fortunately, these cases can generally be detected by the preliminary Western blot and light microscopy screening recommended earlier. Harsh or careless processing for immuno-EM may also cause displacement of the antigen from the normal location in the cell, thus producing a signal at locations not predicted by immunofluorescence or biochemical characterization. Again, most of these artifacts can be detected by proper control experiments.

We routinely use a mixture of ovalbumin, BSA, and Tween 20 in phosphate-buffered saline (PBS, "standard IEM block," see later). When this mixture is not effective, we employ instead glycine, fish gelatin, or both. We also sometimes add purified mannan to the blocking mixture (recommended by Ben Glick, University of Chicago) to avoid a particularly common problem in yeast: the nonspecific localization of both affinity-purified and, especially, unfractionated serum antibodies to the cell wall or other heavily glycosylated materials in the cell. This background is sometimes not observed at the light level because the cell wall is removed for immunofluorescence light microscopy.

Controls

Microscopy, by its nature, tends to be used as a qualitative method, despite the fact that robust quantitative techniques are available for immuno-EM and should be applied whenever possible.[13,32] The qualitative nature of most EM experiments makes it even more imperative that microscopists do basic controls necessary to show that the immunolocalization results are meaningful. As indicated earlier, antibodies should have been well characterized with Western blots and immunofluorescent light microscopy prior to using them in immuno-EM experiments. Assuming that this has been done, with controls appropriate to those methods, a few essential controls remain to be conducted at the EM level. Many of these are particularly easy with yeast, and make this organism, despite its small size and other difficulties, an attractive subject for morphological study with affinity reagents.

Tests for Specificity of Primary Antibody. If the protein antigen is encoded by a nonessential gene, then the best and most logical control is immuno-EM examination of a deletion mutant. If the gene encoding the protein target is an essential gene, then a mutant can nevertheless sometimes be found that encodes a protein that exhibits a greatly reduced affinity for the antibody (e.g., $sec4\text{-}8^{33}$). One can screen for this phenotype in Western blots and by immunofluorescence light microscopy. A complementary approach to finding evidence for antibody specificity is to examine, by immuno-EM, strains that have been genetically altered to overproduce the antigen and observing a suitable increase in the antibody localization signal(s). It should be noted that often these methods indicate that a particular antibody has a specific signal over a background of relatively nonspecific localization (or even some specific localization to alternative structures). In such cases the deletion and/or overproduction controls can allow one to interpret immunolocalization nevertheless. Finally, a purified antigen, if available, can also be incubated with the antibody (or other affinity molecule) and then immuno-EM reactions done. The degree of depletion of the signal is a measure of specificity of the antibody.

Tests for Specificity of Secondary Antibody. With indirect detection methods it is necessary to also control for the specificity of the secondary antibody (or protein A) to which the gold particles are attached. This is done easily by eliminating the primary antibody from the immunolocalization protocol. If the secondary goldconjugate is specific for the primary antibody, then its application in the absence of primary antibody should give little or no staining. This control should be done every time, without fail, because secondary gold conjugates degrade with age. This degradation can produce free gold particles as well as aggregates of immunogold conjugates that can stick to sections.

Occasionally, the affinity of the detection molecule for its target needs to be confirmed. This can be done using dot-blot assays. If testing the affinity of immunoreagents, then purified immunoglobulins (IgG or IgM) or normal serum obtained

---

[32] M. Clark, *Methods Enzymol.* **194,** 608 (1991).
[33] B. Goud, A. Salminen, N. Walworth, and P. Novick, *Cell* **53,** 753 (1988).

from the organism that the immunogold conjugate is directed against is applied to nitrocellulose and probed with the secondary immunogold conjugate. If employing nonimmunodetection methods, then purified reactive material is blotted to nitrocellulose and probed with the affinity molecule–gold conjugate. The gold signal is then detected by amplification using silver enhancement.[34]

Positive Controls. Finding no signal is, unfortunately, not a rare event. Any primary antibody–gold conjugate combination that has previously given good, reproducible immunolocalization results should routinely be used as a positive control for the entire process.

Double-Label Immunoelectron Microscopy

Double Labeling on Same Surface of a Section. In order to carry out successful double labeling, it is necessary to have two specific antibody reagents and also two distinct secondary reagents labeled with alternative sizes of gold particles. It is particularly important that the secondary reagents used to localize one primary antibody not react with the other primary antibody, its secondary antibody–gold conjugate, or with complete complexes containing both. This problem is usually surmounted by the use of affinity-purified, primary antibodies obtained from different organisms. For example, a rabbit primary antibody is used in combination with a primary antibody from a guinea pig or mouse. These two primary antibodies are localized and detected using different secondary gold conjugates directed against immunoglobulins of the organisms in which the primary antibodies were raised. To discriminate between the localization of the different antibodies, the secondary antibodies are conjugated to different sized gold particles. We generally use 10- and 15-nm gold conjugates in the double-label experiments. We avoid using 5-nm gold conjugates because it is difficult to visualize them against the dense cytoplasm (stained ribosomes are appropriately 5 nm in size) (see Figs. 3A and 3B). We also avoid 20-nm gold conjugates because their large size appears to produce a decrease in signal, which is most likely due to steric hindrance.

Thus, in the example just given, we might use goat antirabbit 15 nm gold (GaRab-15) and a goat anti-mouse 10 nm gold (GaGP-10) (see Fig. 3C). To guard further against false colocalization due to the antibodies adsorbing to each other, all the antibodies are usually first cross-adsorbed against normal serum of the opposite species. If control experiments demonstrate that the antibodies used really do not cross-react, then primary antibody incubations can be done together, followed (after washing) by detection using a mixture of the secondary gold conjugates.

However, it will generally be the case that the primary antibodies of interest will have been raised in rabbits. In such cases, two different rabbit antibodies will

[34] M. Moeremans, G. Daneels, A. Van Dijck, G. Langanger, and J. De Mey, *J. Immunol. Methods* **74**, 353 (1984).

need to be localized, and a different approach to prevention of cross-localization is needed. One way to prevent cross-adsorption of the immunoreagents is to carry out the two labeling reactions serially. After the first primary and secondary gold antibodies have been applied, the complexes formed are modified (blocked). Two common blocking methods have been published: one involves application of protein A or Fab fragments that "coat" the first primary antibody, immunogold complex.[35] In the other, aldehydes are applied to cross-link and block (presumably by denaturation) the first primary antibody–immunogold complex.[36] We have not attempted the protein A or Fab blocking method but have successfully used the aldehyde blocking method using the protocol given later.[11]

With all multiple labeling techniques that label the same surface of the section, the possibility must be taken seriously that the localization of the first antibody–immunogold complex will sterically hinder access to the other antigen. Therefore, it is important to reverse the sequential order of the labeling experiments, as well as to localize the antigens singularly.

Double Labeling Using Two Different Surfaces. Two different methods can be used to avoid the problems inherent in conducting multiple localizations on the same surface of the section; in one of these we label opposite faces of the same section and in the other we label the adjacent faces of serial sections.

The first method is accomplished by floating the grid and sections on immunolocalization reagents rather than submerging the grid and sections in the reagents.[37] If this method is not done carefully, antibodies applied to one side will come in contact with the opposite side of the section, thus giving erroneous localization results. An additional problem with using the opposite sides of the same section is that localized antigens can be separated by as much as 80 nm of embedded cellular material. This separation introduces a significant loss of resolution that can greatly complicate interpretation.

The second method uses serial sections for the localization of two or more antigens and it avoids all the problems inherent in methods that apply antibodies to the same or opposite sides of the same section.[13] This method, which we call "adjacent-face double localization," uses two immediately sequential sections (80 to 100 nm thick, silver to gold interference color) that are picked up on two separate, 200-mesh nickel grids. These sections are cut to be almost large enough to cover the entire 3.05-mm grid. Each section is then subjected to a standard (single), immunogold localization but each with a different primary antibody. Only one side of each of the sections is exposed to the antibody; the side that is exposed to the antibody on each of the two sections is the side or "face" that was contiguous with

---

[35] H. J. Geuze, J. W. Slot, P. A. van der Ley, and R. C. Scheffer, *J. Cell Biol.* **89,** 653 (1981).
[36] I. L. van Genderen, G. van Meer, J. W. Slot, H. J. Geuze, and W. F. Voorhout, *J. Cell Biol.* **115,** 1009 (1991).
[37] M. Bendayan, *J. Histochem. Cytochem.* **30,** 81 (1982).

the other section prior to sectioning. To accomplish this, the first section is picked up by touching the grid to the section from above; the second section is picked up by submerging the grid in the water of the sectioning knife and picking the section up from below. These two different ways of picking up the sequential sections allow the adjacent face of both sections to be comparably exposed to subsequent antibody incubations and staining procedures. Once the two adjacent faces are recovered onto separate grids, they can be treated with the same reagents without fear of cross-labeling. We prefer to use the same secondary antibodies labeled with the same size gold particles (generally 10 nm) to minimize differences in detection of the two antigens of interest. The immunogold-localized section pairs are then stained with uranyl acetate and lead citrate as described later.

Each of the adjacent-face double localization sections is examined separately in the electron microscope. Subsequent analysis is facilitated greatly by the use of a digital camera. Low magnification images are obtained, and areas of the adjacent-face pair that are visible on both grids are identified and marked. High magnification digital images of cell sections having localization to proteins A and B are then acquired separately. The digital image pairs are merged using Photoshop software (Adobe Systems Inc., San Jose, CA) and examined for colocalization of anti-A and anti-B antibodies. It should be noted that this technique can eliminate considerable observer bias because the colocalization is not evident until the final merging step. We find it convenient to present images as triptychs, with the adjacent faces printed flanking a merged image in which the gold particles from each face represented in different symbols or colors (see Fig. 4).

The following comments should be kept in mind when examining adjacent-face double localization results. As with all the localization methods presented here, immunolocalization occurs after the cells have been chemically fixed, dehydrated, embedded in resin, and sectioned (postembedding immunolocalization). Therefore, although structures can be visualized throughout the cell section, only antigens that are within 5 nm of either side of the cell section surface are accessible to antibody. Accordingly, in the adjacent-face double localization technique, colocalization of two proteins will be restricted to an area of approximately 10 nm (approximately 5 nm on each section) sandwiched between two cell sections. Thus, it is possible to observe localization of an antigen on a cell section in which there is no visible structure only to find the structure clearly visible in the sequential section. This result suggests that in the first section the structure was just "grazed" during sectioning; proteins are present on the surface of the section but not enough of the structure is present to visualize. Conversely, it is possible to observe a structure that in previous experiments had localized a specific antibody, which now shows no localization of that antibody. In this case, observation of the adjacent section shows no structure, suggesting that the structure in the first section was not exposed at the surface of the adjacent face and instead extended in the opposite direction. Thus, in the "adjacent-face" double localization technique it is expected

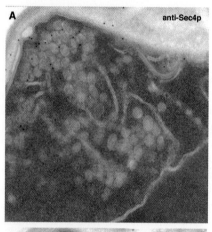
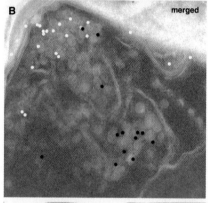
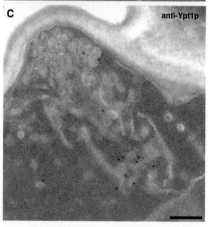

that occasionally, colocalization will not be observed when expected for the trivial reason that the structure and/or antigen is not present within the narrow 10-nm area of the adjacent faces. Conversely, when colocalization does occur it can be inferred that the antigens are within 10 nm or less of each other.

Correlative Studies

It is often necessary to correlate protein localization with structural information. However, because immuno-EM uses weak fixation in combination with low contrasting stains, it does not produce the level of preservation and resolution needed for structural studies. Therefore, when a protein is found to be located, using immuno-EM, in a structure or organelle not previously identified or only minimally characterized, it becomes necessary to provide a more detailed characterization. Although tedious, this is done by splitting cultures into two and processing the cells for both immuno-EM and structural EM. The method applied most commonly to yeast for structural EM is the one developed by Byers and Goetsch[29]; this method produces well-fixed cells that are stained with OsO_4, the heavy metal that imparts considerable contrast to ultrastructure. Figure 5 illustrates the difference between the two methods both in terms of the appearance of fine structure and in the type of information each technique can provide.

Standard Protocol for Obtaining and Immunolabeling Yeast Thin Sections

Grid Preparation and Ultramicrotomy

Once cells have been embedded in resin, thin sections are cut and mounted on grids for use in immunoreactions and observation in the microscope. Thin sectioning, or ultramicrotomy, is difficult to learn and requires some time to acquire skill. Therefore, it is recommended that you use the local EM facility staff to do the sectioning or to train you. Regardless of who does the sectioning, the following observations and recommendations should be kept in mind.

FIG. 4. Adjacent-face, double localization (see text for details) of Ypt1p and Sec4p using two serial cell sections of a *sec8-4* temperature-sensitive mutant. (A) Image shows accumulation of late secretory vesicles; polyclonal anti-sec4p antibodies (rabbit) label the vesicles that accumulate in the bud. (C) The adjacent face localization of polyclonal anti-Ypt1p antibodies (rabbit) to the next serial section; note that Ypt1p is located on the secretory vesicles located in the mother cell. Both anti-Sec4p and anti-Ypt1p localizations were detected with a goat anti-rabbit IgG–10-nm gold secondary. (B) The merged image of (A) and (C) showing the double label of anti-Sec4p (white dots) and anti-Ypt1p (black dots). (B) Merged and processed image using Photoshop software (Adobe, Inc.); A and C were digitally adjusted for contrast and brightness only. Bars: 0.1 μm.

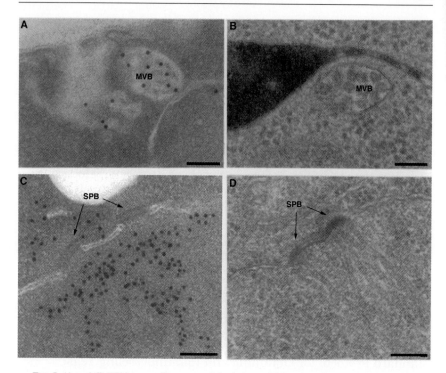

FIG. 5. (A and C) Wild-type cells processed for immuno-EM as described in the text. (B and D) Wild-type cells processed for structural EM using the method of Byers and Goetsch, *Methods Enzymol.* **194,** 602 (1991). (A and B) Images of late endosomes are shown. (A) Note that the endosome is well labeled (10 nm gold) with antibodies directed against the pheromone receptor, Ste2p, but that its fine structure is not well preserved and only low contrast profiles of internal membranes are evident. (B) In contrast, the late endosome is well preserved and is filled with well-stained, small vesicles; its morphology is similar to that of a multivesicular body (MVB). (C and D) Images of duplicate spindle pole bodies (SPB). (C) Note that the spindle is well labeled with antibodies directed against tubulin (10 nm gold) but that a microtubule fine structure is not evident. Note also that SPBs are slightly electron dense and that the nuclear membrane is negatively stained. (D) In contrast, the SPBs and microtubules are well preserved and contrasted and the nuclear membrane is positively stained. Bars: 0.1 μm.

1. Use a sharp, 35° diamond knife, (Diatome, EMS, Inc., Fort Washington, PA). LR White sections cut with a 35° knife angle appear to produce better cell morphology than those cut with a 45° knife.

2. For good resolution, sections should be silver to gray in interference color, approximately 50–70 nm thick.

3. Do not use heat pens or chloroform vapors to spread thin sections. LR White sections are hydrophilic and will, on their own, spread out on the water surface.

4. Prior to picking up sections, they should be grouped together using an eyelash attached (nail polish is a good adhesive) to a sharpened wood applicator

stick. Be sure there are enough sections to cover the grid. Sections can be picked up from above by gently pressing the grid down onto the sections. Press hard enough to slightly indent the water surface without breaking the surface tension. Do not submerge the grid. With the sections trapped beneath the grid, gently roll the grid off the water surface; do not pull the grid straight up off the water because this can damage sections. Reverse tweezers (Ted Pella, Inc., 510-EMX) are recommended because they always hold onto the grid and you do not have to slide an O ring up the tweezers.

5. Because phosphate buffer reacts with copper, grids made of nickel or gold are recommended. Nickel grids can become magnetized and, therefore, nonmagnetizing forceps are useful. Also, magnetized grids can cause astigmatism during EM examination, requiring frequent correction. We routinely use nickel grids with little difficulty but have a demagnetizer available. Grids of 200–300 mesh provide adequate support for the LR White sections; coarser mesh grids will require a support film made of formvar or carbon.[7] We routinely use 300 mesh hexagonal mesh nickel grids purchased from Polysciences.

6. To prevent the loss of sections from grids during antibody incubations it is strongly recommended that the grids be made "sticky" prior to picking up sections; see recipe given later.[1] We prepare 100 sticky grids at a time and store them on hardened #50 Whatman (Clifton, NJ) filter paper in a covered glass petri dish (do not use plastic; static electricity will cause the grids to fly to the top and sides of the dish).

7. After picking up sections, grids are placed on hardened #50 Whatman filter paper in a glass petri dish. When placing the grid on the filter paper, it is important to let go of the grid when the water on the grid is absorbed into the filter paper. The movement of the water into the paper will pull the grid away from the tweezers; try not to drag the grid across the filter. If the grid jumps up the forceps due to capillary action, simply float the grid off on a clean drop of water and then pick up the grid and try again. Alternatively, grids can be left, held in the tweezers and covered, until the grid has dried, approximately 30 min, and then stored in a grid box. Sections mounted on grids and stored in a grid box should be used within a week. Sections up to 1 month old can be used; however, the quality of the ultrastructure, as well as the stability of the sections under the electron beam, is noticeably compromised with older sections.

Immunolabeling

Immunolabeling of yeast cell sections has been well described and our method is essentially unchanged from what is recommended by Wright and Rine.[6] However, for convenience, we have outlined here the immunoreactions for immuno-EM of yeast.

Immunolabeling of thin sections is done in a humidity chamber. Our humidity chamber is a large, covered plastic container that has several damp paper towels placed in the bottom. Immunolocalizations are done in the chamber either directly on a clean sheet of Parafilm or in the wells of a spot plate (Coors, VWR, West Chester, PA). We prefer spot plates for incubations. All incubations are done at room temperature.

1. Two to three drops (approximately 200 µl) of block are delivered into the spot plate wells through a Millex-GS 0.22-µm filter (Millipore, Inc., Bedford, MA) unit attached to a disposable 10-ml syringe. Grids are wetted on each side by touching the surface of the block and then submerged by slowly slicing the grid into the blocking solution. This helps prevent damage to the sections as the grid breaks the surface tension. Sections are blocked in standard immuno-EM block (see later) for 15 min.

2. Grids are removed from the blocking solution, and excess block is absorbed by carefully touching the grid to a damp Kimwipe.

3. Grids are then gently submerged in 20 µl of diluted primary antibody. Grids are typically incubated for 1 hr at room temperature.

4. Grids are removed from the primary antibody and are washed for 5 min in each of three separate wells containing approximately 0.2 ml of wash solution (PBST, see later); 15 min total wash time. There is no need to blot grids dry during this step.

5. Grids are removed from the last wash, and excess PBST is absorbed by carefully touching the grid to a damp Kimwipe.

6. Grids are then submerged in 25 µl of secondary antibody. Incubate for 1 hr at room temperature. Typically, the secondary immunogold conjugate is diluted to between 1 : 50 and 1 : 100 in standard IEM block.

7. After secondary antibody incubation, the grids are moved through three washes of fresh PBST wash. Grids are then immediately rinsed by slowly slicing the grid (about 10 slices at approximately 1 slice/sec) through about 500 ml of high-purity water (Milli-Q or doubly distilled H_2O). Grids can now be placed to dry on Whatman #50 hardened filter paper in a glass petri dish or postfixed and stained as described later. For best results, it is recommended that postlocalization fixation and staining be done immediately following immunolocalization (see later).

Aldehyde Blocking Method: Used for Double Labeling on Same Section

In the aldehyde blocking method the first primary antibody and gold-labeled secondary are applied as described earlier. The cell sections are then washed in PBST and incubated in 4% formaldehyde, 0.1% glutaraldehyde, and 40 mM potassium phosphate, pH 6.7, for 15 min at room temperature. The immunolocalized, blocked cell sections are then washed five times (5 min each time) in 50 µl of PBST. Finally, the cell sections are incubated with the second, primary antibody followed by immunogold secondaries and postfixed and stained with uranyl acetate and lead citrate as described earlier. It should be noted that this technique does not work for all antibodies. The immunoreactivity of the epitopes recognized by the second, primary antibody may be partially or completely eliminated using this blocking procedure. Thus, extensive control experiments may be necessary to determine the optimal concentration of aldehydes needed to adequately block unwanted adsorption but not block localization of the second, primary antibody.

Postlocalization Fixation and Electron-Dense Staining

1. To prevent the loss of antibody localization during low and high pH, electron-dense staining (uranyl acetate and lead citrate, respectively), sections are postfixed in a small amount (ca. 100 μl) of 8% glutaraldehyde for 15–30 min. This postembedding fixation cross-links the antibodies to the section. Post-embedding fixation also appears to provide more extensive fixation of minimally fixed cellular components, especially the endoplasmic reticulum (see earlier discussion).

2. After postlocalization fixation, grids are rinsed immediately by slowly "slicing" (about 20 times; approximately once per second) through 500 ml of high-purity water as described earlier.

3. Grids are then blotted with a Kimwipe and incubated in a drop of 2% aqueous uranyl acetate (2% uranyl acetate can be stored for 1 month at 4°, covered with foil. We store ours in a disposable 10-ml syringe and deliver it through a 0.2-μm syringe filter (Millex-GS, Millipore). Sections can be stain in uranyl acetate for 30 to 60 min at room temperature; longer staining tends to increase the electron density of the cytoplasm, thereby decreasing resolution. Uranyl acetate staining can be done on Parafilm or in a spot plate.

4. Following uranyl acetate staining, sections are washed by slowly slicing the grids 20–30 times through high-purity water as described earlier. After washing, grids are placed on #50 Whatman filter paper in a glass petri dish and allowed to dry for 15–30 min.

5. After drying, sections can be stained with lead citrate (also called Reynolds' lead[38]). Lead citrate imparts considerable contrast but can make the cytoplasm grainy (compare Figs. 3B to 3C). Also, lead staining, if not done carefully, can produce lead carbonate precipitate. Grainy lead staining can look very similar to gold particles 5 nm or smaller (see, for example, Fig. 2A). Additionally, prolonged incubation in lead citrate stain, pH 12, can damage sections. Therefore, it can be advantageous to examine sections prior to lead staining.

To avoid the formation of lead carbonate precipitate, lead staining is done in a semiclosed chamber in the presence of sodium hydroxide. The chamber is made by covering a glass plate (like those used for small gel boxes) with Parafilm on which a glass petri dish cover is placed. Under the cover, we place about 5 g of sodium hydroxide pellets. The first few drops of lead solution are delivered, using a syringe and a 0.2-μm filter, into the hydroxide pellets and then a droplet

[38] E. Reynolds, *J. Cell Biol.* **17**, 208 (1963).

of approximately 100 μl is placed on the Parafilm next to the pellets. The grid is placed in the droplet for 30 sec. During staining we keep the grid held in the tweezers and hold the petri dish cover slightly above the staining area to minimize exposure to air. One should also hold one's breath as CO_2 will cause the formation of lead carbonate. After 30 sec the grid is removed from the droplet and is washed immediately by slicing (described earlier) in fresh Milli-Q water or recently boiled and cooled doubly distilled H_2O. For more contrast, wash only 10 slices (approximately 1 sec/slice) but this will also impart a granularity to the cell section. Longer washes (>20 slices) will give less contrast, especially on thinner sections (<60 nm) but will impart a less grainy appearance. Washed grids are then placed on #50 Whatman filter paper in a glass petri dish and are allowed to dry overnight prior to examination in the electron microscope. Tweezers must be rinsed in water and wiped dry before staining the next grid.

Alternatively, lead acetate or vanadium can be used as electron-dense stains. Lead acetate stain can be used at near neutral pH and does not produce the granularity of lead citrate. Vanadium stain has little to no granularity but provides less contrast than either lead citrate or acetate. Vanadium staining[32] is particularly useful when using gold conjugates 5 nm and smaller (e.g., nanogold).

Sticky Grid Solution

4 ml 0.5% Formvar (Ted Pella, Inc.; final concentration of Formvar is 0.08% w/v)

21 ml of a 24 : 1 (v/v) dichloroethane : chloroform solution

Keep sticky grid solution tightly sealed. Grids can be placed in the solution, removed with tweezers, and placed, singularly, on #50 Whatman filter paper. Work with sticky grid solution and reagents in the hood; carefully wipe tweezers clean with a Kimwipe and ethanol.

Lead Citrate Stain

Lead citrate can be prepared using several different methods.[18] The method we use is as follows:

0.004 g Lead citrate (Polysciences, Inc.) per 1 ml staining solution needed (0.4% solution). We typically make 5 ml.

20 μl of 5 N sodium hydroxide per ml of 0.4% staining solution. NaOH should be no more than a week old.

Mix using vigorous vortexing until the lead citrate goes into solution. The lead solution can be taken up in a disposable syringe that is then fitted with a 0.2-μm filter. The lead solution can be stored in the syringe (remove filter) for 1–2 weeks.

Wash Buffer and Blocks

PBST: 140 mM NaCl, 3 mM KCl, 8 mM Na$_2$HPO$_4$, 1.5 mM KH$_2$PO$_4$, 0.05% Tween 20, pH 7.4
Standard block: PBST, 0.5% (w/v) ovalbumin (Sigma), 0.5% (w/v) BSA (Sigma)
Fish gelatin (0.5% w/v) (BBInternational, gel#10) in PBS or PBST
Glycine (0.15%) block in PBS or PBST
Mannan (0.1 mg/ml) (Sigma) block in PBST or standard block. A 1-mg/ml stock solution can be made and stored at 4° with 0.1% (w/v) sodium azide (Ben Glick, University of Chicago)

[5] Electron Tomography of Yeast Cells

By EILEEN T. O'TOOLE, MARK WINEY, J. RICHARD MCINTOSH, and DAVID N. MASTRONARDE

Introduction

Electron microscopy (EM) is a useful method for the study of yeast cells and is complementary to a genetic analysis. EM provides high-resolution structure data about wild-type cells and can document structural defects in mutant strains.[1,2] A number of EM techniques have been used to study the complex three-dimensional (3D) arrangements of organelles in yeast. Examples include the use of freeze-fracture replicas to reveal the surface topology of nuclear envelopes and cytoplasmic membrane systems,[3–5] as well as the use of serial sections to reconstruct entire mating factor-arrested cells,[6] to describe the 3D geometry of the mitotic spindle,[7,8] and to document the 3D distribution of nuclear pore complexes in wild-type[9] and mutant strains.[10] More recently, electron tomography has been used to describe

[1] B. Byers, in "Molecular Genetics in Yeast" (D. von Wettstein, J. Friis, M. Kielland-Brandt, and A. Stenderup, eds.), p. 119. Munksgaard, Copenhagen, 1981.
[2] B. Byers and L. Goetsch, *Cold Spring Harb. Symp. Quant. Biol.* **38,** 123 (1974).
[3] H. Moor and K. Muhlethaler, *J. Cell Biol.* **17,** 609 (1963).
[4] E. G. Jordan, N. J. Severs, and D. H. Williamson, *Exp. Cell Res.* **104,** 446 (1977).
[5] O. Necas and A. Svoboda, *Eur. J. Cell Biol.* **41,** 165 (1986).
[6] M. Baba, N. Baba, Y. Ohsumi, K. Kanaya, and M. Osumi, *J. Cell Sci.* **94,** 207 (1989).
[7] R. Ding, K. L. McDonald, and J. R. McIntosh, *J. Cell Biol.* **120,** 141 (1993).
[8] M. Winey, C. L. Mamay, E. T. O'Toole, D. N. Mastronarde, T. H. Giddings, K. L. McDonald, and J. R. McIntosh, *J. Cell Biol.* **129,** 1601 (1995).
[9] M. Winey, D. Yarar, T. H. Giddings, and D. N. Mastronarde, *Mol. Biol. Cell* **8,** 2119 (1997).
[10] N. Gomez-Ospina, G. Morgan, T. H. Giddings, B. Kosova, E. Hurt, and M. Winey, *J. Struct. Biol.* **132,** 1 (2000).

the 3D architecture of the yeast spindle pole body (SPB)[11] and forming mitotic spindles.[12]

Electron tomography has proven to be a useful method for obtaining 3D structure data to describe complex cellular substructures.[13] Tomography is based on a series of tilted images, usually collected from a comparatively thick (0.2–1 μm) section or from isolated organelles, to generate a 3D reconstruction using back projection algorithms.[14] This method is like a "CAT" scan of a cell, where the result is a computer-generated reconstruction that can be computationally sectioned and viewed in any orientation at 5–10 nm spatial resolution. In contrast to serial section reconstruction, where multiple thin (50–70 nm) sections have been used to reconstruct cellular objects that extend over many micrometers (e.g., an anaphase mitotic spindle[15,16]), tomography is best suited for structures whose dimensions change significantly, such that standard thin sections cannot reconstruct them accurately. Tomographic reconstruction permits the viewing of computer-generated "slices" that are much thinner than could ever be cut physically with a microtome. This approach is therefore ideal for structures that have a complex 3D geometry, such as arrays of cytoskeletal elements or convoluted membrane systems. This chapter describes the steps necessary to prepare yeast cells for tomography and provides examples that illustrate the procedures for calculating tomographic reconstructions of organelles *in situ* and of whole cell profiles.

Equipment Requirements and Software Packages

Microscope Requirements

An intermediate voltage microscope (IVEM) operating at 200–400 kV or a high-voltage electron microscope (HVEM) operating at 750–1000 kV is necessary for collecting tilt series data from thick (0.2–1 μm) sections. The increased accelerating voltages of these microscopes reduce the scattering cross section of the beam electrons, allowing them to penetrate the thicker specimens with less inelastic and plural scattering, thus improving image quality. This feature of high-energy beams becomes particularly important when the specimen is tilted up to 60–70°, where the section thickness doubles or nearly triples, relative to the electron beam. More limited questions can be addressed by tomography using thinner sections

---

[11] E. Bullitt, M. P. Rout, J. V. Kilmartin, and C. W. Akey, *Cell* **89**, 1077 (1997).
[12] E. T. O'Toole, M. Winey, and J. R. McIntosh, *Mol. Biol. Cell* **10**, 2017 (1999).
[13] J. Frank (ed.), "Electron Tomography," p. 399. Plenum, New York, 1992.
[14] J. Frank and M. Radermacher, *in* "Advanced Techniques in Biological Electron Microscopy" (J. Koehler, ed.), p. 1. Springer-Verlag, Berlin, 1986.
[15] D. N. Mastronarde, K. L. McDonald, R. Ding, and J. R. McIntosh, *J. Cell Biol.* **89**, 1457 (1993).
[16] K. L. McDonald, E. T. O'Toole, D. N. Mastronarde, M. Winey, and J. R. McIntosh, *Trends Cell Biol.* **6**, 235 (1996).

(e.g., 50 nm) with lower voltage EMs. In the United States, there are several NIH-supported national research resources that have the expertise and equipment to do routine tomography (Boulder Laboratory for 3-D Fine Structure, University of Colorado, Boulder, http://bio3d.colorado.edu; Wadsworth Center, Albany, NY, http://www.wadsworth.org; and the National Center for Microscopy and Imaging Research, University of California, San Diego, http://www-ncmir.ucsd.edu). These facilities are set up to help investigators collect and analyze data, and the reader is encouraged to contact the web pages of these facilities to obtain more information.

Equipment for Digital Image Capture

The images can be collected digitally using either standard EM film or a high-resolution charge-coupled device (CCD) camera. If the images are collected on film, the film optical density must be scanned and converted to digital form for further analysis. Our method has been to place the negative on a motor-controlled light box and digitize the image with a high-quality CCD camera.[17] Generally, a pixel size corresponding to 1–3 nm at the plane of the specimen is suitable for tomographic studies of yeast. If images are collected directly in digital form on the microscope, the pixel size may need to be smaller than when working from film because of the limited resolution of microscope CCD cameras.

Software Packages

There are several software packages available to generate, display, and analyze tomographic data. Our laboratory has developed the IMOD software package, which contains all of the programs needed for calculating tomograms, displaying the reconstructions, and modeling image data.[18] The Imod viewing program uses the MRC image file format and is convenient because multiple images can be stored in a single file, or "image stack." The Imod viewing program allows an investigator to read in an image stack and then step or "movie" through the series of images. The program can also be used to model features within a data set and to display 3D models. In addition, the package contains roughly 85 programs for 3D analysis, including measurement and display-enhancement features. The IMOD package was originally developed to run on a Unix platform using SGI computers with 24-bit graphics, but has now been ported to run on a PC under Linux. Executable versions of the IMOD programs are freely available on our website (http://bio3d.colorado.edu), and the details of their use will be discussed in examples given later. The web site also provides a detailed guide for tomographic

---

[17] M. S. Ladinsky, J. R. Kremer, P. S. Furcinitti, J. R. McIntosh, and K. E. Howell, *J. Cell Biol.* **127**, 29 (1984).
[18] J. R. Kremer, D. N. Mastronarde, and J. R. McIntosh, *J. Struct. Biol.* **116**, 71 (1996).

reconstruction that describes how to use specific programs and how to troubleshoot problems that arise. Once the IMOD package is installed, the operator can run the programs with a series of command files that contain all of the information needed to initiate the programs that align tilt series data and calculate a tomographic reconstruction.

Similar software packages have been developed and made available, including the SPIDER, WEB, and STERECON packages developed at the Wadsworth Center (Albany, NY),[19–21] the SUPRIM package developed at the University of Texas (Austin, TX),[22] and several image processing and display software packages developed and used at the National Center for Microscopy and Image Research (San Diego, CA).[23–25] Although there are operational differences in the software packages used to generate, display, and analyze tomographic reconstructions, the basic steps required to create a tomographic reconstruction are similar in all of them; these are outlined in Table I. The details of each step and examples illustrating their use in yeast cell biology are discussed later.

Specimen Preparation for Electron Tomography

Because the purpose of tomographic study in yeasts or in any cell is to obtain high-quality 3D fine structural data, the initial fixation of the cells for electron microscopy is a critical step. Artifacts introduced by poor fixation will have a serious negative effect on the quality of the final reconstruction. We have found that high-pressure freezing, followed by freeze substitution, results in excellent preservation of cellular fine structure. The reader is referred elsewhere in this volume for details of the method.[26] We have prepared the specimens imaged in this chapter for EM by high-pressure freezing, followed by freeze substitution in 3% glutaraldehyde (v/v) and 0.1% uranyl acetate (w/v) in acetone at $-90°$ for 3 days and then low temperature embedded in Lowicryl HM20. Because this method results in a pellet of cells that are embedded in random orientations, it is advisable to cut serial sections in order to find suitable cell profiles or images of specific organelles. Serial thick (200–500 nm) sections are cut using a microtome and are collected onto slot grids coated with 0.7% (w/v) Formvar. Because of the increased thickness of

---

[19] J. Frank, M. Radermacher, P. Penczek, J. Zhu, Y. Li, M. Ladjadj, and A. Leith, *J. Struct. Biol.* **116**, 190 (1996).
[20] M. Marko and A. Leith, *J. Struct. Biol.* **116**, 93 (1996).
[21] B. F. McEwen and M. Marko, *Methods Cell Biol.* **61**, 81 (1999).
[22] J. P. Schroeter and J. P. Bretaudiere, *J. Struct. Biol.* **116**, 131 (1996).
[23] D. Hessler, S. J. Young, and M. H. Ellisman, *NeuroImage* **1**, 55 (1992).
[24] D. Hessler, S. J. Young, and M. H. Ellisman, *J. Struct. Biol.* **116**, 113 (1996).
[25] G. A. Perkins, C. W. Renkin, J. Y. Song, T. G. Frey, S. J. Young, S. Lamont, M. E. Martone, S. Lindsey, and M. H. Ellisman, *J. Stuct. Biol.* **120**, 219 (1997).
[26] K. McDonald and T. Müller-Reichert, *Methods Enzymol.* **351**, [6], 2002 (this volume).

TABLE I
STEPS INVOLVED IN CREATING TOMOGRAPHIC RECONSTRUCTIONS OF YEAST

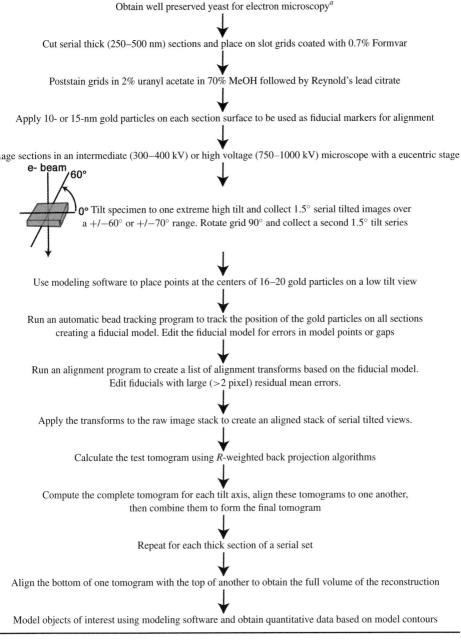

Obtain well preserved yeast for electron microscopy[a]
↓
Cut serial thick (250–500 nm) sections and place on slot grids coated with 0.7% Formvar
↓
Poststain grids in 2% uranyl acetate in 70% MeOH followed by Reynold's lead citrate
↓
Apply 10- or 15-nm gold particles on each section surface to be used as fiducial markers for alignment
↓
Image sections in an intermediate (300–400 kV) or high voltage (750–1000 kV) microscope with a eucentric stage
↓
Tilt specimen to one extreme high tilt and collect 1.5° serial tilted images over a +/−60° or +/−70° range. Rotate grid 90° and collect a second 1.5° tilt series
↓
Use modeling software to place points at the centers of 16–20 gold particles on a low tilt view
↓
Run an automatic bead tracking program to track the position of the gold particles on all sections creating a fiducial model. Edit the fiducial model for errors in model points or gaps
↓
Run an alignment program to create a list of alignment transforms based on the fiducial model. Edit fiducials with large (>2 pixel) residual mean errors.
↓
Apply the transforms to the raw image stack to create an aligned stack of serial tilted views.
↓
Calculate the test tomogram using R-weighted back projection algorithms
↓
Compute the complete tomogram for each tilt axis, align these tomograms to one another, then combine them to form the final tomogram
↓
Repeat for each thick section of a serial set
↓
Align the bottom of one tomogram with the top of another to obtain the full volume of the reconstruction
↓
Model objects of interest using modeling software and obtain quantitative data based on model contours

[a] See Ref. 26.

the sections used for tomography compared to standard electron microscopy, the samples must be poststained for longer periods of time. We have found that staining with 2% (w/v) uranyl acetate in 70% (v/v) methanol for 10 min followed by Reynold's lead citrate for 5 min provides good contrast and stain penetration.

The final step in preparing specimens for electron tomography is the addition of colloidal gold particles to each surface of the section. This is done by placing a drop (~20 µl) of 10 or 15 nm colloidal gold suspension (BBInternational) on top of the grid for several minutes, dipping the grid in distilled H_2O, and then repeating the procedure on the other side of the grid. Gold particles that adhere to the two surfaces serve as fiducial markers for subsequent image alignment.[27] A thin film of carbon may also be evaporated on the grids to stabilize the samples during electron microscopy.

Data Collection

The thick sections can first be imaged in a conventional transmission microscope operating at 100 kV to find sections that contain specific organelles or to identify a cell in a particular stage of the growth and division cycle. Low magnification overview maps can be collected to facilitate finding the particular cell in the higher voltage microscopes, where contrast is greatly reduced. For data collection in an IVEM or HVEM, the grid is placed in a tilting specimen holder and the goniometer of the microscope stage is adjusted to permit eucentric tilting by adjusting the tilt axis so it lies in the plane of the specimen and passes through the region of interest. This adjustment is useful because it decreases the amount of time the microscope operator spends focusing and positioning the specimen during data collection. The specimen is then tilted to 60 or 70°, and serial tilted views are collected to the opposite extreme tilt at intervals of 1–2°, depending on the resolution desired in the tomogram. For dual-axis tomography, the grid is rotated in the specimen holder by 90° and a second tilt series is acquired.

A microscope equipped to collect a tilt series automatically is very useful for facilitating image capture and increasing throughput. Several laboratories have developed procedures for microscope automation and papers describing their use have been published.[28–31] The Boulder 3D laboratory has developed a program for semiautomated data collection procedure that drives Gatan's Digital Micrograph software to collect images from a 1024 × 1024 pixel CCD camera. The program

---

[27] M. C. Lawrence, in "Electron Tomography" (J. Frank, ed.), p. 39. Plenum, New York, 1992.
[28] K. Dierksen, D. Typke, R. Hegerl, A. J. Koster, and W. Baumeister, *Ultramicroscopy* **40,** 71 (1992).
[29] A. J. Koster, H. Chen, J. W. Sedat, and D. A. Agard, *Ultramicroscopy* **46,** 207 (1992).
[30] M. B. Braunfeld, A. J. Koster, J. W. Sedat, and D. A. Agard, *J. Microsc.* **174,** 75 (1994).
[31] J. C. Fung, W. Liu, W. J. de Ruijter, H. Chen, C. K. Abbey, J. W. Sedat, and D. A. Agard, *J. Struct. Biol.* **116,** 181 (1996).

also controls essential microscope functions such as focus and tilt so it can collect single frame or montaged images of large cellular profiles and store the serial images in MRC format. Generally, a magnification of 12,000 on the microscope, corresponding to a pixel size of 1.4 nm, is used.

Alignment of Serial Tilts and Tomographic Reconstruction

Our standard method for alignment of serial images first begins by reading the image stack containing the raw, serial tilts into Imod, and one section at low tilt is displayed. The positions of 10–20 gold particles are then marked by placing a model point in the center of each gold particle (Fig. 1a, open circles). An automatic bead-tracking program is then used to track and model the corresponding images of each gold particle through all of the images of the tilt series. The resulting computer-generated "fiducial" model is then edited manually to correct any small errors in the placement of the model point or any gaps in the data set. Figure 1a and video sequence 1a (available on our web site, http://bio3d.colorado.edu/MIE.html) show images of a 200-nm-thick section through a small yeast bud with the positions of 16 gold particles marked (Fig. 1a, open circles; video sequence 1a, green circles). Gold particles that are attached to the top and bottom surfaces of the section appear

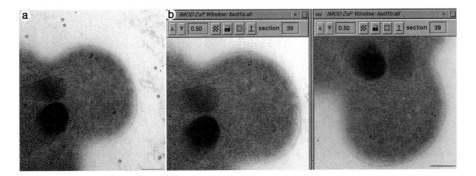

FIG. 1. Serial, tilted views of a single thick section are used as raw data for tomographic reconstructions. (a) An image of a 200-nm-thick section through a portion of an emerging bud from a diploid yeast cell. Gold particles placed on the top and bottom surfaces of the section are used as fiducial markers for alignment because their position can be located in every view. Approximately 10–20 gold particles are modeled [open circles around the 15-nm gold (a)]. Images are collected every 1.5° over a ±60 or ±70° range. For best results, a dual-axis approach is applied, where the grid is rotated 90° and a second tilt series is acquired (see Fig. 1b). Supplementary video material for (a) and (b) can be found at our website, http://bio3d.colorado.edu/MIE.html. Video sequence 1a shows 78 serial tilted views of the 200-nm-thick section shown in (a) with green model points marking the position of 16 gold particles used as markers for alignment. Video sequence 1b shows the complete, aligned serial tilt series from each axis. Note there is little fine structure resolved in any one view because there is so much cellular material superimposed in a 200-nm-thick section. This is particularly true at high tilts, where section thickness has doubled relative to the electron beam axis. Bar: 200 nm.

to move in opposite directions relative to each other as one moves through the tilt series (video sequence 1a).

The serial, tilted images are then aligned by the program Tiltalign, which uses the positions of the gold particles and a least-squares approach to solve for tilt angles, shifts, rotations, magnification changes, and section distortion.[32] The transforms created by Tiltalign are then applied to image data, resulting in an aligned stack. Figure 1b shows images from aligned stacks of a small yeast bud taken from a tilt series obtained from tilting around the Y (vertical) axis and the same bud rotated 90° and then tilted again about the vertical axis. Video sequence 1b (available on our web site, http://bio3d.colorado.edu/MIE.html) shows movies of the aligned tilt series containing 78 serially tilted views separated by 1.5° over a ±60° range. Note that little fine structure can be resolved in these thick sections in any one view because there is so much cellular material superimposed within the volume of the thick section.

The next step is to compute the tomographic reconstruction using the density information in the aligned tilt series. The program we use employs an R^*-weighted back projection algorithm.[33] This step is the most computationally intensive in the whole process and results in reconstructions that can be quite large (>500 megabytes). Moreover, the reconstructed section is often not flat or centered improperly within the volume of the data set. To save time, it is useful first to run the program on a small (10 pixel) slice of the aligned stack to identify any adjustments that need to be made to make the reconstruction flat and centered, and to fit within the smallest volume possible. The procedures in the IMOD package include command files to generate three samples of the reconstruction and analyze them to determine the shifts and rotations that must be made to keep the section level and centered. They also measure the thickness of the reconstruction. With these adjustments, the final reconstruction is calculated from the full-sized, aligned tilt series.

For electron tomography of sectioned biological material, there is a limitation in the range of tilts that one can collect, generally up to ±60 or ±70°. This is due to several factors such as the specimen support grid occluding the image at high tilt, the physical stops in the microscope that prevent the specimen rod from hitting pole pieces and apertures, and the difficulty in obtaining well-focused images of thick sections at tilt angles greater than 60–70°. These limitations in the range of tilt angles that can be sampled result in a missing wedge of information, which leads to a decrease in resolution of the tomogram.

Several laboratories have worked to overcome or at least minimize the impact of this missing wedge. One approach is conical tilting[34] and others include combining information from tilt series taken about two orthogonal axes, also known as

---

[32] P. K. Luther, M. C. Lawrence, and R. A. Crowther, *Ultramicroscopy* **24**, 7 (1988).
[33] P. F. C. Gilbert, *Proc. R. Soc. Lond. B Ser. Biol. Sci.* **182**, 89 (1972).
[34] M. Radermacher, *J. Electron Micros. Tech.* **9**, 359 (1988).

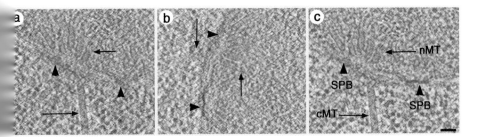

FIG. 2. Benefits of dual-axis tomography. Slices taken from tomograms of duplicated side-by-side SPBs from tilt series collected about two orthogonal axes (a and b) can be very different. The cytoplasmic and nuclear microtubules [(a) arrows] are well resolved in the tilt series taken about the first axis, yet the SPBs are poorly resolved [(a) arrowhead]. When the grid is rotated 90° and a second tilt series is collected, SPBs in the resulting tomogram are well resolved [(b) arrowheads], yet the microtubules show up poorly. The combined tomogram (c) shows features at all orientations well resolved. Bar: 50 nm.

dual-axis tomography.[35,36] Our laboratory has developed an improved method of dual-axis tomography, where tomograms are computed separately from tilt series taken about two orthogonal axes and then the two tomograms are aligned to each other and combined to achieve a single tomogram.[37] This dual-axis approach results in resolution that is almost isotropic, which is valuable, especially for extended features that are perpendicular to a tilt axis.

Figure 2 shows an example of a budding yeast cell that contains duplicated, side-by-side SPBs and associated nuclear and cytoplasmic microtubules. The tomogram calculated from tilt series collected about the first axis shows good resolution of nuclear and cytoplasmic microtubules (Fig. 2a, arrows), yet the microtubules are poorly resolved (Fig. 2b, arrows) in the tomogram calculated from the tilt series collected after rotating the grid 90°. Similarly, features that are perpendicular to the cytoplasmic microtubules, such as the SPB, are better resolved in the tomogram calculated from the second tilt series (Figs. 2a and 2b, arrowhead). Figure 2c shows the combined tomogram where both microtubules and SPBs are well resolved. This illustrates that the dual-axis approach is the method of choice for a tomographic study of complex biological material.

Modeling Image Data

Cellular features in the dual axis tomogram can be modeled, and the resulting 3D model can be displayed for study and analysis (Fig. 3, and supplementary

[35] K. A. Taylor, M. C. Reedy, L. Cordova, and M. K. Reedy, *Nature* **310,** 285 (1984).
[36] P. Penczek, M. Marko, K. Buttle, and J. Frank, *Ultramicroscopy* **60,** 393 (1995).
[37] D. N. Mastronarde, *J. Struct. Biol.* **120,** 343 (1997).

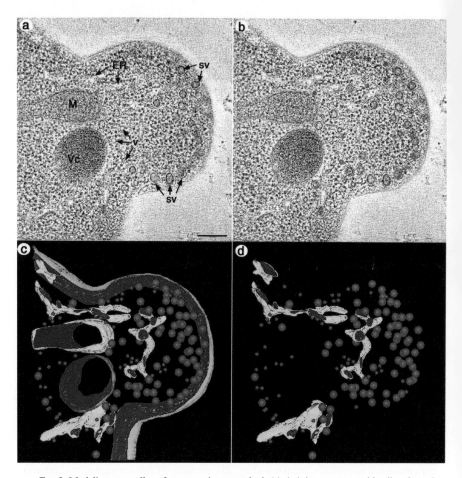

FIG. 3. Modeling organelles of an emerging yeast bud. (a) A 1.4-nm tomographic slice through a portion of a small yeast bud. The tomographic slice is viewed parallel to the plane of the original 200-nm-thick section. Organelles such as ER, mitochondrion, vacuole, and secretory vesicles are clearly resolved. Bar: 200 nm. (b) Organelles were modeled by depositing model points along the membranes of the plasma membrane (green), ER (yellow), a mitochondrion (light blue), and a portion of the vacuole (purple) forming contours that defined the boundaries of the organelle in any given slice. Vesicles were modeled by depositing a single point on the tomographic slice that contained the maximum diameter of the vesicle. Such points were displayed as scattered points with a sphere size set for the particular kind of vesicle (blue, red spheres). (c and d). A three-dimensional model created by surface rendering of contour data. The plasma membrane of the bud is shown in green and encapsulates a bud filled with secretory vesicles (blue, red), ER (yellow), a mitochondrion (light blue), and a portion of the vacuole (purple). (d) Individual objects can be turned on or off to display the 3D relationship of particular organelles to each other, such as the vesicles (blue, red) relative to ER (yellow). Supplementary video material for (a–d) can be found at our web site, http://bio3d.colorado.edu/MIE.html. Video sequence 3a shows a movie of 55 serial 1.4-nm-thick tomographic slices. The bud is filled with organelles that can be clearly resolved. In video sequence 3b, modeling software was used to place model points to define contours of specific organelles. The convoluted nature of the ER (yellow) in this bud is particularly evident when moving through the tomographic volume. Video sequences 3c and 3d are rotating 3D models of the organelles modeled in this tomographic reconstruction.

video material available at our website, http://bio3d.colorado.edu/MIE.html). The tomographic slices shown in Figs. 3a and 3b are 1.4 nm thick and are viewed parallel to the plane of the original 200-nm section. Organelles such as secretory vesicles (sv), a mitochondrion (M), endoplasmic reticulum (ER), and a portion of the vacuole (Vc) can be resolved clearly in these slices (Fig. 3a). Video sequence 3a is a movie of 55 serial, 1.4-nm-thick tomographic slices from the top surface of the original thick section, through to its bottom surface. The 3D relationships of organelles and their arrangement in the bud are particularly evident when moving slice by slice through the entire tomographic volume.

Objects of interest are modeled separately and are made up of a series of stacked contours or a set of scattered points. In the example shown in Fig. 3b, the different organelles within the cell are represented by different graphic objects, each designated by a unique color. The boundaries of each organelle are traced by hand to create a contour, and a new contour is created on each successive tomographic slice in which the organelle can be found. Video sequence 3b is a movie of 55 serial, 1.4-nm-thick tomographic slices with model contours marking the positions of the plasma membrane (green), the ER (yellow), a mitochondrion (light blue), a portion of the vacuole (pink), and two classes of vesicles that have different sizes (marked blue and red). The elaborate membrane system of the ER (yellow) in this emerging bud would be difficult to model with accuracy using other reconstruction methods because the contours of the membrane change significantly over very short distances.

Once the contours of all the objects of interest are traced, the resulting model can then be viewed as a 3D projection of the entire modeled volume and modified for viewing in several ways. Contour data can be "meshed" to create a skin over each modeled object (Figs. 3c and 3d and video sequences 3c and 3d). Small, regular objects like vesicles can also be displayed as spheres whose diameters are set to correspond to the diameter of the object of interest (Figs. 3b–3d, red and blue vesicles). The full model may be displayed to show all objects (Fig. 3c) or only selected objects, facilitating the study of the 3D relationships among a particular subset of the objects in the model (Fig. 3d, ER relative to the vesicles). Video sequences 3c and 3d show the 3D models rotating 360° about a vertical axis. They illustrate the polarity of the organelle distribution in the emerging bud.

Finally, quantitative information, such as the number of particular objects, their surface areas, volumes, length distributions, and nearest neighbor distances, can be determined from model contour data using companion programs such as Imodinfo. This information is useful for testing ideas about structural relationships among organelles within the cell in 3D.

Tomographic Reconstructions of Larger Cellular Volumes

When tomography is used to study large intracellular compartments, such as endomembrane systems or cytoskeletal arrays, it is often best to reconstruct as large

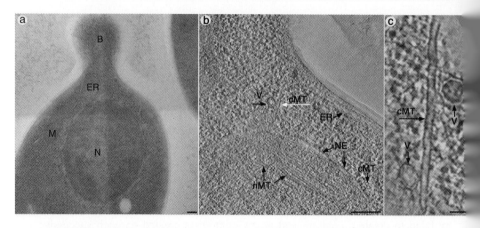

Fig. 4. Tomographic reconstruction of a large cellular area. (a) Four adjacent image frames (1024 × 1024 pixels each) were collected using a pixel size of 2.1 nm with automated image capture software developed in our laboratory. The total area captured equals 4.3 × 4.3 μm. The pixels that overlapped between adjacent frames were blended, the tilt series was aligned using local alignment methods, and a tomogram was calculated. Bar: 200 nm (b) The resulting tomogram shows fine structure detail when panned at high magnification. Individual microtubules in the nucleus (nMT) and cytoplasm (cMT) can be detected as well as vesicles (v), endoplasmic reticulum (ER), and the double membrane of the nuclear envelope. Bar: 200 nm. (c) The Imod "slicer" tool was used to extract a piece of image data that contained an oblique cytoplasmic microtubule (white arrow). The orientation of the slice was then rotated 9° about the horizontal axis to bring the entire length of the microtubule into one view. Bar: 50 nm. Supplementary video material for (a) and (b) is available on our web site, http://bio3d.colorado.edu/MIE.html. Video sequence 4a is a movie of 78 aligned serial tilt series from the blended image stack. Video sequences 4b and 4b′ are movies of the resulting tomographic reconstruction shown at low magnification (video sequence 4b) and viewed at higher magnification with image viewing software (video sequence 4b′). Note that the trajectory of the oblique cytoplasmic microtubule [marked by the white arrow in (b)] can be followed to the bud neck when stepping through the tomographic volume.

a volume as possible. To image large cellular areas with high, uniform resolution, multiple image frames or "montages" must be assembled. Such images of large cellular areas or even entire cell profiles can also be used for tomographic analysis, using minor modifications of the procedures outlined earlier. First, the edges of the montaged pieces must be "blended" to reconcile any differences in intensity or position between the pixels that overlap between adjacent frames of the images. Figure 4a shows an image of a montaged cell captured using four, 1024 × 1024 pixel frames, corresponding to an area of 4.3 × 4.3 μm from this small-budded yeast cell. Note that there are small differences in image intensity among the four frames, yet the edges between the frames show gradual transitions.

When imaging comparatively large areas (>2 × 2 μm), we have found that there are distortions in the resulting images that are not homogeneous over the area sampled. These distortions are likely due to differential thinning and/or shrinkage of the section under the action of the electron beam or from slight bending of the section on the Formvar film, all of which can lead to a poor global alignment. To solve this problem, the tilt alignment program has been modified to use a subset of fiducial markers to solve for local alignments, which are then taken into account by the back projection program. Video sequence 4a shows a movie of an aligned, blended tilt series. The nucleus, bud, and most of the cell is represented in this montage.

Details of the cellular fine structure can be clearly resolved in tomograms calculated from montaged images when panning the tomogram at higher magnification. For example, the double membrane of the nuclear envelope, the nuclear microtubules of the mitotic spindle (nMT), vesicles (v), and cytoplasmic microtubules (cMT) are clearly resolved (Figs. 4b and 4c). Supplementary video material for Fig. 4 shows movies of 40, 2.11-nm tomographic slices of the whole montaged cell (video sequence 4b) and viewed with a zoom tool at higher magnification to show an area of the cell that contains the mitotic spindle and cytoplasmic microtubules (video sequence 4b'). The trajectory of the oblique cytoplasmic microtubule identified in Fig. 4b (cMT, arrow) can be followed up to the bud neck in this movie. The high density of spindle microtubules in the nucleus is also evident.

The Imod software developed in our laboratory provides an additional tool, the "slicer" window, which allows an investigator to rotate the orientation of a particular tomographic slice to view a structure from any chosen angle. For example, the tomographic slice shown in Fig. 4b contains an oblique cytoplasmic microtubule (cMT). The orientation of the tomographic slice was then rotated 9° about the horizontal axis to sample the entire microtubule axis in one view (Fig. 4c, cMT). The close associations between vesicles and this cMT can be appreciated more easily when viewed in this manner.

Higher resolution, dual-axis tomograms can also be calculated from montaged images and can yield a tremendous amount of structural data from larger cellular volumes. Figures 5a and 5b and video sequences 5a and b show tomographic slices from a 1.5×2.9-μm area of a cell that has just undergone cytokinesis. Portions of the two resulting cells are shown at the top and bottom of the image, respectively. The cells are filled with ribosomes, secretory vesicles, ER, and Golgi. Electron-dense cell wall material can be seen deposited between the adjacent cells. When the tomogram is viewed at higher magnification (Fig. 5b), details of the electron-dense secretory vesicles (SV), a budded cisterna of the Golgi membrane (G), and the plasma membranes of the two adjacent cells (arrows) can be clearly resolved. When moving slice by slice through the tomographic volume (video sequence 5b), vesicles of different sizes can be detected as well as the contours of ER and Golgi that are in close proximity to the electron-dense secretory granules.

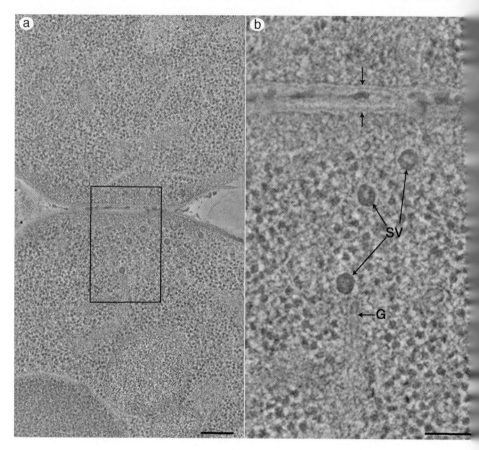

FIG. 5. Dual-axis tomography of montaged images. (a) A 1.4-nm-thick tomographic slice through a cell that has just undergone cytokinesis. Portions of the two resulting cells are at the top and bottom of the image, respectively. The total cellular area represented in the tomogram is 1.5 × 2.9 μm. Bar: 200 nm. (b) The tomogram can be viewed at higher magnification using a zoom tool in the image viewing software [boxed area in (a)] revealing a tremendous amount of cellular fine structure. Individual, electron-dense secretory vesicles (SV), a Golgi cisterna containing a coated bud (G), and the individual leaflets of the plasma membranes of the adjacent cells can be seen (arrows). Bar: 100 nm. The 3D fine structure is best appreciated when viewing supplementary video material (found at our web site, http://bio3d.colorado.edu/MIE.html). Video sequences 5a and 5b show movies of 50, 1.4-nm-thick tomographic slices through the volumes represented in (a) and (b). The cells are filled with ribosomes, and membrane systems such as ER and Golgi can be detected. Secretory vesicles filled with electron-dense material similar to the material deposited between the plasma membranes of the adjacent cells can be detected.

Finally, even larger volumes can be reconstructed by calculating tomograms from serial thick sections.[38,39] The serial tomograms can then be knitted together to obtain high-resolution structure information about large cellular volumes. This procedure has been applied in a tomographic study describing the reconstruction of a $3 \times 4 \times 1.2$-μm region of the Golgi apparatus in a pancreatic β cell.[40] The potential disadvantage of this method is the apparent loss (up to 25–40 nm) of material that occurs between serial thick sections, which sometimes makes alignment and modeling across these gaps difficult.

Conclusion

Due to recent technical advances in specimen preparation and image analysis, electron tomography is now a routine method with which to study the 3D fine structure of cellular subsystems. Tomographic reconstruction results in 3D image data with resolutions that significantly exceed those obtained by serial section reconstruction. In addition, the technique enables one to slice and view an organelle in many orientations, giving a more comprehensive understanding of complex biological structures. When combined with a mutant analysis of yeast, tomography will likely provide new structural insights to help define structure–function relationships in this organism.

Acknowledgments

The authors thank Mary Morphew and Kent McDonald for the preparation of yeast for electron microscopy and for advice on freeze substitution, the laboratory of David Drubin for providing yeast cells for the emerging bud studies, Andrew Staehelin for use of a high-pressure freezer, and Mark Ladinsky for organizing the supplementary video material on the laboratory web page. This work was supported by Grant RR-00592 to J. R. McIntosh from the National Center for Research Resources of the National Institutes of Health (NIH) and by NIH Grant GM59992 to M. Winey.

---

[38] G. E. Soto, S. J. Young, M. E. Martone, T. J. Deernick, S. Lamont, B. Carragher, K. O. Hama, and M. H. Ellisman, *NeuroImage* **1,** 230 (1994).

[39] M. S. Ladinsky, D. N. Mastronarde, J. R. McIntosh, K. E. Howell, and L. A. Staehelin, *J. Cell Biol.* **144,** 1135 (1999).

[40] B. J. Marsh, D. N. Mastronarde, K. F. Buttle, K. E. Howell, and J. R. McIntosh, *Proc. Natl. Acad. Sci. U.S.A.* **98,** 2399 (2001).

[6] Cryomethods for Thin Section Electron Microscopy

By KENT MCDONALD and THOMAS MÜLLER-REICHERT

> By means of snap-freezing and/or glycerol impregnation, yeast cells have been frozen-fixed so as to preserve the life of the organisms. The utilization of this "ideal" fixation in conjunction with an improved freeze-etching technique... has yielded a great deal of new information about fine structure in baker's yeast
> H. Moor and K. Mühlethaler (1963)[1]

> If the freeze-substitution fixation technique were consistently applied to all kinds of yeasts, it would be a great advance, because several problems that are induced by chemical fixation in the preparation of yeast cells for electron microscopy are excluded
> M. Baba and M. Osumi (1987)[2]

Introduction

The current vigorous activity in yeast genomics research has meant that researchers are looking at yeast cells through the light microscope (LM) with unprecedented intensity. When this is combined with ever-better LM technology and reporter molecules, the result is an abundance of information about the distribution of important molecules in the yeast cell. For a subset of these observations, the resolving power of even the best LMs is insufficient to determine the exact spatial relationship of the molecule of interest to particular yeast organelles. For example, we can use LM to map proteins to the spindle pole body (SPB) with certainty, but we cannot always tell if they are on the inner or outer face, or both. A mutant missing a known cell wall protein may affect the growth properties of the cells, but is the wall thinner or is it altered in some more subtle way? These are the types of questions more suited to the resolving power of the electron microscope (EM). This chapter presents information needed to get state-of-the-art preservation of yeast cell ultrastructure for both phenotype characterization and EM immunolabeling.

Checkpoint

Before we get to the detailed protocols, let us briefly review what it means to do EM in terms of time, money, and equipment. Then, we will consider whether you need state-of-the-art ultrastructure or if more routine processing methods might work for you. This section is primarily for the person who has not done EM before

---

[1] H. Moor and K. Mühlethaler, *J. Cell Biol.* **17,** 609 (1963).
[2] M. Baba and M. Osumi, *J. Electr. Micros. Tech.* **5,** 249 (1987).

or for the person who is collaborating with an EM investigator but who may not realize the time and effort required to achieve these goals.

Equipment Resources

First, you should know what equipment is available for use. Just because there is a microscope in your department or institution does not mean that you will be able to do EM. The microscope itself is only one of a number of pieces of specialized EM equipment that you will need. For thin-sectioning projects, here is a minimal list: a fume hood that can be used with fixatives; an oven set to 60° for resin polymerization; a microtome for cutting thin sections; a diamond knife, or glass knife maker; and a tranmission electron microscope. The age and condition of the last three items can determine how quickly and smoothly you will get results. The Reichert–Jung or Leica microtomes are good, especially the Ultracut E or later models. If you use an RMC microtome, try to find a Model 6000 or later. Porter–Blum MT-2 microtomes work well, but the lighting arrangement makes learning more difficult.

Sectioning will be the most difficult and time-consuming technique to learn. It requires developing hand–eye coordination skills that can only be learned with repetition and practice. Learning to cut routine thin sections well enough will take several weeks. While it is essential to start out practicing with a glass knife, the sooner you can begin using a diamond knife, the better (a new diamond knife will cost $1000–2000).

An older electron microscope can take perfectly good images if it has been well maintained. Microscopes 10 years old or less should be satisfactory.

Human Resources

More critical and harder to find than the equipment you need will be the people who can show you how to use it. There may be a core EM facility at your institution where you can be trained. Otherwise, you will be dependent on others to show you how to section and how to use a microscope.

Financial Resources

Electron microscopy can be expensive. You will need to buy a diamond knife and there may be charges for using and/or maintaining the equipment. Finally, you should count on taking at least 1 to 2 months of full-time work to achieve even a modest EM goal if you are working alone and have not done EM before.

Safety Considerations

There are many toxic chemicals used in preparing samples for electron microscopy. Lead, uranium, osmium, arsenic, and cyanide are only the most commonly

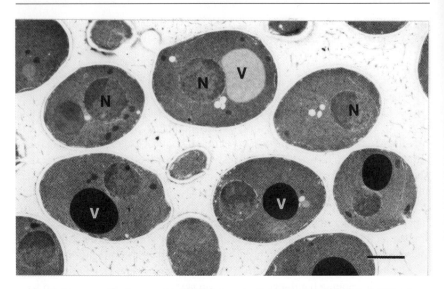

FIG. 1. Low-magnification overview of budding yeast cells in a thin section. Both nuclei (N) and vacuoles (V) have smooth, rounded countours indicating minimal distortion during fixation and embedding. Compare these shapes with those of nuclei and vacuoles in conventionally prepared cells in the literature. Note also the nearly uniform texture of the vacuolar contents, although not all have the same density. These cells were cryofixed by HPF, freeze substituted in 2% osmium tetroxide plus 0.1% uranyl acetate in acetone at $-90°$ in a homemade device [J. Z. Kiss and K. L. McDonald, *Methods Cell Biol.* **37,** 311 (1993)], and embedded in Epon–Araldite resin. Sixty-nanometer sections were poststained with 2% aqueous uranyl acetate for 5 min and lead citrate for 3 min. Bar: 2 μm.

used. Consult a basic EM textbook such as Bozzola and Russell[3] and read the chapter on safety. If you are unfamiliar with any chemical, consult the material safety data sheets that are available for it. In general, you should always wear gloves when handling any EM chemical, work in a fume hood, and wear protective clothing. If you are working with cryogens such as liquid nitrogen, you should wear eye protection and consult the article by Sitte *et al.*[4]

Choosing the Right Methods of Specimen Preparation

When starting an EM study, the first question you will need to answer is: What is the best way to prepare samples to see what I want to see? The key to the answer is knowing what level of resolution is needed. To address this, think about how you will present the final image. For example, if you need to show the whole cell (Figs. 1 and 2) to demonstrate the distribution of endomembranes and their

---

[3] J. J. Bozzola and L. D. Russell, "Electron Microscopy," 2nd Ed. Jones and Bartlett, Sudbury, MA, 1999.

[4] H. Sitte, K. Neumann, and L. Edelmann, *in* "Cryotechniques in Biological Electron Microscopy" (R. Steinbrecht and K. Zierold, eds.), p. 285. Springer-Verlag, Berlin, 1987.

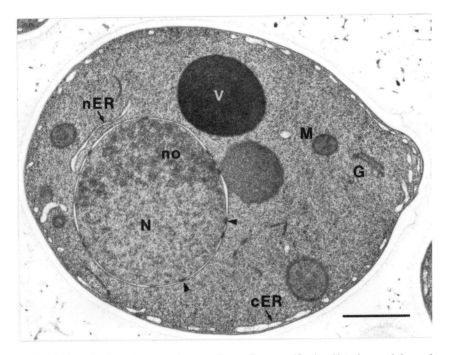

FIG. 2. Example of a well-preserved yeast cell at medium magnification. Note the round shape of the nucleus (N) and vacuole (V), plus details in the nucleus such as nuclear pores (arrowheads) and the distinct texture of the nucleolar region (no). Endomembranes such as Golgi (G), mitochondria (M), nucleus-associated endoplasmic reticulum (nER), and cortical endoplasmic reticulum (cER) are easily visualized. Finally, note the overal uniform texture of the nucleoplasm and cytoplasm. Cells were prepared by HPF, freeze substituted in 2% osmium tetroxide plus 0.1% uranyl acetate in acetone at $-78°$ by the dry ice method, and embedded in Spurr's–Epon resin mixture. Fifty nanometer sections were stained for 4 min with aqueous uranyl acetate and for 2 min in lead citrate. Bar: 0.5 μm.

relationship to other organelles, then only low to moderate resolution will be required. However, if you are interested in morphometry, such as the spacing to the nearest nanometer between spindle microtubules,[5,6] or are doing three-dimensional (3D) reconstructions[7,8] or tomography,[9] then you will need to prepare samples in such a way that the highest level of structural preservation is achieved. Most projects will fall somewhere in between these extremes.

[5] M. Winey, C. L. Mamay, E. T. O'Toole, D. N. Mastronarde, T. H. Giddings, K. L. McDonald, and J. R. McIntosh, *J. Cell Biol.* **129,** 1601 (1993).
[6] R. Ding, K. McDonald, and J. R. McIntosh, *J. Cell Biol.* **120,** 141 (1993).
[7] M. Baba, N. Baba, Y. Ohsumi, K. Kanaya, and M. Osumi, *J. Cell Biol.* **94,** 207 (1989).
[8] M. Baba, K. Takeshige, N. Baba, and Y. Ohsumi, *J. Cell Biol.* **124,** 903 (1994).
[9] E. O'Toole, M. Winey, J. R. McIntosh, and D. N. Mastronarde, *Methods Enzymol.* **351,** [5], 2002 (this volume).

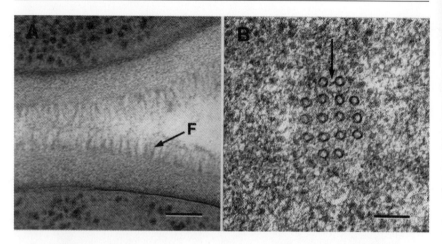

FIG. 3. Examples of high-magnification views of yeast. (A) When cells are fixed without spheroplasting, it is possible to see details in the cell wall such as the fine filaments (F) making up the outer layer. Cells processed as in Fig. 1. Bar: 100 nm. (B) Cross section through a bundle of central mitotic spindle microtubules in *S. pombe*. Cryofixation preserves the material between these square-packed MTs, including possible cross-bridge molecules (arrow). Cells prepared as in Fig. 1. Bar: 100 nm.

For low and moderate resolution studies, you can use any of a number of published protocols of a type called "conventional" methods. These are variations on the basic methods developed by electron microscopists over the years using immersion fixation and room temperature diffusion chemistry to prepare cells for EM observation. The advantage of this approach is that the reagents and equipment needed are readily available and familiar to anyone who has done EM before and are easily learned by anyone just beginning.

To preserve ultrastructure for high-resolution study (Figs. 3–5) or postembedding immunolabeling (Fig. 7), you will probably want to use one of the cryofixation methods, followed by low-temperature dehydration or so-called freeze substitution. The advantage of this approach is that molecules remain in close approximation to where they were in the living cell, i.e., they are not extracted or grossly rearranged as they can be by conventional methods. For illustrations and discussion of why this is so, see Kellenberger[10] and Steinbrecht and Müller.[11] The disadvantages to using cryomethods are that they are unfamiliar to most EM people and may require specialized equipment that is not always readily available. In the

[10] E. Kellenberger, *in* "Cryotechniques in Biological Electron Microscopy" (R. A. Steinbrecht and K. Zierold, eds.), p. 35. Springer-Verlag, Berlin, 1987.

[11] R. A. Steinbrecht and M. Müller, *in* "Cryotechniques in Biological Electron Microscopy" (R. A. Steinbrecht and K. Zierold, eds.), p. 149. Springer-Verlag, Berlin, 1987.

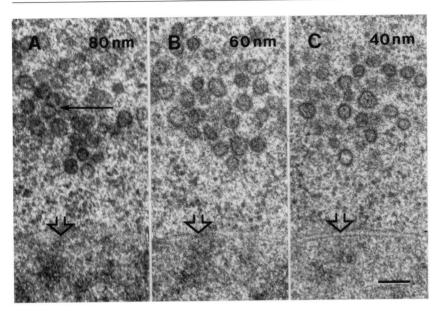

FIG. 4. Illustration of how section thickness can influence visualization of cell details. Adjacent sections were cut at 80, 60, and 40 nm thickness. In thicker sections (A), the small vesicles are superimposed on each other, making it difficult to see clear outlines of individual vesicles. As the sections get thinner (B and C), single vesicle contours become easier to see. Likewise, membranes that are not perfectly perpendicular to the plane of the section, such as the nuclear envelope shown here (arrows), are easier to detect in thinner sections. Cells were prepared as in Fig. 1. Bar: 100 nm.

section that follows, we will list and/or describe the best available conventional and cryopreparation methods for yeast cells.

Conventional Methods

If you decide that conventional methods will work for you, you should survey the literature to see if someone has previously worked out detailed protocols for visualizing what you are hoping to see. If you find such a method and you like the results, then try duplicating it. If it does not work for you, then contact the authors, if possible, and see if anything crucial was left out in the methods section of their article. That is essential for reproducing results.

If you cannot find a published article that deals specifically with your interest, consider using one of these listed below.

1. B. Byers and L. Goetsch, *Methods Enzymol.* **194**, 602 (1991): In the previous volume in this series Breck Byers and Loretta Goetsch provide all the details needed to prepare cells for EM using routine specimen preparation methods.

This protocol is especially useful for visualizing spindle microtubules, spindle pole bodies, and nexin filaments. Although there are no illustrations of yeast EM in this article, good examples can be seen in other Byers references cited therein.

2. R. Wright, M. Basson, L. D'Ari, and J. Rine, *J. Cell Biol.* **107,** 101 (1988): This is the original reference to a method also described in the article by Byers and Goetsch cited earlier. It uses an osmium tetroxide–potassium ferrocyanide solution to enhance membrane contrast. Cytoskeletal features are not well visualized.

3. P. Y. Goh and J. Kilmartin, *J. Cell Biol.* **121,** 503 (1993): A slight modification of the basic Byers and Goetsch protocol, using an acetone step following ethanol dehydration and acetone–resin mixtures for infiltration instead of going straight into Spurr's resin or ethanol–Spurr's mixtures. Results of this approach plus a further modification of the lead citrate poststaining method are illustrated in Donaldson and Kilmartin.[12] This method specially useful for SPBs and cytoskeletal elements.

4. C. A. Kaiser and R. Schekman, *Cell* **61,** 721 (1990): This protocol uses potassium permanganate to enhance membrane contrast and is especially useful for visualizing components of the endomembrane system. Other cell components are not well preserved or easy to visualize.

5. M. G. Heiman and P. Walter, *J. Cell Biol.* **151,** 719 (2000): This paper includes another variation on the permanganate strategy for visualizing membranes.

Depending on your goals, one of the methods just given should suffice for any low- to medium-resolution EM study or a high-resolution study if you are looking at a robust organelle, such as the SPB or nexin filaments. Making modifications of basic methods using acetone instead of ethanol for dehydration or Epon–Araldite instead of Spurr's resin probably would not be harmful and may even give better results. Do not be afraid to experiment if you are not fully satisfied with your first results.

Cryofixation Methods

Background

The most compelling argument for the superiority of cryofixation for EM is the fact that, after freezing, yeast cells can be warmed and resume normal growth.[1,13] This means that the ultrastructure of the yeast cell has been cryoimmobilized in its native state. If the frozen cells are given minimal processing, such as fracturing, coating with metal, and observed as a replica in the electron microscope, the resulting images may be as close as we can get to observing native cell structure

---

[12] A. D. Donaldson and J. V. Kilmartin, *J. Cell Biol.* **132,** 887 (1996).
[13] I. Erk, G. Nicolas, A. Caroff, and J. Lepault, *J. Microsc.* **189,** 236 (1998).

at EM resolution. These images can then serve as the "gold standard" by which all other specimen preparation methods for yeast can be judged.

The excellent preservation of fine structure following freezing first shown by Moor and Mühlethaler[1] may be less impressive to the yeast community because they are freeze-fracture images. Freeze fracture is excellent for membrane studies but lacks the convenience and familiarity of thin sections and is also less suitable for cytoskeletal ultrastructure. Using low-temperature methods, Hereward[14] was able to produce very good thin sections of yeast cells that were rapidly frozen, freeze substituted, and embedded in resin. The paper by Tanaka and Kanbe[15] on *Schizosaccharomyces pombe* was the first comprehensive study of yeast using thin sections, followed soon by Baba and Osumi's[2] work on *Klockera* sp. and *Saccharomyces cerevisiae*. The excellent images in the latter paper show budding yeast ultrastructure in thin sections. In the years since then, methods have improved and become more automated and commercially available. The next section discusses some options for fast freezing of yeast cells.

Freezing Procedures

There are five common ways to do fast freezing: plunge, impact, spray, double propane jet, and high pressure. For the purposes of this chapter, we will only discuss plunging, double propane jet, and high-pressure freezing. The interested reader can learn more about freezing theory and practice by consulting Echlin,[16] Robards and Sleytr,[17] Gilkey and Staehelin,[18] or Steinbrecht and Zierold.[19] Fast freezing can be done with homemade, inexpensive equipment, but if you are able to purchase a commercial machine, you will gain something in convenience and reproducibility.

Because the details of how to freeze are so varied depending on the device and are well-covered elsewhere, we will not devote much space to that aspect here. Instead, we will discuss only the advantages and disadvantages of each type of freezing as well as how to prepare cells for any of these types. We will then pick up the detailed processing procedure from the point of having a frozen sample stored in liquid nitrogen.

Plunge Freezing. All the early studies of fast frozen yeast mentioned earlier used plunge freezing, and from the mid-1980's to the present, Baba and

---

[14] F. V. Hereward, *Protoplasma* **87**, 8 (1976).
[15] K. Tanaka and T. Kanbe, *J. Cell Sci.* **80**, 253 (1986).
[16] P. Echlin, "Low Temperature Microscopy and Analysis." Plenum, New York, 1992.
[17] A. W. Robards and U. B. Sleytr, *in* "Practical Methods in Electron Microscopy" (A. M. Glauert, ed.), Vol. 10. Elsevier, Amsterdam, 1985.
[18] J. C. Gilkey and L. A. Staehelin, *J. Electron Microsc. Tech.* **3**, 177 (1986).
[19] R. A. Steinbrecht and K. Zierold (eds.), "Cryotechniques in Biological Electron Microscopy." Springer-Verlag, Berlin, 1987.

colleagues have used this method to get well-preserved yeast morphology[7] and immunolabeling[8] in thin sections. Their equipment is probably homemade according to the recommendations of Elder et al.[20] However, if you choose to make your own plunge freezing device, we recommend using the instructions provided by Howard and O'Donnell[21] or consulting one of the references in Echlin[16] or Robards and Sleytr.[17] These books contain much valuable information on how to optimize the conditions for successful plunge freezing. There are also commercial machines available such as the Leica KF80 or the the Bal-Tec TFD 010.

In this technique, yeast cells are concentrated and put between small (around 3 mm) metal disks held together by forceps or some other type of clamp and then plunged rapidly into the cryogen cooled to about −185° with liquid nitrogen. Sometimes cells are spread in a thin layer on a plastic film for plunging.[22] Propane is the most often used cryogen for immersion cooling and can be obtained from any hardware store. Once frozen, the cells can be stored in liquid nitrogen for an indefinite time.

The main advantage of this technique is its simplicity and low cost. The disadvantage is that only very small volumes of cells can be frozen at a time, and some of these will be lost in subsequent processing. Samples from many separate freezing runs can be pooled for freeze substitution and embedding.

Double Propane Jet Freezing. This equipment design takes advantage of the fact that when a sample is cooled from both sides at once with jets of cryogen, a threefold increase in freezing rate over plunge freezing is obtained.[16] Samples are put between two metal holders, which are then placed into a commercial machine, the Bal-Tec JFD030 [Technotrade International, Manchester, NH (U.S.); Bal-Tec AG, Liechtenstein (non-U.S.)], where they are simultaneously hit from both sides with cooled liquid propane. Then the sample is automatically dropped into a well of cooled propane for temporary storage until it is transferred out for further processing.

If you are only planning to cryofix yeast cells, the Bal-Tec JFD 030 propane jet freezer is the best way to do so. It is relatively inexpensive and very well suited to the reliable freezing of yeast. You will have many more well-frozen cells per run than with plunge freezing, much of the operation is automated, and quality control will be improved.

High-Pressure Freezing (HPF). The most versatile freezing device currently available is the Bal-Tec HPM 010 high-pressure freezer. Leica also markets a high-pressure freezer called the EMPact. It is essential for freezing large (up to several hundred micrometers) cells or tissues without cryoprotectant and is convenient

---

[20] H. Y. Elder, C. C. Gray, A. G. Jardine, J. N. Chapman, and W. H. Biddecombe, *J. Microsc.* **126,** 45 (1982).
[21] R. J. Howard and K. L. O'Donnell, *Exp. Mycol.* **11,** 250 (1987).
[22] R. W. Ridge, *J. Electron Microsc.* **39,** 120 (1990).

for working with larger volumes of cells for each freezing run. All of the samples shown in this chapter (Figs. 1–7) were prepared by HPF. Cells are loaded into cylindrical specimen holders 2 mm wide and 100–200 μm deep,[23] although yeast cells can be well frozen in specimen holders of different design with a fixed depth of 600 μm.[5,6,9,24] For a general discussion of HPF, see Moor[25] and Kiss and Staehelin,[26] and for details of how to do the procedure, see McDonald.[23]

Disadvantages of the high-pressure freezer are cost and space. The Bal-Tec is large and must be used in a large or well-ventilated room, although the EMPact is smaller and can be used in a smaller space. Although it seems the EMPact has the advantage, it should be noted that it has yet to be proven as versatile as the Bal-Tec machine for a wide variety of sample sizes and shapes. If you are only going to work with yeast, then the Leica EMPact should work well,[27] but the volume of well-frozen sample may not be much more than what you can obtain using the Bal-Tec JFD030 propane jet freezer.

Preparing Cells for Freezing

Materials Needed

50 ml of cells
Controlled temperature shaker
Millipore (or similar) 15-ml suction filtration apparatus (Fisher)
Whatman #1 filter paper, 2.5-cm circles (Fisher)
0.4-μm polycarbonate filter (Fisher)
Toothpicks or wooden sticks sharpened and flattened out at one end to use as scraper
Fine forceps (Ted Pella)
Vacuum source
Freezing apparatus of your choice

Procedure

1. Have the cells growing in liquid culture at early log phase (OD 0.2–0.4). If they have to be in stationary phase or are growing on agar slants, you should be aware that they will probably not fix as well.
2. Concentrate the cells by gentle suction filtration onto a filter, taking care not to let them dry out.

---

[23] K. L. McDonald, *Methods Mol. Biol.* **117**, 77 (1999).
[24] S. Craig, J. C. Gilkey, and L. A. Staehelin, *J. Microsc. (Oxford)* **48**, 103 (1987).
[25] H. Moor, in "Cryotechniques in Biological Electron Microscopy" (R. A. Steinbrecht and K. Zierold, eds.), p. 175. Springer-Verlag, Berlin, 1987.
[26] J. Z. Kiss and L. A. Staehelin, in "Rapid Freezing, Freeze Fracture, and Deep Etching" (N. J. Severs and D. M. Shotton, eds.), p. 89. Wiley-Liss, New York, 1995.
[27] D. Studer, *Microsc. Microanal.* **5**(Suppl. 2: Proceedings), 432. Springer-Verlag, New York, 1999.

3. If the cells seem to dry out too quickly place the filter pad with cells on YPD medium in 1% agar in a small petri dish[28] and cover. This will keep them hydrated.
4. Scrape a small amount from the filter to load into the holder chosen for freezing. We find that a toothpick works well for this.
5. To get the fastest heat transfer and the greatest depth of good freezing, use the smallest convenient volume in the sample holder to be frozen. Also, reducing the liquid phase of the paste to a minimum will improve the yield of well-frozen cells.
6. Freeze cells by your chosen method.

Preparing Fixatives for Freeze Substitution

Fixatives for freeze substitution (FS) are usually made by dissolving either glutaraldehyde and/or osmium in an organic solvent. Low (0.05–0.2%) concentrations of glutaraldehyde are used for immunolabeling studies, and osmium tetroxide at 1–2% is usually best for preservation of the high-resolution ultrastructure. Acetone is the organic solvent of choice, although it is worth experimenting with other solvents such as methanol, diethyl ether, or even combinations of solvents. We also include a little uranyl acetate (0.1% or more) in these mixtures because it enhances membrane contrast. This section gives one recipe that deviates from this basic formula of glutaraldehyde for immunolabeling and osmium for morphology, and a later section (Resin Infiltration and Embedding) emphasizes how nontypical combinations of fixatives and resins can give excellent results for specific organelles.

Materials Needed

 Pure acetone (we prefer 100-ml bottles of EM grade; various EM vendors as listed in the Appendix)
 Pure methanol (use a new bottle)
 One (1) gram crystalline osmium tetroxide (various EM vendors)
 Uranyl acetate crystals (various EM vendors)
 10% Glutaraldehyde in acetone (Electron Microscopy Sciences)
 LN_2 (liquid nitrogen) in 4 liter dewar
 50-ml conical tube with screw cap
 Vial rocker
 Aluminum foil
 Sonicator
 Nalgene cryovials (Nalge)
 Nalgene cryovial rack (Nalge)

---

[28] M. G. Heiman and P. Walter, *J. Cell Biol.* **151,** 719 (2000).

#2 Pencil
Repeater pipettor to dispense up to 2-ml aliquots
Styrofoam box large enough to accept Nalgene cryovial rack

To Make up 50 ml of Fixative

Remember! You must wear double gloves for this procedure and work well inside in a fume hood that draws well. Osmium in acetone is very volatile!

1. Label, with the #2 pencil, about 35 Nalgene 2.0-ml cryovials as "2.1-Os-UA," which stands for 2% osmium tetroxide plus 0.1% uranyl acetate. For the glutaraldehyde fixative, label the vial "2.1 G-UA," which stands for 0.2% glutaraldehyde plus 0.1% uranyl acetate. If you are using a freeze substitution solvent other than acetone, indicate that also.

2. Put vials with tops off in vial rack.

3. Make up stock solution of 5% uranyl acetate in methanol by adding 0.1 g of uranyl acetate to 2 ml pure methanol, cover with foil, and put on a rocker to dissolve. It should dissolve in about 10 min, but if it does not, just let it rock longer. Different sources of uranyl acetate crystals seem to dissolve at different rates.

4. Sonicate a 1-g OsO_4 ampule for about 10 sec to loosen crystals from glass.

5. Fill a 50-ml conical tube with 45 ml pure acetone.

6. Add 1-g OsO_4 crystals (or 1 ml of a 10% solution of glutaraldehyde in acetone) to the 45 ml of acetone, cover, and mix until completely dissolved.

7. Add 1 ml 5% uranyl acetate stock to OsO_4–acetone, add pure acetone to 50 ml, and mix well.

8. Use a repeater pipettor to dispense 1.5 ml into each cryovial. Work quickly, but not carelessly. Cap cryovials as soon as they are filled.

9. Add LN_2 to a Styrofoam box to a level that will cover cryovials in the rack.

10. Slowly lower rack into LN_2. Vials must remain upright so fixative will remain in the bottom of the vial.

11. When completely frozen, transfer vials to the LN_2 storage device until ready to use.

Müller-Reichert and Antony Freeze Substitution Cocktail

A fixative cocktail developed by one of us (T.M-R.) and Claude Antony (Curie Institute in Paris) utilizes glutaraldehyde, osmium tetroxide, and uranyl acetate dissolved in acetone as a fixative for cells embedded in Lowicryl HM20 for immunolabeling.[29] Prepare in the same way as described earlier with a final concentration of 0.1% glutaraldehyde, 0.25% uranyl acetate, and 0.01% osmium

---

[29] T. Müller-Reichert, A. Ashford, C. Antony, and A. Hyman, *Microsc. Microanal.* **6**(Suppl. 2: Proceedings), 300. Springer-Verlag, New York, 2000.

tetroxide. In this particular study, the low concentration of osmium added membrane contrast but did not interfere with immunolabeling (Fig. 7). Be aware that higher concentrations of osmium may block UV polymerization. This formulation may not work for every primary antibody, probably because the osmium blocks the antibody–antigen reaction.

Storing Samples between Freezing and Freeze Substitution

After freezing the sample, you may not be ready to proceed directly to freeze substitution and may want to store your samples. One way is to put them into a cryovial and place them in a LN_2 refrigerator or dewar storage system. What we prefer to do is transfer them directly into the frozen freeze substitution fixative cocktail, labeling the tube with all necessary information.

Materials Needed

1. LN_2 in a 4 liter dewar
2. Small (approximately 20 cm^2 and 7–8 cm deep) Stryofoam box
3. Soft (e.g., #2) pencil
4. Cryovials that you can write on with a pencil, e.g., Nalgene cryovials (Nalge). These should not be immersed in liquid nitrogen (LN_2), but we have been using them for years without problems. We keep the tops slightly loose, which allows LN_2 to escape. If you are at all concerned about this safety aspect, choose a vial type designed for LN_2 immersion and put small holes in the top to allow LN_2 to vent.
5. A cryovial rack such as a Nalgene cryovial rack (Nalge). The tubes interlock with the rack so you can take off the top without having to hold the vial in liquid nitrogen. Other vial types also have racks with interlocks.
6. Cryogloves can be purchased from any scientific supplier such as Fisher or VWR. We use Neoprene diver gloves obtained from a diving supply store. They provide good insulation but are form fitting and allow you to do more delicate tasks than with large cryogloves.
7. Large forceps (Electron Microscopy Sciences) with broad and strong serrated jaws.
8. Fine tweezers (Ted Pella). Glue a piece of foam or use self-adhesive Velcro on the outside where you grip them for insulation against cold.
9. Frozen freeze substitution fixative in cryovials (see "Preparing Fixatives for Freeze Substitution").
10. Several small plastic caps such as those from a glass scintillation vial or similar types.
11. Liquid nitrogen storage system. We use a dewar and cane system.
12. Canes for cryovial storage in the dewar.

Procedure

1. Place the cryovial rack in a Styrofoam box and fill with LN_2 so the rack is just covered.
2. Transfer frozen sample in the specimen holder from the freezing device to a Styrofoam box using LN_2 in the small plastic cap.
3. Transfer freeze substitution fixative vial from fixative storage in LN_2 to the Styrofoam LN_2 box.
4. Holding the fixative vial with the large forceps, use the soft pencil to label the vial with information about the sample.
5. Place the vial in the rack, take off the top, and then place both top and bottom under LN_2.
6. Precool the tips of the fine tweezers in the LN_2, pick up the sample from the plastic cap, and transfer it *under LN_2* to the fixative vial. Do not let it warm up at this or any other stage prior to freeze substitution. Make sure the sample is on top of the fixative and not stuck to the side of the vial by static charges, which can happen easily.
7. Put the fixative vial/sample in the rack, making sure that LN_2 in the vial covers the sample.
8. Screw the top on the vial.
9. Set vial aside in box or in a separate LN_2 dewar flask.
10. Repeat steps 2–9 as necessary until all samples are in appropriate fixative vials. If you do not know yet how you want to fix the samples, they can be put in an empty cryovial for storage and transferred to fixative vials later using this same setup.
11. Transfer labeled vials to a LN_2 storage system until you are ready to do freeze substitution.

Freeze Substitution

Once the samples are frozen, you will need to dehydrate them and fix them before embedding in resin. Freeze substitution is dehydration and fixation at low temperature, typically $-78°$ to $-90°$. There is a huge literature about freeze substitution theory and practice, which is summarized in Echlin.[16] This section covers some of the equipment and processing options for freeze substituting yeast cells.

Dry Ice Method

This is an easy and inexpensive option for freeze substitution that has one important advantage over other methods, i.e., the ability to automatically and constantly agitate the samples during the substitution process. The disadvantage is that you cannot hold intermediate cold temperatures such as $-50°$ for low-temperature embedding in Lowicryl HM20.

Materials Needed

5–10 lbs dry ice in slices
Styrofoam box approximately 22 cm deep by 15 cm wide by 20 cm long (inner dimensions)
Aluminum block with 13-mm holes (Fisher)
Digital thermometer (Cole-Parmer Instrument Co., 625 East Bunker Court, Vernon Hills, IL 60061-1844)
Type T thermocouple probe (Omega Engineering, Inc., One Omega Drive, Box 4047, Stamford, CT 06907)
Rotary shaker
Elastic cord
Cryotube with hole in the cap to admit the thermocouple probe
Acetone

Procedure

1. Load a Styrofoam box two-thirds full with dry ice. Break the slices or, better yet, cut them with a handsaw into pieces that just fit the Styrofoam box. By having the box deeper than wide, you keep more dry ice under the aluminum block for longer as it sublimates. It may take some trial and error to see how much dry ice you need for the time you choose to do the substitution.

2. Into one of the holes in the block place a cryotube filled with acetone into which has been inserted a thermocouple probe through the vial cap.

3. Put the aluminum block on the ice to cool. Cover the top of the block with aluminum foil to prevent frost buildup on the surface. Arrange the block so the cryotubes are vertical, horizontal, or somewhere in between. Horizontal will give the greatest mixing during agitation.

4. When the acetone has cooled to about $-78°$ (it usually takes about 1 hr), transfer HPF-frozen samples in fixative from liquid nitrogen storage into the holes of the block.

5. Fill the remainder of the box with dry ice, covering the top of the aluminum block and the vials with aluminum foil.

6. Put the lid on the box and place the whole assembly onto the rotary shaker. (You must have a shaker that will take this much weight.)

7. Use elastic cord to tie down the Styrofoam box on the shaker. You may also have to tie down the rotary shaker so it does not "walk" off the bench top.

8. Start the shaker and rotate at a rate of at least 100 rotations per minute.

9. Shake continuously until the ice has sublimed and the temperature of the vials has risen to the appropriate temperature for the kind of resin you intend to use for infiltration and embedding. The entire process from the initial transfer of samples onto dry ice to warm up should take 3–4 days. This is a conservative estimate, and you may find you can shorten the time for your samples.

10. For epoxy embedding such as Epon or Epon–Araldite or for methacrylate resins such as LR White, warm to room temperature, rinse, and embed as usual (see "Resin Infiltration and Embedding").

11. For low-temperature embedding, find a way to hold the samples at low temperature for several days, perhaps in a low-temperature freezer. Then follow the manufacturer's instructions for resin : solvent exchange and polymerization.[30]

Low-Temperature Freezers

In many laboratories there are $-80°$ freezers used for storing antibodies and other proteins. Before you try to do freeze substitution in such a freezer, get the permission of everyone who has samples there. Osmium tetroxide is a potent fixative in vapor form. While there is less chance of osmium vapors escaping at low temperatures, it is still a potential hazard. If you try this method, you may have to find dedicated fixative freezers. In any case, put the samples in a sealed container. See the article by Howard and O'Donnell (1987) for further discussion and ideas for using low temperature freezers for freeze substitution.

Materials Needed

1. A $-80°$ (or cooler, down to $-90°$ if you can) freezer. A chest type works better than an upright if you have a choice.

2. A $-20°$ freezer. Most freezer compartments of regular refrigerators are about this temperature.

3. Aluminum block with holes of a diameter that the cryovials fit snugly in them. For Nalgene cryovials (Nalge), 13-mm holes work well. Have a shop drill holes in a block of aluminum for you, or you can use the dry block heater modules available from various vendors (e.g., Fisher) for a block with twenty 13-mm holes.

4. A plastic container (e.g., Tupperware, Rubbermaid) that will hold the aluminum block with vials.

5. Drierite or similar indicator desiccant.

Procedure

1. Put the aluminum block in a plastic container with a little Drierite in the bottom and then place in the $-80°$ freezer to precool. Do this several hours before you are ready to start freeze substitution. If you are in a hurry, you can cool the block in liquid nitrogen.

2. Transfer the fixative cryovials containing the sample from liquid nitrogen storage to the cooled block in the freezer. Leave them in the freezer for 2–3 days. Agitate the block manually at least several times a day during this time and for all subsequent steps at different temperatures.

[30] G. R. Newman and J. A. Hobot, "Resin Microscopy and On-Section Immunocytochemistry." Springer-Verlag, Berlin, 1993.

3. Transfer the block/vials to the $-20°$ freezer and leave for 12 hr.
4. Transfer the block/vials to a $4°$ refrigerator for 2 hr.
5. Transfer the block/vials to a fume hood until they reach room temperature. Do not open before or the cooled vials will act as a cold trap and water will condense in the acetone. The samples are now ready for resin infiltration (see "Resin Infiltration and Embedding").

Leica AFS Device

The Leica AFS automatic freeze substitution device is a convenient instrument for doing freeze substitution. You can set three separate temperatures for one run, plus the rate of warm up between temperatures. A big advantage is the ability to hold any temperature for low-temperature embedding. It also has a provision for UV illumination of an oxygen-free chamber for low-temperature polymerization of resins. The main disadvantage is the lack of any mechanism for agitating the samples to enhance the action and exhange of chemicals.

Materials Needed

30–35 liters liquid nitrogen
AFS machine

Procedure

1. Program the AFS according to the manufacturer's instructions. There are many options for time and temperature for freeze substitution. For yeast, we set the first temperature to $-90°$ for 72 hr, warm up to $-20°$ at a rate of $10°/hr$, and hold at $-20°$ for 12 hr. Warm up to room temperature ($20°$) at $10°/hr$.
2. Fill the dewar with liquid nitrogen. We do this directly from a 160-liter LN_2 tank, pouring the LN_2 into the AFS through a large funnel. It helps to cover the top of the AFS with lab bench paper to prevent the cold nitrogen gas from chilling the top too much.
3. Start the program. This can be done while you are filling the dewar.
4. When the temperature in the working chamber reaches $-90°$, transfer your samples from LN_2 storage to the AFS chamber.
5. Agitate the samples in the working chamber by hand as often is convenient over the entire time of the program.
6. At the end of the program, when the samples are at room temperature, they are ready for resin infiltration (see next section).

Resin Infiltration and Embedding

From the point where samples have been freeze substituted and warmed to room temperature, processing is the same as conventional methods as discussed in any basic EM text.[3] For yeast fixed in osmium tetroxide, we typically embed in an Eponate 12–Araldite resin (Figs. 1, 3, and 4) or a mixture of Spurr's and Epon resins (Fig. 2). There is a mistaken impression in some yeast papers that you need to use Spurr's low-viscosity resin to get adequate resin infiltration. We find that this is not the case following freeze substitution and advise against using Spurr's by itself because it is less beam stable than Eponate12–Araldite for high magnification work.

If we are going to do immunocytochemistry on sections, then we usually embed in LR White resin. There are other options available such as LR Gold and the Lowicryl resins, which can give better morphology and/or immunolabeling. Space does not permit an adequate description of these low-temperature embeddng methods, but the interested reader can find detailed instructions for using these resins in Newman and Hobot[30] and in the manufacturer's instruction sheets. These resins are available from various EM vendors and come with instructions.

In some cases, we have found it useful to combine a "hard" fixation with high (1% or more) concentrations of glutaraldehyde with a low-temperature resin such as Lowicryl HM20. We find this gives good preservation of spindle pole body (SPB) structure as well as other organelles (Fig. 5). The Boulder Lab for 3-D Fine Structure has adopted this method for use in tomographic analysis of yeast SPBs and spindles.[9]

Materials Needed for Epoxy Resins

    EM-grade acetone in an unopened 100-ml bottle (various EM vendors)
    Pure ethanol (various vendors), preferably in an unopened bottle
    2-ml Eppendorf tubes
    Standard laboratory microcentrifuge
    Plastic transfer pipettes, fine-tipped, 4–7 ml volume
    500-ml plastic jars for waste disposal
    Fine-tipped needle or forceps (Ted Pella)
    Plastic (e.g., Tripour) beakers, 100–250 ml (various EM vendors)
    Stir bar, 1/2 inch
    Magnetic stirrer
    Embedding resins: (a) Eponate 12–Araldite 502 (sold as a kit by Ted Pella), (b) Eponate 12 kit with DMP-30 (Ted Pella), and (c) low viscosity "Spurr" kit (Ted Pella)
    BEEM embedding capsules, size 00 (various EM vendors) (optional)
    BEEM capsule Holder for size 00 (various EM vendors) (optional)
    Embedding oven set to 60°
    Vacuum desiccator
    Rotary mixer (various EM vendors)

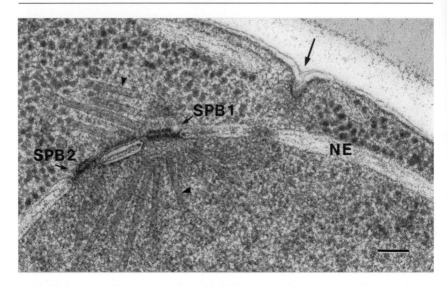

FIG. 5. By varying fixative and resin combinations, visualization of certain organelles can be enhanced. We employed a fixative usually used for morphology (1% glutaraldehyde) with a resin normally used for immunocytochemistry (Lowicryl HM20) to get improved visualization of the spindle pole bodies (SPB1 and 2) and associated microtubules (arrowheads). This method also seems to reveal the details around plasma membrane invaginations (arrow) better than others we have tried. The ribosome-free area just under the invagination may represent an actin patch. Antibody staining of lightly fixed material in Lowicryl HM20 sections with antiactin antibodies shows specific labeling in these areas [K. L. McDonald, *Methods Mol. Biol.* **117,** 77 (1999)]. Cells were cryofixed by HPF, freeze substituted in 1% glutaraldehyde plus 0.1% uranyl acetate at −78° by the dry ice method, except they were warmed to −35° in a Pelco UVC2 Cryo Chamber (Ted Pella, Inc.), infiltrated with Lowicryl HM20 resin according to the manufacturer's recommendations, and the resin UV polymerized for 1 day at −35° and for 2 days 20°. Seventy-nanometer sections were poststained for 4.5 min in 1.5% uranyl acetate in 70% methanol and for 2.5 min in lead citrate. Bar: 100 nm.

Disposable latex laboratory gloves, acetone-resistant (see scientific catalog for grades of gloves and chemical resistance properties if unsure)
15- and 50-ml conical plastic tubes

Procedure for Epoxy Resin

1. Make up resin *without accelerator* (to be added later). Put a stir bar in a 100-ml Tripour or other disposable plastic beaker and place the beaker on a top-loading lab balance with a capacity of at least 100 g and an accuracy to 0.1 g or more. We list two of the epoxy recipes routinely used.

Epon–Araldite mix: 6.2 g of Eponate 12, 4.4 g Araldite 502, 12.2 g of DDSA (dodecenyl succinic anhydride) (accelerator), and 0.8 ml of BDMA (benzyldimethylamine)

Epon–Spurr's mix: If you decide to use this resin formulation, you should make up each of the resins separately and then when they are well mixed, pour the Spurr's into the Epon beaker and mix again until homogeneous.

Spurr's		Epon	
ERL 4206	10 g	Eponate 12	25 g
DER 736	6 g	DDSA	13 g
NSA (accelerator)	26 g	NMA	12 g
DMAE	24 drops	DMP-30	32 drops

Place on a magnetic stirrer and stir gently until all components are mixed, taking care not to mix in too much air. Air bubbles can be removed in a vacuum desiccator. The total volume of resin needed will depend on the number of samples you have, so multiply or divide the recipe accordingly. If you use 2-ml tubes for each sample, you will need a total of about 9 ml resin per sample tube for all resin solution changes.

2. In a fume hood (important!) and wearing gloves, carefully rinse out the fixative solution with at least three 5-minute changes of pure acetone. Be careful not to let the vial and sample dry out; always leave a little liquid in the bottom to cover the sample each time you rinse. Dispose of the fixative–acetone mixture in an appropriately labeled plastic waste container, which should be kept in the fume hood. Place the vials on a mixer or rotator between changes.

3. Transfer the samples from cryovials to labeled Eppendorf tubes by pouring from one to the other, or any method in which the sample does not get exposed to air.

4. With a fine-tipped needle, or forceps, remove the sample from the holder in which it was frozen. We do this to get maximum exposure of the cells to all the subsequent solution changes, but it is not absolutely necessary and you can leave the cells on the holder. Baba and Osumi[2] leave the specimen and holder intact and only separate them after polymerization in resin.

5. Infiltrate with acetone : resin (without accelerator) mixtures and pure resin according to the following schedule:

Mixture	Time
3 acetone : 1 resin	30–60 min
2 acetone : 2 resin	60 min
1 acetone : 3 resin	120 min
Pure resin	30–60 min
Pure resin	Overnight
Pure resin	30–60 min

All steps should be done on a rotator for thorough mixing. After removing the samples from the freezer holders, spin down the cells in a microfuge between each step. The times given are minimum times and leaving cells for a longer period is not harmful.

6. Make up enough *resin plus accelerator* for the samples. (Work in a fume hood, as the accelerator is quite volatile and the fumes are noxious if not toxic. This is especially true for the Spurr's resin components.) If embedding in BEEM capsules, each capsule takes about 0.75 ml of resin. The accelerator is measured out volumetrically and you will need to add 0.8 ml of BDMA (benzyldimethylamine) for each 22.8 g of Epon–Araldite resin (see step 1 given earlier). If you are making up the Epon–Spurr's mixture, add 24 drops DMAE to the Spurr's resin components (already mixed well) and 32 drops DMP-30 to the Epon components. Use a plastic transfer pipette for the drops. Never use glass pipettes for this, or any other fluid transfer, as there are microscopic glass shards in the pipettes which will end up in your polymerized resin. When sectioning, these can ruin the diamond knife. Mix the resin thoroughly on a magnetic stirrer and then degas in a vacuum desiccator.

7. Replace the resin without accelerator with the resin plus accelerator. If you are pelleting the samples, be sure to resuspend the pellet in the new resin. Otherwise, you may not get infiltration of the accelerator and the cells will not polymerize properly. Mix for a minimum of 4 hr.

8. Spin down samples. The resin (especially Epon–Araldite) may be quite viscous and may require longer spin times than for resin without accelerator.

9. Put the Eppendorf tube in a 60° oven for 48 hr to polymerize. When cured, cut off the tip containing the cells and remount on a blank stub for sectioning. If you prefer to use BEEM capsules, transfer pieces of pellet to capsules fill with resin, place in a BEEM capsule holder, and polymerize for 48 hr.

Materials Needed for LR White

> LR White, hard grade. Various EM vendors (we prefer Ted Pella, because the catalyst is supplied separately)
> Gelatin capsules, size 00 (various EM vendors)
> Gelatin capsule holder (various EM vendors) to hold gelatin caps upright while in the oven

Procedure for LR White

1. If not already done, add catalyst to LR White and mix well for several hours. We usually add all the catalyst to the whole bottle and then keep it in a refrigerator at 4° where it will be good for at least a year.

2. Rinse cells with pure acetone as in step 2 for epoxy resins.

3. Rinse three times for 5 min each with pure ethanol.

4. Make up a 1 : 1 mixture of pure ethanol and LR White in sufficient volume for the number of samples you have. Mix for at least 5 min.

5. Transfer samples to Eppendorf tubes without exposure to air and remove from freezer holders as in step 4 for epoxy resins.
6. Change sample solution to a 1 : 1 mixture of ethanol : LR White and put on a rotator for at least 1 hr or longer.
7. Rinse with pure LR White twice for 30 min each and then add LR White and leave overnight on a rotator.
8. Degas the appropriate volume of LR White in a vacuum desiccator for 10–15 min and then rinse cells with the degassed resin.
9. Spin cells down into a hard pellet and then transfer the pieces of pellet to gelatin capsules filled with degassed LR White. Add tops to gelatin capsules, making them as filled with resin as you can because too much oxygen in the capsule will inhibit resin polymerization. You will need to be wearing gloves for this operation.
10. Place capsules in a 50° oven for 2 days. We find it convenient to put the capsules in an aluminum block that has been drilled out to hold the bottom portion of the gelatin capsule. This may help in promoting even polymerization, but is not essential.

Sectioning

We typically cut 40- to 50-nm sections of yeast that are processed by cryomethods. The cells are denser than they would be if processed by conventional methods that typically extract much more of the cytoplasm. Thinner sections are particularly useful for imaging microtubules, but in our experience, all components image a little crisper than in thicker sections (Fig. 4). Regular thin sections in the 70- to 80-nm range are useful for low-magnification images because they tend to have more contrast. A useful exercise is to cut sections of 40, 60, and 80 nm thickness and put them on the same grid. Repeat until you have 5–10 grids. Then use these to experiment with different poststaining (next section) conditions. There is usually a combination of thickness and poststain that gives the most satisfying images, depending on the structures being studied.

Yeast are so small relative to most cells that you can get many in even the smallest section area. To try serial sections, trim the block face so that it is much wider than it is high, e.g., 250 μm wide by 50 μm high. Then, it is relatively easy to find the same cell in adjacent sections.

Poststaining

In all thin section imaging, the final contast is mostly due to the heavy metals used in specimen preparation and/or poststaining (sometimes called counterstaining). In freeze substitution, we typically add osmium and/or uranyl acetate and then more uranium and lead in poststaining. Thinner sections have less heavy metal than thick ones from the same block, so they benefit by poststaining procedures

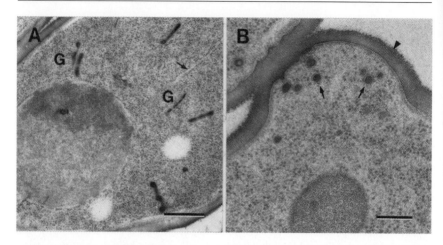

FIG. 6. Examples of yeast cell sections treated with Hamamoto's tannic acid poststain. Both Golgi membranes (A) and bud vesicles (B) show increased contrast with this method. The endoplasmic reticulum (arrow, A), however, does not show enhanced staining. Cell walls (arrowhead, B) are also selectively stained. Cells were cryofixed by HPF, freeze substituted with the Müller–Reichert–Antony cocktail in a Leica AFS machine at $-90°$, warmed to room temperature, and embedded in LR White. Fifty-nanometer sections were poststained for 3 min in 1% aqueous tannic acid, rinsed well with distilled H_2O, and then stained with 2% aqueous uranyl acetate for 8 min and lead citrate for 3 min. Bar: 0.5 μm (A) and 0.25 μm (B). Micrographs courtesy of Susan Hamamoto, University of California, Berkeley.

that give more contrast. We typically poststain with 2% (w/v) uranyl acetate in 70% methanol instead of aqueous uranyl acetate because it gives more contrast. Stain for 4–5 min in uranyl acetate and for 2–3 min in lead citrate. Longer times simply make the cells darker and may even reduce the relative contrast.

*Method for Tannic Acid Poststaining**

Tannic acid can be used as a poststain to bring up membrane contrast on selected organelles such as Golgi membranes, vesicles in the developing bud, and the cell wall (Fig. 6). In cells prepared by fast freezing, freeze substituted in the Müller-Reichert and Antony cocktail, and embedded in LR White, grids are stained for about 3 min in 1% aqueous tannic acid (Mallinkrodt) before regular poststaining in uranyl acetate and lead citrate. The extent to which this method works with other freeze substitution cocktails and other types of resin has not been determined. Here is the method as currently used.

---

* This section has been prepared by Susan Hamamoto (Schekman Laboratory, University of California, Berkeley, CA) based on a method she has developed.

1. Prepare 1% tannic acid in water. Mix until completely dissolved.
2. Float grid, section side down, on a droplet of 1% tannic acid for 3–5 min. Use filmed, carbon-coated grids. If tannic acid touches the side of the grid that is not carbon coated or if you do not use carbon-coated grids, the tannic acid will stick nonspecifically as a film over the entire grid surface.
3. Rinse thoroughly with water.
4. Poststain with uranyl acetate and lead citrate as usual.

Microscopy

At this point there is little you can do to improve the image, although you can improve contrast by using smaller objective apertures and working at lower than normal accelerating voltages, e.g., 60 kV instead of 80 or 100 kV. However, you can still *undo* some of what you have gained by how you record the image. Today's microscopes tend to come with CCD cameras for digital imaging and it is very easy to capture images compared to the traditional method of recording on 3.25 by 4-inch film stock. We encourage you to avoid using the CCD for primary imaging if you intend to publish images of your cells. If you are just counting organelles or measuring contours, then the low-resolution CCD image will be fine. However, even the best CCDs cannot match the 10,000 pixels per inch resolution of EM film.[31] Furthermore, you have a large area of recorded image with each piece of film equivalent to about 10 gigabytes of digital data. If you need to have a digital image, there are easy ways to convert the negative, or a small portion of it to a digital file. Space does not permit more discussion of this issue here, but we strongly encourage you to read the article by Heuser[31] before recording EM images directly by CCD.

When using the microscope, you should assume that all the images you take will be for publication. This is not true, of course, but it will help you decide the magnifications to use and how to orient the cell or organelle in the frame of the negative. Film images can be enlarged up to three or four times in the darkroom without a noticeable loss of image quality. For this reason, it is better to take a lower magnification view than trying to fit everything into the frame at the highest possible magnification. Taking a low *and* a high magnification image is often useful (Fig. 7). When in doubt, take extra pictures. It is usually easier and more cost effective to sort through extra negatives to find what you need than to go back to the microscope and hope to find the same cell again.

Figure 7 illustrates EM localization data using immunogold. The same immunolabeled mitotic spindle is printed at low (Fig. 7A) and high (Fig. 7B) magnification. With excellent reproduction and a trained eye, it is possible to see the gold labels on the spindle in the low magnification view (Fig. 7A); but most people, even

[31] J. Heuser, *Traffic* **1**, 614 (2000).

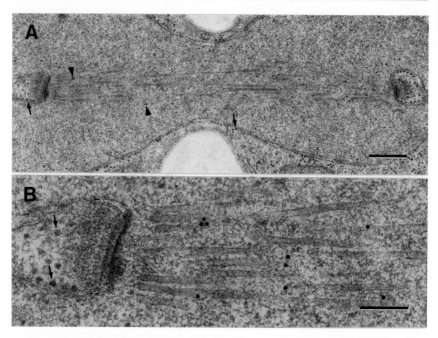

FIG. 7. A longitudinal section through a cdc20 mitotic spindle showing immunolocalization of a kinetochore protein (Ndc10p) with a 10-nm gold secondary. At low magnification (A) it is difficult to distinguish gold particles (arrowheads) from ribosomes (arrows). However, at high magnification (B) the difference is obvious. The arrows point to ribosomes; gold particles are associated with spindle MTs. Cells were cryofixed by HPF, freeze substituted in a Leica AFS at $-90°$ for 3 days, and infiltrated with Lowicryl HM20 resin at $-50°$ over 2 days. Resin polymerization was by UV irradiation at $-50°$ in the AFS for 2 days, followed by 2 days UV irradiation at $20°$. Sixty-nanometer sections were immunolabeled with antibody to Ndc10p and 10-nm gold secondary antibody, poststained for 4 min in 2% uranyl acetate in 70% methanol and for 2.5 min in lead citrate. Bar: 0.25 μm (A), and 100 nm (B).

if they can see the gold, may not be able to tell if the dark dots behind the left SPB and elsewhere in the cell (arrows in Fig. 7A) are gold particles or ribosomes. In an appropriately enlarged view (Fig. 7B), the differences are clear. Both low- and high-magnification views for context and clarity are important. Heiman and Walter[28] give a good example of how to arrange low-, medium-, and high-magnification views of ultrastructure in a minimum of space.

Judging Image Quality

Having invested weeks of work in preparing samples and then finally being able to look at them on the microscope, you will be faced with the task of deciding whether what you are seeing is well preserved or not. How do you know the difference between good and bad fixation? Are there artifacts, either of conventional or cryopreparation? One way to address this issue is to ask an experienced electron microscopist to help you, and another is to consult some references. For general information, consult the "Interpretation of Micrographs" section in Bozzola and Russell,[3] Chapter 1 in Maunsbach and Afzelius,[32] and Chapter 2 in Griffiths.[33]

For specific interpretation of yeast micrographs, we offer the following suggestions.

1. Look at the shape of the nucleus. Freeze-fracture images of yeast consistently reveal a round nucleus and vacuoles,[1] except in areas where they abut each other or if the vacuoles are fragmented. Likewise, images of yeast prepared by cryofixation, freeze substitution, and embedding for sectioning show profiles of nuclei and vacuoles that are smooth and round (Figs. 1, 2, and 6).[2,7,13,14] In general, this is not true for yeast prepared by conventional methods. For membrane preservation, use the shape of the nuclear envelope as a guide to the amount of collapse and distortion suffered in sample preparation. It is reasonable to assume that other membrane components of the cell have suffered similar fates.

2. Look at the shape and texture of the vacuole. In well-fixed cells (Figs. 1 and 2),[2] the vacuole is usually a rounded organelle with contents of uniform density unless there is a lot of autophagic activity.[8] In many published images of conventionally fixed cells, the vacuoles are an irregular mixture of black and white areas. The black area shows coagulated and condensed vacuole contents and the white area shows where the contents were before they collapsed.

3. Look at the shape of the microtubules. Microtubules in well-fixed cells should look straight or gently curved as we know them to be from light microscopy of living cells or cryo-EM.[34] Figures 5 and 7 show straight MTs from mitotic spindles. Published images of fast-frozen, freeze-substituted spindle MTs show even longer MTs that are not obviously distorted. Some MT-related organelles such as the SPB are not good indicators of fixation problems because they are structurally very compact. SPBs are often visualized quite well by conventional methods.[35–37]

4. Look for extraction of the ground cytoplasm. The cytoplasm of cells is packed with proteins and does not contain a lot of empty space. In sections of well-preserved cells, the ground cytoplasm should have an even density without large (bigger than a ribosome) white spaces. White spaces in electron micrographs of cytoplasm usually mean that something has been lost from that space. Selective extraction can be very useful in enhancing contrast of the organelle of interest, but

---

[32] A. B. Maunsbach and B. A. Afzelius, "Biomedical Electron Microscopy." Academic Press, San Diego, 1999.
[33] G. Griffiths, "Fine Structure Immunocytochemistry." Springer-Verlag, Berlin, 1993.
[34] E. M. Mandelkow, E. Mandelkow, and R. A. Milligan, *J. Cell Biol.* **114,** 977 (1991).
[35] B. Byers and L. Goetsch, *J. Bacteriol.* **124,** 511 (1975).
[36] M. P. Rout and J. V. Kilmartin, *J. Cell Biol.* **111,** 1913 (1990).
[37] A. D. Donaldson and J. V. Kilmartin, *J. Cell Biol.* **132,** 887 (1996).

it is a problem for immunolabeling you want to retain as much of the native cell structure as possible, including soluble proteins.

About EM Immunolabeling of Resin Sections

Space does not permit any extended discussion of labeling resin sections for EM. This topic is well covered in Newman and Hobot,[30] Griffiths,[33] and Kellenberger and Hayat,[38] and specifically for yeast in Clark.[39] Mulholland and Botstein[40] discuss similar procedures for labeling LR White sections. We have explained how to get yeast cells embedded in LR White. Whether sections of these cells will label with immunogold or some other secondary depends almost entirely on the reactivity of the primary antibody with fixed proteins at the section surface. The fact that antibodies may work well on Western blots or at the light microscope level is, unfortunately, no guarantee that they will work well on sections for EM. In addition to the information in the aforementioned references, our protocols for on-section (sometimes called postembedding) labeling as shown in Fig. 7 can be found in McDonald et al.[41] Finally, if the images of cells embedded in LR White are not satisfactory, use a low temperature embedding medium such as Lowicryl HM20, as shown in Figs. 5 and 7. See Newman and Hobot[30] for detailed protocols for Lowicryl embedding.

Appendix : Vendor Information

1. General EM supplies:
    a. Ted Pella, Inc.
       P.O. Box 492477
       Redding, CA 96049-2477]
       Tel: 800-237-3526 (USA),
       800-637-3526 (CA),
       800-243-7765 (Canada),
       Fax: 916-243-3761
    b. Electron Microscopy Sciences
       321 Morris Road
       Box 251
       Fort Washington, PA 19034
       Tel: 800-523-5874,
       215-646-1566
       Fax: 215-646-8931

---

[38] E. Kellenberger and M. A. Hayat, in "Colloidal Gold: Principles, Methods and Applications, Vol. 3," (M. A. Hayat, ed.), p. 1, Academic Press, San Diego, 1991.
[39] M. W. Clark, *Methods Enzymol.* **194,** 608 (1991).
[40] J. Mulholland and D. Botstein, *Methods Enzymol.* **351,** [4], 2002 (this volume).
[41] K. L. McDonald, D. J. Sharp, and W. Rickoll, in "Drosophila: A Laboratory Manual," (W. Sullivan, M. Ashburner, and S. Hawley, eds.), 2nd Ed., p. 245. Cold Spring Harbor Press, Cold Spring Harbor, NY, 2000.

c. PLANO
      W. Plannet GmbH
      Ernst-Befort-Straße 12
      D-35578 Wetzlar, Germany
      Tel: ++49 6441 9765 0
      Fax: ++49 6441 9765 65
   d. Science Services
      Landshuter Allee 116
      D-80637 Munich, Germany
      Tel: ++49 89 15980 280
      Fax: ++49 89 15980 282
2. Cryopreparation equipment:
   a. Technotrade International
      7 Perimeter Road
      Manchester, NH 03103-3343
      Tel: 603-622-5011,
      800-875-3713
      Fax: 603-622-5211
   b. Bal-Tec A.G.
      FL 9496 Balzers
      Liechtenstein
      Tel: +75 388 12 12
      Fax: +75 388 12 60
   c. Leica Mikrosysteme GmbH
      Hernalser Hauptstrasse 219
      Postfach 95
      1170 Vienna
      Austria
      Tel: +43 1 486 990
      Fax: +43 1 486 1579
   d. Leica Microsystems, Inc.
      111 Deer Lake Rd.
      Deerfield, IL 60015
      Tel: 847-405-0123,
      800-248-0123
      Fax: 847-405-0300

Section II

Biochemistry

[7] Vacuolar Proteases and Proteolytic Artifacts in *Saccharomyces cerevisiae*

By ELIZABETH W. JONES

Introduction

The yeast *Saccharomyces cerevisiae* contains a large number of proteases that are located in various compartments (e.g., cytosol, vacuole, mitochondria, endoplasmic reticulum, and Golgi complex) and membranes (e.g., vacuole, endoplasmic reticulum, Golgi complex, and plasma) of the cell. These include endoproteinases, carboxypeptidases, aminopeptidases, and dipeptidylaminopeptidases.[1–6] Some of the proteases pose significant impediments to the analysis of biochemical processes and/or purification of proteins and can generate artifacts concerning the activity, structure, and, even, intracellular location of proteins.

Of the many cellular proteases, lumenal vacuolar proteases probably comprise the major source of problems. Found soluble within the vacuole are endoproteinases A and B (PrA and PrB), carboxypeptidases Y and S (CpY and CpS), aminopeptidase I (ApI; called LAPIV by Trumbly and Bradley[7]), and aminopeptidase ApY (formerly yscCo or ApCo).[1–3] Salient characteristics of the vacuolar proteases are summarized in Table I. Also listed are the relevant structural genes and inhibitors for the enzymes. Of the vacuolar proteases, protease B (PrB) is thought to be the greatest source of protease problems for several reasons. Some time ago John Pringle compiled a list of protease artifacts, most of which were caused by PrB.[8] The pH optimum of PrB is near neutrality, unlike the acidic pH optimum of PrA, the other major vacuolar endoproteinase. Conditions that activate PrB and free it from its cytosolic inhibitor I_B [heat and sodium dodecyl sulfate (SDS), see later][8–13] are

[1] E. W. Jones, G. C. Webb, and M. A. Hiller, *in* "The Molecular and Cellular Biology of the Yeast Saccharomyces: Cell Cycle and Cell Biology" (J. Pringle, J. Broach, and E. Jones, eds.), p. 363. Cold Spring Harbor Laboratory Press, Cold Spring Harbor, NY, 1997.
[2] E. W. Jones and D. G. Murdock, *in* "Cellular Proteolytic Systems" (A. Ciechanover and A. Schwartz, eds.), p. 115. Wiley, New York, 1994.
[3] H. B. Van den Hazel, M. C. Kielland-Brandt, and J. R. Winther, *Yeast* **12**, 1 (1996).
[4] T. Achstetter, C. Ehmann, and D. H. Wolf, *Arch. Biochem. Biophys.* **207**, 445 (1981).
[5] T. Achstetter, C. Ehmann, and D. H. Wolf, *Arch. Biochem. Biophys.* **226**, 292 (1983).
[6] T. Achstetter, O. Emter, C. Ehmann, and D. H. Wolf, *J. Biol. Chem.* **259**, 13334 (1984).
[7] R. Trumbly and G. Bradley, *J. Bacteriol.* **156**, 36 (1983).
[8] J. R. Pringle, *Methods Cell Biol.* **12**, 149 (1975).
[9] E. W. Jones, *Genetics* **85**, 23 (1977).
[10] D. H. Wolf and C. Ehmann, *FEBS Lett.* **92**, 121 (1978).
[11] G. S. Zubenko, A. P. Mitchell, and E. W. Jones, *Proc. Natl. Acad. Sci. U.S.A.* **76**, 2395 (1979).
[12] J. Schwenke, *Anal. Biochem.* **118**, 315 (1981).
[13] R. E. Ulane and E. Cabib, *J. Biol. Chem.* **251**, 3367 (1976).

TABLE I
VACUOLAR PROTEASES[a]

Enzyme	Abbreviation	Structural gene	Type	Inhibitors
Proteinase A	PrA	PEP4	Aspartic protease; endoproteinase	Pepstatin
Proteinase B	PrB	PRB1	Serine protease-subtilisin family; endoproteinase	DFP, PCMB, PMSF, Hg^{2+} chymostatin, antipain
Carboxypeptidase Y	CpY	PRC1	Serine carboxypeptidase	DFP, PMSF, TPCK, PCMB
Carboxypeptidase S	CpS	CPS1	Metallo(Zn^{2+})-carboxypeptidase	EDTA
Aminopeptidase I	ApI	LAP4	Metallo(Zn^{2+})-aminopeptidase	EDTA, PCMB, nitrilotriacetic acid, bestatin
Aminopeptidase Y	ApY	APE3	Metallo(Co^{2+})-aminopeptidase	EDTA, Zn^{2+}, bestatin

[a] DFP, Diisopropyl fluorophosphate; PMSF, phenylmethylsulfonyl fluoride; PCMB, 4-chloromercuribenzoic acid; TPCK, L-1-p-tosylamino-2-phenylethylchloromethyl ketone; EDTA, ethylenediaminetetraacetic acid.

built into most biochemical analyses. Finally, from the fact that strain EJ101, which on retest (by Jones, upon receipt from Dave Engelke, University of Michigan) proved to be of genotype α his1 prb1-1122 prc1-126, not α trp1 prb1-1122 pep4-3 prc1-126,[14] has proved to have significant utility for purposes of enzyme purification, the author infers that elimination of PrB but not PrA activity can eliminate some protease problems.

An additional source of protease problems is Zymolyase, as commercial preparations of Zymolyase contain substantial amounts of a protease that will catalyze hydrolysis of Azocoll (E. W. Jones, unpublished, 1985). Users of procedures that employ spheroplasts will need to bear this in mind.

This chapter describes conditions and procedures that affect the levels and activities of the vacuolar proteases and presents both genetic and biochemical methods for coping with protease problems. For many purposes, a combination of a mutant strain and an inhibitor cocktail may provide the optimum solution.

Effects of Growth Stage, Medium Composition, and Genotype on Enzyme Levels

Levels of activity of all of the vacuolar proteases except CpS increase as cells approach stationary phase.[5,6,15-20] For PrB, enzyme activity and antigen are

[14] R.-J. Lin, A. J. Newman, S.-C. Cheng, and J. Abelson, *J. Biol. Chem.* **260,** 14780 (1985).
[15] T. Saheki and H. Holzer, *Biochim. Biophys. Acta* **384,** 203 (1975).
[16] A. Klar and H. Halvorson, *J. Bacteriol.* **124,** 803 (1975).
[17] J. Frey and K. H. Röhm, *Biochim. Biophys. Acta* **527,** 31 (1978).
[18] D. H. Wolf and C. Ehmann, *FEBS Lett.* **91,** 59 (1978).
[19] T. Achstetter, C. Ehmann, and D. H. Wolf, *Biochem. Biophys. Res. Commun.* **109,** 341 (1982).
[20] C. M. Moehle, M. W. Aynardi, M. R. Kolodny, F. J. Park, and E. W. Jones, *Genetics* **115,** 255 (1987).

undetectable in log-phase cells growing on YEPD.[20,21] A small increase in activity occurs at the diauxic plateau. The largest increase, to a level at least 100 times that of log-phase cells, occurs as the cells enter stationary phase.[20,22] These increases as a function of growth stage may reflect release from glucose or cAMP repression, for all of the enzymes save CpS are expressed at higher levels when acetate or lactate serves as a carbon source or when cAMP levels fall (ApY was not tested).[7,15,17,22–26] For CpY, ApI, ApY, and PrB, increased enzyme levels are correlated with increased levels of mRNA.[22,24,26,27]

Levels of all of the vacuolar proteases increase on provision of a poor nitrogen source such as valine or proline[23,28,29] or Cbz-Gly-Leu for CpY and CpS (PrB did not respond; other enzymes were not tested).[18] Nitrogen starvation, under conditions conducive or not to sporulation, results in increased levels of proteases (CpS was not examined).[5,16,23,30,31] *snf2* and *snf5* mutations[32,33] result in high, constitutive levels of PrB activity, whether or not glucose is present in the medium.[23,25] Levels of the other proteases were not examined in the mutants. *snf1*, *snf3*, *snf4*, *snf6*, and *hex2* mutations were without effect on PrB levels.[23,25] Some investigators have suggested that the presence of peptone in the medium induces higher levels of vacuolar proteases. What evidence there is suggests the reverse, for protease levels on minimal medium seem higher than on complex media.[17,23,34] No systematic study has been reported, however.

Activation of Proteases

When cells are broken open, PrA, PrB, and CpY, at least, complex with their corresponding polypeptide inhibitors to form inactive complexes.[35–42] Addition of

[21] V. L. Nebes and E. W. Jones, unpublished observations (1989).
[22] C. M. Moehle, Ph.D. Thesis, Carnegie Mellon University, Pittsburgh, Pennsylvania (1988).
[23] R. J. Hansen, R. L. Switzer, H. Hinze, and H. Holzer, *Biochim. Biophys. Acta* **496,** 103 (1977).
[24] B. Distel, E. J. M. Al, H. F. Tabak, and E. W. Jones, *Biochim. Biophys. Acta* **741,** 128 (1983).
[25] C. M. Moehle and E. W. Jones, *Genetics* **124,** 39 (1990).
[26] D. Tadi, R. Hasan, F. Bussereau, E. Boy-Marcotte, and M. Jacquet, *Yeast* **15,** 1733 (1999).
[27] J. L. DiRisi, V. R. Iyer, and P. O. Brown, *Science* **24,** 278 (1997).
[28] D. J. Klionsky, L. M. Banta, and S. D. Emr, *Mol. Cell. Biol.* **8,** 2105 (1988).
[29] C. M. Moehle, C. K. Dixon, and E. W. Jones, *J. Cell Biol.* **108,** 309 (1989).
[30] H. Betz and U. Weiser, *Eur. J. Biochem.* **62,** 65 (1976).
[31] R. R. Naik, V. Nebes, and E. W. Jones, *J. Bacteriol.* **279,** 1469 (1997).
[32] L. Neigeborn and M. Carlson, *Genetics* **108,** 845 (1984).
[33] E. Adams, L. Neigeborn, and M. Carlson, *Mol. Cell. Biol.* **6,** 3643 (1986).
[34] T. R. Manney, *J. Bacteriol.* **96,** 403 (1968).
[35] H. Betz, H. Hinze, and H. Holzer, *J. Biol. Chem.* **249,** 4515 (1974).
[36] T. Saheki, Y. Matsuda, and H. Holzer, *Eur. J. Biochem.* **47,** 325 (1974).
[37] T. Saheki and H. Holzer, *Biochim. Biophys. Acta* **615,** 187 (1980).
[38] E. P. Fischer and H. Holzer, *Biochim. Biophys. Acta* **615,** 187 (1980).
[39] J. F. Lenney and J. M. Dalbec, *Arch. Biochem. Biophys.* **129,** 407 (1969).
[40] J. F. Lenney, *J. Bacteriol.* **122,** 1265 (1975).
[41] J. F. Lenney, *J. Biol. Chem.* **221,** 919 (1956).
[42] J. F. Lenney and J. M. Dalbec, *Arch. Biochem. Biophys.* **120,** 42 (1967).

SDS to crude extracts activates PrB, as does increasing the temperature.[8–13] The process of sample preparation (boiling) for denaturing electrophoresis activates PrB in the run up to maximum temperature, providing the opportunity for proteolysis and artifact generation. Protease B was initially reported not to stain with Coomassie Brilliant Blue. Freezing of the PrB-containing sample prior to its incorporation and heating in SDS-containing sample buffer resulted in staining of PrB polypeptide with Coomassie, presumably because freezing inactivated PrB and thus protected it from autodigestion (E. W. Jones, unpublished, 1977). The effects can be seen readily by comparing the protein signatures of wild-type and *prb1* strains or of extracts of wild-type cells with and without preincubation with phenylmethylsulfonyl fluoride (PMSF) prior to denaturation. The most striking effects are seen for the largest polypeptides. For a more extensive discussion of the contribution of denaturation to accelerated proteolysis, see Pringle.[8]

Protease Inhibitor Cocktails

Table II gives a number of different "cocktails" that have been used in various purification procedures. In several cases, cocktails were used even though the starting strain was BJ2168 or BJ926, strains that have greatly reduced levels of PrA, PrB, CpY, and ApI [as well as RNase(s) and the repressible alkaline phosphatase] because they carry *pep4-3*, *prb1-1122*, and *prc1* mutations. In the various cocktails, EDTA will obviously inhibit metalloproteases; mercurials will inhibit PrB and CpY. Pepstatin A will inhibit carboxylproteases, including PrA.[40] Chymostatin inhibits some serine proteases, including PrB,[40] and PMSF reacts covalently with active site serines of serine proteases,[43,44] such as PrB[13,45] and CpY.[46] 4-(2-Aminoethyl)-benzenesulfonyl fluoride (AEBSF) is an alternative to PMSF; it is highly soluble in aqueous solution and is more stable in solution than PMSF. The target(s) for leupeptin is unclear, for although it inhibits some serine and thiol proteases, it does not inhibit PrA, PrB, or CpY.[40,45] The target for benzamidine is likewise unclear. It inhibits trypsin, plasmin, and thrombin,[47–49] apparently through its resemblance to arginine.[50] No protease with trypsin-like specificity has been described in yeast. Data presented by Achstetter *et al.*[6] supply possible candidates, however. Benzamidine might significantly improve polypeptide integrity even when the enzyme source is a multiply protease-deficient strain

---

[43] A. M. Gold and D. Fahrney, *Biochemistry* **3**, 783 (1964).
[44] A. M. Gold, *Biochemistry* **4**, 897 (1965).
[45] E. Kominami, H. Hoffschulte, and H. Holzer, *Biochim. Biophys. Acta* **661**, 124 (1981).
[46] R. Hayashi, Y. Bai, and T. Hata, *J. Biochem.* **77**, 1313 (1975).
[47] J. W. Ensink, C. Shepard, R. J. Dudl, and R. H. Williams, *J. Clin. Endocrinol. Metab.* **35**, 463 (1972).
[48] S. L. Jeffcoate and N. White, *J. Clin. Endocrinol. Metab.* **38**, 155 (1974).
[49] M. Mares-Guia and E. Shaw, *J. Biol. Chem.* **240**, 1579 (1965).
[50] M. Krieger, L. M. Kay, and R. M. Stroud, *J. Mol. Biol.* **83**, 209 (1974).

TABLE II
REPRESENTATIVE INHIBITOR COCKTAILS[a]

Cocktail	Strain Normal	Strain Deficient	Source
0.1 mM EDTA 5 mM 2-mercaptoethanol 10% Me$_2$SO 10 mM NaHSO$_3$ 1 mM phenylmethylsulfonyl fluoride (PMSF)	Wild type	—	Badaracco et al.[b]
1 mM EDTA 1 mM PMSF 2 μg/ml pepstatin A 1 mM EGTA 1 mM benzamidine 1 μg/ml leupeptin	Wild type (A364A)	BJ926 (called PEP4D[c])	Jong et al.[c]
10 mM EDTA 2 mM benzamidine 1 mM PMSF	—	BJ926 (called PEP4D[c])	Johnson et al.[d]
1 mM EDTA 1 mM PMSF 1 μg/ml leupeptin 1 μg/ml pepstatin	Wild type	—	Olesen et al.[e]
1 mM EDTA 1 mM PMSF 1 mM benzamidine or 1 μg/ml pepstatin A	Wild type (X2180-1A)	"pep4-3"	Davis et al.[f]
1 mM EDTA 1 mM PMSF 2 μM pepstatin A 0.6 μM leupeptin	—	BJ926	Lue and Kornberg[g]
1 mM EDTA 1 mM PMSF 2 μM pepstatin A 0.6 μM leupeptin 2 μg/ml chymostatin 2 mM benzamidine	—	BJ926	Lue et al.[h]
1 mM ε-aminocaproic acid 5 μg/ml aprotinin 1 mM p-aminobenzamidine 1 μg/ml chymostatin 5 μg/ml pepstatin 250 μM PMSF 50 μM p-Chloromercuri-phenylsulfonic acid	—	BJ2168	Aris and Blobel[i]

[a] The strains used are indicated as normal or (protease) deficient. Protease-deficient strains (made by us and sent to the Yeast Genetics Stock Center and then to the ATCC) for which genotypes are given in the listing at the end of the chapter are identified by BJ numbers.
[b] G. Badaracco, L. Capucci, P. Plevani, and L. M. S. Chang, J. Biol. Chem. **258,** 10720 (1983).
[c] A. Y. S. Jong, R. Aebersold, and J. L. Campbell, J. Biol. Chem. **260,** 16367 (1985).
[d] L. M. Johnson, M. Snyder, L. M. S. Chang, R. W. Davis, and J. L. Campbell, Cell **43,** 369 (1985).
[e] J. Olesen, S. Hahn, and L. Guarente, Cell **51,** 953 (1987).
[f] T. N. Davis, M. S. Urdea, F. R. Mesiarz, and J. Thorner, Cell **47,** 423 (1986).
[g] N. F. Lue and R. D. Kornberg, Proc. Natl. Acad. Sci. U.S.A. **84,** 8839 (1987).
[h] N. F. Lue, A. R. Buchman, and R. D. Kornberg, Proc. Natl. Acad. Sci. U.S.A. **86,** 486 (1989).
[i] J. P. Aris and G. Blobel, J. Cell Biol. **107,** 17 (1989).

like BJ926, and although the author cannot document the finding, she includes it for its possible utility in designing experiments. Possibly ε-aminocaproic acid, aprotinin, and *p*-aminobenzamidine also are targeted to these enzymes, for they inhibit some of the same serine proteases as are inhibited by benzamidine. PrB is not inhibited by aprotinin (Trasylol).[45] The soybean trypsin inhibitor inhibits PrB, but at concentrations 100 times that shown in Johnson *et al.*[51] What its target here might be is unknown. EGTA will inhibit Ca^{2+}-dependent proteases, including the *KEX2* protease.[52] $NaHSO_3$, which is known to inhibit PrA and PrB,[42] is thought to enhance histone stability in nuclear preparations.[49,53]

Premade protease inhibitor cocktails are available commercially, some specifically formulated for yeast extracts (Roche Applied Science, Indianapolis, IN, Sigma, St. Louis, MO; Boehringer Ingelheim, Ridgefield, CT; Calbiochem, La Jolla, CA). The author has never used any of these prepared cocktails and cannot comment on their utility. When using inhibitors, it is important that they be included in all buffers at all steps. Because proteases can be cryptic through association with their inhibitors and because the purification steps may aid dissociation, protease activity can be generated during a purification. (See the final warning section for *in vitro* maturation of the PrB precursor present in *pep4* mutant extracts as a source of PrB activity during purification.) For PMSF, there is the added complication that the compound is quite unstable, particularly if the pH of the buffer is above 7.[54] If use of a protease-deficient mutant and/or use of PMSF is precluded for some reason, fairly extensive analyses of potential inhibitors are available.[40,42,45]

Protease-Deficient Mutants

Most of the genetic analysis has concentrated on the two endoproteinases, PrA and PrB, and the two carboxypeptidases, CpY and CpS. Plate tests or microtiter well tests have been developed that allow one to test directly for activity of PrB, CpY, CpS, and aminopeptidases and indirectly for activity of PrA in colonies. This chapter first presents plate and well tests for assaying protease activities of colonies, followed by assays that allow quantitation of activity levels in cell-free extracts. Guidelines for designing useful protease-deficient strains and a list of strains that have been sent to the Yeast Genetics Stock Center are provided (see "Comments") A cautionary note in the form of a summary of the problems that have been known to surface at least once during the use of protease-deficient strains is also given.

[51] L. M. Johnson, M. Snyder, L. M. S. Chang, R. W. Davis, and J. L. Campbell, *Cell* (*Cambridge, Mass.*) **43**, 369 (1985).
[52] R. S. Fuller, A. Brake, and J. Thorner, *Proc. Natl. Acad. Sci. U.S.A.* **86**, 1434 (1989).
[53] J. Thorner, personal communication (1989).
[54] G. T. James, *Anal. Biochem.* **86**, 574 (1978).

Genetic Analyses

Requirement of Tests

Several requirements must be met to succeed in assaying activities of particular intracellular enzymes in colonies. The first condition is that the enzyme gain access to the externally supplied substrate. This is accomplished either by permeabilizing the cells in a colony with a solvent or by establishing conditions that result in the lysis of cells. The second requirement is that, where lysis is employed, conditions be established that free the enzyme from its naturally occurring, intracellular, polypeptide inhibitor.[1,2] This is accomplished, for PrB, by including SDS in the overlay. The third requirement is to find a substrate or condition that tests for one enzyme activity only. How this is accomplished for each enzyme will be given in the procedure for that enzyme.

Protease B

Protease B activity in colonies can be assayed using an overlay test. Initially, we developed a procedure for cells grown on YEPD plates (20 g Difco Bacto-peptone, 10 g Difco yeast extract, 20 g dextrose, 13–20 g agar, according to brand, per liter) that necessitated use of a lysis mutation to cause the release of intracellular protease B from cells. We have since realized that growth of cells on YEPG plates (20 g Difco Bacto-peptone, 10 g Difco yeast extract, 50 g glycerol, 13–20 g agar, according to brand, per liter) obviates use of a lysis mutation, as some lysis occurs when cells are grown on this medium. The substrate for PrB is particulate Hide Powder Azure (HPA Calbiochem, La Jolla, CA). Hide Powder Azure is cowhide to which the dye Remazol Brilliant Blue has been conjugated. PrB is the only protease in *S. cerevisiae* that catalyzes cleavage and solubilization of this substrate.[55] Various investigators have reported lack of success with the following procedure. In the author's past experience, Calbiochem has changed the formulation of products such as Hide Powder Azure (HPA) or Azocoll (see enzymatic assay for PrB later) and will not reveal current or past procedures or formulations. If the HPA does not work, try Azocoll. The author has not worked out a plate test using Azocoll, but the procedure should be similar to that for HPA.

HPA Overlay Test for PrB Activity[11,20]

Principle. Protease B, which is freed from cells by lysis and from its inhibitor by the SDS present in the overlay, solubilizes the particles of Hide Powder Azure in the overlay, uncovering the colony, and surrounding it with a clear halo. Mutant colonies remain covered.

---

[55] R. E. Ulane and E. Cabib, *J. Biol. Chem.* **249**, 3418 (1974).

Reagents

SDS, 20% (w/v) in 0.1 M Tris–HCl, pH 7.6

Cycloheximide, sterile solution at 5 mg/ml

Penicillin G–streptomycin: Use a sterile solution containing 5000 units/ml penicillin base and 5000 μg/ml streptomycin base. This inhibits bacterial growth and allows viewing of the test plates for more than 1 day.

0.6% agar, molten, held at 50°

Hide Powder Azure (Calbiochem) is pulverized by homogenization (VirTis homogenizer) of 200 ml of a slurry [100 mg/ml 95% (v/v) ethanol] in a 500-ml flask for 5 min at 40,000 rpm or by sonication of a slurry of the same proportions. Aliquots containing 50–100 mg are transferred to sterile 13 × 100-mm tubes, centrifuged (1650g for 5 min at room temperature), and the supernatants are discarded. The pellets are washed with 2.5–3 ml of sterile water, repelleted, and the supernatants discarded.

Procedure. Add 0.2 ml cycloheximide solution, 0.2 ml penicillin–streptomycin solution, and 0.1 ml of 20% SDS to a Hide Powder Azure pellet, vortex, and then add 4 ml of molten agar. Vortex to mix and resuspend particles. Streaks or replica plates of cells grown for 2 days at 30° on YEPG agar are overlaid with the molten agar cocktail, with pouring along the length of the stripes rather than across. After the agar solidifies, the plates are incubated at 34–36° for 8 hr to 2 days, depending on the properties of the strains. (Some strains lyse well and are difficult to score at later times.) Plates can (and should) be incubated upside down, so long as the medium will absorb the moisture in the overlay. It is important that a seal *not* form between the lid and the base of the petri dish. If a seal forms, the test simply does not work for unknown reasons. Even wild-type strains will fail to form a proper halo. Thus, the use of freshly poured plates is not advised. One can drape a doubled Kimwipe over the lid opening and then close over the plate (not touching the agar, of course), in effect lining the lid with the Kimwipe to absorb moisture.

Utility. The HPA overlay test works very well for following mutations in the PrB structural gene, *PRB1*, and the *pbn1-1* mutation[56] and less well for pleiotropic mutations such as *pep* mutations. The *pep4::HIS3* insertion mutation present in strains such as BJ3501 and BJ3505, the BJ5400 series, and the BJ5600 series can be scored very easily in this test, as *pep4::HIS3* causes a much tighter Prb$^-$ phenotype than the *pep4-3* mutation.[57] In using *prb1* mutations or in constructing strains, be advised that although *prb1* homozygotes sporulate, the asci may be very small (the size of a normal spore) and that superimposition of heterozygosity for *pep4* in such *prb1* homozygotes may prevent sporulation.[58]

[56] R. R. Naik and E. W. Jones, *Genetics* **149**, 1277 (1998).
[57] C. A. Woolford and E. W. Jones, unpublished observations (1988).
[58] G. S. Zubenko and E. W. Jones, *Genetics* **97**, 45 (1981).

Carboxypeptidase Y

Carboxypeptidase Y activity in colonies can be assessed by using an overlay test that relies on the esterolytic activity of the enzyme. The substrate is *N*-acetyl-DL-phenylalanine β-naphthyl ester (APE), cleavage of which, in colonies anyway, is catalyzed only by CpY. (PrB, when overproduced, will also give a red colony in the assay.)

APE Overlay Test for CpY Activity[9]

Principle. Dimethylformamide present in the initial overlay permeabilizes cells on the surface of colonies. CpY within cells catalyzes cleavage of the ester. The product β-naphthol reacts nonenzymatically with the diazonium salt Fast Garnet GBC to give an insoluble red dye: Cpy^+ colonies are red, whereas Cpy^- colonies are yellow or pink.

Reagents

N-Acetyl-DL-phenylalanine β-naphthyl ester (Sigma, St. Louis, MO): Make a solution of 1 mg/ml dimethylformamide
0.6% Agar, molten, held at 50°
Fast Garnet GBC (Sigma)
0.1 *M* Tris–HC1, pH 7.3–7.5

Procedure. Replica plate strains to thick YEPD plates (40–45 ml/100-mm plate). Grow 3 days at 30°. To form the overlay mix, add 2.5 ml of the ester solution to 4 ml molten agar in a 13 × 100-mm tube. Vortex or cover with Parafilm and invert three or four times until the schlieren pattern disappears. After the bubbles exit, pour the contents over the surface of colonies or stripes (along, not across, stripes; colonies must be covered). After 10 min (or after the agar is hard), carefully flood the surface of the agar with 4.5–5 ml of a solution of Fast Garnet GBC (5 mg/ml 0.1 *M* Tris–HC1, pH 7.3–7.5). Do not tear the agar overlay during the flooding. The Fast Garnet GBC solution must be made immediately before use, as diazonium salts are very unstable in solution. We use Fast Garnet GBC from Sigma and store it in the freezer. Watch the color develop and pour off the fluid when Cpy^+ colonies turn red (a few to several minutes). If color development takes longer than 5–10 min, use a fresh bottle of Fast Garnet GBC. Do the test at room temperature. The color is not stable, but is more stable if the plates are placed in the cold in the dark. Although diazonium salts other than Fast Garnet GBC could, in principle, be used, Fast Garnet GBC is usually preferred because it is not a zinc salt and, thus, is less inhibitory to enzyme activity.

Utility. The APE test has wide utility for following many mutations that reduce protease activity so long as CpY activity is among the activities reduced as a consequence of the mutation. The APE test can be used for following mutations in the CpY structural gene, *PRC1*, and for following pleiotropic *pep* mutations, including *pep4-3*.[9] If an *ade1* or *ade2* mutation is segregating in the cross, the red pigmentation problem can be circumvented by growing cells on YEPG or on YEPD supplemented with 100 μg/ml adenine sulfate. Petites give aberrant phenotypes in the test (but see well test later).

PEP4 is the structural gene for the PrA precursor.[59,60] PrA activity is required either directly or indirectly, through its requirement for maturation of PrB, for proper maturation of several vacuolar hydrolases, including PrB, CpY, one or more RNase species, ApI, ApY, and the repressible species of alkaline phosphatase.[61-64] We have cloned and sequenced several of the pleiotropic *pep4* mutations (including *pep4-3*), as well as the allelic *pra* mutations[65] that are not fully pleiotropic.[66] Pleiotropic *pep4* mutations such as *pep4-3* usually prove to be nonsense mutations that totally eliminate protease A activity and greatly reduce levels of all hydrolase activites, including CpY, that require PrA activity for maturation.[65] For this reason, one can follow many mutations that eliminate protease A activity (the pleiotropic *pep4* mutations) by means of the APE test. Indeed, the more devastating the effect of the *pep4* mutation on protease A activity, the easier the mutation is to follow by the APE test. Of course, *pep4* mutations, because of their effects on hydrolase maturation, result in reduction or elimination of several protease activities simultaneously, including PrA, PrB, CpY, ApI, and ApY (but not CpS), as well as RNase and alkaline phosphatase activity.[62,64]

In working with pleiotropic *pep4* mutations (in crosses or in constructing alleles), be advised that *pep4* homozygotes do not sporulate[58] and that *pep4* segregants show phenotypic lag.[67] Strains showing phenotypic lag will appear to be

---

[59] C. A. Woolford, L. B. Daniels, F. J. Park, E. W. Jones, J. N. Van Arsdell, and M. A. Innis, *Mol. Cell. Biol.* **6,** 2500 (1986).

[60] G. Ammerer, C. Hunter, J. Rothman, G. Saari, L. Valls, and T. Stevens, *Mol. Cell. Biol.* **6,** 2490 (1986).

[61] B. A. Hemmings, G. S. Zubenko, A. Hasilik, and E. W. Jones, *Proc. Natl. Acad. Sci. U.S.A.* **78,** 435 (1981).

[62] E. W. Jones, G. S. Zubenko, and R. R. Parker, *Genetics* **102,** 665 (1982).

[63] B. Mechler, M. Müller, H. Müller, and D. H. Wolf, *Biochem. Biophys. Res. Commun.* **107,** 770 (1982).

[64] T. Yasuhara, T. Nakai, and A. Ohashi, *J. Biol. Chem.* **269,** 13644 (1994).

[65] E. W. Jones, C. A. Woolford, C. M. Moehle, J. A. Noble, and M. A. Innis, in "Proceedings UCLA Symposium, Cellular Proteases and Control Mechanisms" (T. E. Hugli, ed.), p. 141. A. R. Liss, New York, 1989.

[66] E. W. Jones, G. S. Zubenko, R. R. Parker, B. A. Hemmings, and A. Hasilik, in "Alfred Benzon Symposium" (D. von Wettstein, J. Friis, M. Kielland-Brandt, and A. Stenderup, eds.), Vol. 16, p. 182. Munksgaard, Copenhagen, 1981.

[67] G. S. Zubenko, F. J. Park, and E. W. Jones, *Genetics* **102,** 679 (1982).

Pep$^+$, even though genetically they carry the *pep4* allele. The lag is physiological in origin (the activity perdures), not genetic, and is apparently a reflection of the ability of active PrB packaged within the spore to continue to activate its own as well as the CpY precursor. This positive feedback loop results in the continued production of CpY (detected in the APE test) long after division would have diluted away active PrA. To circumvent phenotypic lag, streak spore clones for single colonies on YEPD plates (quadrants suffice). Do the APE test. Stab negative colonies through the agar overlay and make a new master plate. Then carry out the usual analyses for other markers. If this procedure is followed, the overlays should be made with sterile solutions and the colonies should be stabbed soon after the test, for the cocktail will kill cells.

Well Test for CpY Activity[59]

Principle. Dimethylformamide present in the solution permeabilizes cells. Cleavage of the amide bond in *N*-benzoyl-L-tyrosine *p*-nitroanilide (BTPNA) to give the yellow product *p*-nitroaniline is catalyzed only by CpY. This test works for petites as well as for grandes and is unaffected by *ade1* and *ade2* mutations.

Reagents

N-Benzoyl-L-tyrosine *p*-nitroanilide (Sigma): Make a solution of 2.5 mg/ml dimethylformamide.
0.1 *M* Tris–HCl, pH 7.5. Use a sterile solution.

Procedure. Mix 4 volumes of buffer to 1 volume BTPNA solution. Distribute 0.2 ml into wells in a 96-well microtiter test plate. Cells are transferred into the solution by rotating an applicator stick in the fluid after dipping the sterile stick into a colony grown on a YEPD plate. (Alternatively, a 48-prong replicator can be used, taking care to adjust individual wells for colonies that may not transfer effectively with the technique.) Cover the wells and incubate overnight at 34–37°. Cpy$^+$ cells give yellow fluid in the wells; Cpy$^-$ cells produce no color.

Utility. This test can be used to follow *prc1* mutations as well as pleiotropic *pep* mutations, such as *pep4*, that result in CpY deficiency. This test is reliable for petite and grande colonies.

Carboxypeptidase S

Carboxypeptidase S activity in colonies can be assessed by using a well test that incorporates a coupled assay that detects the release of free leucine from the blocked dipeptide carbobenzoxyglycyl-L-leucine (Cbz-Gly-Leu). CpY, which also catalyzes this cleavage, is inactivated by preincubation with PMSF, which reacts covalently to inactivate serine proteases such as CpY.

Principle. The principle of the coupled assay, devised by Lewis and Harris[68] and adapted by Wolf and Weiser,[69] is shown in the following reactions:

$$N\text{-Cbz-Gly-Leu} + H_2O \xrightarrow{\text{carboxypeptidase}} \text{leucine}$$

$$\text{Leucine} + O_2 \xrightarrow{\text{L-amino-acid oxidase}} \text{keto acid} + NH_3 + H_2O_2$$

$$H_2O_2 + o\text{-dianisidine} \xrightarrow{\text{peroxidase}} \text{oxidized dianisidine}$$

Oxidized dianisidine is dark brown in color.

The amino acid generated by cleavage must be a substrate for the L-amino acid oxidase employed. Typically snake venoms serve as sources. For *Crotalus adamanteus* (eastern diamondback rattlesnake) venom, leucine, isoleucine, phenylalanine, tyrosine, methionine, and tryptophan are good substrates; arginine, valine, and histidine are poor substrates; and the other amino acids are not oxidized. We have successfully used venoms from *Crotalus atrox* (western diamondback rattlesnake) and *Bothrops atrox* (a viper). Cps$^+$ colonies give brown fluid in the wells; Cps$^-$ cells do not. Cells are permeabilized and CpY is inactivated by preincubation of cells in a solution containing Triton X-100 and PMSF.

Reagents

0.1% Triton X-100, made 1 mg/ml in PMSF
0.2 M potassium phosphate, pH 7.0
50 mM MnCl$_2$
N-Cbz-Gly-Leu (Sigma)
Horseradish peroxidase, type I (Sigma)
L-Amino acid oxidase, type VI (Sigma; actually crude dried venom from *C. atrox*) or type II (Sigma; dried venom from *B. atrox*)
o-Dianisidine dihydrochloride (Sigma)

Procedure. Cells are transferred into 50 μl/well of the Triton X-100/PMSF solution in a 96-well microtiter test plate by applicator stick or multiprong replicator (see CpY procedure) after growth on a YEPD plate. Cover and let sit 2 hr at room temperature. Add to each well 150 μl of the substrate mix made in the following proportions: 1 ml buffer, 10 μl MnCl$_2$, 3.22 mg Cbz-Gly-Leu (final concentration, 10 mM), 0.2 mg peroxidase, 0.4 mg amino acid oxidase type VI or 1.2 mg type II, and 0.4 mg o-dianisidine dihydrochloride. Cover and incubate at 37° for 17–18 hr. (We have not investigated other concentrations of venoms but know that these work.)

---

[68] W. H. P. Lewis and H. Harris, *Nature* (*London*) **215**, 351 (1967).
[69] D. H. Wolf and U. Weiser, *Eur. J. Biochem.* **73**, 553 (1977).

Utility. This test appears to work well for following mutations that result in CpS deficiency, whether or not the strain lacks CpY activity. We have used it particularly to follow *dut1-1*,[70] a mutation that might be allelic to *cps1*.[71]

Protease A

We have found no general plate test to directly assess protease A activity that is satisfactory, although preliminary tests indicate that it may be possible to adapt the fluorescence-based assay given in the biochemistry section of this chapter for a well test. (A plate test necessitating that strains carry a lysis mutation has been described.[72]) However, because total loss of function for protease A results in failure to activate a set of vacuolar hydrolase precursors, including that of CpY, many mutations in the protease A structural gene, *PEP4*, can be followed using the indirect test (APE test, see under carboxypeptidase Y) that detects the esterolytic activity of CpY, the processing and activation of which is dependent on protease A. This latter test is satisfactory for following the pleiotropic *pep4* mutations (Pra$^-$Prb$^-$Cpy$^-$...) that are of most utility in biotechnological applications.

Aminopeptidases

Trumbly and Bradley[7] isolated mutants defective in one or more aminopeptidases by sequential mutagenesis. An agar overlay test was used that incorporated leucine β-naphthylamide (LBNA) as a substrate. Cleavage releases β-naphthylamine. In fairly extensive but unpublished work, the author isolated and partially characterized mutants by similar procedures but used several different amino acid β-naphthylamides as substrate.[73] Both procedures are described.

Principle. Dimethylformamide present in the overlay permeabilizes cells on the surface of colonies. Aminopeptidases catalyze cleavage of the β-naphthylamide. The product β-naphthylamine reacts nonenzymatically with the diazonium salt Fast Garnet GBC to give an insoluble red dye. *Caution:* β-Naphthylamine is a carcinogen.

Trumbly and Bradley Method[7]

Reagents

L-Leucine β-naphthylamide (Sigma): Make a solution of 50 mg/ml dimethylformamide
0.8% molten agar, held at 50°
Fast Garnet GBC (Sigma)

---

[70] G. S. Zubenko, Ph.D. Thesis, Carnegie Mellon University, Pittsburgh, Pennsylvania (1981).
[71] D. H. Wolf and C. Ehmann, *J. Bacteriol.* **147,** 418 (1981).
[72] B. Mechler and D. H. Wolf, *Eur. J. Biochem.* **121,** 47 (1981).
[73] E. W. Jones, unpublished observations (1973–1979).

Procedure. Replica plate or streak strains on YEPD plates. Grow 2–3 days. Mix LBNA solution, agar, and Fast Garnet GBC in the proportions of 0.02 ml, 1 ml, and 1 mg, respectively, and cover colonies with the solution. Red color should develop within 30 min and be stable for 2–3 days.

Jones Method[73]

Reagents

DL-Methionine β-naphthylamide (Sigma), L-leucine β-naphthylamide (Sigma), DL-alanine β-naphthylamide (Sigma), threonine β-naphthylamide (Bachem, Inc., King of Prussia, PA): Make solutions of 3 mg naphthylamide/ml dimethylformamide
0.6% Molten agar, held at 50°
Fast Garnet GBC (Sigma)
0.1 M Tris–HCl, pH 7.3–7.5

Procedure. Replica plate or streak strains on YEPD plates. Grow 3 days at 30°. Mix 4 ml of molten agar with 2.5 ml of naphthylamide solution in a 13 × 100-mm tube. Vortex until the schlieren pattern disappears. After the bubbles exit, pour the contents over the surface of colonies or stripes (along, not across, the stripes; colonies must be covered). After 10 min (or after the agar is hard), carefully flood the surface of the agar with 4.5–5 ml of a solution of Fast Garnet GBC (5 mg/ml 0.1 M Tris–HCl, pH 7.3–7.5). Do not tear the agar overlay during flooding. The Fast Garnet GBC solution must be made immediately before use, as diazonium salts are very unstable in solution. We use Fast Garnet GBC from Sigma and store it in the freezer. Watch the color develop and pour off the fluid when wild-type colonies are red (several minutes to 1 hr, depending on the substrate and/or genotype). If color development takes more than 5–10 min for the wild type, replace the bottle of Fast Garnet GBC. The test is done at room temperature. Fast Garnet GBC is the preferred diazonium salt because it does not contain zinc, which inhibits some enzyme activities.

Utility. When leucine β-naphthylamide is used as substrate, activity in colonies is apparently the sum of four different aminopeptidase activities.[7] Whether segregation of the *lap* mutations other than *lap1* (*lap2-lap4*)[7] can be followed in the presence of wild-type alleles for the other three genes has not been determined.

Analysis of cleavage patterns for several naphthylamides or nitroanilides across column profiles or in activity gels for mutant and wild-type extract indicates that different enzymes have different cleavage specificities.[5,17,19,73] In principle, one amino acid naphthylamide or nitroanilide might be used to follow segregation of one mutation and a second amide to separately follow segregation of a second mutation. Both the periplasmic aminopeptidase II (LAPI) and ApY catalyze cleavage of lysine amides[19,64]; only the latter requires Co^{2+} to catalyze the reaction. ApI (LAP4) is essentially inactive against lysine amides, but is very active against leucine amides. Obviously, the particular substrate for each must be chosen judiciously. Because the relationship to *lap1-lap4* of the mutations the author isolated and studied is not known, the author cannot provide more guidance for substrate choice. For wild-type colonies, the rates of cleavage of naphthylamides decline in the order methionine > alanine > leucine > threonine. The last three are better for scoring segregation of mutations than methionine β-naphthylamide.

The structural genes for vacuolar ApI (*LAP4*) and ApY (*APE3*) have been cloned and sequenced.[74,75] It seems likely that plate or well tests could now be devised, although recombinant methods may render them moot.

Biochemical Analyses

Growth of Cells and Preparation of Extracts

For most assays of cell-free extracts for protease activities we employ extracts prepared according to the following protocol. Cells are grown to stationary phase in YEPD at 30° with vigorous shaking (usually 48–52 hr for our conditions of inoculation), harvested by centrifugation, washed once with distilled water, and resuspended in 2 ml of 0.1 M Tris–HCl, pH 7.6, per gram of cells. The cells are broken (3 min) with 0.45-mm glass beads [40 : 60 to 50 : 50 (v/v) glass beads to cell suspension] in a Braun homogenizer (Braun, Melsungen, Germany) without CO_2 cooling. After centrifugation for 30 min at 35,000g in the cold, the supernatant is removed to a fresh tube and placed on ice.

Enzymatic Assays

Protease A

Two assays for protease A activity are described; the first is based on the release of peptides from hemoglobin and the second is a fluorescence assay based on cleavage of a peptide.

Principle. The most commonly used assay for protease A activity measures the release of tyrosine-containing acid-soluble peptides from acid-denatured hemoglobin. Protease A is apparently the only protease to catalyze the reaction at acid pH. The procedure is based on that of Lenney *et al.*[76]

---

[74] M. Nishizawa, T. Yasuhara, T. Nakai, Y. Fujiki, and A. Ohashi, *J. Biol. Chem.* **269**, 13651 (1994).
[75] Y.-H. Chang and J. A. Smith, *J. Biol. Chem.* **264**, 6979 (1989).
[76] J. Lenney, P. Matile, A. Wiemken, M. Schellenberg, and J. Meyer, *Biochem. Biophys. Res. Commun.* **60**, 1378 (1974).

Reagents

2% Acid-denatured hemoglobin: Dissolve 2.5 g hemoglobin (Sigma) in 100 ml distilled water; dialyze against three changes of 3 liters of water in the cold. Bring the pH to 1.8 with 1 N HCl. After 1 hr of incubation with stirring at 35°, bring the pH to 3.2 with 1 M NaOH and adjust the volume to 125 ml. Aliquots can be stored frozen for years.
0.2 M glycine hydrochloride, pH 3.2
1 N perchloric acid
0.5 M NaOH
2% (w/v) Na_2CO_3 in 0.1 M NaOH
1% (w/v) $CuSO_4 \cdot 5H_2O$
2% Sodium or potassium tartrate
Folin and Ciocalteau's phenol reagent (diluted 1 : 1 with water)

Procedure. The reaction mixture consists of 2 ml of a hemoglobin solution (prepared by mixing equal volumes of the 2% hemoglobin, pH 3.2, described earlier, and 0.2 M glycine, pH 3.2) and 0.1 ml of cell-free extract (1–2 mg protein/ incubation) with incubation at 37°. At 0, 15, and 30 min, 0.4-ml samples are removed to 0.2 ml of 1 N perchloric acid on ice, and the tubes are shaken briefly. After centrifugation at 1650g for 5 min at room temperature, 0.1 ml of each sample is removed to 0.1 ml of 0.5 M NaOH. Tyrosine-containing peptides in the neutralized 0.2-ml sample are determined with the Folin reagent according to Lowry *et al.*[77] To each 0.2-ml sample add 1 ml of a reagent consisting of 2% Na_2CO_3 in 0.1 M NaOH, 1% $CuSO_4 \cdot 5H_2O$, and 2% sodium or potassium tartrate (100 : 1 : 1) (mix just before use). Incubate at room temperature for at least 10 min. Add 0.1 ml of the diluted phenol reagent and vortex immediately. After 30 min, determine the absorbance at 750 nm.

Definition of Unit and Specific Activity. One unit corresponds to 1 μg tyrosine per minute. The A_{750} of 1 μg tyrosine is 0.058 in the Lowry assay performed as described. Using the change in absorbance for a 30-min incubation, the conversion to micrograms Tyr/minute/milligram protein is made by the following calculation:

$$\frac{\Delta A_{750}}{30 \text{ min}} \times \frac{1}{0.058} \times \frac{2.1}{0.1} \times \frac{6}{4} \times \frac{1}{0.1 \text{ ml (mg protein/ml extract)}}$$
$$= (181)(\Delta A_{750})/\text{min/mg protein}$$

An abbreviated protease A assay can be used for segregants of crosses known to be segregating a *pra* mutation, where distinction between + and − is all that

---

[77] O. Lowry, N. Rosebrough, A. Farr, and R. Randall, *J. Biol. Chem.* **193**, 265 (1951).

is sought. It can also be used for mutant screens if a 0-min point is added. Fifty microliters of extract is added to 1 ml of the hemoglobin solution (equal volumes of 2% hemoglobin and 0.2 M glycine, pH 3.2). Incubate at 37° for 30 min. Remove a sample to perchloric acid and work up as described earlier.

An alternate, much more sensitive assay for protease A is available. We find it more satisfactory than the hemoglobin assay with respect to linearity, reproducibility, etc. It was developed for assaying renin[78] but has been used for assaying protease A.[79] We have adapted it for use in crude extracts.

Principle. Protease A will catalyze cleavage at the Leu-Val bond of the octapeptide N-succinyl-L-arginyl-L-prolyl-L-phenylalanyl-L-histidyl-L-leucyl-L-leucyl-L-valyl-L-tyrosine-7-amido-4-methylcoumarin. After removal of valine and tyrosine residues by aminopeptidase M, the fluorescence of 7-amino-4-methylcoumarin can be determined at 460 nm after excitation at 380 nm. PMSF is included to covalently react with and inactivate PrB, a serine protease that could also catalyze cleavage of the peptide.

Reagents

McIlvaine's buffer, 0.2 M Na$_2$HPO$_4$, 0.1 M citric acid; adjust to pH 6.0 with NaOH or HCl

N-Succinyl-L-arginyl-L-prolyl-L-phenylalanyl-L-histidyl-L-leucyl-L-leucyl-L-valyl-L-tyrosine-7-amido-4-methylcoumarin (Sigma, molecular weight 1300), 0.325 mg/ml dimethylformamide (0.25 mM)

7-Amino-4-methylcoumarin (Sigma, molecular weight 175): Make a 0.4-mg/ml solution and serially dilute this 1/100 × 1/40 (1/4000) to make a stock solution for the standard curve

PMSF (Sigma), 0.2 M in 95% (v/v) ethanol (34.8 mg/ml)

Aminopeptidase M (Sigma), 0.5 mg/ml in McIlvaine's buffer (we have not explored whether other apparently more active preparations will work, as the original protocol for renin calls for only 50 mU^{24})

Procedure. Mix 90 or 85 μl McIlvaine's buffer, pH 6.0, 10 μl peptide solution, and 10 or 15 μl crude extract (pretreat 1 ml of extract with 5 μl of 0.2 M PMSF for 2 hr at room temperature). Incubate for 15 min at room temperature. Immerse the tube in boiling water for 5 min to stop the reaction. Cool. Add 5 μl aminopeptidase M. After a 90-min incubation at room temperature, add 1.61 ml buffer (to dilute the reaction 15-fold) and centrifuge the tubes at 1650g for 5 min. Remove the supernatant and determine the fluorescence in a fluorimeter, with excitation at 380 nm and emission at 460 nm. A standard curve is constructed for

---

[78] K. Murakami, T. Ohsawa, S. Hirose, K. Takada, and S. Sakakibara, *Anal. Biochem.* **110**, 232 (1981).
[79] H. Yokosawa, H. Ito, S. Murata, and S.-I. Ishii, *Anal. Biochem.* **134**, 210 (1983).

7-amino-4-methylcoumarin. The diluted stock (0.1 μg/ml) is mixed with buffer as follows and the fluorescence is determined.

Stock (μl)	Buffer (ml)	Total (nmol)
350	1.375	0.2
175	1.550	0.1
88	1.637	0.05
44	1.681	0.025
22	1.703	0.0125
11	1.714	0.00625

For experimental samples, read nanomoles in the sample from the standard curve (the volumes for the standard and the experimental sample are the same). The samples and standards must be read in exactly the same way (same slit width, etc.). Prepare the standard curve the same day.

Definition of Unit and Specific Activity. We have been using 1 unit to equal 1 nmol/min and the specific activity to be units per milligram protein.

Crude extracts for this assay procedure are made as follows. The volume of cells sampled is 2500 ml/Klett units (A_{600} of 1.0 = 40–45 Klett units). Pellet the cells, wash with water, and freeze the pellet. Resuspend the pellet in 1.5 ml of 0.1 M Tris–HCl, pH 7.6, and transfer into a small Braun homogenizer tube (40–50% full of glass beads). Add buffer to the top (fill tube completely) to prevent foaming. Homogenize for 3 min at room temperature. Transfer to a long centrifuge tube and centrifuge for 20 min at 25,000g in the cold. Transfer the supernatant to a 1.5-ml microcentrifuge tube. (The small homogenizer tubes are about 35 mm high and are cut down from 12 × 75-mm tubes.)

Preliminary evidence suggests that this assay can be adapted for use in a well test for genetic analyses.

Protease B

Protease B is an endoproteinase that will solubilize particulate substrates such as Hide Powder Azure and Azocoll. It is apparently the only enzyme in the yeast cell that can do so.[80] Calbiochem has changed the formulation of Azocoll and will not release details about past or current formulations. This can result in a dramatic change in the PrB specific activity for a wild-type strain (E. W. Jones, unpublished, 1980).

Principle. Protease B catalyzes hydrolysis of the peptide bonds in Azocoll, resulting in release of the trapped red dye. Absorbance of the dye is read at 520 nm.[11]

---

[80] E. Juni and G. Heym, *Arch. Biochem. Biophys.* **127,** 89 (1968).

Reagents

Azocoll (Calbiochem; now available in two mesh sizes: <50 mesh >100 mesh; the latter is probably preferred for solution assays)
1% Triton X-100
0.1 M Tris–HCl, pH 7.6
20% SDS in 0.1 M Tris–HCl, pH 7.6

Procedure. Use 20 mg Azocoll for each 0.54 ml of solution made by mixing the three listed solutions in the following proportions: 0.125 ml Triton X-100, 0.375 ml buffer, and 40 μl SDS solution. With a wide-bore pipettor (cut 1/4 inch off a 1-ml pipette tip), transfer 0.375 ml of the suspension to a tube. Two-tenths milliliter of extract of a suitable dilution is added, and the tube is placed in a 37° constant temperature block. At 1-min intervals each tube is removed and shaken gently (do not vortex) to resuspend the Azocoll and is then replaced in the block. Avoid leaving Azocoll on the tube walls. At the end of the 15-min incubation, the tubes are plunged into ice and 3.5 or 2 ml of ice-cold distilled water is added. Tubes are immediately centrifuged for 3–5 min at 1650g, and the supernatants are removed to fresh tubes. The absorbance of these supernatants is relatively stable. Absorbance is read at 520 nm. Dilutions are chosen such that the kinetics are linear with time and protein concentration. Best results are obtained if the ΔA_{520} is less than 0.3. When comparing different strains, the extracts should be diluted such that the protein concentrations are similar for all strains. For extracts made from stationary-phase wild-type cells, about 0.1 mg extract protein/assay is appropriate. Correction is made for a blank lacking extract. A 0-min time point is needed for *ade2* mutant strains.

Definition of Unit and Specific Activity. One unit of protease B activity is defined as a change in absorbance at 520 nm of 1.0 per minute for the 0.74-ml reaction mixture as assayed at 37°. For a reaction stopped with 3.5 ml of water and run for 15 min, the conversion to units/mg protein is

$$\frac{\Delta A_{520}}{15 \min} \times \frac{4.24}{0.74} \times \frac{1}{0.2 \text{ ml (mg protein/ml extract)}} = (1.91)(\Delta A_{520})/\text{mg protein}$$

Carboxypeptidase Y

Carboxypeptidase Y will catalyze cleavage of esters, amides, and peptides. An assay based on its amidase activity can be employed for kinetic analyses[81] or modified to a fixed time point assay.[9]

Principle. Carboxypeptidase Y will catalyze the hydrolysis of *N*-benzoyl-L-tyrosine *p*-nitroanilide to give the yellow product *p*-nitroaniline. Production can be followed by absorbance at 410 nm.

---

[81] S. Aibara, R. Hayashi, and T. Hata, *Agric. Biol. Chem.* **35,** 658 (1971).

Reagents

N-Benzoyl-L-tyrosine p-nitroanilide (Sigma) (6 mM): Dissolve 2.43 mg in 1 ml dimethylformamide
0.1 M Tris–HCl, pH 7.6
1 mM HgCl$_2$
20% (w/v) SDS in 0.1 M Tris–HCl, pH 7.6

Procedure. One-tenth milliliter of 6 mM BTPNA in dimethylformamide is added to a tube containing 0.40 ml of 0.1 M Tris–HCl, pH 7.6, and 0.1 ml of extract at 37°. After 30 min, 1.5 ml of 1 mM HgCl$_2$ is added to stop the reaction. If the extract being assayed has low activity (as is typical for wild-type strains), 0.2 ml of 20% SDS, pH 7.6, is added and, after vortexing, the tubes are incubated at 70° until solubilization of the protein, as evidenced by clearing, ensues. Absorbance at 410 nm is determined. Permeabilized cells can be used as an enzyme source in this assay. We have used cells permeabilized with 0.1–0.2% Triton X-100 or with 10–20% dimethylformamide.

Definition of Unit and Specific Activity. One unit of activity corresponds to 1 μmol p-nitroaniline produced per minute, assuming a molar absorbance of 8800. Corrections for absorbance due to substrate and protein are made. The conversion to units/mg protein is

$$\frac{\Delta A_{410}}{30\,\text{min}} \times \frac{2.3}{8800} \times \frac{10^3}{0.1\,\text{ml}\,(\text{mg protein/ml extract})} = (0.087)(\Delta A_{410})/\text{mg protein}$$

Carboxypeptidase Y levels can also be determined using the kinetic assay described for carboxypeptidase S, but using Cbz-Phe-Leu as the substrate, as 95% of the hydrolytic activity toward this peptide is apparently due to CpY.[69]

Carboxypeptidase S

Carboxypeptidase S will catalyze hydrolysis of dipeptides that are blocked at the amino terminus.

Principle. Free leucine released from a peptide by CpS catalysis is oxidized by L-amino acid oxidase. Reduction of the product hydrogen peroxide by horseradish peroxidase is coupled to the oxidation of o-dianisidine, yielding brown oxidized dianisidine. The absorbance at 405 nm is followed. The kinetic assay was developed by Wolf and Weiser.[69]

Reagents

0.2 M potassium phosphate buffer, pH 7.0, containing 0.5 mM MnCl$_2$
L-Amino acid oxidase type I (Sigma)
Horseradish peroxidase, type I (Sigma)

N-Carbobenzoxyglycyl-L-leucine, Cbz-Gly-Leu (Sigma) (20 m*M*), 6.44 mg/ml 0.2 *M* potassium phosphate buffer, pH 7.0 [or carbobenzoxy-L-phenylalanyl-L-leucine, Cbz-Phe-Leu (Sigma) (15 m*M*) in 0.2 *M* potassium phosphate buffer, pH 7.0, for CpY]
o-Dianisidine dihydrochloride (Sigma) 2 mg/ml water
0.2 *M* PMSF in 95% ethanol

Procedure. A solution is made in the proportion 1 ml phosphate buffer-$MnCl_2$, 0.25 mg L-amino acid oxidase, and 0.4 mg peroxidase. To 0.5 ml of this is added 0.5 ml of the peptide solution followed by 50 μl of the *o*-dianisidine solution and 50 μl of dialyzed extract. The mixture is incubated at 25°, and the absorbance is followed at 405 nm. To render the assay specific for CpS when Cbz-Gly-Leu is the substrate, extracts are preincubated for 2 hr at 25° with 0.1 m*M* PMSF to inactivate CpY. Use Cbz-Phe-Leu as the substrate if CpY activity is to be measured with the assay.

Definition of Unit and Specific Activity. One unit corresponds to the production of 1 nmol L-leucine per minute; 0.1 μmol of leucine corresponds to a change in absorbance of 0.725 for this procedure. Specific activity is expressed as nanomoles L-leucine/minute/milligram extract protein.

Aminopeptidases

Quantitative assays for the aminopeptidase activity of extracts exist but do not distinguish among the several activities.[5,17,19,64] Little has been reported on the aminopeptidase activities of crude extracts in wild-type or mutant strains. An alternative procedure using the fluorescence of 7-amino-4-methylcoumarin from amino acid MCAs has been described but was not employed on crude extracts of wild-type cells.[64]

Principle. Aminopeptidase will catalyze the hydrolysis of amino acid *p*-nitroanilides to give the yellow *p*-nitroaniline. Production is monitored by absorbance at 405 nm.[5]

Reagents

Amino acid *p*-nitroanilide, 2 m*M* in 4 m*M* H_2SO_4
0.2 *M* Tris–HCl, pH 7.5

Procedure. Add extract to buffer. Start the reaction by mixing this with an equal volume of the substrate solution and incubate at 37°. Follow absorbance at 405 nm. Frey and Röhm[17] add Zn^{2+} to 50 μ*M* to activate ApI. Achstetter *et al.*[5] and Frey and Röhm[17] used various amino acid *p*-nitroanilides (PNA), Trumbly and Bradley[7] used leucine PNA, and Chang and Smith[75] used methionine PNA.

Utility. A great many aminopeptidases are present in cells. No substrate differentiates among them in crude extracts.[5,17] The difference between the activity toward leucine PNA measured in the presence of 50 μM Zn^{2+} and that measured in the presence of only 1 nM Zn^{2+} (by use of nitrilotriacetic acid)[17] is taken as the activity of ApI.[17]

Comments

Designing the Most Useful Protease-Deficient Strain

In all known cases the use of a protease-deficient strain has eased protease problems. Strains bearing the pleiotropic *pep4-3* (UGA) mutation have greatly reduced, but (except for PrA) not zero, levels of PrA, PrB CpY, and ApI [as well as RNase(s) and the repressible alkaline phosphatase]. Strains such as EJ101 (α *his1 prb1-1122 prc1-126*) that carry *prb1-1122* have also been used successfully in some purifications. The author recommends use of a double mutant that carries mutations in both the *PEP4* gene and the *PRB1* gene. Whether a mutation in the CpY gene, *PRC1*, is needed has not been determined, but, in most cases, it probably would not be required.

We constructed and lodged a number of strains in the Yeast Genetics Stock Center (University of California, Berkeley, CA). They have been transferred to the American Type Culture Collection [10801 University Blvd., Manassas, VA 20110-2209 (www.atcc.org)]. A number of the genotypes listed on their web site are incorrect or incomplete. Listed below are the correct genotypes.

ATCC number	Jones number	Genotype
208274	BJ1984	α 20B-12 *MATα trp1 pep4-3 gal2*
208275	BJ1991	*MATα leu2 trp1 ura3-52 prb1-1122 pep4-3 gal2*
208276	BJ1995	*MATα leu2 trp1 ura3-52 prb1-1122 pep4-3 gal2*
208277	BJ2168	*MAT**a** leu2 trp1 ura3-52 prb1-1122 pep4-3 prc1-407 gal2*
208278	BJ2407	*MATα/MAT**a** leu2/leu2 trp1/trp1 ura3-52/ura3-52 prb1-1122/prb1-1122 prc1-407/prc1-407 pep4-3/pep4-3 gal2/gal2*
208279	BJ926	*MATα/MAT**a** trp1 +/+/ his1 prc1-126/prc1-126 pep4-3/pep4-3 prb1-1122/prb1-1122 can1/can1 gal2/gal2*
208280	BJ3501	*MATα pep4::HIS3 prb1-Δ1.6R his3-Δ200 ura3-52 can1 gal2*
208281	BJ3505	*MATα pep4::HIS3 prb1-Δ1.6R his3-Δ200 lys2-801 trp1-Δ101 (gal3) ura3-52 gal2 can1*
208282	BJ5457	*MATα ura3-52 trp1 lys2-801 leu2Δ1 his3Δ200 pep4::HIS3 prb1-Δ1.6R can1 GAL*
208283	BJ5458	*MATα ura3-52 trp1 lys2-801 leu2Δ1 his3Δ200 pep4::HIS3 prb1-Δ1.6R can1 GAL*
208284	BJ5459	*MAT**a** ura3-52 trp1 lys2-801 leu2Δ1 his3Δ200 pep4::HIS3 prb1-Δ1.6R can1 GAL*

208285	BJ5460	MATa ura3-52 trp1 lys2-801 leu2Δ1 his3Δ200 pep4::HIS3 prb1-Δ1.6R can1 GAL
208286	BJ5461	MATa ura3-52 trp1 lys2-801 leu2Δ1 his3Δ200 pep4::HIS3 prb1-Δ1.6R can1 GAL
208287	BJ5462	MATα ura3-52 trp1 leu2Δ1 his3Δ200 pep4::HIS3 prb1-Δ1.6R can1 GAL
208288	BJ5464	MATα ura3-52 trp1 leu2Δ1 his3Δ200 pep4::HIS3 prb1-Δ1.6R can1 GAL
208289	BJ5465	MATa ura3-52 trp1 leu2Δ1 his3Δ200 pep4::HIS3 prb1-Δ1.6R can1 GAL
208290	BJ5626	MATα/MATa ura3-52/ura3-52 trp1/+ +/leu2Δ1 his3Δ200/his3Δ200 pep4::HIS3/pep4::HIS3 prb1-Δ1.6R/prb1-Δ1.6R can1/can1 GAL/GAL
208291	BJ5627	MATα/MATa ura3-52/ura3-52 trp1/+ +/leu2Δ1 his3Δ200/his3Δ200 pep4::HIS3/pep4::HIS3 prb1-Δ1.6R/prb1-Δ1.6R can1/can1 GAL/GAL
208292	BJ5628	MATα/MATa ura3-52/ura3-52 leu2Δ1/+ +/trp1 his3Δ200/his3Δ200 pep4::HIS3/pep4::HIS3 prb1-Δ1.6R/prb1-Δ1.6R can1/can1 GAL/GAL
208293	BJ5407	MATa ura3-52 trp1 lys2-801 leu2-Δ1 his3-Δ200 GAL Control for BJ5459, ATCC# 208284
208294	BJ5405	MATα ura3-52 trp1 lys2-801 leu2-Δ1 his3-Δ200 GAL Control for BJ5458, ATCC# 208283

The *prb1-1122* and *pep4-3* alleles are nonsense mutations (UAA[82] and UGA,[65] respectively). The *pep4::HIS3* mutation is an insertion of a *Bam*HI fragment bearing *HIS3* into the *Hin*dIII site in *PEP4*.[59] The *prb1-Δ1.6R* mutation is a deletion of a 1.6-kb *Eco*RI fragment internal to the *PRB1* gene.[20] The *trp1-Δ101* deletion inactivates the adjacent GAL3 gene.[83]

Cautions to Users of Protease-Deficient Strains

The *pep4-3* and *prb1-1122* mutations are nonsense mutations (UGA and UAA, respectively). Investigators employing nonsense suppressor-bearing plasmids must bear this in mind. *pep4-3*-bearing strains accumulate suppressors of the *pep4* mutation on continued subculturing.[62,84,85] Stress in the form of overproduction of toxic proteins may exacerbate the problem and result in selection of revertants[84] or reactivation of phenotypic lag.[84,85]

While attempting to purify the 40-kDa precursor to PrB (see Moehle *et al.*[29] for the PrB maturation pathway) from a *pep4*-bearing strain, Moehle[22] found that the precursor became activated to PrB of mature size during one column purification

[82] G. S. Zubenko, A. P. Mitchell, and E. W. Jones, *Genetics* **96**, 137 (1980).
[83] P. A. Hieter, personal communication (1988).
[84] J. R. Shuster, A. Randolph, and C. George-Nacimento, *J. Cell Biochem.* **9C**(Suppl.), 111 (1985).
[85] E. W. Jones, unpublished observations (1984–1988).

step. We have no information on whether the activation was autocatalytic or due to the activity of another protease. Nevertheless, it seems clear that strains of the *pep4 PRB1* genotype that contain PrB precursor at levels comparable to levels of PrB in wild-type strains may not be ideal starting strains for biochemical analyses and/or purifications, as the PrB precursor constitutes a potential reservoir for the production of PrB during a purification. This provides an additional reason for using *prb1 pep4* double mutants.

[8] Analysis of the Size and Shape of Protein Complexes from Yeast

By SCOTT C. SCHUYLER and DAVID PELLMAN

Introduction

The era of gene discovery in budding yeast ended a few years ago with the complete sequence of the yeast genome.[1] Use of the genome sequence and all of the remarkable tools generated from it are now the common currency of everyday life in yeast laboratories. The idea that the genome sequence would change the way we do genetics was easy to anticipate. What was perhaps less obvious was the degree to which the genome sequence would revolutionize biochemical experiments. New methods in mass spectrometry and informatics, now combined with many novel methods for epitope tagging, affinity chromatography, and high-throughput proteomics, have created a boom industry in the discovery and characterization of protein complexes.

There is a proud history of biochemistry in yeast dating back to the early studies on metabolism. Because of a new emphasis on analyzing protein complexes, many of the "classic" biochemical approaches have come to the fore. These methods, especially when used in combination with more recently developed procedures, are powerful experimental tools for analyzing the physical properties of protein complexes. This chapter focuses on methods for estimating the molecular weight and shape of soluble protein complexes. We review in practical terms the preparation of native extracts, size-exclusion chromatography (gel filtration), and velocity sedimentation (sucrose gradients). We also briefly comment on commonly used methods to detect protein–protein interactions such as polymerase chain reaction (PCR)-based epitope tagging of proteins followed by coimmunoprecipitation. We anticipate an increasingly wide use of these methods by the community of yeast researchers.

[1] A. Goffeau, B. G. Barrell, H. Bussey, R. W. Davis, B. Dujon, H. Feldmann, F. Galibert, J. D. Hoheisel, C. Jacq, M. Johnston *et al., Science* **274**, 546 (1996).

It is worthwhile saying a few words about the circumstances in which yeast researchers will find a need to use these methods. In laboratories where protein complexes are being purified based on a biochemical activity, the protocols described here will be of use to beginning students. However, many experiments with yeast start with the identification of a protein through a genetic screen, by homology, or by an interesting gene expression pattern. In this case, one of the first issues to arise is whether the protein of interest acts alone or as part of a macromolecular complex. The answer to this question can have an important impact on how one interprets genetic data. For example, a protein identified by genetic interactions with tubulin would be viewed differently if it were found to be a monomeric microtubule-binding protein versus being part of a large motor protein complex. One simple and information-rich experiment is to compare the size of pure recombinant protein with that of the protein from a native yeast extract. A second issue is that many proteins have multiple functions and are components of distinct protein complexes. The methods reviewed in this chapter can be used to dissect the partitioning of proteins into different complexes. These methods become particularly informative when used in combination with mutational studies. A third issue is that the discovery of many new protein complexes will invariably lead to studies on the structure of these complexes. The methods described here are critical for the preparation of proteins for structure determination. One benchmark for the suitability of a protein preparation for crystallization is whether it runs as a single discrete peak by size-exclusion chromatography. Finally, many protein complexes are dynamic, and the ability to determine the size of the complex under different physiologic conditions is important for understanding many aspects of cell signaling and protein regulation.

Native Yeast Extracts

Here we review two different methods for making native yeast extracts. The N_2(liquid) method tends to be more convenient for large-scale extract preparations. Glass bead lysis is quicker and can be used on a small scale, which is more convenient when preparing a large number of samples. For most uses, it is important to keep the extracts concentrated and to avoid proteolysis by use of protease inhibitors (see later). In order to preserve the integrity of protein complexes, it is important to use lysis solutions that are well buffered for pH and do not contain too much salt or detergent.

Liquid Nitrogen Lysis

In this protocol, cells in lysis buffer are rapidly frozen in N_2(liquid) and can then be stored for an extended period at $-80°$.[2] In the following method, a low

[2] K. B. Kaplan and P. K. Sorger, in "Protein Function: A Practical Approach" (T. Creighton, ed.). IRL Press, Oxford, 1996.

salt buffer, which should maintain protein–protein interactions, is used for lysis. A typical yield is 3–5 ml of extract at a concentration of 5–15 mg/ml.

1. Grow cells (500–1000 ml) to an OD_{600} of 0.2–0.5. Harvest cells by centrifugation for 10 min at 4° at 13,000g [e.g., JA-14 rotor at 9500 rpm (Beckman Instruments, Palo Alto, CA)]. Pour off the supernatant.

2. Wash cell pellets once in 50 ml of 50 mM HEPES–NaOH (pH 7.4), 150 mM NaCl. This removes excess medium and ensures that the pH is near 7.4.

3. Resuspend cell pellets in an equal volume of lysis buffer [50 mM HEPES–NaOH (pH 7.4), 150 mM NaCl, 1 mM phenylmethylsulfonyl fluoride (PMSF), 1× protease inhibitors mix]. We use PMSF combined with protease inhibitors such as the "Complete Mini-EDTA Free" protease inhibitor mix (Roche, Indianapolis, IN). A stock of 100 mM PMSF should be made fresh in 100% ethanol. When resuspending the cell pellet, slurries of less than an equal volume do not yield good lysis. Conversely, do not make the extracts too dilute by adding too much lysis buffer. An equal volume mix usually yields a 10-mg/ml extract. Detergents, salts, inhibitors, and other small molecules should be included if necessary (see later). Freeze the cell slurry by dropping the cell suspension from the tip of a pipette directly into N_2(liquid) drop by drop and store at −80° until use. The slurry forms little frozen beads in the N_2(liquid). Do not thaw and refreeze the mix.

4. Grind frozen cell beads with a mortar and pestle (50–100 strokes) into a fine powder that is kept cold in N_2(liquid). It is important to precool and keep the mortar, pestle, and a spatula cool. For 0.5–1 liter cultures, grind by hand in a mortar/pestle (Coors, Golden, CO); for larger volumes, grind the cells using a stainless-steel blender chilled with N_2(liquid). Do not use ball-bearing blenders because N_2(liquid) can freeze and crack the ball bearings. Use carbon-brushed blenders such as those made by Waring (see Fischer, Pittsburgh, PA). One can check the lysis efficiency by looking at the cells in a microscope.

5. Collect powder with a cold spatula into a 15-ml conical tube and spin for 15 min at 4° at 2000g to concentrate [e.g., an RT-6000D tabletop centrifuge at 3000 rpm (Sorvall, Newtown, CT)]. Transfer the whole lysate and thick pellet to 1.5-ml Eppendorf tubes and spin for 10 min at 4° at 10,000g to make a low-speed supernatant (e.g., top speed in a microfuge). Transfer the supernatant to a new Eppendorf tube. This is the low-speed extract.

6. To make a high-speed supernatant, spin for 1 hr at 4° at 100,000g [e.g., a RP100 AT3-168 rotor in an RC M120 miniultracentrifuge at 51,000 rpm (Sorvall) or a TLA 120.1 rotor in an Optima TLX Ultracentrifuge at 50,000 rpm (Beckman Instruments)].

Glass Bead Lysis

In this protocol, cells are lysed by vortexing them in the presence of lysis buffer and glass beads.[3] This method is faster than the mortar and pestle N_2(liquid) method and is easier to use with multiple small samples. Using this method, 1 ml of extract at a concentration of 10–20 mg/ml can routinely be obtained.

1. Grow cells (100–200 ml) to an OD_{600} of 0.2–0.5. Harvest cells by centrifugation for 10 min at 4° at 13,000g [e.g., a JA-14 rotor at 9500 rpm (Beckman Instruments)]. Pour off the supernatant.

2. Resuspend pellet in remaining liquid and transfer to a 2-ml screw cap tube. Make sure these tubes (Fischer, PA) fit into the Mini-BeadBeater (Biospec Products, Bartlesville, OK). If the volume is too big (more than 1.5 ml), split the sample into multiple tubes.

3. Spin down screw cap tubes in an Eppendorf centrifuge for 1 min at 4° at 10,000g (13,000 rpm). Aspirate away all of the medium. At this point, if you want to freeze down the pellet, wash twice with ice-cold lysis buffer and store at $-80°$. To proceed with the lysis, after washing twice with lysis buffer, spin for 1 min at 13,000 rpm (10,000g) at 4° and aspirate away lysis buffer.

4. Resuspend the cell pellet in 300 μl ice-cold lysis buffer + protease inhibitors (see earlier discussion). Add an equal volume of prechilled acid-washed glass beads (Sigma, St. Louis, MO) with a diameter of $\sim$500 μm. Place tubes in an ice–water slurry.

5. Make sure the tubes are tightly capped. In a cold room, set the Mini-BeadBeater (Biospec Products) to 20-sec pulses at a speed of "50". Beat for 20 sec and then place the tubes on ice for 1 min to cool down in between pulses of bead beating. Repeat five times so there is a total of 100 sec of beating. One can check the lysis efficiency by looking at the cells in a microscope.

6. To separate extract from beads, use a thin pipette tip. However, we have obtained better recovery using the following centrifuge-based method: (a) cut the cap off an Eppendorf tube, (b) invert the 2-ml screw cap tube that contains the beads and extracts, and tap it a few times to get its contents off the bottom of the screw cap tube, (c) use a needle to poke a hole at the very bottom of the screw cap tube, and (d) put the screw cap tube, sitting right on top of the capless Eppendorf tube, into the 15-ml conical tube so that the Eppendorf tube can collect the extract. Spin the 15-ml tube for 3 min at 4° at 2000g [e.g., a tabletop centrifuge at 3000 rpm (Sorvall)]. Carefully remove screw cap tubes with a forceps. Take out the Eppendorf tube with the extract.

7. Transfer extract to a new Eppendorf tube. Spin for 15 min at 4° at 10,000g.

8. Transfer the supernatant to a new tube. This is the low-speed extract.

9. To make a high-speed supernatant, spin for 1 hr at 4° at 100,000g [e.g., a RP100 AT3-168 rotor in an RC M120 miniultracentrifuge at 51,000 rpm (Sorvall) or a TLA 120.1 rotor in an Optima TLX Ultracentrifuge at 50,000 rpm (Beckman Instruments)].

---

[3] E. Harlow and D. Lane, "Using Antibodies: A Laboratory Manual." Cold Spring Harbor Laboratory Press, Cold Spring Harbor, NY, 1999.

Troubleshooting

One of the main challenges for biochemical manipulations is maintaining the integrity of the protein complex in extracts. It is important to work at 4° and to include protease inhibitors in the lysis buffer. If the protein to be studied is present in very small amounts, then it may be necessary to start with a large number of cells and scale the amount of extract to the absolute amount of the protein. It may also be necessary to include small molecules such as phosphatase inhibitors in order to maintain complex structure. Commonly used phosphatase inhibitors include 1 mM NaVO$_4$ (sodium-orthovanadate), 5 mM sodium-pyrophosphate, 10 mM sodium-β-glycerophosphate, and 10 mM sodium-fluoride.

Another issue when making extracts is that of efficient extraction and solubility. Proteins vary dramatically in their solubility from being essentially insoluble (<10 μg/ml) to very soluble (>300 mg/ml). Key variables that affect the solubility of a protein include pH, ionic strength, temperature, and the polarity of the solvents.[4]

Once cells are lysed it is important to remove any unlysed cells or large particles by centrifugation. Making a low-speed supernatant is sufficient most of the time, but the supernatant may still be viscous and contain large particles. It is preferable to use a high-speed spin to clarify the extract; however, this may lead to pelleting of the protein of interest. Increasing the salt or detergent concentrations can promote solubility, but may also disrupt protein complex integrity.

Finally, it may be necessary to supplement the lysis buffer with salts (many enzymes bind to metals such as Cu^{2+}, Zn^{2+}, Ca^{2+}, Co^{2+}, and Ni^{2+}) and other small molecules (such as ATP) in order to maintain protein–protein interactions.

Estimating Protein Complex Molecular Weight and Shape

Proteins in yeast vary in molecular mass from $\sim$3 to $\sim$470 kDa. Most proteins in yeast have molecular masses in the range of 10–150 kDa.[5] Protein complexes of course may be much larger. For example, the yeast 20S proteasome core particle is composed of 28 proteins with a combined molecular mass of $\sim$670 kDa.[6] The following methods apply to either monomeric proteins or protein complexes.

Regardless of molecular size and quaternary structure, proteins and protein complexes have a variety of shapes. Protein shapes range from globular (compact and spherical) to extended (rod shaped). Most proteins tend to be globular, but those containing α-helical coiled-coil motifs commonly form extended rods.[7] Estimates

---

[4] D. R. Marshak (ed.), "Strategies for Protein Purification and Characterization." Cold Spring Harbor Laboratory Press, Cold Spring Harbor, NY, 1996.

[5] M. C. Costanzo, M. E. Crawford, J. E. Hirschman, J. E. Kranz, P. Olsen, L. S. Robertson, M. S. Skrzypek, B. R. Braun, K. L. Hopkins, P. Kondu *et al., Nucleic Acids Res.* **29**, 75 (2001).

[6] M. Bochtler, L. Ditzel, M. Groll, J. Hartmann, and R. Huber, *Annu. Rev. Biophys. Biomol. Struct.* **28**, 295 (1999).

[7] K. Beck and B. Brodsky, *J. Struct. Biol.* **122**, 17 (1998).

of the molecular weight of a protein and its shape in a native complex are made from measuring its hydrodynamic properties in solution.

Two hydrodynamic values that can be determined experimentally from the methods outlined later are the Stokes radius (a) from gel filtration and the sedimentation coefficient (s) from sucrose gradients. The Stokes radius is the radius of a hypothetical sphere that has the same hydrodynamic properties as the protein or protein complex. Stokes radii are usually given in nanometers. The sedimentation coefficient of a protein is equal to the rate of movement of the protein through a centrifuge tube divided by the centrifugal force and reflects the molecular weight and shape of the protein. Sedimentation coefficients are usually given in Svedberg (S) units. One Svedberg is 1×10^{-13} sec.

In addition, the Stokes radius can be correlated with the shape of a protein or complex. In a commonly used approach, one models the protein(s) as a prolate ellipsoid.[8] In a prolate ellipsoid, the up/down (a_1) and forward/back (a_2) axes are small and equal to each other, whereas the left/right (b) axis is larger ($a_1 = a_2 < b$). A large axial ratio ($a:b$) suggests that the protein(s) has an extended rod shape.

Gel Filtration

Analytical gel filtration is a common experimental method used to measure the Stokes radius of a protein or complex. A gel filtration column contains porous beads, and the movement of a protein into and out of the small pores during a column run is influenced by the size and shape of the protein. This is illustrated by comparing two monomeric proteins of the same molecular weight, where one is spherical and the other is rod shaped. During gel filtration the spherical protein, which has a small Stokes radius, will diffuse more readily into the small pores in the bead matrix and thus explore a larger volume and elute later from the column. A rod-shaped protein has a larger Stokes radius and does not explore as much of the solution volume inside the beads. Relative to the globular protein of the same mass, the protein with a rod shape will appear to be bigger by gel filtration because the protein explores less volume and elutes earlier.

In general, there are two kinds of gel filtration columns: preparative and analytical. Large volumes can be loaded onto preparative columns, but the resolution is poor. Analytical columns have high resolution, but work best with small load volumes. Manufacturers usually recommend that one does not load more then 5% of the total column volume. However, a smaller load volume gives higher resolution, and optimal results are obtained with loads of only 1 or 2% of the total column volume.

---

[8] An ellipsoid is a solid of which all the plane sections normal to one axis are circles and all the other plane sections are ellipses. Prolate means lengthened in the direction of a polar diameter, or growing or extending in width. See Ref. 16.

TABLE I
GLOBULAR MOLECULAR WEIGHT MARKERS FOR CALIBRATING ANALYTICAL GEL FILTRATION
COLUMN AND STANDARDS FOR SUCROSE GRADIENTS[a]

Protein	Stokes radius (nm)	Sedimentation coefficient (S)	M
Thyroglobulin	8.5	19.4	349 kDa × 2 = 698 kDa
Ferritin	6.1		
Catalase	5.2	11.3	51 kDa × 4 = 206 kDa
Aldolase	4.8		
Yeast Alcohol Dehydrogenase	4.6	7.4	37 kDa × 4 = 150 kDa
Bovine serum album	3.5	4.3	66 kDa × 1 = 66 kDa
Cytochrome c	1.0	1.9	12 kDa × 1 = 12 kDa

[a] From Pharmacia Biotech and Sigma. Some of the markers are themselves multisubunit protein complexes (values are from the manufacturer).

To determine the Stokes radius, one must first determine a K_d, the distribution coefficient. K_d can be thought of as the percentage of the solution volume inside the beads explored by a protein relative to the total solution volume inside of the beads. Siegel and Monty[9] found that the cube root of K_d, $(K_d^{1/3})$, is linear with the Stokes radii of proteins. K_d is defined as

$$K_d = [(V_e - V_o)/(V_t - V_g - V_o)] \tag{1}$$

where V_e is the volume that the sample protein elutes in, V_o is the void volume (the solution volume outside of the beads), V_t is the total volume in the column, and V_g is the volume in the column occupied by the bead matrix.

Before an experiment, the column should be equilibrated in a low salt buffer to ensure that protein–protein interactions are not disrupted. An analytical column can be calibrated with globular markers of known Stokes radii (see Table I). Calf thymus DNA can be used to determine the void volume (V_o) of the column, and acetone is used to determine the total solution volume. The volume in the column occupied by the bead matrix (V_g) is determined by subtracting the total column volume (provided by the manufacturer) from the total solution volume (determined by acetone). To generate the linear calibration curve for your column, determine the elution volume (V_e) for each marker and then plot the Stokes radii of the markers versus their $K_d^{1/3}$.

Experimental Protocol

Commonly used low-pressure analytical gel filtration columns include Superose 6 and Superose 12 (Pharmacia Biotech, Piscataway, NJ) or SE-1000 (Bio-Rad, Hercules, CA). These columns have different molecular weight ranges,

[9] L. M. Siegel and K. J. Monty, *Biochim. Biophys. Acta* **112**, 346 (1996).

where the molecular weight information provided by the manufacturer is for globular proteins or complexes. If there is no estimate of the size of the protein or complex of interest, one has to try several different columns. These columns are not gravity flow columns and should be used in combination with pumps such as the "FPLC Basic" (Pharmacia Biotech) or "BioLogic HR" (Bio-Rad) systems. One may also use gravity flow columns, which are much less expensive. However, the results may not be as reproducible when compared to pump-driven columns.

1. All work should be done on ice, in a refrigerator, or a cold room at 4°.
2. Equilibrate the column with 3 column volumes of buffer [e.g., a 50 mM HEPES–NaOH (pH 7.4), 150 mM NaCl solution]. The column running buffer should be matched with the yeast extraction buffer. We typically do not add protease inhibitors to our column running buffer; however, one may wish to add these (i.e., if you find that proteolysis is a problem during the column run) or other small molecules. Be sure to follow the manufacturer's guidelines for flow rate and back pressure limits.
3. Load the sample onto the column, but save a portion of the load for analysis after the column run. Typical load volume limits are 2% of the total column volume. Avoid overloading as it creates broad elution profiles and decreases the overall column resolution. Follow the recommendations made by the manufacturer regarding sample preparations and load limits. Samples should not be too concentrated (70 mg/ml or less) or viscous. Samples should be filtered or spun at a high speed to remove any large particles before they are loaded onto the column. Load high-speed supernatant extracts when possible. When using a filter, save a sample of prefiltered extract and compare it with the filtered extract to check if the protein(s) stuck to the filter. To minimize protein loss during filtration, use a polyvinylidene difluoride (PVDF)-based 0.45-μm sterile syringe filter that has reduced protein binding (e.g., Millex-HV filters, Millipore, Bedford, MA).
4. After loading, run 1–2 full column volumes of buffer through the column at a slow and continuous flow rate. Use a flow rate of 0.1–0.5 ml/min for a Superose 6 or SE-1000 column. Slower flow rates tend to give better column resolution.
5. Collect fractions during the entire column run for analysis. Collect 0.5- to 1.0-ml fractions from a 15-ml column [e.g., using an automated Model 2128 fraction collector (Bio-Rad)].
6. Assay the load sample and column fractions for the protein of interest by biochemical activity or Western blotting. The maximum(s) of the peak(s) of biochemical activity should be used as an average value for the elution volume. Using Eq. (1) and the elution volume (V_e), determine the $K_d^{1/3}$ value for the protein of interest. Then, determine the Stokes radius for the protein based on the standard curve for the column. This Stokes radius value is used for estimating the molecular weight and shape of the protein in solution (see later). Analytical columns should be stable and peak elution volumes should be reproducible. Multiple runs should be performed to ensure reproducibility.

Troubleshooting

Keys to good analytical gel filtration are quality sample preparation and slow and even column runs. The extract preparations outlined earlier should be sufficient for quality sample preparation.

One potential problem is that the small sample load volume (~250 μl of extract for a 15-ml column) necessary for good column resolution may lead to a loss of biochemical activity. For example, the activity could be a Western blot signal, which may become too dilute after the column run. As such, it may be necessary to concentrate the fractions before analysis. One can precipitate the fractions [using trichloroacetic acid (TCA), see steps 4–7 in the denaturing lysis protocol outlined later] before loading them onto sodium dodecyl sulfate–polyacrylamide gel electrophoresis (SDS–PAGE) gels for Western blot analysis to ensure a robust signal.

The goal of the methods outlined for analytical gel filtration and sucrose gradients (see later) is to estimate the molecular weight and shape of a protein complex. These values can be changed in solution by the association of proteins with detergents. Thus, it is best to avoid detergents when preparing yeast extracts for analysis. However, you may find that you get poor extraction in the absence of detergent or poor solubility of the protein complex. In this case, it may be necessary to try extracting at a higher salt concentration or including detergent. If detergent is included, try to use it at the lowest level possible and do not exceed the critical micelle concentration.

Sucrose Gradients

Sucrose gradients are used to determine sedimentation coefficients. Both the molecular size and the shape of the protein affect how it moves through solution during centrifugation. Massive globular proteins sediment further into the gradient compared to smaller globular proteins. Also, an extended rod-shaped protein has a larger frictional coefficient compared to that of a compact sphere of equal molecular weight. Therefore, a globular protein sediments faster than a protein with an extended rod shape.

The movement of a protein or a complex through a sucrose gradient is determined by a combination of several forces. The sedimenting force on a particle is equal to the mass (m) multiplied by the centrifugal field $\omega^2 r$, where ω is the angular velocity of the rotor (in radians per second) and r is the radius or distance of the particle from the axis of rotation. As the particle travels further away from the axis of rotation, the centrifugal force increases. Opposing the sedimentation force are (i) flotation force, (ii) frictional resistance, and (iii) diffusion. The flotation force is equal to $m\omega^2 r v \rho$, where v is the partial specific volume (i.e., the volume displaced by 1 g of sedimenting particles) and ρ is the density of the solution. The net sedimentation force is equal to $[m\omega^2 r(1 - v\rho)]$. Frictional resistance against a particle moving through solution is equal to (fv), where f is the frictional coefficient and v is the particle velocity. Finally, diffusion is the result of random thermal motions of particles in all directions and leads to the equal dispersion of particles in solution.

Experimental Protocol

To measure the sedimentation coefficient of a protein or complex, it is standard to use either 5–20% (for proteins or complexes from 1S to 20S) or 10–40% (for proteins or complexes from 1S to greater than 20S) sucrose gradients spun in swinging bucket rotors, which have a constant volume-to-radius ratio. These sucrose gradients are nearly isokinetic, which means that a particle of a given mass moves at a constant velocity through the gradient. The velocity is constant because the increase in centrifugal force, as the particle moves further from the axis of rotation, is balanced by an increase in density and viscosity of the solution. It is also common to use a 5–20% glycerol gradient, which has a very similar density profile as a 5–20% sucrose gradient. Glycerol has the added advantage that many enzyme activities and protein complexes are stabilized in it. During centrifugation the maximal rotor speeds (rpm) and the shortest durations should be used because this minimizes peak broadening by diffusion. This is particularly a concern for 5–20% gradients because they are less viscous and are disrupted more easily by diffusion.

It is necessary to have molecular weight standards with known sedimentation coefficients to construct a standard curve (Table I). Always have one separate centrifuge tube for the standards during each experiment. Do not mix the molecular weight standards in with a yeast extract because *in vitro* interactions may cause the standards and/or the protein(s) to sediment abnormally.

As with gel filtration chromatography, it is important to prepare a quality sample. Viscous and particulate extracts should be avoided, as these can lead to streaming effects: particles that are not free to diffuse may be aberrantly dragged down into the gradient by other larger particles. Do not load too much sample onto the gradient, and volume load limits should be followed (Table II[10]). It is also important that the load sample is not too dense or highly concentrated, as this can lead to broadening of the peaks and poor gradient resolution. When the sample density exceeds the initial density in the gradient, mixing effects can occur. Diffusion of large protein particles in a highly concentrated load sample is slowed by particle–particle interactions. The small particles that make up the density gradient, such as sucrose, diffuse and tend to mix rapidly into the less mobile sample and destroy the linearity of the gradient. Use extracts that are at a concentration of 5 mg/ml or less for a 5–20% gradient.

---

[10] O. W. Griffith, "Techniques of Preparative, Zonal and Continuous Flow Ultracentrifugation." Beckman Instruments Inc., Palo Alto, CA, 1986.

TABLE II
PARAMETERS FOR RUNNING SUCROSE GRADIENTS[a]

Rotor	Sample volume (ml)	Gradient density (%)	Revolutions per minute
SW 65 Ti	0.2	5–20	65,000
	0.2	10–40	65,000
SW 60 Ti	0.2	5–20	60,000
	0.2	10–40	60,000
SW 55 Ti	0.2	10–40	55,000
SW 50.1	0.2	5–20	50,000
	0.2	10–40	50,000
SW 41 Ti	0.5	5–20	40,000
	0.5	10–40	40,000
SW 40	0.5	10–40	40,000

[a] All are Beckman swinging bucket rotors (Beckman Instruments). Information on recommended sample volume loads can be obtained from the manufacturer (O. W. Griffith, "Techniques of Preparative, Zonal and Continuous Flow Ultracentrifugation." Beckman Instruments Inc., Palo Alto, CA, 1986).

1. All work should be done on ice, in a refrigerator, or a cold room at 4°. Precool the centrifuge and rotor before spinning.

2. Prepare a stock solution of a high percentage (40–60% w/v) sucrose and filter sterilize it. Measure the refractive index of this stock solution to determine the percentage sucrose accurately (see step 7). Prepare one low (5 or 10) and one high (20 or 40) percent solution, both with a final concentration of 50 mM HEPES–NaOH (pH 7.4) 150 mM NaCl. As with gel filtration, we recommend that the buffer used to make the extracts is the same as the buffer used for pouring the gradients. Again, avoid detergents and high salt solutions.

3. Pour sucrose gradients (5–20% or 10–40%). The total gradient volume depends on the rotor and centrifuge tubes. Follow the manufacturer's recommendation on the type of tube and the maximum volumes allowed for any given tube. Gradients should be poured slowly and carefully to avoid mixing. Use a solution flow rate of 1 ml/min or less. To obtain a high degree of reproducibility from tube to tube and experiment to experiment, use one of the automated dual pump systems, such as the "FPLC Basic" (Pharmacia) or "BioLogic HR" (Bio-Rad). Gradients made by simple gravity flow mixing chambers are less reproducible, although they are much less expensive. Regardless of the device used, the tubing leading into the centrifuge tube should be placed on the side of the centrifuge tube. This allows the drops to roll down the side rather then drip into the center of the centrifuge tube, which would disrupt the gradient. Do not let tubes sit unused for an extended time (more than 2 hr), as diffusion will destroy the gradients.

4. Load the recommended volume of extract onto the top of the gradient (Table II). Save an aliquot of material as a loading control for analysis after centrifugation. If multiple gradients are to be compared, be sure to load an equal volume on top of each one so that the total solution volume from tube to tube remains the same. This is also true for the molecular weight marker tube. Do not disrupt the gradient. Load the tube with either a pipette tip or a needle resting against the side of the tube at a slow rate, drop by drop. Once the centrifuge tubes are loaded, balance each one with lysis buffer one drop at a time. When prepared properly, the tubes should be nearly balanced to begin with.

5. Spin at 4° at or near the top rotor speed. Use the rotor speeds provided by the manufacturer (Table II). Use slow acceleration and deacceleration rates to avoid disrupting the gradient. Although optimal times are determined empirically, anywhere from 6 to 16 hr should be adequate for average samples. Sufficient run times should be used to ensure good separation within the gradient, but extended run times should be avoided as they can lead to broadening of the peaks by diffusion. On inspection after the run, the molecular weight marker tube should have one red band (cytochrome c) near the top and green band (catalase) clearly separated down into the tube.

6. Collect fractions at 4°. Carefully insert a glass capillary tube connected to a peristaltic pump (Bio-Rad). Collect at a flow rate of 1 ml/min from the bottom of the tube. One can also poke a hole in the bottom of the tube and collect the drops. Either way, be sure to scrape off any pellet at the bottom of the tube, which contains very large complexes that may have quickly sedimented to the bottom. It is convenient to collect 0.5- to 1-ml fractions.

7. Confirm the linearity of the gradients by refractometry following the manufacturer's protocol (#ABBE-3L, Thermo Spectronic, Rochester, NY). Plot fraction number versus percentage sucrose. The plot should show that the gradients are nearly linear. The slopes of the gradients should be the same from one tube to the next.

8. For molecular weight marker fractions it is sufficient to run an SDS–PAGE gel followed by Coomassie staining. Remember that some markers are multisubunit complexes, which are denatured when run through SDS–PAGE gels (Table I). Generate a graph of fraction number versus sedimentation coefficient (in Svedberg units). The slope of this graph should be linear.

9. Assay load sample, pellet, and fractions for the protein of interest. The fraction(s) that contains the maximum(s) of the peak(s) of biochemical activity, such as a signal on a Western blot, should be used as the fraction value(s) when determining the sedimentation coefficient. Using the molecular weight standard curve, estimate the sedimentation coefficient for your protein. This is only an estimate and it is usually reported without any decimal places. Sedimentation coefficient values for standard markers are usually measured under standard conditions of 20° and in pure water ($s_{20,w}$). Although the protocol outlined here recommends using nonstandard conditions, the markers and the complexes in the extracts should be

affected in the same way by the difference in temperature and density. As such, one can present the measured sedimentation coefficient of a protein as an $s_{20,w}$ value.

Troubleshooting

The small sample load volume (200–500 μl of extract) necessary for good gradient resolution may lead to a loss of biochemical activity or Western blot signal. As such, it may be necessary to perform a concentrating step on the fractions before analysis. If the protein of interest is being assayed by Western blotting, you can TCA precipitate the fractions before loading them onto SDS–PAGE gels (see steps 4–7 in the denaturing extract protocol outlined later).

Finally, the density of most proteins is between 1.3 and 1.4 g/cm^3. However, proteins containing large amounts of phosphate (phosvitin, 1.8 g/cm^3) or lipid moieties (β-lipoprotein, 1.03 g/cm^3) are substantially different in density compared to the average protein. This difference in density can affect the calculation of molecular weight[4] (see later).

Hydrodynamic Calculations

Protein complex molecular weight is estimated using the method of Siegel and Monty.[9] The equation that relates molecular weight (M) with the Stokes radius and sedimentation coefficient is

$$M = [(6\pi \eta N a s)/(1 - \upsilon\rho)] \qquad (2)$$

where N is Avogadro's number (6.022 × 10^{23}), a is the Stokes radius in centimeters, and s is the sedimentation coefficient (in Svedberg units or 1 × 10^{-13} sec).[9] Standard values for the density of water ($\rho = 1$ g/cm^3) and buffer viscosity ($\eta = 0.01005$ g/cm sec, or 0.01005 poise) can be used. Water has a viscosity close to that of 0.01 poise, as do solutions that are made mostly of water (i.e., lysis buffers). It should be noted that these commonly used values for density and viscosity are approximations based on the assumption that the experiment is performed under standard conditions of 1 atm of pressure at 20° in pure water.[10] Here we are outlining a very simple method for estimating the molecular weight and it is sufficient to use these values with the assumption that the nonstandard conditions shall have the same uniform effect on the molecular weight markers and proteins in the extract. The average specific volume (υ) of a protein can be estimated using the method of Cohn and Edsall[11] (see Perkins[12] for a more recent discussion). Calculate the average specific volume using Table III[11,12] by adding up the values for the given

---

[11] E. J. Cohn and J. T. Edsall, "Proteins, Amino Acids and Peptides." Reinhold, New York, 1943.
[12] S. J. Perkins, *Eur. J. Biochem.* **157**, 169 (1986).

TABLE III
VOLUMES AND PARTIAL SPECIFIC VOLUMES (v) OF AMINO ACIDS[a]

Amino acid	Volume $\times 10^{-3}$ nm^3	cm^3/g or ml/g
Ile	168.9	0.90
Phe	187.9	0.77
Val	141.4	0.86
Leu	168.9	0.90
Trp	228.5	0.74
Met	163.1	0.75
Ala	87.2	0.74
Gly	60.6	0.64
Cys	106.7	0.63
Tyr	192.1	0.71
Pro	122.4	0.76
Thr	117.4	0.70
Ser	91.0	0.63
His	152.4	0.67
Glu	141.4	0.66
Asn	117.4	0.62
Gln	142.4	0.67
Asp	114.6	0.60
Lys	174.3	0.82
Arg	181.3	0.70

[a] The overall average partial specific volume for all 20 amino acids is 0.724 cm^3/g or an average density of 1.38 g/cm^3 [E. J. Cohn and J. T. Edsall, "Proteins, Amino Acids and Peptides." Reinhold, New York, 1943; S. J. Perkins, *Eur. J. Biochem.* **157**, 169 (1986)].

amino acid sequence in the complex and dividing by the total number of amino acids. If the primary amino acid sequence of the protein or protein complex is not known, use the average value of 0.724 cm^3/g. It is important to stress that this method for calculating molecular weight is only an estimate with a typical systematic error of about ±20% of the calculated value.[10,13–15]

The Stokes radius can be correlated with the shape of a protein or complex after making an assumption about the percentage hydration of the protein(s) and

[13] J. Steensgaard, S. Humphries, and S. P. Spragg, in "Preparative Centrifugation: A Practical Approach" (D. Rickwood, ed.). IRL Press, Oxford, 1992.
[14] M. Potschka, *Anal. Biochem.* **162**, 4 (1987).
[15] C. M. Field, O. al-Awar, J. Rosenblatt, M. L. Wong, B. Alberts, and T. J. Mitchison, *J. Cell Biol.* **133**, 605 (1996).

modeling the protein(s) as a prolate ellipsoid. This does not mean that the complex literally has an ellipsoid shape, but rather that the hydrodynamic behavior of the complex matches the predicted hydrodynamic behavior of an ellipsoid with a given axial ratio. In fact, the geometric shapes of proteins are often quite irregular. Nonetheless, if a protein is found to have a large axial ratio, it suggests a rod-like shape versus a globular shape.

To determine the axial ratio, first determine the translational frictional coefficient (f) from the measured Stokes radius (a) and then compare it with the translational frictional coefficient (f_0) of a perfect sphere with the same molecular weight (M) as your complex[13,16]:

$$f = 6\pi \eta a = [M(1 - \upsilon\rho)]/(Ns)] \qquad (3)$$

$$f_0 = 6\pi \eta [(3\upsilon M)/(4\pi N)]^{1/3} \qquad (4)$$

$$(f/f_0) = a/[(3\upsilon M)/(4\pi N)]^{1/3} \qquad (5)$$

The P hydrodynamic shape function provides a relationship between the translational frictional coefficient ratio (f/f_0) and the axial ratio ($a:b$). Determine the P function value for the protein complex by

$$P = (f/f_0) [(\omega/\upsilon\rho) + 1]^{-1/3} \qquad (6)$$

Typical hydration values (ω) are 0.35–0.40, and υ and ρ are the same as in Eq. (2).[13,16,17] Table IV contains relationships between the P hydrodynamic shape function value and axial ratios ($a:b$) when assuming a prolate ellipsoid shape.[13,16]

PCR-Based Epitope Tagging

We briefly comment on PCR-based epitope tagging of proteins and provide one example of a simple coimmunoprecipitation protocol. Western blotting of an epitope-tagged protein is a simple approach for following the behavior of a protein through the procedures outlined in the previous sections. Coimmunoprecipitation from native yeast extracts is a standard approach for identifying protein–protein interactions.[18] Coimmunoprecipitations can also be performed on gel filtration or sucrose gradient fractions to determine if the cofractionation of proteins in fact reflects a complex between the proteins of interest.

---

[16] S. E. Harding and H. Colfen, *Anal. Biochem.* **228,** 131 (1995).
[17] A. M. Taylor, J. Boulter, S. E. Harding, H. Colfen, and A. Watts, *Biophys. J.* **76,** 2043 (1999).
[18] E. M. Phizicky and S. Fields, *Microbiol. Rev.* **59,** 94 (1995).

TABLE IV
RELATIONSHIP BETWEEN P VALUE AND AXIAL RATIO[a]

P value	Axial ratio
1.0000	1.0000
1.0009	1.1000
1.0031	1.2000
1.0063	1.3000
1.0103	1.4000
1.0149	1.5000
1.0201	1.6000
1.0256	1.7000
1.0315	1.8000
1.0377	1.9000
1.0440	2.0000
1.1130	3.0000
1.1830	4.0000
1.2500	5.0000
1.3140	6.0000
1.3750	7.0000
1.4340	8.0000
1.4900	9.0000
1.5430	10.000
1.9960	20.000
2.3590	30.000
2.6710	40.000
2.9500	50.000
3.2050	60.000
3.4420	70.000
3.6640	80.000
3.8740	90.000
4.0740	100.00

[a] Assuming a prolate ellipsoid of revolution [J. Steensgaard, S. Humphries, and S. P. Spragg, in "Preparative Centrifugation: A Practical Approach" (D. Rickwood, ed.). IRL Press, Oxford, 1992; S. E. Harding and H. Colfen, *Anal. Biochem.* **228**, 131 (1995)].

Several elegant and rapid PCR-based methods take advantage of homologous recombination in yeast to generate a variety of epitope-tagged proteins. Tags can be placed anywhere in an open reading frame (ORF),[19,20] as well as at the NH_2

[19] B. L. Schneider, W. Seufert, B. Steiner, Q. H. Yang, and A. B. Futcher, *Yeast* **11**, 1265 (1995).
[20] M. Knop, K. Siegers, G. Pereira, W. Zachariae, B. Winsor, K. Nasmyth, and E. Schiebel, *Yeast* **15**, 963 (1999).

and COOH termini.[21-29] Epitope tags are most often placed at the COOH terminus because one-step NH_2-terminal tagging methods remove the native promoter. Some of these tagging systems take advantage of selectable markers that are unique and do not share homology with *Saccharomyces cerevisiae* or *Schizosaccharomyces pombe* genes. This is beneficial because it increases the frequency of the desired recombination event. With the continued identification of novel selectable markers, it is quite likely that several new PCR-based tagging cassettes shall soon become available.[30]

Once a gene is tagged, it is necessary to confirm that the protein is expressed and that the tagged protein is functional. The following simple denaturing lysis method is used for rapidly screening through potential recombinant clones or confirming the expression of an epitope-tagged protein.

Denaturing Lysis

This protocol is a convenient and rapid way to isolate denatured whole cell yeast extracts from small culture volumes for SDS–PAGE and Western blot analysis.[31]

1. Grow cells (5 ml) to an OD_{600} of 0.2–0.5. Harvest the cells by centrifugation for 10 min at 4° at 13,000g (i.e., top speed in a microfuge). Resuspend the cell pellet in 600 μl of doubly distilled H_2O at 4°. As an aside, when collecting closely spaced time points, one can skip the water wash and add 100 μl of lysis buffer directly to 600 μl of cells in medium.
2. Add 100 μl of lysis buffer (1.85 M NaOH, 7.4% 2-mercaptoethanol). Incubate sample on ice for 15 min.
3. Spin for 10 min at 4° at 13,000g (i.e., top speed in a microfuge).
4. Place supernatant, which should have a volume of about 700 μl, into a new tube and then add 42 μl of 100% TCA, which is a 6% final solution. Mix and incubate on ice for 15 min.

---

[21] A. Wach, A. Brachat, C. Alberti-Segui, C. Rebischung, and P. Philippsen, *Yeast* **13**, 1065 (1997).
[22] M. S. Longtine, A. McKenzie III, D. J. Demarini, N. G. Shah, A. Wach, A. Brachat, P. Philippsen, and J. R. Pringle, *Yeast* **14**, 953 (1998).
[23] J. Bahler, J. Q. Wu, M. S. Longtine, N. G. Shah, A. McKenzie III, A. B. Steever, A. Wach, P. Philippsen, and J. R. Pringle, *Yeast* **14**, 943 (1998).
[24] O. Puig, B. Rutz, B. G. Luukkonen, S. Kandels-Lewis, E. Bragado-Nilsson, and B. Seraphin, *Yeast* **14**, 1139 (1998).
[25] M. D. Krawchuk and W. P. Wahls, *Yeast* **15**, 1419 (1999).
[26] G. Rigaut, A. Shevchenko, B. Rutz, M. Wilm, M. Mann, and B. Seraphin, *Nat. Biotechnol.* **17**, 1030 (1999).
[27] A. De Antoni and D. Gallwitz, *Gene* **246**, 179 (2000).
[28] A. Brachat, N. Liebundguth, C. Rebischung, S. Lemire, F. Scharer, D. Hoepfner, V. Demchyshyn, I. Howard, A. Dusterhoft, D. Mostl *et al., Yeast* **16**, 241 (2000).
[29] S. Honey, B. L. Schneider, D. M. Schieltz, J. R. Yates, and B. Futcher, *Nucleic Acids Res.* **29**, E24 (2001).
[30] A. L. Goldstein and J. H. McCusker, *Yeast* **15**, 1541 (1999).
[31] M. P. Yaffe and G. Schatz, *Proc. Natl. Acad. Sci. U.S.A.* **81**, 4819 (1984).

5. Spin down for 10 min at 4° at 13,000g and remove the supernatant. Wash the pellet with 1 ml ice-cold acetone. This helps remove acids and salts.
6. Spin down for 10 min at 4° at 13,000g and air dry the pellet.
7. Resuspend pellet in SDS–PAGE protein sample buffer. The TCA pellet can be difficult to resuspend, and it may be necessary to work the pellet into solution with a pipette tip. If the protein sample buffer turns yellow, add 2 M Tris–base that has not been adjusted for pH, 1 μl at a time, until it turns blue again. Be sure to add an equal amount of Tris–base to each sample as the extra salt can cause the samples to run differently on the SDS–PAGE gel.

Coimmunoprecipitation

Coimmunoprecipitation (co-IP) is a simple and effective method for detecting or confirming a physical interaction between two proteins.[3,18,32] There are a wide variety of protocols for co-IPs, and here we present one sample protocol. The amount of extract or column and/or gradient fraction to be used and the amount of antibody necessary for efficient and effective immunoprecipitation can vary widely and has to be determined empirically for each experiment. A starting point is to use 1 μl for polyclonal antibodies (serum solutions usually have 0.5 mg/ml or less of the specific antibodies of an antigen or 50 μl of tissue culture supernatant (supernatants usually have 0.05 mg/ml or less of the specific antibody of an antigen) or 0.5 μl of ascites fluid (ascites fluid usually has 0.5 mg/ml or less of the specific antibody of an antigen with 2 mg of total protein per reaction (200 μl from a 10-mg/ml extract[3]).

1. Add the appropriate amount of antibody to the extract or fraction and incubate on ice at 4° for 1 hr to allow binding of antibody to epitopes. Spin extract + antibody for 30 sec at 4° at 10,000g to remove any aggregates.
2. Prepare protein A (Sigma) or protein G-Sepharose beads as recommended by the manufacturer in a 1 : 1 slurry in immunoprecipitation buffer.
3. Add protein A-Sepharose beads (1/10 of the volume of the extract) and tumble on a rotator for 1 hr at 4° (i.e., a Barnstead/Thermolyne Labquake Shaker/Rotisserie, Dubuque, IA).
4. Pellet beads for 30 sec at 4° at 10,000g. Save a fraction of the supernatant to determine if the immunoprecipitation depleted the protein of interest from the extract. Aspirate the supernatant with a bent 20-gauge needle to ensure a slow flow and preserve the pellet (we bend the needle twice to make an "N" shape). Wash five times with at least 1 volume of lysis buffer + protease inhibitors.
5. After the last wash, cut off the end of a pipette tip to make the opening larger and pipette the beads into a new Eppendorf tube. This will reduce the amount of nonspecific proteins, which are stuck to the sides of the original tube.

---

[32] P. A. Kolodziej and R. A. Young, in "Guide to Yeast Genetics and Molecular Biology" (C. Guthrie and G. R. Fink, ed.). Academic Press, San Diego, 1991.

6. Resuspend the bead pellet in protein sample buffer plus 2-mercaptoethanol and boil for 5 min (or heat to 65° for 20 min). Run an SDS–PAGE gel including some or all of these controls: lysates of untagged strains, antibody itself, bead-only control, antibody-only control, post-IP supernatant, and low- and high-speed supernatants of extracts. The samples can also be frozen at $-20°$ prior to Western blotting. Run only half of each sample and save the other half in case there is a gel disaster.

Conclusion

We have described in detail how to estimate the molecular weight and shape of soluble proteins from yeast. These estimates, when coupled with epitope tagging and coimmunoprecipitation, give insight into the physical properties and molecular makeup of protein complexes. These estimates can also be very informative when pure recombinant protein is compared with the native protein. The strength of these methods lies in the fact that knowledge of only a single biochemical activity, even as simple as a Western blot, is sufficient to gain some insight into molecular composition and structure. These methods may also be applied to biological complexes with mixed classes of components such as protein–RNA complexes. Finally, the results of size and shape measurement are quantitative and thus can be compared directly with previously determined experimental values. Such comparisons can be very powerful and can provide insight into molecular structure and function.

Acknowledgments

We thank an astute reviewer and T. Wolkow for very helpful suggestions and critically reading the manuscript. We also thank M. Longtine and T. Mitchison for helpful conversations, J. Huang and L. Lee for outlines of the coimmunoprecipitation and glass bead lysis protocols, and Y. L. Juang for the outline of the denaturing lysis protocol.

[9] Purification of Glutathione S-Transferase Fusion Proteins from Yeast

By AVITAL A. RODAL, MARA DUNCAN, and DAVID DRUBIN

Introduction

Purification of affinity-tagged recombinant proteins from prokaryotic hosts is an efficient method to obtain milligram quantities of protein for binding and activity studies. Unfortunately, eukaryotic proteins often are not expressed or are not soluble in prokaryotic hosts, making it necessary to use a eukaryotic expression system such as baculovirus, which can be time-consuming and require access to tissue culture facilities. Yeast provides an opportunity to express large amounts of recombinant protein in a eukaryotic host using microbiological media. However, difficulties with the expression level of recombinant protein, toxicity of the recombinant protein, proteases, and yield from the affinity purification step have prevented researchers from taking advantage of yeast as an expression host. The choice of an appropriate host strain, induction media, and expression plasmid can overcome the expression level and protease problems, and the use of an ion-exchange resin before the affinity purification step increases recombinant protein yield, presumably by removing an inhibitor of binding to the affinity resin.

Materials and Methods

To alleviate the problem of contaminating proteases, a protease-deficient strain (BJ2168 MATa *leu2 trp1 ura3-52 prb1-1122 pep4-3 gal2*)[1] is used. For maximal expression of the recombinant protein, one can use a 2μ plasmid [pEG(KT)][2] bearing the *GAL* promoter, GST (glutathione *S*-transferase), the URA3 marker for selection, and the *leu2-d* allele for plasmid amplification. This allele of LEU2 is compromised for function so that the plasmid needs to be in very high copy in order to complement leucine auxotrophy. Finally, the induction is carried out in rich medium, which increases expression of the recombinant protein 10-fold over the amount produced in synthetic medium. Recombinant proteins bind poorly to glutathione agarose resin directly from yeast lysates, presumably due to an inhibitor of binding to the affinity resin. Use of an ion-exchange resin before the affinity purification step partially overcomes this problem.

Induction of Recombinant Protein

For a 4-liter culture (synthetic complete and "drop out" media are described elsewhere)[3]:

1. Inoculate 200 ml of SC-URA,LEU (synthetic complete media lacking uracil and leucine) containing 2% (w/v) dextrose with cells expressing the recombinant protein in pEG(KT). Grow to late log phase (OD 1.5–3).
2. Add the inoculum to 4 liter SC-URA,LEU containing 2% (w/v) raffinose and grown to an OD of 0.8–1.2 at 30° with good aeration. Achieving this OD can take up to 2 days because the cells are growing on a poor carbon source.
3. Add 40 g of yeast extract, 80 g Bacto-peptone, and 400 ml 20% (w/v) galactose to the culture. It is not necessary to sterilize these reagents. Incubate for 4–16 hr at 30°.

[1] E. W. Jones, *Methods Enzymol.* **194,** 428 (1991).
[2] D. A. Mitchell, T. K. Marshall, and R. J. Deschenes, *Yeast* **9,** 715 (1993).
[3] D. Burke, D. Dawson, and T. Stearns, "Methods in Yeast Genetics: A Cold Spring Harbor Laboratory Course Manual." Cold Spring Harbor Laboratory Press, Cold Spring Harbor, NY, 2000.

4. Spin down cells at 4000g for 5 min at room temperature (5000 rpm in a Sorvall GSA rotor), wash once with 250 ml water, and resuspend in 50 ml water. Drip slowly in 50- to 100-μl drops into liquid nitrogen. These frozen pellets can be stored indefinitely at $-80°$.
5. Lyse pellets in a Waring blender (five rounds of blending give good lysis) with liquid nitrogen.[4] Do not allow the powder to thaw. Powder can be stored indefinitely at $-80°$.

Initial Purification of Recombinant Protein

We have used both phosphate-buffered saline (PBS) and 20 mM HEPES, pH 7.5, 50 mM KCl, 1 mM EDTA successfully as base buffers for this purification.

1. Quickly thaw lysed, frozen powder in 2× PBS with protease inhibitors to a final concentration of 1× PBS [inhibitor cocktail: 1 mM phenylmethylsulfonyl fluoride (PMSF) and 0.5 μg/ml each of antipain, leupeptin, pepstatin A, chymostatin, and aprotonin].
2. Spin at 25,000g for 15 min at 4° (12,000 rpm in a Sorvall GSA rotor).
3. Collect the supernatant and spin it at 300,000g at 4° (64,000 rpm for 60 min in a Beckman Type 70Ti rotor or 80,000 rpm for 20 min in a Beckman TLA100.3 rotor). Transfer the supernatant to a fresh tube. When removing the supernatant from this high-speed spin, be careful not to get any of the cloudy lipids at the top or at the bottom of the tube.
4. Run 10 μl of the low-speed supernatant (25,000g) and high-speed supernatant (300,000g) on an SDS–PAGE gel to test solubility and expression of the recombinant protein. Stain the gel with Coomassie blue if the fusion protein is highly expressed and runs clear of most yeast proteins (above 60 kDa) or use an anti-GST Western blot if not (Fig. 1).
5. Load the supernatant directly onto a cation- or anion-exchange resin (whichever the recombinant protein is likely to bind; GST binds Q resins at pH 7.5) equilibrated in PBS. We load 25 ml of high-speed supernatant expressing GST-Abp1p onto a Q Hi-Trap column (Pharmacia, Piscataway, NJ) and run a 50-ml gradient from 100 to 700 mM KCl at 1 ml/min (Fig. 2). Fractions are analyzed by Coomassie-stained SDS–PAGE or by anti-GST Western blot (see earlier discussion).

Affinity Purification of GST Fusion Protein

1. Pool the fractions containing recombinant protein and dilute to 1× PBS. Load pooled fractions onto a 1-ml GST HiTrap column (Pharmacia). Wash with 10 column volumes of 1× PBS.

---

[4] B. L. Goode, J. J. Wong, A. C. Butty, M. Peter, A. L. McCormack, J. R. Yates, D. G. Drubin, and G. Barnes, *J. Cell Biol.* **144,** 83 (1999).

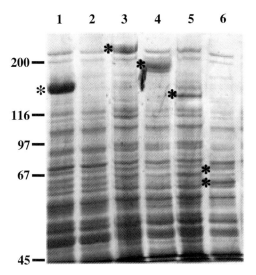

FIG. 1. Coomassie-stained gel of high-speed supernatants prepared from cells expressing various GST fusion proteins. Lane 1, GST Sla2p; lane 2, GST; and lanes 3–6, fragments of GST Pan1p. The recombinant protein is marked with an asterisk.

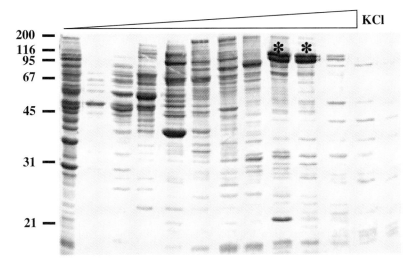

FIG. 2. Twenty milliliters of high-speed supernatant prepared from cells expressing GST-Abp1p was loaded onto a 5-ml Q HiTrap column (Pharmacia). Ten-microliter samples of every other 1.5-ml fraction (eluted from 200 to 700 mM KCl) were fractionated by SDS–PAGE and stained with Coomassie blue. GST-Abp1p is the 95-kDa doublet marked with an asterisk.

2. Elute in 2 ml 50 mM Tris, pH 8.0, 10 mM glutathione.
3. Concentrate the protein in Centricon (Amicon, Danvers, MA) devices to 100–300 μl. Add 1.9 ml of the final desired buffer. Repeat the concentration and buffer dilution twice. Concentrate to final desired protein concentration and drop freeze in small aliquots into liquid nitrogen.

General Comments

We have not had any problems using this method to purify proteins that are lethal on overexpression, including Abp1p and Duo1p. Although cells are dying during the induction, significant amounts of recombinant protein are produced. Because some proteins are sensitive to longer (>4 hr) induction times, it may be important to test protein expression at different time points during the induction. The expression level varies from approximately 0.01 to 0.2 mg recombinant protein/ml high-speed supernatant, and all proteins tested have been soluble. The yield from these preparations is approximately 10% of the expressed protein; most of the loss is at the affinity purification step. We have not been successful in dialyzing out the putative inhibitor of binding to glutathione resins. We have successfully cleaved GST from GST-Aip1p bound to glutathione agarose using the thrombin site present in pEG(KT) and standard techniques,[5] but this approach has not worked for other more thrombin-sensitive proteins. To request plasmids or yeast strains described here, please contact Dr. Bruce Goode at goode@brandeis.edu.

[5] A. A. Rodal, J. Tetrault, P. Lappalainen, D. G. Drubin, and D. C. Amberg, *J. Cell Biol.* **146**, 1251 (1999).

[10] Protein- and Immunoaffinity Purification of Multiprotein Complexes

By DOUGLAS R. KELLOGG and DANESH MOAZED

Introduction

Virtually all proteins must interact with other proteins to carry out their activities within the cell. Protein affinity chromatography allows one to exploit this fundamental property to identify proteins that function together as components of multiprotein complexes. Affinity chromatography is similar to genetics in that it offers a highly specific and powerful means of identifying proteins that participate in a biological activity. Affinity chromatography offers an advantage over genetics in that it can identify proteins that would be difficult or impossible to identify by genetic screens (e.g., proteins that carry out redundant functions or proteins that play a broad role in cellular activities that can be missed in specific genetic screens). In addition, genetic approaches to identifying interacting proteins can be biased by particular mutant alleles or preconceived ideas about the functions of proteins. Affinity chromatography is not biased by function and can therefore lead one in new and interesting directions.

This chapter discusses two methods for identifying interacting proteins by affinity chromatography. In the first method, a protein affinity column is made by linking relatively large amounts of a purified protein to a column matrix. A crude extract is then passed over the column and specifically bound proteins are eluted with a buffer containing high salt or denaturants such as urea. In the second approach, endogenous protein complexes are purified using an immunoaffinity column made by linking antibodies that recognize a specific protein to a column matrix.

Purification of Interacting Proteins Using Protein Affinity Chromatography

Protein affinity chromatography using proteins that are expressed in bacteria provides a powerful method for the identification of interacting proteins. This method has been used to isolate proteins that bind to the gene N transcription antiterminator of bacteriophage lambda (λ),[1] bacteriophages T4 proteins essential for DNA replication,[2] actin and microtubule-binding proteins,[3–5] RNA polymerase C-terminal tail-binding proteins,[6] Sir2- and Sir4-binding proteins,[7,8] and proteins that associate with cell cycle control proteins.[9–11] One primary advantage of affinity chromatography is that the high concentration of purified protein immobilized on the column will allow the isolation of proteins that have a weak affinity for the protein of interest. A second advantage of this approach is that column chromatography can be performed using buffer conditions that are independent of initial cell

[1] J. Greenblatt, P. Malnoe, and J. Li, *J. Biol. Chem.* **255**, 1465 (1980).
[2] T. Formosa, R. L. Burke, and B. M. Alberts, *Proc. Natl. Acad. Sci. U.S.A.* **80**, 2442 (1983).
[3] K. G. Miller and B. M. Alberts, *Proc. Natl. Acad. Sci. U.S.A.* **86**, 4808 (1989).
[4] D. R. Kellogg, C. M. Field, and B. M. Alberts, *J. Cell Biol.* **109**, 2977 (1989).
[5] R. V. Aroian, C. Field, G. Pruliere, C. Kenyon, and B. M. Alberts, *EMBO J.* **16**, 1541 (1997).
[6] C. M. Thompson, A. J. Koleske, D. M. Chao, and R. A. Young, *Cell* **73**, 1361 (1993).
[7] D. Moazed and A. D. Johnson, *Cell* **86**, 667 (1996).
[8] A. F. Straight, W. Shou, G. J. Dowd, C. W. Turck, R. J. Deshaies, A. D. Johnson, and D. Moazed, *Cell* **97**, 245 (1999).
[9] D. R. Kellogg, A. Kikuchi, T. Fujii-Nakata, C. W. Turck, and A. W. Murray, *J. Cell Biol.* **130**, 661 (1995).
[10] R. Altman and D. Kellogg, *J. Cell Biol.* **138**, 119 (1997).
[11] C. W. Carroll, R. Altman, D. Schieltz, J. R. Yates, and D. Kellogg, *J. Cell Biol.* **143**, 709 (1998).

lysis conditions. Thus, lysis may be carried out in the presence of high concentrations of salt that may be required for the efficient solubilization of proteins of interest, which can be removed from the extract prior to chromatography. The overall experiment involves three steps. First, the protein chosen as an affinity probe (and a suitable control protein) must be purified and immobilized on a solid matrix to construct an affinity column. Second, a whole cell extract is prepared and passed over the columns. Third, specifically bound proteins are identified by mass spectroscopy analysis. These steps must be followed by experiments designed to test the *in vivo* significance of the identified interactions (discussed later).

Choosing Proteins for Construction of Affinity Matrix

Affinity chromatography experiments generally begin with a protein that one would like to learn more about. This is usually a protein that has a known function in a biological process but its mode of action is not fully understood. The most important factor is the ability to purify relatively large quantities of the protein. An easy way to do this is by overexpression of the protein as a fusion with glutathione *S*-transferase (GST) in *Escherichia coli*.[12] We have successfully used many different GST fusions expressed in *E. coli* for our experiments, but in principle, any source that can produce milligram quantities of the protein can be used. In general, 3–6 mg of protein is required for construction of an affinity column. A general method for the purification of GST fusion proteins is described later.

Controls

In order to distinguish proteins that are specifically retained on affinity columns, a control column is always run in parallel with the column containing the protein of interest. In the case of GST fusion proteins, any other GST fusion protein that is not functionally related to the protein of interest could be used as a control. We usually try to use GST fusion proteins of similar size: one as the control column and one as the experimental column. GST alone is not an ideal control because it is not very sticky and elution profiles of GST columns are not an accurate representation of nonspecifically bound proteins. If available, an ideal control is the same protein carrying a point mutation or a truncation that disrupts a specific function *in vivo*.

Purification of GST Fusion Proteins

GST fusion proteins are purified based on the method described by Smith and Johnson.[12] The following is an adaptation of this protocol that has worked well in our laboratories for the purification of several yeast proteins involved in cell cycle regulation or gene silencing.[7,11]

---

[12] D. B. Smith and K. S. Johnson, *Gene* **67,** 31 (1988).

1. Inoculate 4 liters of 2× YT media[13] containing 75 µg/ml ampicillin with 100–200 ml of overnight culture. We use three 2.8-liter flasks containing 1.3 liters each.

2. Grow the cultures at room temperature until the OD reaches 0.8 to 1, and then add isopropylthiogalactoside (IPTG) to 0.1 mM and continue incubating the cultures at room temperature for another 2–3 hr. Many proteins are more soluble when expressed at room temperature. To speed things up, we sometimes start the cultures at 37°, and by the time they reach an OD of 1.0 they usually have cooled to room temperature.

3. Pellet the cells at room temperature by spinning at 5000 rpm for 20 min (Sorvall) and rapidly scrape them out of the bottle and freeze in liquid nitrogen. The frozen chunks of cell paste can be stored at −80°.

4. We have found that the degradation of many proteins can be prevented by using a rapid lysis procedure that gets the extract into a buffer that prevents proteolysis as rapidly as possible. To do this, grind the frozen cells under liquid nitrogen in a mortar and pestle until you have a fine powder (5–10 min). Transfer the cell powder to a cold beaker and allow the cell powder to warm at room temperature for about 5 min or until the powder is just starting to thaw around the edges of the beaker. (This avoids the formation of ice crystals when the buffer is added.) Add approximately 5 volumes of room temperature phosphate-buffered saline (PBS) that contains 0.5% (v/v) Tween 20, 1 mM phenylmethylsulfonyl fluoride (PMSF), and 1 M NaCl. Immediately resuspend the cell powder by stirring with a spatula and then add a stir bar and stir in a cold room for a few minutes. (For proteins where degradation is not a problem, skip the grinding step and resuspend the frozen cell pellet in the aforementioned lysis buffer containing 200 µg/ml lysozyme.) Sonicate for about a minute to reduce viscosity. The PMSF is added immediately before resuspending the cells because it is unstable in water. All of the following purification steps are carried out at 4°.

5. Add dithiothreitol (DTT) to 10 mm. Spin the lysate for 60 min at 35,000 rpm in a Beckman 50.2 Ti rotor.

6. Load the supernatant onto a 5- to 10-ml glutathione agarose column (Sigma, St. Louis, MO) over a period of 2–4 hr.

7. Wash the column with 50–100 ml of wash buffer (PBS containing 0.05% Tween 20, 0.5 mM DTT, and 0.25 M KCl). Monitor the effluent from the column (Bradford assay) to make sure that there is no protein still washing off at the end of the wash step. At the end of the wash step, wash the column with 2 column volumes of wash buffer without Tween 20.

8. Elute the column with 50 mM Tris, pH 8.1, containing 0.25 M KCl and 5 mM reduced glutathione. We do the elution by pipetting 1/7 column volume aliquots

---

[13] J. Sambrook, E. F. Fritsch, and T. Maniatis, "Molecular Cloning: A Laboratory Manual." Cold Spring Harbor Laboratory Press, Cold Spring Harbor, NY, 1989.

of elution buffer directly onto the top of the column bed. We then let the aliquot flow through into a plastic tube and advance to another fraction. The fractions are assayed, and the peak fractions are pooled and dialyzed extensively into 50 mM HEPES, pH 7.6, 50–300 mM KCl, and 30% (v/v) glycerol (the glutathione must be removed prior to coupling to Affigel 10.) The yield of GST should be approximately 100 mg, whereas the yield of GST fusions is usually much lower.

Preparation of Affinity Columns

We usually make 1- to 2-ml affinity columns by cross-linking 3–6 mg of protein/ml of Affi-Gel 10 (Bio-Rad, Hercules, CA).

1. Protein should be at a concentration of 1–5 mg/ml in 50 mM HEPES, pH 7.6, 100–500 mM KCl, and 30% (v/v) glycerol. The amount of buffer, salt, or glycerol is not critical. What is important is that the pH is between 7.5 and 8 and that there are no primary amines or sulfhydryls in the buffer.

2. Affi-Gel 10 is stored in 2-propanol at -20. Before using Affi-Gel 10, remove the bottle from the freezer and allow it to warm to room temperature. This prevents water from condensing in the bottle (water inactivates Affi-Gel 10).

3. Transfer the desired quantity of Affi-Gel to a Buchner funnel (Pyrex, 30 ml) assembled on a vacuum line. Apply gentle or intermittent vacuum and do not allow the gel to dry. Wash the resin three times with ice-cold water (each wash is with at least two to three times the volume of the resin) and one with a small amount of coupling buffer. The active groups on Affi-Gel 10 hydrolyze in water, so the washes should be completed as rapidly as possible (20 min or less).

4. Using a spatula, transfer the moist gel to a fresh tube and add the protein solution. Mix immediately. Do the coupling in the smallest possible tube. Affi-Gel 10 sticks to surfaces and much of it will be lost if a larger than necessary tube is used for coupling. Mix using an end-over-end mixer at 4° for 2–4 hr. Monitor the rate of the coupling reaction by spinning down the beads and removing a 25-μl sample of the supernatant. Do a Bradford assay on each time point, and when 25 μl looks like 5 μl of the starting reaction mix, you have reached approximately 80% coupling and the reaction should be stopped (overcoupling kills some proteins). We usually take time points at 5, 10, and 20 min and at 20-min intervals thereafter. The rate of coupling varies considerably, depending on the protein. Some proteins are completely coupled within 5 min, some take more than an hour, and some do not couple at all. Proteins that do not couple well can often be made to couple by adding 50–100 mM MgCl$_2$ to the reaction. This minimizes charge repulsion between negatively charged proteins and the column matrix.

5. Stop the reaction by adding 1 M ethanolamine, pH 7.5, to a final concentration of 50 mM. Leave on ice for 1 hr or overnight to block any residual active groups.

6. Pour the coupled resin into a small column. We usually use a 5- to 10-cm Bio-Rad column with a 1-cm internal diameter or the Bio-Rad Bio-Spin columns.

7. Pack and wash the columns using a peristaltic pump at a flow rate of ~20 ml per hour. The maximum cross-sectional flow rate for the Affi-Gel 10 resin is about 25 cm/hr. For a column with a 1-cm internal diameter, flow rate = cross-sectional flow rate (cm/hr) × ID (cm^2) = 25 × 0.79 = 20 ml/hr.

8. Wash each column with 10 column volumes of 50 mM HEPES (pH 7.6), 250 mM KCl, and 1 mM DTT. Before using the column, wash it with whatever elution conditions that you will be using. Most protein affinity columns can be stored for prolonged periods of time at −20° after they have been equilibrated with a buffer containing 50% (w/v) glycerol. For shorter storage periods of up to 2 weeks, columns can be stored at 4° with a buffer containing 0.05% azide.

Preparation of Yeast Extracts for Affinity Chromatography

Obtain 10–15 g of yeast cells from a protease-deficient strain (e.g., BJ2168)[14] grown to log phase in rich medium. Resuspend cells in 50 ml of lysis buffer [buffer L: 50 mM HEPES–KOH, pH 7.6, 5% glycerol 1 mM MgCl$_2$, 1 mM EGTA, 1 mM DTT] containing 500 mM KCl, 0.5% nonidet p-40 (NP-40), and the following protease inhibitors, 1 mM PMSF, 1 μg/ml leupeptin, 1 μg/ml pepstatin, and 1 μg/ml bestatin. PMSF is added to this buffer immediately before lysis from a 100 mM stock made in 100% methanol. Break cells by grinding under liquid nitrogen in a mortar and pestle (see later) or by using a BeadBeater (Biospec).

When using a BeadBeater, leave all BeadBeater components in the cold room overnight. We use the 89-ml BeadBeater chamber filled with glass beads until the screw that holds the blade in place is covered with beads. The cell suspension is then added to the chamber containing the glass beads. A glass rod or plastic pipette is used to remove air bubbles that are trapped in the beads, and care is taken to minimize air that is trapped in the chamber when the cap is screwed on. The lysis chamber is then assembled with an ice-water chamber, and cells are broken using 12–15 pulses of bead beating each lasting 15 sec with 2 min in between pulses to allow for cooling. The crude extract is centrifuged at 20,000g for 10 min (SS34 rotor) followed by 100,000g for 1 hr. Prior to the latter spin, the salt concentration in the extract is reduced to 150–275 mM by dilution with lysis buffer lacking KCl. However, if the desired binding proteins are thought to be soluble at lower salt concentrations, the initial lysis can be performed at 150–275 mM KCl.

Running Affinity Columns and Elution of Bound Proteins

1. Load the columns in parallel at a flow rate of about 6–8 ml per hour (for 1-ml columns). It takes about 8–9 hr to load. We usually do this step overnight. To monitor how well the column is performing and also how much extract is required to saturate the binding sites on the column, take flow-through samples

---

[14] E. W. Jones, *Methods Enzymol.* **194,** 428 (1991).

in the beginning and the end of the experiment. The first flow through equals the first flow through with the same protein concentration as the load. The last flow through equals the flow through near the end of the experiment. A bovine serum albumin (BSA) precolumn can be used to reduce the background of nonspecific proteins as described in Kellogg and Alberts.[15] However, it is possible to omit this step if the binding proteins are relatively abundant.

2. Wash the columns with 5–40 column volumes of buffer L containing 0.5% NP-40 and the same KCl concentration as the extract (150–250 mM KCl). The amount of wash buffer to use is determined empirically. Proteins that interact with low affinity can be lost by extensive washing, but too little washing can sometimes give a high background. A good starting point is to wash with 10–15 column volumes.

3. Wash the columns with 5 column volumes of buffer L containing the same KCl concentration as in step 2 (but without NP-40).

4. Elute each column using a salt gradient going from 0.3 to 1.0 M KCl (see Altman and Kellogg).[10] The gradient is made by pipetting 400-μl aliquots of elution buffer directly onto the top of the column, with each aliquot increasing in KCl concentration by 50 mM. Collect 400-μl fractions in plastic tubes. Some interacting proteins remain bound to the column after the 1 M KCl elution step. These proteins can be eluted from the column using 2.5 M urea in the elution buffer containing 150 mM KCl.[7,8]

5. Wash columns with 10 column volumes of buffer L containing 0.05% NaN$_3$ and store at 4° or equilibrate with buffer L containing 50% glycerol for storage at −20°.

Gel Electrophoresis and Identification of Bound Proteins

1. Precipitate the proteins present in each fraction with trichloroacetic acid (TCA). To each 400-μl fraction add 40 μl 100% TCA, mix, leave on ice for 15 min, and spin in a microfuge at 4° for 15 min. Discard the supernatant, spin again for a few seconds, remove the last bit of supernatant, and resuspend in 50 μl SDS sample buffer. Leave at room temperature for 1 hr and then heat to 85° for 10 min.

2. Load 10 μl on an 8.5% SDS–polyacrylamide gel. Visualize proteins by silver or Coomassie blue staining.

3. Proteins that are specifically bound to the experimental column and not the control column are then excised from the gel and identified by mass spectroscopy analysis. Approximately 10–100 ng of protein is sufficient for identification by most mass spectroscopy facilities. A silver staining protocol compatible with mass spectroscopy analysis is described by Morrissey.[16] However, it is advisable to

---

[15] D. R. Kellogg and B. M. Alberts, *Mol. Biol. Cell* **3**, 1 (1992).
[16] J. H. Morrissey, *Anal. Biochem.* **117**, 307 (1981).

consult the mass spectrometry facility that you intend to use for their preferred staining protocol.

Purification of Endogenous Multiprotein Complexes by Immunoaffinity Chromatography

Protein affinity chromatography utilizing proteins expressed in bacteria has provided a powerful means of identifying interacting proteins. However, many proteins can be difficult or impossible to purify from *E. coli*, and proteins purified from heterologous expression systems lack posttranslational modifications that may be necessary for protein–protein interactions. Ideally, one would like to be able to rapidly and specifically purify endogenous multiprotein complexes because these are assembled under native conditions and therefore provide information that is most relevant to the *in vivo* functions of the protein of interest.

Immunoaffinity chromatography has provided a good approach to isolating endogenous multiprotein complexes.[15,17–19] In this approach, immunoaffinity beads are made by binding antibodies that recognize the protein of interest to protein A beads. The beads are then incubated with crude extract to allow binding of the protein complex to the antibody. After washing the beads with buffer, the protein complex is eluted. In some cases, associated proteins have been eluted with a high salt buffer, leaving the original protein still bound to the antibody on the column. In other cases, the immunoaffinity beads have been made using an antibody raised against a C-terminal or N-terminal peptide, allowing competitive elution of the protein complex with excess peptide. Competitive elution with peptide has the strong advantage of allowing highly specific elution of the entire complex under gentle conditions. This results in a considerably lower background and allows one to purify an intact complex for functional studies. In some cases, it has been possible to identify associated proteins after elution with denaturing conditions; however, this often results in a high background that can make analysis difficult or impossible.

Generally Applicable Protocol for Immunoaffinity Purification of Multiprotein Complexes

A problem with immunoaffinity chromatography is that one must generate a new antibody for each protein complex to be purified. In addition, many antibodies raised against synthetic peptides do not recognize native proteins. To make

---

[17] Y. Zheng, M. L. Wong, B. Alberts, and T. Mitchison, *Nature* **378**, 578 (1995).
[18] C. M. Field, O. al-Awar, J. Rosenblatt, M. L. Wong, B. Alberts, and T. J. Mitchison, *J. Cell Biol.* **133**, 605 (1996).
[19] C. M. Field, K. Oegema, Y. Zheng, T. J. Mitchison, and C. E. Walczak, *Methods Enzymol.* **298**, 525 (1996).

immunoaffinity chromatography more generally applicable, we have developed a protocol that allows single-step immunoaffinity purification of proteins tagged with three copies of the HA epitope. Because this is a commonly used epitope tag, the same methods and reagents can be used to purify numerous different protein complexes. Proteins are eluted from the immunoaffinity column with an HA dipeptide (i.e., a synthetic peptide that includes a tandem repeat of the standard HA peptide). The use of a dipeptide is essential for obtaining quantitative elution of 3× HA-tagged proteins, due most likely to high-avidity binding of anti-HA antibodies to triple HA repeats. We have now used this method to purify six different multiprotein complexes in our laboratory, including complexes that assemble in response to cell cycle-dependent posttranslational modifications. In addition, we have analyzed all of these complexes by mass spectrometry, and we know which proteins appear in the elutions from multiple columns and are likely to represent background bands. More importantly, when we identify new proteins that have never been observed binding to other immunoaffinity columns, we have confidence that the proteins are binding due to specific interactions with the HA-tagged protein. Each multiprotein complex that is purified therefore provides an additional control for purifications of other multiprotein complexes. Similar approaches for affinity purification of protein complexes have been described elsewhere, including a two-step purification scheme that employs a combination of protein A and a calmodulin-binding peptide, called the TAP tag.[20]

A detailed protocol for immunoaffinity purification of 3× HA-tagged protein complexes from yeast cells is presented. This protocol is based on previous work using antipeptide antibodies to purify protein complexes from *Drosophila* and *Xenopus*.[17,18]

Generation of Anti-HA Peptide Antibodies

Anti-HA antibodies are generated by immunizing rabbits with an HA peptide conjugated to keyhole limpet hemocyanin (KLH). We initially use a peptide that includes the last half of the standard HA peptide followed by a full peptide to ensure that we obtain antibodies that recognize the juncture between HA peptides in tandem repeats (peptide sequence: CPDYAGYPYDVPDYAG, the cysteine is included to allow coupling to KLH via the sufhydryl group). It should also work to use an HA dipeptide so that one can use the same peptide to generate the antibody and to elute proteins from the antibody. The antibody is affinity purified using a column constructed with a purified GST-2× HA fusion protein. We generally obtain approximately 20 mg of purified antibody from each rabbit. Commercially available monoclonal anti-HA antibodies (12CA5 and HA.11) can also be used for

---

[20] G. Rigaut, A. Shevchenko, B. Rutz, M. Wilm, M. Mann, and B. Seraphin, *Nature Biotech.* **17,** 1030 (1999).

affinity purification experiments. However, we have found that the commericially available antibodies are prohibitively expensive.

Preparation of Immunoaffinity Beads

Immunoaffinity beads are prepared by binding 0.45 mg of antibody to 0.45 ml of protein A beads (Bio-Rad). Binding is carried out in the presence of PBS containing 0.05% Tween 20 for 1–3 hr at room temperature or overnight at 4°. Binding is carried out in a 1.6-ml tube. The presence of Tween 20 minimizes sticking of the beads to plastic or glass surfaces. After binding, the beads are washed twice with extract buffer (see later).

Controls

The ideal control for these experiments is an identical strain that does not carry an HA-tagged protein. We have also used a column made from a nonspecific antibody as a control. The use of a column constructed with a nonspecific antibody has several advantages. First, one needs to grow only one strain for each experiment. Second, one reduces the amount of anti-HA antibody needed. However, we have found that there can be background bands that bind to anti-HA affinity columns, but not to control columns constructed with anti-GST antibodies, depending on the batch of antibody. In our experience, these include Ydj1 and Cdc48. All binding interactions should therefore be verified in experiments using identical columns and strains carrying untagged proteins as a control.

Choice of Extract and Wash Buffers

The choice of the extract and wash buffers used for affinity chromatography experiments is critical and often must be determined empirically. We generally start by using buffers of physiological ionic strength (150–200 mM) containing 50 mM HEPES–KOH, pH 7.6, and 100–150 mM KCl. In many cases, the salt concentration can be increased without disrupting specific protein interactions, resulting in a lower nonspecific background. Yeast extracts contain highly active phosphatases that rapidly dephosphorylate proteins, which will disrupt protein interactions that are dependent on phosphorylation. In recent experiments, therefore, we have replaced the KCl in the extract and wash buffers with 100 mM β-glycerophosphate and 50 mM NaF, which act as good phosphatase inhibitors. This has proven to be crucial for the purification of multiprotein complexes that are dependent on cell cycle-specific phosphorylations.

Immunoaffinity Purification of 3× HA-Tagged Proteins from Yeast Cells

1. Obtain approximately 10–15 g of cells carrying the 3× HA-tagged protein. We usually grow 4–6 liters of culture to an OD of 1.0. The cells are pelleted,

resuspended in 50 mM HEPES–KOH, pH 7.6, and then pelleted again in a 50-ml conical tube. The cells are either frozen directly in the tube on liquid nitrogen or a hole is made in the bottom of the tube with an 18-gauge syringe needle and a stream of cell paste is extruded directly into the liquid nitrogen using a plunger from a 60-ml syringe. The latter method makes a "spaghetti" that is easier to grind. Frozen cells may be stored indefinitely at $-80°$.

2. Break open cells by grinding for 30 min in a mortar and pestle under liquid nitrogen to obtain a fine powder the consistency of flour. An initial grinding can be done for several minutes in a coffee mill, which reduces the total grinding time in the mortar and pestle by approximately 10 min. The coffee mill should be prechilled by grinding dry ice. A motorized mortar and pestle can also be used.

3. Transfer the powder to a 50-ml beaker prechilled with liquid nitrogen. Allow the powder to warm for approximately 5 min until the powder around the edges of the beaker is just beginning to thaw. This avoids the formation of ice crystals when the extract buffer is added. Resuspend the powder in 25 ml of extract buffer at room temperature.

50 mM HEPES–KOH, pH 7.6
100 mM β-glycerolphosphate
50 mM NaF
1 mM MgCl$_2$
1 mM EGTA
5% Glycerol
0.25% Tween 20
1 mM PMSF

The PMSF is added from a 100 mM stock made in 100% ethanol and is added to the extract buffer immediately before resuspending the cells. After adding the extract buffer, immediately stir the powder into solution with a spatula. Stir for another 5–10 min with a magnetic stir bar in a cold room, taking care to avoid foaming. Spin the extract at 10,000g for 5 min, followed by 100,000g for 1 hr. Take 10-μl samples of the crude extract and the supernatants from both spins. Store on ice. Carry out all of the following steps at 4°.

4. Add the supernatants from tagged and untagged strains to 0.45 ml of protein A beads that have 0.45 mg of polyclonal anti-HA bound in 15-ml conical tubes. Prepare the beads by prebinding the antibody to the beads for at least 1 hr at room temperature in 1.5 ml PBS containing 0.1% Tween 20.

5. Mix for 2–3 hr on a rotator. Pellet beads and take a 10-μl sample of each supernatant.

6. Wash twice with 15 ml of extract buffer with 25 mM NaF and without PMSF.

7. Transfer to a column and wash with 5 ml of extract buffer. Washes are carried out by pipetting 1-ml aliquots of buffer onto the column and letting them wash through by gravity. Before eluting, wash with 1 ml of extract buffer without

Tween 20. If there is a possibility that you will be sequencing the purified proteins, wear gloves for all of the following steps.

8. Make 1 ml of elution buffer: 50 mM HEPES–KOH, pH 7.6, 100 mM β-glycerophosphate, 1 mM MgCl$_2$, 1 mM EGTA, 5% glycerol, and 0.5 mg/ml HA dipeptide. Add 0.25 ml of elution buffer and collect the flow through in a 1.6-ml tube. Incubate for 30 min and then add another 0.25 ml of elution buffer. Allow to incubate overnight. Elute with two more aliquots of elution buffer, allowing each aliquot to incubate on the column for 30 min. Wash the last aliquot of elution buffer off the column with 0.25 ml of elution buffer without peptide. You should have a total of five fractions for each column.

9. Take 10 μl of each fraction and dilute into 90 μl of 1× sample buffer. In addition, dilute the crude extract and supernatant fractions into 90 μl of sample buffer. Heat all of the samples at 100° for 5 min and load 15 μl of each on an SDS–polyacrylamide gel. Use this gel for a Western blot and probe with an antibody against the protein or against HA.

10. Pool fractions 2–5 and precipitate proteins by the addition of trichloroacetic acid to 10%. Resuspend the precipitates from each column in 50 μl of sample buffer and load 15 μl/lane for a Coomassie blue-stained gel. Also load samples of the crude extract and supernatants for comparison.

Determining *in Vitro* Protein Interactions Relevant *in Vivo*

Once interacting proteins have been identified by affinity chromatography, an important next step is to gain additional information to help confirm that the proteins functionally interact *in vivo*. False positives may arise, for example, from the mixing of proteins in whole cell extracts, which are localized to separate compartments *in vivo*. Protein interactions that can only be disrupted with high salt (e.g., >0.5 M) or denaturing conditions are most likely to interact *in vivo*, and it is perhaps best to focus efforts on these first. Genetic interactions can provide perhaps the strongest evidence that two proteins that interact *in vitro* also interact *in vivo*. For example, if loss of function of an interacting protein causes a mutant phenotype that is similar or identical to the mutant phenotype caused by loss of function of the protein used for affinity chromatography, it is likely that the two proteins functionally interact *in vivo*. Similarly, unique genetic interactions between the two interacting proteins would provide strong evidence for an *in vivo* interaction. In the case of proteins isolated by standard protein affinity chromatography, it is important to show that the two proteins coimmunoprecipitate with each other in crude extracts. Finally, another criteria that can be used is to determine whether interacting proteins are colocalized within the cell.

[11] In Vitro DNA Replication Assays in Yeast Extracts

By PHILIPPE PASERO and SUSAN M. GASSER

Introduction

Nuclear extracts from budding yeast cells synchronized in S phase support the synthesis of DNA *in vitro*. Two assays for semiconservative DNA replication are described here. The first uses intact yeast nuclei from cells arrested in G_1 phase as the template. The second monitors the replication of bacterially produced supercoiled plasmid. S-phase nuclear extracts are supplemented with nucleotides and an energy-regenerating system, and the efficiency of DNA replication is monitored by the substitution of newly synthesized DNA with a heavy nucleotide derivative, BrdUTP, followed by density gradient analysis. In addition, neutral-neutral two-dimensional (2D) gel analyses of replication intermediates and/or immunodetection of newly synthesized DNA in replication foci can be used to monitor DNA synthesis.

Yeast genetics, the study of SV40 (simian virus 40) viral replication, and an *in vitro* replication assay based on *Xenopus* oocyte extracts have led to significant progress in identifying the factors that catalyze eukaryotic DNA replication (reviewed in Refs. 1–4). Nonetheless, to understand the regulation and molecular mechanisms of site-specific initiation of DNA replication, it is essential to have efficient systems that reconstitute DNA replication *in vitro*. To complement the powerful replication assay based on *Xenopus* oocyte extracts, the budding yeast *Saccharomyces cerevisiae* is particularly useful. Yeast cells have a normal mitotic cell cycle, are readily synchronized, and are amenable to both molecular and classical genetics. In addition, origins of replication are well defined and a large number of conditional mutations in components of the replication machinery are available. Finally, yeast is suitable for a number of well-established techniques that precisely map sites at which DNA replication initiates.[5–7]

Criteria for physiological DNA replication go beyond a simple monitoring of nucleotide incorporation into high molecular weight DNA. To differentiate between repair and replicative DNA synthesis, one must demonstrate complete complementary strand synthesis and dependence on the replicative DNA

[1] D. Coverley and R. A. Laskey, *Annu. Rev. Biochem.* **63,** 745 (1994).
[2] J. F. Diffley, *Genes Dev.* **10,** 2819 (1996).
[3] P. Pasero and E. Schwob, *Curr. Opin. Genet. Dev.* **10,** 178 (2000).
[4] H. Takisawa, S. Mimura, and Y. Kubota, *Curr. Opin. Cell Biol.* **12,** 690 (2000).
[5] B. J. Brewer and W. L. Fangman, *Cell* **51,** 463 (1987).
[6] J. A. Huberman, L. D. Spotila, K. A. Nawotka, S. M. el-Assouli, and L. R. Davis, *Cell* **51,** 473 (1987).
[7] A. K. Bielinsky and S. A. Gerbi, *Science* **279,** 95 (1998).

TABLE I
PROTEINS REQUIRED FOR DNA REPLICATION IN YEAST NUCLEAR EXTRACTS[a]

Function	sc plasmid[b]	G_1-phase nuclei[f]
ORC	Not required[c]	Required (cis)[g]
Cdc6	Not required[c]	Required (cis)[h]
MCMs	Required[d]	Required (cis)[d]
Pol α/primase	Required[e]	ND
Pol δ	Required[e]	ND
Dna2	Required[e]	ND
Rad52	Not required[e]	ND
Cdc28	Required[c]	Required (trans)[i]
Cdc7	Not required[c]	Required (cis)[i]
Dbf4	ND	Required (cis)[i]

[a] Using various mutant strains to prepare S-phase nuclear extracts or G_1-phase template nuclei, one could test which activities are required for semiconservative DNA replication *in vitro*.
[b] Results from the soluble system using supercoiled plasmid as a template. Requirements reflect the necessity of having the activity in the S-phase extract.
[c] From B. P. Duncker, P. Pasero, D. Braguglia, P. Heun, M. Weinrich, and S. M. Gasser, *Mol. Cell Biol.* **19**, 1226 (1999).
[d] P. Heun, Ph.D. Thesis, University of Lausanne (2000).
[e] From D. Braguglia, P. Heun, P. Pasero, B. P. Duncker, and S. M. Gasser, *J. Mol. Biol.* **281**, 631 (1998).
[f] Results from the nuclear replication assay using late G_1-phase nuclei. In this case, the activities mentioned might be required either in the S-phase extract (in *trans*) or in the template nuclei (in *cis*).
[g] From P. Pasero, D. Braguglia, and S. M. Gasser, *Genes Dev.* **11**, 1504 (1997).
[h] P. Pasero and S. Gasser, unpublished observation, 1998.
[i] From P. Pasero, B. P. Duncker, E. Schwob, and S. M. Gasser, *Genes Dev.* **13**, 2159 (1999).

polymerases δ, ε, and pol α/primase. This last is the only polymerase uniquely implicated in the *de novo* initiation of DNA replication (reviewed in Refs. 8–10). Bona fide DNA replication should also show a requirement for correct cell cycle coordination and should be dependent on the origin recognition complex (ORC), Cdc6p, and MCM proteins for origin-specific initiation.[4,9] The first three of these criteria have been achieved with a soluble replication assay using a supercoiled template as substrate.[11] All four have been demonstrated when G_1 phase yeast nuclei are used as a template in yeast nuclear extracts[12] (Table I).

[8] J. L. Campbell, M. Budd, C. Gordon, A. Jong, D. Sweder, A. Oehm, and M. Gilbert, *Basic Life Sci.* **40**, 463 (1986).
[9] A. Dutta and S. P. Bell, *Annu. Rev. Cell Dev. Biol.* **13**, 293 (1997).
[10] M. Foiani, G. Lucchini, and P. Plevani, *Trends Biochem. Sci.* **22**, 424 (1997).
[11] D. Braguglia, P. Heun, P. Pasero, B. P. Duncker, and S. M. Gasser, *J. Mol. Biol.* **281**, 631 (1998).
[12] P. Pasero, D. Braguglia, and S. M. Gasser, *Genes Dev.* **11**, 1504 (1997).

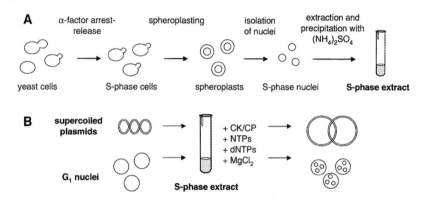

FIG. 1. Yeast systems for *in vitro* DNA replication. (A) Nuclear extracts for replication assays are prepared by ammonium sulfate extraction of nuclei isolated from cells synchronized in S phase with the α factor block release procedure (see protocol). (B) Either yeast nuclei isolated from cells arrested in G_1 or naked DNA (supercoiled plasmids) are added to S-phase extracts as a template for the replication reaction. The reaction is generally performed for 90 min at 25° in the presence of nucleotides, ribonucleotides, and an energy regeneration system (CP, creatine phosphate; CK, creatine kinase). Incorporation of derivatized nucleotides at discrete nuclear foci is monitored by immunofluorescence and confocal microscopy. Semiconservative DNA synthesis is quantitated by density substitution, and the appearance of replication intermediates is followed by 2D gel analysis.

As shown in Fig. 1, both replication assays are based on the use of nuclear extracts prepared from cells synchronously traversing the S phase. Nuclear templates can be efficiently replicated *in vitro*, such that 15–20% of the input genomic DNA migrates with the density of heavy–light (H–L) DNA on CsCl density gradients following BrdU substitution, indicative of complete second-strand synthesis (Figs. 2A and 2D). In this system, the template shows a cell cycle-dependent competence for initiation, an S-phase promoting factor (Cdc28/Clb5 kinase) is required in *trans*, and the subsequent initiation step is both origin specific and ORC dependent.[12] In contrast, the semiconservative replication of supercoiled plasmid is far less efficient, with less than 1% of the input plasmid recovered in the H–L peak after substitution with BrdUTP. It is likely that the initiation step limits the replication efficiency, for ORC is not able to assemble into a prereplication complex containing Cdc6p and MCMs, in an S-phase extract. Nonetheless, semiconservative replication can be detected at a low level, and requires an intact, supercoiled plasmid as template (see Fig. 2A). It is possible that local unwinding of the double helix due to torsional stress allows ORC-independent initiation in this assay, as Rad52-mediated recombination events are not required. Nonetheless, the reaction is cell cycle dependent and requires the activity of a Cdk1/B-type cyclin complex for efficient DNA synthesis.[13]

---

[13] B. P. Duncker, P. Pasero, D. Braguglia, P. Heun, M. Weinreich, and S. M. Gasser, *Mol. Cell Biol.* **19**, 1226 (1999).

This chapter provides detailed protocols for the preparation of yeast nuclear extracts from S-phase cells, the isolation of template nuclei, the replication assay itself, using either plasmid DNA or nuclei as the template, and assays for the detection of replication foci and for fully substituted H–L DNA after replication in the presence of BrdUTP.

Strains and Reagents

Yeast Strains and Media

A variety of strain backgrounds can be used for the isolation of nuclei for either extracts or template DNA, yet optimal strains are those that can be efficiently synchronized and readily converted to spheroplasts (see later). We have observed that strains lacking all three major vacuolar proteases are useful when mastering techniques, although this is not essential if the protocol is done quickly and care is given to keep the extract on ice. The phenotype of *pep4*-deficient cells can be tested regularly by use of a simple chromogenic assay.[14] To eliminate DNA synthesis due to recombination in the soluble assay, the *RAD52* gene can be disrupted in the starting strain.[11] Finally, to eliminate background due to replication of mitochondrial DNA, some strains were rendered rho° by growth in ethidium bromide.[12] The *Saccharomyces cerevisiae* strain GA-59 (*MAT*a, *leu2, trp1, ura3-52, prb1-1122, pep4-3, prc1-407, gal2*) is used as our standard wild-type strain and standard yeast media are used.[15]

Synchronization

The pheromone α factor efficiently arrests *MAT*a cells in mid-G_1 phase prior to start (see detailed protocol later). We have observed that cells that form a shmoo shape after a long α factor arrest do not spheroplast well, and therefore we usually optimize the pheromone concentration and the length of the arrest for each strain to minimize this problem. Fluorescence-activated cell sorting (FACS) analysis confirms that a block of one generation time is generally sufficient to get a homogeneous G_1 arrest. Disruption of the *BAR1* gene facilitates the arrest in G_1, yet in the absence of the Bar1 protease, it is more difficult to prevent schmoo formation. For release into S phase, α factor is removed by filtration or washing in fresh medium. Alternatively, 50 μg/ml of pronase is added to the medium to degrade α factor. In this case, the pH of the medium should be adjusted to 7.0 with the addition of sodium phosphate buffer. At 25°, cells generally enter S phase 30 min after the addition of pronase. To make extracts and nuclei from different stages of the cell cycle, conditional *cdc* mutants are a convenient tool. Thermosensitive strains are used to arrest cells in G_1 phase (*cdc4*-1), at the G_1/S transition (*cdc7*-1), or in

---

[14] E. W. Jones, *Genetics* **85**, 23 (1977).
[15] M. D. Rose, F. Winston, and P. Hieter, "Methods in Yeast Genetics." Cold Spring Harbor Laboratory Press, Cold Spring Harbor, NY, 1990.

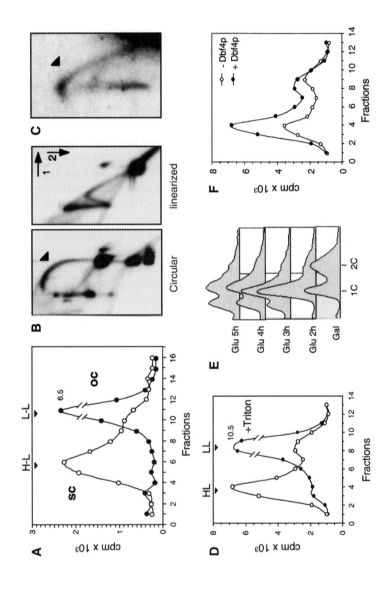

M phase (*cdc28*-1N or *cdc16*-1) after a 4-hr shift to the restrictive temperature. Synchronization in G_1 phase prior to start can be performed with *cln1 cln2 cln3 GAL-CLN3* cells by shifting for 3 hr to glucose medium or with wild-type cells by arresting in G_1 with α factor and spheroplasting in the presence of pheromone. Post start G_1 nuclei and extracts are prepared after α factor arrest with wild-type or *clb5 clb6* cells, but in this case, pheromone is not maintained during spheroplasting. This allows cells to progress through late G_1 while spheroplasting proceeds. For elongation assays with nuclei, we have found it preferable to isolate S-phase nuclei from cells that are released from an α factor block and allowed to enter S phase in the presence of hydroxyurea (200 mM). G_2-phase nuclei are also conveniently isolated by releasing α factor-arrested cells into 15 μg/ml nocodazole in media containing 1% dimethyl sulfoxide (DMSO).

FIG. 2. Analysis of *in vitro* replication products. (A) Density substitution analysis of plasmid DNA replicated *in vitro*. Incorporation of [α-^{32}P]dCTP into DNA migrating as either H–L or L–L DNA was monitored by scintillation counting. Use of negatively supercoiled pH4ARS (sc) as a template results in a strong H–L peak, whereas use of mung bean nuclease-nicked pH4ARS (oc) results only in repair synthesis. (B) Two-dimensional gel analysis of plasmid DNA replication in nuclear extracts. A strong bubble arc is detected on the gel before linearization of the plasmid (arrow head), which is consistent with bidirectional elongation from a single origin on the plasmid. However, this bubble is almost completely lost after digestion. A typical Y arc, indicative of the presence of replication forks, is observed, in addition to a major spot corresponding to fully replicated plasmids. Moreover, a "tail" of small (<1N) labeled fragments is visible in the bottom right of the gel. Because the only difference between the two panels is restriction digestion after the replication reaction, we assume that replication bubbles are broken, either by a nicking activity or by physical breakage during plasmid recovery. The plasmid pH4ARS, corresponding to the H4 ARS fragment cloned into a pGEM backbone, was used here as a template. (C) Two-dimensional gel analysis of a multicopy vector (pRS424) replicated in G_1 nuclei incubated in an S-phase nuclear extract. Genomic DNA was purified after the replication assay and was digested with *Eco*RI, which cuts opposite to the ARS element. The gel was transferred and hybridized with the vector backbone. Adapted with permission from P. Pasero, D. Braguglia, and S. M. Gasser, *Genes Dev.* **11,** 1504 (1997). (D) Density substitution analysis of genomic DNA replicated *in vitro*. In a standard reaction, corresponding to late G_1-phase nuclei incubated in an S-phase extract, most of the newly synthesized DNA (measured as cpm of [α-^{32}P]dCTP incorporated) is found at the heavy–light (H–L) position of the gradient (○). However, upon addition of 0.1% Triton X-100 in the reaction mixture, only light–light (L–L) DNA is observed (●), corresponding to path repair. (E) Analysis of the DNA replication block in cells depleted of Dbf4p, the regulatory subunit of the Cdc7 kinase, which is essential for progression into S phase. The *DBF4* gene was placed under the control of a galactose-inducible promoter. *GAL-DBF4* cells were grown exponentially on galactose-containing medium and were shifted to a glucose-containing medium to block *DBF4* expression. Analysis of DNA content by FACS indicates that cells are arrested in G_1 2 hr after the addition of glucose. A peak at 0.5 C corresponding to cells undergoing a reductional anaphase is detected after 3 hr on YPD. (F) Nuclei were prepared from *GAL-DBF4* cells arrested in G_1 by Dbf4p depletion as shown in D or from wild-type cells arrested in G_1 with α factor. Their ability to support semiconservative replication was monitored in standard reactions in a wild-type S-phase extract. Profiles of the radioactivity recovered after buoyant density gradient analysis are shown for wild-type nuclei (+Dbf4p, ●) or for Dbf4p-depleted nuclei (○). The images were adapted with permission from D. Braguglia, P. Heun, P. Pasero, B. P. Duncker, and S. M. Gasser, *J. Mol. Biol.* **281,** 631 (1998) and from P. Pasero, B. P. Duncker, E. Schwob, and S. M. Gasser, *Genes Dev.* **13,** 2159 (1999).

Spheroplasting

Because it is essential that spheroplasting is rapid and efficient for the preparation of nuclei and extracts, we use our own laboratory stock of lyticase, isolated from an *Escherichia coli* strain that overexpresses the *Oerskovia xanthineolytica* β-1,3-glucanase gene under the control of the *lacUV5* promoter.[16] Purification is carried out according to Shen *et al.*,[16] with modifications described in Braguglia *et al.*[11] Alternatively, commercially available lyticase (Roche Research Diagnostics, Rotkreuz, Switzerland) can be used in combination with a small amount of a cruder extract of lysing enzyme (Zymolyase 20T, Seikagaku Corp., Tokyo, Japan). In the case of α factor block release procedures, spheroplasting is normally completed within 15 min to prevent cells from advancing through the cell cycle. For G_1-phase extracts or nuclei, α factor is kept present during spheroplasting to prevent further progression through the cell cycle.

Preparation of Yeast S-Phase Nuclear Extract

For the preparation of nuclear extracts, crude nuclei are isolated and extracted by the addition of $(NH_4)_2SO_4$ as described in the following detailed protocol. Synchronization of yeast cells and spheroplasting are the most critical steps for the quality of the extract. These steps should be optimized on small-scale cultures for each individual strain before the preparation of the extract.

Method

1. Cell Culture. Inoculate 8 liters of 2% yeast extract, 1% bactopeptone, 2% dextrose (YPD) with a saturated preculture and grow overnight at 30° (or at the appropriate permissive temperature) to a density of 0.5 to 1×10^7 cells/ml. Harvest cells by centrifugation (5 min at 4300g at room temperature). Wash twice with water to eliminate the Bar1 protease. This is not necessary if the strain is *bar1*.

2. Synchronization. Resuspend in 2 liters prewarmed YPD, pH 5.0, and add α factor to about 1×10^{-7} *M* for *BAR1* cells or 2×10^{-8} *M* for *bar1* cells (the optimal concentration must be determined empirically for each strain). Check the G_1 arrest (90–100% unbudded cells) under the microscope after at least one generation time (90 to 120 min). Harvest the cells (4300g for 5 min at room temperature), resuspend in 2 liters fresh, prewarmed YPD, and follow the appearance of small buds, indicative of entry into S phase (30 to 45 min after release). Generally, cells are taken when 50% of the population has a small bud.

3. Spheroplasting. Harvest the cells (4300g for 5 min at room temperature) and weigh the pellet. Resuspend in 100 m*M* EDTA–KOH, pH 8.0, 10 m*M* dithiothreitol (DTT) (10 ml/g cells). Incubate for 10 min at 30° with gentle shaking and harvest cells (4300g for 5 min at room temperature). Resuspend the cells in

[16] S. H. Shen, P. Chretien, L. Bastien, and S. N. Slilaty, *J. Biol. Chem.* **266**, 1058 (1991).

TABLE II
BUFFERS FOR PREPARATION OF NUCLEI AND NUCLEAR EXTRACTS[a]

10× buffer A		Buffer B	
200 mM	Tris-HCl, pH 7.4	100 mM	Tris–acetate, pH 7.9
800 mM	KCl	50 mM	potassium acetate
80 mM	EDTA–KOH, pH 7.4	10 mM	MgSO$_4$
5 mM	spermidine	20%	glycerol
2 mM	spermine	2 mM	EDTA–KOH
		3 mM	DTT
Breakage buffer		0.5 mM	PMSF
0.25×	buffer A	300 μg/ml	benzamidine
18%	Ficoll	1 μg/ml	pepstatin A
1%	thiodiglycol	2 μg/ml	antipain
1%	Trasylol (aprotinin)	0.5 μg/ml	leupeptin
0.5 mM	PMSF		
300 μg/ml	benzamidine	Buffer C	
1 μg/ml	pepstatin A	20 mM	HEPES, pH 7.6
2 μg/ml	antipain	10 mM	MgSO$_4$
0.5 μg/ml	leupeptin	10 mM	EGTA
100 μg/ml	TPCK	20%	glycerol
50 μg/ml	TLCK	5 mM	DTT
		0.5 mM	PMSF

[a] The concentration of buffer A in the breakage buffer is 0.25×, which means a 40× dilution of the 10× stock solution. These buffers must be prepared before the experiment, with the exception of buffer A, which is stable for a few months at 4° after filtration. Protease inhibitors are added just before the extraction. PMSF, benzamidine, and pepstatin A stocks are made in methanol; antipain and TLCK in water; TPCK in ethanol; and leupeptin in 15% methanol and 25% DMSO.

10 ml/g YPD/1.1 M sorbitol and add 40 U/ml of lyticase (β-glucanase, see earlier discussion) and 0.2 mg/ml Zymolyase 20T (Seikagaku Corp., Tokyo, Japan).[17] Incubate at 30° with gentle shaking (90 rpm) for about 15 min or until cells observed through a phase-contrast microscope become dark and round shaped. As the cell wall is degraded, cells stick together to form aggregates. The progression of spheroplasting should be checked frequently by light microscopic observation. When complete, harvest the spheroplasts at 1000g for 5 min at room temperature and resuspend in an equal volume (10 ml/g) of YPD/1.1 M sorbitol, 0.5 mM phenylmethylsulfonyl fluoride (PMSF). Incubate for 10 min at 30° for reactivation of cellular metabolism (optional).

4. *Isolation of nuclei.* After centrifugation of the culture at 1000g for 5 min, resuspend the pellet in 40 ml (80 ml if more than 10 g of cells) of ice-cold breakage buffer (see Table II for buffer composition). Dounce on ice (about 20 strokes) in

[17] J. M. Verdier, R. Stalder, M. Roberge, B. Amati, A. Sentenac, and S. M. Gasser, *Nucleic Acids Res.* **18**, 7033 (1990).

a 40-ml glass Dounce with the tightest available pestle. Check the lysis under the microscope and repeat douncing until less than 20% of the cells remain intact. Spin at 5000g for 12 min at 4° in two 40-ml Sorvall polycarbonate tubes (or four tubes for 80 ml) in a HB4 rotor. Recover the supernatant and repeat centrifugation to eliminate unbroken spheroplasts. Recover the supernatant and spin at 4° for 15 min at 24,000g (HB4 rotor). Resuspend the pellet of crude nuclei by pipetting gently but thoroughly in one pellet volume of ice-cold buffer B. At this point, nuclei should be checked by both phase microscopy and 4'-6-diamidino-2-phenylindole-2 HCl (DAPI) fluorescence to ensure that there are very few mitochondria and unbroken spheroplasts remaining. Adjust nuclei to a concentration of about 110 OD_{260}/ml. The optical density of nuclei is determined by disrupting 1–10 μl of nuclei in 1 ml 1% sodium dodecyl sulfate (SDS) and reading the absorption at 260 nm. OD_{260}/ml is a convenient way to standardize nuclear concentration. For haploid strains, 1 OD_{260} correspond, roughly to 10^8 nuclei.

5. Extraction of nuclear proteins. Add chilled and saturated $(NH_4)_2SO_4$ (4.1 M, pH 7.1) to a final concentration 0.9 M. Stir gently on ice for 30 min. Sediment undissolved proteins at 215,000g for 30 min at 4° in a SW 50.1 rotor. Recover the supernatant and slowly dissolve 0.35 g/ml of ultrapure $(NH_4)_2SO_4$ plus 10 μl of 1 M KOH per gram of $(NH_4)_2SO_4$ in the supernatant to attain 75% $(NH_4)_2SO_4$ saturation. Incubate at 4° for 30 min with gentle stirring. Sediment precipitated proteins at 165,500g for 30 min at 4° in a SW 50.1 rotor (Beckman). Discard the supernatant and resuspend the pellet slowly, pipeting in a minimum volume of buffer C (usually 1–2 ml if starting from 8 × 10^{10} cells). Additional protease inhibitors can be added if strains are not deficient for vacuolar proteases.

6. Dialysis and storage. Dialyze for 12–15 hr against 500 ml of buffer C at 4°. Clarify the extract by centrifugation in an Eppendorf 5415C centrifuge for 10 min at 12,000 rpm at 4°. Discard the pellet, which generally contains nuclear remnants. Check the protein concentration of the extract, which should be between 10 and 20 mg/ml (Bradford assay). Freeze and store 50-μl aliquots in liquid nitrogen. Nuclear extracts are stable for several months when stored in liquid nitrogen, but should not be frozen and thawed repeatedly.

Preparation of Crude Nuclei

For the *in vitro* replication of genomic DNA, we use as template crude nuclei identical to those isolated for extract preparation except that they are synchronized in G_1 phase (see detailed protocol later). In contrast, nuclei prepared on Percoll gradients[18] are poor templates for replication, probably because they lack essential components that diffuse out during long centrifugation steps in hypotonic buffers (P. Pasero, unpublished observations, 1998). We have shown that the integrity of

[18] B. B. Amati and S. M. Gasser, *Cell* **54**, 967 (1988).

the nuclear envelope, which can be checked using fluorescent dextran molecules,[19] is essential for semiconservative DNA replication.[12] Nuclei permeabilized with Triton X-100 show a high repair activity, but very little semiconservative DNA replication (Fig. 2D). A rapid isolation of intact nuclei is therefore crucial for the preparation of both active nuclear extracts and template nuclei. Nuclei are stable for a week at $-20°$ and for at least a month in liquid N_2, although after longer periods of time, G_1-phase nuclei lose their competence for initiation of DNA replication. This loss of competence may result from a dissociation of the prereplicative complex, as no signs of proteolysis are detectable on Western blots (P. Pasero, unpublished observations, 1998).

Method

1. Cell culture. Inoculate 2 liters of YPD with a saturated preculture and grow to a density of 1×10^7 cells/ml at $30°$ or the appropriate permissive temperature. Synchronize the cells in G_1 phase with α factor and prepare spheroplasts as described earlier.

2. Cell fractionation. Lyse the spheroplasts in 20 ml of ice-cold breakage buffer as described earlier. After centrifugation at $5000g$ for 12 min at $4°$ in two 40-ml Sorvall polycarbonate tubes (HB4 rotor), recover the supernatant and transfer it into 40-ml Sorvall tubes containing a 1-ml cushion of $0.25\times$ buffer A ($10\times$ buffer A diluted 40-fold), 50% glycerol, and with the addition of all protease inhibitors and the antioxidant, thiodiglycol, as in the breakage buffer. Centrifugation at $24,000g$ for 15 min at $4°$ (HB4 rotor) allows one to eliminate the supernatant above the glycerol cushion.

3. Storage. Resuspend nuclei carefully in the $0.25\times$ buffer A/glycerol buffer and check for the presence of contaminants (e.g., whole cells, mitochondria, cell debris) under the microscope. Store at $-20°$ in or in liquid nitrogen at a concentration of 100 OD_{260}/ml.

Replication Assays with Plasmid DNA

Replication of naked DNA in yeast S-phase extracts is monitored by the incorporation of radiolabeled nucleotide in supercoiled plasmid. We usually prepare plasmids without nicks by a standard alkaline lysis method and purify them either over Qiagen columns (Qiagen Inc., Basel, Switzerland) or by banding on CsCl gradients.[20] In the latter case, the DNA is never exposed to light (UV or natural) in the presence of ethidium bromide, and all ethidium bromide is removed by

---

[19] R. Peters, *J. Biol. Chem.* **258**, 11427 (1983).
[20] J. Sambrook, E. F. Fritsch, and T. Maniatis, "Molecular Cloning." Cold Spring Harbor Laboratory Press, Cold Spring Harbor, NY, 1989.

isopropanol extraction. DNA is stored in small aliquots to avoid freeze–thaw cycles. A functional S-phase extract must have a highly active Cdk1/cyclin B kinase. Extracts should be tested for kinase activity using histone H1 as the substrate.[13,21]

Replication Assay

Prepare 25 μl of standard reaction mixture containing 2.5 μl of 10× replication buffer (120 mM HEPES–NaOH, pH 7.6, 48 mM MgCl$_2$, 3 mM EDTA–NaOH, pH 7.6, 6 mM DTT), 300 ng of supercoiled plasmid DNA, an energy regeneration system composed of 40 mM creatine phosphate, 0.125 mg/ml creatine kinase, 100 μM of dATP, dGTP, and dTTP, 2 μM of dCTP, 0.2 μCi/μl of [α-^{32}P]dCTP (Amersham, Piscataway, NJ), 200 μM of each ATP, CTP, GTP, and UTP, and 40–60 μg of nuclear protein extract. The optimal amount of a given extract should be determined by titration. For CsCl gradient analysis, and 200 μM of the heavy base analog BrdUTP in place of dTTP. Incubate for 90 or 180 min at 25° (or another relevant temperature in the case of thermosensitive extracts). Stop the reaction by adjustment to 0.3 M sodium acetate, pH 5.2, 0.1% SDS, 5 mM EDTA sodium, pH 8.0, and 100 μg/ml proteinase K and incubate for 30 min at 37°. Add tRNA carrier, extract the DNA with phenol/chloroforme, precipitate, and wash with 1 ml of 70% ethanol. Dry the pellet and resuspend in 20 μl distilled water. Remove unincorporated label with a Qiaquick column (Qiagen). Run one-fourth of each reaction on a 0.8% agarose gel and stain with ethidium bromide to verify equal recovery of DNA.

Cesium Chloride Density Gradient Centrifugation

Because repair synthesis usually results in short stretches of substituted DNA, a rigorous distinction between repair and replicative DNA synthesis relies on density gradient centrifugation following substitution with a heavy nucleotide analog. This is the method we recommend for monitoring replication *in vitro*. After the replication reaction and removal of free label, plasmid DNA (or genomic DNA) is mixed with a CsCl solution to a final density of 1.7176 g/ml at 25° ($\eta = 1.7176$). The gradients are generated at 36,000 rpm in a fixed-angle T1270 rotor (Sorvall, Zurich, Switzerland) for 40 hr at 20°. Fractions of 250–300 μl are collected from the bottom of the gradient with a capillary pipette, counted in a Packard liquid scintillation counter, and the refractive index of every second fraction is read. The gradients typically span densities from 1.6742 ($\eta = 1.3970$) to 1.7936 ($\eta = 1.4080$) g/ml, and H–L DNA is usually recovered at refractive index values between 1.4035 and 1.4040, while the unsubstituted DNA peaks at values between 1.4000 and 1.4010. Sedimentation of the light–light (L–L) and H–L DNA was standardized by elongating a primed M13 template with Sequenase (Roche Research

---

[21] S. Moreno, J. Hayles, and P. Nurse, *Cell* **58**, 361 (1989).

Diagnostics, Rotkreuz, Switzerland) incorporating either BrdUTP or dTTP. Typical examples of the gradient profiles are shown in Fig. 2A. The relative height of the L–L peak depends entirely on the integrity of the template. If plasmids are damaged or nicked, the L–L peak represents a majority of the incorporated label.[11]

DpnI Resistance Analysis of Plasmid DNA Replication in Vitro

Conversion of methylated bacterial plasmid DNA to a hemimethylated or fully unmethylated form can be achieved by *de novo* complementary strand synthesis. This is assayed by simple restriction enzyme digestion of the radiolabeled replication products using the restriction enzymes *Dpn*I and *Nde*II (an isoschizomer of *Mbo*I), which cleave either fully methylated or entirely unmethylated GATC sequences, respectively. Another isoschizomer, *Sau*3AI, cuts both. Thus, the conversion of a fully methylated template to hemimethylated DNA by semiconservative replication correlates with an acquired resistance to *Dpn*I (while remaining sensitive to *Sau*3AI and resistant to *Nde*II, see Ref. 22). For this analysis, replication products are extracted following a standard reaction performed in the presence of a radioactive nucleotide. The plasmid is linearized with a restriction enzyme that is not methylation sensitive and is subsequently digested with *Nde*II, *Sau*3AI, or *Dpn*I for 1 hr at 37° prior to gel electrophoresis on a 1% agarose gel.[22] *Nde*II and *Sau*3AI digests are performed in the recommended buffers, whereas the *Dpn*I digest uses buffer H (Roche Research Diagnostics) adjusted to 150 mM NaCl. A fourfold excess of unreplicated plasmid is added to the restriction reaction to monitor the efficiency of *Dpn*I digestion. After electrophoresis, agarose gels are dried, and label incorporation is quantified on a PhosphorImager (Molecular Dynamics, Sunnyvale, CA) or equivalent. The relative efficiency of semiconservative replication as compared to localized patch repair is calculated by comparing incorporation into *Dpn*I-resistant DNA vs *Dpn*I-sensitive DNA. Under optimal conditions, up to 86% of [α-^{32}P]dCTP incorporated into a supercoiled plasmid is *Dpn*I resistant. This is in good agreement with the fraction of label recovered at the H–L peak in CsCl gradients.

Two-Dimensional Gel Electrophoresis

Two-dimensional gel electrophoresis is performed on plasmids replicating *in vitro* essentially as described by Brewer and Fangman[5] and Friedman and Brewer[23] for *in vivo* replication, except that a low level of radioactivity is included in the reaction. The gel is dried on Whatman (Clifton, NJ) 3MM paper and exposed by standard methods. When nonlinearized plasmids are analyzed on 2D gels, a clear bubble arc extending from 1n- to 2n-sized circular molecules can be

---

[22] J. A. Sanchez, D. Marek, and L. J. Wangh, *J. Biol. Chem.* **267**, 25786 (1992).
[23] K. L. Friedman and B. J. Brewer, *Methods Enzymol.* **262**, 613 (1995).

detected (Fig. 2B). However, this bubble arc usually disappears almost completely after linearization of the vector, suggesting that the replication bubble is broken during the restriction digestion step. The hypersensitivity of our *in vitro*-replicated products to breakage may reflect a lack of polymerase coordination at the replication fork and/or the presence of residual single-stranded DNA regions that are susceptible to nicks after deproteinization. In contrast, replication bubbles are detected readily in our hands when plasmids are replicated in the nuclear replication system (Fig. 2C).

Replication Assays with Nuclei

We have shown that the replication assay using isolated nuclei as template obeys cell cycle controls, as G_1- but not G_2- or M-phase nuclei initiate DNA replication in the presence of an S-phase nuclear extract.[12] It should be noted that nuclei arrested in G_1 phase prior to start ($\Delta cln1$-3 or α factor arrest, with α factor present during the spheroplasting step) do not replicate very efficiently, presumably because they lack essential components synthesized after start, such as Cdc6p or Dbf4p.[24] However, when nuclei are spheroplasted in the absence of α factor, they appear to progress sufficiently through late G_1 phase during the cell wall degradation to be fully competent for initiation without reaching S phase. A lack of entry into S phase can be checked by 2D gel analysis of the endogenous 2-μm plasmid.[12] To arrest cells in late G_1 phase, $cdc4$-1 cells can also be used, although they do not spheroplast well at 37° because they form very elongated buds.

Replication Assay

Mix 10^7 freshly prepared G_1-phase nuclei (corresponding to approximately to 100 ng of genomic DNA or 0.1 OD_{260} with nuclei monitored after lysis in 0.1% SDS) with 25 μl of a reaction mixture containing 2.5 μl of 10× replication buffer (see earlier discussion), 40 mM creatine phosphate, 0.125 mg/ml creatine kinase, 100 μM of dATP, dGTP, and dTTP, 2 μM dCTP, 0.2 μCi/μl of [α-^{32}P]dCTP (Amersham), 200 μM of each NTPs, and 40–60 μg of S-phase nuclear extract. For density substitution experiments, replace dTTP with 200 μM BrdUTP. For immunofluorescence experiments, replace dTTP with 20 μM digoxigenin-11–dUTP (Roche Research Diagnostics) and omit [α-^{32}P]dCTP. Incubate for at least 30 min at 25° or another relevant temperature. Stop the reaction with the addition of 25 μl of 10 mM Tris–HCl, pH 7.0, 0.1% SDS, 5 mM EDTA, 100 μg/ml proteinase K and incubate for 60 min at 37°. For 2D gel analysis, precipitate the DNA after the proteinase K treatment. For immunofluorescence experiments, the reaction is stopped with the addition of 4% paraformaldehyde (see later).

---

[24] P. Pasero, B. P. Duncker, E. Schwob, and S. M. Gasser, *Genes Dev.* **13,** 2159 (1999).

Cesium Chloride Density Gradient Centrifugation

Density shift experiments are performed as described earlier for plasmid DNA. After the replication assay and deproteinization, genomic DNA is sheared by sonication into fragments of 2–5 kb. Because of variation in sonicators, the conditions for shearing must be determined empirically by estimating the size of sonicated genomic DNA on agarose gels. The free label is removed with Qiaquick columns (Qiagen, Basel, Switzerland) after sonication. Quantitation of semiconservative replication is done by integrating the peak of H–L DNA and converting the cpm into moles of dCTP incorporated (Figs. 2D and 2E).

Two-Dimensional Gel Electrophoresis

Two-dimensional gel electrophoresis is carried out on genomic DNA as described previously. Gels are transferred to nitrocellulose and hybridized with the relevant probes.[12] It should be noted that the amount of genomic DNA present in a standard reaction (about 100 ng) does not allow the study of single-copy chromosomal origins, although initiation at origins present on multicopy plasmids can be detected readily when the reaction is scaled up 10- to 20-fold (Fig. 2C).

Immunofluorescence and Confocal Microscopy

Following the replication assay in the presence of digoxigenin(DIG)–dUTP, yeast nuclei are fixed for 60 min at 4° through the addition of 4% paraformaldehyde to the reaction mixture and spotted on a microscope slide. Air-dried slides are then fixed for 6 min in prechilled ethanol ($-20°$) and 1 min in prechilled acetone ($-20°$) and are incubated for 30 min at room temperature in phosphate-buffered saline (PBS), 0.1% Triton-X100 before the addition of the primary antibody. A fluorescein isothiocyanate (FITC)-coupled anti-DIG F(ab) fragment or a monoclonal antibody directed against DIG is used to detect DIG–dUTP incorporation. Slides are incubated for 1.5 hr at 37°, washed six times for 5 min in PBS, 0.1% Triton X-100 and mounted in 1× PBS, 50% (v/v) glycerol, 24 μg/ml 1,4-diazabicyclo[2.2.2]octane (DABCO) with 1 μg/ml ethidium bromide. In our laboratory, confocal microscopy is performed on a Zeiss Axiovert 100 microscope (Zeiss Laser Scanning Microscope 410 or 510) with a 63× Plan-Apochromat objective (1.4 oil). High-resolution microscopy is necessary to visualize replication foci in haploid yeast nuclei. *De novo* incorporation of nucleotides occurs at roughly 20 replication centers (Fig. 3A). Because there are about 400 origins of replication per haploid genome, this may mean that there are up to 20 origins per cluster. These foci colocalize to a large extent with immunostaining for ORC, although the staining for Dbf4p is found in regions that are not yet replicated (see Fig. 3B).[24] The quantitation and techniques used to produce these images are discussed elsewhere.[12]

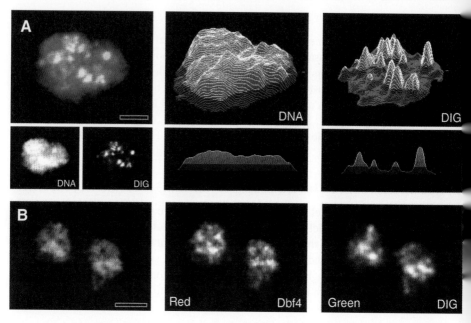

FIG. 3. (A) Replication foci are detected in isolated yeast nuclei on incorporation of DIG–dUTP *in vitro*. G$_1$-phase wild-type nuclei were incubated in a wild-type S-phase extract for 30 min in the presence of 20 μm DIG–dUTP. After fixation with paraformaldehyde, genomic DNA is stained with ethidium bromide, and DIG–dUTP is detected on slides with an anti-DIG monoclonal antibody and an anti-mouse IgG coupled to DTAF. A high-resolution image of a single yeast nucleus obtained with a Zeiss LSM 410 confocal microscope with a 63× Plan-Apochromat objective (1.4 oil) is shown. Surface topography profiles of red (DNA) and green (DIG) channels are calculated with Carl Zeiss LSM 3.95 software. About 15 replication foci appear as individual peaks above the background level of fluorescence (blue). The threshold is determined for green and red channels in a 50 × 50 pixel area containing no nuclei. Bar: 200 nm. (B) A similar reaction using G$_1$ nuclei prepared from Dbf4–Myc cells was performed to localize Dbf4p relative to the sites of incorporation of DIG–dUTP. The replication assay was stopped by the addition of paraformaldehyde, and both sites of DIG–dUTP (green) incorporation and of Dbf4–Myc (red) localization were detected by immunofluorescence. Bar: 2 μm. Adapted with permission from P. Pasero, D. Braguglia, and S. M. Gasser, *Genes Dev.* **11**, 1504 (1997) and from P. Pasero, B. P. Duncker, E. Schwob, and S. M. Gasser, *Genes Dev.* **13**, 2159 (1999).

Chromatin-Binding Assay

A simple chromatin fractionation assay has been developed in yeast in order to monitor the assembly of the prereplication complex on origins.[25,26] We have adapted the assay to follow the fate of chromatin-bound proteins in isolated yeast nuclei during the *in vitro* replication reaction.[24] The replication reaction is mixed

[25] S. Donovan, J. Harwood, L. S. Drury, and J. F. Diffley, *Proc. Natl. Acad. Sci. U.S.A.* **94**, 5611 (1997).
[26] C. Liang and B. Stillman, *Genes Dev.* **11**, 3375 (1997).

with an equal volume of 0.25× buffer A (see Table II), supplemented with 1% thiodiglycol, 1% Trasylol, 0.5 mM PMSF, and 2% Triton X-100. Lysis of nuclei occurs within a few seconds, and the chromatin-associated fraction is recovered by centrifugation in an Eppendorf centrifuge for 15 min at 13,500 rpm at 4°. Total protein concentration is determined for chromatin-bound and soluble fractions by a Bradford assay.

Concluding Remarks

The assays described here provide the basis for biochemical fractionation of both extracts and nuclei to determine the minimal components required for a properly controlled origin-specific initiation of DNA replication. The use of mutant extracts can be very powerful. The replication of supercoiled plasmid is not yet optimized for the study of origin-dependent initiation, as the prereplication complex is unable to assemble in S-phase extracts.[27] Seki and Diffley[28] have shown that a prereplicative complex containing ORC, Cdc6p, and Mcm2-7 proteins can assemble *in vitro* on *ARS1* multimers, provided that the extract used presents a low Cdk activity and a high amount of Cdc6p. This is a very encouraging step toward the development of a fully soluble origin-specific replication system. The nuclear replication system is perfectly suited to study how nuclear organization and compartmentation might facilitate and regulate origin-dependent initiation. Using this system, we have shown that yeast origins are assembled into replication foci,[12] as it is the case in metazoans.[29] We have also shown that the Dbf4/Cdc7 complex, an essential S-phase promoting kinase, associates with the prereplication complex at origins in G_1 nuclei and is displaced from replication foci on initiation.[24] The subnuclear organization of origins appears to affect the timing of replication,[30] although the mechanism controlling the distribution of replication origins within the nucleus is still poorly understood. Coupled with yeast genetics, these two biochemical assays should allow new insights into the regulation of eukaryotic DNA replication.

Acknowledgments

We acknowledge and thank Diego Braguglia, whose Ph.D. Thesis (1995, University of Lausanne) provided the basis for all the assays described here. We also thank Bernard Duncker and Patrick Heun for optimization of the CsCl gradient and of the *Dpn*I assay, respectively, for helpful collaboration, and for stimulating discussions. P.P. thanks ARC, EMBO, and the Roche Research Foundation for support in the Gasser laboratory. This research was supported by the Swiss National Science Foundation, the Swiss Cancer League, and the Human Frontiers Science Program.

---

[27] S. Piatti, T. Bohm, J. H. Cocker, J. F. Diffley, and K. Nasmyth, *Genes Dev.* **10**, 1516 (1996).
[28] T. Seki and J. F. Diffley, *Proc. Natl. Acad. Sci. U.S.A.* **97**, 14115 (2000).
[29] R. Berezney, D. D. Dubey, and J. A. Huberman, *Chromosoma* **108**, 471 (2000).
[30] P. Heun, T. Laroche, M. K. Raghuraman, and S. M. Gasser, *J. Cell Biol.* **152**, 385 (2001).

[12] Yeast Pre-mRNA Splicing: Methods, Mechanisms, and Machinery

By SCOTT W. STEVENS and JOHN ABELSON

History of Pre-mRNA Splicing

Prior to the discovery of intervening sequences (introns) present in eukaryotic viral messenger RNA (mRNA),[1,2] it was thought that the sequence of mature mRNA was colinear with that of the gene from which it was transcribed. The baroque process by which these extraneous sequences are removed came to be known as pre-mRNA splicing. The genes for dozens of protein factors involved in the removal of pre-mRNA introns have been noted in the deeply rooted eukaryote *Giardia lamblia*,[3,4] indicating that introns are present from the simpler eukaryotes to humans. Curiously, although the majority of trans-acting factors involved in the pre-mRNA splicing reaction are highly conserved and present throughout Eukarya, the number and complexity of introns present in the genomes of eukaryotes mirror the complexity of the organism. Yeast have only a few hundred introns, generally limited to one intron per intron-containing gene; only a few yeast genes have more than one intron. Human genes have multiple introns yielding tens or hundreds of thousands of introns coded for in the human genome.

As introns were discovered in the human adenovirus, the mechanism of the pre-mRNA splicing reaction was initially studied using extracts from human cells.[5,6] An *in vitro* extract to study splicing soon followed the discovery of yeast introns.[7–9] Because human cells are not easily amenable to genetic studies, the promise of yeast genetics in the study of the components of the splicing machinery was quickly realized.[10–14] Not to be completely outdone in the realm of

---

[1] L. T. Chow, J. M. Roberts, J. B. Lewis, and T. R. Broker, *Cell* **11**, 819 (1977).
[2] S. M. Berget, C. Moore, and P. A. Sharp, *Proc. Natl. Acad. Sci. U.S.A.* **74**, 3171 (1977).
[3] A. G. McArthur, H. G. Morrison, J. E. Nixon, N. Q. Passamaneck, U. Kim, G. Hinkle, M. K. Crocker, M. E. Holder, R. Farr, C. I. Reich, G. E. Olsen, S. B. Aley, R. D. Adam, F. D. Gillin, and M. L. Sogin, *FEMS Microbiol. Lett.* **189**, 271 (2000).
[4] http://www.mbl.edu/Giardia/
[5] N. Hernandez and W. Keller, *Cell* **35**, 89 (1983).
[6] R. A. Padgett, S. F. Hardy, and P. A. Sharp, *Proc. Natl. Acad. Sci. U.S.A.* **80**, 5230 (1983).
[7] D. Gallwitz and I. Sures, *Proc. Natl. Acad. Sci. U.S.A.* **77**, 2546 (1980).
[8] R. Ng and J. Abelson, *Proc. Natl. Acad. Sci. U.S.A.* **77**, 3912 (1980).
[9] R. J. Lin, A. J. Newman, S. C. Cheng, and J. Abelson, *J. Biol. Chem.* **260**, 14780 (1985).
[10] A. J. Lustig, R. J. Lin, and J. Abelson, *Cell* **47**, 953 (1986).
[11] P. G. Siliciano and C. Guthrie, *Genes Dev.* **2**, 1258 (1988).
[12] U. Vijayraghavan, M. Company, and J. Abelson, *Genes Dev.* **3**, 1206 (1989).
[13] D. Frank, B. Patterson, and C. Guthrie, *Mol. Cell. Biol.* **12**, 5197 (1992).
[14] S. M. Noble and C. Guthrie, *Genetics* **143**, 67 (1996).

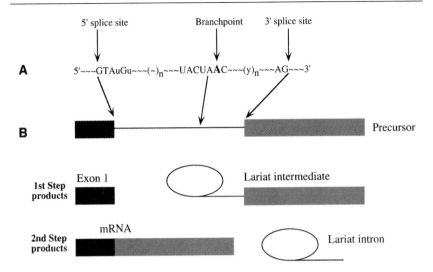

FIG. 1. Structure of a yeast intron and of the splicing reaction products. (A) The conserved 5' splice site, branchpoint, and 3' splice site sequences are delineated. Highly conserved nucleotides are presented in uppercase letters, and nucleotides, which are less conserved, are shown in lowercase letters. The branchpoint adenosine residue is highlighted in boldface type. (B) Schematic representation of the chemical steps of the pre-mRNA splicing reaction. The pre-mRNA precursor is processed into two species during the first step of the reaction, exon 1 and the lariat intermediate. In the second step of the reaction, the lariat intron is excised and the two exons are linked convalently.

biochemistry, discoveries regarding the apparatus, which came to be known as the spliceosome,[15] and the corresponding subcomplexes known as small nuclear ribonucleoprotein particles (snRNPs) were made with near simultaneity in the yeast and human systems. We present here a review of the methods for the study of the mechanisms and machinery of the yeast pre-mRNA splicing system.

Precursors and Products

To faithfully remove sequences from what will become the mature message, intronic sequences must be defined precisely to preserve the reading frame. Frequent translation of improperly spliced mRNA would be catastrophic for the cell. In Fig. 1A, the signals that define the exon (expressed sequence) and intron (intervening sequence) are delineated. In yeast, there exist three nearly invariant sequences, which are also conserved with respect to their location in the linear sequence. The 5' splice site consists of a nearly invariant GU, the guanosine

---

[15] E. Brody and J. Abelson, Science **228**, 963 (1985).

nucleotide defining the first nucleotide of the intron. The nearly invariant sequence termed the branchpoint is UACUAAC. The adenosine in boldface type is the branchpoint nucleotide, which participates in the chemistry of the pre-mRNA splicing reaction (see later). The final sequence in yeast, which defines the boundaries of the intron, is the 3' splice site. In yeast, it is always AG, where the guanosine represents the very last nucleotide of the intron. Additionally, a tract of pyrimidine (U and C) residues frequently exists between the branchpoint sequence and the 3' splice site. In human cells, these sequences are less stringently conserved with a branchpoint sequence that is often highly variable. A bioinformatic analysis of all intron-containing genes in yeast has been presented elsewhere.[16,17]

In Fig. 1B, a schematic of the products of each of the chemical steps of the splicing reaction is presented. It should be noted that the diagram presented is a simplified version of events not meant to represent the intricate and dynamic interactions of the pre-mRNA with the trans-acting factors. Although the signals are more divergent in humans, the chemistry of the pre-mRNA splicing reaction is conserved through evolution. The products of the first step of splicing are the 5' exon (black bar in Fig. 1B) and the lariat intermediate, which contains the intron in the form of the branched lariat and the 3' exon (gray bar in Fig. 1B). The second step of splicing produces the mature mRNA and the lariat intron (Fig. 1B). A number of excellent reviews detailing these mechanisms have been published previously.[18–20]

Spliceosome and SnRNPs

The entity that effects the removal of introns was termed the spliceosome by Brody and Abelson.[15] The spliceosome is the macromolecular machine consisting of the pre-mRNA, nuclear pre-mRNA-binding proteins, the snRNPs, and non-snRNP or "exchangeable" factors, which include a number of ATP-dependent RNA helicases. The spliceosome has been characterized by glycerol gradient sedimentation as a 40S (yeast[15]) or 60S (human[21]) complex. Although their roles in pre-mRNA splicing were not definitively proven until the mid-1980s,[22,23] spliceosomal snRNPs were identified due to their great abundance first based on their

[16] P. J. Lopez and B. Séraphin, *RNA* **5**, 1135 (1999).
[17] M. Ares, L. Grate, and M. H. Pauling, *RNA* **6**, 1138 (1999).
[18] C. Guthrie, *Science* **253**, 157 (1991).
[19] M. J. Moore, C. C. Query, and P. A. Sharp, in "Splicing of Precursors to mRNA by the Spliceosome," (R. F. Gesteland and J. F. Atkins, eds.), p. 303. Cold Spring Harbor Laboratory Press, Cold Spring Harbor, NY, 1993.
[20] J. P. Staley and C. Guthrie, *Cell* **92**, 315 (1998).
[21] P. J. Grabowski, S. R. Seiler, and P. A. Sharp, *Cell* **42**, 345 (1985).
[22] D. L. Black, B. Chabot, and J. A. Steitz, *Cell* **42**, 737 (1985).
[23] D. L. Black and J. A. Steitz, *Cell* **46**, 697 (1986).

RNA component[24] and subsequently based on their composition as ribonucleoproteins.[25,26]

What was to become a workhorse for snRNP biochemical preparation in mammals and later in yeast was the development by Lührmann and colleagues[27] of the anti-2,2,7-trimethylguanosine (α-TMG) cap antibody. The U1, U2, U4, and U5 snRNAs (but not the U6 snRNA) each contain a unique trimethylated cap structure to which this antibody efficiently binds with the ability to selectively elute with free 7-methylguanosine nucleoside.

The purification of snRNPs from yeast has revealed remarkable similarities as well as remarkable differences in the protein complement from the respective snRNPs. The methods and results of the purification of snRNPs from yeast will be discussed later. Three groups nearly simultaneously developed two-step affinity chromatography techniques in yeast for snRNP purification, each with its own advantages and disadvantages.

Yeast as a Model System

The ability to study the splicing reaction *in vitro* has been achieved with both yeast and human (HeLa) extracts. The benefit in using human extracts has, until recently, been the ability to study purified splicing complexes and components. The benefit of using yeast has been largely the ability to perform genetic manipulation of the proteins and snRNAs and then to test by classical genetic techniques for intra- and extragenic mutants to gain insight on the interactions in the spliceosome. Additionally, the complete genome of the yeast *Saccharomyces cerevisiae* has been known since 1996.[28] This allows many analyses previously not possible, such as definitive database homology searches, and the ability to unambiguously identify the genes for purified proteins by mass spectrometry.

In Vitro Splicing Extract Preparation from Saccharomyces cerevisiae

Dounce Extracts

There are two common methods of extract preparation. We will present protocols for each of these methods, and enumerate the advantages and disadvantages of

---

[24] R. Weinberg and S. Penman, *J. Mol. Biol.* **38**, 289 (1968).
[25] M. R. Lerner and J. A. Steitz, *Proc. Natl. Acad. Sci. U.S.A.* **76**, 5495 (1979).
[26] M. R. Lerner, J. A. Boyle, J. A. Hardin, and J. A. Steitz, *Science* **211**, 400 (1980).
[27] R. Lührmann, B. Appel, P. Bringmann, J. Rinke, R. Reuter, S. Rothe, and R. Bald, *Nucleic Acids Res.* **10**, 7103 (1982).
[28] A. Goffeau, B. G. Barrell, H. Bussey, R. W. Davis, B. Dujon, H. Feldmann, F. Galibert, J. D. Hoheisel, C. Jacq, M. Johnston, E. J. Louis, H. W. Mewes, Y. Murakami, P. Philippsen, H. Tettelin, and S. G. Oliver, *Science* **274**, 546 (1996).

each. The first method was that used to demonstrate yeast splicing *in vitro* and, in the experience of many laboratories, yields the greatest splicing activity, although it is often quirky and may require some optimization. The yeast strain used will depend on the requirement of the experiment. For general use extracts, we use the strain BJ411 (aka EJ101 from Elizabeth Jones, Carnegie Mellon University, Pittsburgh, PA), which is a protease-deficient strain; however, active extracts can be made from virtually any strain.

YPD (per liter): 10 g Yeast extract, 20 g Bacto-peptone, and 20 g Glucose per liter
SB: 1 M Sorbitol, 50 mM Tris, pH 7.8, and 10 mM MgCl$_2$
Zymolyase buffer: 20 mM K$_2$PO$_4$ pH 7.4, and 5% Glucose
Buffer A: 10 mM KCl, 10 mM HEPES, pH 7.9, at 25°, 1.5 mM MgCl$_2$, 0.5 mM Dithiothreitol (DTT). Optional: Phenylmethylsulfonyl fluoride (PMSF) to 0.5 mM, leupeptin and pepstatin each to 1 μg/ml
Buffer D: 20 mM HEPES, pH 7.9, 50 mM KCl, 0.2 mM EDTA, pH 8.0, and 20% (v/v) Glycerol

Growth of the strain is performed at the required temperature (generally 30°) in the required medium (generally YPD) to an optical density of 3.0 at 600 nm. Active extracts are generally made from cells grown to OD 2.0 to 4.0; one should avoid further growth for the best results. Two liters of growth medium generally yields approximately 5 ml of splicing extract. Cells are pelleted at 5000g for 5 min, and the resulting pellet is washed once in doubly distilled H$_2$O. Measure the wet mass of the cell pellet after washing with water. Resuspend cells in 2 ml SB containing 30 mM DTT per gram of wet cell mass. Incubate at room temperature for 15 min. Pellet cells at 5000g for 5 min. Resuspend cells in 1.5 ml SB containing 3 mM DTT per gram of wet cell mass. Remove a 200-μl pre-enzymatic treatment sample aliquot to determine the extent of cell lysis. Add 100 μl Zymolyase solution [20 mg/ml Zymolyase (Seikagaku Corp., Tokyo, Japan) in Zymolyase buffer] per 10 ml cell resuspension. Incubate at 30° with gentle shaking for 30 min. Remove 10 μl cell suspension at 10-min intervals. Add 10 μl prelysis suspension and 10 μl from each time point to separate tubes containing 1 ml of 10% sodium dodecyl sulfate (SDS). Vortex and measure the optical density at 800 nm. When the ratio of optical density of the Zymolyase digestion sample to that of the predigestion sample is <0.25, it is time to proceed. Cells at this point have been spheroplasted and are rather fragile. Spin spheroplasts at 3000g for 5 min at 4°. Resuspend the spheroplasts very gently in 1.5 ml of ice-cold SB containing 3 mM DTT per gram wet cell mass. Spin washed spheroplasts 3000g for 5 min at 4°. Resuspend spheroplasts in 1.4 ml buffer A per gram wet cell mass. In a prechilled (4°) 40-ml Dounce homogenizer, using the tightest available pestle (the fit is critical and may require testing several pestles), add the resuspended spheroplasts. Slowly perform 10 strokes up and down in the cold, preferably submerging the Dounce homogenizer in an ice water bath. Pipette the suspension into a chilled small glass beaker containing a stir bar, noting the volume transferred. Slowly add 0.11 (1/9) volume ice-cold 2 M KCl to the suspension with stirring. Although splicing activity was shown to require Na^+, K^+, or NH_4^+ ions, potassium is the generally accepted cation for the splicing reaction. Reichert and Moore[29] have shown that there may be a benefit to changing the anion to glutamate in human splicing extracts, although it is not known if this effect is beneficial when using yeast extracts. Gently stir this homogenate on ice for 30 min. Spin at 17,000 rpm in a Sorvall SS34 rotor for 30 min at 4°. Avoiding the pellet, as well as the lipoprotein material at the top of the tube, pipette the supernatant into a chilled, appropriately sized (e.g., Beckmann Ti-60) ultracentrifuge tube. Spin at 100,000g (37,000 rpm in Ti-60) for 1 hr. Using a Pasteur pipette, carefully remove the middle phase. The appearance of the material in the tube will vary widely with the strain used; however, again one wants to avoid anything that has pelleted as well as the lipoprotein material at the top of the tube. Generally, one-fourth to one-fifth of the spun volume is recovered and for optimal splicing activity, purity is more critical than quantity. Dialyze the extract at 4° against 2 × 2 liters of buffer D for 1.5–2 hr each with stirring. We generally use dialysis tubing or a cassette (Pierce, Rockford, IL) with a molecular weight cutoff of 10,000. Transfer the extract into chilled microcentrifuge tubes, filling them to the top. Spin in a chilled (4°) microcentrifuge for 10 min at top speed to pellet precipitated material. Remove the supernatant and aliquot into screw cap microcentrifuge tubes in manageable portions (usually ∼500 μl) and freeze in liquid nitrogen. Store at −80°. Splicing extract can survive repeated freeze thaws; however, it is wise to realiquot into portions appropriate to your experimental needs and refreeze.

Liquid Nitrogen Method

Although extract prepared by a Dounce homogenizer is generally superior for *in vitro* splicing experiments, activity from batch to batch can vary widely. A more consistent method for the preparation of splicing extract is achieved by grinding or blending in liquid nitrogen. For small-scale preparations (<8 liters of culture), one should use an appropriately sized ceramic mortar and pestle. For larger preparations, one uses an appropriately sized laboratory blendor. Batch-to-batch variation using this method is greatly reduced; however, some extracts exhibit more precursor degradation and/or reduced activity when compared to a dounced extract. Some hints to increase the chances of getting active extracts include (i) using very pure water (double distilled, deionized water–diethyl pyrocarbonate treatment is not necessary), (ii) making sure that all reagents, pipettes, and so on used are prechilled to 4°, and (iii) making sure that all tubes are cleaned thoroughly and, when possible, autoclaved.

[29] V. Reichert and M. J. Moore, *Nucleic Acids Res.* **28**, 416 (2000).

Buffer A: 10 mM KCl, 10 mM HEPES, pH 7.9, at 25°, 1.5 mM MgCl$_2$, and 0.5 mM DTT. Optional: PMSF to 0.5 mM, leupeptin and pepstatin each to 1 µg/ml
Buffer AGK: 10 mM HEPES, pH 7.9, 200 mM KCl, 1.5 mM MgCl$_2$, 0.5 mM DTT, and 10% glycerol

An advantage to this method is the ability to freeze large amounts of the cell suspension for use when needed. The cells are grown and pelleted as in the dounce method. Cultures as large as 200 liters have been processed with great success as follows. One has the ability at this point to resuspend the cell pellet in a minimal buffer (such as buffer A) or as in a method outlined in Ansari and Schwer,[30] which allows one to resuspend with the buffer in which you will perform the subsequent manipulations (e.g., buffer AGK). A benefit in using a minimal buffer is the ability to later optimize other conditions, such as salt extracting the homogenate with different ions, use of different additives, and so on. When one is resuspending fermenter-sized cell masses, use of a minimal buffer is recommended.

For the mortar and pestle method, one should thoroughly clean and dry the ceramic materials to avoid breakage on addition of the liquid nitrogen. Resuspend the cells in 7.5 ml of buffer AGK per 2 liters of cell culture unless one desires to use a buffer of lower or higher complexity. The cell suspension is then slowly extruded through a syringe (without a needle) into liquid nitrogen, forming small pellets of frozen cells. At this stage, the cells can be frozen at −80° for later use or processed directly. To homogenize the cell suspension, the cell pellets are dispensed into an appropriately sized mortar prechilled with liquid nitrogen. It is important to keep the cells under liquid nitrogen throughout the procedure, but not have too much. Using the pestle, grind the cells until they are a fine powder, frequently adding liquid nitrogen. This can take up to 30 min of grinding to achieve an appropriate degree of homogenization. One may monitor cell breakage microscopically by diluting the sample in sterile water. Compare the ground sample to that of a pregrinding aliquot: broken cells appear duller and the presence of cellular debris indicates breakage. Although this method is adequate for preparing splicing extract, the proportion of broken cells is smaller than that realized using the dounce method. After the grinding is completed, thaw the cells at 4° in a small beaker containing a stir bar. If one has used the AGK buffer, stir the cells gently for 30 min after they have completely melted. Others have reported thawing rapidly at 30°. Although we prefer to avoid temperatures greater than 4°, it is not clear that this affects the activity of the extracts. If using a minimal buffer, add the appropriate components (e.g., KCl to 200 mM see dounce method) and continue stirring for 30 min. From this point on, low- and high-speed spins and dialysis are performed as in the dounce procedure.

[30] A. Ansari and B. Schwer, *EMBO J.* **14**, 4001 (1995).

To homogenize in a blender, one must choose the appropriately sized blender cup. A 1-liter blender can homogenize up to 150 g of cell suspension. A 2-liter blender can homogenize up to 350 g of cell suspension. Measure the cell mass in grams. Resuspend the cell mass thoroughly in 0.5–1.0 ml per gram of cell mass of the desired buffer. Extrude into liquid nitrogen as described earlier. Cells have been successfully frozen for up to 2 years at $-80°$ with no adverse effects. An appropriately sized portion of frozen cells is poured into a chilled (generally in a cold room) blender of the appropriate size. One should have a substantial amount of liquid nitrogen on hand. Blending 350 g of cell suspension will require ∼6 liters of liquid nitrogen. One should be wary of two important parameters in this procedure. The first is the mess that may be made by this method. At the beginning of the blending, some yeast dust can be released if the N_2 is not added quickly enough. The second is to become familiar with the blender: speed, capacity, and length of the blending are all critical. Too much N_2 added at once can result in the ejection of material out the top of the blender. Once the blending has begun, continue to add more liquid nitrogen, maintaining a thick custard-like consistency. Fifteen minutes of blending at the highest speed is generally sufficient. After blending, the homogenate can be placed at $-80°$ until needed. The processing of this homogenized material is as in the mortar and pestle method after the grinding step, adjusting the volumes accordingly if larger amounts of cells have been homogenized. The activity of splicing extracts can vary greatly from preparation to preparation.

In Vitro Assays for Yeast Pre-mRNA Splicing

Preparation of Pre-mRNA Precursor

The preparation of an extract capable of performing the yeast pre-mRNA splicing reaction was first demonstrated by Lin *et al.*[9] Using a synthetic pre-mRNA derived from a fragment of the *ACT1* or *CYH2* gene, it was shown that the RNA species outlined in Fig. 1B were resolvable on a denaturing polyacrylamide gel. Others have used constructs derived from other genes (mainly ribosomal protein genes), but for simplicity, the *ACT1* gene fragment will be addressed here.

RNA loading buffer: 8.3 M urea, 0.05% bromphenol blue, and 0.05% xylene cyanol
RNA elution buffer: 0.3 M sodium acetate, pH 5.3, 1 mM EDTA, and 0.1% SDS

To prepare the synthetic precursor, the purified restriction enzyme-digested plasmid (here *Eco*RI-digested pSPACT) is transcribed with the appropriate RNA polymerase (here SP6 RNA polymerase). The resulting RNA species is purified on a denaturing polyacrylamide gel to ensure its homogeneity, as well as to purify the RNA from contaminating DNA and ribonucleotides.

1 μg EcoRI-digested pSPACT
2 μl 10× SP6 reaction buffer
2 μl each ATP, GTP, and CTP (5 mM)
1 μl UTP (1 mM)
5 μl [^{32}P]UTP 3000 Ci/mmol
20 U SP6 RNA polymerase
Optional: 20 U RNasin (Promega, Madison, WI)
Doubly distilled H$_2$O to 20 μl

Incubate at 37° for 2 hr. Add 1 μl RNase-free DNase and incubate for 15 min at 37°. Add 80 μl RNase-free doubly distilled H$_2$O and 100 μl phenol. Vortex well. Spin in a microcentrifuge at room temperature for 5 min at 14,000 rpm. Extract aqueous phase with 100 μl chloroform and spin again. To the aqueous phase, add 10 μl 3 M sodium acetate and 250 μl ice-cold ethanol. Vortex well and store at −20° or −80° for at least 1 hr. Some investigators use a carrier such as tRNA or glycogen to aid in the precipitation of nucleic acids, although under the transcription conditions outlined earlier, we have not found this necessary. Spin in a chilled microcentrifuge at 14,000 rpm for 10 min. Remove supernatant and wash RNA pellet twice with 80% (v/v) ethanol. Dry sample briefly in a Speed-Vac. Resuspend RNA pellet in 20 μl of RNA loading buffer. Heat to 65° for 5 min and electrophorese on a prerun (250 V for 30 min) 5% (w/w) polyacrylamide gel (19:1 acrylamide: bisacrylamide) made to 8.3 M urea, 1× TBE (urea concentrations reported in the literature range from 6 to 8.3 M). An 8 cm × 8 cm × 1 mm minigel will suffice for this procedure. Electrophorese the sample until the xylene cyanol dye has just exited the gel. Expose the gel to film, taking care to ensure orientation and location of the well of interest. Align the gel with the autoradiograph and carefully excise the gel band and dice into small pieces. Add 300–500 μl RNA elution buffer to the gel material and rock/shake/nutate for 1 hr at room temperature. If RNA degradation due to RNases is a problem, one may include an equal volume of phenol in the extraction step. Carefully gather the contents of the closed microcentrifuge tube into the cap side. Slice the bottom of the microcentrifuge tube off to allow the material to exit efficiently. Place the tube right side up in a Quik-Sep polypropylene column (QS-P, Isolab Inc. Akron, OH) that has been placed into a Falcon 2059 tube. Centrifuge the contents through the separator at top speed in a clinical centrifuge for 5 min. Collect the flow through and phenol extract twice, chloroform extract once, and ethanol precipitate with 2.5 volumes 95% (v/v) ethanol. Wash the RNA pellet with 80% ethanol and resuspend the dried pellet in 50 μl RNase-free water. Measure the radioactivity of 1 μl by Cerenkov counting.

In Vitro Pre-mRNA Splicing Assay

Once the pre-mRNA precursor and the extract are prepared, it is now possible to perform an *in vitro* splicing reaction. This protocol is a slightly modified version of the one developed by Lin *et al.*[9]

5× SpB: 300 mM KH$_2$PO$_4$, pH 7.0, 10 mM ATP, pH 7.0, 16 mM MgCl$_2$, 15% polyethylene glycol (PEG) 8000, and 5 mM spermidine
PK buffer: 20 mM Tris, pH 8.0, 50 mM EDTA, pH 8.0, 1% SDS, and 1 mg/ml proteinase K
RNA extraction buffer: 50 mM sodium acetate, pH 5.3, 1 mM EDTA, 0.1% SDS, and 30 μg/ml tRNA

A 5-μl splicing reaction is set up as follows and can be adjusted in volume accordingly.

1 μl 5× SpB
2 μl splicing extract
1 μl (2000–20,000 cpm/μl) ^{32}P-labeled pre-mRNA (2–20 fmol cold pre-mRNA)
1 μl doubly distilled H$_2$O

Mix thoroughly with a pipette and incubate at 23° for 30 min; the temperature and length of incubation may vary depending on the experimental conditions. Add 1 μl PK buffer and incubate at 40° for 20 min. Add 200 μl RNA extraction buffer and 200 μl phenol (pH 5.3) : chloroform : isoamyl alcohol (25 : 24 : 1). Vortex thoroughly (1–3 min) and spin in a microcentrifuge for 5 min at top speed. Repeat extraction of the aqueous phase with an equal volume of phenol. Chloroform extract the aqueous phase. To the aqueous phase, add 26 μl 3 M sodium acetate and 775 μl 95% ethanol. Vortex vigorously and incubate at −20° overnight. Spin for 10 min in a microcentrifuge at 4°, wash the pellet with 80% ethanol, and dry briefly in a Speed-Vac. Resuspend in 10 μl RNA loading buffer, heat sample at 65° for 5 min and load into an appropriately sized 7% polyacrylamide gel (29 : 1), 1× TBE, 8.3 M urea. Generally, a 10 cm × 15 cm × 0.4 mm gel is used for this purpose. Preelectrophorese the polyacrylamide gel for 30 min at 300 V. Run samples until the xylene cyanol dye reaches the bottom. Expose gel to film or phosphorimager screen overnight. Depending on the strength of the signal, it may take more or less time to see the products of the splicing reaction.

The results of a wild-type yeast *in vitro* splicing reaction are shown in Fig. 2. Listed are the pre-mRNA (starting material), the product mRNA, the lariat intermediate, the lariat intron, and the exon 1 fragment. Even though the lariat intermediate and the lariat are shorter in length than the precursor, they migrate more slowly in a polyacrylamide gel due to the presence of the branched nucleotide structure. On the right-hand side of Fig. 2, the commonly used schematics for the different products are presented.

Substantial precursor degradation can result from a number of causes. Most commonly, the precursor RNA is degraded during the purification step. Including RNase inhibitors during transcription, phenol in the gel extraction, and/or using DEPC-treated water for RNA resuspension can often fix this problem. If one

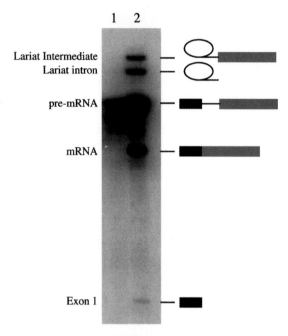

FIG. 2. Yeast *in vitro* pre-mRNA splicing of the *ACT1* gene fragment. Lane 1 is the precursor pre-mRNA mock treated in the absence of extract. Lane 2 is *ACT1* pre-mRNA, which has undergone an *in vitro* pre-mRNA splicing reaction as described in the text. The names of the RNA species of this reaction are presented on the left-hand side. Typical schematic representations of the RNA species are presented on the right-hand side.

determines that the RNA is intact on addition to the extract, there may be problems with the extract that are difficult to diagnose. Often, remaking the extract and taking care to avoid pitfalls listed (see earlier discussion) are the only ways to alleviate this problem. Adding RNase inhibitors to the splicing reaction generally has no effect on the integrity of the pre-mRNA.

In Vivo Yeast Pre-mRNA Splicing Assay

Occasionally it is necessary to determine the pre-mRNA splicing activity of a strain under a series of conditions such as varying the growth temperature or induction or repression of genes. One may also wish to assay pre-mRNA splicing under conditions that change over the course of time. There are two common ways to address the splicing of pre-mRNA *in vivo*. The first to be used was a standard Northern blotting procedure using total RNA and probing for a mRNA/pre-mRNA to monitor for the presence of intron.[12] This procedure is fairly time-consuming and can often be technically quirky. A less labor-intensive and highly sensitive method

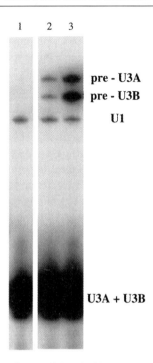

FIG. 3. Assay for the splicing of yeast U3A and U3B snoRNAs *in vivo*. Samples were processed as described in the text. Lane 1 represents a wild-type culture. Lane 2 is a sample derived from a strain grown at 30°, which has the *SNU66* gene deleted. Lane 3 is a sample derived from the *SNU66* deletion strain grown at the nonpermissive temperature of 16°. A splicing defect is shown in that pre-U3A- and pre-U3B-sized primer extension products accumulate in both lane 2 and lane 3. The U1 band is an internal loading control to assure that samples have been processed equivalently.

is to use primer extension analysis. A commonly assayed RNA in *S. cerevisiae* is the U3 small nucleolar RNA (snoRNA), which contains a pre-mRNA-like intron[31] and is very stable, thus making it easier to assay than less stable pre-mRNAs. There are two genes for this snoRNA termed U3A and U3B (also *snR17A* and *snR17B*). The mature snoRNAs from these two genes are identical in size; however, the products of the following reaction allow for differentiation of the unspliced message from the two gene products. In Fig. 3, the products of a primer extension reaction are shown. In a wild-type strain, the mature U3 is the only species present (lane 1). On removing the gene for a nonessential protein involved in splicing, a pre-mRNA splicing defect is detected (lane 2) and is further exacerbated by growth at lower temperatures, as evidenced by the increased accumulation of the pre-U3A and pre-U3B size primer extension products (lane 3). The U1 band is an optional

[31] E. Myslinski, V. Ségault, and C. Branlant, *Science* **247**, 1213 (1990).

internal loading control using an oligonucleotide designed to produce a primer extension product distinguishable from the experimental products. One drawback to this method is the difficulty in resolving a block to the second step of splicing. In the *in vitro* splicing assay, accumulation of the lariat intermediate and first exon with no product or lariat indicates a block in the second step of pre-mRNA splicing. In *S. cerevisiae*, the first exon is generally very short and is often similar in size to the distance between the 3' splice site and the branchpoint adenosine (S.W. Stevens, unpublished observations, 2000). This leads to a primer extension product too similar to the primer extension product of the precursor to resolve these two species by polyacrylamide gel electrophoresis. Choosing several highly expressed pre-mRNAs with these primer extension product sizes calculated to be resolvable did not allow for the reliable detection of a second step splicing block (S.W. Stevens, unpublished observations, 2000) using a strain with a temperature-sensitive mutation in a second step splicing factor. This indicates that the *in vitro* splicing assay is often more appropriate for assaying second step splicing blocks.

The method described here is a hybrid variation of previously published methods.[32,33] To perform an *in vivo* splicing assay, one must first purify total RNA from cells grown to the specifications dictated by the experiment. Remember to also include appropriate controls, such as a wild-type culture. Cells are grown in the appropriate medium to an optical density (absorbance at 600 nm) of 1.0. Generally, a 10-ml culture will suffice for this application; the volumes given here should be adjusted accordingly if using larger or smaller volumes. Cells are pelleted by centrifugation ($5000g$ for 5 min at 4°), washed once with ice-cold RNase-free water, and repelleted. Resuspend cell pellet in 1 ml 50 mM sodium acetate, pH 5.3, 10 mM EDTA by rapid vortexing. Transfer to a fresh 15-ml polypropylene conical tube. Add 100 μl 10% SDS and 1.2 ml phenol (equilibrated to pH 5.3) preheated to 65°. Vortex vigorously for 30 sec and place in a 65° water bath for 1.5 min. Alternately, if a shaking water bath is available, vigorous shaking during this step will increase the recovery of full-length RNA. Cycle through the vortexing and 65° incubation steps three times. Chill the sample rapidly to room temperature by shaking in an ice water bath. Centrifuge the tube(s) for 10 min at $3000g$ at room temperature in a clinical centrifuge. Spinning at lower temperatures will result in difficulty resolving the aqueous and organic phases. Repeat the extraction of the aqueous phase in a new tube with an equal volume of phenol at 65° as just described until there is no longer an interphase of cell debris. Extract the aqueous phase once with an equal volume of chloroform. Transfer the aqueous phase in 300-μl volumes to microcentrifuge tubes. Add 30 μl 3 M sodium acetate and 900 μl ice-cold ethanol. Mix well and store at $-20°$ overnight. Spin at 4° for 10 min at top speed in a microcentrifuge and wash once with 500 μl 80% ethanol.

---

[32] M. I. Zavanelli and M. Ares, *Genes Dev.* **5**, 2521 (1991).
[33] J. A. Wise, *Methods Enzymol.* **194**, 405 (1991).

Dry briefly in a Speed-Vac and resuspend the total RNA in 50 μl RNase-free water. Calculate the concentration of RNA by measuring the optical density of a 1 : 50 or 1 : 100 dilution in RNase-free water by standard methods.[34]

5× RNA hybridization buffer: 1.5 M NaCl, 50 mM Tris–Cl, pH 7.5, and 10 mM EDTA, pH 8.0

1.25× reverse transcriptase buffer: 1.25 mM each dNTP, 12.5 mM DTT, 12.5 mM Tris–Cl, pH 8.0, and 7.5 mM MgCl$_2$

For analysis of the splicing of U3A and U3B snoRNAs, the oligonucleotide 5′-CCAAGTTGGATTCAGTGGCTC-3′ is used. The optional internal loading control using an oligonucleotide designed for the U1 snRNA utilizes 10 fmol of a DNA oligonucleotide with the sequence 5′-GAATGGAAACGTCAGCAAACAC-3′ (this may be mixed with the U3 oligonucleotide throughout the following experiment). One hundred femtomoles of the U3 oligonucleotide is labeled with [^{32}P]ATP for each experimental lane required.[34] Mix 1–5 μg total RNA (in no more than 3 μl), 1 μl 5× RNA hybridization buffer, and 1 μl ^{32}P-labeled oligonucleotide(s) in a microcentrifuge tube. Heat to 75° for 1 min. Chill rapidly in an ice–water bath. Add 25 μl 1.25× reverse transcriptase buffer and 1 μl (5–10 U) avian myeloblastosis virus (AMV), reverse transcriptase. Incubate the primer extension reaction for 30 min at 37°. Best results are achieved by degrading the RNA by adding 1 μl 2 N NaOH and incubating for 1 hr at 65°, but one may proceed without this treatment. Adjust the volume to 100 μl with doubly distilled H$_2$O and extract once with 100 μl phenol and once with 100 μl chloroform. Add 10 μl 3 M sodium acetate and 250 μl ice-cold ethanol. Store at $-20°$ for 1 hr to overnight. Pellet the primer extension products by spinning in a microcentrifuge for 10 min at 4°. Wash the pellets with 500 μl ice-cold 80% ethanol and dry briefly in a Speed-Vac. Resuspend in 5 μl RNA loading buffer. Heat for 5 min at 65° and load into an appropriately sized (15 cm × 15 cm × 0.75 mm) 7% polyacrylamide gel (19 : 1) 8.3 M urea in 1× TBE. Electrophorese samples until the xylene cyanol has migrated two-thirds the length of the gel. Dry the gel and expose to film or a phosphorimager screen for 1 hr to overnight.

In Vivo Splicing Reporter Assays

In addition to the primer extension analysis outlined earlier, one can assay for the splicing of a reporter pre-mRNA *in vivo*. The first such construct reported was that of a β-galactosidase fusion, which contained a pre-mRNA intron from the *RP51* gene.[35] This construct uses β-galactosidase activity to semiquantitatively

---

[34] J. Sambrook, E. F. Fritsch, and T. Maniatis, "Molecular Cloning: A Laboratory Manual." Cold Spring Harbor Laboratory Press, Plainview, NY, 1989.

[35] J. L. Teem and M. Rosbash, *Proc. Natl. Acad. Sci. U.S.A.* **80**, 4403 (1983).

measure the splicing activity of a particular strain. Variations of this construct were also designed for assaying the illegitimate export of unspliced introns.[36]

Another reporter system described uses an *ACT1* intron insertion into the *CUP1* gene of *S. cerevisiae*.[37] The *CUP1* gene confers resistance to copper concentrations that are lethal to *CUP1*-deficient strains. The resistance to copper ions is proportional to the expression of the *CUP1* gene. Measurable defects in pre-mRNA splicing are therefore assayed by determining the maximal copper concentration allowable for growth of the strain. Although these reporters are not as sensitive in detecting splicing defects as the primer extension method, they are alternative methods that allow for subsequent genetic selections *in vivo*.

Native Gel Analysis of Splicing Complexes

Early in the study of the splicing reaction, three groups identified several pre-mRNA splicing complexes, which were represented by bands migrating on specially formulated polyacrylamide gels, each with a slightly different composition.[38–40] We concentrate here on the two systems used to study pre-mRNA splicing complexes in yeast, each with its own advantages.

Stop buffer: 60 mM KH$_2$PO$_4$, pH 7.0, 3 mM MgCl$_2$, 3% PEG-8000, 8% glycerol, and 0.01% bromphenol blue

To prepare the splicing reactions to be analyzed by native polyacrylamide gel electrophoresis, standard splicing reaction conditions are used (see earlier discussion); however, variations can be performed such as using oligonucleotide-directed ablation of snRNAs.[41,42] Add an equal volume of stop buffer and 2 μg heparin/μl of splicing extract. Some protocols do not call for heparin; however, if you notice aggregation of material in the wells, you may wish to try heparin or tRNA at similar concentrations to eliminate this effect.

Native gel analysis of the early events in splicing has been described by Pikielny and colleagues.[39] This system has been optimized to observe the migration of a species termed the "commitment complex." This complex is the first association of splicing machinery with the pre-mRNA in the form of a U1 snRNP/pre-mRNA complex[43,44] and is stable to dilution with competitor pre-mRNA thus being

---

[36] P. Legrain and M. Rosbash, *Cell* **57,** 573 (1989).
[37] C. F. Lesser and C. Guthrie, *Genetics* **133,** 851 (1993).
[38] M. M. Konarska and P. A. Sharp, *Cell* **46,** 845 (1986).
[39] C. W. Pikielny, B. C. Rymond, and M. Rosbash, *Nature* **324,** 341 (1986).
[40] S. C. Cheng and J. Abelson, *Genes Dev.* **1,** 1014 (1987).
[41] P. Fabrizio, D. S. McPheeters, and J. Abelson, *Genes Dev.* **3,** 2137 (1989).
[42] D. S. McPheeters, P. Fabrizio, and J. Abelson, *Genes Dev.* **3,** 2124 (1989).
[43] S. W. Ruby and J. Abelson, *Science* **242,** 1028 (1988).
[44] B. Séraphin and M. Rosbash, *Cell* **59,** 349 (1989).

committed to the splicing reaction. The recipe for resolving commitment complex is as follows: 3% polyacrylamide (60:1), 0.5% agarose, 0.5× TBE, 1 mM EDTA.[39] This gel is run at 4° for up to 12 hr at 100 V in 0.5× TBE. A 20-cm-long, 0.5-mm-thick gel allows for sufficient resolution of these complexes. A gel recipe that allows better resolution of larger yeast splicing complexes such as the prespliceosome and the several species of spliceosome has been described by Cheng and Abelson[40] and is as follows: 4% polyacrylamide (80:1) in 0.5× TAE. The consistency of these types of gels is very glue-like and can be difficult to handle. Best results are achieved when using plates that have been soaked in 1 N NaOH, washed thoroughly in distilled water, and one plate well silanized. This gel is run at 25 V/cm for 4 hr in 0.5× TAE buffer. After the gel has been run, the plates are separated and the gel can be transferred carefully to Whatman (Clifton, NJ) 3 MM paper or wrapped in Saran wrap directly on the glass plate for exposure to film or a phosphorimager screen. Although the procedure is technically challenging, one can probe for the presence of a protein or an RNA after transferring to an appropriate membrane for Western or Northern blotting. One should use an unlabeled pre-mRNA substrate in the splicing reaction for this purpose and keep in mind that the low percentage of acrylamide and the low amount of cross-linking agent in these gels result in a gel with substantially less integrity than the ones generally used for this purpose.

Figure 4 shows a gel that resolves prespliceosome and spliceosomal complexes. In lane 1 (Fig. 4), the prespliceosome, or B complex, accumulates due to the oligonucleotide ablation of U6 snRNA.[41] A wild-type splicing reaction is shown in lane 2 (Fig. 4) and contains spliceosome, or A complex, as well as B complex. Lane 3 (Fig. 4) has been treated with EDTA, which resolves two A complexes: A1 and A2. The names and composition of the complexes are nicely described by Cheng and Abelson.[40]

Glycerol Gradient Analysis of snRNPs and Spliceosomes

In addition to native gel electrophoresis, one can resolve snRNPs and splicing complexes in glycerol gradients of varying compositions. Using a radioactive pre-mRNA, it was determined by glycerol gradient sedimentation that the yeast spliceosome is a 40S complex.[15] It was later determined that individual snRNPs from yeast sediment at discrete locations under certain salt conditions: 18S for the U1 snRNP, 12S for the U2 snRNP, and 25S for the U4/U6·U5 snRNP.[45–47]

Pouring a consistent gradient is more an art than a science. To ensure the reproducible sedimentation of material, it is crucial that one maintain consistency in the

---

[45] P. Fabrizio, S. Esser, B. Kastner, and R. Lührmann, *Science* **264**, 261 (1994).
[46] S. W. Stevens and J. Abelson, *Proc. Natl. Acad. Sci. U.S.A.* **96**, 7226 (1999).
[47] A. Gottschalk, G. Neubauer, J. Banroques, M. Mann, R. Lührmann, and P. Fabrizio, *EMBO J.* **18**, 4535 (1999).

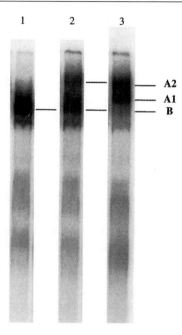

FIG. 4. Native gel analysis of yeast pre-mRNA splicing complexes. Samples prepared as described in the text were analyzed by native gel electrophoresis. Lane 1 shows accumulation of the prespliceosome or B complex. Lane 2 is a wild-type splicing reaction and shows spliceosome, or A complex as well as B complex. Lane 3 is a pre-mRNA splicing reaction treated with EDTA, which allows the resolution of two types of A complex: A1 and A2.

method. Although gradient formers can perform the job well, we have found that a layering technique gives the most reliable results. To analyze splicing complexes, a 10–40% gradient is most appropriate. To sediment splicing snRNPs, a 10–30% gradient works nicely. Making fresh material each time yields better results and can be made in 10- or 20-ml batches relatively quickly. Each solution is adjusted with the appropriate buffer (20 mM HEPES, pH 7.9), salt (50–250 mM monovalent ion; 1.5 mM MgCl$_2$), and additives such as NP-40 (to 0.01%), DTT (to 0.5 mM), and protease inhibitors (e.g., leupeptin, pepstatin, and PMSF). In a SW41 tube, calculate appropriate volumes for an 11-ml gradient (a four-step gradient will use 2.75 ml for each step, etc.). The gradient is layered *carefully* in four to six steps starting with the highest percentage of glycerol at the bottom and proceeding to the top with solutions of successively less glycerol. The tube is covered with plastic wrap and is set aside upright to diffuse at room temperature for 4 hr, at which time it is transferred to 4° for 1–3 hr. The sample being separated must be diluted or dialyzed to <10% glycerol (8% works nicely) before layering on top. For the highest resolution, it is best to not apply more than 0.5 ml of the sample, although up to 1 ml of the sample can be separated effectively. Although the centrifugation conditions will vary, sedimentation is generally carried out at 4°. For resolution of spliceosomal snRNPs (or other similarly sized complexes), centrifuge samples in a Beckman SW40 or SW41 rotor at 29,000 rpm for 24 hr. To resolve larger spliceosomal complexes, centrifuge at 20,500 rpm for 16 hr at 4°. Although it requires more time, endeavor to turn the centrifuge brake off as braking can decrease the resolution of the gradient. Gradient fractionation can be achieved in two ways, the first is the easier and requires an apparatus. One such device delivers the gradient material through the bottom of the tube by displacing volume at the top with an inert substance such as mineral oil. Another from Labconco uses a peristaltic pump and can both pour and fractionate gradients using one machine. Although these methods are rapid ways to fractionate, one runs the risk of contaminating fractions with material that may have pelleted at the bottom of the tube or from mixing in the tubing using the pump-type device. To avoid these pitfalls, one may also manually fractionate using an automatic pipette by removing 200- to 500-μl fractions from the top. With practice, a gradient can be reproducibly fractionated in this manner in about 5 min. Gradient fractions may be used to collect material for biochemical analyses or to analyze the RNA or protein components by extraction and silver staining or Northern or Western blotting the resulting material.

Analysis of Splicing Complexes by Immunoprecipitation

The assignment of a splicing protein as one that is snRNP associated can be achieved by analysis of the material that is immunoprecipitated using antibodies or epitope tags. This can also be used to determine the temporal association of a non-snRNP protein with the splicing machinery during the splicing reaction.

After generating antibodies or an epitope-tagged version of the protein, the extract is prepared by one of the aforementioned procedures. Adjust the salt and other additives to the appropriate levels, preferably by dialysis. Incubate 100–500 μl of splicing extract with 10–50 μl settled bed volume of affinity media that has been prewashed with a compatible buffer. Incubate by gently rocking or on a rotating wheel for 1–3 hr at 4°. Gently pellet the affinity matrix in a microcentrifuge at 5000 rpm for 1 min. Wash with an appropriate buffer 4× with 1 ml of a buffer similar to that used for immunoprecipitation.

If one wishes to analyze the coimmunoprecipitated RNA, resuspend the pelleted affinity matrix in 100 μl of 20 mM Tris, pH 8.0, 0.1% SDS, 100 μg/ml proteinase K. Incubate at 37° for 1 hr. Phenol : chloroform : isoamyl alcohol (25 : 24 : 1) extract this solution, including the beads. Remove the aqueous phase and chloroform extract again. To the aqueous phase, add 10 μl 3 M sodium acetate and 250 μl ice-cold ethanol. Store at −20° for 1 hr. Pellet the nucleic acid by microcentrifuging at top speed for 10 min at 4°. Wash pellet with 1 ml ice-cold 80%

ethanol. If you wish to radiolabel the resulting material with [^{32}P]pCp, protocols have been published previously in this series.[48] To analyze by Northern blotting, resuspend the nucleic acid in 10 μl RNA loading buffer and electrophorese in a 5% polyacrylamide gel (19 : 1) 8.3 M urea, 1× TBE until the bromphenol blue dye reaches the bottom. Transfer RNA to the appropriate support and probe for the RNA of interest using standard methods.[34] A method to assay for the presence of an RNA by primer extension has also been described.[49]

To analyze the proteins coimmunoprecipiated, incubate the washed pelleted affinity matrix in one bead volume 1× SDS loading buffer (50 mM Tris, pH 6.8, 100 mM DTT, 2% SDS, 0.1% bromphenol blue, 10% glycerol). Boil samples for 3 min. Spin at 14,000 rpm for 5 min at room temperature in a microcentrifuge. Load 10–30 μl of this solution on an appropriate (a 4–20% gradient gel resolves most proteins) SDS–polyacrylamide minigel and electrophorese at 25 mA until the bromphenol blue dye migrates to the bottom. Blot the gel onto an appropriate support and perform a Western blot using detection methods for the coimmunoprecipitating protein(s) of interest using standard methods.[34]

Affinity Purification Techniques

Three different groups using similar but technically unique methods have achieved the purification of spliceosomal snRNPs and splicing complexes. In all cases, two affinity steps were needed to achieve highly pure samples. The U1 snRNP was the first to be purified from yeast by the laboratory of Reinhardt Lührmann.[50,51] This method used a combination of α-TMG (see earlier discussion) affinity chromatography to isolate all of the TMG- containing snRNPs followed by Ni-NTA chromatography, taking advantage of a polyhistidine epitope engineered into an integral U1 snRNP protein.

The next snRNP to be purified was the U4/U6·U5 tri-snRNP by the authors[46] and concurrently by the Lührmann laboratory.[47] In Stevens and Abelson, a dual-purpose epitope was inserted at the carboxyl terminus of the SmD3 polypeptide. Because SmD3 is an integral component of U1, U2, and tri-snRNPs, it was an efficient method for their purification. The dual-purpose epitope is composed of a monoclonal antibody epitope preceded by a polyhistidine sequence.[52] Using a peptide-elutable antibody affinity matrix followed by Ni-NTA chromatography, the tri-snRNP was then separated from the other snRNPs by glycerol gradient

[48] T. E. England, A. G. Bruce, and O. C. Uhlenbeck, *Methods Enzymol.* **65**, 65 (1980).
[49] B. Séraphin, *EMBO J.* **14**, 2089 (1995).
[50] G. Neubauer, A. Gottschalk, P. Fabrizio, B. Séraphin, R. Lührmann, and M. Mann, *Proc. Natl. Acad. Sci. U.S.A.* **94**, 385 (1997).
[51] A. Gottschalk, J. Tang, O. Puig, J. Salgado, G. Neubauer, H. V. Colot, M. Mann, B. Séraphin, M. Rosbash, R. Lührmann, and P. Fabrizio, *RNA* **4**, 374 (1998).
[52] S. W. Stevens, *Methods Enzymol.* **318**, 385 (2000).

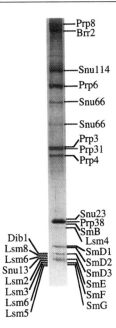

FIG. 5. Yeast U4/U6·U5 tri-snRNP purified by affinity chromatography and glycerol gradient sedimentation. The yeast tri-snRNP was purified as described in the text. The identity of the proteins was determined by mass spectrometry microsequencing.

ultracentrifugation (see earlier discussion). A demonstration of the affinity chromatography of the tri-snRNP is shown in Fig. 5. This preparation was derived from a CHP^{52}-tagged Brr2 protein. Proteins were identified by mass spectrometry and have been reported previously.[46,47]

The U2 snRNP was partially purified in the laboratory of Bertrand Séraphin[53] using the tandem affinity chromatography (TAP) technique.[54] This method uses the IgG-binding domain of the *Staphylococcus aureus* protein A as a first affinity step. Because there is not an efficient, nondestructive affinity elution step for this complex, a tobacco etch virus (TEV) protease cleavage site was engineered preceding the protein A module. After cleavage with TEV protease, the protease and other contaminants are removed by immobilized calmodulin chromatography in the presence of calcium ions by virtue of a calmodulin-binding peptide engineered upstream of the TEV cleavage site. Elution from this matrix is achieved by chelation of the calcium ions using EGTA.

[53] F. Caspary, A. Shevchenko, M. Wilm, and B. Séraphin, *EMBO J.* **18,** 3463 (1999).
[54] G. Rigaut, A. Shevchenko, B. Rutz, M. Wilm, M. Mann, and B. Séraphin, *Nature Biotechnol.* **17,** 1030 (1999).

Each of the aforementioned techniques has advantages and disadvantages. For example, α-TMG and polyoma chromatography requires that one has the monoclonal antibody hybridoma or purchase the relatively expensive resin commercially. Additionally, α-TMG chromatography is only applicable to complexes containing that epitope naturally. The TAP protocol is more economical but has potential drawbacks if a protein in the desired complex is cleaved by TEV protease cleavage or if the protein/complex is sensitive to the removal of divalent ions. These affinity techniques can be applied to other non-snRNP complexes. The sequenced genome is of great utility in this matter when coupled with the power of mass spectrometry peptide sequencing. A variety of methods and different machines are available as well as commercial and academic resources. Descriptions of the mass spectrometry sequencing methods have been published elsewhere.[55-57]

The exciting next stage of yeast pre-mRNA splicing research lies in the structural determination of the components of the spliceosome. Using the methods outlined here, it is now possible to perform cryoelectron microscopy of yeast snRNPs and various spliceosome assembly intermediates with the ultimate goal of atomic scale information by X-ray crystallography.

Acknowledgments

We acknowledge the efforts of the pre-mRNA splicing community for developing these methods. We are also grateful to Dan Ryan for the contribution of Fig. 3 and to Christine Guthrie for strains. Work in the Abelson laboratory is supported by a grant from the NIH (GM32627). S.W.S. was supported by a fellowship from the American Cancer Society (PF 4447).

[55] D. C. Stahl, K. M. Swiderek, M. T. Davis, and T. D. Lee, *J. Am. Soc. Mass Spectrom.* **7**, 532 (1996).
[56] M. T. Davis and T. D. Lee, *J. Am. Soc. Mass Spectrom.* **8**, 1059 (1997).
[57] A. J. Link, J. Eng, D. M. Schieltz, E. Carmack, G. J. Mize, D. R. Morris, B. M. Garvik, and J. R. Yates, *Nature Biotechnol.* **17**, 676 (1999).

[13] Analysis and Reconstitution of Translation Initiation *in Vitro*

By KATSURA ASANO, LON PHAN, THANUJA KRISHNAMOORTHY, GRAHAM D. PAVITT, EDITH GOMEZ, ERNEST M. HANNIG, JOSEPH NIKA, THOMAS F. DONAHUE, HAN-KUEI HUANG, and ALAN G. HINNEBUSCH

Introduction

Translation initiation is the rate-limiting step in protein biosynthesis, and alteration of the initiation factors by covalent modification, such as phosphorylation, or by mutation can have dramatic effects on the rate of protein synthesis. The budding yeast *Saccharomyces cerevisiae* provides an ideal system to investigate structure–function relationships for conserved eukaryotic translation initiation factors (eIF) by combining powerful genetic tools with biochemical analysis of cell-free extracts and purified factors. Earlier biochemical studies with mammalian cell extracts revealed the following model for the steps involved in the eukaryotic initiation pathway. First, the (80S) ribosome is dissociated into 40S and 60S subunits, and free 40S subunits are stabilized by interaction with the multisubunit factor eIF3. The initiator tRNA^{Met} (Met-tRNA$_i^{Met}$) in a ternary complex (TC) with the three-subunit factor eIF2 and GTP, then binds to the 40S subunit, producing the 43S preinitiation complex. mRNA bound to eIF4F (m^{7}G-cap-binding complex) then joins the assembly to produce the 48S complex. On recognition of an AUG triplet by Met-tRNA$_i^{Met}$, in a process known as scanning, the eIF5 stimulates hydrolysis of GTP bound to eIF2, triggering ejection of all eIFs to produce the 40S initiation complex. The 60S ribosomal subunit joins with the 40S preinitiation complex, and the resulting 80S initiation complex can enter the elongation phase of translation. Because only the GTP-bound form of eIF2 binds Met-tRNA$_i^{Met}$ and because the eIF2 ejected from the ribosome is stably bound to GDP, the GDP must be exchanged with GTP by a guanine nucleotide exchange factor eIF2B. There is evidence that eIF3 stimulates the recruitment of TC and the eIF4F–mRNA complex to the 40S ribosome and also has a role in positioning eIF5 with respect to the TC in the 48S complex. The single-subunit factors eIF1 and eIF1A have been implicated in TC binding, scanning, and AUG recognition (for a review, see Refs. 1 and 2).

---

[1] J. W. B. Hershey and W. C. Merrick, *in* "Translational Control of Gene Expression" (N. Sonenberg, J. W. B. Hershey, and M. B. Mathews, eds.), p. 33. Cold Spring Harbor Laboratory Press, Cold Spring Harbor, NY, 2000.

[2] A. G. Hinnebusch, *in* "Translational Control of Gene Expression" (N. Sonenberg, J. W. B. Hershey, and M. B. Mathews, eds.), p. 185. Cold Spring Harbor Laboratory Press, Cold Spring Harbor, NY, 2000.

Some of the individual reactions in the initiation pathway can be assayed in yeast whole cell extracts (WCEs) or by using the relevant purified factors in model assays. These assays were originally developed for mammalian eIFs and have been adapted for use with budding yeast. Assays in yeast WCEs frequently make use of mutant strains with a temperature-sensitive lesion in the factor of interest. The mutant extract is heat treated and analyzed for biochemical defects that can be complemented by addition of the purified wild-type factor. In other cases, the wild-type factor is depleted *in vivo* by repressing its synthesis and WCEs lacking the factor are examined. This chapter describes assays using both WCEs and purified eIFs. For the latter, we focus on formation of the TC, the recycling of eIF2-GDP to eIF2-GTP by eIF2B, and stimulation of GTP hydrolysis in the TC by eIF5.

Assays Using Cell-Free Extracts

Yeast WCEs have been used to assay different activities of the eIFs.[3-8] Here we describe the use of a single extract to assay the overall rate of protein synthesis with a luciferase reporter mRNA and the ability of the endogenous eIFs to deliver Met-tRNA$_i^{Met}$ and mRNA to the 40S ribosome.

Preparation of Whole Cell Extracts

Cell extracts with high activity have been prepared from several *S. cerevisiae* strains by a gentle disruption of cells.[6] Low molecular weight components, including amino acids, nucleotides, and inhibitors, are removed from the extracts by gel filtration using a Sephadex G-25 column. Fractions in the void volume contain active protein factors, ribosomes, and tRNAs competent for *in vitro* protein synthesis on addition of amino acids, with either endogenous or exogenous mRNAs. Strains with a temperature-sensitive mutation in a given factor should be grown at the permissive temperature to avoid secondary effects of the mutation on eIF expression *in vivo*.

Materials

Buffers. These include buffer WCE-A [30 mM HEPES (pH 7.4), 100 mM potassium acetate, 2 mM magnesium acetate, 2 mM dithiothreitol (DTT)], with fresh DTT added each time, and buffers WCE-AM, WCE-AP, or WCE-AMP,

---

[3] P. J. Hanic-Joyce, R. A. Singer, and G. C. Johnston, *J. Biol. Chem.* **262,** 2845 (1987).
[4] N. Iizuka, L. Najita, A. Franzusoff, and P. Sarnow, *Mol. Cell Biol.* **14,** 7322 (1994).
[5] P. Danaie, B. Wittmer, M. Altmann, and H. Trachsel, *J. Biol. Chem.* **270,** 4288 (1995).
[6] S. Z. Tarun and A. B. Sachs, *Genes Dev.* **9,** 2997 (1995).
[7] L. Phan, X. Zhang, K. Asano, J. Anderson, H. P. Vornlocher, J. R. Greenberg, J. Qin, and A. G. Hinnebusch, *Mol. Cell Biol.* **18,** 4935 (1998).
[8] S. K. Choi, J. H. Lee, W. L. Zoll, W. C. Merrick, and T. E. Dever, *Science* **280,** 1757 (1998).

which are identical to WCE-A but contain 8.5% w/v mannitol, 0.5 mM phenylmethylsulfonyl fluoride (PMSF), or both, respectively.

Sephadex G-25 Column. Swell a 8.5-g Sephadex G-25 superfine gel (Amersham Pharmacia Biotech, Piscataway, NJ) overnight in buffer WCE-A and decant the buffer to achieve a ∼70% slurry. Swirl and pour into a 33 × 1.5-cm column in the cold room. If bubbles occur in the resin, invert the column and wait for the resin to settle again. There should be as ∼10-cm space above the resin for sample loading. For long-term storage, the resin is equilibrated with buffer WCE-A containing 0.03% w/v sodium azide.

Procedures

1. Grow cells in 2 liters of yeast extract-peptone-dextrose (YPD) from Sorvall (Newtown, CT) or synthetic complete (SC) medium[9] in a 6-liter flask to an OD_{600} of 3–5 and harvest in six 400-ml centrifuge bottles by centrifugation at 5000 rpm in a GSA rotor (4000g) for 5 min at 4°.

2. Resuspend all the cells in the six bottles into 15 ml of buffer WCE-AM and combine them in one of the six bottles. The total cell wet weight should be >5 g. Wash the cells three more times using 10 ml buffer WCE-AM for each wash.

3. Resuspend cells in 1.5 volumes of buffer WCE-AMP and transfer to a 50-ml disposable plastic conical tube. Add six times the cell weight of cold acid-washed glass beads (0.5 mm diameter from BioSpec Products, Bartlesville, OK).

4. Break the cells by shaking the tube manually in the cold room. Hold the tube tight with its long axis perpendicular to your forearm. Shake vigorously through an arc of ∼2 ft, at two strokes/sec for 1 min and cool for 1 min on ice. Repeat this shaking/cooling protocol four more times.

5. With a Pipetman, remove as much cell suspension as possible and transfer to a 40-ml centrifuge tube. Centrifuge for 6 min at 18,000 rpm in a Sorvall SS-34 rotor (38,000g), at 4°.

6. Remove the supernatant and centrifuge again under the same conditions as in step 5. Carefully remove the supernatant, avoiding lipids at the top and cell debris at the bottom.

7. Load up to 3.75 ml of sample on a Sephadex G-25 column preequilibrated with 50 ml of buffer WCE-AP. After the sample enters the resin, carefully load 15 ml buffer WCE-AP without disturbing the top of the resin. Allow the buffer to flow under gravity at a rate of ∼2 drops/sec and start collecting 0.5-ml fractions in microfuge tubes when the yellow sample reaches the bottom of the column. Peak fractions will appear slightly opaque.

8. Measure the OD_{260} of a 100-fold dilution of each fraction and pool fractions of $OD_{260} = 90$ or higher. Quickly freeze 100-μl aliquots in liquid nitrogen and store them under liquid nitrogen.

[9] F. Sherman, *Methods Enzymol.* **194**, 13 (1991).

Note. Inexperienced researchers tend to make dilute extracts in step 6 and lose track of the sample during the column fractionation in step 7. Extracts of final OD_{260} <90 are inactive in assays described later. To avoid this, it is important to start with >5 g of cells.

Preparation of ^{3}H-Labeled Met-tRNA$_i^{Met}$

^{3}H-Labeled Met-tRNA$_i^{Met}$ is synthesized using purified *Escherichia coli* methionyl-tRNA synthetase (MetRS) from [^{3}H]methionine and the tRNA$_i^{Met}$ that is present in commercially available crude yeast tRNA. Note that *E. coli* MetRS will aminoacylate the initiator but not elongator tRNA^{Met} in total yeast tRNA.[10] The [^{3}H]Met-tRNA$_i^{Met}$ is separated from unincorporated [^{3}H]methionine by gel filtration through a Sephadex G-25 column.

Purification of MetRS

Plasmid. pQE60MRS5 encodes *E. coli* hexahistidine (His$_6$)-tagged MetRS with the size of ~70 kDa.[11]

Buffers. These include buffer MRS-1 [20 mM Tris–HCl (pH 8.0), 100 mM NaCl], 2 mg/ml lysozyme, 100 mM PMSF, buffers MRS-2 and MRS-3 (as buffer 1 but contains 10 and 100 mM imidazole, respectively), and storage buffer [20 mM imidazole hydrochloride (pH 7.5), 150 mM KCl, 50% (v/v) glycerol, 10 mM 2-mercaptoethanol].

Procedures. XL1-Blue (Stratagene, La Jolla, CA) carrying pQE60MRS5 is grown at 37° in 500 ml 2×YT medium supplemented with 100 μg/ml ampicillin and 15 μg/ml tetracycline to $OD_{600} = 0.8$, and expression of the recombinant protein is induced is with isopropylthiogalactoside (IPTG, 0.5 mM, final) for 2 hr and purified as described.[12] Cells are harvested by centrifugation and resuspended in the fresh mixture of 20 ml buffer MRS-1, 1 ml lysozyme, and 0.5 ml PMSF. The suspension is incubated for 20 min and subjected to a single freeze–thaw cycle, with rapid freezing in a dry ice–ethanol bath followed by rapid thawing. Two hundred and fifty microliters of DNase I (2.5 mg/ml in 100 mM MgCl$_2$) is added and incubated for a further 20 min. The lysate is clarified by ultracentrifugation at 100,000g for 1 hr at 4° and applied onto a column of Talon Sepharose Co^{2+} (5 ml bed volume) (Clontech, Palo Alto, CA), which has been preequilibrated by washing twice with 50 ml of buffer MRS-1. The recommended flow rate is between 0.5 and 1 ml/min. After the lysate enters the resin, the column is washed sequentially with 50 ml buffer MRS-1 and 50 ml buffer MRS-2. The bound protein is eluted with 10 ml of buffer MRS-3 into 500-μl fractions. Protein concentration

[10] U. L. RajBhandary and H. P. Ghosh, *J. Biol. Chem.* **244**, 1104 (1969).
[11] V. Ramesh, S. Gite, and U. L. RajBhandary, *Biochemistry* **37**, 15925 (1998).
[12] V. Ramesh, S. Gite, Y. Li, and U. L. RajBhandary, *Proc. Natl. Acad. Sci. U.S.A.* **94**, 13524 (1997).

is determined by the Bradford assay, and peak fractions are combined, dialyzed overnight against storage buffer, and stored in aliquots at $-70°$ after rapid freezing in liquid nitrogen.

tRNA and Reaction Components

The starting material is 1000 U (52.6 mg) of baker's yeast tRNA (Sigma, St. Louis, MO type X) dissolved in 8 ml water (final $A_{260} = 125$) and stored at $-20°$. Reaction components include 0.5 M creatine phosphate (CP), 1000 U/ml (3.2 mg/ml) creatine phosphokinase (CPK), 1000 U/ml myokinase, 100 μM methionine, 40 mM ATP (pH 7.0; adjusted with NaOH), reaction buffer [200 mM potassium phosphate buffer (pH 7.0), 22 mM magnesium acetate, 2 mM DTT], and [^{3}H]methionine (70–80 mCi/μmol, 5 mCi/ml).

Pilot Assay of [^{3}H]Met-tRNA$_i^{Met}$ Synthesis by Purified MetRS

Buffers. These include TCA-1 (10% TCA, 1.5 g/liter methionine, 6 mM 2-mercaptoethanol), TCA-2 (10% TCA, 6 mM 2-mercaptoethanol), and Hokin's reagent (94% ethanol, 6% acetic acid, 0.08% 10 N NaOH).

Procedure
1. Using different amounts of purified MetRS ranging from 0.5 to 1.5 μg, compose a series of 100-μl reactions each containing 5 μl CP, 2.5 μl CPK, 2.5 μl myokinase, 37.5 μl reaction buffer, 4.2 μl 40 mM ATP, 1.66 μl [^{3}H]methionine (5 mCi/ml), 6.66 μl of 100 μM unlabeled methionine, and 20 μl baker's yeast tRNA.
2. Incubate each reaction mixture at 37° for 1 hr, withdrawing 10-μl aliquots at regular time intervals and spotting them on Whatman No. 1 filter paper disks.
3. Immerse the filters sequentially for 10 min each in 200 ml of ice-cold TCA-1, twice in 150 ml of ice-cold TCA-2, and once in Hokin's reagent, rinse in ether, and air dry.
4. Count the dried filters in liquid scintillation fluid.
5. Analyze data to determine the amount of MetRS that will be required to aminoacylate all of the tRNA$_i^{Met}$ present in 8 ml of the tRNA solution.

Preparative Scale Synthesis and Partial Purification of [^{3}H]Met-tRNA$_i^{Met}$

Buffers and Size Standards. Buffers include cacodylate buffer (0.1 mM cacodylic acid adjusted to pH 5.5 with HCl), 2 M potassium acetate adjusted to pH 5.5 with glacial acetic acid, and storage buffer [10 mM sodium acetate, 1 mM DTT (pH 5.0)]. Standards are Blue dextran (4 mg/ml, blue) and vitamin B_{12} (2.5 mg/ml, pink) dissolved in cacodylate buffer.

Sephadex G-25 Column. On day 1, allow ~100 g of Sephadex G-25 to swell overnight in cacodylate buffer to produce an ~500 ml bed volume. On day 2, degas the resin, autoclave, and allow to cool. Pack the resin in a 76 × 2.6-cm column in the cold room until the bed volume reaches 400 ml. Equilibrate the column with 3–5 column volumes of cacodylate buffer at a flow rate of 1.25–2.5 ml/min. On day 3, calibrate the column by loading a mixture of blue dextran and vitamin B_{12} and eluting with cacodylate buffer while collecting 5- to 6-ml fractions immediately after loading. Blue dextran should elute between fractions 28 and 37 in the void volume; vitamin B_{12} should elute between fractions 53 and 68. After vitamin B_{12} comes off completely, wash the column with 1 column volume of cacodylate buffer.

Procedure

1. A cocktail of 9.33 ml reaction buffer, 1.04 ml 40 mM ATP, 0.9 ml CP, 0.227 ml CPK, 0.340 ml of myokinase, and 2.0 ml of [^{3}H]methionine (5 mCi/ml) is mixed in a 50-ml conical tube and incubated at 37° for 5 min. One thousand units of baker's yeast tRNA (8 ml) and the appropriate amount of purified MetRS are added to the cocktail and incubated for 45 min at 37°. Ten microliters of the cocktail is withdrawn at appropriate times and spotted on filter disks for scintillation counting.

2. Mix the remainder of the cocktail with 0.226 ml of 2 M potassium acetate and extract with 25 ml of water-saturated phenol for 20 min at room temperature. Centrifuge for 15 min at 3000 rpm in a Beckman J-6 type centrifuge (1000g) at 4°. Remove the top aqueous layer and transfer to a 50-ml conical tube. Add 4–5 ml of water to the lower phenol layer, mix, and centrifuge as described earlier. Remove the top aqueous layer and combine it with the aqueous layer from the first extraction.

3. Apply the crude [^{3}H]Met-tRNA$_i^{Met}$ solution to the equilibrated Sephadex G-25 column. Once the sample enters the resin, connect the column to a continuous supply of cacodylate buffer at a flow rate of 1.25–2.5 ml/min and collect 5-ml fractions. Mix 5 μl from alternating fractions with an appropriate scintillation fluid for aqueous samples and count in a scintillation counter. The first peak of radioactivity will contain the [^{3}H]Met-tRNA$_i^{Met}$ and the second will contain unincorporated [^{3}H]methionine.

4. Fractions containing the [^{3}H]Met-tRNA$_i^{Met}$ are pooled in a lyophilization flask, shell frozen (in liquid nitrogen), and lyophilized for ~2 days to completion. Alternatively, the sample can be placed in a 250-ml Corning flask, covered with Parafilm containing small punctures, shell frozen, and lyophilized under vacuum (e.g., in a Speed-Vac with the tube holder removed). The lyophilized sample is resuspended in 3–5 ml of storage buffer, and a 2-μl aliquot is counted in liquid scintillation fluid. Aliquots of 25 μl are stored at $-80°$.

5. The Sephadex G-25 column is washed thoroughly to remove all radioactivity.

Preparation of mRNAs

The mRNAs used in the assays described later are synthesized with the Ampli Scribe transcription kit (Epicentre Technologies, Madison, WI) or the mMessage mMachine kit (Ambion, Valencia, CA), followed by purification with the RNAeasy spin column (Qiagen, Austin, TX), all as recommended by the manufacturers.

Luciferase mRNA. m^7G-capped luciferase mRNA (capped *LUC* mRNA), used to measure the overall rate of protein synthesis in WCEs, is synthesized by T7 RNA polymerase using plasmid pRG166,[8] linearized with *Sma*I, as the template.

MFA2 mRNA. Radiolabeled *MFA2* mRNA is used to assay mRNA binding to 40S subunits in WCEs. Its small size allows unbound and 40S ribosome-bound molecules to be separated by sucrose gradient-velocity sedimentation.[6] [^{32}P]*MFA2* mRNA is synthesized with SP6 RNA polymerase from plasmid pAS225,[6] linearized with *Pst*I, in the presence of [α-^{32}P]UTP, as recommended by the vendor. This plasmid encodes a poly(A) tail following the *MFA2*-coding region for the synthesis of polyadenylated *MFA2* mRNA. The mRNA can be capped or uncapped as desired, as uncapped poly(A) mRNA can bind to 40S ribosomes, albeit with lower efficiency than the capped species.[6]

Translation of Luciferase mRNA

To measure the overall rate of protein synthesis by endogenous factors in a WCE, capped *LUC* mRNA is added and luciferase synthesis is assayed by measuring the light emitted from an aliquot of the reaction mixture after combining with the appropriate luciferase substrates. We do not treat the extracts with nuclease to digest endogenous mRNAs in an effort to maintain physiological conditions of competition among mRNAs for limiting amounts of eIFs and ribosomes, and also to minimize the effects of nuclease action on ribosome function.

If the effect of a Ts mutation on translational activity is under study, the extract is incubated at 37° for 5 min prior to the addition of *LUC* mRNA. WCEs prepared from some wild-type strains (e.g., KAY8[13]) are sensitive to this heat treatment and lose most of their *LUC* mRNA translational activity; however, they retain activity for binding of Met-tRNA$_i^{Met}$ and poly(A) mRNA binding to 40S ribosomes. Apparently, a factor(s) required at a step following 48S complex formation is sensitive to heat treatment in such wild-type strains.

Materials

Reagents. These include buffers TR-A [120 mM HEPES (pH 7.4), 9 mM ATP, 1.2 mM GTP, 300 mM creatine phosphate (freshly prepared)], TR-B [120 mM HEPES (pH 7.4), 360 mM potassium acetate, 24 mM magnesium acetate, 0.5 mM

---

[13] K. Asano, L. Phan, J. Anderson, and A. G. Hinnebusch, *J. Biol. Chem.* **273**, 18573 (1998).

amino acid mixture (Promega, Madison, WI), 9 mM DTT], 4 mg/ml creatine phosphokinase (CPK, Sigma) in 50% glycerol (stored at $-20°$), RNasin ribonuclease inhibitor (40 units/μl; Promega). Buffers TR-A and TR-B are stored in aliquots at $-80°$ and can be used for up to 1 year; however, repeated freeze–thaw cycles should be avoided to prevent precipitation of essential components.

Procedure

1. Assemble 30 μl of a 2× reaction mixture by combining 5 μl each of solution TR-A and TR-B, 4 μl CPK, 0.4 μl RNasin, and 2 μg capped *LUC* mRNA. Combine the 2× mixture with 30 μl of the cell extract prepared as described earlier (heat treated when necessary) and incubate at 26° for 1 hr.
2. Aliquots (7.5 μl) are withdrawn at 10-min intervals, diluted in 22 μl distilled water, and quick frozen on dry ice.
3. After all time points are taken, thaw the diluted samples and measure the luciferase activity in 20 μl of each sample using luciferase assay reagents (Promega) and a Monolight 3010 (Pharmingen, San Diego, CA) or equivalent luminometer.

Note. To test for complementation of a defect in *LUC* mRNA translation in a mutant extract, first measure the amount of endogenous factor present in the isogenic wild-type extract by immunoblotting and then add the purified factor to the reaction at step 1 in an amount equivalent to that of the endogenous factor.

Initiator tRNA^{Met} and mRNA-Binding Assays

The ability of eIFs to transfer Met-tRNA$_i^{Met}$ or mRNA to 40S ribosomes can be assayed in WCEs by incubating exogenous radiolabeled Met-tRNA$_i^{Met}$ or mRNA with the extract, resolving the reaction by sucrose gradient velocity sedimentation, and scintillation counting of gradient fractions to detect cosedimentation of labeled molecules with the 40S subunits. To increase the abundance of 48S preinitiation complexes containing Met-tRNA$_i^{Met}$ and mRNA, the nonhydrolyzable GTP analog (GMPPNP) is added to these reactions. Ternary complexes produced in the presence of GMPPNP can bind to 40S ribosomes, but the hydrolysis of GTP, with the subsequent release of eIFs and joining of 60 subunits to the 48S complexes, is blocked. By labeling Met-tRNA$_i^{Met}$ and mRNA with different radioisotopes, the binding of both molecules to 40S ribosomes can be assayed in a single reaction tube.

Materials

Reagents. These include 7.5 and 30% sucrose solutions prepared in gradient buffer [20 mM Tris–HCl (pH 7.5), 100 mM KCl, 1 mM MgCl$_2$] used for making sucrose gradients, 6× binding buffer [120 mM HEPES (pH 7.4), 24 mM magnesium acetate, 9 mM ATP, 300 mM creatine phosphate, 20 mM DTT], 100 mM GMPPNP (Sigma), CPK and RNasin (as in the luciferase synthesis assay), radiolabeled [^{3}H]Met-tRNA$_i^{Met}$, and [^{32}P]*MFA2* mRNA.

Sucrose Gradients. Using a syringe fitted with a blunt-ended needle, add 6 ml of the 7.5% sucrose solution to a 12.5-ml polyallomer centrifuge tube (14 × 89 mm, Beckman, Fullerton, CA). Carefully load 6 ml of the 30% sucrose solution beneath the 7.5% sucrose solution in the tube. After capping the tube, generate a 7.5–30% linear sucrose gradient by rotation on a Gradient Master (Biocomp, Fredericton, NB, Canada). Leave the gradient at 4° for more than 1 hr (but not longer than 18 hr) before loading the sample.

Met-tRNA$_i^{Met}$ and mRNA-Binding Assay

1. Assemble fresh 2× reaction buffer by combining 33.3% 6× binding buffer, 1.6% 100 mM GMPPNP, 3.3% CPK, 0.6% RNasin, [^{3}H]Met-tRNA$_i^{Met}$ (0.77 μCi per reaction), [^{32}P]poly(A) *MFA2* mRNA (2.0 μCi per reaction), and the appropriate amount of water. Mix 20 μl of the 2× reaction buffer with 20 μl cell extracts, prepared as described earlier (heat treated when necessary), followed by incubation at 26° for 20 min.
2. Terminate the reaction by adding 6 μl of 3% formaldehyde and chilling on ice.
3. Carefully layer the reaction mixture on a 7.5–30% sucrose gradient.
4. Centrifuge at 41,000 rpm in a Beckman SW41 rotor (288,000g) for 5 hr at 4°.
5. Separate the gradient into twenty 0.6-ml fractions using an ISCO gradient fraction collector, scanning continuously at 254 nm with the range set to $A = 0.2$. Typically, peaks present in fractions 12, 16, and 18 indicate the positions of 40S, 60S, and 80S ribosomes, respectively.
6. Dilute 0.2 ml of each fraction in 1 ml water and mix with 10 ml of appropriate liquid scintillation fluid for counting ^{3}H and ^{32}P radioactivity.

Assays Using Purified eIFs

Although conventional column chromatography has been applied to purify yeast eIFs, these preparations require many steps and have a low yield of active factors.[14–16] Here we describe procedures allowing more rapid purification of factors using affinity-tagged proteins and the corresponding affinity resins. To purify multisubunit complexes by this approach, the epitope tag is appended to the gene encoding one subunit of the complex, and genes for all of the subunits are

---

[14] M. F. Ahmad, N. Nasrin, A. C. Banerjee, and N. K. Gupta, *J. Biol. Chem.* **260**, 6955 (1985).
[15] A. M. Cigan, J. L. Bushman, T. R. Boal, and A. G. Hinnebusch, *Proc. Natl. Acad. Sci. U.S.A.* **90**, 5350 (1993).
[16] T. Naranda, S. E. MacMillan, and J. W. B. Hershey, *J. Biol. Chem.* **269**, 32286 (1994).

introduced into the same yeast strain on one or two high-copy plasmids. The overexpressed subunits form a complex *in vivo* containing the tagged subunit, which can be purified from the WCE using the appropriate affinity matrix. In purifying eIF2, eIF2B, and eIF3, we make use of polyhistidine-tagged subunits and Ni^{2+} affinity resins, as the latter are relatively inexpensive and easy to handle and can be regenerated for repeated purifications. A Ni^{2+} affinity purification generally must be followed by one or two additional purification steps, as yeast contains many proteins that copurify with the polyhistidine-tagged proteins of interest (e.g., native SNF1 contains 12 consecutive histidine residues at its N terminus).

We have also employed the FLAG antibody resin to affinity purify eIFs tagged with the FLAG epitope (DYKDDDDK) expressed from a high-copy plasmid. In terms of purity, this method is superior to Ni^{2+} affinity purification of polyhistidine-tagged proteins, and a single affinity purification step using anti-FLAG resin can yield a FLAG-tagged protein of ~50–70% purity, as described later. However, anti-FLAG resin is expensive and sometimes gives low yields due to problems with binding or elution. Double affinity purification of a complex containing one subunit tagged with polyhistidine and the other tagged with FLAG has provided the best yield and purity. An example of this technique is described for the purification of eIF3.

Purification of eIF2

Purification of polyhistidine-tagged eIF2 has been conducted successfully.[17–20] Because eIF2B binds tightly to eIF2, eIF2B will copurify with a fraction of eIF2 through conventional chromatographic separations. However, this interaction can be completely disrupted by NaCl/KCl concentrations of 500 mM and above. Consequently, our procedure uses KCl concentrations of 500 mM or above to ensure that eIF2 is purified free of eIF2B. The same holds true for the purification of eIF2B (see later).

Materials

The eIF2-Overexpressing Yeast Strain. We constructed strain GP3511 [*MATα- leu2-3,-112 ura3-52 ino1 gcn2Δ pep4::LEU2 sui2Δ HIS4-lacZ* pAV1089 (*SUI2 SUI3 6xHis-GCD11 URA3*)] for purification of eIF2.[18] Its salient features include (i) pAV1089, a high-copy plasmid bearing *SUI2*, *SUI3*, and *6xHis-GCD11*, encoding eIF2α, eIF2β, and eIF2γ bearing a hexahistidine affinity tag,

---

[17] F. L. Erickson and E. M. Hannig, *EMBO J.* **15**, 6311 (1996).
[18] G. D. Pavitt, K. V. A. Ramaiah, S. R. Kimball, and A. G. Hinnebusch, *Genes Dev.* **12**, 514 (1998).
[19] E. Gomez and G. D. Pavitt, *Mol. Cell Biol.* **20**, 3965 (2000).
[20] J. Nika, W. Yang, G. D. Pavitt, A. G. Hinnebusch, and E. M. Hannig, *J. Biol. Chem.* **275**, 26011 (2000).

respectively. This plasmid elevates expression of eIF2 about 10-fold above the native level. (ii) A deletion of *GCN2* prevents phosphorylation of eIF2α so that the purified eIF2 can be isolated in its unphosphorylated state and phosphorylated *in vitro* only if required. (iii) A deletion of *PEP4*, encoding the vacuolar protease A, serves to reduce proteolysis during purification.[21] (iv) Because *SU12* is essential, the chromosomal *sui2*Δ allele allows cells to maintain pAV1089 without selecting for the *URA3* nutritional marker; hence, cells can be cultured in rich medium.

Buffers. These include IF2-LYS [20 mM Tris–HCl (pH 7.5), 500 mM KCl, 10 mM imidazole, 10% (v/v) glycerol, 0.1 mM MgCl$_2$, 1 mM PMSF, 1 μg/ml pepstatin, 1 μg/ml leupeptin, 1 μg/ml aprotinin, 5 mM 2-mercaptoethanol, 100 μM Na$_3$VO$_4$, 50 mM NaF, and 10 μM GDP], IF2-ELU (identical to IF2-LYS but contains 500 mM imidazole and only 150 mM KCl), IF2-HS1 (same as IF2-LYS but lacks imidazole and Na$_3$VO$_4$ and contains 150 mM KCl and 5 mM NaF), IF2-HS2 and IF2-HS3 (same as IF2-HS1, but with 250 and 450 mM KCl, respectively), IF2-MQ1 [20 mM Tris–HCl (pH 7.5), 100 mM KCl, 10% glycerol, 0.1 mM MgCl$_2$, 1 mM DTT, 0.1 mM PMSF], and IF2-MQ2 and IF2-MQ3 (same as IF2-MQ1, but with 200 and 300 mM KCl, respectively).

Ni-NTA Resin. Eight milliliters of Ni-NTA resin (Qiagen) is added to the top of a bottle-top filtration unit (Nalgene, Rochester, NY) and vacuum is applied to remove the storage buffer. The vacuum is disconnected and ~50 ml buffer IF2-LYS is added and swirled to wash the resin. The vacuum is reapplied and the washing procedure is repeated two more times. The "dry" resin is added immediately to the whole cell extract in step 4.

Procedures

1. Grow strain GP3511 in 10 liters of YEPD medium in five 6-liter Erlenmeyer flasks at 30°, 200 rpm, to an A_{600} of ~8. All remaining steps are performed at 4°.

2. Cells (~100 g wet weight) are harvested by centrifugation at 4200g for 10 min, washed with ice-cold deionized water, and resuspended in buffer IF2-LYS at 2 ml/g wet weight cells.

3. Cells are lysed in a BeadBeater (Biospec Products) as follows. The cell suspension is combined with an equal volume (approximately 175 ml) of 0.5-mm acid-washed glass beads in a 350-ml stainless-steel chamber (Biospec Products) with an ice-water jacket, five cycles of 45 sec each, with 1-min cooling intervals. The lysate is transferred to 250-ml centrifuge bottles and centrifuged two times at 13,000g for 30 min each to remove the glass beads and insoluble cell debris.

4. The supernatant from step 3 is combined with ~5 g of washed Ni-NTA resin in a 250-ml centrifuge bottle and incubated with gentle mixing on a nutator at 4° for 2 hr.

---

[21] E. W. Jones, *Methods Enzymol.* **194**, 428 (1991).

5. Centrifuge at 5000g for 10 min in a fixed angle rotor to collect the resin and remove the supernatant by pipetting. Resuspend the resin in 10 ml of buffer IF2-LYS and transfer to a 15-ml conical tube.

6. Collect the resin by centrifuging at ~1000g for 5 min at 4°. Wash the resin three times with 10 ml of IF2-LYS, as described earlier, and resuspend in 5 ml of elution buffer, IF2-ELU. After mixing for 1 hr at 4°, collect the resin by centrifugation as described earlier and reserve the supernatant. Repeat the elution step and combine the two 5-ml eluates.

7. The eluate is dialyzed overnight against 1 liter of IF2-HS1 buffer. Following dialysis, insoluble material is removed by centrifugation at 100,000g for 30 min. The dialyzate is then applied to a 10-ml column of heparin Sepharose (AP Biotech) equilibrated in IF2-HS1 buffer. The column is washed with 4 column volumes of IF2-HS1 and then with 4 column volumes of IF2-HS2, and the eIF2 is eluted using IF2-HS3 buffer and collected in 2.5-ml fractions. A flow rate of 2 ml/min is maintained throughout.

8. Fractions containing eIF2 are identified by Bradford assay or by assaying TC formation (see later) and combined. The contaminating proteins remaining at this step do not interfere with the activities of eIF2; hence the preparation can be dialyzed and concentrated in storage buffer, as in step 13 later, or purified further as described next.

9. Fractions from step 9 are dialyzed overnight against 1 liter of buffer IF2-MQ1 and loaded on a 1-ml column of HitrapQ (AP Biotech) equilibrated in buffer IF2-MQ1. The column is washed with 5 column volumes of IF2-MQ1 and with 5 column volumes of IF2-MQ2 buffer, and the eIF2 is eluted with IF2-MQ3 buffer, collecting 0.3-ml fractions. All solutions are applied to the HitrapQ column manually using 5-ml disposable syringes according to the manufacturers instructions.

10. Fractions containing eIF2 are identified by SDS–PAGE or by assaying TC formation, and the appropriate fractions are pooled.

11. The pooled fractions can be dialyzed or concentrated using a Centricon 30 spin concentrator (Amicon, Danvers, MA) according to the manufacturers instructions to reduce the KCl concentration to 100 mM prior to storage. Aliquots, 50 μl, are stored at $-80°$. The eIF2 preparations are stable when stored at $-80°$ and remain active for over 1 year. Yields of up to 3 mg of eIF2 from 10 liters of cells have been obtained.

Notes. Erickson and Hannig[17] recommend adding an ammonium sulfate precipitation step after cell breakage. The lysate (in 100 mM KCl) is brought to 75% ammonium sulfate, and the material that precipitates is resuspended in ~45 ml nickel-NTA buffer (500 mM KCl) and dialyzed for 2 hr against 2 liters of the same buffer. This step concentrates the protein in a smaller volume. It may also be advantageous to centrifuge this dialyzate at 200,000g for 2 hr before applying to the nickel column to remove ribosomes (S. Rippel and E. M. Hannig, unpublished observations, 2000).

Purification of eIF2B: Method I

Materials

The eIF2B-Overexpressing Strain. Strain H2649 (WY2479; *MATα leu2-3 leu2-112 trp1-Δ63 ura3-52 prb1-1122 pep4-3 gcd6Δ gcd7Δ::hisG gal2*) harbors two high-copy number plasmids, p1871 (*GCD2 GCD7 GCN3 URA3*) and p2337 [*GCD-1-6xHis GCD6 LEU2*], that allow for overexpression of the eIF2B heteropentamer with the GCD1 subunit bearing a hexahistidine affinity tag. *PEP4* and *PRB1* encode vacuolar proteinases A and B, respectively. *GCD1, GCD2, GCD6*, and *GCD7* are single-copy essential genes encoding the γ, δ, ε, and β subunits of eIF2B, respectively. *GCN3* is a nonessential gene encoding the eIF2Bα subunit. Chromosomal null alleles of *gcd6* and *gcd7* allow for the stable maintenance of p2337 and p1871, respectively, in the absence of nutritional selection. This permits growth of the strain in rich media (YEPD) for purposes of purification.

Buffers. They include I2B-LYS1 [75 mM Tris–HCl (pH 7.5), 100 mM KCl, 1 mM Na$_2$EDTA, 12.5 mM 2-mercaptoethanol, 1 μg/ml pepstatin A, 1 μg/ml leupeptin, 1 μg/ml aprotinin, 0.5 mM PMSF], I2B-A [20 mM Tris–HCl (pH 7.5), 0.1 mM MgCl$_2$, 150 mM KCl, 25 $\mu$$M$ GDP, 10% (v/v) glycerol, 6.25 mM 2-mercaptoethanol, 1 μg/ml pepstatin A, 1 μg/ml leupeptin, 1 μg/ml aprotinin, 0.5 mM PMSF], I2B-B (same as I2B-A but contains 1 M KCl), I2B-C (same as I2B-A but contains 600 mM KCl), I2B-HS1 [20 mM Tris–HCl (pH 7.5), 150 mM KCl, 0.1 mM MgCl$_2$, 25 $\mu$$M$ GDP, 1 mM DTT, 10% glycerol, 0.5 μg/ml pepstatin A, 0.5 μg/ml leupeptin, 0.5 μg/ml aprotinin, 0.1 mM PMSF], I2B-HS2 and -HS3 (same as I2B-HS1 but contains 300 and 600 mM KCl, respectively), I2B-SIZ [20 mM Tris–HCl (pH 7.5), 600 mM KCl, 0.1 mM MgCl$_2$, 25 $\mu$$M$ GDP, 1 mM DTT], storage buffer [20 mM Tris–HCl (pH 7.5), 100 mM KCl, 0.1 mM MgCl$_2$, 1 mM DTT, 50% glycerol].

Procedures

1. A preculture of strain H2649 is grown in YEPD at 30°, 300 rpm, to saturation ($\sim$18–24 hr). The preculture is diluted 1 : 50 (this can vary, depending on time constraints) into 12 liters of fresh YEPD (typically, 2 liters in each of six 6-liter Erlenmeyer flasks), and the cultures are then grown at 30°, 300 rpm, to an A_{600} of 4–8. (All remaining steps are performed at 4°.)

2. Cells ($\sim$150–175 g wet weight) are harvested by centrifugation, washed with 1 liter of ice-cold water, and resuspended in 100 ml I2B-LYS.

3. Cells are lysed using a BeadBeater (Biospec Products) as follows. The cell suspension is combined with an equal volume (approximately 175 ml) of 0.5-mm

acid-washed glass beads in a 350-ml stainless-steel chamber (Biospec Products). Cells are lysed for four cycles of 1 min breaking/1 min resting; the chamber is cooled continuously using an ice-water jacket. The lysate is removed and the beads are rinsed twice using 50 ml (per rinse) of I2B-LYS. The pooled lysate is clarified by centrifugation for 20 min at 13,000g.

4. The supernatent from step 3 is centrifuged for 2 hr at 200,000g. The pellet, which contains ribosomes, is resuspended in 100 ml I2B-C and stirred for 30 min on ice. Ribosomes are pelleted as described earlier, and the supernatent (which contains eIF2B) is dialyzed overnight against 4 liters of I2B-A. The dialysate is clarified by centrifugation for 15 min at 100,000g.

5. Phosphocellulose chromatography. Cellulose phosphate (P11, Whatman) is precycled as recommended by the manufacturer and equilibrated extensively against I2B-A prior to use. The cleared dialyzate is then applied to a 1.5 × 23-cm cellulose phosphate column (40 ml bed volume) at a flow rate of 100 ml/hr. The column is washed using 10 column volumes of I2B-A and is then developed using a 500-ml 150 mM to 1 M linear KCl gradient using I2B-B. Ten-milliliter fractions are collected. eIF2B activity typically elutes as a broad peak between 300 and 720 mM KCl. Activity is detected as described later and corresponds with eIF2Bε protein detected by Western blotting.[20] Active fractions are pooled, and the imidazole concentration is adjusted to 10 mM by the addition of 1 M imidazole (pH 7.5).

6. Nickel-NTA agarose chromatography. Ni-NTA resin (Qiagen) is prepared in I2B-C containing 10 mM imidazole as described earlier for purification of eIF2. A 15-ml bed volume of prewashed resin is added to the pooled active fractions from step 5. The suspension is mixed gently for 1 hr and is then transferred to a 1.5-cm (inner diameter) column. Beads are packed by gravity flow and washed using 10 column volumes of I2B-C containing 10 mM imidazole and 100 μM GDP. Bound protein is eluted using the same buffer containing 200 mM imidazole (collect 3-ml fractions) at a flow rate of 100 ml/hr. Protein-containing fractions are pooled and dialyzed overnight against 2 liters of I2B-HS1.

7. Heparin-Sepharose CL-6B chromatography. Insoluble material is cleared by centrifugation for 15 min at 100,000g and applied to a heparin-Sepharose CL-6B column (1 × 12 cm; 10 ml bed volume; Amersham Pharmacia Biotech) that has been precycled according to the manufacturer's instructions. The column is washed using 10 column volumes of I2B-HS2 buffer and is then developed using a 50-ml linear 300–600 mM KCl gradient with I2B-HS3, collecting 3-ml fractions at the flow rate of 25 ml/hr. The eIF2B typically elutes between 460 and 530 mM KCl. Fractions containing eIF2B, as identified by SDS–PAGE or activity, are pooled and spin concentrated to a volume of 0.5 ml (Centri-Plus 100; Amicon).

8. Superdex 200 HR10/30 chromatography. Concentrated material from step 7 is applied to a Superdex 200 HR10/30 FPLC column (Amersham Pharmacia Biotech) that has been precycled according to the manufacturer's instructions.

The column is developed in 2 column volumes of I2B-SIZ at a flow rate of 0.75 ml/min, and 0.5-ml fractions are collected. Fractions containing eIF2B (the predominant peak) are identified by SDS–PAGE, pooled, and dialyzed overnight against 1 liter to storage buffer. Protein is stored in small aliquots at $-70°$. Typical yields from 12 liters of cells (A_{600} 4–8) are up to 1 mg of protein that is 80% pure. Protein can be frozen/thawed for at least two cycles without significant loss of activity.

Schedule for eIF2B Purification

Day 1. Harvest and break cells (steps 2 and 3), harvest ribosome-containing pellet, wash with high salt buffer, pellet ribosomes, and dialyze supernatant overnight (step 4).
Day 2. Phosphocellulose chromatography (step 5), pool fractions, Ni-NTA chromatography (step 6), pool fractions, and dialyze overnight.
Day 3. Heparin-Sepharose chromatography (step 7), pool and concentrate fractions, Superdex chromatography (step 8), pool fractions, and dialyze overnight.
Day 4. Quantitate, aliquot, and store protein.

Purification of eIF2B: Method II

Levels of eIF2B expression are higher in a newly constructed yeast strain carrying the same high-copy plasmids encoding eIF2B subunits that are present in strain H2649 described earlier. This feature permits purification of eIF2B in a single step by Ni^{2+} affinity chromatography, with a yield of $\sim$1 mg eIF2B at $\sim$80% purity.[19] The new strain (GP3667) does not contain chromosomal deletions of any essential eIF2B subunit genes and, thus, selection must be maintained for the auxotrophic markers on the two plasmids during growth of the cells. Salient features of this purification scheme include the use of high salt (1 M KCl) and detergent (0.1% Triton X-100) in the binding buffer and stringent washes of the Ni-NTA-bound material using 40 mM imidazole prior to elution.

Materials

Yeast Strain. Strain GP3667 (*MATα leu2-3 leu2-112 trp1-Δ63 ura3-52 gcn2Δ GAL2+*) transformed with plasmids pAV1494 (*LEU2 GCN3 GCD7 GCD2* in high copy) and pAV1533 (*URA3 GCD6 GCD1-FLAG-6His* in high copy) encoding all five subunits of yeast eIF2B.

Buffers. These include I2B′-LYS [20 mM Tris–HCl (pH 7.5), 1 M KCl, 10 mM imidazole, 5% (v/v) glycerol, 3 mM MgCl$_2$, 5 mM 2-mercaptoethanol, 5 mM NaF, 0.1% Triton X-100, and complete EDTA-free protease inhibitors (Roche)], I2B′-WSH (same as I2B′-LYS buffer but containing 40 mM imidazole and omitting

Triton X-100), I2B'-ELU (same as before only with 500 mM imidazole), storage buffer [20 mM Tris–HCl (pH 7.5), 100 mM KCl, 5% (v/v) glycerol, 1 mM PMSF, 0.7 μg/ml pepstatin, 1 μg/ml leupeptin, 1 μg/ml aprotinin, 5 mM 2-mercaptoethanol, 5 mM NaF].

Procedures

1. The transformant of GP3667 carrying pAV1494 and pAV1533 is grown in 2.4 liters of SC medium lacking uracil, leucine, isoleucine, and valine in four 2-liter Erlenmeyer flasks at 30°, 200 rpm, to an A_{600} of ~2.5–4.5. All remaining steps are performed at 4°.

2. Cells (~10 g wet weight) are harvested by centrifugation as described earlier, washed with 50 ml ice-cold deionized water, and resuspended in 2 ml/g of I2B'-LYS.

3. Cells are lysed with ~10 ml volume of acid-washed glass beads (0.5 mm diameter) by vortexing in 50-ml conical tubes for five cycles of 1 min each with 1-min cooling intervals on ice. Cell lysates are cleared by centrifugation once at 5000g for 10 min to pellet beads and cellular debris followed by a second step at 30,000g for 30 min to obtain a clear lysate. The lysate is incubated with 1 g Ni-NTA agarose resin and is preequilibrated with I2B'-LYS for 3 hr at 4° with mixing.

4. Resin is collected by low-speed centrifugation (2000g) at 4° and washed once in I2B'-LYS and twice in I2B'-WSH, 5 ml per wash.

5. eIF2B complexes are eluted by two 15-min incubations in 1 ml I2B'-ELU, collecting the resin as described earlier between elutions.

6. The two eluates are combined and dialyzed in storage buffer. Aliquots are stored at $-80°$.

Purification of eIF3: Method I

Yeast eIF3 purified by different combinations of conventional column chromatography contained nine different polypeptides, including degradation products of certain eIF3 subunits,[16,22] or only five different subunits.[5] Affinity purification of eIF3 directed against a polyhistidine-tagged form of the PRT1 subunit yielded a five subunit complex, similar to that described by Danaie *et al.*,[5] containing stoichiometric amounts of eIF3 subunits TIF32, NIP1, His$_6$-PRT1, TIF34, and TIF35, and substoichiometric amounts of eIF5.[7] This last protocol is described next, in which eIF3 containing His$_6$-PRT1 is purified from a ribosome salt wash by Ni affinity purification and gel filtration.

[22] J. W. B. Hershey, K. Asano, T. Naranda, H. P. Vornlocher, P. Hanachi, and W. C. Merrick, *Biochimie* **78**, 903 (1996).

Materials

Yeast Strain. Yeast strain LPY201 [*MATa leu2-3, -112 prt1::kanMX pLPY101 (PRT1-His URA3)*] is employed, in which chromosomal *PRT1* is replaced with the *kanMX* marker encoding kanamycin resistance.[7] Plasmid pLPY101, encoding polyhistidine-tagged PRT1, is stably maintained during growth of this strain in rich medium because *PRT1* is essential.

Buffers. These include IF3-LYS [20 mM Tris–HCl (pH 7.5), 100 mM KCl, 10% (v/v) glycerol, 5 mM MgCl$_2$, 0.1 mM EDTA, 7 mM 2-mercaptoethanol, 1 mM PMSF, 1× complete protease inhibitor], IF3-NI1 [20 mM Tris–HCl (pH 7.5), 350 mM KCl, 10% (v/v) glycerol, 5 mM MgCl$_2$, 1 mM PMSF, 20 mM imidazole 1× EDTA-free complete protease inhibitor], IF3-NI2 (same as IF3-NI1 but with 250 mM imidazole), and IF3-SIZ [20 mM Tris–HCl (pH 7.5), 75 mM KCl, 1 mM PMSF, 10% (v/v) glycerol].

Procedure

1. Strain LPY201 is grown in 10 liters of YEPD medium in four 6-liter Erlenmeyer flasks at 30°, 200 rpm, to an A_{600} of ∼4, yielding ∼4 g/liter of cell wet weight. The yield can be more than doubled (∼10 g/liter) by growing cells in a 10-liter bench-top fermenter (New Brunswick, Edison, NJ) at 30°, 400 rpm, to an A_{600} of ∼10. All remaining steps are performed at 4°.

2. Cells (∼100 g wet weight) are harvested by centrifugation as described earlier, washed with 10 cell volumes of ice-cold deionized water, and resuspended in IF3-LYS at 1 ml/g wet weight cells.

3. Cells are lysed by two passages through a French press at 19,000 psi. Cell debris is removed by centrifugation at 17,000g and then at 25,000g for 15 min each, and the final supernatant is centrifuged at 200,000g for 2 hr to pellet the ribosomes. The ribosome pellet is resuspended in 25 ml of IF3-NI1 and centrifuged at 200,000g for 2 hr, yielding the ribosome salt wash (RSW).

4. The RSW is mixed with 0.75 ml of a 50% slurry of Ni-silica resin (Qiagen), equilibrated in IF3-NI1, in a 50-ml conical tube, and incubated overnight with gentle rocking at 4°.

5. Centrifuge at 1000g for 5 min to collect the Ni-silica resin, remove the supernatant by pipetting, and wash the resin three times with 10 ml of IF3-NI1 in a 15-ml conical tube. Elute the bound proteins by resuspending the resin in 0.75 ml of IF3-NI2, transferring to a 1.5-ml microfuge tube, and mixing on a nutator at 4° for 5 min. Collect the resin by centrifuging as described earlier and repeat the elution step once more. Combine the two 0.75-ml eluates and concentrate to ∼0.3 ml using a Centricon-10 spin column (Amicon).

6. An aliquot of 0.2ml of the concentrated eluate is injected onto a Sepharose-6 sizing column (1 × 25 cm) of an FPLC system (Pharmacia Biotech) preequilibrated

in IF3-SIZ. The proteins are resolved into 0.3-ml fractions in IF3-SIZ at a flow rate of 0.3 ml/min.

7. Fractions containing the eIF3–eIF5 complex (generally numbers 49–57) are pooled and concentrated to 100 µl volume (~1 mg/ml of complex) using a Centricon-10 spin column (Amicon), and 20-µl aliquots are stored in liquid nitrogen. The eIF3 remains active for over 1 year. A typical yield is 50–100 µg of the highly purified eIF3–eIF5 complex from 100 g of cells.

Purification of eIF3: Method II

To improve the yield of eIF3, we constructed a yeast strain overexpressing all five subunits of the eIF3 complex from two high-copy plasmids. The overexpressed PRT1, TIF34, and TIF35 subunits are tagged with His_6, hemagglutinin (HA) epitope, and FLAG epitope, respectively, although only the His_6 and FLAG tags are exploited for affinity purification. Because eIF3 is produced in excess of ribosomes in this strain, we use a postribosomal supernatant (PRS) rather than RSW as the starting material for affinity chromatography. The PRS is prepared in a buffer containing 750 mM KCl, which significantly reduces the interaction of eIF3 and eIF5 without impairing eIF3 activity; hence, eIF3 can be purified free of eIF5.

Materials

Yeast Strain Overexpressing eIF3. We employ protease-deficient diploid strain LPY87 [*MATa/MATα ura3-52/ura3-52 trp1/trp1 leu2-Δ1/leu2-Δ1/his3-Δ200/his3-Δ200 pep4::HIS4/pep4::HIS4 prb1-Δ1.6/prb1-Δ1.6 can1/can1 GAL^+* (pLPY-PRT1His-TIF34HA-TIF35Flag) (pLPY-TIF32-NIP1)][23] containing high-copy plasmids encoding all five eIF3 subunits. LPY87 is grown in SC medium lacking uracil and leucine to maintain selection for the plasmids.

Buffers. These include IF3-NI4 (same as IF3-NI1 but contains 750 mM KCl and 0.5 mM 2-mercaptoethanol), IF3-NI5 (same as IF3-NI4 but contains 100 mM KCl), IF3-NI6 (same as IF3-NI5 but with 250 mM imidazole), and IF3-FL (1× TBS, 10% glycerol, 1 mM PMSF, 1× complete protease inhibitor).

FLAG Resin and Peptide. Both of these are from Sigma. The resins were prepared according to the manufacturer prior to use.

Procedure

1. Strain LPY87 is grown in 10 liters of SC medium without uracil and leucine (SC-UL) in four 6-liter Erlenmeyer flasks at 30°, 250 rpm, to an A_{600} of ~4, which yields about 4 g/liter of cell wet weight. All remaining steps are performed at 4°.

[23] L. Phan, L. Schoenfeld, L. Valasek, K. H. Nielsen, and A. G. Hinnebusch, *EMBO J.* **20**, 2954 (2001).

2. Cells (~40 g wet weight) are harvested by centrifugation and washed with ice-cold deionized water, as described earlier, and resuspended at 1 ml/g of cells in IF3-NI4.

3. Cells are lysed in the French press as described earlier, cell debris is removed by centrifugation at $17,000g$ and $25,000g$ for 15 min each, and the ribosomes are collected by centrifugation at $200,000g$ for 1 hr. Discard the ribosome pellet and save the supernatant for the next step.

4. The supernatant is incubated with 0.75 ml of a 50% slurry of Ni-silica resin in a 50-ml conical tube overnight with gentle rocking at $4°$.

5. Centrifuge at $1000g$ for 5 min to collect the resin. Remove the supernatant and wash the resin with 10 ml of IF3-NI4 in a 15-ml conical tube by inverting repeatedly for 5 min at $4°$. Collect the resin as described earlier and repeat the washing procedure two more times with IF3-NI4 and once more with IF3-NI5. Elute the bound proteins as described previously three times with 1 ml of IF3-NI6 in a 1.5-ml microfuge tube. Pool the eluates (~3 ml) in a 50-ml conical tube and add 27 ml of IF3-FL.

6. Add 0.5 ml of FLAG resin to the eluate from step 5 and rock gently at $4°$ for 2 hr. Centrifuge at $1000g$ for 5 min to pellet the resin. Remove the supernatant without disturbing the resin by pipetting slowly, starting from the top of the tube, and leaving the last 2–3 ml with the pellet. Wash the pellet three times in 10 ml of IF3-FL in a 15-ml conical tube by inverting repeatedly for 5 min at $4°$ and collecting the resin as described earlier. Elute bound proteins by resuspending the resin in 0.5 ml of IF3-SIZ containing 25 pmol of FLAG peptide and transferring the mixture to a 1.5-ml microfuge tube. Mix on a nutator for 2 hr or overnight at $4°$ and collect the resin by centrifuging in a microfuge for 1 min at $4°$. Repeat the elution step and pool the two eluates (~1 ml). Concentrate to ~0.5 ml using a Centricon-10 spin column (Amicon)

7. Separate the concentrated elute on a Sepharose-6 column using FPLC, concentrate the pooled fractions containing eIF3, and store aliquots in liquid nitrogen, all as described earlier. The typical yield is 50–100 μg from 40 g of cells.

Purification of FLAG-Tagged eIF5

Here we describe a one-step affinity purification of FLAG-tagged eIF5 from yeast WCE. Yeast eIF1, eIF1A, eIF4B, eIF4E, eIF4G, eIF5, eIF5B, and all subunits of eIF2, eIF2B, and eIF3 are all intact in extracts prepared as described later; hence, it could serve as starting material for the purification of essentially any FLAG-tagged eIFs. Because FLAG–eIF5 purified from bacteria was equally active as that purified from yeast,[24] we also describe affinity purification of FLAG–eIF5 from bacterial WCE.

[24] K. Asano, A. Shalev, L. Phan, K. Nielson, J. Clayton, L. Valasek, T. F. Donahue, and A. G. Hinnebusch, *EMBO J.* **20,** 2326 (2001).

Materials

Yeast Strain Overexpressing FLAG–eIF5. KAY39 [*MATa leu2-3, -112 ura3-53 trp1-Δ63 gcn2Δ tif5Δ::hisG YEpTIF5 (TIF5-FL LEU2 2μ)*].[25] Because *TIF5* is essential and chromosomal *TIF5* is deleted in KAY39, YEpTIF5, encoding C-terminally FLAG-tagged eIF5, is stably maintained in rich medium.

Escherichia coli Strain Overexpressing FLAG–eIF5. KAB330 is a transformant of BL21 (DE3) (Novagen, Madison, WI) carrying pT7-TIF5,[26] encoding C-terminally FLAG-tagged yeast eIF5 under the T7 promoter, and the DE3 prophage encoding T7 RNA polymerase under an IPTG-inducible promoter. Thus, expression of FLAG–eIF5 is induced by the addition of IPTG.

Buffers. These include IF5-A [20 mM Tris–HCl (pH 7.5), 100 mM KCl, 5 mM MgCl$_2$, 0.1 mM EDTA, 7 mM 2-mercaptoethanol, 5 mM NaF, 1 mM PMSF, 1× complete protease inhibitor, 1 μg/ml pepstatin, 1 μg/ml leupeptin, 1 μg/ml aprotinin] for purification from yeast WCE and buffer PBS-C [140 mM NaCl, 2.7 mM KCl, 10.1 mM Na$_2$HPO$_4$, 1.8 mM KH$_2$PO$_4$ (pH 7.3), 1× complete protease inhibitor] for purification from bacterial WCE.

Procedure for Purification of FLAG–eIF5 from Yeast

1. Grow strain KAY39 in 1 liter of YEPD medium at 30° to an OD$_{600}$ of 1–2. Collect cells in four 400-ml bottles by centrifuging at 5000 rpm in a GSA rotor (4000g) for 5 min at 4°.

2. Resuspend the pellet in each bottle with 5 ml of ice-cold sterile water, and combine the cells from two bottles in a 15-ml conical tube. Centrifuge at 2000 rpm with a Beckman J-6 swinging centrifuge (500g) for 3 min at 4° and remove as much supernatant as possible using a Pipetman. The volume of each cell pellet should be >1 ml.

3. Resuspend cells in 1.5 cell volumes of buffer IF5-A and add 1 cell volume of acid-washed glass beads (Sigma, 425–600 μm in diameter). Vortex the tubes in the cold room eight times for 30 sec each, with 30-sec intervals on ice.

4. Centrifuge at 2000 rpm with the J-6 centrifuge for 3 min at 4°, transfer the supernatant to 1.5-ml sterile microfuge tubes, and centrifuge at >12000 rpm in a microfuge in the cold room for 10 min. Carefully recover the supernatant using a Pipetman, avoiding the upper lipid layer, and transfer to a clean 1.5-ml microfuge (the WCE fraction).

5. Determine the protein concentration of the WCE by the Bradford assay. It should be >10 mg/ml.

---

[25] K. Asano, T. Krishnamoorthy, L. Phan, G. D. Pavitt, and A. G. Hinnebusch, *EMBO J.* **18,** 1673 (1999).

[26] K. Asano, J. Clayton, A. Shalev, and A. G. Hinnebusch, *Genes Dev.* **14,** 2534 (2000).

6. Distribute the WCE containing 10 mg of total protein to microfuge tubes each containing 100 μl of M2 anti-FLAG affinity resin (Sigma) and incubate at 4° for 2 hr with gentle rocking.
7. Spin in a microfuge at 2000 rpm for 2 min at 4° to collect the resin, add 1 ml of ice-cold buffer IF5-A, and invert the tubes several times. Repeat this washing step three more times.
8. Elute the FLAG–eIF5 from each tube with 200 μl IF5-A containing 10% glycerol and 400 μg/ml of FLAG peptide (Sigma) at room temperature for 20 min with rocking. Spin at 2000 rpm for 2 min at room temperature, and reserve the supernatant as eluate E1. Repeat the elution step and obtain eluate E2. Typically, E1 contains ~100 μg/ml FLAG–eIF5 of 50–70% purity, as revealed by Coomassie staining of the sample separated by SDS–PAGE. E2 contains ~4-fold less FLAG–eIF5. The total yield is ~70–120 μg from the 1-liter culture. Store aliquots at $-80°$.

Note. When stored for long periods (>6 months) and thawed, the sample contains a white precipitate. Do not try to remove the precipitate by centrifugation; instead, remove the sample by pipetting while avoiding the precipitate.

Procedure for Purification of FLAG-eIF5 from Bacteria

1. Grow KAB330 in 50 ml LB medium supplemented with 100 μg/ml ampicilin to OD_{600}~1.0 at 25°. Add IPTG to a final concentration of 0.1 mM. Continue incubation overnight.
2. Collect cells in two 40-ml centrifuge bottles by centrifugation at 7000 rpm in a GSA rotor (8000g) for 5 min. Suspend cells in each bottle in 2 ml PBS-C and combine them in one of the two bottles.
3. Disrupt cells by sonication. Placing the sample-containing bottles on ice, sonicate for 2 min, and turn off for 30 sec. Repeat for a total of 6 min of sonication.
4. Add 20% Triton X-100 to 1% and mix on a nutator at 4° for 30 min.
5. Spin the tube at 18,000 rpm in a Sorval SS-34 rotor (38,000g) for 20 min at 4°. Collect supernatant as cell lysate.
6. Incubate cell lysate with anti-FLAG affinity resin, using 1 ml of lysate per 100 μl bed volume, at 4° for 2 hr with gentle rocking.
7. Wash the resin and elute FLAG–eIF5 as described in steps 7 and 8 of "Purification of FLAG–eIF5 from Yeast," except that PBS-C and PBS-C containing 10% glycerol and 400 μg/ml of FLAG peptide are used for washing and elution, respectively.

Note. This method yields FLAG–eIF5 of ~70–80% purity, and less precipitates occur in this preparation than in the sample purified from yeast after long-term storage.

Purification of Yeast 40S Ribosomes

The following is a modified version of methods described previously,[27,28] in which ribosomes are pelleted by high-speed centrifugation and then resolved into 40S and 60S subunits by sucrose gradient-velocity sedimentation. Yeast strains TD28[28] and LPY201 (described earlier) have been used as the source of ribosomes.

Buffers. These include RIB-LYS [20 mM Tris–HCl (pH 7.2), 50 mM NH$_4$Cl, 12 mM MgCl$_2$, 1 mM DTT], RIB-CSH [20 mM Tris–HCl (pH 7.2), 500 mM KCl, 5 mM MgCl$_2$], RIB-A [50 mM Tris–HCl (pH 7.5), 800 mM KCl, 12 mM magnesium acetate, 20 mM 2-mercaptoethanol], RIB-B [20 mM Tris–HCl (pH 7.5), 100 mM KCl, 5 mM MgCl$_2$, 0.1 mM EDTA, 1 mM DTT, 50% glycerol], dilution buffer (20 mM 2-mercaptoethanol, 12 mM magnesium acetate).

1. Cells are grown in 4 liters of YEPD in two 6-liter Erlenmeyer flasks at 30°, 250 rpm, to an A_{600} of ~0.8–1.0. All subsequent steps are carried out at 4°.

2. The culture is divided into six 1-liter centrifuge bottles, and the cells are cooled immediately by adding ice to the bottles to bring the volume to ~1 liter. The cells (2–3 g wet weight) are harvested by centrifugation 12,000g, washed once with ice-cold water, and once with RIB-LYS. After the final wash, the cell pellet is resuspended in 2 cell volumes of RIB-LYS (4–6 ml).

3. Lyse cells by two passages through a French press at 19,000 psi using the small press chamber (5 ml volume) with a piston diameter of 1/2 inch. Cell debris is removed by centrifugation at 17,000g for 10 min. Centrifuge again if necessary to remove all cell debris.

4. Layer 6 ml of the supernatant in step 3 on a 6-ml 10% sucrose cushion prepared in RIB-CSH contained in a 12.5-ml tube for the SW41 rotor and centrifuge at 41,000 rpm for 2 hr.

5. Remove the supernatant by pipetting and gently wash the ribosome pellet by swirling the tube with 1 ml of RIB-CSH to remove all the white debris. Repeat the washing until only a gold-colored pellet remains.

6. Dissolve the pellet in RIB-A (~2 ml) to make a final dilution of A_{260}/ml = 400.

7. Add 100 mM puromycin to a final concentration of 1 mM and incubate at 30° for 15 min.

8. Prepare six 15–25% sucrose gradients in buffer RIB-A in centrifuge tubes for the SW28 rotor (37 ml). Layer 200 μl (~75 A_{260} unit) of the sample from step 7 onto each gradient. Do not overload the gradient. Centrifuge the gradients at 25,000 rpm (15°) for 12 hr.

9. Fractionate the gradients using an ISCO fraction collector equipped for continuous monitoring of A_{254}. Separately pool the fractions corresponding to the

---

[27] J. R. Warner and C. Gorenstein, *Methods Cell Biol.* **20**, 45 (1978).
[28] H. Huang, H. Yoon, E. M. Hanning, and T. F. Donahue, *Genes Dev.* **11**, 2396 (1997).

more rapidly sedimenting halves of the 40S and 60S peaks, as these are more likely to contain intact ribosomes. Dilute each of the samples with an equal volume of dilution buffer and centrifuge at 55,000 rpm at 4° in a Beckman 70.1 Ti rotor for 12–14 hr. A small white translucent pellet, ~3 mm in diameter, containing the ribosomes will form at the bottom of the tube.

10. Carefully discard the supernatant and gently layer 200 μl of buffer RIB-B over the pellet to wash. Slowly pipette out the wash without disturbing the pellet and discard. Dissolve the pellet in RIB-B (15–20 μl/tube) by shaking on ice and pipetting slowly up and down. Measure the A_{260} and store as 20-μl aliquots in liquid nitrogen. The typical yield is about 20 A_{260} units of ribosome.

Note. Step 7 may be optional.

Initiation Factor Assays

Here we describe three different assays using purified components: (i) formation of ternary complex (TC) by eIF2, (ii) stimulation of GTP hydrolysis in a model 48S complex by eIF5 (GAP function), and (iii) guanine-nucleotide exchange on eIF2-GDP by eIF2B. We also describe assays for the ability of purified eIF3 to restore binding of Met-tRNA$_i^{Met}$ and mRNA to 40S ribosomes in a *prt1-1* WCE.

Ternary Complex Formation Assay

In this assay, the TC of purified eIF2, GTP, and [^{3}H]Met-tRNA$_i^{Met}$ is bound to a nitrocellulose filter, dried, and counted by liquid scintillation. Because eIF2 binds with higher affinity to GDP than GTP and becuase commercial preparations of GTP contain significant levels of GDP, a GTP-regenerating system is included in the reaction. The assay is based on one used extensively for TC formation with mammalian eIF2 (see, e.g., Refs. 29,30) and can be used to assess the activity of eIF2 during purification in heparin Sepharose column fractions. With some variations the assay can be adapted easily for other purposes.

Assay Components. Prepare two 3-ml aliquots of TC assay mix [20 mM Tris–HCl (pH 7.5), 100 mM KCl, 5% (v/v) glycerol, 1 mM DTT, 3.2 mM phosphoenol pyruvate, 50 U pyruvate kinase, 200 μg/ml CPK, 2 mM MgCl$_2$, 20 pmol [^{3}H]Met-tRNA$_i^{Met}$]. CPK is used as a carrier; others have successfully used BSA instead. Pyruvate kinase and phosphoenol pyruvate are included for GTP regeneration. The exact amount of Met-tRNA$_i^{met}$ required will depend on the eIF2 preparation, but 20 pmol [^{3}H]Met-tRNA$_i^{Met}$ at 3000 Ci/mmol should suffice (prepared as described earlier). Three milliliters of TC assay mix is sufficient for assaying 16 column

[29] C. G. Proud and V. M. Pain, *FEBS Lett.* **143,** 55 (1982).
[30] S. T. Wong, W. Mastropaolo, and E. C. Henshaw, *J. Biol. Chem.* **257,** 5231 (1982).

fractions collected during eIF2 purification and should be scaled as required. The wash buffer is 20 mM Tris–HCl, (pH 7.5), 100 mM KCl, 5 mM MgCl$_2$.

1. To one 3-ml aliquot of TC assay mix (mix A) add GTP to 0.2 mM and to the other (mix B) add an equal volume of water.
2. Set up two series of tubes, one with 180 μl TC-mix A and one with 180 μl mix B. Add 20 μl of each column fraction to one tube from each series A and B. Incubate at 21° for 20 min.
3. Stop the reaction with the addition of 2.5 ml ice-cold wash buffer and filter through nitrocellulose filters (Whatman type WCN 0.45-μm pore size or equivalent) using a vacuum-sampling manifold (Millipore, Bedford, MA). Rinse each assay tube with two further volumes of wash buffer and filter the washes through the same filter as the sample.
4. Filters are baked dry in an oven (80°) or under a heat lamp until completely dry, placed in scintillation vials with appropriate scintillation fluid, and counted. By subtracting the counts obtained with mix B from those obtained with mix A, fractions with GTP-dependent eIF2 ternary complex formation can be identified.

eIF2 GTPase Stimulation Assay

The purified eIF5 can stimulate hydrolysis of GTP present in a model 48S complex containing the TC (eIF2/[γ-^{32}P]GTP/Met-tRNA$_i^{Met}$), rAUG triplet, and the 40S ribosome. The rate of GTP hydrolysis is monitored by measuring the rate of production of free labeled inorganic phosphate resulting from the hydrolysis of [γ-^{32}P]GTP. This assay was established for mammalian eIF5, eIF2, and 40S subunits[31] and modified for use with the corresponding yeast factors and ribosomes.[28] Stimulation of GTP hydrolysis in the TC by eIF5 should be dependent on the presence of the rAUG triplet and 40S ribosomes.

Solutions. These include preequilibration buffer PE [20 mM Tris–HCl (pH 7.5), 5 mM MgCl$_2$, 100 mM KCl, 0.1% Nonidet P-40], PE-G (same as PE but contains 1 mM DTT and 10 μM GTP), Blue dextran solution (4 mg/ml in PE, filtered) to use as a marker in the Sepharose CL-6B columns, initiation complex buffer 2× IC [40 mM Tris–HCl (pH 7.5), 200 mM KCl, 2 mM DTT, 50 μM GTP, 6.4 mM phosphoenol pyruvate, 30 U/ml pyruvate kinase, 200 μg/ml creatine phosphokinase, 0.2% Nonidet P-40], 0.3 mM solution of rAUG triribonucleotide (NBI) in water, and 115 mM MgCl$_2$ for the 48S complex formation. For assaying the production of free [^{32}P] phosphate, the following solutions are needed: 20 mM tungstosilicic acid dissolved in 10 mM sulfoic acid (TS), 2 mM KH$_2$PO$_4$, 5% ammonium molybdate dissolved in 2 M sulfuric acid (AM), and a 1-to-1 mixture of benzene and isobutyl alcohol (BI mix).

[31] A. Chakrabarti and U. Maitra, *J. Biol. Chem.* **21,** 14039 (1991).

Sepharose CL-6B column. Approximately 5 ml of CL-6B resin (Pharmacia) is transferred to a 50-ml conical tube and washed batchwise with 40 ml buffer PE three times. The resin is collected by low-speed centrifugation (2000 rpm, 2 min) in a Beckman J6 centrifuge or equivalent (500g). Resuspend the washed resin in 1 ml PE and load ~1.5 ml in three disposable polystyrene columns (0.7 × 4.0 cm) (Pierce). The resin columns should be 3.5 cm or higher after settling overnight at 4°. All the fractionation steps are done at 4°. Preequilibrate the column with 3 ml PE and load 23 μl of Blue dextran. After the dye enters the resin, carefully load 1.5 ml PE without disturbing the resin and collect fractions of two drops on a sheet of 3MM paper. The dye should elute between fractions 16 and 20 (void volume). Wash the resin with 3 ml PE for stock at 4°. If the resin is used immediately for purification of the model 48S complex, wash the resin instead with 3 ml PE-G.

1. Form TC in a 20-μl reaction containing 10 μl 2× IC, [γ-^{32}P]GTP (5000 Ci/mmol, 10 Ci/liter), 3 pmol [^{3}H]Met-tRNA$_i^{Met}$, and 1 μg purified eIF2. Incubate the reaction at 37° for 5 min. The [^{3}H]Met-tRNA$_i^{Met}$ and purified eIF2 are prepared as described earlier.
2. Add 1 μl each of purified 40S ribosomes (0.2 A_{260} unit) (prepared as described earlier), rAUG triplet, and 115 mM MgCl$_2$. Continue incubation at 37° for 4 min.
3. Load the entire reaction onto the Sepharose CL-6B column equilibrated with PE-G at 4°.
4. To elute the 48S complex, load 1.5 ml of PE-G to the column and collect two-drop fractions in 36 microfuge tubes.
5. Measure the amount of ^{32}P in each tube using a Geiger counter. The model 48S complex elutes at the void volume, followed by a much larger peak containing free TC and [γ-^{32}P]GTP.
6. Fractions containing the radiolabeled 48S complex are pooled and used as the substrate for assaying the GAP function of eIF5. Determine the amount in picomoles of the ^{32}P-labeled 48S complex based on scintillation counting. Dilute the pooled 48S complex with PE-G to a final concentration of ~1 pmol/ml.
7. Place 250 μl of diluted 48S complex in a 1.5-ml microfuge tube and equilibrate to room temperature. Add 0.5 to 3 μg of purified eIF5. Continue incubation for 20 min to 1 hr.
8. Withdraw 50-μl aliquots from the reaction at intermittent times (e.g., 0, 2.5, 5, 10, 20 min) and mix sequentially with 50 μl TS, 50 μl AM, and 100 μl 2 mM KH$_2$PO$_4$. The terminated mixture is incubated at 37° for 1 min and is extracted with 250 μl BI mix by vortexing for 10 sec followed by a 1-min spin in a microcentrifuge at room temperature.
9. Take 200 μl from the upper phase of the extracted sample and mix it with 8 ml of appropriate scintillation fluid for counting of ^{32}P.

Guanine-Nucleotide Exchange Assay

The assay for yeast eIF2B function described here is a modified version of one originally developed for the mammalian factor.[32,33] In the absence of eIF2B, eIF2-[^{3}H]GDP is stable even in the presence of excess unlabeled guanine nucleotide.[15,18,20] On the addition of eIF2B, [^{3}H]GDP is released from eIF2 and replaced with unlabeled nucleotide. The amount of [^{3}H]GDP remaining bound to eIF2 is assessed by filter binding of the eIF2–[^{3}H]GDP complex and scintillation counting. We have shown that eIF2Bε alone can perform nucleotide exchange in this assay, in place of the eIF2B holocomplex.[19] GDP rather than GTP is used as the unlabeled nucleotide because eIF2 has a higher affinity for GDP than GTP and because GTP does not stably bind eIF2 in the absence of Met-tRNA$_i^{Met}$.[17]

Reagents. These include binary complex buffer BC [20 mM Tris–HCl (pH 7.5), 100 mM KCl, 2 mg/ml CPK, 10% (v/v) glycerol, 5 mM NaF, 1 mM DTT, 0.1 mM EDTA] with the CPK added from a 10-mg/ml stock in 50% glycerol, the same wash buffer employed for the "TC Formation Assay" described earlier, and 0.5 mM unlabeled GDP.

Procedure

1. Add eIF2 purified as described earlier (typically 50 pmol, 5–10 μg) and [^{3}H]GDP (20 pmol) to buffer BC and incubate at 21° for 10–12 min to form the eIF2–[^{3}H]GDP binary complex. If assessing the effects of eIF2α phosphorylation on the subsequent exchange reaction catalyzed by eIF2B, ATP (0.1 mM) and the appropriate eIF2α kinase can be added either before or after the addition of [^{3}H]GDP without any apparent effect on the rate or final amount of binary complex formed.[18]

2. The eIF2–[^{3}H]GDP binary complex is stabilized by the addition of MgCl$_2$ to 3 mM and continued incubation at 21° for 4 min, and then transferred to ice. The binary complexes are stable on ice for at least several hours.

3. To assay eIF2B activity, the preformed eIF2-[^{3}H]GDP complexes, the eIF2B, and unlabeled GDP are each equilibrated at 10° for 4 min. The final reaction volume and the amount of eIF2B added will vary with each preparation of eIF2 and eIF2B. With our preparations, we employ 0.5 μg eIF2B or less, added in a volume of 25 μl or less. The assay is initiated by the addition of 4 μl GDP and then eIF2B to the binary complexes. Aliquots are removed at regular time intervals over a 6- to 9-min reaction so that the total volume of the aliquots taken does not exceed 75% of the total reaction volume (i.e., 4 × 15-μl aliquots for an 80-μl total reaction) and diluted into 2.5 ml wash buffer on ice. Zero time points are taken before the addition of eIF2B. Note that the exchange assay is conducted at 10° because the

---

[32] R. L. Matts and I. M. London, *J. Biol. Chem.* **259**, 6708 (1984).
[33] A. G. Rowlands, R. Panniers, and E. C. Henshaw, *J. Biol. Chem.* **263**, 5526 (1988).

stability of the eIF2–[^{3}H]GDP complex is reduced at higher temperatures, and the resulting eIF2B-independent release of GDP interferes with the assessment of eIF2B function.

4. Samples are filtered through nitrocellulose filters (Whatman, type WCN) using a vacuum sampling manifold (Millipore). The reaction tubes are rinsed twice with 2.5 ml wash buffer, and the washes are filtered through the same filters. The filters are baked and counted in a scintillation counter as described in step 4 of the "TC Formation Assay."

Note. Note that when using eIF2 purified according to Erickson and Hannig[17] and eIF2B purified as described earlier, Nika *et al.*[34] employed conditions for assaying eIF2B that differed somewhat from those described earlier, particularly with respect to magnesium concentration and temperature.

Assay of Purified eIF3 for Rescue of Met-tRNA$_i^{Met}$ and mRNA Binding in Heat-Inactivated prt1-1 Extracts

Yeast *PRT1* encodes the essential 90-kDa subunit of eIF3.[16] The temperature-sensitive *prt1-1* mutation is known to reduce Met-tRNA$_i^{met}$ and mRNA binding to 40S ribosomes in extracts heat treated at 37° using the assays "Initiator tRNA and mRNA Binding Assays" described earlier. This defect can be complemented by the addition of purified eIF3.[5,7] Thus, Met-tRNA$_i^{Met}$ and mRNA binding assays conducted using cell extracts can be used to measure the activity of purified eIF3 for these two functions. [Note that an alternative assay for yeast eIF3 has been described[16] in which the purified factor is tested for the ability to substitute for mammalian eIF3 in promoting first peptide bond formation (measured as methionylpuromycin synthesis) in an assay containing several other purified mammalian eIFs and mammalian ribosomes.]

Materials. These include WCEs prepared from yeast strains H1676 (*MATa part 1-1 ade 1 leu2-3,-112 ura3-52*)[7] and isogenic *PRT1* strain LPY201, described earlier, and all assay components described earlier.

Procedure. Heat treat the WCEs for 5 min at 37° and return to ice. Assemble the binding reactions in the presence of ∼1 μg purified eIF3 and analyze the reactions as described earlier.

Acknowledgments

G.D.P was supported by a career development fellowship from the Medical Research Council (UK). E.M.H. was supported by a grant from the American Cancer Society (RPG-97-061-01-NP). T.F.D. was supported by a grant from the National Institutes of Health (GM32263).

[34] J. Nika, S. Rippel, and E. M. Hannig, *J. Biol. Chem.* **276**, 1051 (2001).

[14] Assaying Protein Ubiquitination in *Saccharomyces cerevisiae*

By JEFFREY D. LANEY and MARK HOCHSTRASSER

Covalent modification of target proteins by the polypeptide ubiquitin (Ub) is involved in a wide array of cellular processes, ranging from cell cycle progression and receptor-mediated endocytosis to endoplasmic reticulum-associated degradation and cell-type specification.[1-4] The best understood function of ubiquitination is to tag protein substrates for destruction, with a polymeric chain of Ub molecules being required to target the substrate to the 26S proteasome for hydrolysis. In addition to this common role in protein degradation, a number of examples of nonproteolytic functions for Ub attachment also exist. This chapter describes several different methods to determine whether Ub modifies a particular protein *in vivo*. These assays can be adapted to ask if the substrate is attached to Ub chains and to ascertain the topology (linkages) of the Ub monomers in these polymeric chains. Finally, methods for determining which enzymes are required for substrate ubiquitination are discussed.

Ubiquitination of Protein Substrate *in Vivo*

Three different methods for detecting the modification of target proteins by Ub are described. Most of the assays rely on immunoprecipitation of the substrate; therefore, antibodies capable of precipitating the protein of interest are a prerequisite. One drawback of the protocols is that these critical reagents are sometimes not available. However, with the advent of polymerase chain reaction (PCR)-based techniques for epitope tagging any open reading frame, a protein of interest can be engineered to carry any one of a variety of different antigenic determinants.[5] These epitope-tagged proteins can then be monitored with commercially available monoclonal antibodies whose immunoprecipitation capabilities are well established. It should be noted, however, that modifying a protein by epitope addition can lead to artifactual ubiquitination and degradation by the Ub–proteasome pathway, even though these tagged versions can complement the null allele.[6]

---

[1] A. Hershko and A. Ciechanover, *Annu. Rev. Biochem.* **67**, 425 (1998).
[2] D. M. Koepp, J. W. Harper, and S. J. Elledge, *Cell* **97**, 431 (1999).
[3] L. Hicke, *Trends Cell Biol.* **9**, 107 (1999).
[4] J. L. Brodsky and A. A. McCracken, *Semin. Cell Dev. Biol.* **10**, 507 (1999).
[5] M. E. Petracek and M. S. Longtine, *Methods Enzymol.* **350**, 445 (2002).
[6] C. Schauber, L. Chen, P. Tongaonkar, I. Vega, D. Lambertson, W. Potts, and K. Madura, *Nature* **391**, 715 (1998).

All of the following assays evaluate the modification state of the protein of interest in cell lysates. Because deubiquitinating activity is often high in such extracts, precautions must be taken to guard against deubiquitination of the substrate. Therefore, it is essential to include an inhibitor of deubiquitinating enzymes (Dubs) in all lysis buffers. Since these enzymes have a cysteine at their active site, the sulfhydryl alkylating agent N-ethylmaleimide (NEM) is commonly used. A highly specific Dub inhibitor, ubiquitin aldehyde, is available commercially (Affiniti Research Products Ltd., Devon, UK) but is rather expensive. This inhibitor may be helpful in circumstances when the use of NEM is not practical.

Immunoprecipitation of Target Protein from Cell Lysates Followed by Anti-Ub Immunoblotting

A standard protocol to assay protein ubiquitination by immunoprecipitation and immunoblotting is outlined, and an example of results from a typical experiment is shown in Fig. 1.

1. Grow a 10-ml culture of cells to mid-log phase ($OD_{600} = 0.5$–1.0).
2. Collect cells by centrifugation ($1200g$, 5 min, room temperature) and wash once with 10 ml sterile water.
3. Centrifuge cells as before and prepare a lysate from the pelleted cells. Although extracts can be prepared successfully in a number of ways, we have found that lysis with glass beads in 100% ethanol containing freshly dissolved NEM effectively preserves Ub–protein conjugates and is both rapid and easy. For ethanol–glass bead lysis, the cell pellet is resuspended in 500 μl ice-cold ethanol containing 50 mM NEM (Sigma, St. Louis, MO) and $\sim$300 μl acid-washed glass beads (425–600 μm size, Sigma) are added. The samples are vortexed continuously for 5 min at room temperature, and the liquid (which is cloudy from precipitated protein) is removed to a new tube (do not centrifuge the sample, the glass beads will settle rapidly). Wash the glass beads twice with 500 μl fresh ethanol/NEM and pool the liquid (total $\sim$1.5 ml). Dry the samples completely in a Speed-Vac without heating. The protein precipitates are resuspended in 100 μl sodium dodecyl sulfate (SDS) buffer [1% (w/v) SDS, 45 mM Na-HEPES, pH 7.5] by vortexing and heating at 100° for 5 min. Dilute the sample with 1 ml Triton lysis buffer [1% (v/v) Triton X-100, 150 mM NaCl, 50 mM Na-HEPES, pH 7.5, 5 mM Na-EDTA] containing 10 mM NEM [diluted from a freshly prepared 1 M stock in dimethyl sulfoxide (DMSO)] and protease inhibitors [complete protease inhibitor cocktail tablets (Roche Molecular Biochemicals, Indianapolis, IN) supplemented with 10 μg/ml pepstatin]. Remove any remaining precipitate by centrifugation at 4° for 15 min at 14,000g.
4. Immunoprecipitate the protein of interest from the cell lysates. Transfer normalized volumes of extract to fresh tubes (extracts are typically normalized according to culture OD_{600}, but normalization based on total protein concentration

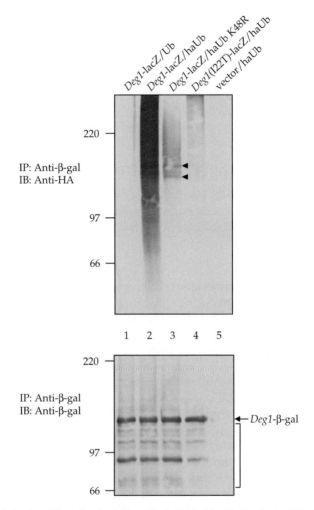

FIG. 1. Evaluation of Deg1-β-galactosidase (Deg1-β-gal) ubiquitination *in vivo*. Ubiquitination was assayed by immunoprecipitating Deg1-β-gal from ethanol/NEM lysates with anti-β-galactosidase antibodies (rabbit polyclonal, ICN), followed by immunoblotting with anti-HA antibodies (top: mouse monoclonal 16B12, Covance Research Products) or anti-β-galactosidase antibodies (bottom). Extracts were made from cells expressing Deg1-β-gal and Ub (lane 1), Deg1-β-gal and HA-tagged Ub (lane 2), Deg1-β-gal and HA-tagged Ub-K48R (lane 3), Deg1-β-gal carrying the I22T mutation and HA-tagged Ub (lane 4), or the empty YEplac195 vector and HA-tagged Ub (lane 5). The I22T mutation is located within the Deg1 portion of the protein and has been shown previously to stabilize a number of different Deg1-containing proteins [P. R. Johnson, R. Swanson, L. Rakhilina, and M. Hochstrasser, *Cell* **94**, 217 (1998)]. Overexpression of haUb-K48R, a Ub chain terminator, inhibits the formation of highly ubiquitinated forms of Deg1-β-gal and leads to the relative overaccumulation of short chain Ub–Deg1-β-gal conjugates (arrowheads). The anti-HA immunoreactivity that extends below the position of full-length Deg1-β-gal represents ubiquitinated degradation products of Deg1-β-gal (marked by a bracket). To the left are the molecular masses (in kilodaltons) of mass standard proteins.

or cell count may be preferable). Add an appropriate amount of antibody and incubate at 4° for 1–4 hr with end-over-end rotation. After incubation, add protein A-agarose (Repligen, Needham, MA) and rotate end over end for 1 hr at 4°. Protein G beads should be substituted when using antibodies that are not bound effectively by protein A.[7] For epitope-tagged proteins, anti-tag antibodies covalently coupled to beads are available commercially and convenient for precipitations. Collect the immunoprecipitates by brief centrifugation (<5 sec at top speed in a microcentrifuge) and wash the beads three times with Triton lysis buffer supplemented with 0.05% (w/v) SDS.

5. After completely removing the wash buffer from the beads with a Pipetman, add SDS–PAGE sample buffer [5× : 30% (v/v) glycerol, 10% (w/v) SDS, 0.3 M Tris–HCl, pH 6.8, 0.5 M dithiothreitol (DTT), 0.1 mg/ml bromphenol blue] and place in a boiling water bath for 3 min. Load the samples on a SDS–PAGE gel, electrophorese, and transfer to PVDF or nitrocellulose membrane. Note that Western transfer of high molecular mass Ub conjugates may be more efficient from a low percentage polyacrylamide gel.

6. After blocking nonspecific binding sites, incubate the membrane with a dilution of an anti-Ub monoclonal antibody (e.g., clone P4D1, Covance Research Products, Berkeley, CA). Alternatively, the anti-Ub monoclonal antibody sold by Zymed Laboratories (South San Francisco, CA) has been reported to give good results with yeast Ub–protein conjugates as well.[8] The immunoblot signal is developed with ECL (enhanced chemiluminescence), ECF (enhanced chemifluorescence) (Amersham-Pharmacia Biotech) or ^{125}I-labeled reagents according to the manufacturer's instructions. Using ^{125}I-labeled or chemifluorescent detection reagents allows for the ubiquitinated species to be quantitatively analyzed.

Constructs that express N-terminally epitope-tagged Ub (Table I) in cells can also be used in these types of experiments (Fig. 1). Endogenous Ub need not be deleted from these cells; overproduction of the tagged Ub is usually sufficient. In this case, the immunoblot signal is developed with an anti-tag antibody. Using these constructs also generates a useful negative control: cells expressing untagged Ub should be free of signal (Fig. 1, top, lane 1). A number of other control experiments can be performed to substantiate any positive results. Although it is not always possible (e.g., with essential proteins), cells lacking the antigen used to immunoprecipitate the target should be assayed in parallel (Fig. 1, lane 5). With epitope-tagged substrates, the corresponding strain expressing the untagged protein is routinely used. Assays with mutant strains in which Ub activation is

---

[7] E. Harlow and D. Lane, "Using Antibodies: A Laboratory Manual." Cold Spring Harbor Laboratory Press, Cold Spring Harbor, NY, 1999.
[8] N. W. Bays, R. G. Gardner, L. P. Seelig, C. A. Joazeiro, and R. Y. Hampton, *Nature Cell Biol.* **3**, 24 (2001).

TABLE I
PLASMIDS FOR OVEREXPRESSION OF MUTANT AND EPITOPE-TAGGED Ub[a]

Ub allele[b]	Marker gene	Reference
Ub	TRP1, LYS2, or URA3	20, 24, 34
Ub-K29R	LYS2	21
Ub-K48R	TRP1 or LYS2	15, 24
Ub-K63R	LYS2	21
Ub-no-K	URA3	23
Ub-K48R G76A	TRP1 or LYS2	24, 25
mycUb	TRP1 or URA3	14, 20
mycUb-K29R	URA3	20
mycUb-K48R	TRP1 or URA3	20, 25
mycUb-K63R	URA3	20
mycUb-no-K	URA3	20
mycUb-K48R G76A	TRP1	25
haUb	TRP1	15
haUb-K48R	TRP1	15
His$_6$Ub	TRP1 or LYS2	35, 36

[a] All plasmids are based on a 2-μm vector that contains the Ub-coding sequences under the control of the copper-inducible CUP1 promoter.

[b] Ub-no-K, ubiquitin with all seven lysines mutated to arginine; mycUb, ubiquitin N-terminally tagged with the myc epitope; haUb, ubiquitin N-terminally tagged with the influenza hemagglutinin (HA) epitope; His$_6$Ub, ubiquitin N-terminally tagged with a hexahistidine sequence.

impaired or Ub levels are depleted (e.g., cells carrying the $uba1$-2[9] or $doa4\Delta$[10] mutations) should also produce a decrease in signal. Further experiments could include analysis of cells lacking the enzymes suspected to be required for substrate ubiquitination and/or experiments with cells carrying target protein mutations that are suspected to inhibit Ub attachment (Fig. 1, lane 4). Of course, these latter experiments require substantial information about the system being studied, usually gleaned from prior analysis of degradation of the target protein.[11]

Immunoblot signals obtained with these experiments are often smears of immunoreactivity (Fig. 1). It is possible to increase the signal from these immunoblots by overexpressing Ub, the target protein, or both. A danger to this approach is that abnormally high levels of substrate (and possibly of Ub) may lead to substrate ubiquitination that does not occur normally. The sensitivity of anti-Ub immunoblots may also be improved by boiling or autoclaving the membrane after Western

[9] R. Swanson and M. Hochstrasser, FEBS Lett. **477,** 193 (2000).
[10] S. Swaminathan, A. Y. Amerik, and M. Hochstrasser, Mol. Biol. Cell **10,** 2583 (1999).
[11] D. Kornitzer, Methods Enzymol. **351,** [37], 2002 (this volume).

transfer, which presumably denatures and exposes buried epitopes.[12] Finally, a note of caution: the use of anti-Ub antibodies for these immunoblots may cross-react and detect protein modification by a Ub-like protein instead of Ub,[13] as a protein highly related to Ub, Rub1, exists in the yeast proteome.

Immunoprecipitation from Radiolabeled Extracts

Another method to detect Ub modification of a target protein involves immunoprecipitation of the protein with anti-target antibodies from lysates of radiolabeled cells expressing wild-type Ub or epitope-tagged Ub.[14,15] The precipitates are then examined for minor species that migrate more slowly than the target protein in SDS–PAGE gels. To confirm that these higher molecular mass species contain the target protein, parallel immunoprecipitation experiments with cells that do not express the target should be compared. To provide evidence that the slowly migrating species contain Ub, the immunoprecipitated material from extracts of cells expressing high levels of epitope-tagged Ub or untagged Ub is examined. The molecular mass of the presumptive Ub–protein conjugates should increase when untagged Ub is replaced with epitope-tagged Ub.[14,15] Alternatively, the immunoprecipitate can be boiled in SDS buffer, diluted as before in Triton lysis buffer, and reprecipitated with an antibody to the epitope tag of Ub. The precipitated material from this second precipitation should comigrate with the higher molecular weight species from the initial immunoprecipitation. Furthermore, these species should be absent from serial immunoprecipitations of extracts of cells expressing untagged Ub.[15]

To prepare a lysate of radiolabeled cells for immunoprecipitation, a 5-ml exponentially growing culture ($OD_{600} = 0.5–1.0$) is harvested by centrifugation (1200g, 5 min, room temperature), washed with 1 ml minimal media (0.67% w/v yeast nitrogen base without amino acids, 2% w/v glucose), transferred to a 1.5-ml tube, and washed again with the same media. Cells are resuspended in 200 μl of SD-Met media[16] and placed at 30°. Pulse labeling is initiated by the addition of 100–150 μCi of Tran^{35}S-label (ICN, Costa Mesa, CA) and continued for 5–10 min, with intermittent vortexing to keep cells in suspension. The cells are collected by centrifugation at 14,000g for 10 sec, resuspended in 50 μl water, and lysed by adding 50 μl of 2% (w/v) SDS/90 mM Na-HEPES, pH 7.5/30 mM (DTT) and

---

[12] P. S. Swerdlow, D. Finley, and A. Varshavsky, *Anal. Biochem.* **156**, 147 (1986).

[13] A. R. Willems, S. Lanker, E. E. Patton, K. L. Craig, T. F. Nason, N. Mathias, R. Kobayashi, C. Wittenberg, and M. Tyers, *Cell* **86**, 453 (1996).

[14] M. J. Ellison and M. Hochstrasser, *J. Biol. Chem.* **266**, 21150 (1991).

[15] M. Hochstrasser, M. J. Ellison, V. Chau, and A. Varshavsky, *Proc. Natl. Acad. Sci. U.S.A.* **88**, 4606 (1991).

[16] F. Sherman, G. R. Fink, and J. B. Hicks, "Methods in Yeast Genetics." Cold Spring Harbor Laboratory Press, Cold Spring Harbor, NY, 1986.

heating at 100° for 5 min. The lysate is diluted with 1 ml of Triton lysis buffer (see earlier discussion) and cleared by centrifugation at 14,000g for 15 min. Immunoprecipitations are usually performed with extract volumes that are normalized according to the amount of radiolabeled protein present, typically determined by the amount of radioactivity precipitated with 10% (v/v) trichloroacetic acid (TCA). A 10-μl sample of each extract is spotted onto a filter paper disk (740E, Schleicher and Schuell, Keene, NH) and allowed to air dry. The disks are dipped into 10% TCA and blotted face up on a paper towel. This procedure is repeated with fresh 10% TCA, and the filters are soaked in 10% TCA at 100° for 5 min. After blotting, the disks are washed twice in 10% TCA (as before), followed by two washes in 100% ethanol. The filter disks are dried (air dried or dried under a heat lamp), placed in vials containing scintillation fluid, and counted in a scintillation counter.

Affinity Purification of Ubiquitinated Proteins from Cells Expressing His$_6$-Ubiquitin

A third method to assay the ubiquitination of a protein of interest utilizes hexahistidine (His$_6$)-tagged Ub as a purification reagent for Ub–protein conjugates.[13,17] Outlined here is a method from Kaiser et al.[17]

1. Cells expressing His$_6$-tagged Ub are lysed in buffer C [6 M guanidinium–HCl/50 mM phosphate buffer, pH 8.0/10 mM Tris–HCl, pH 8.0/300 mM NaCl/ 5 mM NEM (from a freshly prepared stock)/1 mM phenylmethylsulfonyl fluoride (PMSF)/2 μg/ml of aprotinin, leupeptin, and pepstatin] with glass beads. Extraction with 6 M guanidinium-hydrochloride/NEM appears to effectively preserve Ub-modified species.

2. Centrifuge the lysate at 4° for 15 min. Combine the clarified extract with Ni^{2+}-NTA-agarose beads (50 μl of beads per 2.75 mg total protein) and add imidazole to a final concentration of 10 mM. Incubate with end-over-end rotation for 4 hr at 4°.

3. Wash the beads with 1 ml buffer C + 20 mM imidazole, 1 ml buffer D (same as buffer C, but with the 6 M guanidinium-hydrochloride replaced by 8 M urea) + 20 mM imidazole, and twice with 1 ml buffer D adjusted to pH 6.0 + 20 mM imidazole. Wash the beads well with the urea-containing buffers, as it is critical to remove all traces of guanidinium that will precipitate in the presence of SDS.

4. Elute the bound material from the beads by boiling in urea sample buffer [2× : 8 M urea, 4% (w/v) SDS, 0.125 M Tris–HCl, pH 6.8, 0.2 M DTT] and analyze by SDS gel electrophoresis and immunoblotting with antibodies directed against the target protein.

[17] P. Kaiser, K. Flick, C. Wittenberg, and S. I. Reed, *Cell* **102,** 303 (2000).

A positive result with extracts expressing His_6-tagged Ub and the target protein is validated by the absence of an immunoblot signal with extracts of cells lacking the protein of interest and/or with lysates of cells lacking His_6-tagged Ub.

A direct comparison of the three different methods for detecting Ub modification has not been reported. Therefore, it is unclear if a negative result from one assay necessarily means that substrate ubiquitination cannot be detected by an alternative method. In fact, our unpublished results indicate that the ubiquitinated forms of the Matα2 repressor, which can be detected readily in immunoprecipitations of radiolabeled extracts, cannot be visualized by the immunoprecipitation/anti-Ub immunoblotting approach. If modification of a protein substrate by Ub is suspected, multiple methods should therefore be attempted. Furthermore, the results of these different assays may be visually quite different, ranging from a smear of immunoreactivity to a distinct ladder of bands (or a single band) above the unmodified species, depending on the substrate being analyzed and the detection method that is used.

Detection of Ubiquitin Chains Conjugated to the Protein Substrate

Efficient targeting of ubiquitinated proteins to the 26S proteasome for proteolysis requires the formation of a poly-Ub chain on the substrate protein.[18,19] The individual Ub subunits in these chains are linked to one another through the carboxyl-terminal glycine (G76) of one Ub joined covalently to the side chain of a lysine (K48) in another Ub molecule (a K48-linked chain).[18] In addition, Ub chains with alternative amide linkages (through K29 or K63) also exist in cells and appear to function in a number of other cellular processes.[20–22] Ub need not operate solely through polymeric chains. For example, some endocytosis events are triggered by monoubiquitination of the substrate protein.[23]

To address whether a protein of interest is modified by a Ub polymer and to determine the topology of the Ub monomers in these chains, the assays described in the previous section can be performed with strains overexpressing various Ub mutants (Table I). The mutations substitute an arginine residue for different lysines in the Ub sequence; they do not alter the ability of the mutant Ub to be conjugated to proteins but rather inhibit its ability to form multi-Ub chains normally linked through particular lysine residues. In total, there are seven lysines in the Ub sequence. Although each of these lysine residues could potentially serve as a linkage

---

[18] V. Chau, J. W. Tobias, A. Bachmair, D. Marriott, D. J. Ecker, D. K. Gonda, and A. Varshavsky, *Science* **243,** 1576 (1989).
[19] J. S. Thrower, L. Hoffman, M. Rechsteiner, and C. M. Pickart, *EMBO J.* **19,** 94 (2000).
[20] T. Arnason and M. J. Ellison, *Mol. Cell Biol.* **14,** 7876 (1994).
[21] J. Spence, S. Sadis, A. Haas, and D. Finley, *Mol. Cell. Biol.* **15,** 1265 (1995).
[22] J. Galan and R. Haguenauer-Tsapis, *EMBO J.* **16,** 5847 (1997).
[23] J. Terrell, S. Shih, R. Dunn, and L. Hicke, *Mol. Cell* **1,** 193 (1998).

site for a multi-Ub chain, evidence only exists for chains with linkages involving three of them (K29, K48, and K63).[20,21] The K48 residue of Ub is essential for viability, whereas cells expressing Ub mutants carrying the other lysine substitutions are viable.[21,24] Therefore, to assay for K48-linked chains, overexpression of the K48R mutant (or Ub-no-K) must be performed in the presence of endogenous wild-type Ub. Because the other lysine residues are not essential, those mutations may be assayed in a strain whose sole source of Ub in the cell is the mutant version.[21]

To ascertain if a substrate is conjugated to a multi-Ub chain with K48 linkages, for example, one can overexpress Ub-K48R in cells and compare the ubiquitination pattern to that in cells overexpressing wild-type Ub (see Fig. 1). Similar experiments can be performed with other Ub mutant constructs to determine if chains with alternative linkages form on the protein of interest. If overexpression of the chain-terminating mutant Ub suppresses the formation of a multi-Ub chain on the protein, it will be visible as a shift from more highly ubiquitinated forms to conjugates with fewer Ub molecules attached (Fig. 1, compare lanes 2 and 3). This effect is sometimes more pronounced in cells overexpressing a doubly mutant Ub construct in which the chain-terminating Lys-to-Arg substitution is combined with a G76A mutation. This latter mutation inhibits deubiquitination, so that once a Ub chain is capped with the doubly mutant Ub, it is irreversibly modified because the capped chain cannot be disassembled efficiently.[24,25] As with the Ub-K48R single mutant, an increase in the monoubiquitinated species at the expense of polyubiquitinated forms should be observed with this double mutant.

The $doa4\Delta$ mutation may also be useful for these types of experiments. Cells lacking the Doa4-deubiquitinating enzyme grow fairly robustly, yet contain low levels of Ub.[10] Consequently, a number of Ub-dependent processes are inhibited in $doa4$ mutants.[10,22,23,26] By reducing the amount of competing wild-type ubiquitin, the $doa4\Delta$ mutant background can be used to examine if overexpression of the different chain-terminating Ub mutations can still promote the formation of multi-Ub chains on the substrate of interest. If, for example, the K63R mutation cannot rescue chain formation, this suggests that the substrate is modified by K63-linked Ub chains. Using this approach, such alternative Ub chains have been linked to the endocytosis of certain membrane proteins.[22]

A common question is whether the formation of a multi-Ub chain on a substrate is required for degradation of that protein. To address this, the effect of overexpressing a chain-terminating Ub mutant (e.g., Ub-K48R) on the degradation of the substrate[11] and the formation of Ub chains is examined. A correlation between a

---

[24] D. Finley, S. Sadis, B. Monia, P. Boucher, D. Ecker, S. Crooke, and V. Chau, *Mol. Cell. Biol.* **14**, 5501 (1994).
[25] R. R. W. Hodgins, K. S. Ellison, and M. J. Ellison, *J. Biol. Chem.* **267**, 8807 (1992).
[26] F. Papa and M. Hochstrasser, *Nature* **366**, 313 (1993).

defect in the degradation of the substrate and a reduction of poly(Ub) chains on the protein is evidence that formation of a Ub chain on the substrate is necessary for its degradation.

Gene Products Required for Substrate Ubiquitination

Determining what genes are required for a particular process *in vivo* requires genetic analysis. Indeed, mutant screens have been quite useful in defining the requirements for a number of Ub-dependent processes. Because Ub conjugation is commonly associated with protein degradation, screening for mutations that stabilize a protein of interest can lead to the identification of proteins required for substrate ubiquitination.[27-29] To address whether any of these genes function in ubiquitination of the substrate, the mutant strains may be assayed with any of the methods described earlier. Alternatively, a more directed approach for identifying genes required for Ub modification is possible: sifting through a panel of strains deficient for the known genes encoding Ub-conjugating and -ligating enzymes using the ubiquitination assays outlined earlier may identify necessary factors. This is not a small task, however, as there are 11 different Ub-conjugating enzymes (E2s) and a greater number of Ub-ligating enzymes (E3s).

Cell-free reactions have also aided in defining the requirements for substrate ubiquitination in yeast. Biochemical systems that recapitulate certain Ub-dependent reactions have been established, and *in vitro* ubiquitination has been observed in extracts made from either whole yeast cells or spheroplasts.[30,31] These methods provide yet another assay to determine whether a gene product is involved in substrate modification by Ub. Extracts made from mutant strains can be assayed to determine their competence to promote substrate ubiquitination *in vitro*. Results from such experiments can provide evidence that the gene product is required for ubiquitination of a particular protein, but as with the genetic experiments outlined earlier, the assays cannot establish whether that product is directly involved in the modification reaction. Reconstitution of the ubiquitination reaction using purified components is necessary for this conclusion.[32,33]

---

[27] M. Hochstrasser and A. Varshavsky, *Cell* **61**, 697 (1990).
[28] S. Irniger, S. Piatti, C. Michaelis, and K. Nasmyth, *Cell* **81**, 269 (1995).
[29] R. Y. Hampton, R. G. Gardner, and J. Rine, *Mol. Biol. Cell* **7**, 2029 (1996).
[30] W. Zachariae and K. Nasmyth, *Mol. Biol. Cell* **7**, 791 (1996).
[31] R. Verma, Y. Chi, and R. J. Deshaies, *Methods Enzymol.* **283**, 366 (1997).
[32] D. Skowyra, K. L. Craig, M. Tyers, S. J. Elledge, and J. W. Harper, *Cell* **91**, 209 (1997).
[33] R. M. Feldman, C. C. Correll, K. B. Kaplan, and R. J. Deshaies, *Cell* **91**, 221 (1997).
[34] D. J. Ecker, M. I. Khan, J. Marsh, T. R. Butt, and S. T. Crooke, *J. Biol. Chem.* **262**, 3524 (1987).
[35] M. Hochstrasser, unpublished observations.
[36] R. Ling, E. Colon, M. E. Dahmus, and J. Callis, *Anal. Biochem.* **282**, 54 (2000).

[15] Vesicle Budding from Endoplasmic Reticulum

By YUVAL SHIMONI and RANDY SCHEKMAN

Introduction

Protein transport along the secretory pathway has historically been evaluated by molecular characterization of covalent modifications on transported proteins, lipids, and carbohydrates and by the fractionation and evaluation of membrane compartments. In a previous volume in this series on yeast genetics and molecular and cell biology, which was written over a decade ago, protocols for organelle fractionation, membrane association, and analysis of marker enzymes, along with other protocols for the analysis of protein transport, were described in detail.[1] Since that time, progress has focused on the evaluation of transport events in cell-free systems that combine isolated membranes and purified cytosolic proteins. This chapter details the procedures necessary to evaluate polypeptide translocation and protein sorting into transport vesicles using purified components.

Cell-free reactions that reproduce transport steps between organelles in the secretory pathway are used to identify the proteins involved in particular steps and mechanistic aspects of this process.[2] The first cell-free reaction that achieved intercompartmental transport was established with permeabilized mammalian cells,[3] and proteins involved in vesicle formation and membrane fusion were isolated and their role was elucidated by reconstitution. However, the discovery of the full range of gene products involved and establishing their cellular roles is limited in the mammalian system. Our effort began with a genetic approach through the isolation of temperature-sensitive yeast secretory (*sec*) mutants.[4] Genetic and morphological analysis defined the genes required for vesicle formation and fusion in the endoplasmic reticulum (ER) to Golgi limb of the secretory pathway,[5] and a temperature-dependent Sec protein defect was reconstituted *in vitro*.[6] This allowed a detailed characterization of the transport process[7] and isolation of the proteins necessary for vesicle budding using biochemical complementation of the transport defect.[6,8] One outcome of this approach was the identification of a novel coat protein complex, COPII, comprising the cytosolic proteins required for vesicle

---

[1] A. Franzusoff, J. Rothblatt, and R. Schekman, *Methods Enzymol.* **194**, 662 (1991).
[2] N. K. Pryer, L. J. Wuestehube, and R. Schekman, *Annu. Rev. Biochem.* **61**, 471 (1992).
[3] E. Fries and J. E. Rothman, *Proc. Natl. Acad. Sci. U.S.A.* **77**, 3870 (1980).
[4] P. Novick, C. Field, and R. Schekman, *Cell* **21**, 205 (1980).
[5] C. A. Kaiser and R. Schekman, *Cell* **61**, 723 (1990).
[6] D. Baker, L. Hicke, M. Rexach, M. Schleyer, and R. Schekman, *Cell* **54**, 335 (1988).
[7] M. F. Rexach and R. W. Schekman, *J. Cell Biol.* **114**, 219 (1991).
[8] L. Hicke, T. Yoshihisa, and R. W. Schekman, *Methods Enzymol.* **219**, 338 (1992).

formation.[9] Other requirements for coat assembly, including nucleotide dependence and the GTP-dependent cycle of COPII assembly and dynamics, emerged from studies of the purified proteins.[10,11] Components involved in vesicle targeting, docking, and fusion with the target organelle were also discovered using these approaches, and a cell-free reaction that reconstitutes a full cycle of transport and retrieval between the ER and the Golgi has been established.[12]

Secretory proteins comprising soluble and membrane-bound cargo molecules are assembled in the ER and sorted from resident ER proteins by their capture into transport vesicles that are directed to the Golgi apparatus. Vesicle formation and cargo packaging have been reconstituted with yeast membranes and soluble proteins in what is referred to as a budding assay. Reconstitution of vesicle budding requires three cytosolic components: Sar1p (21 kDa), a small GTPase belonging to the *RAS* superfamily, a ~400-kDa heterodimeric complex containing Sec23p (85 kDa), the Sar1p-specific GTPase-activating protein (GAP), and Sec24p (105 kDa) and a larger heterooligomeric complex of ~700 kDa composed of two WD-40 motif-containing protomers, Sec13p (34 kDa) and Sec31p (150 kDa). In the presence of nucleotides and a source of ER membranes, these three components form a proteinaceous coat, called COPII, on the membrane and drive the formation of vesicles containing cargo proteins and excluding ER resident proteins. In yeast, Sec24p has two functional nonessential homologues that copurify with Sec23p, termed Iss1p and Lst1p. A putative Sec23p homolog [open reading frame (ORF) : YHR035$_W$] was also identified in the yeast genome, but the nature of this gene product and its associations have not been reported.

We have compiled here our current protocols for the preparation of donor ER membranes and procedures for the isolation of the cytosolic coat components. We describe their use in the *in vitro* biochemical analysis of vesicle formation and the packaging of soluble and membrane proteins into COPII vesicles.

Preparation of Donor Membranes for Budding Assay

Either one of two membrane preparations, microsomes or semi-intact cells (SICs), is typically used in the budding assay as a source of ER.[13,14] Microsomes are a subcellular preparation enriched with ER membranes (but still contaminated with Golgi and other cellular membranes). SICs are perforated yeast spheroplasts (yeast cells stripped of their cell wall). Microsomes take longer to prepare but may

---

[9] C. Barlowe, L. Orci, T. Yeung, M. Hosobuchi, S. Hamamoto, N. Salama, M. F. Rexach, M. Ravazzola, M. Amherdt, and R. Schekman, *Cell* **77,** 895 (1994).
[10] T. Yoshihisa, C. Barlowe, and R. Schekman, *Science* **259,** 1466 (1993).
[11] C. Barlowe and R. Schekman, *Nature* **365,** 347 (1993).
[12] A. Spang and R. Schekman, *J. Cell Biol.* **143,** 589 (1998).
[13] L. J. Wuestehube and R. W. Schekman, *Methods Enzymol.* **219,** 124 (1992).
[14] M. F. Rexach, M. Latterich, and R. W. Schekman, *J. Cell Biol.* **126,** 1133 (1994).

be beneficial, e.g., if high background budding levels are obtained with SICs and in morphological analysis. Although cruder than microsomes, SICs are advantageous when packaging of a newly synthesized cargo protein is to be examined because they can be made easily from pulse labeled cells (see later discussion). In such a case, protein analysis is confined to nascent protein (i.e., mostly ER localized) and therefore additional purification of the ER can be avoided. Importantly, the integrity of the ER and other subcellular organelles is largely preserved in SICs, whereas most cytosolic proteins are lost. When a cleaner source of ER membranes is required, either microsomes or nuclei preparations are used. The nuclear envelope is contiguous with the rough ER and because nuclei preparations are relatively free of other organelles and membranes they can serve as an uncontaminated source of ER as is evident morphologically.[15] Liposomes of a defined phospholipid composition have also been used as a membrane source in budding assays.[16] The use of liposomes in vesicle formation and cargo sorting has been described elsewhere.[17]

Reagents and Buffers

YP: 1% (w/v) Bacto yeast extract, 2% (w/v) Bacto-peptone (Difco Laboratories Inc., Detroit, MI)

YPD: 1% (w/v) Bacto yeast extract, 2% (w/v) Bacto-peptone (Difco), 2% (w/v) glucose

SD: 0.67% (w/v) yeast nitrogen base with ammonium sulfate, 2% (w/v) glucose, and required nutrients

Spheroplasting buffer: 0.7 M sorbitol, 10 mM Tris–HCl (pH 7.4), 1 mM dithiothreitol (DTT), 20 mM NaN$_3$

Lysis buffer: 0.4 M sorbitol, 0.15 M potassium acetate, 20 mM HEPES (pH 6.8), 2 mM magnesium acetate, 0.5 mM EGTA

JR lysis buffer: 0.2 M sorbitol, 50 mM potassium acetate, 2 mM EDTA, 20 mM HEPES (pH 7.4), 1 mM DTT, 1 mM phenylmethylsulfonyl fluoride (PMSF)

Regeneration buffer: 0.7 M sorbitol, 0.75% (v/v) YP, 1% glucose

B88: 20 mM HEPES, pH 6.8, 250 mM sorbitol, 150 mM potassium acetate, 5 mM magnesium acetate

Sucrose solutions: 20 mM HEPES, pH 7.4, 50 mM potassium acetate, 2 mM EDTA, 1.2 or 1.5 M sucrose (as required), 1 mM DTT, and 1 mM PMSF (added prior to use)

[15] S. Y. Bednarek, M. Ravazzola, M. Hosobuchi, M. Amherdt, A. Perrelet, R. Schekman, and L. Orci, *Cell* **83**, 1183 (1995).
[16] K. Matsuoka, L. Orci, M. Amherdt, S. Y. Bednarek, S. Hamamoto, R. Schekman, and T. Yeung, *Cell* **93**, 263 (1998).
[17] K. Matsuoka and R. Schekman, *Methods* **20**, 417 (2000).

Preparation of Semiintact Cells

The cell wall is a rigid structure that defines the morphology of the yeast cell and preserves its osmotic integrity. Cell wall removal ("spheroplasting") and lysis rupture the plasma membrane but not intracellular membranes and release the cytosolic content. This treatment can therefore be used as an initial step in the preparation of various subcellular organelles (such as the microsomal preparation described later). Because spheroplasts are capable of regenerating cell wall polymers, they can also be used in cell wall biogenesis studies. Perforation of the spheroplasts to generate SICs also provides access (for exogenous macromolecules) to the ER membranes and makes them an easy and convenient source of donor membranes for the budding assay.

SICs may be prepared from either pulse radiolabeled yeast cells (i.e., for the analysis of packaging of newly synthesized cargo proteins, which can be genetically tagged for this purpose; see "The COPII Vesicle Packaging Assay" later) or from unlabeled cells. Unlabeled cells can be loaded with a radioactive tracer cargo (e.g., the α-factor precursor protein). Here we describe the preparation of both labeled and unlabeled SICs. Several preparations are often done in parallel (e.g., for the comparison of packaging of a newly synthesized cargo protein to that of a radiolabeled cargo introduced into the ER by posttranslational translocation). The procedure consists of an optional metabolic labeling of the cells followed by spheroplast formation[1] and perforation in lysis buffer.

Procedure. Freshly plated cells are inoculated in a small volume of YPD and are grown at 30° overnight. The next day the cells are transferred to 100 ml YPD starting from an OD_{600} of 0.1. Growth is followed and after $\sim$6 h at an OD_{600} of 0.5–1 (early to midlog phase) the cells are harvested (5 min at 4000g, in a Sorvall GSA rotor) and washed three times in minimal (SD) medium without methionine (Met) and cysteine (Cys). Cells are resuspended in 15–30 ml of the just-described medium (at $\sim$5 OD_{600}/ml) and are distributed into three disposable tubes (e.g., 50-ml polypropylene conical tubes) and shaken for another 15 min at 30°. This amino acid deprivation is done to improve incorporation of the radiolabeled amino acids during pulse labeling. Two culture tubes are labeled with 1 mCi (100 μl) of ^{35}S-Promix (1200 Ci/mmol, Amersham Pharmacia Biotech, Piscataway, NJ) for 3 min (the duration of pulse labeling may need optimization and the shortest time allowing good detection should be used) at 30° and one is left unlabeled. Metabolic activity is stopped by the addition of an equal volume of 40 mM ice-cold sodium azide (to 20 mM final). The cells are incubated on ice for 10 min and washed with 10 ml ice-cold 20 mM sodium azide. Cells are resuspended in 1 ml of a freshly made 0.1 M Tris–HCl (pH 9.4), 10 mM DTT, 20 mM NaN$_3$ solution, transferred to 1.5-ml Eppendorf tubes, and incubated for 10 min at room temperature. This incubation loosens the outer mannoprotein layer, which promotes the activity of the lytic enzyme on the glucan layer of the cell wall. Cells are then sedimented in a tabletop centrifuge at 10,000 rpm for 30 sec, and the pellet is resuspended in 1 ml spheroplasting buffer. An aliquot (from the unlabeled tube) is diluted 1 : 200 in H_2O and saved (for determination of the efficiency of spheroplast formation, see later). Lyticase[18] is added at ~3600 U/ml to each tube with the cell suspension, and all tubes are placed in a roller device and rotated for 30 min at 30°. (Commercial zymolase may be also used for cell wall removal.) Following this incubation, an aliquot (from the unlabeled tube) of lyticase-treated cells is diluted 1 : 200 in H_2O and both 1 : 200 cell dilutions of lyticase-treated and -untreated cells are read at OD_{600} to determine the extent of cell lysis. The OD_{600} reading of lyticase-treated cells should be less than 10% of the initial one. Spheroplast formation is crucial for obtaining efficient vesicle budding and is best when cells grow under optimal conditions (temperature and media; see later). Each one of the tubes containing the spheroplasts is divided into two equal volumes and is centrifuged briefly in a tabletop centrifuge. Pellets are rinsed in 1 ml lysis buffer, centrifuged again, and resuspended in 100 μl lysis buffer on ice. Labeled and unlabeled SICs are aliquoted as convenient for the number of reactions to be performed in each experiment (see "COPII Vesicle Packaging Assay" later). Aliquots are frozen over liquid nitrogen vapor for 1 hr (this is done by placing the tubes in an Eppendorf tube holder that is fixed on top of an ice bucket containing liquid nitrogen in the cold room) and stored at $-80°$ until use. Slow freezing is important to prevent organelle rupture and preserves spheroplasts intact.

Newly synthesized proteins are often detected by epitope tagging their genes in an expression system. If the tagged gene is placed under an inducible promoter, an induction period must be included prior to labeling. For example, for a gene controlled by a *GAL* promoter, a glucose depletion step done by a period of growth in a raffinose-containing medium followed by galactose induction should be included. The length of glucose depletion, which serves to prevent repression of the promoter, and galactose induction should be determined experimentally. Minimal medium containing galactose and lacking Met and/or Cys can be used to induce gene expression concomitantly with specific amino acid deprivation, just prior to pulse labeling.

This procedure can be scaled up for the preparation of larger amounts of unlabeled SICs (i.e., omitting Met/Cys starvation and [^{35}S]Met/Cys pulse-labeling steps). For a scaled-up preparation of SICs, see "Preparation of Microsomes." In this case, an additional step of resuspending the spheroplasts (after the lyticase treatment) in regeneration buffer may be performed (this may improve budding activity), and the spheroplasts are shaken very gently for 30 min at 30° and then harvested and treated with lysis buffer as described.

Preparation of Microsomes

When cells are homogenized, the ER is fragmented and reseals into microsomal vesicles with their interior being biochemically equivalent to the luminal space of

[18] J. H. Scott and R. Schekman, *J. Bacteriol.* **142**, 414 (1980).

the ER. The following microsomal preparation procedure is described as it is done in our laboratory for use in budding assays. However, microsomal preparations can serve in various studies on ER function and biochemistry because they are purified in a form that is capable of protein and lipid synthesis, as well as protein glycosylation.

The preparation of microsomes begins with the formation of spheroplasts essentially as described earlier. The spheroplasts are then lysed osmotically and homogenized, and a microsomal membrane fraction enriched in ER membranes is obtained following differential centrifugation and sucrose gradient fractionation.

Procedure. Cells (1 liter) are grown overnight in YPD at 30° and harvested at an OD_{600} reading of ~1. Pellets are resuspended to 100 OD_{600} units/ml in freshly made 0.1 M Tris-HCl (pH 9.4), 10 mM DTT, and incubated at room temperature for ~10 min. Cells are harvested again and resuspended to 100 OD_{600} units/ml in lyticase buffer and a small aliquot is kept aside. Lyticase (~50 units/OD_{600} cells) is added, and cells are incubated in a shaking water bath at 30° for 30 min. Spheroplast formation is verified by comparing OD_{600} readings of lyticase-treated to -untreated cells as described earlier. Spheroplasts are chilled on ice for ~2 min and harvested (at 5000 rpm for 10 min in a GSA rotor, 4°). Care should be taken when removing the supernatant because the spheroplast pellet does not adhere to the tube as strongly as do pellets from untreated cells. Spheroplast pellets are resuspended gently in 2XJR lysis buffer (without DTT or PMSF) to ~250 OD_{600} units/ml, centrifuged again (at 10,000 rpm for 5 min in an SS34 rotor, 4°), and resuspended in 2XJR lysis buffer (without DTT or PMSF) to ~1000 OD_{600} units/ml and frozen at −80°.

Lysed spheroplasts are thawed in an ice water bath, and an equal volume of cold water plus DTT and PMSF (each to 1 mM final) is added. All subsequent steps are done quickly and on ice for best results. Spheroplasts are disrupted with 10 strokes of a motor-driven Potter–Elvehjem homogenizer. Cell breakage is assessed with the aid of a microscope. The homogenate is centrifuged (at 3000 rpm for 10 min in an SS34 rotor at 4°), and the low-speed supernatant is collected (avoiding the pellet) and centrifuged at 15,000 rpm for 10 min in an SS34 rotor at 4°. The microsomal pellet is resuspended gently in B88 at 2500 OD_{600} units/ml using a Dounce homogenizer (about 7 strokes). The homogenate (1–1.5 ml) is loaded onto a 2-ml step gradient composed of 1 ml each of 1.2 and 1.5 M sucrose solutions (see earlier) in a "thick wall" tube and centrifuged at 40,000 rpm for 1 hr in an SW55 Ti rotor at 4°. The upper volume and vacuole band (at the load/1.2 M sucrose interface) is aspirated, and the ER-enriched microsomal fraction (at the 1.2/1.5 M sucrose step interface) is collected with a Pasteur pipette. Microsomes are diluted with 4–5 volumes of B88 and centrifuged at 15,000 rpm for 10 min in an SS34 rotor at 4°. The microsome pellet is resuspended in a small volume (about equal to the volume of the pellet) of B88 with gentle homogenizing (using a 1-ml Dounce homogenizer). The OD_{280} of a 1:100 dilution in 2% SDS is measured using a quartz cuvette and adjusted with B88 to ~40 (which is equivalent to ~8 mg/ml protein). Aliquots (~50 µl) are frozen in liquid nitrogen and stored at −80°.

Preparation of Cytosol

Cytosol, the fraction of the cytoplasm that excludes membrane-bounded organelles, can be prepared for many purposes ranging from the purification of soluble cellular components to studies on metabolism. The amount of luminal content from various organelles and membrane-associated contamination in the cytosol preparation and the preservation of activity of cytosolic components depend on many factors. These include the conditions and method used (see later) and, importantly, the composition of the buffer used for extraction during homogenization. For example, in the presence of sucrose, ATP is hydrolyzed in yeast cytosol.[19] For this reason, in situations where cytosol is to be used in an energy-requiring reaction, we use sorbitol in the lysis buffer and supplement reactions with ATP and an ATP-regenerating system. The extraction buffer should be tailored for the optimal extraction and solubilization of the desired components and for the preservation of their activity. Cytosol is frequently regarded as a "physiological" mixture in terms of its composition and the relative concentration of its constituents. Therefore, it is used many times in initial experiments to demonstrate the effect of a cytosolic component in a particular assay. Growth conditions, the strain used, and various treatments of the cells prior to preparation can all be tested or manipulated. Cytosol can also serve as a crude source of COPII components. The requirement for a cytosolic component in vesicle budding and cargo packaging activities can be tested by using cytosol prepared from various mutant strains. Cytosol of various sources can also be supplemented with purified components and tested in the budding assay.

There are several ways to prepare yeast cytosol. We describe here a quick and simple procedure based on liquid nitrogen cell lysis. Additional protocols have been described such as the preparation of membrane-free cytosol[14] or a preparation employing glass beads for cell lysis.[13]

Procedure. Yeast cells (3 liter) are grown to midlog phase in the appropriate medium and temperature, depending on the strain used. The cells are harvested by centrifugation (4000g for 5 min at 4°) and washed once with water and once with B88. The cell pellet is resuspended in 2 ml B88, and cell droplets are pipetted into liquid nitrogen, allowing small frozen cell pellets to form. The cell pellets are ground with a pestle in a mortar placed on dry ice. Best lysis is achieved under liquid nitrogen and not when it has all evaporated. If this procedure is scaled up (e.g., ~10-fold or more), it may be more practical to grind the frozen cell pellets in a blender (Waring Corp., New Hartford, CT) under liquid nitrogen. The cells are ground up for 10–15 min with a mortar and pestle until a very fine powder is formed. The cell powder is thawed in an ice/water bath. At this stage, protease

[19] W. G. Dunphy, S. R. Pfeffer, D. O. Clary, B. W. Wattenberg, B. S. Glick, and J. E. Rothman, *Proc. Natl. Acad. Sci. U.S.A.* **83**, 1622 (1986).

inhibitors (e.g., PMSF) are added and 1 mM (final) each of DTT and ATP may be stirred in (optional). The cell lysate is precleared by centrifugation for 10 min at 10,000g (e.g., at 8000 rpm in a GSA rotor), and the supernatant is centrifuged for 1 hr at 100,000g (32,000 rpm in a Beckman 70Ti rotor). The supernatant (cytosol) is collected, carefully avoiding the top lipid layer and the pellet. Protein concentration is determined (typically ∼10 mg/ml) and aliquots are stored at −80°.

Purification of COPII Components

A combination of genetic and biochemical approaches have led to the identification of a set of proteins that in the presence of GTP and ATP satisfies the requirement for cytosol in the cell-free production of vesicles that bud from the ER.[9,20] These proteins (Sar1p and the Sec23p and Sec13p complexes) form the COPII coat.[9] In addition to being biochemically required for vesicle formation from the ER, these proteins actually coat the transport vesicles formed. Another coat complex that functions early in the secretory pathway termed COPI or coatomer[21] mediates anterograde and reterograde transport between the ER and the Golgi and within the Golgi.[15,22] Unlike coatomer, COPII does not form a cytosolic complex, but instead only forms in the context of vesicle formation. Order-of-addition experiments showed that GTP-bound Sar1p is the first to bind to the membrane followed by the Sec23p and finally the Sec13p complexes, which complete the coat.[16] In the presence of nonhydrolyzable analogs of GTP, vesicles form but cannot deliver their contents to the Golgi complex and COPII remains tightly associated with the vesicles.[9] Two functional homologues of Sec24p have been isolated as part of Sec23p complexes.[23–25] More details regarding the development of the procedures for the purification of each one of the COPII proteins are given later.

COPII components, prepared as described later, yield a reproducible pattern that can be recognized following separation on SDS–PAGE. A 10% SDS–PAGE separation profile of each component is given in Fig. 1.

Sar1p Preparation

Sar1p has been purified in the laboratory from yeast and bacteria. Initially, Sar1p was purified from a yeast strain (YPH500) containing a plasmid with *SAR1*

---

[20] N. R. Salama, T. Yeung, and R. W. Schekman, *EMBO J.* **12**, 4073 (1993).
[21] V. Malhotra, T. Serafini, L. Orci, J. C. Shepherd, and J. E. Rothman, *Cell* **58**, 329 (1989).
[22] L. Orci, M. Stamnes, M. Ravazzola, M. Amherdt, A. Perrelet, T. H. Sollner, and J. E. Rothman, *Cell* **90**, 335 (1997).
[23] T. Kurihara, S. Hamamoto, R. E. Gimeno, C. A. Kaiser, R. Schekman, and T. Yoshihisa, *Mol. Biol. Cell* **11**, 983 (2000).
[24] K. J. Roberg, M. Crotwell, P. Espenshade, R. Gimeno, and C. A. Kaiser, *J. Cell Biol.* **145**, 659 (1999).
[25] Y. Shimoni, T. Kurihara, M. Ravazzola, M. Amherdt, L. Orci, and R. Schekman, *J. Cell Biol.* **151**, 973 (2000).

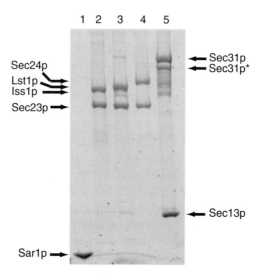

FIG. 1. SDS–PAGE migration pattern of purified COPII components. Purified COPII components were resolved by 10% SDS–PAGE and SYPRO Red stained. Lanes 1–5 were loaded with Sar1p, Sec23/Iss1p, Sec23/Lst1p, Sec23/24p, and Sec13/31p, respectively. Sec31p* indicates a degradation product of Sec31p that is present in the Sec13/31p preparation.

controlled by a *GAL1* promoter.[26] In that protocol, a clarified lysate from induced cells was applied onto a Sephacryl S-100 column followed by loading the peak fractions on a DEAE-Sepharose CL-4B and eluted peak fractions were used. Later, Sar1p purification was simplified by expressing a recombinant glutathione *S*-transferase (GST)–Sar1p fusion protein in *Escherichia coli*.[9] Because we currently use Sar1p purified from *E. coli* in our budding assays, only this protocol is described.

Reagents and Buffers

B88*: 20 mM HEPES, pH 6.8, 250 mM sorbitol, 150 mM potassium acetate, 5 mM magnesium acetate, 5 μM GDP

TBS*: 50 mM Tris, pH 7.4, 150 mM NaCl, 5 μM GDP, 5 mM MgCl$_2$

TBST*: TBS + 0.1% Tween 20, 5 μM GDP, 5 mM MgCl$_2$

TCB*: 25 mM Tris, pH 8, 5 mM CaCl$_2$, 250 mM potassium acetate, 5 μM GDP, 5 mM MgCl$_2$

Glutathione-agarose (Sigma)

---

[26] C. Barlowe, C. d'Enfert, and R. Schekman, *J. Biol. Chem.* **268**, 873 (1993).

* All buffers for Sar1p purification are supplemented with 5 μM GDP (to help maintain Sar1p in its soluble, GDP-bound form) and the Tris buffers also contain 5 mM MgCl$_2$ (for stabilization of the nucleotide-bound form of Sar1p).

Procedure. An overnight stationary culture of *E. coli* strain RSB1091 (expressing the recombinant Sar1p–GST fusion protein) is used to inoculate 2 liters of LB-ampicillin. Cells are grown at 30° (this generally helps improve solubility and prevent accumulation in inclusion bodies) to an OD_{600} of ~0.7, induced by adding isopropylthiogalactoside (IPTG) to a final concentration of 0.1 mM and grown for an additional 2 hr. Cells are sedimented by centrifuging at 5000 rpm for 15 min in a GS3 rotor, washed once in 200 ml TBS (pH 7.4), and centrifuged at 6000 rpm for 10 min in a GSA rotor. Pellets are resuspended in 20 ml TBS, transferred to SS34 Oakridge tubes, and frozen and stored at −80°.

The frozen cells are thawed, lysozyme is added to a final concentration of 2 mg/ml, and the cells are incubated at 30° for 25 min with occasional mixing. Concurrently, 2.5 ml glutathione-agarose gel (which was preswelled overnight in TBST at 4°) is washed three times. Washes are done in a 50-ml conical tube by adding buffer, inverting several times to suspend the gel beads, centrifuging briefly in a clinical centrifuge, and removing the wash buffer. After incubation, Triton X-100 is added to the cells to 1% final concentration. The cells are sonicated three times for 30 sec using a small tip at ~60 W with chilling on ice between each sonication. The cells are then centrifuged twice at 12,500 rpm for 5 min each time in an SS34 rotor. The supernatant is combined with the washed glutathione-agarose gel in a 50-ml conical tube, and the mixture is rotated at 4° for 1 hr. Following centrifugation at 4° and removal of the supernatant, the gel is washed three times with TBST, once with TBS, once with TCB, and then transferred into a short column and washed with 25 ml of TCB at 4°. A thrombin solution [2.5 units (Roche) dissolved in 1.5 ml TCB] is added, the top and bottom of the column are sealed, and the suspension is incubated in a 25° water bath for 1 hr with occasional mixing (making sure that the gel gets thoroughly resuspended). Sar1p cleaved from the column matrix is collected into a 15-ml conical tube. B88 (2 ml) is added to the top of the column and all liquid is collected again in a fresh tube. This is repeated once more with 2 ml B88 and then twice with 1 ml B88. The protein concentration is assayed (Bio-Rad, Hercules, CA) and the protein-containing fractions are combined (to ~0.5–1 mg/ml) and saved. PMSF is then added to a final concentration of 2 mM, and the samples are centrifuged in a microcentrifuge for 1 min to remove any precipitate. The protein concentration of the supernatant is reassayed and aliquots are frozen in liquid nitrogen and kept at −80°. This procedure should yield several milligrams of protein.

The uncleaved Sar1–GST fusion protein also supports vesicle budding and cargo packaging (but requires a nonhydrolyzable analog of GTP rather than GTP).[27] For purification of Sar1–GST, the aforementioned protocol is used except that elution is done in a final volume of 3–5 ml of TBS, 0.1% Triton X-100, 10 mM glutathione. The eluted fusion protein may be further dialyzed

[27] M. J. Kuehn, J. M. Herrmann, and R. Schekman, *Nature* **391,** 187 (1998).

against 2× 1 liter of 20 mM HEPES, pH 6.8, 1 mM magnesium acetate, 150 mM potassium acetate.

Purification of Sec23p Complexes: Sec23/24p, Sec23/Iss1p, and Sec23/Lst1p

Sec23p associates tightly with Sec24p to form a 400-kDa complex.[8] Two nonessential but functional homologs of Sec24p termed Iss1p and Lst1p, which, like their essential counterpart, are also tightly associated with Sec23p, have been described.[23–25]

In the time since it was first reported,[8,28] purification schemes for Sec23p complexes have been modified and improved. The initial purification[28] took advantage of tight binding properties of Sec23p to both anion- and cation-exchange columns at neutral pH. This procedure yielded a complex that complemented *sec23* mutant cytosol but was inactive in the budding assay due to proteolysis of the Sec24p component. A newer purification method was developed[29] employing a strain that overproduces both subunits of the complex from two 2μ vectors. The inclusion of EGTA in the buffers and a more rapid purification prevented Sec24p proteolysis. Purification was facilitated by modification of the *SEC24* gene to encode six histidine residues at the C terminus of Sec24p. This modification does not affect functionality of the complex in the budding assay. The most recent published protocol consists of cytosol preparation, ammonium sulfate precipitation, DEAE-FF column separation, Ni-NTA column separation, and fractionation on a Mono S column.[29] The current procedure for the purification of Sec23/24p described later is largely based on the aforementioned protocol. A modification of this procedure is currently employed in the laboratory only for the purification of Sec23/Iss1p and Sec23/Lst1p complexes and requires cloning of both genes of the complex under a galactose-inducible promoter. Whereas this protocol calls for an intermediate shift to a raffinose-containing medium, it offers several advantages. First, culture (and column) volumes needed to obtain a given yield are reduced greatly. Second, both components of the complex are expressed at about the same levels, eliminating the need to purify the Sec23 complex from the large amounts of Sec24p (or Sec24p homologue) monomers that also bind to the Ni-NTA column generated in the procedure not involving galactose induction. A third advantage is that fewer column separation steps are required to obtain a highly pure complex, helping to increase yield and to reduce the overall purification time and cost. The quicker purification probably also helps preserve the activity of the complex.

Both the noninducible (Sec23/24p) purification scheme and the procedure employed with the GAL constructs (for purification of Sec23/Iss1p and Sec23/Lst1p)

---

[28] L. Hicke, T. Yoshihisa, and R. Schekman, *Mol. Biol. Cell* **3,** 667 (1992).
[29] T. Yeung, T. Yoshihisa, and R. Schekman, *Methods Enzymol.* **257,** 145 (1995).

as they are currently performed in the laboratory are described. This latter procedure should also be amenable for Sec23/24p purification.

Reagents and Buffers

HSLB (high salt lysis buffer): 0.75 M potassium acetate, 50 mM HEPES, 0.1 mM EGTA, 10 or 20% (w/v; see note below) glycerol (pH adjusted to 7.0 with 5 M KOH)

Protease inhibitors final concentrations: 1 μM leupeptin, 1 μM pepstatin A, 1 mM ε-aminocaproic acid, 0.5 mM PMSF. These are included in all buffers and are added just prior to use

Reducing agent final concentration: 1.4 mM 2-mercaptoethanol

Glass beads: 0.5 mm diameter (Biospec Products, Bartlesville, OK)

BeadBeater (Biospec Products)

Wash buffer I: 0.75 M potassium acetate, 50 mM MES (pH 6.3), 0.1 mM EGTA, 10% (w/v) glycerol, 40 mM imidazole

Wash buffer II: 0.5 M potassium acetate, 50 mM HEPES (pH 7.0), 0.1 mM EGTA, 10% (w/v) glycerol, 40 mM imidazole

Ni elution buffer: 0.5 M potassium acetate, 50 mM HEPES (pH 7.0), 0.1 mM EGTA, 10% (w/v) glycerol, 400 mM imidazole

Desalting buffer: 0.5 M potassium acetate, 50 mM HEPES, (pH 7.0), 0.1 mM EGTA, 10% (w/v) glycerol, 100 mM imidazole

Buffer A: 0.5 M potassium acetate, 50 mM HEPES (pH 7.0), 0.1 mM EGTA, 10% (w/v) glycerol

Buffer B: 0.8 M potassium acetate, 50 mM HEPES (pH 7.0), 0.1 mM EGTA, 10% (w/v) glycerol, 50 mM imidazole

B-II: 0.75 M potassium acetate, 50 mM 2-(N-morpholino)ethanesulfonic acid, 0.1 mM EGTA, 20% (w/v) glycerol, 40 mM imidazole (pH adjusted to 6.3 with 5 M KOH)

B-III: 0.75 M potassium acetate, 50 mM HEPES, 0.1 mM EGTA, 0.25 M sorbitol, 20% (w/v) glycerol, 40 mM imidazole (pH 7.0)

B-IV100, B-IV150, B-IV250, and B-IV500: same as B-III except that they contain 100, 150, 250, and 500 mM imidazole, respectively

Note that for the Sec23/Iss1p and Sec23/Lst1p purifications, sorbitol, as well as 20% glycerol, is used (instead of 10% as for Sec23/24p). The pH of all buffers should be adjusted at 4°.

Procedure for Sec23/24p Purification. Strain RSY1069[29] is grown at 30° in 6 liters of minimal medium (SD) supplemented with 20 mg/liter of the following: Ade, His, Lys, and Trp (Sigma). Cells are harvested (5 min at 4000g in a Sorvall GS3 rotor) at late log phase (this should yield ~25 g cells). Cells are washed in cold distilled water and are resuspended in ~4 ml HSLB, frozen in liquid nitrogen, and stored at −80° until used.

The cells are partially thawed in an ice-cold water bath and are suspended with ~50 ml of HSLB (supplemented with protease inhibitors and reducing agent) to a final volume of 70 ml. The chamber of a chilled BeadBeater is half-filled with chilled glass beads and is then topped off with the cell suspension and the outer jacket is filled with ice. Cells are disrupted by five 1-min periods of agitation followed by 2-min intervals of chilling. The lysate is recovered and the beads are washed (to increase recovery) with 20 ml HSLB (supplemented with the protease inhibitors and reducing agent). The combined lysate (~80 ml) is centrifuged at 10,500 rpm (~13,000g) for 10 min in an SS34 rotor (Sorvall/DuPont, Wilmington, DE). The supernatant is centrifuged at 40,000 rpm (~185,000g) for 75 min in a 45Ti rotor (Beckman, Palo Alto, CA). The supernatant is collected carefully, avoiding the floating lipids and the turbidity near the pellet.

The supernatant (~50 ml) is loaded onto a 5-ml nickel-nitriloacetic acid (Ni-NTA) agarose column (Qiagen, Valencia, CA), which was preequilibrated with HSLB containing protease inhibitors and reducing agent. The column is washed successively with 50 ml of wash buffer I, 10 ml of wash buffer II, and Sec23/24p is eluted with 20 ml of 400 mM imidazole. Fractions (~1 ml) are collected with the aid of a fraction collector and tested with a protein assay reagent (Bio-Rad), as well as by SDS–PAGE. Peak fractions containing the Sec23p complex are pooled and kept on ice overnight. The column may be regenerated and stored with 20% ethanol.

The pool fraction from the Ni-NTA column is loaded on a 50-ml G-25 column (preequilibrated with desalting buffer). After all of the load has entered the column, more desalting buffer is added and ~1-ml fractions are collected. A total of ~30 fractions are collected (depending on the size of the column used), and peak fractions are pooled following a Bradford assay. It may be possible to skip this desalting step and load the Ni-NTA pool directly on the DEAE, depending on the size of the DEAE column, the protein concentration, and the concentration of imidazole (e.g., if a gradient of 100–400 mM imidazole rather than a single high concentration of imidazole is applied).

The G-25 pool is loaded onto a 5-ml DEAE column that was preequilibrated with 10 ml buffer A. If some of the Sec23/24p complex precipitates at this point, it should be centrifuged at 15,000 rpm for 10 min in an SS34 rotor before loading on the DEAE column. The column is then washed with 10 ml buffer A. Because the Sec24p monomer is more abundant than the Sec23/24p complex, the monomeric form could be recovered (if needed) by saving the flow through of the load and the first wash. The Sec23/24p complex is eluted with 10 ml buffer B and ~0.5-ml fractions are collected. The peak samples are determined by the Bradford assay and pooled. Aliquots of the Sec23/24p complex are made in convenient volumes and frozen and kept at $-80°$. The total protein yield is typically ~0.5 mg. The homogeneity and quality of the preparation may be tested by SDS–PAGE.

Procedure for Sec23/Iss1p and Sec23/Lst1p Purification. Strain RSY620 (*MAT* a *leu2-3,112 ura3-52 ade2-1 trp1-1 his3-11,15PEP4::TRP1*) is transformed with pTKY9, which is a pGAL426GAL1 2μ *URA3* plasmid containing *SEC23* (*NcoI–Hin*dIII) between the *GAL1* promoter and the *CYC1* terminator and either one of the following: pTKY7 (a pGAL426GAL1 2μ *LEU2* plasmid containing a hexahistidine (His$_6$)-tagged version of *ISS1* between the *GAL1* promoter and the *CYC1* terminator) or pTKY12 (a pGAL426GAL1 2μ *LEU2* plasmid containing a His$_6$-tagged version of *LST1* between the *GAL1* promoter and the *CYC1* terminator). Transformant cells are grown at 30° in SC-Ura-Leu (2% glucose) to early stationary phase, washed with water, and then used to inoculate 6 liters of SC-Ura-Leu (2% raffinose) at an initial OD$_{600}$ of 0.005. To ensure sufficient utilization and depletion of intracellular glucose (which inhibits the *GAL* promoter), the cells are grown at 30° for 1 day to an OD$_{600}$ of ~1.2. Galactose is then added to a final concentration of 0.2%, and growth is continued for ~5 hr, to an OD$_{600}$ of ~2.7 for the co-overproduction of Sec23p and either Iss1 or Lst1p. The cells are harvested, washed twice with water, and kept frozen at −80° until use. About 25 g of cells (wet weight) is obtained from a 6-liter culture.

The cells are thawed and lysis and centrifugation steps are done exactly as described for the purification of Sec23/24p. The ~185,000g supernatant (~50 ml) is loaded onto a 10-ml Ni-NTA superflow (or agarose) column (Qiagen, Valencia, CA), which was preequilibrated with HSLB containing protease inhibitors and a reducing agent. The column is then washed successively with 90 ml of B-II, 20 ml of B-III, and 35 ml each of B-IV100, B-IV150, B-IV250 (Sec23/Lst1p elutes), and B-IV500 (Sec23/Iss1p elutes). During elution, ~1-ml fractions are collected with the aid of a fraction collector and later tested with a protein assay reagent (Bio-Rad), as well as by SDS–PAGE. Peak fractions containing the Sec23p complex are pooled and aliquots are frozen in liquid nitrogen and stored at −80°. Typical yields range from 1 to 5 mg purified complex (when starting from 25 g of wet cells).

Sec13/31p Preparation

Initially, a construct with the mouse *DHFR* gene fused to the 3′ coding terminus of *SEC13* was employed for affinity purification of the Sec13/31p complex by using methotrexate-agarose chromatography.[30] Expression of the fusion protein from a multicopy vector complemented the null allele of *SEC13* but resulted mainly in overproduction of monomeric inactive Sec13p.[30] Nevertheless, the purified Sec13–dhfr/31p complex was active in a budding assay performed with all purified components[20] and was used for the subsequent cloning of *SEC31*.[31] To

---

[30] N. K. Pryer, N. R. Salama, R. Schekman, and C. A. Kaiser, *J. Cell Biol.* **120,** 865 (1993).
[31] N. R. Salama, J. S. Chuang, and R. W. Schekman, *Mol. Biol. Cell* **8,** 205 (1997).

increase the yield of the purified Sec13/31p complex, we found it desirable to produce equivalently high expression levels of both *SEC13* and *SEC31* gene products. Because Sec31p is expressed at lower levels than Sec13p, *SEC31* was chosen as the tagged partner to minimize the amount of monomer in the final purified preparation. His tagging was employed instead of *DHFR* to avoid potential structural perturbations in the fusion protein. Strain RSY1113, which harbors the 2μ plasmid pNS3141 containing both *SEC13* and *SEC31-HIS*, is now routinely used. The following protocol results in higher yields of protein complex and includes fewer purification steps than that described before.[20]

Reagents and Buffers

B88 + EDTA: 0.15 M potassium acetate, 20 mM HEPES (pH 6.8), 5 mM magnesium acetate, 0.25 M sorbitol, 1 mM EDTA

Protease inhibitors at final concentrations: 1 μM leupeptin, 1 μM pepstatin A, 0.5 mM PMSF, and 1.4 mM 2-mercaptoethanol are added to all buffers

Resuspension buffer: 50 mM HEPES (pH 7.0), 0.1 mM EGTA, 10% (v/v) glycerol

Equilibration buffer: 0.5 M potassium acetate, 50 mM HEPES (pH 7.0), 0.1 mM EGTA, 10% glycerol

Wash buffer: 0.5 M potassium acetate, 50 mM HEPES (pH 6.8), 0.1 mM EGTA, 10% glycerol, 40 mM imidazole (check pH at 4°)

Elution buffer: 50 mM potassium acetate, 20 mM HEPES (pH 7.4), 0.1 mM EGTA, 10% glycerol, 200 mM imidazole (check pH at 4°)

Buffer A: 0.4 M potassium acetate, 20 mM HEPES (pH 7.4), 0.1 mM EGTA, 10% glycerol

Buffer B: 0.7 M potassium acetate, 20 mM HEPES (pH 7.4), 0.1 mM EGTA, 10% glycerol

Procedure. A 10-liter culture of RSY1113 (*his3-11,15, trp1-1, MATα*) containing pNS3141 is grown at 30° to an OD$_{600}$ of 2–3. Cells ($\sim$50 g wet weight) are harvested, resuspended in 100 ml of B88 + EDTA containing protease inhibitors, and lysed by agitation in a BeadBeater as described for the Sec23/24p preparation. The lysate is centrifuged at 13,000g (e.g., at 9000 rpm in a GSA rotor) for 15 min at 4°, and the supernatant is collected and centrifuged at 40,000 rpm ($\sim$185,000g) for 90 min in a 45Ti rotor (Beckman). The supernatant is collected, carefully avoiding the pellet and floating lipids, and stirred on ice. Powdered ammonium sulfate is added slowly to 35% of saturation (i.e., 0.208 g/ml) and stirring is allowed to continue for $\sim$30 more min. Precipitated proteins are centrifuged at 9000 rpm for 15 min at 4° in a GSA rotor, the supernatant is discarded, and the pellets are left overnight on ice.

Pellets are resuspended in 5 ml of resuspension buffer and homogenized with a Dounce homogenizer. Insoluble material is removed by centrifugation at 15,000 rpm for 10 min in an SS34 rotor at 4° and the supernatant is collected. After adjusting the salt concentration to ∼0.5 M (determined by conductivity measurement) and the final pH to ∼7.0, the supernatant is loaded onto a preequilibrated 3-ml Ni-NTA column, which is then washed with 30 ml wash buffer. Elution is performed with 10 ml elution buffer, and ∼20 fractions of ∼0.5 ml are collected. Loading, washing, and elution are all done at an even flow rate of ∼0.5 cm/min. Protein peak fractions are determined by performing a Bradford protein assay and peak samples are pooled (∼4 ml).

The pooled samples are adjusted to 0.4 M salt using 5 M potassium acetate and are then applied to a 1-ml Mono Q column (Pharmacia, Piscataway, NJ), which was preequilibrated with 10 ml buffer A (0.4 M potassium acetate) using a Pharmacia FPLC. After loading the sample, the column is washed with 10 ml of buffer (A and B mix) at 0.45 M potassium acetate. Then, the complex is eluted with a 1-ml gradient from 0.45 to 0.7 M potassium acetate using buffers A and B followed by 10 ml of buffer B (at 0.7 M). Samples (0.5 ml) are collected after the wash. Peak fractions (typically eluting at ∼0.55 M potassium acetate) are determined by a Bradford assay and the protein quality is assessed by SDS–PAGE. Peak fractions (∼1 mg protein) are pooled, frozen in liquid nitrogen, and stored at −80°. Both Ni-NTA and Mono Q columns can be washed, regenerated, and stored with 20% ethanol for additional usage.

COPII Vesicle Packaging Assay

Packaging of small soluble proteins, such as the yeast pheromone α-factor precursor, can be assessed by producing radiolabeled protein (by *in vitro* transcription/translation) and posttranslationally translocating it into ER membranes prior to the initiation of vesicle budding[6] (see Fig. 2A). However, this approach is not possible for membrane proteins and proteins that translocate into the ER cotranslationally. Alternatively, cargo proteins are labeled in yeast cells during a short radioactive pulse period and membranes are prepared and used in the budding assay (Fig. 2B). A simple fractionation step permits separation of the COPII vesicles that bud from the donor membranes *in vitro*. The cargo protein of interest is then immunoprecipitated for quantification. Obtaining efficient packaging depends on many factors (e.g., quality of donor membranes, cytosolic components, nucleotides, buffer system, the concentration of the various components, temperature, time), all of which have to be optimized. The specificity of the budding reaction can be assessed by its dependence on each one of the required factors; nucleotide dependence serves as a typical control. The following procedure describes parallel analysis of the [^{35}S]Met-labeled glycosylated prepro-α factor

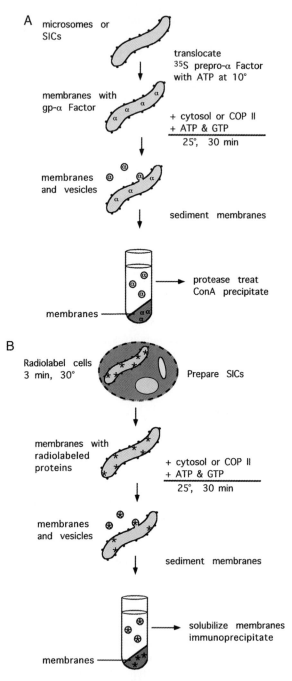

FIG. 2. Schematic presentation of the budding assay: A diagram outlining the COPII packaging assay for (A) a posttranslationally translocated cargo (e.g., α-factor precursor protein) and (B) a pulse-labeled protein cargo.

([^{35}S]gpαF) posttranslationally translocated into unlabeled membranes and a cargo protein (e.g., epitopically tagged plasma membrane ATPase, Pma1p) in membranes that were prepared from pulse-labeled cells. Packaging of [^{35}S]gpαF is quantified by scintillation counting, whereas metabolically labeled proteins are isolated by immunoprecipitation and the packaging efficiency is determined by Phosphor-Imager quantification of bands separated on SDS–PAGE. The assay consists of three steps and is summarized in Fig. 2.

Reagents and Buffers

[^{35}S]gpαF: preparation is done using a standard procedure[6]; a coupled transcription/translation kit such as TNT (Promega Corp., Madison, WI) can be used for small-scale preparations

10× ATP plus ATP regeneration system mix: 10 mM ATP, 500 μM GDP-mannose, 400 mM creatine phosphate, and 2 mg/ml creatine phosphokinase

Con A buffer: 500 mM NaCl, 1% Triton X-100, 20 mM Tris–HCl, pH 7.5, 2 mM NaN$_3$

IP buffer: 150 mM NaCl, 1% Triton X-100, 0.1% SDS, 15 mM Tris–HCl, pH 7.5, 2 mM NaN$_3$

Urea buffer: 2 M urea, 200 mM NaCl, 1% Triton X-100, 100 mM Tris–HCl, pH 7.5, 2 mM NaN$_3$

Tris/NaCl buffer: 50 mM NaCl, 10 mM Tris–HCl, pH 7.5, 2 mM NaN$_3$

Procedure

Step I. Translocation: Microsomes or SICs from unlabeled or pulse-labeled yeast (which may express an epitopically tagged version of the protein of interest) are used as the source of donor membranes. Aliquots of donor membranes are thawed (the number of aliquots to be used should be calculated according to the number of budding reactions to be performed; see later). SICs are resuspended in "low acetate" B88 (containing 50 mM instead of 150 mM potassium acetate) and are incubated on ice for 2 min, and potassium acetate is added to 150 mM and the SICs are centrifuged (1 min at 10,000 rpm in a microcentrifuge). This low osmotic support B88 wash is done only with SICs (i.e., not with microsomes) to maximize spheroplast lysis. Donor membranes (microsomes or "low acetate" washed SICs) are washed (1 min at 10,000 rpm in a microcentrifuge) twice with B88 (see earlier discussion) and [^{35}S]gpαF is imported into the unlabeled donor membranes by posttranslational translocation,[14] whereas the pulse-labeled SICs are mock treated (mock treatment can reduce nonspecific/background budding signals). The concentration of membranes in the translocation reaction is the equivalent of ∼60 OD$_{600}$ cells/ml with SICs and 3 mg/ml with microsomes. The translocation reaction is initiated by mixing the washed membranes with [^{35}S]gpαF (using ∼5 μl translation product per each budding reaction) and the ATP plus ATP regeneration system mix (from a 10× stock) consisting of (final concentrations) 1 mM ATP, 50 μM GDP-mannose, 40 mM creatine phosphate, and 200 μg/ml creatine phosphokinase. The translocation mixture (consisting of donor membranes, [^{35}S]gpαF, and ATP plus ATP regeneration mix) is incubated for 25 min at 10° (translocation occurs at low temperature, whereas vesicle budding requires a higher temperature) and chilled for 5 min on ice. A urea (2.5 M final in B88) wash step is optional at this point for the removal of membrane-tethered constituents, including prebound COPII components. This washing step may impede packaging of some cargo proteins (e.g., HA-tagged Pma1p) and therefore should be tested for the protein analyzed. The urea wash is performed by resuspending the membranes in 450 μl of B88, adding 450 μl of 5 M urea (made in B88) and incubating on ice for 5 min. Finally, membranes are resuspended and washed three times with 1 ml B88 in a microcentrifuge.

Step II. Budding. [^{35}S]gpαF-loaded membranes and pulse-labeled SICs are distributed into two parallel sets of tubes for the budding step. The final concentration of SICs to be used in the budding assay is the equivalent of ~15 OD$_{600}$ cells/ml (a typical reaction volume using labeled SICs is 150 μl). Microsomes are used at 0.25 mg/ml (e.g., ~0.06 OD$_{280}$ units of microsomes per each 50-μl budding reaction). ATP plus ATP regeneration system (see earlier discussion) and guanine nucleotides (Boehringer Mannheim, GmbH, Germany), GTP, or a nonhydrolyzable GTP analog, e.g., GMP-PNP (in reactions where preservation of the coat is desired), are added to the reaction mixtures in B88 at a final concentration of 0.1 mM. Yeast cytosol (2 mg/ml final) or purified COPII components are added to initiate vesicle budding. Standard concentrations of purified COPII components are 20 μg/ml Sar1p, 20 μg/ml Sec23/24p, and 40 μg/ml Sec13/31p. Following an incubation at 25° for 30 min, reactions are chilled on ice for 5 min and aliquots representing the "total" are transferred to fresh tubes. The rest is sedimented in a microcentrifuge (12,000g, 4 min), which retains COPII vesicles in the medium speed supernatant (MSS), whereas the heavy donor membranes sediment to a pellet.

Step III. Detection. In order to remove [^{35}S]gpαF, which is associated on the cytosolic face (not in the lumen) of the donor membranes and vesicles, MSS and total fractions from [^{35}S]gpαF-loaded SICs are treated with trypsin (250 μg/ml final) for 10 min on ice followed by quenching by the addition of the trypsin inhibitor (250 μg/ml final). Following denaturation with SDS (1% final) for 5 min at 95°, 1 ml concanavalin A (Con A) buffer and 100 μl of 20% Con A-Sepharose (Amersham Pharmacia Biotech, Uppsala, Sweden) are added, and precipitation samples are incubated in a rolling device for 2 hr at room temperature. Reactions are then washed once in 1 ml of each of the following buffers: IP buffer, urea buffer, Con A buffer, and finally Tris/NaCl buffer. Pellets are resuspended in 1% SDS, and bound [^{35}S]gpαF is quantified in a scintillation counter. The packaging efficiency is calculated as the ratio of Con A-bound [^{35}S]gpαF detected in the MSS to that present in the total fraction.

MSS and total fractions from pulse-labeled SICs are solubilized in 1% SDS and heated for 5 min to aid solubilization (caution: some membrane proteins aggregate if warmed at high temperature). The samples are then diluted (10-fold) with IP buffer not containing SDS and supplemented with protease inhibitors and immunoprecipitated (using antibodies directed against the protein of interest) overnight at 4°. Immunoprecipitates are washed twice with IP buffer and once each with urea, Con A, and Tris/NaCl buffers (see earlier discussion). In some cases, it may be necessary to omit urea and Con A steps because some antibody–antigen complexes are disrupted by these washes. A preclearing step may be done prior to immunoprecipitation by omitting the antiserum from the immunoprecipitation mixture, and following an incubation the supernatant is transferred to fresh tubes for immunoprecipitation. Protein sample buffer is added to the washed immunoprecipitates and samples are separated on SDS–PAGE and the gel is fixed and dried. The packaging efficiency of the cargo of interest is determined following exposure of the gel to a PhosphorImager (STORM 860; Molecular Dynamics, Sunnyvale, CA) and quantitation of the metabolically labeled protein bands in the MSS and total reactions.

Establishing and performing the budding assay can serve to test COPII (or cytosol)-dependent packaging of essentially any potential cargo molecule into ER-derived vesicles. Mutations or truncations can be made in YFP (your favorite protein) in order to identify novel functional domains that may be required for its packaging. Various factors (e.g., proteins, lipids, ions) can be tested for their requirement or effect on vesicle formation and cargo packaging. The requirement for a particular soluble or membrane protein for vesicle budding or for the packaging of a specific cargo molecule can be tested using cytosol or membranes from which the protein has been depleted. Yeast mutants are a powerful tool.[32] For example, if the gene for the protein tested is identified, then cytosol or membranes can be prepared from null cells (if the gene is nonessential) or from mutant cells grown under restrictive conditions (for a conditional gene). It is also possible to preincubate the membranes or cytosol (depending on the nature of the inspected protein) at the restrictive conditions prior to performing the assay or to perform the budding assay under restrictive conditions in comparison to the optimal conditions. Another option is to titer into the budding assay antiserum made against the tested protein. Not only depletion of one or more proteins but also overexpression, as well as the effects of competition, synergism, or cooperation and other synthetic interactions, can be tested relatively easily in yeast. The effect of a potential regulator of activity of the COPII components or an accessory component required for

---

[32] M. F. Rexach and R. W. Schekman, *Methods Enzymol.* **219**, 267 (1992).

the packaging of a specific cargo[33] or a suspected COPII homologue can also be examined. Cytosol (or purified COPII components) can be prepared from a mutant and used in the budding assay instead of wild-type cytosol or a strain where the factor has been depleted/inhibited can be used. Ultimately, if the factor can be provided in an active purified form, reconstitution of COPII function, cargo packaging, and vesicle budding can be tested by titrating it into the depleted/inhibited assay.[25,34,35] Morphological, biochemical, and biophysical studies on ER-derived vesicles generated *in vitro* have revealed interesting mechanistic insights. Electron microscopic analysis of COPII vesicles generated in the budding assay revealed differences in the morphology (size) of COPII vesicles, as well as in cargo packaging efficiency depending on the coat composition indicating that different subtypes of COPII vesicles may form in the cell.[25] Immunoisolation and density gradient fractionation of ER-derived vesicles generated with cytosol have revealed two populations of vesicles based on the class of cargo proteins being sorted.[36] A recent real-time kinetics assay sheds light on the dynamics and transit intermediates in COPII formation.[37] Evidently, the COPII budding mechanism is universal and general, as the equivalent mammalian components can initiate vesicle budding and cargo packaging from a mammalian source of donor membranes.[38] The budding assay described here may be adapted to assay vesicle budding from other organelles.

Note added in proof. Protocols for the preparation of COPII components were given in previous volumes of *Methods in Enzymology* for the publication of Sar1p,[39] Sec23/24p,[29] and Sec13/31p.[40]

Acknowledgments

We thank past and present members of the Schekman laboratory for the work described here and for providing a most pleasant working environment. We are grateful to P. Malkus, F. Jiang, and especially R. Lesch for critical reading of the manuscript and expert opinion and to M. Kuehn for Fig. 2. This work was supported by a postdoctoral fellowship from the Cancer Research Fund of the Damon Runyon-Walter Winchell Foundation, DRG-1469 to Y.S. Work in the Schekman laboratory is supported by the Howard Hughes Medical Institute.

[33] J. M. Herrmann, P. Malkus, and R. Schekman, *Trends Cell Biol.* **9**, 5 (1999).
[34] M. J. Kuehn, R. Schekman, and P. O. Ljungdahl, *J. Cell Biol.* **135**, 585 (1996).
[35] W. T. W. Lau, R. W. Howson, P. Malkus, R. Schekman, and E. K. O'Shea, *Proc. Natl. Acad. Sci. U.S.A.* **97**, 1107 (2000).
[36] M. Muñiz, P. Morsomme, and H. Riezman, *Cell* **104**, 313 (2001).
[37] B. Antonny, D. Madden, S. Hamamoto, L. Orci, and R. Schekman, *Nature Cell Biol.* **3**, 531 (2001).
[38] M. Aridor, J. Weissman, S. Bannykh, C. Nuoffer, and W. E. Balch, *J. Cell Biol.* **141**, 61 (1998).
[39] K. Kimura, T. Oka, and A. Nakano, *Methods Enzymol.* **257**, 41 (1995).
[40] W. J. Belden and C. Barlowe, *Methods Enzymol.* **329**, 438 (2001).

[16] Mapping Phosphorylation Sites in Proteins by Mass Spectrometry

By WENYING SHOU, RATI VERMA, ROLAND S. ANNAN,
MICHAEL J. HUDDLESTON, SUSAN L. CHEN, STEVE A. CARR,
and RAYMOND J. DESHAIES

Introduction

Many intracellular pathways are regulated by protein phosphorylation. To understand how protein phosphorylation controls a pathway, two experiments are required (Fig. 1). The common starting point is to mutate the gene that encodes the relevant protein kinase (a *trans* mutant) (Fig. 1B). There are four limitations to this experiment. First, the large number of protein kinases expressed in eukaryotic cells can make it difficult to identify the responsible enzyme. The budding yeast genome, for example, codes for 120 different protein kinases.[1] Second, homologous protein kinases can act redundantly to regulate a process. For example, the cyclin-dependent protein kinases Pho85 and Srb10 both phosphorylate Gcn4 and target it for ubiquitin-dependent proteolysis.[2] Third, most protein kinases have many substrates; therefore, a protein kinase mutant may display many phenotypes that obscure the pathway of interest. Fourth, from the characterization of protein kinase mutants alone, it can be difficult to decipher the exact mechanism of regulation and untangle primary and secondary effects.

The second experiment required to establish that phosphorylation of a given protein plays a causal role in a process under study is the evaluation of the effect of nonphosphorylatable mutations in the candidate substrate (*cis* mutations) (Fig. 1C). If one can establish that a *cis* mutant in substrate Y (Fig. 1C) and a *trans* mutant in protein kinase X (Fig. 1B) both prevent the execution of process Z, then it is very likely that phosphorylation of Y by X activates Z.

A confounding problem for "closing the circle" by the analysis of *cis* mutants is that it is often very difficult to map phosphorylation sites in substrate proteins. Many phosphoproteins are expressed in low abundance and can be recovered in only picomole quantities. In addition, most phosphoproteins are phosphorylated on more than one site and phosphorylation of any given site is often substoichiometric.

The serious constraints of low phosphopeptide yield and stoichiometry make it essential to have analytical methods that can preferentially detect and analyze

---

[1] H. Zhu, J. F. Klemic, S. Chang, P. Bertone, A. Casamayor, K. G. Klemic, D. Smith, M. Gerstein, M. A. Reed, and M. Snyder, *Nature Genet.* **26,** 283 (2000).
[2] Y. Chi, M. J. Huddleston, X. Zhang, R. A. Young, R. S. Annan, S. A. Carr, and R. J. Deshaies, *Genes Dev.* **15,** 1078 (2001).

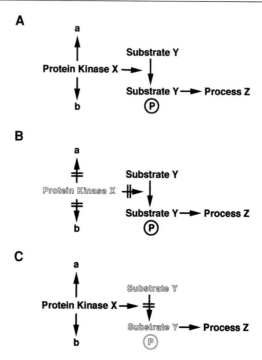

FIG. 1. The significance of protein phosphorylation can be tested in two complementary experiments. (A) Hypothesis: phosphorylation of substrate Y by protein kinase X leads to process Z, where kinase X also phosphorylates additional substrates such as a and b. Either a loss-of-function mutation in kinase X (B) or a nonphosphorylatable mutation in substrate Y (C) should block process Z (blockage is indicated by double lines).

phosphopeptides. By far the most common approach is to isolate a phosphorylated substrate protein from ^{32}P-labeled cells and separate tryptic digests of the labeled protein by thin-layer chromatography to reveal phosphopeptides.[3-6] One can then use a combination of secondary digestion with other proteases, deductive logic, and, in some cases, direct protein sequence to assign sites of phosphorylation.[7] In practice, this is not a trivial exercise. It is especially unappealing to contemplate the isolation of a low abundance substrate from large cultures of ^{32}P-labeled cells.

[3] K. L. Gould and K. L. Nurse, *Nature* **342,** 39 (1989).
[4] W. J. Boyle, P. van der Geer, and T. Hunter, *Methods Enzymol.* **201,** 110 (1991).
[5] T. Moll, G. Tebb, U. Surana, H. Robitsch, and K. Nasmyth, *Cell* **66,** 743 (1991).
[6] J. M. Sidorova, G. E. Mikesell, and L. L. Breeden, *Mol. Biol. Cell* **6,** 1641 (1995).
[7] W. J. Boyle, T. Smeal, L. H. Defize, P. Angel, J. R. Woodgett, M. Karin, and T. Hunter, *Cell* **64,** 573 (1991).

Mass spectrometry (MS), however, is ideally suited to the direct identification of protein phosphorylation sites. Phosphopeptides present in mixtures can be sequenced at the femtomole level without the need for extensive purification. A variety of mass spectrometry-based approaches have been employed to map phosphorylation sites in proteins.[8-13] A great advantage of these methods is that they do not require prior labeling of the target protein with ^{32}P. Thus, it is possible to isolate rare proteins from large-scale cultures for phosphopeptide mapping studies. Regardless of the method used to map phosphorylation sites, it is imperative that the native phosphorylation state of the target protein be preserved during isolation. This chapter describes general strategies for isolating phosphoproteins from budding yeast cells by drawing reference to two specific examples: the S-Cdk inhibitor Sic1 and the mitotic exit inhibitor Net1. We then provide an overview of the mass spectrometric methods used to map specific phosphorylation sites in these proteins.

General Strategies for Protein Isolation

Strain Design

Complete phosphorylation site mapping by mass spectrometry typically requires approximately 5–200 pmol of purified protein (depending on the complexity and stoichiometry of phosphorylation) preserved in its native phosphorylation state. To obtain reasonable amounts of material [Your favorite phosphoprotein (Yfp)] that is maximally phosphorylated, the following molecular genetic manipulations can be performed with the strain from which Yfp is to be isolated.

1. Yfp can be expressed from an inducible promoter (e.g., *GAL1,10*). A short 1- to 3-hr pulse of galactose may be enough to yield sufficient amounts of Yfp. Constitutive overproduction of Yfp, however, could overwhelm its cognate protein kinase[14] or could circumvent other cellular regulatory systems, resulting in nonphysiological phosphorylation of the substrate. If a known kinase is being tested, then it could also be overproduced.

---

[8] P. Cohen, B. W. Gibson, and C. F. Holmes, *Methods Enzymol.* **201**, 153 (1991).
[9] J. D. Watts, M. Affolter, D. L. Krebs, R. L. Wange, L. E. Samelson, and R. Aebersold, *J. Biol. Chem.* **269**, 29520 (1994).
[10] X. Zhang, C. J. Herring, P. R. Romano, J. Szczepanowska, H. Brzeska, A. G. Hinnebusch, and J. Qin, *Anal. Chem.* **70**, 2050 (1998).
[11] G. Neubauer and M. Mann, *Anal. Chem.* **71**, 235 (1999).
[12] M. C. Posewitz and P. Tempst, *Anal. Chem.* **71**, 2883 (1999).
[13] R. S. Annan, M. J. Huddleston, R. Verma, R. J. Deshaies, and S. A. Carr, *Anal. Chem.* **73**, 393 (2001).
[14] G. Alexandru, F. Uhlmann, K. Mechtler, M.-A. Poupart, and K. Nasmyth, *Cell* **105**, 459 (2001).

2. Determine physiological states under which Yfp is maximally phosphorylated. It may be beneficial to use mutants that stabilize the phosphorylation state of the protein. For example, if the counteracting phosphatase is known, Yfp can be isolated from a cell deficient in phosphatase activity.

3. Isolate Yfp from strains that can be conditionally inactivated for the kinase and compare the phosphorylation pattern with that of Yfp isolated from wild-type cells.

Substrate and Experimental Design

The purification of Yfp is facilitated by the introduction of one or more affinity tags at the N or C terminus of the protein. Several commonly used epitopes that have proven to be effective include the hexahistidine (His_6), HA, Myc, FLAG, polyoma, and "ZZ" (IgG-binding domain of protein A) tags. These epitopes can be combined in various permutations (e.g., His_6-HA) to enable consecutive affinity purification steps.[15,16] Regardless of the tagging strategy, it is our bias that it is critical to inactivate phosphatases to prevent dephosphorylation of the target protein during purification. In both cases described, the extract was denatured prior to affinity purification of the target protein.

Growing and Harvesting Cells

Growth. Cells are grown to exponential phase ($OD_{600} = 0.5$) before further manipulations, such as a temperature shift to 37° to impose a *ts* block or induction of substrate or protein kinase expression with galactose.

Harvesting. After cells have reached the desired stage, they are chilled and harvested as quickly as possible to preserve the phosphorylation status of Yfp. We routinely fill 1-liter centrifuge bottles halfway with ice, place them on ice, and pour culture directly into the bottles with vigorous shaking. The cultures are centrifuged at 4° for 5–10 min at 4000g to pellet cells. Pelleted cells are resuspended rapidly with either ice-cold water or 25 mM Tris, pH 7.5, containing a phosphatase and protease inhibitor cocktail described later, transferred to 50-ml conical screw-cap tubes, and sedimented in a clinical centrifuge at 5000 rpm for 5 min at 4°. It is important to work as quickly as possible during the cell harvest to minimize dephosphorylation.

Phosphatase and Protease Inhibitor Cocktails

Phosphatase and protease inhibitor cocktails are typically used as a prophylactic measure during lysis and purification to help preserve Yfp in its native

[15] R. Verma, R. S. Annan, M. J. Huddleston, S. A. Carr, G. Reynard, and R. J. Deshaies, *Science* **278**, 455 (1997).
[16] G. Rigaut, A. Shevchenko, B. Rutz, M. Wilm, M. Mann, and B. Seraphin, *Nature Biotechnol.* **17**, 1030 (1999).

state. Protease inhibitor cocktails consist of combinations of 5 mM EDTA (stock, 0.5 M, pH 8), 2 mM EGTA (stock, 0.2 M, pH 8), 0.2 mM 4-(2-aminoethyl)benzene sulfonyl fluoride hydrochloride (AEBSF: frozen stock, 100 mM in water), 25 μg/ml aprotinin (frozen stock, 10 mg/ml), 1 mM benzamidine (frozen stock, 1 M), 1 mM phenylmethylsulfonyl fluoride (PMSF, stock, 100 mM in 100% 2-propanol), 5–10 μg/ml pepstatin, leupeptin, chymostatin [frozen stock, 5–10 mg/ml in dimethyl sulfoxide (DMSO)]. Phosphatase inhibitor cocktails consist of 10 mM NaF (stock, 0.5 M), 60 mM β-glycerolphosphate (stock, 1 M, pH 7.5, store at 4°), 10 mM sodium pyrophosphate (stock, 100 mM), 2 mM sodium orthovanadate (stock, 200 mM; for method of preparation, see Ref. 17), and 3 μM microcysteine-LR (frozen stock, 300 mM in DMSO).

Three methods can be used to monitor how well the phosphorylation state of a protein is preserved throughout purification. Functional assays are the most reliable, e.g., phosphorylated but not unphosphorylated Sic1 can serve as a substrate for SCF^{Cdc4} (see later). However, functional assays are also the most demanding because they require that the proteins remain competent after purification. If a phosphorylated protein migrates differently from its unphosphorylated counterpart in an SDS–polyacrylamide gel, then gel mobility can be used to track the degree of phosphorylation (see later). If these two methods failed, then the last resort would be to label yeast cells with [^{32}P]phosphate and to use the radioactivity of the protein (normalized against the level of the protein) as an indicator.[15]

Isolation of Sic1 and Net1 for Mapping Sites of Phosphorylation *in Vivo*

Sic1

Sic1 is an S phase-specific Cdk Clb5/Cdc28 (S-Cdk) inhibitor that has to be degraded for cells to enter S phase.[15,18] Following its ubiquitination by the SCF^{Cdc4}/Cdc34 pathway, Sic1 is recognized by the 26S proteasome and degraded.[19] Although it was known that G1-Cdk is required for both ubiquitination and degradation of Sic1 and that Sic1 is phosphorylated by G1-Cdk,[18,20,21] it was not known whether phosphorylation of Sic1 itself (as opposed to phosphorylation of some other protein) triggered its ubiquitination and degradation. To address this key issue, we sought to construct a "nonphosphorylatable" mutant of Sic1. Sites at which G1-Cdk phosphorylated Sic1 *in vitro* were mapped by nano-electrospray

---

[17] D. J. Brown and J. A. Gordon, *J. Biol. Chem.* **259**, 9580 (1984).
[18] E. Schwob, T. Bohm, M. D. Mendenhall, and K. Nasmyth, *Cell* **79**, 233 (1994).
[19] R. Verma, S. Chen, R. Feldman, D. Schieltz, J. Yates, J. Dohmen, and R. J. Deshaies, *Mol. Biol. Cell* **11**, 3425 (2000).
[20] B. L. Schneider, Q. H. Yang, and A. B. Futcher, *Science* **272**, 560 (1996).
[21] R. Verma, R. M. Feldman, and R. J. Deshaies, *Mol. Biol. Cell* **8**, 1427 (1997).

tandem mass spectrometry (nano-ESMS/MS),[13] and a mutant (Sic1-Δ3P) lacking a subset of the identified sites was constructed.[15] Sic1-Δ3P was not ubiquitinated by SCF^{Cdc4}/Cdc34 *in vitro* and was stable *in vivo,* suggesting that G1-Cdk enabled the G$_1$/S transition by phosphorylating Sic1, thereby targeting it for ubiquitination and degradation via the SCF^{Cdc4}/Cdc34 pathway.

To confirm our hypothesis, we sought to demonstrate that the same residues on Sic1 that are phosphorylated by G1-Cdk *in vitro* are also phosphorylated *in vivo.* At the time that we set out to map the sites of *in vivo* phosphorylation on Sic1 using nano-ESMS/MS, there was little precedent for using this technique to map phosphorylation sites on proteins isolated from their native environment. Thus, to maximize our likelihood of achieving success, we took great care to ensure that sufficient amounts of Sic1 were isolated to enable a thorough analysis, that the isolated Sic1 was of high purity, and that the native phosphorylation state of Sic1 was preserved as much as possible. To achieve these goals, we implemented the following experimental design.

1. Sic1 was transiently overexpressed from a chromosomally integrated *GAL* promoter-driven cassette to yield enough material for analysis, but not so much as to overwhelm the cell.

2. The expressed Sic1 was tagged at the C terminus with a bipartite hemagluttinin–hexahistidine (HA-His$_6$ epitope). This tandem epitope provided a key advantage because hexahistidine binds Ni-NTA even in the presence of strong denaturants. Thus, we were able to prepare the cell lysate and conduct the first purification step in the presence of 6 *M* guanidinium hydrochloride, which is expected to inactivate all phosphatases and proteases in the extract. However, single-step purification on Ni-NTA rarely yields material of sufficient purity unless the protein is expressed at very high levels, and therefore the HA domain of the tandem epitope was critical because it enabled consecutive affinity purification steps. Following these two steps, Sic1 was sufficiently pure to be submitted directly to mass spectrometric analysis without further fractionation by SDS–PAGE (Fig. 2A).

3. Sic1^{HAHis6} was expressed in a *cdc34-2* strain held at the nonpermissive temperature. Cdc34 is the E2 enzyme that mediates the ubiquitination and degradation of phosphorylated Sic1. Mutant *cdc34^{ts}* cells accumulate high levels of G1-Cdk activity,[20,22] but are unable to degrade Sic1[18] and thus accumulate the phosphorylated protein.

Detailed Procedure. RJD 1044 (*GAL-SIC1^{HAHis6} cdc34-2*) cells are grown in 6 liters of YP plus raffinose at 24° to an optical density (OD$_{600}$) of 0.5. Sic1^{HAHis6} synthesis is induced by the addition of galactose to 2%, and after 3 hr, the culture is shifted to 37° for 3 hr. After harvesting (as described earlier), frozen cells are

[22] M. Tyers, *Proc. Natl. Acad. Sci. U.S.A.* **93,** 7772 (1996).

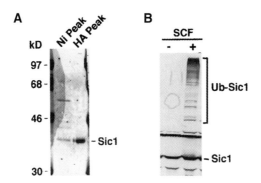

FIG. 2. Purification of phospho-Sic1 from yeast cells. (A) Analysis of purity of the Sic1 preparation. Aliquots of eluates from Ni-NTA and 12CA5/protein A resins were resolved by SDS–PAGE and visualized by Coomassie blue staining. (B) Evaluation of the ability of purified phospho-Sic1 to serve as a substrate of the ubiquitin pathway. An aliquot of purified phospho-Sic1 was incubated with ATP, E1, and E2 (Cdc34) in the absence or presence of SCF^{Cdc4} ubiquitin ligase activity (supplied by yeast extract, see Ref. 21). Reaction products were visualized by immunoblotting with anti-Sic1.

ground to powder in liquid nitrogen.[23] The cell powder (40 g) is thawed in five volumes of denaturing lysis buffer (DLB: 100 mM sodium phosphate, 10 mM Tris, pH 8.0, and 6 M guanidine hydrochloride), and the resulting slurry is stirred for 30 min at 24° and then centrifuged at 26,000g for 15 min at 4° in a Sorvall SS34 rotor. The supernatant is mixed with 2 ml of Ni-NTA resin (Qiagen, Valencia, CA) for 40 min at 24°, after which the beads are washed twice with 20 ml DLB, twice with 100 mM sodium phosphate, pH 5.9, 10 mM imidazole, and 2 M urea, and twice with 25 mM Tris, pH 8.0, 500 mM NaCl, and 0.2% Triton X-100. Bound proteins are eluted with 6 ml of buffer containing 250 mM imidazole, 50 mM Tris, pH 8.0, and 250 mM NaCl. The eluate is supplemented with NaCl (500 mM final), Triton X-100 (0.2%), and the protease and phosphatase inhibitor cocktail described earlier and is incubated with 0.5 ml 12CA5 resin (anti-HA antibody covalently cross-linked to protein A beads) for 45 min at 4°. The beads are collected by centrifugation and washed twice with binding buffer and three times with 10 mM Tris, pH 6.8. Sic1^{HAHis6} (Fig. 2A) is eluted with 1.5 ml 0.1% trifluoroacetic acid and processed for electrospray mass spectrometry as described later. To confirm that the functionally relevant phosphorylation state of Sic1 is preserved throughout isolation, we demonstrated that the purified material is competent to serve as a substrate for ubiquitination by the SCF^{Cdc4}/Cdc34 pathway (Fig. 2B).

Net1

Net1 is a subunit of the nucleolar RENT complex. The disassembly of the RENT complex in late anaphase of the cell cycle culminates in the release of

---

[23] R. Verma, Y. Chi, and R. J. Deshaies, *Methods Enzymol.* **283**, 366 (1997).

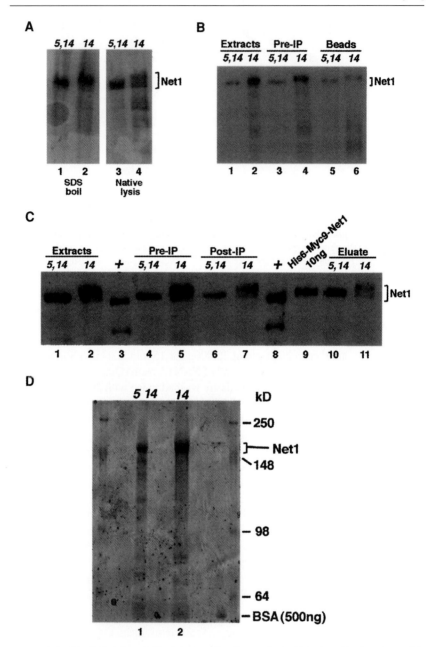

FIG. 3. Purification of phospho-Net1 from yeast cells. (A–C) Net1-Myc9 purifications from wild-type (+), *cdc5 cdc14* (*5,14*), or *cdc14* (*14*) strains were analyzed by SDS–PAGE followed by Western blot using 9E10 antibodies (against the Myc epitope). (A) The degree of phosphorylation of Net1

the protein phosphatase Cdc 14 from the nucleolus, which then triggers the exit from mitosis.[24–26] The disassembly mechanism is unknown, but we hypothesize that it involves phosphorylation of Net1 because Net1 accumulates in a highly phosphorylated form in a $cdc14^{ts}$ mutant that is unable to exit mitosis.[24] Moreover, the release of Cdc14 from Net1 requires the action of three protein kinases (Cdc5, Cdc15, and Dbf2), suggesting that protein phosphorylation may play a direct role in the release mechanism.

To address whether phosphorylation of Net1 regulates disassembly of the RENT complex, we sought to determine the phosphorylation state of Net1 in a $cdc14^{ts}$ mutant. We specifically selected the $cdc14^{ts}$ mutant because the Cdc14^{ts} protein is released from Net1 in late mitosis with normal kinetics in $cdc14^{ts}$ cells,[27] but because the phosphatase is inactive, the cells remain arrested in late mitosis with dispersed Cdc14. As a control, we sought to determine sites of Net1 phosphorylation in $cdc14^{ts} cdc5^{ts}$ cells because preliminary evidence suggested that Cdc5 is required for the disassembly of the RENT complex. Thus, a comparison of the sites at which Net1 is phosphorylated in these two mutant strains might highlight amino acids whose phosphorylation correlates with release of Cdc14 from Net1.

Detailed Procedure. To isolate Net1 in a manner that preserves its native phosphorylation state, we pursued two alternative approaches. We first tried nondenaturing lysis by grinding frozen cells in liquid nitrogen and thawing the cell powder in lysis buffer comprised of 20 mM HEPES, pH 7.2, 0.5 M NaCl, 2 mM dithiothreitol (DTT), 0.5% (v/v) Triton X-100, protease inhibitors, and phosphatase inhibitors. Unlike the SDS boiling method,[24] in which cells are boiled in SDS sample buffer to achieve fast inactivation of all proteins prior to lysis (Fig. 3A, lanes 1 and 2), the native lysis method fails to preserve the phosphorylation state of Net1 isolated from $cdc14$ cells (Fig. 3A, compare lane 4 with lane 2).

Because Net1 is dephosphorylated rapidly in native cell extract, we sought to purify it from denatured cell extract. However, the strategy described earlier

---

[24] W. Shou, J. H. Seol, A. Shevchenko, C. Baskerville, D. Moazed, S. Z. W. Chen, J. Jang, A. Shevchenko, H. Charbonneau, and R. J. Deshaies, *Cell* **97**, 233 (1999).
[25] A. F. Straight, W. Shou, G. J. Dowd, C. W. Turck, R. J. Deshaies, A. D. Johnson, and D. Moazed, *Cell* **97**, 245 (1999).
[26] R. Visintin, E. S. Hwang, and A. Amon, *Nature* **398**, 818 (1999).
[27] S. L. Jaspersen and D. O. Morgan, *Curr. Biol.* **10**, 615 (2000).

(as judged by its reduced mobility on SDS–PAGE) was preserved less well in native lysis than in the SDS boil lysis method. (B and C) Two independent samples of phospho-Net1 prepared from yeast cells using denaturing lysis method (see text for details). (C) Wild-type extracts (+) were prepared by the SDS boiling method. Net1 phosphorylation was the most extensive in $cdc14$ and the least extensive in wild-type cells (compare lanes 3–5 in C). (D) Eluate of phospho-Net1 from (C) was concentrated, fractionated on SDS–PAGE, and evaluated by Coomassie blue staining.

for Sic1 is deemed unsuitable because we consider it unlikely that the large Net1 protein would efficiently refold on removal of denaturant. Instead, we decided to isolate Net1 from SDS-denatured cells. A beaker containing 180 ml of H_2O is brought to boiling. Cell pellets (cells are harvested as rapidly as possible; see prior section) from a 4.5-liter culture that is shifted to 37° for 3 hr (final $OD_{600} = 1$) are resuspended in ~100 ml ice-cold H_2O and added to the boiling water bath in 20-ml aliquots with each addition being initiated when the water bath reaches boiling. This ensures that the temperature of the bath never drops below 90°. After all cells are processed, the bath is boiled for 3 more min and allowed to cool. The boiled cells are then harvested by centrifugation for 10 min at 5000 rpm in a clinical centrifuge. Cell pellet aliquots are frozen in liquid N_2 and stored at −80°.

One pellet (~5 ml, from ~1100 OD_{600}) is thawed on ice, resuspended in 1.5 volume of lysis buffer (100 mM Tris, pH 7.5, 200 mM NaCl, 10 mM DTT, 2% SDS, plus protease and phosphatase inhibitors), and boiled immediately for 3 min. Aliquots (320 μl) of cell suspension are distributed in 2-ml flat-bottom microtubes (USA Scientific, Ocala, FL), acid-washed 0.5-mm glass beads (200 μl per tube of Sigma, St. Louis, MO) are added, and tubes are vortexed at 4° in a multibead vortexer (six pulses of 3 min each, with 3-min intermissions). When glass bead lysis is carried out in a single large centrifuge tube, we find the lysis to be very inefficient. After vortexing, all tubes are boiled for 2 min to ensure that SDS has completely dissolved all extractable proteins.

The following steps (until after the addition of IP buffer) are carried out at room temperature to prevent precipitation of SDS. Supernatants are pooled into 1.7-ml microcentrifuge tubes, centrifuged at 14,000 rpm for 10 min, transferred to a 50-ml tube, and sonicated with a microtip at room temperature (Branson Sonifier 450, VWR, San Dimas, CA, six cycles of five pulses per cycle at #5 output control and 50% duty cycle). Sonication appears to reduce high molecular weight contaminants, possibly by shearing DNA into small fragments. Sonicated extracts are supplemented with 3 volumes of IP buffer (50 mM Tris, pH 7.5, 1% Triton X-100, 0.5 M NaCl, and protease inhibitors) ("extracts" in Figs. 3B and 3C). Triton forms micelles that absorb SDS not bound to proteins, thereby eliminating free SDS and creating an environment conducive for antibody–antigen interaction. Pansorbin cells (100 μl; Calbiochem, La Jolla, CA) are incubated with extracts for 20 min to preabsorb proteins that bind nonspecifically to protein A, and samples are then centrifuged at 160,000g (40,000 rpm in a Beckman Ti60 rotor) for 20 min at 12°. The supernatant (~30 ml) ("pre-IP" in Figs. 3B and 3C) is transferred to two 15-ml tubes, with each tube containing 0.32 ml of 9E10 resin (antimyc antibody cross-linked to protein A beads). The antibody beads are incubated with extracts for 1 hr at 4°, sedimented, resuspended in ice-cold wash buffer (20 mM HEPES, pH 7.2, 0.5% Triton X-100, 0.5 M NaCl, 1 mM DTT), and distributed to seven 0.6-ml tubes. An aliquot of post-IP extracts ("post-IP" in Fig. 3C) is set aside. The beads are washed seven times with 0.5 ml wash buffer/tube, transferred to fresh tubes (to avoid contaminants adsorbed to the side of the tube), and washed three more times ("beads" in Fig. 3B). The beads are then washed three times with 2 mM Tris, pH 8.8, 1 mM DTT, and eluted in 2 mM Tris, pH 8.8, 0.5 mM DTT, 0.1% SDS at 100° for 3 min. The supernatant is collected, and elution is repeated. The eluates are pooled ("eluate" in Fig. 3C), lyophilized, resuspended in 30 μl of SDS sample buffer, fractionated on a 7.5% SDS–polyacrymide gel, stained by Coomassie Brilliant Blue (0.1% Coomassie R-250, 20% methanol, 0.5% acetic acid in water), and destained in 30% MeOH in water (Fig. 3D). Samples that are set aside throughout the preparation are fractionated by SDS–PAGE and evaluated by immunoblotting with the antimyc antibody. As shown in Fig. 3C, ~70% of Net1-myc is recovered in the immunoprecipitation step (compare lanes 4,5 with 6,7). In the first preparation (Fig. 3B), the phosphorylation state of Net1 is apparently preserved throughout isolation. In the second preparation, Net1 from *cdc5 cdc14* is upshifted with respect to that from wild-type cells, suggesting that the basal phosphorylation of Net1 is also preserved in *cdc5 cdc14* (compare Fig. 3C, lanes 3 and 4). However, ~50% of the Net1 isolated from *cdc14* cells collapses into a lower molecular weight species after the elution step (Fig. 3C, lane 11), presumably as a result of dephosphorylation. Note that the relative phosphoshifts are difficult to see in the Coomassie-stained gel (Fig. 3D) due to overloading. The source of variability in our preparations is unknown, but smaller scale preparations that require less time may better preserve phosphorylation states. This procedure generates ~3 μg (~20 pmol) of Net1-Myc9 (Fig. 3D). Mass spectrometric analysis of Net1 samples excised from the SDS–polyacrylamide gel reveals up to 20 sites of phosphorylation, which will be reported elsewhere.

Multidimensional Electrospray Mass Spectrometry-Based Phosphopeptide Mapping Method

Before any effort is made to determine the site of phosphorylation on a protein, the modification is usually first localized to a peptide from an enzymatic digest. A major challenge in phosphopeptide mapping is to isolate or identify the phosphorylated peptides from the usually overwhelming amount of nonphosphorylated peptides present in the digest. Although the phosphopeptide need not be purified prior to sequencing by MS/MS, it is necessary to identify which peptide in the mixture to sequence.

Earlier we reported on a method utilizing nanoelectrospray MS and a technique called precursor ion scanning, which takes advantage of the fact that in the mass spectrometer, under experimentally controllable conditions, phosphopeptides undergo facile loss of phosphate from serine, threonine, and tyrosine residues to produce a PO_3^- ion with m/z 79.[28] During a precursor scan, the mass spectrometer is set to detect only those peptides that fragment to yield the diagnostic m/z 79 ion, thus allowing the selective detection of phosphopeptides. This high degree of selectivity allows the precursor scan to identify phosphopeptides that are present as very minor components of the sample. Once the phosphopeptides have been identified, the MS can be switched to the positive ion mode and the peptides sequenced by tandem MS.

Although this approach can successfully identify and sequence phosphopeptides from unfractionated mixtures, there are limitations. First, as the phosphorylation state of the protein becomes more complex, the least abundant phosphopeptides may go undetected. Second, large phosphoproteins can yield an overwhelming number of unmodified peptides that make it difficult to sequence the phosphopeptides even after they have been identified by the precursor ion scan. This difficulty stems from the unavoidable problems of a wide dynamic range of peptide abundance, ion suppression effects, charge state overlap, and the possible need to sequence many phosphopeptides in a single sample. Studies of highly phosphorylated proteins have illustrated the advantages of employing a separation step to fractionate the sample or enrich the phosphopeptide pool prior to site-specific analysis.

In 1993, we reported on a liquid chromatography (LC) MS-based method that used phosphopeptide-specific marker ions to selectively detect phosphopeptides in complex mixtures.[29] As components of a proteolytic digest elute from a high-performance liquid chromatography (HPLC) column into the MS, they are subjected to collision-induced dissociation conditions in the ion source. Peptides that contain phosphorylated residues fragment to produce highly diagnostic $[PO_2]^-$ and $[PO_3]^-$ marker ions at m/z 63 and m/z 79, respectively. Monitoring for these two marker ions permits selective collection of LC fractions that contain phosphopeptides (as well as coeluting nonphosphorylated peptides). The marker ion elution profile is analogous to the output from an HPLC radioactivity detector or the autoradiogram from a two-dimensional phosphopeptide map, but the MS-based method does not rely on $^{32/33}P$ labeling.

Combining these two orthogonal MS techniques—both of which selectively detect phosphopeptide-specific marker ions—with MS-based peptide sequencing yielded a multidimensional analytical method, which is highly selective for phosphopeptides.[13] The overall strategy is outlined in Fig. 4. In the first step, the proteolytic digest of a protein is subjected to reversed phase (RP)-HPLC and collected into fractions. On-line monitoring of the phosphopeptide-specific marker

[28] S. A. Carr, M. J. Huddleston, and R. S. Annan, *Anal. Biochem.* **239**, 180 (1996).
[29] M. J. Huddleston, R. S. Annan, M. F. Bean, and S. A. Carr, *J. Am. Soc. Mass Spectrom.* **4**, 710 (1993).

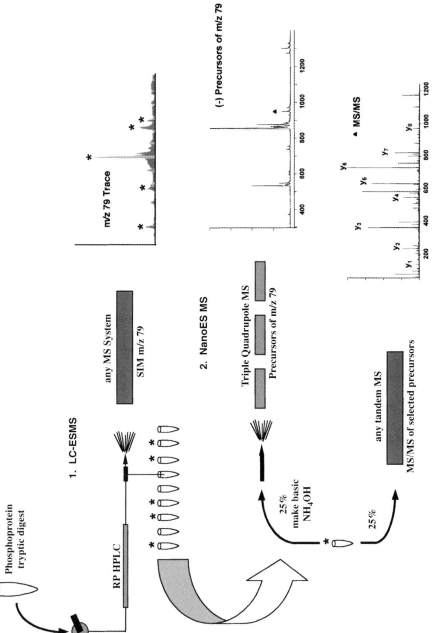

FIG. 4. Schematic diagram of the multidimensional electrospray–mass spectrometry method for mapping phosphorylation sites in proteins.

ion m/z 79 by ESMS in the negative ion mode identifies those fractions that contain phosphopeptides [those marked with an asterisk (*)]. This HPLC step greatly simplifies the complexity of phosphopeptide mixtures. In the second dimension of the analysis, the molecular weights of the phosphopeptides in each collected fraction are determined by precursor scans for m/z 79 in the negative ion mode. Finally the phosphopeptides are sequenced by positive-ion nanoelectrospray tandem MS. This strategy, which we have developed and refined, is particularly well suited to phosphoproteins that are phosphorylated with widely varying stoichiometry at many sites.

Analysis of Sic1 and Net1 Phosphorylation by Mass Spectrometry

Sic1

Sic1 is of sufficient purity that it can be analyzed directly from 0.1% TFA elution from anti-HA beads without the need for further purification by SDS–PAGE. Approximately 70% of the purified Sic1 solution is lyophilized to remove the acid. The sample is then reconstituted in 20 μl of 8 M urea and vortexed. The urea concentration is reduced to 2 M by the addition of 80 μl of 50 mM NH$_4$HCO$_3$, pH 8.5. Sic1 proves resistant to typical in-solution tryptic digestion conditions, finally succumbing to a 16-hr incubation at an enzyme : substrate ratio of 1 : 7. Sic1 contains no cysteines residues, thus reduction and alkylation are not performed. Half of the digest is loaded onto a 1-mm C$_{18}$ trap cartridge installed in place of a sample loop on the HPLC injector. After washing with 200 μl of 0.1% TFA–5% acetonitrile, the peptides are backflushed off the trap cartridge onto a 0.5 mm × 15 cm C$_{18}$ Reliasil (Michrom BioResources, Inc., Auburn, CA) HPLC column with an acetonitrile : water : TFA gradient run at 20 μl/min. The column eluate is split 4 : 1 after the UV detector, with 4–5 μl/min going to the mass spectrometer and the remainder going to a fraction collector taking half-minute fractions.

Electrospray mass spectra are acquired on a modified PerkinElmer (Norwalk, CT) Sciex API-III atmospheric pressure ionization triple quadrupole tandem mass spectrometer using the standard articulated, pneumatically assisted nebulization electrospray probe. The mass spectrometer, operating in the negative ion mode, is optimized to produce and detect m/z 63 and 79 product ions (PO$_2^-$ and PO$_3^-$) produced by CID in Quad 0, the high-pressure region located between the sampling cone and the quadrupole mass filter. The yield of the m/z 63 and 79 product ions is maximized by the application of -350 V (m/z 63) and -300 V (m/z 79) to the orifice of the API III as has been described.[29,30] The MS is operated in a single ion-monitoring mode for enhanced sensitivity.

The summed LC-ESMS ion trace for m/z 63 and 79 from the Sic1 sample shows more than 12 phosphopeptide-containing peaks, which are collected in

[30] S. A. Carr, R. S. Annan, and M. J. Huddleston, *Methods Enzymol.*, in press (2002).

30 fractions. To simplify the subsequent analysis, we pool adjacent fractions from clusters of phosphopeptide peaks and identify the phosphopeptides in each pool using a precursor scan for m/z 79. Precursor scans are acquired in the negative ion mode on the same PE Sciex API III as described previously, but using a nanoES source designed and built at the EMBL.[31] To enhance detection of the phosphopeptides, an aliquot of each fraction is made basic prior to precursor ion analysis as follows: one-fourth to one-half of a desired fraction is lyophilized and reconstituted in 50/50 methanol : water containing 5% concentrated ammonium hydroxide (30% by weight), and 1–2 μl of the sample is introduced into the nanoES needle.[28] After the molecular weights of the phosphopeptides are determined from the m/z 79 precursor ion scan, the mass spectrometer is switched into the positive ion mode. When sufficient signal is present, the phosphopeptides are sequenced from the same sample loading by acquiring a full-scan CID product ion spectrum (MS/MS) on a chosen multiply charged precursor. In some cases, an acidic aliquot of the sample in 50 : 50 (v/v) methanol water with 5% (v/v) formic acid is used for MS/MS sequencing of the phosphopeptides.

Precursor ion scan analysis and tandem MS of the pooled fractions reveal that Sic1 phosphorylation is localized on three tryptic peptide sequences: amino acids 2–8, 14–50, and 54–84. The 2–8 sequence appears to be quantitatively monophosphorylated on Thr-5, whereas the latter two peptides are differentially phosphorylated (0-2 and 0-3 phosphates, respectively), with the lowest phosphorylation state being the most abundant. We sequenced the monophosphorylated form of the 14–50 peptide and the 54–84 peptide by MS/MS and determined that the major sites of phosphorylation are Thr-33 and Ser-76, respectively. Tandem mass spectrometry of the 54–84 peptide failed to distinguish between Thr-75 and Ser-76, and assignment of the phosphorylation site in this case was based on the fact that Ser-76 is in a Cdk consensus sequence.

Because the purification protocol just described yields purified Sic1 in solution, we are able to use nanoESMS to determine the overall phosphorylation state of the intact protein.[13] To facilitate the nanoES, we pass the remaining protein (approximately 30% of the original material) over a 0.5-mm C_{18} HPLC column and collect the Sic1 peak. The solvent is removed by lyophilization, and the protein is reconstituted in 4 μl of 50 : 50 methanol : water with 5% formic acid and is loaded into the nanoES needle. The distribution of phosphorylation on Sic1 corresponds to between 1 and 6 mol of phosphate per mole of Sic1, with quadruply and quintuply phosphorylated Sic1 being the predominant forms,[13] and suggests that Sic1 is phosphorylated by Cln/Cdc28 in a distributive manner *in vivo*.

The physiological significance of the three major *in vivo* phosphorylation sites determined by mass spectrometry is confirmed by analysis of a mutant in which Thr-5, Thr-33, and Ser-76 are substituted with Ala.[15] When this triple mutant is

---

[31] M. Wilm and M. Mann, *Int. J. Mass Spectrom. Ion Proc.* **136,** 167 (1994).

analyzed by nanoES, the mass spectrum shows that the most abundant phosphorylation state decreases from 4–5 mol of phosphate (observed on wild-type Sic1) to 2 mol of phosphate.[13] These data suggest that these three sites are indeed major phosphoacceptor sites *in vivo*.

Whenever it is possible to generate sufficient amounts of purified protein (10–50 pmol), we advocate making a molecular weight measurement on the intact protein. The intact molecular weight provides the total number of moles of phosphate added to the protein and the relative distribution of protein molecules with different numbers of phosphate on them. As was demonstrated here for Sic 1, this information can be useful for interpreting the phosphopeptide mapping and mutagenesis data.

Net1

Proteins derived by SDS–PAGE require special preparation protocols to ready the sample for MS analysis. Elution of proteins from acrylamide gels is inefficient at best and is a hopeless proposition for a large protein like Net1. The most straightforward approach to preparing gel-fractionated proteins for MS analysis is direct digestion of the protein in the gel.[32] The Coomassie-stained Net1 band (Fig. 3D, lane 2) is excised from the gel with a razor blade, and the gel piece is washed in a 1.5-ml Eppendorf tube with several changes of 50:50 acetonitrile:50 mM NH_4HCO_3 until the Coomassie blue is completely extracted. The wash solvent is discarded and replaced with 50 μl of fresh 50 mM NH_4HCO_3, pH 8.5, and 5 μl of 45 mM DTT. The sample is heated for 1 hr at 38°. After cooling to room temperature, 5 μl of 100 mM iodoacetamide is added, and the sample is allowed to react for 1 hr at room temperature in the dark. After 1 hr, the gel piece is washed for 1 hr with 50:50 acetonitrile:50 mM NH_4HCO_3 and is then covered with acetonitrile and allowed to shrink until it turns white. The acetonitrile is removed, and the gel piece is dried in a Speed-Vac. The gel piece is rehydrated with 15 μl of 50 mM NH_4HCO_3 containing 500 ng of modified trypsin. After the gel piece is completely reswollen, an additional 50 μl of 50 mM NH_4HCO_3 is added to cover the slice, and the sample is incubated overnight at 38°. After overnight digestion, the NH_4HCO_3 solution is removed and transferred to a fresh 200-μl Eppendorf tube. The gel piece is washed once with an additional 50 μl of fresh 50 mM NH_4HCO_3 and once with 50 μl of 5% formic acid. Both washes are combined with the original NH_4HCO_3 solution, and 25–50% of the sample is injected onto a C_{18} trap cartridge, which is installed in place of the sample loop on an HPLC injector as described earlier. Following a wash of the trap with 200 μl 0.1% TFA, the Net1 digest is analyzed using the multidimensional MS-based phosphopeptide mapping strategy described previously.

[32] J. Rosenfeld, J. Capdevielle, J. Guillemot, and P. Ferrara, *Anal. Biochem.* **203**, 173 (1992).

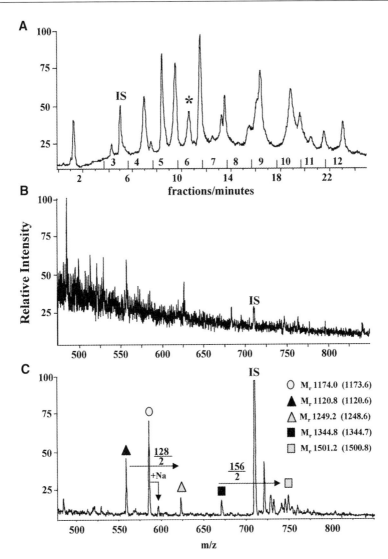

FIG. 5. Multidimensional electrospray–MS phosphopeptide mapping of Net1. A Coomassie blue-stained band was excised from the gel (Fig. 3D, lane 2) and digested *in situ* with trypsin. Five hundred femtomoles of phosphopeptide internal standard was added to the digest prior to injection onto the HPLC. (A) First-dimension phosphopeptide-specific LC-ESMS trace (m/z 79). The phosphopeptide internal standard peak is marked as "IS." Twelve fractions were taken for further analysis, and 100 fmol of internal standard was added to each fraction. (B) NanoES (−) ion full-scan MS spectrum of fraction 6 from the LC-ESMS run in (A). (C) NanoES (−) ion precursor scan for m/z 79 of fraction 6. Five phosphopeptides with listed molecular weights were detected.

The much smaller amount of protein typically available from a gel-derived sample necessitates some changes in the analysis. The sensitivity of the method as described for Sic1 is limited by the HPLC column diameter. Because ESMS (at flow rates above ca. 50 nl/min) acts like a concentration-sensitive detector, smaller internal diameter HPLC columns provide higher sensitivity analyses. However, the requirement to split the HPLC flow postcolumn to allow collection of fractions imposes a practical limitation on the HPLC column diameter that can be used. By switching the MS to a micro ion spray source flowing at 0.2–0.5 μl/min, we are able to use a 180 μm i.d. HPLC column flowing at rates of 3–4 μl/min. By setting up an approximately 10 : 1 postcolumn split, we are able to collect 3.5-μl/min fractions. Preliminary data obtained using this setup indicate that sensitivity in the first-dimension analysis is 20- to 50-fold higher than the 0.5-mm i.d. column used for the Sic1 analysis (Zappacosta, unpublished results, 2001). The smaller HPLC column diameter also required us to use a smaller diameter C_{18} trap cartridge (300 μm i.d.).

The m/z 79 trace for Net 1 from $cdc14^{ts}$ $cdc5^{ts}$ cells is shown in Fig. 5A. The Net1 digest is spiked with 500 fmol of a phosphopeptide standard (marked IS) prior to analysis. The peak heights of the Net 1 phosphopeptides relative to the internal standard allow an estimate of the overall level of phosphorylation present in the sample. Ten fractions (collected by hand) labeled 3–12 in Fig. 5A were analyzed using precursor ion scans and tandem MS as described earlier. No changes were required in this part of the analysis to accommodate the lower sample amounts. The negative ion MS and the precursor scan for m/z 79 for fraction 6 are shown in Figs. 5B and 5C, respectively. The precursor ion scan shows five phosphopeptides present in this fraction, whereas none of these peptides are detectable in the negative ion scan. The sequences, which have been assigned to these molecular weights, were verified by tandem MS; in each case the specific phosphorylated residue was established. In all, 20 unique phosphorylation sites were determined from 22 different phosphopeptides.

[17] Identification of Yeast Proteins by Mass Spectrometry

By ALEXANDRE V. PODTELEJNIKOV and MATTHIAS MANN

Yeast provides a unique model system that enables elegant and rapid genetic analyses in addition to extremely powerful biochemical studies. Mass spectrometry has further improved the toolbox of biochemical strategies by providing rapid and high throughput identification of proteins in yeast. In fact, many of the new proteomic approaches are being established first in yeast because *Saccharomyces cerevisiae* was the first fully sequenced eukaryotic genome,[1] a precondition to rapid protein database identification by mass spectrometry.

The proteomic revolution is still continuing and is based on the large-scale identification of proteins, similar to rapid DNA sequencing, which fueled the genomics revolution. Today, any yeast protein that can be purified so as to be silver stained can be identified rapidly by mass spectrometry. This is opening the way for large-scale yeast protein interaction studies (through large numbers of immunoprecipitations of tagged yeast proteins) and the study of quantitative changes in the yeast proteome. It is also providing us with the knowledge of the composition of various yeast organelles and other cellular compartments. This chapter outlines the principles and practical protocols involved in the mass spectrometric identification of yeast proteins.

Introduction to Mass Spectrometry

This chapter covers the basics of mass spectrometric techniques and concepts for yeast protein identification. Mass spectrometers have the ability to generate ions from macromolecules in the gas phase and to separate them according to their mass-to-charge ratios (m/z). Mass spectrometry is a versatile technique that enables a researcher to study different compounds ranging from single atoms to whole viruses. The relevant mass spectrometric equipment used depends on the goals of the investigation. The wide variety of mass spectrometers available commercially can be distinguished by the way ions are produced (ion sources) and the way the ions are separated (mass analyzers). The most commonly used ion sources for biological applications are matrix-assisted laser desorption and ionization (MALDI) and electrospray (ES) ionization. Mass analyzers resolve ions in either electric or magnetic fields according to m/z and generate signals on the detector. Different types can operate with very different physical principles, but their common feature is that they all assign mass-to-charge values to the measured ions. For biological applications, the most applicable mass analyzers include time-of-flight (TOF), quadrupole, ion trap, and Fourier transform ion cyclotron resonance (FTICR).

The quality of data varies from one method to another depending on the type of analyzer (see Table I). Among the most important characteristics of data are mass accuracy estimated in parts per million (ppm), sensitivity, and resolution. Resolution is typically defined as the full width at half-maximum (FWHM) and can be presented as $RR = M/\Delta M$, where R is the resolution, M is the m/z value of a peak, and ΔM is the width of the peak at half-height.

The high resolving power of modern mass spectrometers makes it possible to distinguish between naturally occurring isotopes. For example, peptides contain

[1] A. B. Goffeau, G. Barrell, H. Bussey, R. W. Davis, B. Dujon, H. Feldmann, F. Galibert, J. D. Hoheisel, C. Jacq, M. Johnston, E. J. Louis, H. W. Mewes, Y. Murakami, P. Philippsen, H. Tettelin, and S. G. Oliver, *Science* **274**, 563 (1996).

TABLE I
MASS SPECTROMETRIC TECHNIQUES FOR PROTEIN IDENTIFICATION

Method	Mass analyzer	Mass range (Da)	Resolution (FHMW)	Mass accuracy (ppm)	LC/MS capabilities	Sequence quality
MALDI TOF	TOF	Unlimited	8000–12,000	5–30	−	+
ES QQQ	Triple quadrupole	3000	1000–2000	100–1000	+++	+++
ES QTOF	Hybrid quadrupole TOF	3000	7000–12,000	5–50	+++	+++
ES IT	Ion trap	2000	1000–2000 Zoom scan up to 4000	100–2000	+++	++
ES FTICR	Ion cyclotron resonance	5000	Up to 1,000,000	1–50	+++	+
MALDI QTOF (o-MALDI)	Hybrid quadrupole TOF	3000	7000–12,000	5–50	−	++

−, poor or not applicable; +, decent; ++, good; +++, excellent.

an average of 60% carbon by weight and 1% of the carbon atoms are ^{13}C instead of ^{12}C. Therefore, peptide mass spectra are composed of a series of peaks that are 1 Da apart. Molecular weight is usually determined by the first or monoisotopic peak. If the resolution is low, the molecular mass is measured by average molecular weight. Generally, the greater the resolving power, the better the mass accuracy provided by the instrument. For example, FTICR instruments can achieve resolution up to one million, providing mass accuracy of up to 1 ppm.

To generate structural information, several stages of mass spectrometry are performed in series. Specific ions are selected by the first mass analyzer and are transmitted to the collision cell. Inside this cell the ions undergo collision-induced dissociation (CID) to yield a set of fragment ions. The resulting fragment ions are detected using another mass analyzer. This technique is known as a tandem mass spectrometry (MS/MS). For MS/MS, mass analyzers are assembled in a linear or an orthogonal arrangement or, alternatively, "trapped" ions are mass isolated and fragmented in the same mass spectrometer.

The most common methods to ionize biological molecules are electrospray and MALDI. Both of them are known as "soft" ionization methods, permitting transfer of the biological compound into the gas phase of the mass analyzer without decomposition.

Principles of MALDI Mass Spectrometry

Laser desorption ionization has been used for years for the analysis of low molecular weight compounds. With the advent of MALDI[2] in 1988, this technique

[2] M. Karas and F. Hillenkamp, *Anal. Chem.* **60**, 2299 (1988).

became applicable to the analysis of biomolecules. MALDI is based on the desorption of matrix–analyte cocrystals after irradiation with a laser pulse. Small UV light-absorbing molecules such as α-cyano-4-hydroxycinnamic acid or 2,5-dihydroxybenzoic acid are used as the matrix. The energy of the pulsed laser beam is absorbed by the matrix, causing the matrix–analyte crystals to vaporize into a vacuum without significant decomposition of the analyte. Ions are formed due to gas-phase reactions, including proton transfer from matrix to analyte, and observed as quasimolecular ions $[M + H]^+$ rather than radicals. MALDI analysis results in single charged ions and it is usually combined with TOF analyzers. Depending on the nature of the compound under investigation, an appropriate matrix should be used. Large numbers of samples are often spotted on metal targets and analyzed by MALDI-TOF in a serial fashion.

Principles of Electrospray Mass Spectrometry

Electrospray (ES)[3,4] is another "soft" ionization method but is based on a completely different physical principle. In ES, a solution containing the analyte is passed through a hypodermic needle at high voltage, which vaporizes or "electrosprays" the solution. Electrically charged droplets evaporate and release multiply protonated analyte ions ($[M + nH]^{n+}$). Unlike the MALDI method, ES allows the rapid and accurate determination of protein molecular weight as the multiple charging phenomenon brings protein ions into the m/z range of typical quadrupole or ion trap mass spectrometers.

The spraying process is less tolerant to salts and detergents than the MALDI process; therefore, a sample purification step is usually necessary in ES analysis. Often, samples are separated chromatographically by high-performance liquid chromatography (HPLC), the effluent of which is directly coupled to an ES source. More efficient electrospray ionization can be achieved using a low flow electrospray source known as nanoelectrospray.[5]

Peptide Fragmentation

During MS/MS, peptides undergo collision-induced dissociation mainly at amide bonds, leading to so-called b ions (charge retention on the N terminus) or y ions (charge retention on the C terminus).[6] The fragmentation scheme and ion types are listed in Fig. 1.

Trypsin is a commonly used protease, which cleaves proteins C-terminal to Arg (R) or Lys (K) residues. Such peptides are usually doubly or triply charged in

[3] M. Yamashita and J. B. Fenn, *J. Phys. Chem.* **88,** 4451 (1984).
[4] M. L. Alexandrov, L. Gall, N. V. Krasnov, V. I. Nikolaev, V. A. Pavlenko, V. A. Shkurov, G. I. Baram, M. A. Grachev, V. D. Knorre, and Y. S. Kusner, *Bioorg. Khim.* **10,** 710 (1984). [In Russian]
[5] M. Wilm and M. Mann, *Anal. Chem.* **68,** 1 (1996).
[6] P. Roepstorff and J. Fohlman, *Biomed. Mass Spectrom.* **11,** 601 (1984).

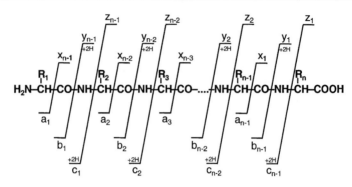

FIG. 1. Scheme of major backbone fragmentation according to P. Roepstorff and J. Fohlman, Biomed. Mass Spectrom. **11**, 601 (1984).

ES and often yield easily interpretable tandem mass spectra. On triple quadrupole-type instruments, y ion series in the high mass range are common, whereas in ion traps, abundant b and y ions are observed. The special structure of the immino acid proline leads to little fragmentation at its C-terminal peptide bond and large fragmentation at its N-terminal bond. The acidic amino acids Asp (D) and Glu (E) are also often sites for preferential fragmentation. The fragment formed can then fragment further in triple quadrupole-type instruments and results in an "internal" fragment ion series.

Liquid Chromatography–Mass Spectrometry (LC-MS)

Electrospray mass spectrometry is ideal for on-line coupling with low flow rate liquid chromatographic separation. Combining LC with mass spectrometry enables desalting, separation, and concentration of biological samples all in the same step. For very complex mixtures, two chromatographic principles with different separation properties can be coupled in series.

The most common and effective way to separate proteins and peptides is to use reversed-phase HPLC. Because the signal strength in electrospray is only dependent on the concentration but independent of flow rate, it is advantageous to use very small columns and very low flow rates (nanoLC). Columns as small as 75 μm with flow rates of 200 to 300 nl/min are commonly used, achieving sensitivities in the low femtomole range of peptides loaded on the column.

In addition to the characterization of peptides, mass spectrometry has proven to be a very useful technique for protein characterization, including accurate molecular weight determination of intact proteins and detection of various post-translational modifications, such as glycosylation and phosphorylation. Usually, these tasks demand somewhat larger amounts of starting biological material (picomoles) and require specialized mass spectrometric approaches. The detection of phosphorylation is described, and a description of these methods can be found elsewhere.[7] This chapter concentrates primarily on yeast protein identification based on proteolytic digestion and mass spectrometric detection of peptide mixtures, as well as on interpretation and evaluation of obtained MS data using methods applied in our laboratory.

Methods for Protein Identification

In principle, proteins can be identified by their molecular mass, the molecular mass of the peptides after sequence specific digestion ("peptide mass fingerprinting") or by the fragmentation of some of the peptides (MS/MS). The first method, identification by protein molecular mass, is usually not practical because the expected molecular mass is shifted subtly by chemical and natural processing events such as oxidation of methionine residues and N- and C-terminal processing. Furthermore, as noted earlier, the analysis of protein molecular mass requires significantly more material than the analysis of peptides. The second method, peptide mass finger printing, is applied after in-gel digestion of protein bands or spots obtained after sodium dodecyl sulfate–polyacrylamide gel electrophoresis (SDS–PAGE), and MALDI is the method of choice. In the third method, tandem mass spectrometry, nanoelectrospray, LC/MS/MS, or MALDI quadrupole-TOF is used to sequence the peptides. LC/MS/MS methods have become sufficiently powerful to allow the direct analysis of peptide mixtures derived from unseparated protein mixtures, completely bypassing gel separation (Fig. 2).

In typical biological experiments, proteins are purified by column chromatography and/or immunopurification methods, often followed by one-dimensional gel electrophoresis for separation by molecular weight. Determination of yeast protein interaction partners after tagging and immunoprecipitation is a good example of a class of experiments that has been made possible by the rapid and sensitive mass spectrometric identification of protein bands. Many multiprotein complexes and signal transduction pathways have been elucidated in this fashion as well.

For very complex mixtures, two-dimensional (2D) PAGE has been used as the referral technique since its introduction in 1975 by O'Farrell.[8] Two-dimensional gel electrophoresis separates proteins according to isoelectric focusing in the first dimension and molecular weight in the second dimension. This technique allows visualization of up to several thousand protein spots and provides information about molecular weight, isoelectric point, and relative abundance of a protein. Experience with large-scale analysis of 2D gels by MS has shown that many of the spots are actually due to minor variations of abundant proteins and that low

---

[7] R. Aebersold, B. Rist, and S. P. Gygi, *Ann. N.Y. Acad. Sci.* **919,** 33 (2000).
[8] P. O'Farrell, *J. Biol. Chem.* **250,** 4007 (1975).

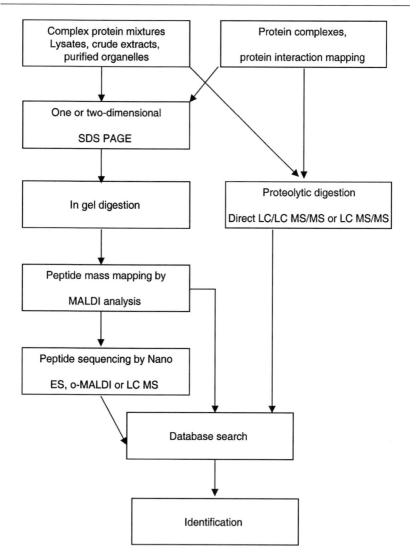

FIG. 2. Schematic of yeast protein identifications by mass spectrometric methods.

abundance proteins are not represented in 2D gels of total yeast cell lysates.[9] Typically only a few hundred different proteins are represented in yeast 2D gels.

To visualize protein bands or spots on polyacrylamide gels, several staining procedures are available. Among them are Coomassie Brilliant Blue, reversed

---

[9] M. Perrot, F. Sagliocco, T. Mini, C. Monribot, U. Schneider, A. Shevchenko, M. Mann, P. Jenö, and H. Boucherie, *Electrophoresis* **20**, 2280 (1999).

staining with zinc/imidazole, a number of fluorescent stains such as SYPRO Ruby, and several silver-stained protocols. Protein bands on polyacrylamide gels stained with colloidal Coomassie or Coomassie blue yield sufficient amounts of yeast proteins for mass spectrometric analysis and identification is straightforward. Coomassie staining has the added advantage of being semiquantitative. Silver staining methods are at least an order of magnitude more sensitive than colloidal Coomassie staining and allow detecting proteins in low femtomole amounts (1–50 ng of a protein). Not all silver-staining protocols are suitable for mass spectrometric identification. For example, using glutaraldehyde as a sensitizing agent leads to very low recovery of peptides due to the cross-linking of proteins with the gel matrix. We recommend using the following silver-staining protocol.[10] To avoid staining artifacts, all solutions need to be of high purity and fresh prepared.

1. Fix the gel with [50:5:45 (v/v/v) methanol : acetic acid : water] for 20–30 min.
2. Rinse the gel with water for 1 hr to remove acid. Extensive washing will eliminate yellowish background usually observed after the development step.
3. Incubate gel with sensitizing solution (0.02% sodium thiosulfate) for 1–2 min.
4. Discard the thiosulfate solution and rinse the gel twice with water (1 min per change).
5. Incubate the gel in 0.1% (w/v) $AgNO_3$ for 20 min at $4°$.
6. Discard the silver nitrate solution and rinse the gel with two changes of water (1 min each).
7. Develop the gel with 0.04% (v/v) formaldehyde in 2% Na_2CO_3. Replace the developing solution when it turns yellow with a fresh one.
8. As soon as sufficient staining has been obtained, discard developer and add 200–300 ml 1% (v/v) acetic acid. Rinse the gel several times with 1% (v/v) acetic acid.
9. Store silver-stained gels in 1% (v/v) acetic acid at $4°$.

In principle, proteins can also be electroeluted from polyacrylamide gels and transferred to nitrocellulose or polyvinylidene difluoride (PVDF) membranes; however, in practice, losses due to elution and removal of SDS usually make mass spectrometric analysis impossible.

As mentioned earlier, mass spectrometry is also quite successful for intact protein analysis, especially for the characterization of recombinant proteins defining various posttranslation modifications and disulfide bonds. However, due to heterogeneity, problems with desalting, and lower sensitivity of mass spectrometers for protein analysis compared to peptide analysis, it is usually more practical to

---

[10] A. Shevchenko, M. Wilm, O. Vorm, and M. Mann, *Anal. Chem.* **68**, 850 (1996).

reduce proteins to peptides.[11] In this step, proteins undergo chemical or proteolytic digestion. Trypsin and endoproteinases Lys-C, Glu-C, and Asp-N are the most commonly used enzymes because of their high sequence specificity. The choice of protease also depends on the size of the resulting peptides and their aggressiveness and stability, as well as their resistance to autoproteolysis. Trypsin is the most popular enzyme for several reasons. It is highly specific for the C terminus of lysine and arginine. On average, tryptic peptides are around 5–20 amino acid residues long, corresponding to the most accurate and sensitive region in modern mass spectrometers. Tryptic peptides also usually carry at least two charges, on N-terminal amino groups and on lysine or arginine located at the C terminus, which makes tandem mass spectra relatively easy to interpret in ES MS/MS.

In-Gel Protein Digestion

1. After washing the gel intensively with water, excise the bands (spots) of interest using a clean, sharp scalpel at the margin of detectable stain. Transfer the gel pieces into an Eppendorf tube or 96-well plate. Wash the gel particles with 100–150 μl of water (5 min). Spin down gel particles and discard the solvent. Add acetonitrile and incubate for 5 min until the gel pieces shrink. Spin down the gel particles and discard the liquid. Dry down gel particles in a vacuum centrifuge.

2. Swell the gel pieces in 10 mM dithiothreitol in 100 mM NH$_4$HCO$_3$ buffer and incubate for 30 min at 56° to reduce the protein. Spin down gel particles and discard the liquid. Shrink the gel pieces with acetonitrile.

3. Replace acetonitrile with 55 mM iodoacetamide in 100 mM NH$_4$HCO$_3$ buffer. Incubate for 20 min at room temperature in the dark. Discard the iodoacetamide solution. Wash the gel pieces with 100–150 μl of 100 mM NH$_4$HCO$_3$ buffer for 15 min. Spin down gel particles and discard the liquid. Shrink the gel pieces with acetonitrile. Dry down gel particles in a vacuum centrifuge.

4. Rehydrate gel pieces in the digestion buffer (50 mM NH$_4$HCO$_3$ containing 5 mM CaCl$_2$) containing 12.5 ng/μl trypsin (or another enzyme). Incubate at 37°. After 15–20 min of incubation, check the samples and add more digestion buffer if needed to cover gel pieces with liquid. After 3–8 hr of incubation, spin down the condensate. Take up a small (0.2–0.5 μl) aliquot of the supernatant for MALDI analysis.

5. Extract peptides by adding 40–50 μl of 5% formic acid. Vortex briefly, incubate for 10 min, spin down the gel pieces, and collect the supernatant. The samples should be stored at −20°.

MALDI Peptide Mass Mapping

MALDI analysis has proven to be very successful in the analysis of peptide mixtures derived from in-solution or in-gel digest of proteins. MALDI peptide mass

[11] R. Beynon and J. Bond, "Proteolytic Enzymes." Oxford Univ. Press, New York, 2001.

mapping is a relatively simple technique based on the detection of usually singly charged ions with high mass accuracy. MALDI MS analysis is fast and consumes small amounts of sample volume. Normally, 1–5% of supernatant is sufficient to obtain high-quality data (roughly 1 fmol of peptide on the target, depending on the instrument). It is also tolerant to buffer constituents. We recommend using the fast evaporation method for sample preparation in MALDI TOF.[12]

1. Prepare a saturate solution of α-cyano-4-hydroxycinnamic acid (HCCA) in acetone. Vortex, centrifuge briefly, and remove supernatant into an Eppendorf tube. Add 2–5% by volume 10% formic acid to a supernatant. Fresh matrix solution should be prepared daily. Precautions should be taken to check the quality of HCCA. In the case of brownish powder, it is recommended to recrystallize the matrix.
2. Dissolve nitrocellulose (Bio-Rad Trans-Blot membrane) in acetone/2-propanol (1 : 1) with a final concentration around 10 g/liter.
3. Mix a nitrocellulose solution with a matrix solution with a 1 : 4 ratio to get Fast Evaporation NitroCellulose (FENC) containing solution.
4. Quickly deposit 0.3 μl of FENC solution on a MALDI target. Because the matrix solution contains mainly acetone, it should evaporate quickly by spreading on the metal surface. No crystals should be detected.
5. Deposit 0.3 μl of 10% formic acid on the thin-layer surface. Immediately load 0.3 μl of the supernatant obtained after in-gel digestion in an acidified water droplet. This will prevent the matrix thin layer from dissolving due to the high pH value of the peptide solution.
6. Keep the samples in air until the target is dry.
7. Wash the thin layer with a deposited sample to remove salts. Deposit 2–5 μl of 10% formic acid and remove it by blowing compressed air or shaking the target. Allow to dry at room temperature. Samples are ready for analysis.

The measured values of peptide masses are compared with values of theoretical protein digests of each protein in large protein databases. Basically, the number of matching peptides is counted within the mass accuracy of the experiment. Ambiguities in MALDI fingerprinting identifications have been minimized by improvements in a mass accuracy to better than 50 ppm achieved using delayed extraction in reflector time-of-flight (TOF) instruments. To calibrate the spectrum from the TOF scale to the mass scale, trypsin autoproteolysis products can be used. Bovine trypsin from Roche Diagnostics yields autolysis products at m/z 805.417, 906.504, 1153.570, and 2163.057. Porcine trypsin from Promega (Madison, WI) undergoes less autolysis, but two peptides with m/z values 842.509 and 2211.104 can be detected and are sufficient for an accurate two-point calibration. Matrix-related

[12] O. Vorm, P. Roepstorff, and M. Mann, *Anal. Chem.* **66**, 3281 (1994).

ions at m/z 855.06, 860.06, and 1066.06 from HCCA can also be useful for internal calibration.

The certainty of protein identification by peptide fingerprinting depends on the number of detected peptides and the mass accuracy. Generally, at least five peptides should match within 30 ppm mass accuracy, and sequence coverage must be better than 15% to call a positive identification.[13] The major limitations for MALDI peptide mapping are low sequence coverage for proteins with a molecular mass of more than 200 kDa or an insufficient number of matched peptides for low molecular weight proteins (less than 20 kDa) where only few peptides can be detected. Furthermore, identification of proteins from species without fully sequenced genomes is also less successful but this does not apply to the yeast system.

As mentioned previously, a small fraction of the protein digest is used by MALDI analysis. The rest of the material can be subjected to MS/MS analysis in case the results of MALDI peptide mapping are uncertain. Sequence analysis was performed previously on nanoelectrospray tandem mass spectrometers or by ES MS/MS analysis on an ion trap or quadrupole TOF instrument. The additional sample handling limited the very high throughput achievable by MALDI analysis alone. In our laboratory, a relatively new method, orthogonal MALDI quadrupole TOF ("o-MALDI"),[14] has superseded both MALDI and the subsequent electrospray sequencing steps in the analysis of yeast proteins. This technique allows the acquisition of MALDI mass spectra at high resolution and mass accuracy combined with the added certainty due to fragmenting selected ions from the peptide mass map.

For o-MALDI, the fast evaporation method does not produce the same sensitivity as for a classical MALDI-TOF. In this case, we recommend using the dried droplet method based on 2,5-dihydroxybenzoic acid (DHB).

1. Prepare a DHB solution (40 g/liter) in 3 : 1 (v/v) acetonitrile : 10% formic acid. Vortex for half a minute, centrifuge briefly, and remove supernatant into an Eppendorf tube.

2. Deposit 0.5 μl of matrix on a MALDI target. Add 0.5 μl of the supernatant obtained after in-gel digestion to a matrix droplet and mix it with an Eppendorf tip to accelerate crystallization.

3. Allow to dry at room temperature. Sample is ready for analysis.

The choice of a matrix will affect the MALDI peptide map. For HCCA, most of the detected peptides will be in the mass range of 800–2500 Da, whereas larger peptides can be detected with the DHB matrix.

---

[13] O. N. Jensen, A. Podtelejnikov, and M. Mann, *Rapid Commun. Mass Spectrom.* **10,** 1371 (1996).

[14] A. V. Loboda, A. N. Krutchinsky, M. Bromirski, W. Ens, and K. G. Standing, *Rapid Commun. Mass Spectrom.* **14,** 1047 (2000).

An example of protein identification by MALDI peptide mapping followed by MS/MS of one selected peptide is shown in Fig. 3. A Coomassie-stained band migrating at an apparent molecular mass of 17 kDa is in-gel digested, and the resulting peptide mixture is subjected to MALDI MS analysis on an orthogonal MALDI quadrupole-TOF instrument (PE-Sciex, Toronto, Canada). Monoisotopic peaks are labeled, and a list of detected masses was submitted to a database search using the database search engine PepSea (MDS-Proteomics). In total, only four masses were matched to yeast peptidylprolyl *cis–trans*-isomerase with an accuracy better than 30 ppm. This was not sufficient for unambiguous identification, and MS/MS data were acquired.

The high mass accuracy of o-MALDI for both peptides ("precursors") and their fragments ("product ions") facilitates the identification of proteins in databases and is limited more by ion statistics than by instrument performance. For MS/MS data, a search algorithm named a "breakpoint search" was used. It is based on the comparison of measured fragment masses to calculated fragment masses followed by scoring. As shown in Fig. 3B, 14 fragment ions were found to correspond to a, b, and y type with a mass accuracy better than 30 ppm. All data processing can be performed automatically, including peak recognition and database searches without human intervention. As a result, the 17-kDa protein was identified as the peptidylprolyl *cis–trans*-isomerase of *Saccharomyces cerevisiae* (Accession number P14832). Precautions should be taken for low femtomole levels of in-gel digested samples because the fragmentation pattern of singly charged peptide ions is more complicated and fragment ions from chemical background can lead to wrong assignments. Very large numbers of samples can be processed by MALDI. Modern MALDI-TOF instruments allow analysis of one sample per minute without human intervention. Taking into account that all open reading frames in the *S. cerevisiae* genome are known, MALDI-TOF is potentially the ideal instrument for the high-throughput identification of yeast proteins derived from SDS–PAGE separation.

Nanoelectrospray Peptide Sequencing

Nanoelectrospray allows a flow rate of about 10–20 nl/min and requires a very small amount of sample. No pumps or chromatographic equipment are required and samples are desalted and concentrated on hand-held microcolumns instead. Nanoelectrospray MS/MS has proven to be a highly sensitive method, but so far it has been difficult to automate, resulting in a relatively low sample throughput.

1. Prepare a microcolumn. Carefully squeeze a GelLoader (Eppendorf) tip on the edge of a tip[15] and pluck it with a short piece of a fused capillary (Fig. 4A).

[15] H. Erdjument Bromage, M. Lui, L. Lacomis, A. Grewal, R. S. Annan, D. E. McNulty, S. A. Carr, and P. Tempst, *J. Chromatogr. A* **826**, 167 (1998).

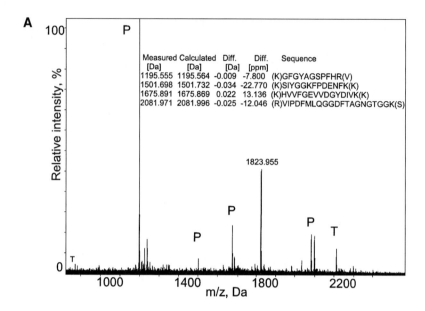

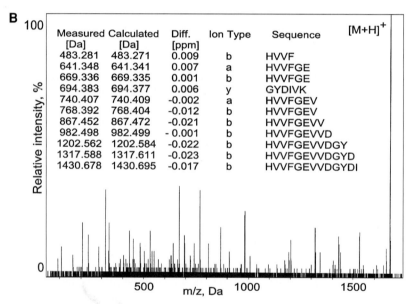

FIG. 3. MALDI peptide map of a 17-kDa yeast protein. (A) A MALDI mass spectrum of unseparated tryptic peptides derived from in-gel digestion was acquired on a quadupole TOF ("o-MALDI") instrument using the dried droplet protocol for sample preparation. The spectrum was calibrated using trypsin autolysis products marked with a "T". Signals marked with "P" correspond to yeast peptidyl-prolyl *cis–trans*-isomerase. The peptide with a m/z 1823.955 was found to be the acetylated N-terminal peptide after evaluation of data. Mass accuracy and sequence of detected peptides are presented in the inset table. (B) MALDI MS/MS of a tryptic peptide with m/z 1675.891 from A. Mass accuracy, type of fragment ions, and sequence of detected fragments are presented in the inset table. The whole sequence of the peptide could be verified by these data.

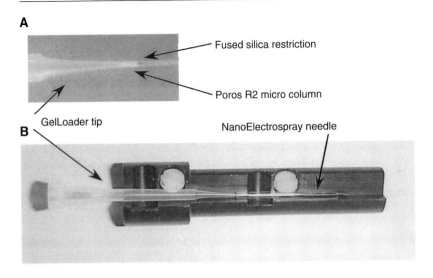

FIG. 4. Sample cleanup setup for nanoelectrospray analysis. (A) Gelloader tip filled with Poros R2 sorbent material using a fused silica capillary as a restrictor. (B) Mounting of microcolumn and nanoelectrospray needle in the MDS Proteomics holder. (C) A table microcentrifuge for gentle spinning down the liquid.

2. Load approximately 100–200 nl of POROS R2 reversed-phase resin (PerSeptive Biosystems, Framingham, MA) in methanol on the microcolumn.

3. Mount the GelLoader tip in a capillary holder, e.g., from MDS Proteomics (Odense, Denmark), and equilibrate the microcolumn with 5 μl of 5% formic acid. Spin down the solvent using a microcentrifuge.

4. Load the sample onto the microcolumn using a table microcentrifuge (Fig. 4C). Remove salts from the microcolumn by elution with 2–3 μl of 5% formic acid. Remove all liquid until the sorbent is dry.

5. Mount a nanoelectrospray needle (MDS-Proteomics or New Objectives, Boston, MA) and align it to the microcolumn in the holder (Fig. 4B). Elute the sample with 0.5–1.5 μl of 50% (v/v) methanol in 5% (v/v) formic acid. Remove the nanoelectrospray needle from the holder and mount it in a nanoelectrospray ion source.

Peptide sequencing projects require the tandem mass spectrometric method. Hybrid quadrupole-TOF, triple-quadrupole, and ion trap instruments are the methods of choice. The last two instruments do not produce high-resolution tandem mass spectra, and interpretation of obtained data can be a challenge, especially for low-level amounts of protein. Here we focus on data interpretation of quadrupole-TOF data. Due to high resolution, it is easy to assign the charge state of a peptide ion. Because tryptic peptides are usually doubly or triply charged, only multiple charged ions are normally selected for MS/MS analysis. Trypsin autolysis products and protein contamination such as keratin also produce multiple charged ions, but they can be excluded based on their precise mass. A list of trypsin and keratin peptide masses can be downloaded from http://www.pil.sdu.dk/ContaTableASMS1999.doc.

Selected precursor ions undergo collisions with an inert gas followed by TOF detection of fragment ions. As shown in Fig. 5, for triply charged peptide ions, singly and doubly charged fragments are detected. During precursor selection, singly charged compounds from chemical background noise also enter the collision cell, producing single charged fragments that complicate data interpretation. It is not very often possible to "read" a complete amino acid sequence from MS/MS data. However, the m/z region above the selected precursor ion only contains fragments from the peptide, as the chemical background is singly charged and cannot contribute to the high m/z region. This fact is exploited in the peptide sequence tag algorithm as follows. Mass differences between large fragment ions (usually y ions in quadruple TOF machines) are taken and used to "spell out" a short amino acid sequence of usually two to three amino acids. This sequence is then combined with the start and end mass of the peptide, yielding a highly specific "probe" to search sequence databases according to the cleavage specificity of the enzyme. Two to three amino acid sequence tags, together with accurate mass information, are sufficient to retrieve unique matches even in very large databases. In general, a sequence tag obtained from a single peptide can unambiguously identify a protein.

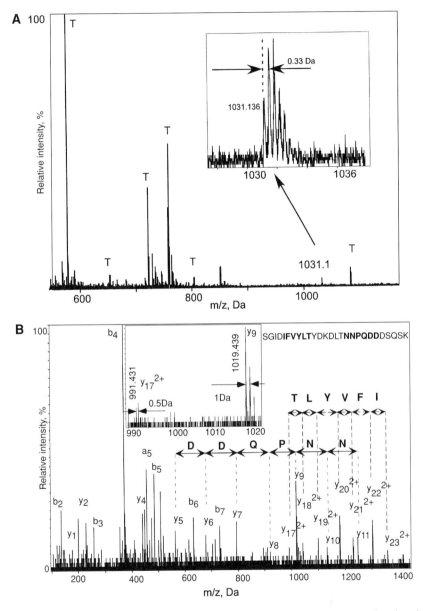

FIG. 5. Peptide sequencing using nanoelectrospray MS/MS. (A) Mass spectrum of an in-gel-digested yeast protein purified by one-dimensional SDS–PAGE. Signals corresponding to trypsin autolysis products are marked with a "T." Triply charged ion at m/z 1031.136 was chosen for MS/MS sequencing (inset). (B) Tandem mass spectrum of a yeast peptide with m/z 1031.136. Two sequence tags were derived from this spectrum and were searched against an nrdb database. Yeast protein MTR3 was unambiguously identified, and the retrieved peptide was confirmed by additional y ion and b ion fragments. (Inset) Isotope spacing for singly charged fragments (1 Da) and doubly charged fragments (0.5 Da)

An example of a fragment of a triply charged peptide with an m/z value of 1031.136 is shown in Fig. 5. In this case, two sequence tags are assigned. Both tags, (564.261)DDQPNN(1247.523) and (1982.862)TLYVFI(2719.271), derived from singly and doubly charged ions correspondingly, match the same peptide from MTR3 (Accession number P48240). When a peptide sequence is retrieved from a database, the assignment should be confirmed by the presence of other y and b ions, which in this case confirms the database hit.

The mass accuracy achieved on quadrupole-TOF instruments is sufficient to distinguish between glutamine and lysine, which have the same nominal mass but have an exact mass difference of 0.036 Da, as well as between phenylalanine and the oxidized form of methionine (0.033 Da). Oxidized methionine can also be detected by a characteristic series of satellite ions (-64.108 Da, loss of CH_3SOH group) in fragmentation spectra.[16]

LC MS/MS Methods

Protein identifications by MALDI peptide mapping or nanoelectrospray sequencing have been extremely successful in the yeast community and have helped solve many challenging biological problems. For example, since the late 1990s, over a dozen new yeast protein–protein complexes have been characterized using the described strategy. The first large-scale analysis of a proteome by mass spectrometric methods was carried out on the yeast proteome[17] using a combination of rapid screening by MALDI followed by nanoelectrospray sequencing when needed. Yeast 2D PAGE reference maps (http://www.expasy.org/ch2d/publi/yeast.html and http://www.ibgc.ubordeaux2.fr/YPM/) have been created and annotated with data obtained by mass spectrometry.

Despite these successes, analysis of the whole "proteome" of yeast by gel-based methods has been difficult for several reasons. Two-dimensional gels have limitations in the range of proteins solubilized and displayed within their coordinate range. Very acidic and basic proteins, as well as high molecular weight and membrane proteins, are often severely underrepresented. The dynamic range of 2D gels is limited, and low abundance proteins are usually not detected in 2D PAGE.[18]

These problems are now being solved using automated LC MS/MS in a datadependent manner. In a typical LC MS/MS experiment, peptides are separated on a reversed-phase column according to their hydrophobicity and are eluted directly into a mass spectrometer. Ion trap and quadrupole TOF mass spectrometers can

---

[16] F. M. Lagerwerf, M. Weert, W. Heerma, and J. Haverkamp, *Rapid Commun. Mass Spectrom.* **10**, 1905 (1996).

[17] A. Shevchenko, O. N. Jensen, A. V. Podtelejnikov, F. Sagliocco, M. Wilm, O. Vorm, P. Mortensen, A. Shevchenko, H. Boucherie, and M. Mann, *Proc. Natl. Acad. Sci. U.S.A.* **93**, 14440 (1996).

[18] S. P. Gygi, G. L. Corthals, Y. Zhang, Y. Rochon, and R. Aebersold, *Proc. Natl. Acad. Sci. U.S.A.* **97**, 9390 (2000).

automatically acquire three to five MS/MS scans on the most intensive ions in every MS spectrum in several seconds, providing sufficient time to acquire MS/MS data even for coeluted peptides. During an LC MS/MS run of 1 hr or so, several thousand peptide fragmentation spectra can in principle be acquired. Sophisticated software can be used, which ensures that the same peptides are not fragmented more than once ("dynamic exclusion").

An example of LC MS/MS analysis of a peptide mixture derived from yeast immunoprecipitation experiment is shown in Fig. 6. The analysis is performed using autosampling equipment combined with a capillary HPLC system (LC Packings, Amsterdam, The Netherlands) coupled to an ion trap mass spectrometer equipped with a nanoLC source. Peptides are eluted from a 75-μm i.d. C_{18} column with a 30-min linear gradient at a flow rate of 250 nl/min. Solvent A is 5% acetonitrile containing 0.4% acetic acid and 0.005% heptafluorobutyric acid (HFBA) as the ion-pairing agent. Solvent B is 80% acetonitrile in 0.4% acetic acid and 0.005% HFBA acid. Results are searched automatically in the yeast database. The analysis is completely automated and routinely allows processing around 50 samples per day on a single instrument.

LC MS/MS is also used in several schemes that allow relative quantitation of proteomic samples. Of two forms of the sample that are to be compared (e.g., yeast in different growth media), one is labeled by a stable isotope reagent and then both forms are analyzed together. In the mass spectrometric analysis, both forms of the peptide behave identically, except that they show different masses. Corresponding peaks can then be quantified. Yeast can be grown on ^{15}N media, which will shift peptide masses by 1 Da for every nitrogen.[19]

An elegant method termed isotope-coded affinity tag (ICAT) combines the stable isotope labeling only of cysteine residues with an affinity label for selective retrieval. When the affinity label is biotin, cysteine-containing peptides can be retrieved easily on avidin-agarose columns. The two labels to be compared for quantification are the normal reagent (normal linker and biotin) and the labeled reagent (d_8 deuterated linker and biotin).[20] Such a reagent is now on the market (PerSeptive Biosystems), and similar schemes, with and without the affinity step, are being developed elsewhere.

LC/LC MS/MS

Protein mixtures derived from cell lysates or purified organelles are still too complex for the LC MS/MS approach and too many peptides will coelute at any

---

[19] Y. Oda, K. Huang, F. R. Cross, D. Cowburn, and B. T. Chait, *Proc. Natl. Acad. Sci. U.S.A.* **96**, 6591 (1999).

[20] S. P. Gygi, B. Rist, S. A. Gerber, F. Turecek, M. H. Gelb, and R. Aebersold, *Nature Biotechnol.* **17**, 994 (1999).

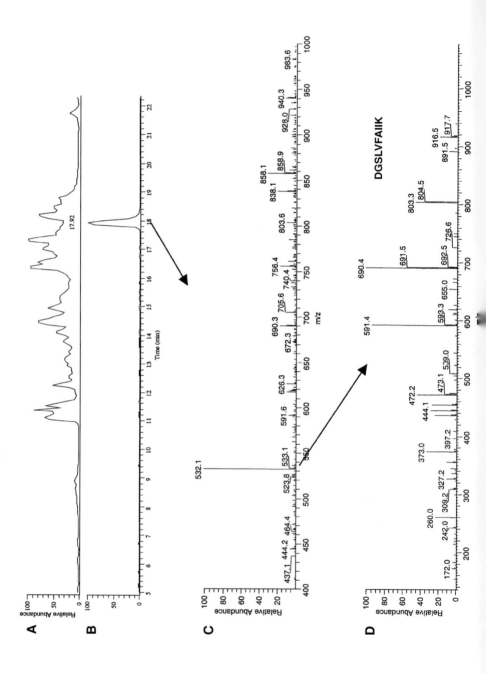

time, resulting in the preferential detection of the abundant proteins. To decrease the level of complexity and extend the dynamic range of the analysis, additional separation steps ahead of the nanoLC MS/MS step can be considered. For example, proteins can be fractionated by 1D PAGE before nanoLC MS/MS of all the bands. Several steps of liquid chromatography can also be considered. For example, strong cation exchange (SCX), which fractionates peptides by their charge state at pH 3, does not have the same resolution as HPLC but provides useful fractionation of peptides. In SCX, peptides are eluted by a salt gradient. Eluted fractions are coupled to nanoLC either directly or off line, which has the advantage of allowing higher organic buffer content and hence better separation by the charge state. Complete integrated LC/LC MS/MS applied on a *S. cerevisiae* proteome enabled the detection of 5540 peptides, leading to the reported identification of 1484 yeast proteins in a single analysis.[21] To perform on-line LC/LC MS/MS, a sophisticated LC setup is usually applied, including several multiport switching valves, an autosampler, and isocratic and binary pumps. The ICAT method mentioned earlier can also be used just to simplify peptide mixtures by retaining only cysteine-containing tryptic peptides.

Database Searching

DNA sequencing projects have collected enormous amounts of sequence information, providing the whole biological community with a very important resource. One of the milestones of the last decade was the sequencing of the first eukaryotic genome *S. cerevisiae*. Unlike the case of mammalian genomes, yeast genes are relatively easy to predict; therefore, the yeast and proteomic communities have had nearly the full inventory of yeast proteins available since 1996.

As mentioned previously, MALDI mass spectrometry of in-gel-digested proteins results in a list of detected ions corresponding to masses of tryptic peptides. A correlation of this set of data to theoretical tryptic digests, derived from every one of the predicted sequence database entries, in principle enables a straightforward identification of every yeast protein.

For tandem mass spectra, the peptide sequence tag described earlier is a powerful way of identifying proteins. It usually requires the manual interpretation of MS/MS data and is therefore time-consuming, making the analysis of thousands of spectra from LC MS/MS runs unrealistic. In other approaches, there is no attempt

[21] M. P. Washburn, D. Wolters, and J. R. Yates, *Nature Biotechnol.* **19**, 242 (2001).

FIG. 6. LC-MS analysis of yeast proteins. (A) Total ion current chromatogram. (B) Selected ion current chromatogram for *m/z* 532.1. (C) Mass spectrum of the peptide with *m/z* 532.1 at an elution time of 17.92 min. (D) MS/MS spectrum for the tryptic peptide (*m/z* 532.1) and its sequence obtained after a database search.

TABLE II
SEQUENCE DATABASE SEARCH PROGRAMS FOR MASS SPECTROMETRIC DATA

Data set	Program	WWW address
MALDI peptide mass maps	PepSea[a]	http://pepsea.protana.com/
	MS-Fit	http://prospector.ucsf.edu/ucsfhtml3.4/msfit.htm
	Mascot	http://www.matrixscience.com/
	ProFound	http://www.proteometrics.com/
Peptide sequence tag	Mascot	http://www.matrixscience.com/
	PepSea[a]	http://pepsea.protana.com/
	MS-Seq	http://prospector.ucsf.edu/ucsfhtml3.4/msseq.htm
Breakpoint search	Sequest[a]	http://fields.scripps.edu/sequest/
	Mascot	http://www.matrixscience.com/
	PepFrag	http://www.proteometrics.com/
	PepSea[a]	http://pepsea.protana.com/
	MS-Tag	http://prospector.ucsf.edu/ucsfhtml3.4/mstagfd.htm

[a] Commercial software.

to interpret mass spectra, but rather the fragmentation pattern is compared to the theoretical fragments calculated for all peptides in the database. In comparison with the sequence tag, these approaches are less specific but are automated more easily. For the future we expect the combination of *de novo* sequencing programs (i.e., programs that interpret the sequence from the spectrum) applied to very short sequence stretches in combination with the peptide sequence tag algorithm to provide the best results both in terms of specificity and automation.

Databases of varying quality can be searched by mass spectrometric data, roughly in the order of protein sequence database, expressed sequence tag database (not usually relevant for yeast), and genomic databases (Table II). As a first step in database searching, we recommend the use of a nonredundant, nonspecies-specific protein database. The reason is that even in yeast experiments it is possible to find proteins from other species such as human, sheep keratins, albumins, fusion proteins such as glutathione transferase (GST) or hemagglutinin (HA) tags, yeast viruses, and different immunoglobulins. Depending on the design of an experiment, *Escherichia coli* contaminations can also be detected. Using the species-specific *S. cerevisiae* database will lead to nonidentification of these proteins or even to misleading assignments.

Analysis of Simple Protein Mixtures Derived from SDS–PAGE Separation

Depending on the complexity of the protein mixtures, sometimes it cannot be completely resolved either by one-dimensional or even by two-dimensional electrophoresis. The iterative algorithm described by Jensen *et al.*[22] for MALDI

[22] O. N. Jensen, A. V. Podtelejnikov, and M. Mann, *Anal. Chem.* **69,** 4741 (1997).

peptide mapping can be very useful in solving such a problem. The principle is to identify the first protein, discard all peaks corresponding to tryptic peptides of that first protein, and then analyze the remaining peaks for the next protein. This procedure is repeated until not statistically relevant hits are found. It is possible to resolve mixtures of several proteins as shown in Fig. 7, where as many as eight proteins were detected in a single band excised from a one-dimensional gel.

In general, MALDI-TOF cannot analyze protein mixtures when components are in less than 10% of the main components. In these cases, tandem mass spectrometry—o-MALDI, nanoelectrospray,[23] or LC/MS/MS—should be used.

Normally for analysis of yeast protein complexes it is sufficient to employ MALDI or o-MALDI analysis. Keratin contamination can cause serious problems in the identification of yeast proteins due to the existence of different parolog forms of that protein. In our experience, if different keratins are still detected after three to four iterative searches, the experiments have to be redone with less contamination.

Data Evaluation

Evaluation of the obtained hits is also an important step. Normally the molecular weight of the identified protein should correlate with the apparent value derived from the migration of the protein on a gel. If this is not the case, but the identification is correct, a reason for the discrepancy can often be found. Sometimes proteins are truncated or modified, leading to dramatic shifts in migration in SDS gel electrophoresis. For example, myristoylation or glycosylation can cause shifts to higher apparent molecular weights. For two-dimensional gels the pI value should also be taken into account. A train of horizontal spots can be evidence of a possible series of phospho groups attached to the protein of interest. For evaluation of obtained results we recommend using the yeast protein database (http://www.proteome.com/databases/index.html), which in our experience is the best annotated yeast database. MIPS (http://speedy.mips.biochem.mpg.de/mips/yeast) and the Stanford *Saccharomyces* genome database (http://genome-www.stanford.edu/Saccharomyces/) are also very useful. In the case of identification of yeast proteins with unknown function, database searching programs can be combined directly with the standard homology and domain searching programs on the web.

Several protein modifications can be detected simultaneously with protein identification. Detailed analysis of mass spectrometric data can lead to detection of the sites of phosphorylation, acetylation, glycosylation, or other posttranslational modifications. For example, the MALDI peptide map of the yeast peptidylprolyl *cis–trans*-isomerase in Fig. 3A contains an intense signal at m/z 1823.55, which fits the N-terminal acetylation noted in its database entry. Assignments of posttranslational modifications on the basis of mass alone can be error prone and should

---

[23] A. Yaron, A. A. Hatzubai, M. Davis, I. Lavon, S. Amit, A. M. Manning, J. S. Andersen, M. Mann, F. Mercurio, and Y. Ben Neriah, *Nature* **396,** 590 (1998).

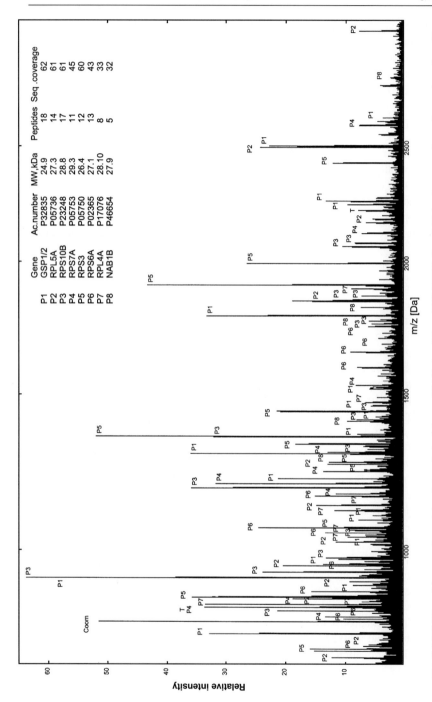

FIG. 7. MALDI peptide map of a protein mixture derived from a 30-kDa band from SDS–PAGE. (Inset) Molecular mass of protein, number of peptides matched, and sequence coverage for each of the eight identified proteins.

be checked by other means such as tandem mass spectrometry. In our example, o-MALDI sequencing of the ion in question did produce the expected fragment data with b ions increased in mass by the acetyl group (42.011 Da) and y ions unchanged.

One more useful feature of MALDI analysis that deserves to be mentioned is that it is possible to evaluate possible truncated forms of a protein. These can be due either to cellular processing events or to proteolysis during cell disruption, which is a particular problem in yeast. Proteolysis can be minimized using standard protocols. Cellular processing can be studied by MS,[24] but detection of the precise truncation site is usually much more difficult than the mere identification of the protein. A large number of peptides need to be sequenced and the peptide in question needs to be in the right mass range for easy MS detection.

Phosphopeptide Analysis

Now that the identification of yeast proteins has become relatively routine, interest is focusing on posttranslational modifications such as phosphorylation or ubiquitination. Sites of phosphorylation are difficult to detect by other methods and many kinase and phosphatase substrates remain to be identified in yeast. Theoretically, it is simple to detect phosphopeptides by mass spectrometry due to the 80-Da mass shifts caused by serine, threonine, or tyrosine bonding to phosphate esters. In reality, detection of phosphopeptides is challenging because phosphopeptides are less well ionized by MS, are of low abundance compared to the nonphosphorylated form, and most of all may be on a peptide that is not prominent in the peptide mass map due to size or other characteristics.

Comparative peptide maps of a single protein can sometimes lead to the detection of modified peptides. An intensive unidentified peak from a MALDI peptide map with a mass increase of 80 Da compared to a predicted peptide from a theoretical digest can be a potential phosphopeptide. In MALDI-TOF, phosphopeptide signals are usually accompanied by high-intensity alkali adducts, $[M + Na]^+$ and $[M + K]^+$.

Another method to identify phosphopeptides by comparative peptide mapping is based on alkaline phosphatase treatment. Peptide mixtures are analyzed by MALDI-TOF before and after phosphatase hydrolysis. The potential phosphopeptide will be assigned for each peptide mass decreased by 80 Da. This approach is very straightforward and can be done on the same sample used to obtain the initial MALDI map but it does not provide sufficient information to localize the phospho group if more than one serine, threonine, or tyrosine is present in the peptide sequence. Structural information can be obtained by the o-MALDI sequencing method.

[24] M. T. Teixeira, S. Siniossoglou, S. Podtelejnikov, J. C. Benichou, M. Mann, B. Dujon, E. Hurt, and E. Fabre, *EMBO J.* **16,** 5086 (1997).

To enrich phosphopeptides from a mixture of peptides, immobilized metal affinity columns (IMAC) can be used. Based on the reversible binding of the phospho group to Fe^{3+}- or Ga^{3+}-containing columns, IMAC technology allows selective binding of phosphopeptides at low pH followed by elution at basic pH. One disadvantage of IMAC is that it is also selective for acidic peptides containing aspartic or glutamic acid.

Electrospray methods are also suitable for phosphopeptide analysis, especially on tandem mass spectrometers where selective ion scans are available. An alternative method to detect phosphopeptides derived from in-gel digestion using triple quadrupole mass spectrometers was described by Neubauer and Mann.[25] It is based on "precursor ion scanning" of peptide mixtures in negative mode. The first quadrupole scans while the third quadrupole is set to a fixed m/z value of 79 corresponding to the loss of PO_3^-. Under CID conditions, only ions having lost PO_3^- would be detected. To obtain MS/MS data, the peptide mixture should be acidified and the spectrum should be obtained in positive mode. Serine and threonine phosphopeptides can be detected by "neutral loss scanning" on triple quadrupole instruments. In this method, the first and the third quadrupoles scan simultaneously to detect a loss of 98 Da due to collision-induced dissociation in the second quadrupole. LC MS/MS data files can also be "interrogated" for the loss of 98 Da from the parent ion, which is very pronounced, especially on ion trap instruments. Modern database-searching algorithms can be instructed to match peptides to a restricted number of sequences, considering all possible sites of phosphorylation. A large number of phosphorylation sites can be assigned provided picomole or Coomassie-stained levels of protein material are available. The combination of IMAC and other enrichment methods with LC/MS/MS may soon allow probing the global phosphorylation state of a yeast cell. First steps in this direction using chemical derivatization of the phosphogroup have been described.[26,27]

Finally, an attractive approach to detect phosphotyrosine-containing peptides has been described.[28] It makes use of the high mass accuracy of modern quadrupole-TOF instruments enabling to distinguish specifically the immonium ion of phosphotyrosine (216.04 ± 0.02 Da) from other fragment ions. Once this ion has been detected, the corresponding precursor ion is fragmented to localize the site of phosphorylation.

Conclusions

For yeast proteins separated by either one-dimensional or two-dimensional SDS gel electrophoresis and visualized by Coomassie staining, MALDI peptide mass mapping is a fast and sufficient method to identify most proteins.

[25] G. Neubauer and M. Mann, *Anal. Chem.* **71**, 235 (1999).
[26] Y. Oda, T. Nagasu, and B. T. Chait, *Nature Biotechnol.* **19**, 379 (2001).
[27] H. Zhou, J. D. Watts, and R. Aebersold, *Nature Biotechnol.* **19**, 375 (2001).
[28] H. Steen, B. Kuster, M. Fernandez, A. Pandey, and M. Mann, *Anal. Chem.* **73**, 1440 (2001).

For low abundance proteins visualized only by silver staining and sometimes for proteins with a molecular mass less than 20 kDa and larger than 200 kDa, tandem mass spectrometry should be used. This method also provides sufficient sequence information to detect and localize posttranslation modifications.

Rapid and sensitive analysis of yeast protein complexes or interacting partners can be achieved by LC/MS approaches.

Complex protein mixtures derived from organelle purification or whole cell lysates should undergo intensive LC/LC MS/MS analysis.

The combination of high-resolution mass spectrometry, data-dependent software, and nanoflow liquid chromatography allows creation of a robust system for the characterization of complex mixtures. Further development of this technology will enable analysis of very low amounts of yeast proteins in one single run, including the detection of different posttranslational modifications, opening new perspectives in an investigation of the yeast system as a model for functional analysis of a complete proteome.

Acknowledgment

The authors acknowledge funding of the Protein Interaction Laboratory of a Center for Experimental Bioinformatics by the Danish National Research Foundation.

Section III
Cell Fractionation

[18] Subcellular Fractionation of Secretory Organelles

By CHRIS A. KAISER, ESTHER J. CHEN, and SASCHA LOSKO

Introduction

The primary function of the *Saccharomyces cerevisiae* secretory pathway is to deliver newly synthesized proteins to growing plasma membrane and vacuoles. The organization of the yeast secretory pathway, composed of the endoplasmic reticulum (ER) and the Golgi complex, is the same as for other eukaryotic cells.[1] The ER is the site where newly synthesized membrane proteins and soluble secretory and vacuolar proteins are folded and receive carbohydrate modifications. Further modifications to glycoproteins take place in the Golgi complex. In addition, the late Golgi represents a major branch point in the pathway where proteins destined for the plasma membrane are segregated from vacuolar proteins.

About one-quarter to one-third of yeast proteins are associated with one or another of the membranes of the secretory pathway. From sequence analysis, more than 1800 of the approximately 6200 *S. cerevisiae* proteins appear to encode proteins with either a predicted N-terminal signal sequence or one or more predicted transmembrane domains. Most of these predicted integral membrane proteins are probably either residents of secretory pathway organelles or are cargo proteins that transit the secretory pathway to be delivered to either the vacuole or the plasma membrane. In addition to integral membrane proteins, there are likely a large number of peripheral membrane proteins bound to the cytoplasmic face of one of the secretory organelles. Fundamental insight into the function of either integral or peripheral membrane proteins can be deduced from information about their intracellular location.

Here we describe a set of cell fractionation protocols used to explore whether a protein might be associated with a membrane compartment and to localize those proteins that are associated with organelles. These procedures complement microscopic localization studies and are usually carried out in parallel. The first step in any localization study is to develop immunological reagents specific for detection of the protein of interest. This is usually done either by raising antibodies to the recombinant yeast protein expressed in *Escherichia coli* or by adding an epitope tag to the yeast gene of interest and verifying that the tagged gene is functional. It is also important to work out methods for detection of the protein of interest by immunoblotting. These preliminaries include finding conditions to stabilize the

---

[1] C. A. Kaiser, R. E. Gimeno, and D. A. Shaywitz, *in* "The Molecular and Cellular Biology of the Yeast *Saccharomyces:* Cell Cycle and Cell Biology" (J. R. Pringle, J. R. Broach, and E. W. Jones, eds.), p. 91. Cold Spring Harbor Laboratory Press, Cold Spring Harbor, NY, 1997.

protein of interest against proteolysis in cell extracts, to solubilize the protein in a gel-loading buffer, and to optimize conditions for electrophoretic transfer to a nitrocellulose membrane for immunoblotting.

Cell fractionation methods depend on a reliable set of marker proteins. The best membrane markers are integral membrane proteins, as their membrane association is stable. In this chapter we have tried to extend the general usefulness of the fractionation procedures using reagents commercially available for the detection of markers. Antibodies for some yeast marker proteins are available commercially from Molecular Probes, Inc. (Eugene, OR). These include monoclonal antibodies to the ER protein Dpm1p, the trans-Golgi/late endosome protein Vps10p, the late endosome protein Pep12p, and the vacuolar membrane protein Vph1p. Two useful marker proteins that can be assayed by enzymatic assay are the GDPase located within cis- and medial-Golgi compartments and Kex2 protease located in the trans-Golgi compartment. We have found the most useful plasma membrane markers to be Pma1p, the plasma membrane H^+-ATPase, and glycosylphosphatidylinositol–anchored proteins such as Gas1p. Antibodies to these plasma membrane proteins are not yet available commercially; however, Pma1p activity can be assayed enzymatically.[2]

Differential Centrifugation

Differential centrifugation can be used to separate a yeast cell extract into crude membrane fractions. Because differential centrifugation is relatively easy, it is usually the first test used to determine whether a protein of interest is associated with membranes. The behavior of a protein in a differential centrifugation experiment can then be used to guide more definitive but more difficult fractionation procedures. Sedimentation of a protein could indicate either association with a membrane-bounded organelle or association with a large proteinaceous complex such as the cytoskeleton. These possibilities can only be distinguished definitively by fractionation on a flotation gradient as described later in this chapter.

Cell extracts used for differential centrifugation are prepared by the gentle lysis of spheroplasts. Gentle lysis is used to maximize the integrity of large organelles and to prevent the release of peripheral membrane proteins that can occur when a more vigorous lysis method such as agitation with glass beads is used. Spheroplasts are prepared by enzymatic digestion of (β1 → 3)-glucan, the major cell wall carbohydrate linkage. (β1 → 3)-Glucanase prepared from the bacteria *Oerskovia xanthineolytica* (also called *Arthrobacter luteus*) is usually known as Zymolyase or Lyticase. Spheroplasts can be lysed efficiently by shearing in a small Dounce homogenizer and fractions can then be separated by centrifugation. It is important to be aware that yeast cell extracts usually have high protease activity, arising from endogenous yeast proteases, as well as contaminating proteases usually present

[2] R. Serrano, *Methods Enzymol.* **157**, 533 (1988).

in Zymolyase preparations.[3] Nevertheless, proteolysis of all but the most labile proteins can be prevented by inclusion of a set of protease inhibitors in the lysis buffer.

Method

Extracts are prepared from 8×10^8 yeast cells, typically prepared by overnight growth of a 100-ml culture to an OD_{600} of 0.4. Cells collected by centrifugation for 5 min at $500g$ are washed once in 10 mM sodium azide, 10 mM potassium fluoride, 50 mM Tris–HCl, pH 7.5, to poison energy-dependent processes and then are incubated in 0.8 ml of 100 mM ethylenediaminetetraacetic acid (EDTA), 0.5% 2-mercaptoethanol, and 10 mM Tris–HCl, pH 7.5, for 20 min at 30° to reduce the disulfide bonds in the cell wall. The cells are then suspended in 0.8 ml S buffer [1.2 M sorbitol, 0.5 mM MgCl$_2$, 40 mM hydroxyethylpiperazine-N'-2-ethanesulfonic acid (HEPES), pH 7.5] and are converted to spheroplasts by the addition of 50 U/OD_{600} Zymolyase 100T (ICN Biomedicals, Costa Mesa, CA) for 45–60 min at 30° (1 U Zymolyase will cause a decrease in OD_{600} of 0.1 in 30 min at 30°). The extent of spheroplast formation can be checked by recording the decrease in optical density when the spheroplast suspension is diluted into 1% Triton X-100. Spheroplasting is complete when the optical density of the suspension in Triton X-100 has decreased to about 10% of that of spheroplasts in S buffer. Spheroplasts are washed once in S buffer and suspended in 1 ml lysis buffer (0.2 M sorbitol, 1 mM EDTA, 50 mM Tris–HCl, pH 7.5). The spheroplasts are lysed by 20 strokes in a Dounce homogenizer with a tightly fitting pestle, with care to avoid introducing bubbles into the lysate. The extent of cell lysis can be checked by phase-contrast microscopy. Unlysed cells and unmanageably large aggregates of organelles are removed by centrifugation at $500g$ for 5 min at 4°.

Lysed spheroplasts are centrifuged at $13,000g$ for 10 min to yield a P13 pellet fraction. An ideal centrifuge is a refrigerated ultracentrifuge (such as an Optima TL Ultracentrifuge, Beckman Instruments, Fullerton, CA) with a rotor such as the TLA100.3 and polyallomer microfuge tubes (Beckman Instruments). Swinging bucket rotors and the corresponding ultracentrifuge tubes can also be used. The supernatant is then centrifuged for 1 hr at 4° at $100,000g$ (55,000 rpm in a TLA100.3 rotor) to generate the P100 pellet and S100 supernatant fractions.

For SDS–PAGE analysis, the P13 and P100 fractions are solubilized in 1× sample buffer [2% SDS, 62.5 mM Tris–HCl, pH 6.8, 10% glycerol, 10 mM dithiothreitol (DTT), 0.1% bromphenol blue]. The cleared lysate and S100 fractions are solubilized by the addition of one-third volume 4× sample buffer. Heating to 100° for 2 min in sample buffer solubilizes most proteins readily. However, very hydrophobic integral membrane proteins will often precipitate if heated to 100° and for these proteins, solubilization is achieved by incubation at 37° for 10 min. Equal

---

[3] J. H. Scott and R. Schekman, *J. Bacteriol.* **142**, 414 (1980).

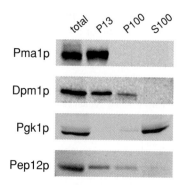

FIG. 1. Differential centrifugation of marker proteins. A wild-type yeast cell culture was converted to spheroplasts, lysed by Dounce homogenization, and fractionated by differential centrifugation. Equal cell equivalents from the crude lysate (total) and P13, P100, and S100 fractions were analyzed by SDS–PAGE and immunoblotting. The following antibodies were used: anti-Pma1p (rabbit polyclonal, gift from R. Kölling, Düsseldorf, Germany), anti-Dpm1p (mouse monoclonal, 5C5-A7, Molecular Probes, Eugene, OR), anti-Pgk1p (mouse monoclonal, 22C5-D8, Molecular Probes), and anti-Pep12p (rabbit polyclonal, E. Chen).

amounts of cell equivalents of each fraction are resolved by SDS–PAGE, transferred to nitrocellulose, and analyzed by immunoblotting using standard methods. An amount of extract corresponding to 0.2–0.4 OD_{600} units is the maximum amount that can be loaded on a standard SDS gel without producing distortions due to overloading.

Figure 1 shows results from a typical differential centrifugation experiment. The plasma membrane (marked by Pma1p) is located almost entirely in the P13 fraction. The ER (marked by Dpm1p), Golgi, and late endosome (marked by Pep12p) are divided between the P13 and P100 fractions.

Technical Notes

Yeast cells can be cultured in either rich or minimal media prior to spheroplasting. Cells grown in minimal medium are usually somewhat resistant to glucanase digestion and require additional Zymolyase for complete spheroplasting. Stationary-phase cells are highly resistant to glucanase digestion and cultures should be harvested while cells are still growing exponentially.

The amount of proteolysis in cell lysates can be minimized by keeping the extract at 4°, using a protease-deficient yeast strain such as $pep4\Delta$[4,5] and inclusion of a protease inhibitor cocktail in the lysis buffer [final concentrations: 1 mM phenylmethylsulfonyl fluoride (PMSF), 0.5 μg/ml leupeptin, 0.7 μg/ml pepstatin, 2 μg/ml aprotinin].

[4] E. W. Jones, *Methods Enzymol.* **194**, 428 (1991).
[5] G. Ammerer, C. P. Hunter, J. H. Rothman, G. C. Saari, L. A. Valls, and T. H. Stevens, *Mol. Cell. Biol.* **6**, 2490 (1986).

Extraction of Proteins from the Particulate Fraction

If a protein of interest sediments with either the P13 or the P100 fraction by differential centrifugation, the biochemical properties of the protein can be further characterized by examining conditions required for extraction from the membrane fraction. An understanding of which treatments can solubilize a protein is valuable for any subsequent biochemical investigations. If the sequence of a protein is known, the presence of significant hydrophobic segments (usually about 20 amino acids in length) in the gene sequence can offer the clearest indication of whether a protein may span the membrane. However, it is not always possible to reliably predict protein topology from primary sequence and an experimental confirmation of the nature of membrane association based on extraction with different agents is desirable.

The following reagents are commonly used to determine the nature of the membrane association of a protein. A sodium carbonate buffer at pH 11.5 will extract most peripheral membrane proteins from membranes without disrupting the integrity of the membranes.[6] Many peripheral proteins can also be extracted by treatment with high salt or a polar chaotropic agent such as urea. Finally, it is necessary to dissolve the membrane lipids with a detergent such as Triton X-100 to solubilize most integral membrane proteins.

Method

Release of a protein from the particulate fraction can be examined by treating spheroplast lysates with 500 mM NaCl, 2.5 M urea, 100 mM sodium carbonate, pH 11.5, or 1% Triton X-100 (v/v). All of these extraction conditions can be tested by preparing twice the amount of cell extract from 1.6×10^9 cells (or 80 OD$_{600}$ units) as described in the protocol for differential centrifugation. The cleared cell extract (0.25 ml) is mixed with an equal volume of lysis buffer containing 1 M NaCl, 5 M urea, 200 mM sodium carbonate, 2% Triton X-100, or no additive, respectively. After incubation for 1 hr at 4°, samples are centrifuged at 100,000g for 1 hr at 4° to separate soluble and particulate fractions. Particulate fractions are suspended in the same buffer and same volume as the corresponding soluble fraction. Then, 0.2 OD$_{600}$ cell equivalents of each fraction (∼10 μl) are diluted 5-fold with 1× sample buffer for SDS–PAGE and immunoblotting.

Technical Notes

Some integral membrane proteins resist detergent solubilization. For example, a population of plasma membrane ATPase is not extracted by Triton X-100.[7,8] The detergent dodecyl-β-D-maltoside has a relatively high critical micell concentration

---

[6] Y. Fujiki, A. L. Hubbard, S. Fowler, and P. B. Lazarow, *J. Cell Biol.* **93**, 97 (1982).
[7] F. Malpartida and R. Serrano, *FEBS Lett.* **111**, 69 (1980).
[8] R. Navarrete and R. Serrano, *Biochim. Biophys. Acta* **728**, 403 (1983).

and is effective for solubilization of refractory membrane proteins.[9] Examples of exceptional peripheral membrane proteins that are not extracted by pH 11.5 are also known.[10]

Tests of different extraction conditions should be performed at 4° to minimize reaggregation, which may occur in the presence of Triton X-100 at higher temperatures.[11]

N-Linked Glycosylation

In the ER lumen, the core oligosaccharide dolichol pyrophosphate (Dol-PP)-GlcNAc$_2$Man$_9$Glc$_3$ can be transferred to the side chain of Asn residues within the sequence Asn-X-Ser/Thr. Thus, the presence of N-linked carbohydrate chains will definitively show that a protein has entered the lumen of the secretory pathway.

To test for N-linked glycosylation, protein extracts can be digested with endoglycosidase H, which cleaves between the GlcNAc residues in the core oligosaccharide, releasing the N-linked carbohydrate, except for a single GlcNAc residue. The release of N-linked carbohydrate typically results in an increase in the electrophoretic mobility of a protein by SDS–PAGE, corresponding to a decrease in apparent mass of 2–3 kDa.

Protein extracts are prepared by suspending 4×10^7 cells in 30 μl 1× sample buffer (see differential centrifugation method given earlier for recipe) and heating to 100° for 2 min to solubilize proteins and inactivate endogenous proteases. Glass beads (0.5 mm diameter, BioSpec Products, Inc., Bartlesville, OK) are added to the meniscus, cells are lysed by vortexing with glass beads for 2 min, extracts are diluted to 0.1 ml with sample buffer, and the extracts are heated again to 100° for 2 min. Then, 30 μl of extract, 3 μl 0.5 M sodium citrate (pH 5.1), and 100 U of EndoH$_f$ (New England Biolabs, Beverly, MA) are mixed and incubated at 37° for 2–16 hr. After digestion, 11 μl of 4× sample buffer is added, and the samples are heated to 100°. The samples are ready for SDS–PAGE and immunoblotting.

Additional modifications of N-linked carbohydrate chains occur in ER and Golgi compartments. One can also test for the presence of extensive ($\alpha 1 \rightarrow 6$)- or ($\alpha 1 \rightarrow 3$)-mannose modifications, which occur in the cis-Golgi and medial-Golgi compartments.[12] Och1p is the primary ($\alpha 1 \rightarrow 6$)-mannosyltransferase located in the cis-Golgi, and Mnn1p initiates ($\alpha 1 \rightarrow 3$)-mannosyltransfer in the medial-Golgi.[13–15]

[9] R. Dunn and L. Hicke, *Mol. Biol. Cell* **12**, 421 (2001).
[10] D. Feldheim and R. Schekman, *J. Cell Biol.* **126**, 935 (1994).
[11] U. Kragh-Hansen, M. le Maire, and J. V. Møller, *Biophys. J.* **75**, 2932 (1998).
[12] T. R. Graham and S. D. Emr, *J. Cell Biol.* **114**, 207 (1991).
[13] K. W. Cunningham and W. T. Wickner, *Yeast* **5**, 25 (1989).
[14] T. R. Graham, M. Seeger, G. S. Payne, V. L. MacKay, and S. D. Emr, *J. Cell Biol.* **127**, 667 (1994).
[15] Y. Nakanishi-Shindo, K. I. Nakayama, A. Tanaka, Y. Toda, and Y. Jigami, *J. Biol. Chem.* **268**, 26338 (1993).

To test for modification in the Golgi, a double immunoprecipitation of the protein can be done. First, 2×10^7 cells grown overnight in minimal medium lacking methionine are radiolabeled with [^{35}S]methionine (New England Nuclear, Boston, MA), and a cell extract is prepared by boiling and glass bead lysis in 20 μl of lysis buffer (60 mM Tris–HCl, pH 6.8, 100 mM DTT, 2% SDS). The extract is diluted with 1 ml immunoprecipitation buffer (IP buffer: 50 mM Tris–HCl, pH 7.4, 150 mM NaCl, 1% Triton X-100) and incubated with the protein-specific antibody for 1–2 hr and then with protein A-Sepharose (Amersham Pharmacia, Piscataway, NJ) for 1 hr. The immunoprecipitates are washed with IP buffer, boiled in lysis buffer, and diluted again with IP buffer. Then, the protein can be reimmunoprecipitated with either the protein-specific antibody or with antibodies that recognize (α1 $\rightarrow$ 6)- or (α1 $\rightarrow$ 3)-mannose linkages. Immunoprecipitates are boiled in SDS–PAGE sample buffer before loading on a gel. The presence of (α1 $\rightarrow$ 6)- or (α1 $\rightarrow$ 3)-mannose linkages would suggest that the protein of interest reached either the cis- or the medial-Golgi, respectively.

Fractionation by Equilibrium Sedimentation

Intracellular membranes can be separated on the basis of their characteristic densities, and cofractionation of the protein of interest with a known membrane marker protein can be examined. In sedimentation gradient fractionation based on Kölling and Hollenberg[16] and Roberg et al.,[17] a cell extract is prepared by agitation with glass beads. Vigorous lysis is needed to break the membranes into small fragments that will exhibit the density characteristic of the organelle from which they were derived. Gentler lysis methods tend to form mixed aggregates of membranes that cannot be well resolved on the basis of density. The membrane extract is layered on a 20–60% sucrose gradient and is centrifuged at 100,000g for 17 hr at 4° to ensure that the membranes reach their equilibrium position in the gradient.

Fractionation by equilibrium sedimentation is ideal for identifying proteins that are located in either the ER or the plasma membrane. The plasma membrane is of relatively high density ($\sim$1.22–1.27 g/ml) and can be resolved well from relatively less dense membranes—vacuole, ER, Golgi, and endosome. The ER membrane has the distinguishing feature that its density differs according to whether Mg^{2+} is present in the cell extract. Mg^{2+} preserves an association of ribosomes with the rough ER membrane so the ER has a high density similar to that of the plasma membrane.[18–20] In contrast, when the extract is prepared in the presence of EDTA,

---

[16] R. Kölling and C. P. Hollenberg, *EMBO J.* **13**, 3261 (1994).
[17] K. J. Roberg, N. Rowley, and C. A. Kaiser, *J. Cell Biol.* **137**, 1469 (1997).
[18] J. M. Lord, T. Kagawa, T. S. Moore, and H. Beevers, *J. Cell Biol.* **57**, 659 (1973).
[19] M. Marriott and W. Tanner, *J. Bacteriol.* **139**, 566 (1979).
[20] C. M. Sanderson and D. I. Meyer, *J. Biol. Chem.* **266**, 13423 (1991).

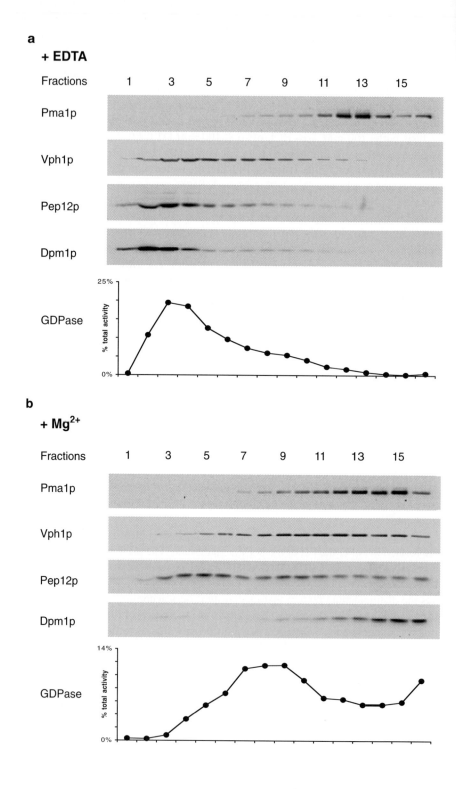

the ER has a low density, similar to that of Golgi and vacuolar membranes. We have found that in the presence of 2 mM Mg^{2+} the ER and Golgi membranes are well resolved (see Fig. 2). Resolution of Golgi and endosomal membranes is considerably more difficult because these membranes have similar properties probably as a result of the extensive exchange of membranes and proteins between these organelles. Resolution of Golgi and endosome membranes requires shallow density gradients or affinity purification. Refined methods for separation of Golgi and endosomal membranes are described by Sipos and Fuller elsewhere in this volume.[20a]

Method

Yeast cells are grown to the exponential phase, and 0.5 to 1 × 10^9 cells are collected by filtration or centrifugation. Cells are washed once in 10 mM sodium azide, 10 mM potassium fluoride, 50 mM Tris–HCl, pH 7.5. For fractionation of membranes in the absence of Mg^{2+}, cells are washed once in STE10 [10% (w/w) sucrose, 10 mM Tris–HCl, pH 7.5, 10 mM EDTA] and suspended in 0.5 ml STE10 with protease inhibitors in a 15-ml conical tube (see differential centrifugation section given earlier for the protease inhibitor cocktail; note that leupeptin will interfere with an assay for Kex2p activity). Glass beads (0.5 mm diameter, BioSpec Products, Inc.) are added to the meniscus and cells are disrupted by vigorous vortexing at top speed for 2 min. After the addition of 1 ml STE10 with protease inhibitors, the extract is transferred to a 1.5-ml microcentrifuge tube and cleared of unlysed cells and cell walls by centrifugation at 500g for 3 min at 4°. Then, 100 μl is removed from the top of a 5-ml, 20–60% sucrose gradient in 10 mM Tris–HCl, pH 7.5, 10 mM EDTA, and 300 μl of the cleared extract is carefully layered on top of the gradient. For fractionation of membranes in the presence of Mg^{2+}, all solutions contain 2 mM Mg^{2+} instead of 10 mM EDTA.

Samples are centrifuged for 17 hr at 4° at 100,000g (with no brake) in a SW50.1 or SW55Ti rotor (Beckman Instruments, Fullerton, CA). Sixteen 330-μl fractions are collected from the top of a gradient using a Pipetman. GDPase and Kex2 activity should be assayed within 1 day of running the gradient; the rest of each fraction can be stored at −20° immediately after collection. For immunoblots, 50 μl of each fraction is mixed with 50 μl 2× sample buffer + EDTA (4% SDS,

[20a] G. Sipos and R. S. Fuller, *Methods Ezymol.* **351** [20], 2002 (this volume).

FIG. 2. Equilibrium sedimentation of membrane proteins. Extracts from 5 × 10^8 wild-type S288c cells (CKY443) grown in YPD were prepared and fractionated in a SW55Ti rotor on 20–60% (w/w) sucrose gradients containing EDTA (a) or Mg^{2+} (b) as described in the sedimentation gradient fractionation procedure. GDPase activity was measured. Proteins were precipitated by TCA, separated by SDS–PAGE on a 9% gel, and transferred by semidry transfer to nitrocellulose membranes. The following antibodies were used: anti-Pma1p (gift from R. Kölling), anti-Vph1p (Molecular Probes), anti-Pep12p (Molecular Probes), and anti-Dpm1p (Molecular Probes).

0.1 M Tris–HCl, pH 6.8, 4 mM EDTA, 20% glycerol, 2% 2-mercaptoethanol, 0.2% bromphenol blue), incubated at 37° for 1 hr to solubilize proteins, and then 50 μl is loaded onto a gel for SDS–PAGE. Alternatively, proteins can be concentrated by trichloroacetic acid (TCA) precipitation: 200 μl of each fraction is diluted to 1 ml with water, deoxycholate is added to 0.0125%, and TCA added to 6%. After incubation for 1 hr on ice, proteins are pelleted by centrifugation and washed with cold acetone. Protein pellets are suspended in 100μl 2× sample buffer + EDTA, and 25 μl of each sample loaded onto a gel for SDS–PAGE.

Technical Notes

We favor the use of continuous density gradients rather than step gradients because step gradients can create artificial protein peaks due to the abrupt difference in densities between steps.

Most soluble proteins peak at the top of the gradient where they were loaded (fraction 1) and trail off in the first few fractions. However, a portion of some abundant soluble proteins may be found in all fractions.

Some useful conversions from Rickwood[21]: 10% (w/w) sucrose is equivalent to 10.38% (w/v); 20% (w/w) sucrose is equivalent to 21.62% (w/v); 30% (w/w) sucrose is equivalent to 33.81% (w/v); and 60% (w/w) sucrose is equivalent to 77.19% (w/v).

A gradient maker (such as one made by BioComp Instruments Inc., New Brunswick, Canada) rotates tubes of sucrose solutions at variable angles, speeds, and times to make continuous gradients. If a gradient maker is unavailable, gradients can be made by the method of Kölling and Hollenberg,[16] in which three sucrose solutions of different densities are layered in a tube and the tube is laid on its side for several hours, forming a continuous gradient by diffusion. The linearity of a gradient can be determined by measuring fraction densities with a refractometer.

GDPase Assay

GDPase in the Golgi cleaves GDP to GMP after GDP has been released from the GDP-mannose precursor. GDPase is localized to the cis- and medial-Golgi compartments.[14,22] The assay (adapted from the method of Ames[23] and Abeijon et al.[24]) is performed with both GDP and CDP. GDP-specific nucleoside diphosphate hydrolysis is determined by subtraction of the CDP hydrolysis from GDP

---

[21] D. Rickwood, "Preparative Centrifugation: A Practical Approach." IRL Press, Oxford Univ. Press, Oxford, 1992.
[22] V. V. Lupashin, S. Hamamoto, and R. W. Schekman, *J. Cell Biol.* **132,** 277 (1996).
[23] B. N. Ames, *Methods Enzymol.* **8,** 115 (1966).
[24] C. Abeijon, P. Orlean, P. W. Robbins, and C. B. Hirschberg, *Proc. Natl. Acad. Sci. U.S.A.* **86,** 6935 (1989).

hydrolysis. In a 96-well plate, 5 μl of each fraction is mixed with 90 μl of GDP reaction buffer in one well and with 90 μl of CDP reaction buffer in another well. The reactions are incubated at 30° for 30 min and are stopped by the addition of 5 μl 10% SDS. The amount of released phosphate is determined by the addition of 150 μl of ascorbic acid/MoO_3 buffer, and the A_{660} is read after incubation at room temperature for 5 to 10 min. GDPase activity per fraction is expressed relative to the total activity in the gradient.

Reaction buffer: 200 mM imidazole-hydrochloride, pH 7.4, 10 mM $CaCl_2$, 0.1% Triton X-100, and 1 mM GDP or CDP. GDP or CDP is usually dissolved in water first before it is added to the reaction buffer.

Ascorbic acid/MoO_3 buffer: A stock solution of 0.42% MoO_3 and 2.86% concentrated H_2SO_4 is stored at 4°. For phosphate determination, a 1 : 6 mixture of 10% ascorbic acid (fresh) and MoO_3/H_2SO_4 solution is used.

Kex2 Assay

Kex2p is an endoprotease in the trans-Golgi that is required for proteolytic processing of pro-α-factor.[12,13] Kex2p cleaves at the carboxyl side of pairs of basic residues and requires Ca^{2+} for activity. Kex2p activity is inhibited by leupeptin and antipain, but not by PMSF or pepstatin.[25,26] In this assay, Kex2p cleaves the substrate QRR-MCA (Boc-Gln-Arg-Arg-7-amino-4-methylcoumarin, Peptides International, Louisville, KY), releasing the fluorescent moiety MCA (7-amino-4-methylcoumarin). This protocol differs from the method of Cunningham and Wickner[13] in that the $CaCl_2$ concentration is increased to 5 mM to compensate for the EDTA present in the gradient fractions. Reactions contain 40 μl of gradient fraction, 0.5 M mannitol, 0.1 M Tris–HCl, pH 7.5, 1% Triton X-100, and 5 mM $CaCl_2$ in a total volume of 200 μl. Reactions are incubated for 1 hr at 30°, boiled for 5 min, and centrifuged at 13,000g for 10 min at room temperature in a microcentrifuge to remove precipitates. Fluorescence is measured in a fluorimeter at excitation 380 nm and emission 460 nm. The standard curve should use a range of MCA concentrations up to 2 nmol/ml.

Fractionation by Equilibrium Flotation

An alternative to the sedimentation gradient is a flotation gradient in which a membrane extract is prepared in high sucrose and loaded at the bottom of a sucrose gradient. After centrifugation to equilibrium, membranous particles will rise into the gradient, whereas protein particles will remain at the bottom of the

---

[25] D. Julius, A. Brake, L. Blair, R. Kunisawa, and J. Thorner, *Cell* **37**, 1075 (1984).
[26] R. S. Fuller, A. Brake, and J. Thorner, *Proc. Natl. Acad. Sci. U.S.A.* **86**, 1434 (1989).

tube, giving definitive evidence for or against membrane localization. A disadvantage of this method is that the plasma membrane-containing fractions are poorly separated from the loaded sample because it is difficult to prepare the membrane sample in a sucrose solution of significantly greater density than the plasma membrane.

Method

A yeast cell extract from 10^9 cells is prepared by lysis with glass beads followed by centrifugation at 500g for 3 min at 4° as described for the sedimentation gradient procedure. A portion of the cleared supernatant (0.6 ml) is layered on top of a step gradient of 1 ml STE20 above a 0.1-ml cushion of 80% (w/v) sucrose, 10 mM Tris–HCl, pH 7.5, 10 mM EDTA and is centrifuged at 100,000g for 1 hr at 4° in a TLS-55 rotor (Beckman Instruments). The membrane fraction concentrated at the interface between 20 and 80% sucrose layers is collected with a pipette and is then mixed with enough 80% (w/v) sucrose, 10 mM Tris–HCl, pH 7.5, 10 mM EDTA solution to bring the sample density equal to the density of the densest solution in the gradient (STE60). This step usually involves dilution of the sample by four or five times and is facilitated by checking the sucrose concentrations with a refractometer. A long needle is then used to carefully load the dense membrane fraction underneath a 30–60% sucrose gradient prepared in a buffer of 10 mM Tris–HCl, pH 7.5, 10 mM EDTA. The gradient is centrifuged for 17 hr at 100,000g at 4° with no brake. Fractions are collected and analyzed as for sedimentation equilibrium fractionation.

Technical Notes

Collecting the membranes on a cushion of dense sucrose solution is preferable to simply pelleting the membranes because it allows the membranes to be resuspended more easily for loading at the bottom of the flotation gradient. Also, pelleting the membranes may cause aggregation, which may prevent their migration upward during the subsequent gradient centrifugation.

Adjusting the density of the sample to that of 60% (w/w) sucrose is difficult to do exactly and usually requires significant dilution of the sample, which is why this procedure begins with twice as much cell extract as with a sedimentation gradient. It is best to avoid the STE20 layer as much as possible when collecting membranes. The sample may float up slightly from the bottom of the tube on loading, but in contrast to a cytosolic marker protein, membrane fractions will nevertheless be able to migrate to a less dense area of the gradient.

Sucrose gradients of 30–60% are used for flotation experiments because suspension of the membranes in a 60% sucrose solution appears to make the membranes slightly more dense overall than if they were loaded in a 10% sucrose solution. Therefore, a 30–60% gradient appears to give maximum resolution of intracellular membranes in a flotation experiment.

A related protocol combines the initial membrane purification on the step gradient with the usual sedimentation gradient procedure, in which the sample is loaded at the top of the gradient. While this method does not provide proof for membrane association like a flotation gradient, it significantly reduces the amount of soluble protein present in the gradient, and it is easier because dilution of the membrane sample to the density of a 10% sucrose solution is easily accomplished.

Plasma Membrane Isolation by Concanavalin A Treatment

Gradient fractionation described earlier is useful for determining the intracellular distribution of a membrane protein, but the need to assay individual gradient fractions makes this method cumbersome in instances where many different strains or growth conditions need to be compared. A useful way to quantify the relative amounts of a protein of interest at the cell surface compared to internal compartments is to chemically modify the plasma membrane in a way that increases its size and density greatly. When intact yeast spheroplasts are treated with the plant lectin concanavalin A (Con A), the lectin will bind mannose residues on cell surface glycoproteins and glycolipids, causing extensive aggregation of the plasma membrane after spheroplast lysis.[16,27,28] The aggregated plasma membranes can then be efficiently separated from other membranes by differential centrifugation.

Method

As in the protocol for differential centrifugation, 40 OD_{600} units of yeast cells are converted to spheroplasts. The spheroplasts are layered on top of 5 ml of 2 M sorbitol, 0.5 mM $MgCl_2$, 40 mM HEPES, pH 7.5, and are centrifuged for 5 min at room temperature at 500g. The spheroplasts are then suspended gently in 5 ml of S buffer, centrifuged for 5 min at room temperature at 500g, and then suspended in 3.2 ml of Con A buffer (1.2 M sorbitol, 1 mM magnesium acetate, 1 mM $CaCl_2$, 1 mM $MnCl_2$, 50 mM Tris–HCl, pH 7.5). The sample is transferred to a 15-ml Corex centrifuge tube, and an equal volume of Con A buffer containing 125 μl of a 5-mg/ml freshly made stock of Con A (Sigma, St. Louis, MO) is added dropwise while gently stirring at room temperature. It is crucial to the success of this method to continue stirring during the addition of Con A to achieve uniform coating of the spheroplasts with Con A. Mixing is continued by rocking the tube for an additional 10 min. The coated spheroplasts are harvested by centrifugation for 5 min at 500g at 4°, washed twice in Con A buffer by centrifugation, and finally suspended in 2 ml of cold lysis buffer (2 mM EDTA, 1 mM DTT, 5 mM HEPES, pH 7.5) containing a protease inhibitor cocktail (see differential centrifugation protocol). All subsequent steps are performed at 4°.

[27] A. Duran, B. Bowers, and E. Cabib, *Proc. Natl. Acad. Sci. U.S.A.* **72**, 3952 (1975).
[28] J. L. Patton and R. L. Lester, *J. Bacteriol.* **173**, 3101 (1991).

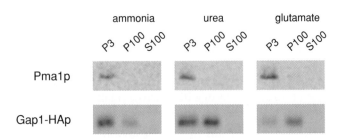

FIG. 3. Cellular distribution of the general amino acid permease, Gap1p, in yeast cells grown on different nitrogen sources. The yeast strain S288c was grown overnight in medium containing ammonia, urea, or glutamate as the sole source of nitrogen. Extracts from 40 OD_{600} units of yeast cells were prepared by Con A treatment of spheroplasts. The fractions were analyzed by immunoblotting and compared to fractionation of the plasma membrane ATPase Pma1p.

The coated spheroplasts are lysed by 20 strokes with a Dounce homogenizer with a tightly fitting pestle. One milliliter of homogenate is transferred to a microcentrifuge tube and centrifuged in a cooled variable speed centrifuge twice for 5 min at 100g. The supernatant is then removed by gentle pipetting. Because some of the Con A-coated plasma membranes will sediment even at low speed, the loose material on top of a tight pellet is collected along with the rest of the supernatant. The supernatant is then centrifuged for 15 min at 3000g to sediment plasma membranes. The resulting pellet (P3), which is highly enriched for plasma membrane, is suspended in 500 μl of storage buffer (20% glycerol, 1 mM DTT, 0.1 mM EDTA, 10 mM Tris–HCl, pH 7.5) for later analysis. The supernatant is then centrifuged for 1 hr at 100,000g (45,000 rpm in a Beckman TLA100.3 rotor). The resulting pellet (P100), which is enriched for intracellular membranes, is also suspended in 500 μl storage buffer. Proteins can be detected either by immunoblotting or by their enzymatic activity (see Fig. 3). Equivalent amounts of each fraction are diluted in two volumes of sample buffer and are resolved by SDS–PAGE and immunoblotting.

Technical Notes

Although this method appears to be straightforward, it takes some practice to get proper fractionation without losing material. Particularly critical steps are the first two clearing spins, because it is easy to lose plasma membrane material even at this low speed. The success of the method can be checked by verifying the separation of known marker proteins.

[19] Plasma Membrane Biogenesis

By AMY CHANG

Introduction

The plasma membrane serves to mediate interaction between the cell and the environment. For an overview of the yeast plasma membrane, including its unique lipid makeup, see van der Rest et al.[1] According to Yeast Protein Database (YPD; http://www.incyte.com/sequence/proteome/index.shtml), there are 232 documented plasma membrane proteins in *Saccharomyces cerevisiae*.[2] This number will undoubtedly increase as work progresses to define an estimated 1900 proteins of unknown function. Plasma membrane proteins fall into numerous classes, including transporters, sensors and receptors, signaling molecules, and landmark proteins. Transporters represent a large group of (polytopic) membrane proteins with multiple membrane-spanning domains that can be further subdivided into (a) ATPases, which couple ATP hydrolysis to transport. ATPases are composed of P-type ATPases, Pma1 and Pmr2, and transporters of the ATP-binding cassette (ABC) family, including Ste6 and pleiotropic drug resistance proteins. (b) Secondary transport proteins derive energy for solute translocation by utilizing electrochemical gradients across the plasma membrane. Numerous secondary transporters are involved in translocating sugars, ions, nucleosides, and amino acids. Many transport activities are modulated in response to extracellular nutrients, which are monitored by cell surface sensors; in several cases in which sensors have been identified, they are structurally similar to the transporters they regulate.[3–7]

A consistent theme at the plasma membrane is protein recruitment and assembly of protein complexes in response to external stimuli. Although the cell surface components that initiate many signal transduction cascades have not yet been defined,[8,9] it is clear that membrane-bound sensors must communicate with

---

[1] M. van der Rest, A. H. Kamminga, A. Nakano, Y. Anraku, B. Poolman, and W. N. Konings, *Microbiol. Rev.* **59**, 304 (1995).
[2] M. C. Costanzo, M. E. Crawford, J. E. Hirschman, J. E. Kranz, P. Olsen, L. S. Robertson, M. S. Skrzypek, B. R. Braun, K. L. Hopkins, P. Kondu, C. Lengieza, J. E. Lew-Smith, M. Tillberg, and J. I. Garrels, *Nucleic Acids Res.* **29**, 75 (2001).
[3] I. Iraqui, S. Vissers, F. Bernard, J.-O. De Craene, E. Boles, A. Urrestarazu, and B. Andre, *Mol. Cell. Biol.* **19**, 989 (1999).
[4] H. Klasson, G. R. Fink, and P. O. Ljungdahl, *Mol. Cell. Biol.* **19**, 5405 (1999).
[5] M. C. Lorenz and J. Heitman, *EMBO J.* **17**, 1236 (1998).
[6] A. M. Marini, S. Soussi-Boudekou, S. Vissers, and B. Andre, *Mol. Cell. Biol.* **17**, 4282 (1997).
[7] S. Ozcan and M. Johnston, *Microbiol. Mol. Biol. Rev.* **63**, 554 (1999).
[8] F. Banuett, *Microbiol. Mol. Biol. Rev.* **62**, 249 (1998).
[9] H. D. Madhani and G. R. Fink, *Trends Genet.* **14**, 151 (1998).

downstream signaling molecules, and the plasma membrane is frequently a target of the signal cascade. During mating, pheromones are sensed by seven transmembrane domain receptors, Ste2 and Ste3, on MATa and MATα cells, respectively.[8] Signaling occurs via a heterotrimeric G protein and recruitment of a MAP kinase complex, leading to remodeling of the cell surface to form a projection (schmoo), among other cellular responses. Similarly, the integrity of the cell wall is monitored and signaled via the protein kinase C (PKC) MAP kinase pathway to downstream targets that include plasma membrane-bound enzymes involved in localized synthesis and remodeling of the cell wall.

The plasma membrane is a polarized organelle. Plasma membrane growth is localized to the bud site during normal cell growth and to the projection in response to mating pheromone. An important class of proteins at the plasma membrane is composed of spatial landmark proteins (including products of *BUD* and *AXL* genes), which are localized to the mother-bud neck. These proteins recruit and assemble protein complexes to initiate bud formation.[10] Similarly, establishment of a polarized actin cytoskeleton is initiated by recruitment of Rho GTPases to specific sites designated by unknown landmark proteins at the plasma membrane,[11] and nuclear movement and spindle orientation are regulated by selective stabilization of astral microtubules with the cell cortex. Polarized secretion is mediated by assembly of a protein complex called the exocyst, which permits the docking of secretory vesicles at the bud site.[12] Major questions that remain unanswered include the identity of certain spatial landmark proteins and the mechanisms by which they assume their specific localization and signal the axis of polarization.

This chapter focuses on protein trafficking as a mechanism for regulating the function of plasma membrane proteins. In this regard, this chapter describes methods that have been found useful for studying cell surface targeting, mislocalization, and degradation of a model plasma membrane protein, Pma1.

Trafficking of Plasma Membrane Proteins

Stability versus Turnover at Plasma Membrane

The plasma membrane is a dynamic organelle. In the face of considerable flux at the cell surface, the plasma membrane ATPase, encoded by *PMA1*, is a paradigm of a stable cell surface protein with a half-life of ~11 hr.[13] In contrast, removal from the plasma membrane represents a mechanism by which numerous cell surface proteins are regulated. Ligand-induced endocytosis of pheromone receptors attenuates signaling. Similarly, downregulation of transport proteins in response

---

[10] J. Chant, *Annu. Rev. Cell Dev. Biol.* **15**, 365 (1999).
[11] D. Pruyne and A. Bretscher, *J. Cell Sci.* **113**, 365 (2000).
[12] F. P. Finger and P. Novick, *J. Cell Biol.* **142**, 609 (1998).
[13] B. Benito, E. Moreno, and R. Lagunas, *Biochim. Biophys. Acta* **1063**, 265 (1991).

to changes in the extracellular environment is brought about by endocytosis and vacuolar degradation of the cell surface proteins. It has been established that ubiquitination of cell surface proteins is a signal for their endocytosis and subsequent vacuolar degradation.[14] Ubiquitination of plasma membrane substrates is dependent on ubiquitin-conjugating enzymes Ubc4 and Ubc5 and the ubiquitin ligase Rsp5.[15] Proteins that have been reported to undergo regulated endocytosis include permeases (Fur4, Gal2, Gap1, Mal61), the pheromone receptors Ste2 and Ste3, and transporters (Zrt1, Ste6, Pdr5, and Sts1).[15] How ubiquitination is regulated in response to the external milieu remains unknown.[14]

Endoplasmic Reticulum Export

All plasma membrane proteins enter the secretory pathway at the endoplasmic reticulum and move via vesicular transport to the Golgi and then to the cell surface. Export from the ER of a number of plasma membrane proteins requires specialized ER resident membrane proteins.[16] Packaging of a family of amino acid permeases into COPII vesicles requires the ER resident Shr3[17,18]; absence of Shr3 results in defective amino acid transport at the cell surface because there is specific accumulation of permeases in the ER. Other examples include the hexose transporters, Hxt1 and Gal2,[19] Pho84p, a high-affinity phosphate transporter,[20] and Chs3, a subunit of chitin synthetase III.[21] It is not understood why ER export of these polytopic membrane proteins requires dedicated ER resident proteins.

Quality Control

Defective plasma membrane proteins are prevented from reaching or residing at the cell surface by two major mechanisms. ER-associated degradation (ERAD) is a mechanism by which misfolded and improperly assembled proteins at the ER are recognized and then removed from the secretory pathway. This occurs by ubiquitination via ER-associated ubiquitin-conjugating enzymes (including Ubc6 and Ubc7), retrotranslocation, and degradation in the cytoplasm by the 26S proteasome.[22,23] In addition to ERAD, defective plasma membrane proteins are

---

[14] J. S. Bonifacino and A. M. Weissman, *Annu. Rev. Cell Dev. Biol.* **14,** 19 (1998).
[15] L. Hicke, *Trends Cell Biol.* **9,** 107 (1999).
[16] J. M. Herrmann, P. Malkus, and R. Schekman, *Trends Cell Biol.* **9,** 5 (1999).
[17] M. J. Kuehn, R. Schekman, and P. Ljungdahl, *J. Cell Biol.* **135,** 585 (1996).
[18] P. O. Ljungdahl, C. J. Gimeno, C. A. Styles, and G. R. Fink, *Cell* **71,** 463 (1992).
[19] P. W. Sherwood and M. Carlson, *Proc. Natl. Acad. Sci. U.S.A.* **96,** 7415 (1999).
[20] W.-T. W. Lau, R. W. Howson, P. Malkus, R. Schekman, and E. O'Shea, *Proc. Natl. Acad. Sci. U.S.A.* **97,** 1107 (2000).
[21] J. A. Trilla, A. Duran, and C. Roncero, *J. Cell Biol.* **145,** 1153 (1999).
[22] J. L. Brodsky and A. A. McCracken, *Trends Cell Biol.* **7,** 151 (1997).
[23] T. Sommer and D. H. Wolf, *FASEB J.* **11,** 1227 (1997).

degraded by delivery to the vacuole. Pma1-7p and Ste2-3p are two examples of defective plasma membrane proteins that fail to move to the cell surface and are instead targeted to the vacuole for degradation.[24,25] It is possible that vacuolar targeting of mutant proteins is mediated by a saturable Golgi-based quality control system, which recognizes misfolded and incompletely assembled proteins. Indeed, delivery to the vacuole is a means by which foreign proteins are eliminated.[26-28] Nevertheless, the idea that delivery to the vacuole is a default pathway has not been excluded, i.e., membrane proteins will move to the vacuole in the absence of a sorting signal.[29] This hypothesis predicts that membrane proteins have plasma membrane-targeting signals, although such signals have not been identified.

Protein Sorting

Protein sorting of newly synthesized protein to the vacuole represents a mechanism for the downregulation of normal plasma membrane proteins in response to nutritional conditions. The heavy metal transporters Smf1p and Smf2p are targeted for vacuolar degradation under metal-replete conditions while they are localized to the plasma membrane under metal starvation conditions.[30] Similarly, sorting of the general amino acid permease, Gap1p, and the high-affinity tryptophan permease, Tat2, to the cell surface or to the vacuole is dependent on the nutrient status of the cell.[31,32]

Plasma Membrane Markers

The list of available plasma membrane markers is ever growing. The plasma membrane [H$^+$]ATPase, encoded by *PMA1*, is probably the most commonly used cell surface marker (see later). Another favorite is Gas1, a protein attached to the plasma membrane by a glycosylphosphatidylinositol (GPI) anchor, which has been suggested to play a key role in cell wall assembly.[33] Gas1 is a useful marker for protein trafficking studies because its transport from ER to Golgi is accompanied by glycosylation and an increase in electrophoretic mobility from an apparent M_r of 105 to 125 kDa.[34,35] The immature and mature forms of Gas1 can also

[24] A. Chang and G. R. Fink, *J. Cell Biol.* **128**, 39 (1995).
[25] D. D. Jenness, Y. Li, C. Tipper, and P. Spatrick, *Mol. Cell. Biol.* **17**, 6236 (1997).
[26] H. Holkeri and M. Makarow, *FEBS Lett.* **429**, 162 (1998).
[27] E. Hong, A. R. Davidson, and C. A. Kaiser, *J. Cell Biol.* **135**, 623 (1996).
[28] B.-Y. Zhang, A. Chang, T. B. Kjeldsen, and P. Arvan, submitted for publication.
[29] S. F. Nothwehr and T. H. Stevens, *J. Biol. Chem.* **269**, 10185 (1994).
[30] X. F. Liu and V. C. Culotta, *J. Biol. Chem.* **274**, 4863 (1999).
[31] T. Beck, A. Schmidt, and M. N. Hall, *J. Cell Biol.* **146**, 1227 (1999).
[32] K. J. Roberg, N. Rowley, and C. A. Kaiser, *J. Cell Biol.* **137**, 1469 (1997).
[33] L. Popolo and M. Vai, *Biochim. Biophys. Acta* **1426**, 385 (1999).
[34] T. L. Doering and R. Schekman, *EMBO J.* **15**, 182 (1996).
[35] C. Sutterlin, T. L. Doering, F. Schimmoller, S. Schroder, and H. Riezman, *J. Cell Sci.* **110**, 2703 (1997).

be resolved by fractionation on Renografin density gradients (described later). Ste2 and Ste3 are popular markers for endocytosis studies because the receptors undergo constitutive as well as ligand-induced internalization, leading to vacuolar degradation.[36] Receptor endocytosis has been assessed by classical pulse-chase analysis by immunoprecipitating from a metabolically labeled cell lysate. Alternatively, an epitope-tagged *GAL1-STE3* construct has been used for Western blotting to detect time-dependent loss of tagged Ste3 after adding glucose to shut off its synthesis.[37–40]

Available Tools for Studying Pma1

Pma1 is a popular membrane marker because it is abundant, representing 10–50% of the total plasma membrane protein, and it is stable at the cell surface.[13] Pma1 is a polytopic membrane protein with 10 transmembrane segments and an apparent $M_r \sim 100$ kDa.[41,42] Pma1 is not glycosylated but it is modified posttranslationally by phosphorylation at serine and threonine residues.[43] Phosphorylation is not a useful marker of Pma1 transit through the secretory pathway because it occurs progressively during the intracellular transport of Pma1, and the modification causes only a subtle mobility shift by SDS–PAGE. Pulse-chase analysis, combined with cell fractionation, indicates that newly synthesized Pma1 arrives at the cell surface by 30 min of chase.[44] Maximal phosphorylation of Pma1 is achieved by 1–2 hr of chase,[43] suggesting that some phosphorylation occurs after residence at the cell surface.

There is a wealth of reagents available for the detection of Pma1. Polyclonal anti-Pma1 antibodies are available from a number of laboratories. Some of the anti-Pma1 polyclonal antibodies were generated using partially purified enzyme from *Neurospora crassa* as antigen, and these antibodies cross-react well with Pma1 from *Saccharomyces cerevisiae*.[45,46] A monoclonal anti-Pma1 antibody has been described as well.[47] All the antibodies work well for Western blotting, immunoprecipitation, and indirect immunofluorescence (IF) (although the polyclonal

---

[36] A. F. Roth, D. M. Sullivan, and N. G. Davis, *J. Cell Biol.* **142**, 949 (1998).
[37] S. R. Gerrard, N. J. Bryant, and T. H. Stevens, *Mol. Biol. Cell* **11**, 613 (2000).
[38] W.-j. Luo and A. Chang, *Mol. Biol. Cell* **11**, 579 (2000).
[39] H. R. Panek, J. D. Stepp, H. M. Engle, K. M. Marks, P. K. Tan, S. K. Lemmon, and L. C. Robinson, *EMBO J.* **16**, 4194 (1997).
[40] R. G. Spelbrink and S. F. Nothwehr, *Mol. Biol. Cell* **10**, 4263 (1999).
[41] M. Auer, G. A. Scarborough, and W. Kuhlbrandt, *Nature* **392**, 840 (1998).
[42] R. Serrano, M. C. Kielland-Brandt, and G. R. Fink, *Nature* **319**, 689 (1986).
[43] A. Chang and C. W. Slayman, *J. Cell Biol.* **115**, 289 (1991).
[44] X. Gong and A. Chang, *Proc. Natl. Acad. Sci. U.S.A.*, **98**, 9104 (2001).
[45] K. M. Hager, S. M. Mandala, J. W. Davenport, D. W. Speicher, E. J. Benz, Jr., and C. W. Slayman, *Proc. Natl. Acad. Sci. U.S.A.* **83**, 7693 (1986).
[46] W.-j. Luo and A. Chang, *J. Cell Biol.* **138**, 731 (1997).
[47] J. P. Aris and G. Blobel, *J. Cell Biol.* **107**, 17 (1988).

antibody should be affinity purified for IF). A hemagglutinin epitope (HA)-tagged *PMA1* in which the epitope is inserted after the second amino acid has been widely used as a cell surface marker.[38,48–50] Localization studies have been performed in strains carrying centromeric plasmids with HA-tagged *PMA1* under the control of a *GAL1*[49] or *MET25* promoter.[38] Expression of tagged Pma1 in addition to chromosomal expression does not appear to affect cell growth. Strains in which chromosomal *PMA1* is replaced by *HA-PMA1* grow normally.[51] Finally, Pma1 can be detected readily by assaying ATPase activity; Pma1 activity is distinguished from that of other ATPases by its pH optimum and its vanadate sensitivity.[43]

Pma1 Mutants

Extensive mutagenesis studies have been carried out with Pma1 in order to understand the structure and function of the physiologically important family of P-type transporters to which Pma1 belongs. Of ~400 single-site mutants that have been published, serious defects in biogenesis (in which less than 50% of the newly synthesized mutant Pma1 is delivered to the cell surface) have been documented for >15%.[52] Many of these mutants are retained in the ER and targeted for ERAD.[48,50,53] Because *PMA1* is essential for cell viability,[42] conditional *pma1* mutants have been isolated that cause temperature-sensitive growth.[24,54] There is no evidence that any of these mutants represents a thermolabile enzyme. Instead, temperature-sensitive growth is the result of (1) mislocalization and degradation of the newly synthesized mutant Pma1 and (2) gradual loss of essential activity from the cell surface. Pma1 may be less susceptible to temperature-sensitive denaturation because it forms an oligomer[41] and associates with ergosterol and sphingolipid-enriched membrane microdomains called lipid rafts.[55]

Separation of Plasma Membrane by Cell Fractionation

Several protocols for the isolation of plasma membrane have been described.[56] Some of these protocols involve making spheroplasts by enzyme digestion of

---

[48] N. D. DeWitt, C. F. Tourinho dos Santos, K. E. Allen, and C. W. Slayman, *J. Biol. Chem.* **273**, 21744 (1998).
[49] S. L. Harris, S. Na, X. Zhu, D. Seto-Young, D. Perlin, J. H. Teem, and J. E. Haber, *Proc. Natl. Acad. Sci. U.S.A.* **91**, 10531 (1994).
[50] A. M. Maldonado, N. de la Fuente, and F. Portillo, *Genetics* **150**, 11 (1998).
[51] M. Ziman, J. S. Chuang, and R. W. Schekman, *Mol. Biol. Cell* **7**, 1909 (1996).
[52] P. Morsomme, C. W. Slayman, and A. Goffeau, *Biochim. Biophys. Acta* **1469**, 133 (2000).
[53] Q. Wang and A. Chang, *EMBO J.* **18**, 5972 (1999).
[54] A. Cid and R. Serrano, *J. Biol. Chem.* **263**, 14134 (1988).
[55] M. Bagnat, S. Keranen, A. Shevchenko, and K. Simons, *Proc. Natl. Acad. Sci. U.S.A.* **97**, 3254 (2000).
[56] B. Panaretou and P. Piper, *Methods Mol. Biol.* **53**, 117 (1996).

the cell wall, coating them with concanavalin A or cationic silica microbeads to stabilize the plasma membrane, followed by lysis. We have employed fractionation on Renografin density gradients to separate plasma membranes from intracellular membranes.[38] The protocol is simple and robust because cell lysis is carried out by vortexing with glass beads (without generating spheroplasts by enzyme digestion). Separation of plasma membrane from other organellar membranes is achieved on one density gradient.

Renografin Density Gradient Protocol

We usually grow cells in synthetic complete medium. (Because some *pma1* mutants are sensitive to the pH of the growth medium, it is important to be aware of pH differences between synthetic medium and YPD.) Roughly 17 OD_{600} units of cells are harvested for each gradient. Wash the cells once with Tris EDTA buffer (50 mM Tris, pH 7.5, 1 mM EDTA). Protease inhibitors are critical during cell lysis. We make a 1000× protease inhibitor cocktail by mixing 0.2 ml aprotinin (10 mg/ml stock in H_2O), 0.2 ml leupeptin (10 mg/ml stock in H_2O), 0.2 ml pepstatin [10 mg/ml stock in dimethyl sulfoxide (DMSO)], 0.4 ml chymostatin (10 mg/ml stock in DMSO), and 1 ml H_2O. Just before use, add 1 mM phenylmethylsulfonyl fluoride (PMSF) (from a 250 mM stock in ethanol). Because of the short half-life of PMSF in aqueous solution, it is important to add protease inhibitors to the buffer just prior to use. Resuspend cells in 0.5 ml Tris EDTA buffer in a microfuge tube and add glass beads two-thirds of the way to the meniscus. Vortex in 30-sec bursts seven times with resting on ice in between each burst. After vortexing, transfer the supernatant to a fresh tube using a 1-ml pipette tip. Wash the glass beads with a small amount of buffer and combine both supernatants so that the final volume of lysate equals 0.5 ml. Centrifuge for 5 min at 400g to remove unbroken cells. Mix 0.5 ml lysate with 0.5 ml Renografin-76 (76% Renografin) and place at the bottom of an SW50.1 centrifuge tube. Overlay with 1 ml each 34 (diluted from 76% with Tris EDTA buffer), 30, 26, and 22% Renografin. Centrifuge overnight (>16 hr) at 40,000 rpm. Renografin-76 has been replaced by a similar product called Renocal-76 (Bracco Diagnostics, Inc., Princeton, NJ). Alternatively, Renografin-60 (60% Renografin) is also available from the same vendor.

After centrifugation, collect 14× 0.35-ml fractions from the top of the gradient with a Pipetman. There is usually no visible pellet at the bottom of the tube. To analyze the distribution of membrane markers following fractionation, dilute each fraction ~10-fold with Tris EDTA buffer (usually 100 μl fraction plus 1 ml buffer) and pellet membranes at 100,000g for 1 hr in a tabletop ultracentrifuge. We have found this step useful because it removes Renografin, which can interfere with subsequent electrophoretic transfer and Western blotting. After centrifugation, there is usually no visible pellet. Remove supernatant, add Laemmli gel sample buffer, and swirl the tubes by hand for 1–2 min. Heat the samples at 37° for 5–10 min before loading gels; 7.5% polyacrylamide gels provide good separation of 100 kDa Pma1 from such membrane markers as Gas1 (125 kDa) and Sec61 (41 kDa) (see later). Do not boil samples because polytopic membrane proteins, e.g., Pma1, are prone to aggregation. In addition, increasing 2–4× SDS concentration in Laemmli sample buffer (to 4–8%) seems to improve the resolution of Pma1 on SDS polyacrylamide gels.

Additional Comments

A number of intracellular markers have been shown to migrate in lower-density fractions on Renografin gradients. The plasma membrane markers Pma1 and Gas1 are usually maximal in fractions 10 and 11 and are well separated from all intracellular membrane markers, which usually migrate in fractions 4–6. Western blots have been used to localize a number of intracellular membrane markers following fractionation. These are Sec61, a 41-kDa ER marker[57]; Pep12, a ~37-kDa membrane marker of Golgi and endosome[58]; and alkaline phosphatase, a ~70-kDa marker of the vacuolar membrane.[59] Commercial antibodies against Pep12 and alkaline phosphatase are available (Molecular Probes, Eugene, OR). Unfortunately, Renografin gradients are not able to separate any of the intracellular membrane markers from each other.

We have used Renografin density gradients to follow the transport of newly synthesized Pma1 either by metabolic labeling cells with [^{35}S]methionine and cysteine or by inducing the synthesis of epitope-tagged Pma1 using a regulated promoter (see later). For metabolic labeling, cells growing exponentially in minimal medium are harvested and resuspended at a density of 1 OD_{600}/ml in fresh minimal medium. Cells are allowed to recover by being shaken for 15 min. Expre^{35}S^{35}S (NEN, Boston, MA) is then added (2 mCi/25 OD_{600} cells). For a pulse-chase format, chase is initiated after a 2- to 5-min incubation with radiolabel by adding an equal volume of synthetic complete medium with an additional 20 mM cysteine and 20 mM methionine. At various times of chase, aliquots are removed and placed on ice in the presence of 10 mM azide. After lysate preparation and before loading Renografin density gradients, small aliquots of lysate (5 μl) are removed to measure TCA-precipitable cpm. Samples from different chase times are normalized to TCA-precipitable cpm. For immunoprecipitation of metabolically labeled Pma1, gradient fractions (up to 100 μl volume) are resuspended in immunoprecipitation buffer (RIPA buffer: 10 mM Tris, pH 7.5, 150 mM NaCl, 2 mM EDTA, 1% Nonidet P-40, 1% deoxycholate, 0.1% SDS).

[57] R. J. Deshaies, S. L. Sanders, D. A. Feldheim, and R. Schekman, *Nature* **349**, 806 (1991).
[58] K. A. Becherer, S. E. Rieder, S. D. Emr, and E. W. Jones, *Mol. Biol. Cell* **7**, 579 (1996).
[59] D. J. Klionsky and S. D. Emr, *EMBO J.* **8**, 2241 (1989).

Visualizing Transport to Plasma Membrane by "Pulse-Chase Immunofluorescence"

To follow the ordered movement of a protein through the secretory pathway, a critical parameter is regulating the synthesis of the protein of interest. In this regard, we have developed a protocol called "pulse-chase immunofluorescence," which allows visualization of transport through the secretory pathway. We have employed a *MET25* promoter to control the synthesis of an epitope-tagged Pma1 protein. Transcriptional activation occurs in the absence of methionine in the medium; in the presence of methionine, there is transcriptional repression. In the classical pulse-chase experiment described earlier, cells are pulsed by metabolically radiolabeling for a brief period. The pulse is rapidly and effectively terminated simply by adding excess cold amino acids to the medium. The precision with which the kinetics of intracellular transport can be measured is dependent on the brevity of the pulse. In contrast to a 2- to 5-min pulse, which is common for metabolic radiolabeling experiments, we use a 30- to 60-min pulse to visualize newly synthesized Pma1 in the "pulse-chase immunofluorescence" protocol. This is necessary to generate an abundant signal easily detectable by indirect immunofluorescence staining. What the "pulse-chase immunofluorescence" method lacks in precision, it makes up for in simplicity. Using this method, we have been able to visualize the kinetically slowest steps in the intracellular transport of Pma1 in different trafficking mutants. We have used this method to visualize the transport pathway of a mutant Pma1 (Pma1-7p) that fails to move to the plasma membrane and instead is mislocalized to the vacuole.[38]

Pulse-Chase Immunofluorescence Protocol

To follow intracellular trafficking of newly synthesized Pma1, we have constructed strains in which *MET25-HA-PMA1* is cloned into a *URA3*-marked Yip (pRS306[60]) and integrated into the genome at *ura3* (after linearizing the plasmid with *Nco*I). Cells are grown overnight in minimal medium supplemented with necessary amino acids and 300 μM methionine. Under these conditions, the background level of HA-Pma1 synthesis is low and undetectable by staining with anti-HA antibody. To initiate HA-Pma1 synthesis, cells (usually 10 OD_{600} units) are harvested, washed with water to remove methionine, and resuspended at a density of 0.5 OD_{600}/ml in fresh minimal medium without methionine. After 30 min, the HA-Pma1 level is high, and by 60 min, HA-Pma1 has reached maximal levels. A "pulse" of 1 hr generates a strong signal detected readily by IF. After the pulse, 5 OD_{600} unit cells are pelleted and resuspended in 5 ml of 0.1 M potassium phosphate, pH 6.5. Cells are fixed by adding 0.6 ml formaldehyde (37% solution).

[60] R. S. Sikorski and P. Hieter, *Genetics* **122**, 19 (1989).

Samples are placed on a nutator overnight at room temperature. To terminate synthesis of HA-Pma1 for the "chase," 2 mM methionine is added. After a 60-min chase, cells are harvested for fixation.

Immunofluorescence Protocol

For indirect immunofluorescence staining of HA-Pma1, we use a standard protocol.[61] Wash fixed cells twice with potassium phosphate buffer, once with 0.1 M potassium phosphate, 1.2 M sorbitol buffer, and resuspend 5 OD$_{600}$ units in 1 ml phosphate–sorbitol buffer. Fixed cells can be stored for up to a week at 4° before staining. We usually process half of the fixed cells, saving the remainder in case spheroplasting conditions or antibody dilutions need to be adjusted. Removal of the cell wall is important to permit antibody accessibility, although overdigestion of the wall can result in poor morphology and loss of antigenicity. Oxalyticase (Enzogenetics, Corvallis, OR) works well for cell wall digestion and is less expensive than Zymolyase.

The optimum oxalyticase concentration for efficient cell wall removal varies depending on strain background and growth conditions. Usually, we test two different concentrations of oxalyticase after adding 1 μl 2-mercaptoethanol (required for oxalyticase activity) to 500 μl cells and dividing the sample into two tubes, e.g., 1.3 and 2.5 μl oxalyticase (1 mg/ml in phosphate–sorbitol buffer) are added to each tube. After 15 min at 30°, wash spheroplasts by adding 1 ml cold phosphate–sorbitol and pelleting at 3600g. Resuspend cells in ~200 μl phosphate–sorbitol buffer. Before placing spheroplasts on slides (masked 10-well slides; ER208WSF from Fisher Scientific), coat slides with drops of 1% (w/v) polyethyleneamine; let drops sit 5 min before washing twice with water. Let dry completely. Allow spheroplasts to adhere to the slides >30 min. Gently aspirate nonadhering spheroplasts. To permeabilize spheroplasts, place slide in methanol at −20° for 6 min, followed by acetone at −20° for 30 sec. Stain cells by incubating overnight at room temperature with anti-HA monoclonal antibody from Covance, Inc. (1 : 500 dilution). Dilute antibodies in phosphate–sorbitol buffer with 2% (v/v) goat serum. Wash three times with buffer containing 0.2% goat serum. Incubate with Cy3-conjugated anti-rabbit secondary antibody (1 : 1000 dilution) (Jackson ImmunoResearch, West Grove, PA) for 2 hr at room temperature. After washing three times, allow slides to dry. Add mounting medium containing 50% (v/v) glycerol in phosphate-buffered saline and a coverslip before visualizing.

Additional Considerations

The nutritional cues on which promoters such as *MET25* or *GAL1* depend are largely influenced by vacuolar metabolism. The regulation of methionine

---

[61] M. D. Rose, F. Winston, and P. Hieter, "Methods in Yeast Genetics: A Laboratory Manual." Cold Spring Harbor Laboratory Press, Cold Spring Harbor, NY, 1990.

utilization and *MET25*-mediated transcription is perturbed in certain *vps* mutants defective in vacuolar biogenesis and vacuole function. Thus, in certain *vps* mutants, one obtains a higher "background" HA signal in the absence of protein induction and less efficient termination of induction. To alleviate a high background problem, it is important that cells are not allowed to overgrow and deplete the supplemented methionine. If necessary, methionine supplementation of the medium can be increased to 600 μM. To ensure that the chase is "tight," cycloheximide (100 μg/ml) can be added to the chase medium in addition to 2 mM methionine.

Other Methods for Assessing Cell Surface Delivery

Steady-State Protein Level

A common effect of mutation of plasma membrane proteins is defective transport to the cell surface with concomitant protein degradation. A direct approach to detect increased turnover of a protein is pulse-chase analysis by metabolic radiolabeling cells. An easier, albeit indirect, method to detect degradation of a plasma membrane marker is to assess the steady-state level of the protein by Western blot. The approach makes the assumption that synthesis of the marker protein remains constant. Nevertheless, immunoblotting is useful especially as a screening tool to compare the stability of multiple mutants.

Temperature-sensitive *pma1* mutants defective in the cell surface delivery of Pma1 display a reduced steady-state level of Pma1 by comparison to that of wild-type cells even at the permissive temperature.[24] The difference between Pma1 levels in mutant and wild-type cells is enhanced by shifting the cells to the restrictive temperature for 3–6 hr.

Western Blot Protocol

For Western blot, cells from a midlog culture are harvested and resuspended at a density of $\sim$200 OD$_{600}$/ml in bead buffer (0.3 M sorbitol, 0.1 M NaCl, 5 mM MgCl$_2$, 10 mM Tris, pH 7.4) with freshly added protease inhibitor cocktail and 1 mM PMSF. Usually, $\sim$50 μl buffer is added to 5–10 OD$_{600}$ units cells and glass beads are added up to the meniscus. Vortex 3× 1 min with resting on ice between each burst. Add $\sim$250 μl bead buffer, vortex, and remove the supernatant with a pipette tip. Wash the beads with an additional small aliquot of bead buffer. Centrifuge the pooled supernatants for 5 min at $\sim$400g to remove unbroken cells from the lysate. Normalize samples to protein content using the Bradford assay (Bio-Rad Laboratories, Hercules, CA) for SDS–PAGE and Western blotting. For Pma1 blots, $\sim$50 μg lysate protein/lane is sufficient. To obtain quantitative results, use ^{125}I-labeled protein A as the secondary reagent ($\sim$1 μCi/blot; Amersham Life Science, Inc., Arlington Heights, IL) and phosphorimaging of the blot (Molecular Dynamics, Sunnyvale, CA). Because normalization of samples is critical, it is advantageous to show that the levels of a control membrane marker, e.g., Sec61p, remain constant.

Additional Considerations

A reduced level of mutant protein by comparison with wild-type protein on Western blots may indicate increased degradation. To address the site of degradation, steady-state protein levels can be compared in strains expressing or deleted of *PEP4*, encoding vacuolar proteinase A. In the absence of Pep4p, vacuolar proteases are inactive and degradation of substrates is inhibited. Thus, the steady-state level of a mutant protein targeted for vacuolar degradation should rise in *pep4*Δ cells. Indeed, the steady-state level of mutant Pma1 targeted for vacuolar degradation is increased in *pep4*Δ cells. Constructs to knockout *PEP4* have been described.[24,62]

If the steady-state level of a mutant protein is decreased because of ERAD, it may be possible to stabilize the protein in a *ubc6*Δ *ubc7*Δ strain background.[63] In a variation on the Western blot assay, Bays and co-workers[63] have been able to detect accumulation of an ERAD substrate in *ubc6*Δ *ubc7*Δ cells by using a green fluorescent protein (GFP) fusion construct and FACS to detect an increase in fluorescence of the GFP-substrate protein. Failure to detect stabilization in *ubc6*Δ *ubc7*Δ cells by assaying steady-state protein level is inconclusive because other ubiquitin-conjugating enzymes, e.g., Ubc1, play roles in ERAD.[64] A more sensitive assay such as pulse-chase analysis by metabolic radiolabeling may reveal partial stabilization in *ubc7*Δ and *ubc1*Δ cells.[64]

Acknowledgments

Research in the author's laboratory was supported by NIH Grant GM58212. The author thanks Jonathan Warner and Peter Arvan for critical reading of the manuscript.

[62] S. F. Nothwehr, N. J. Bryant, and T. H. Stevens, *Mol. Cell. Biol.* **16,** 2700 (1996).
[63] N. W. Bays, R. G. Gardner, L. P. Seelig, C. A. Joazeiro, and R. Y. Hampton, *Nat. Cell Biol.* **3,** 24 (2001).
[64] R. Friedlander, E. Jarosch, J. Urban, C. Volkwein, and T. Sommer, *Nat. Cell Biol.* **2,** 379 (2000).

[20] Separation of Golgi and Endosomal Compartments

By GYÖRGY SIPOS and ROBERT S. FULLER

Introduction

Fractionation and isolation of subcellular organelles have always been a major challenge in cell biology. Membrane-bound compartments are separated on the basis of their differential physical features, including shape, density, and charge. Most frequently, the method of choice is isopycnic centrifugation, in which organelles are separated according to their distinct buoyant densities. Within the past several years, the better availability of immunological reagents has provided an additional approach that makes use of the antigenic properties of the organelles. Immunoisolation has become a powerful tool for identifying and characterizing membrane compartments.

The aim of this chapter is to describe procedures for the fractionation and isolation of Golgi and endosomal membranes. Several important topics in fractionation of yeast secretory organelles, which are not addressed here, are covered elsewhere: secretory vesicles,[1,2] clathrin-coated vesicles,[3] and vacuoles.[4,5] The review of Walworth *et al.*[2] still provides valuable general information about fractionation of yeast membranes. Here, we describe systematic methods for quickly optimizing sucrose step gradients for separating marker proteins, as well as methods for quantitative immunoisolation of organelles with antibody-coated magnetic beads.

Yeast Subcellular Fractionation

After the initial isolation of *sec* mutants and description of the yeast secretory pathway, fractionation of yeast organelles became of particular interest.[6] The resulting pursuit of fractionation methods led to the development of several basic approaches. The first yeast organellar fractionation protocols involved enzymatic removal of the rigid cell wall, allowing spheroplasts to be opened by osmotic lysis and gentle homogenization.[2] Based on assays of their resident enzymes, the major organelles were separated by differential centrifugation and density gradients.

---

[1] C. L. Holcomb, T. Etcheverry, and R. Schekman, *Anal. Biochem.* **166,** 328 (1987).
[2] N. C. Walworth, B. Goud, H. Ruohola, and P. J. Novick, *Methods Cell Biol.* **31,** 335 (1989).
[3] S. C. Mueller and D. Branton, *J. Cell Biol.* **98,** 341 (1984).
[4] C. J. Roberts, C. K. Raymond, C. T. Yamashiro, and T. H. Stevens, *Methods Enzymol.* **194,** 644 (1991).
[5] E. Conibear and T. H. Stevens, *Methods Enzymol.* **351,** [24], 2002 (this volume).
[6] B. Goud, A. Salminen, N. C. Walworth, and P. J. Novick, *Cell* **53,** 753 (1988).

Since then, the same methodology has been in use for various fractionation and localization studies.[7]

The choice of a fractionation procedure depends on the experimental goals. The stringency of a fractionation is necessarily much greater when the goal is purification of an organelle and its protein content than when the goal is to establish a biochemical or phenotypic assay. For example, simple differential centrifugation of a low-speed supernatant of a yeast cell lysate yielding P13 (13,000g pellet), P100 (100,000g pellet), and S100 (100,000g supernatant) fractions has been used with great effectiveness as a sensitive probe for the effects of vacuolar protein sorting (*vps*) mutations on localization of proteins and protein complexes to the vacuole, late endosome (prevacuolar compartment), and late Golgi/TGN (*trans*-Golgi network).[8,9] This is true even though, strictly speaking, such a technique can only achieve a crude separation of soluble proteins (S100) from slowly sedimenting membranes, such as Golgi vesicles (P100), and rapidly sedimenting membranes, such as vacuoles and the endoplasmic reticulum (ER)/nuclear envelope (P13). Late endosome/prevacuolar compartment (PVC) proteins in fact appear to be distributed between P13 and P100 fractions.

In order to achieve separation of endosomal and Golgi membranes, additional steps must be taken. After the removal of cellular debris and bulky, heavier organelles from the crude lysate by medium-speed centrifugation, fractions containing the slowly sedimenting Golgi and endosomal membranes are harvested in a high-speed pellet (HSP). Further fractionation of HSP membranes over an equilibrium sucrose step gradient is able to resolve density differences between Golgi and endosomal markers. For example, the late endosome or PVC was identified as an intermediate organelle fractionating at densities between the heavier Golgi and the lighter vacuolar membranes.[10] In fact, even the distributions of markers for Golgi subcompartments, despite their overlaps, could be clearly distinguished.[11] Highly purified membrane fractions were successfully isolated, with the help of various density gradients, one enriched in *cis*-Golgi enzymes and competent for *in vitro* ER–Golgi transport assays and the other enriched in Kex2 activities characteristic of the late Golgi.[12,13]

In addition to the commonly used sucrose, there are many other alternatives for gradient medium. Nycodenz is an iodinated compound with low osmolality. When following endocytosis of α factor, flotation through a Nycodenz gradient provided sufficient resolution to separate early and late endosomes.[14]

[7] E. Zinser and G. Daum, *Yeast* **11,** 493 (1995).
[8] E. C. Gaynor, S. te Heesen, T. R. Graham, M. Aebi, and S. D. Emr, *J. Cell Biol.* **127,** 653 (1994).
[9] S. E. Rieder and S. D. Emr, *Mol. Biol. Cell* **8,** 2307 (1997).
[10] T. A. Vida, G. Huyer, and S. D. Emr, *J. Cell Biol.* **121,** 1245 (1993).
[11] J. C. Holthuis, B. J. Nichols, S. Dhruvakumar, and H. R. Pelham, *EMBO J.* **17,** 113 (1998).
[12] V. V. Lupashin, S. Hamamoto, and R. W. Schekman, *J. Cell Biol.* **132,** 277 (1996).
[13] E. A. Whitters, T. P. McGee, and V. A. Bankaitis, *J. Biol. Chem.* **269,** 28106 (1994).
[14] L. Hicke, B. Zanolari, M. Pypaert, J. Rohrer, and H. Riezman, *Mol. Biol. Cell* **8,** 13 (1997).

Separation of early (or *cis*) from late (or *trans*) Golgi membranes has also been achieved using immunoisolation procedures. *cis*-Golgi membranes were immunoisolated with monoclonal antibodies against the *myc*-tagged Sed5p and late Golgi with affinity-purified polyclonal antibodies raised against the cytosolic tail of Kex2p.[15,16]

Separating and Distinguishing Golgi and Endosomal Compartments Using Equilibrium Density Gradients

Preparation of Golgi and Endosomal Membrane Fractions

In order to obtain a sufficient yield of intact organelles with well-preserved *in vivo* biochemical activities, appropriate methods for gentle lysis and subsequent homogenization of the cells are required. Such methods must include efficient digestion of the compact yeast cell wall, gentle lysis of the resulting spheroplasts, and disruption of the cell skeleton without releasing proteases from the vacuole. The convenience and reproducibility of *in vitro* assays performed with isolated organelles also rely on easy handling and storage of the spheroplasts and the reproducibility of their preparation. Spheroplasts can be efficiently preserved and stored for weeks by freezing lyticase-treated cells over liquid N_2 vapor. Gently lysing frozen spheroplasts by the combination of thawing, osmotic shock, and homogenization in a ground-glass homogenizer results in reproducible levels of intact subcellular membrane fractions. To maintain intact organellar structures during lysis, we recommend buffers with low ionic strength, slightly acidic pH, and an adjustment to 0.7 M sorbitol after hypoosmotic homogenization.

Protocols

Preparing Frozen Spheroplast Aliquots. This step is adapted from Baker *et al.*[17]

Pretreatment buffer: 150 mM Tris–HCl, pH 9.4, 40 mM 2-mercaptoethanol, 10 mM NaN$_3$, NaF

Spheroplasting buffer: 50 mM HEPES–KOH, pH 6.8, 10 mM NaN$_3$, NaF, 1.2 M sorbitol

Storage buffer: 25 mM HEPES–KOH, pH 6.8, 50 mM potassium acetate, 0.5 M sorbitol

Harvest log-phase yeast cells at room temperature at a density of 1.5–2.0 OD$_{600}$/ml by centrifugation at 5000g for 8 min (Sorwall GS-3 rotor) in the presence of 20 mM NaF and NaN$_3$. Wash cells with distilled water, resuspend them in pretreatment buffer at 50 OD equivalents/ml, and incubate for 10 min at room

---

[15] J. H. Cho, Y. Noda, and K. Yoda, *FEBS Lett.* **469**, 151 (2000).
[16] N. J. Bryant and A. Boyd, *J. Cell Sci.* **106**, 815 (1993).
[17] D. Baker, L. Hicke, M. Rexach, M. Schleyer, and R. Schekman, *Cell* **54**, 335 (1988).

temperature (for the purpose of this chapter, we define an OD equivalent as the amount of cells present in 1 ml of culture grown to an OD_{600} of 1.0). Prepare spheroplasts by incubating 50 OD equivalents/ml cells with 500 units/ml of purified lyticase, in spheroplasting buffer, for 35 min at 30°. Follow the efficiency of spheroplast formation by the decrease of turbidity, due to the lysis of spheroplasts, after aliquots of the cell/spheroplast suspension are diluted 50-fold with distilled water. Collect spheroplasts at ≥90% spheroplasting efficiency by centrifugation at room temperature at 2500g for 5 min. Resuspend the pellet at 250 OD equivalents/ml in ice-cold storage buffer, and freeze 200-μl aliquots over N_2 vapor for 35–40 min. Samples can be stored safely at −80° for at least 2 months.

Comments. Use of dithiothreitol (DTT) for the pretreatment should be avoided because it may interfere later with Kex2 assays. We have used recombinant lyticase[18] purified from an *Escherichia coli* strain supplied by Randy Schekman (Berkeley, CA), using a procedure supplied by the same. To assay lyticase, after incubation at 30° for 30 min, 20 μl of the cell/spheroplast suspension is diluted into 1 ml distilled water, and the OD_{600} is measured. Cell suspension to which no lyticase is added serves as the control. One unit of lyticase activity is sufficient to cause a 10% decrease in OD_{600} in 30 min at 30°.[18] A commercial, purified lyticase (oxalyticase) is also available (Enzogenetics, Corvallis, OR).

High-Speed Membrane Fractions

Lysis buffer: 25 mM HEPES, pH 6.8, 50 mM potassium acetate, 0.2 M sorbitol, 1mM PMSF, 100 μM TLCK, 100 μM TPCK, 2 μM pepstatine A, 5 μM E-64

Membrane buffer: 25 mM HEPES, pH 6.8, 50 mM potassium acetate, 0.7 M sorbitol

Thaw a 200-μl frozen spheroplast aliquot at 25° for 1 min and mix with 400 μl ice-cold lysis buffer, resulting in a sorbitol concentration of 0.3 M. Combine four aliquots (2.4 ml) and homogenize them in an ice-cold ground-glass homogenizer (5 ml, Wheaton, Millville, NJ) with a series of 35 strokes. After homogenization, adjust the sorbitol concentration to 0.7 M by adding 560 μl of 2.4 M sorbitol. Spin lysates in microcentrifuge for 15 min at 15,000g at 4° and pool the clear supernatant fractions in a 2.4-ml Beckman polyallomer tube, leaving behind the substantial pellet of cell debris. Pellet slowly sedimenting membranes in a Beckman TLS55 or comparable rotor at 4° at 55,000 rpm (200,000g) for 30 min. Resuspend the pellet (HSP membranes) in 250–500 μl ice-cold membrane buffer by repeated pipetting.

Sucrose Step Gradients

Despite high osmolality and viscosity, sucrose gradients are still efficient and easily manageable for fractionating subcellular organelles. Equilibrium step

[18] S. H. Shen, P. Chretien, L. Bastien, and S. N. Slilaty, *J. Biol. Chem.* **266**, 1058 (1991).

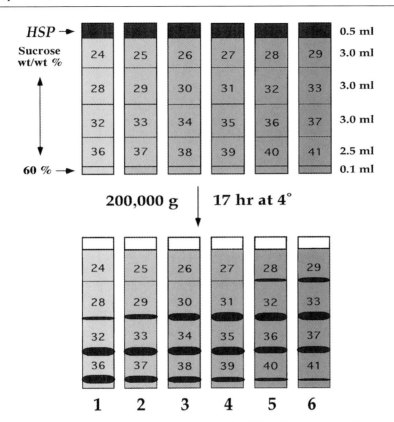

FIG. 1. Running an equilibrium sucrose density matrix. Parallel gradients are set up with volumes and sucrose concentrations as indicated. Two hundred OD equivalents of HSP membranes are loaded on the top of the gradients and centrifuged at 41,000 rpm for 17 hr at 4° using a Beckman SW 41 Ti rotor.

gradients work best because they yield high resolution with minimal dilution of markers and cross-contamination between fractions. Step gradients have additional useful features. Due to the high viscosity of sucrose solutions, step gradients maintain their steps at 4° and membrane fractions accumulate visibly at the interfaces. Because the concentrations of the sucrose steps are well preserved during centrifugation, one may design a simple but systematic set, or matrix, of step gradients that permits one to optimize steps for a particular marker protein (Fig. 1). After running gradients to equilibrium, 250-µl aliquots are selected and analyzed only from the interfaces of the sucrose steps (Fig. 2A).

Figure 2B shows the pattern of fractionation of Anp1p, a cis-Golgi marker, using the matrix of gradients illustrated in Fig. 1. Membranes containing Anp1p clearly fractionated in a consistent, density-dependent manner, allowing the density

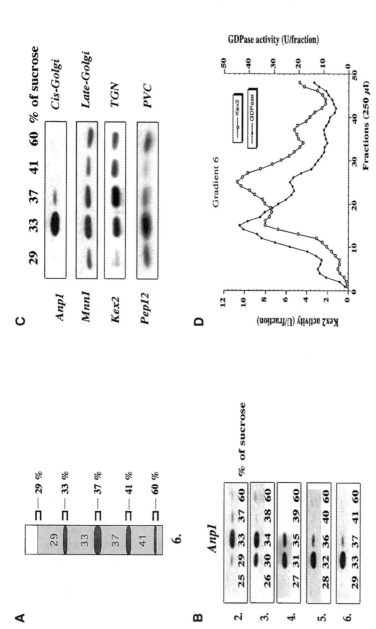

FIG. 2. Analyzing data from step gradients. Fractions (250 µl) were collected manually from the top of the gradients and transferred into a 96-well plate. (A) Samples from the interfaces were analyzed by enzyme assay and Western blotting. Samples are indicated by the concentration of sucrose on the more dense side of the interface. (B) Western blots with the *cis*-Golgi marker Anp1p. Interface samples were taken from gradients 2 through 6 of Fig. 1. (C) Western blots of various Golgi {Anp1p [R. E. Chapman and S. Munro, *EMBO J.* **13**, 4896 (1994)], Mnn1p [T. R. Graham, M. Seeger, G. S. Payne, V. L. MacKay, and S. D. Emr, *J. Cell Biol.* **127**, 667 (1994)], and Kex2p [K. Redding, C. Holcomb, and R. S. Fuller, *J. Cell Biol.* **113**, 527 (1991)]} and endosomal {Pep12p [K. A. Becherer, S. E. Rieder, S. D. Emr, and E. W. Jones, *Mol. Biol. Cell* **7**, 579 (1996)]} markers with interface samples from gradient 6. (D) Fractionation of Kex2p and GDPase on sucrose step gradient 6. Two hundred OD equivalents of HSP were loaded on the gradient. After centrifugation (protocol C), 250-µl fractions were

distribution of the yeast *cis*-Golgi to be determined to a resolution of 1% in terms of sucrose concentration. Moreover, Anp1p can be seen to fractionate predominantly in membranes with a bouyant density greater than that of 29% sucrose and less than that of 33% sucrose. Thus identification of an optimal step gradient (either gradient 2 or 6) requires only running six gradients and assaying only five fractions from each. Markers for late Golgi and endosomal membranes also exhibited consistent and reproducible patterns of fractionation on this gradient matrix (data not shown). However, each of these proteins exhibited a more complicated pattern of fractionation along a broader range of sucrose concentrations, resulting in distributions that overlapped substantially (Fig. 2C). Reasons for this behavior are discussed later.

Running an isopycnic density matrix can also provide additional qualitative information about the membranes. Membranes that are leaky, fragmented, or osmotically unstable may not fractionate at the same concentrations in the parallel gradients. Proteolysis, ionic strength, pH, and the presence of divalent cations can profoundly change the distribution profiles. Our experience is that the use of purified lyticase for spheroplasting, gentle osmotic lysis, subsequent adjustment to isoosmotic conditions with lower salt concentrations, and use of a slightly acidic pH in the lysis buffer were all crucial for keeping membranes intact. Following these parameters, markers fractionated at the expected sucrose concentrations in a consistent fashion.

Harvesting the sample aliquots at the interfaces in the smallest reasonable volumes yields membranes of the highest possible concentration and least cross-contamination. The smallest volume that consistently yields peaks for marker proteins provides the most representative sample of the interface. When GDPase and Kex2 activities were assayed in gradients sampled using 250-μl aliquots, both activities exhibited clear peaks at the interfaces (Fig. 2D). The 250-μl samples are harvested manually using a Gilson P1000 Pipetman (Rainin Instrument Co., Woburn, MA); pipette tips are cut off (to an inside diameter of 2–2.5 mm).

Protocols

Sucrose Step Gradients. Make a 60% (w/w) sucrose stock solution from ultrapure quality sucrose (SigmaUltra; Sigma, St. Louis, MO). Prepare the desired dilutions in 20 mM HEPES–NaOH, pH 6.8. Prior to making the step gradient, cool sucrose solutions on ice. Carefully overlay the sucrose steps in a 12-ml Beckman Ultra-Clear tube, mark the interfaces on the outside of the tubes, and keep the gradients at 4° prior to use. Briefly spin the resuspended HSP membranes for 15 sec at 4° at 15,000g in order to remove insoluble material that would otherwise degrade gradient performance. Then carefully load 200 OD equivalents of HSP membranes on the top of the gradient. Centrifuge gradients at 41,000 rpm for 17–18 hr at 4° using a Beckman SW 41 Ti rotor. Collect sequential 250-μl aliquots from the top of the gradient, store in 96-well plates, and mark the fractions taken at each interface. Use 10 μl of the interface aliquots for the enzyme assays and Western blots.

Comments. For unknown reasons, adjusting the pH of the sucrose solutions with sodium hydroxide rather than potassium hydroxide helped preserve Kex2 activity during centrifugation.

Immunoaffinity Isolation of Organelles Using Magnetic Beads

Integral membrane proteins of the TGN, such as Kex2p, Ste13p, and Vps10p, have been shown to maintain their steady-state localization by cycling between TGN and post-TGN/endosomal compartments.[19–22] Each of these proteins is therefore expected to be present in multiple, distinct membrane compartments. Other data suggest that the yeast Golgi itself consists of four functionally distinct compartments.[23] As a result of this complexity, the actual diversity of Golgi compartments and the physical boundaries between the Golgi and endosomal membrane system are difficult to define by gradient fractionation alone.

Late Golgi and endosomal marker proteins exhibit broad and overlapping distributions in sucrose gradients (Fig. 2C). This result raises two fundamental questions. First, in the case of a single marker protein, to what extent do membranes of different densities represent functionally distinct compartments as opposed to the broad density distribution of a single compartment? Second, in the case of two overlapping marker proteins, to what extent does the overlap represent a presence in the same membrane compartment and to what extent a presence in distinct compartments of comparable density? Both of these questions can be addressed by immunoisolation of membrane compartments followed by analysis of the coisolating markers (coimmunoisolation), which can provide important insights into compartment identities.

Immunoaffinity Isolation of Membranes and Coimmunoisolation of Marker Proteins

Immunoisolation of membranes relies on monoclonal or affinity-purified, polyclonal antibodies that recognize a specific antigen exposed on the cytosolic surface of the membrane compartment of interest. The antigen may correspond to all or part of the native cytosolic domain of a membrane protein[16] or to an epitope tag.[15,24] Antibodies are adsorbed to a solid support, which is then used to isolate the specific membranes.

---

[19] J. H. Brickner and R. S. Fuller, *J. Cell Biol.* **139**, 23 (1997).
[20] N. J. Bryant and T. H. Stevens, *J. Cell Biol.* **136**, 287 (1997).
[21] A. A. Cooper and T. H. Stevens, *J. Cell Biol.* **133**, 529 (1996).
[22] M. J. Lewis, B. J. Nichols, C. Prescianotto-Baschong, H. Riezman, and H. R. Pelham, *Mol. Biol. Cell* **11**, 23 (2000).
[23] W. T. Brigance, C. Barlowe, and T. R. Graham, *Mol. Biol. Cell* **11**, 171 (2000).
[24] P. Rehling, T. Darsow, D. J. Katzmann, and S. D. Emr, *Nature Cell Biol.* **1**, 346 (1999).

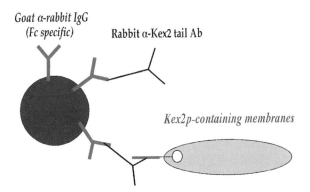

FIG. 3. Immunoisolation of Kex2p-containing organelles. For isolating Kex2p-containing membranes, M-500 subcellular Dynabeads were first coated covalently with affinity-purified goat α-rabbit IgG (Fc specific); the affinity-purified anti-Kex2 tail antibody was then bound as the primary immunoreactive coat.

As an example, we describe a system designed for the isolation of membranes containing the TGN-localized processing protease Kex2p.[25] We have made use of magnetic beads[26,27] as the solid support, both because magnetic beads allow extremely gentle isolation and washing of membranes and because membranes thus isolated can be used immediately (while still bound to the beads) for the analysis of marker protein content. Beads are first coated covalently with the secondary antibodies, affinity-purified Fc-specific goat anti-rabbit IgG. Affinity-purified rabbit anti-Kex2 antibodies, which recognize the distal part of the Kex2p cytoplasmic tail, are then adsorbed to the beads (Fig. 3). After preparing the coated beads, we confirmed that they could bind Kex2p-containing membranes by assaying Kex2 enzymatic activity. Conditions were then optimized for a quantitative isolation of Kex2p-containing membranes so that the degree of coisolation of other marker proteins would represent a reliable measurement of the fraction of those proteins present in Kex2p-containing membranes. (However, we also have found that it is not absolutely required to obtain quantitative isolation of the primary antigen, e.g., Kex2p, in order to obtain quantitative information about coisolation. Use of subsaturating levels of magnetic beads is discussed later.) It is anticipated that the use of other primary antibodies would require similar optimization of ionic strength, pH, and stoichiometry of antibody and beads to membranes. It should also be noted that optimal conditions for the coimmunoisolation of integral and peripheral membrane proteins can differ significantly.

[25] K. Redding, C. Holcomb, and R. S. Fuller, *J. Cell Biol.* **113,** 527 (1991).
[26] J. E. Gruenberg and K. E. Howell, *EMBO J.* **5,** 3091 (1986).
[27] K. E. Howell, R. Schmid, J. Ugelstad, and J. Gruenberg, *Methods Cell Biol.* **31,** 265 (1989).

Efficient isolation of Kex2p-containing membranes requires incubation of coated beads with membranes in 250 mM potassium acetate (Fig. 4A). Under these conditions, we found that the late Golgi integral membrane protein Mnn1p coisolated with Kex2p membranes, but Pma1p, a plasma membrane protein, did not (Fig. 4B). In addition, ER and vacuolar marker proteins also did not coisolate with Kex2p membranes (not shown).

Coimmunoisolation can also provide insight about the localization of peripheral membrane proteins if conditions are optimized to promote retention of the proteins on membranes. Vps35p, a protein that is part of the "retromer" complex required for endosome to Golgi retrieval of TGN membrane proteins, was shown previously to cofractionate with Kex2p in sucrose gradient fractionation.[28] Optimization of ionic strength was crucial to observe the association of Vps35p with immunoisolated Kex2p-containing membranes. As can be seen in Fig. 4C, Vps35p was coisolated with Kex2p in a salt-dependent manner, exhibiting optimal isolation at 150 mM potassium acetate, but being nearly completely stripped off at 250 mM potassium acetate.

Extracting Information from Coimmunoisolation Data

The degree of colocalization of secondary marker proteins in membranes containing the primary marker protein (i.e., the protein to which the primary antibody binds) can be determined quantitatively from a coimmunoisolation experiment once levels of the marker proteins have been calibrated. Calibration is achieved by comparing recovery of the marker protein with a calibration curve obtained by serial dilution of the starting material (i.e., HSP membranes). In the experiment shown in Fig. 4B, the amount of Mnn1p recovered on Kex2p-containing membranes corresponded to that present in 0.125 OD equivalents of cells when 20 μg of anti-Kex2p beads was used. Because the same amount of beads isolated 0.5 OD equivalents of Kex2p (not shown), then the relative coisolation for Mnn1p was 25%, meaning that 25% of the Mnn1p resided in membranes also containing Kex2p. It is important to stress that although it is possible to obtain nearly complete recovery of Kex2p ($\geq$95%) if the level of anti-Kex2p beads is titrated to sufficiently high levels (as measured both by enzymatic activity and by Western blotting), the experiment described here was performed using a subsaturating level of anti-Kex2p beads (in this case, the efficiency of recovery of Kex2p was 20%). Use of subsaturating amounts of beads both reduces nonspecific background and minimizes the use of potentially scarce reagents, namely the beads and the primary antibodies. Control experiments showed that nearly identical coisolation efficiencies were calculated for secondary marker proteins under conditions resulting in substantially different degrees of absolute recovery of Kex2p.

[28] M. N. Seaman, E. G. Marcusson, J. L. Cereghino, and S. D. Emr, *J. Cell Biol.* **137,** 79 (1997).

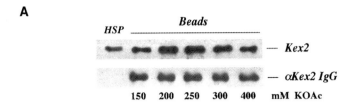

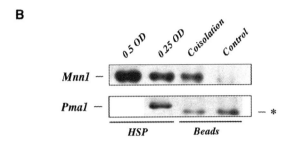

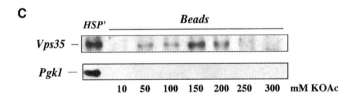

FIG. 4. Coisolation with anti-Kex2 tail antibody-coated beads. (A) Fifty OD equivalents (50 µl) of HSP membranes were coincubated with 300-µg beads in the presence of the indicated concentrations of potassium acetate. Beads were then isolated and washed. The coisolating membranes were solubilized in SDS–PAGE sample buffer and analyzed by Western blotting for Kex2p. Signal intensities were calculated using the NIH Image system. As a control for normalization, the immunoglobulin G (IgG) heavy chain, which also eluted from beads in sample buffer, was detected as a 55-kDa species on the same Western blots. HSP represents 1 OD equivalent of membranes. (B) Twenty-five OD equivalents of HSP membranes were incubated with 200-µg beads in 25 µl buffer for 3 hr. Immunoisolated membranes from 20-µg beads or an identical amount from a control incubation lacking HSP membranes were analyzed by Western blotting for Mnn1p and Pma1p [T. R. Graham, M. Seeger, G. S. Payne, V. L. MacKay, and S. D. Emr, *J. Cell Biol.* **127,** 667 (1994); A. Chang and C. W. Slayman, *J. Cell Biol.* **115,** 289 (1991)]. Known OD equivalents of HSP membranes were included for standardization. Bands indicated with an asterisk (∗) were artifacts observed in the absence of HSP membranes that migrated as 92- to 94-kDa species. (C) Twenty-five OD equivalents of HSP membranes were resuspended in cytosol instead of membrane buffer, yielding HSP' membranes. HSP' membranes were then incubated at 4° with 200-µg beads in the presence of the indicated concentrations of potassium acetate. After 3 hr of incubation, beads were isolated, and the coisolating membranes were analyzed by Western blotting for Vps35p [M. N. Seaman, E. G. Marcusson, J. L. Cereghino, and S. D. Emr, *J. Cell Biol.* **137,** 79 (1997)] and Pgk1p (antibody from Molecular Probes, Eugene, OK).

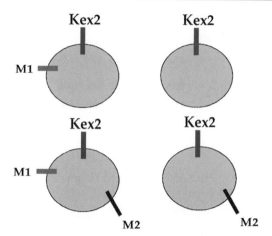

FIG. 5. Theoretical cross-coisolation with Kex2p-containing membranes. When two markers (M1 and M2) are coisolated with Kex2p, there are four possible combinations of these markers within the Kex2p-containing membranes. Isolating membranes using antibodies that recognize M1 and M2 ("cross-coisolation") can be used to evaluate the distribution of these proteins.

The fact that a secondary marker protein coimmunoisolates with the primary marker protein indicates that the proteins colocalize to some degree. If coisolation of the secondary protein is less than 100% efficient, then one may conclude that membranes exist that contain the secondary protein but not the primary protein. For example, the experiment described earlier suggests that 75% of the Mnn1p is present in membranes lacking Kex2p. In order to determine what fraction of the Kex2p is present in membranes that contain Mnn1p, the reciprocal experiment using anti-Mnn1p beads would have to be performed.

If there are several coisolating markers, further "cross-coisolation" experiments may help reveal the degree of compartmental heterogeneity. A simple theoretical model is represented in Fig. 5, where two markers, M1 and M2, coisolated with Kex2p. If anti-M1 and anti-M2 beads both isolate a fraction of Kex2p, but fail to reciprocally coisolate (i.e., anti-M1 beads fail to isolate M2 and anti-M2 beads fail to isolate M1), then one can conclude that at least two compartments exist (Kex2p + M1 and Kex2p + M2; membranes containing only Kex2p, M1, or M2 may also exist). If M1 and M2 do coimmunoisolate reciprocally, then both may share the same compartment with Kex2p. Additional information may be obtained from sequential immunoisolation from one HSP sample using beads charged with different primary antibodies. Systematic pairwise analysis of the other coisolating markers can, in principle, identify precisely the distributions of multiple markers in a system of multiple membrane types.

In the analysis described earlier, one examines the extent to which a secondary marker coisolates with the primary marker from total HSP membranes. A more powerful application is immunoisolation from each interface of a step gradient. From a single experiment, the degree of coisolation of several secondary markers can be determined using enzymatic assays and Western blotting for each interface fraction, yielding a comprehensive colocalization profile.

Protocols

Preparation of Coated Beads. To prepare beads for immunoisolation (Dynabeads M-500 Subcellular; Dynal, Lake Success, NY), follow the manufacturer's instructions carefully. For isolating Kex2p-containing membranes, magnetic beads are first coated covalently with affinity-purified goat anti-rabbit (IgG/Fc fragment specific; Jackson ImmunoResearch, West Grove, PA) and then affinity-purified rabbit anti-Kex2p tail antibodies are adsorbed.[25] Anti-Kex2-coated beads are stable for ∼6 months at 4° using the storage buffer recommended by the manufacturer.

Isolation of Kex2p-Containing Membranes from HSP

Coisolation buffer: 25 mM HEPES, pH 6.8, 200 mM potassium acetate, 0.8 M sorbitol, 5% fetal calf serum, 1 mM EDTA
Washing buffer: 25 mM HEPES, pH 6.8, 200 mM potassium acetate, 0.8 M sorbitol

Prior to immunoisolation, 200-μg-coated beads are washed and equilibrated three times with 400 μl of coisolation buffer. The beads are then mixed with 25 OD equivalents of the HSP membranes resuspended in 25 μl of coisolation buffer. The suspension is incubated for 3 hr at 4°, mixed by pipetting every half hour. The beads are removed magnetically and washed three times at 4° with 400 μl of washing buffer for 5 min each time. Twenty micrograms of beads are used for enzyme assays and Western blots.

Comments. For isolating membranes from sucrose gradient fractions, we recommend using concentrations of sorbitol in the wash buffers that are equimolar to the sucrose concentration in the individual gradient fractions.

GDPase Assay. The enzymatic activity of the integral membrane protein GDPase provides an easily assayed marker that is thought to be localized to early, medial, and late Golgi vesicles.[23,29] The following GDPase protocol is based on the previously described assay; the resulting inorganic phosphate is determined by the methods of Yanagisawa *et al.*[30] and Ames.[31]

---

[29] J. J. Vowels and G. S. Payne, *Mol. Biol. Cell* **9**, 1351 (1998).
[30] K. Yanagisawa, D. Resnick, C. Abeijon, P. W. Robbins, and C. B. Hirschberg, *J. Biol. Chem.* **265**, 19351 (1990).
[31] B. N. Ames, *Methods Enzymol.* **8**, 115 (1966).

The GDPase reaction is initiated by the addition of 10 µl sample (HSP membranes or gradient fraction) to 90 µl of reaction mixture (200 mM imidazole, pH 7.6; 10 mM CaCl$_2$, 0.1% Triton X-100, 2 mM GDP). The reaction is incubated for 20 min at 30° and is stopped by the addition of 200 µl 2% sodium dodecyl sulfate (SDS). ADP (2 mM) must be used to control for nonspecific phosphatase.

For measuring the liberated inorganic phosphate, add 700 µl Ames reagent to the terminated GDPase reaction and incubate for 30 min at 45°. Spin samples for 30 sec in a microcentrifuge at room temperature at 15,000 rpm; measure the absorbance of the supernatant fraction at 660 nm.

Ames reagent: Mix 1 volume of 10% freshly prepared ascorbic acid with 6 volumes of Ames solution (4.2 g ammonium molybdate and 28.6 ml H$_2$SO$_4$ in 1 liter). Solutions containing 10–150 nmol inorganic phosphate are used for calibration. One unit of GDPase produces 1 nmol P$_i$ /min.

Kex2 Assay. Mix 10 µl of sample with 50 µl of reaction mixture [200 mM Bis–Tris, pH 7.25, 1 mM CaCl$_2$, 1% Triton X-100, 5 mM o-phenanthroline, 0.1 mM Boc-Leu-Lys-Arg-AMC (Bachem, Torrance, CA)] and incubate for 15 min at 37°. Stop reactions by adding 800 µl of 125 mM ZnSO$_4$ and 100 µl of saturated BaOH. Vortex well and spin for 30 sec in a microcentrifuge at 15,000 rpm. Measure fluorescence of the supernatant fraction (λ_{ex} = 385 nm, λ_{em} = 465 nm). Use AMC solutions for calibration and determine 1 unit of Kex2 as 1 pmol AMC liberated in 1 min. *o*-Phenanthroline is essential to inhibit an unidentified proteolytic activity that pellets in the HSP and fractionates at a bouyant density equivalent to that of ~29 to 32% sucrose in gradients.

Perspectives

In the case of the Golgi/endosomal system, fractionation and coimmunoisolation of marker proteins can, in principle, provide a comprehensive picture of the distribution of marker proteins in distinguishable types of membranes, yielding information about both compartmental identity and the mechanisms of localization of proteins to compartments. At a practical level, the localization of a novel protein can be assigned rapidly. Uncertainties arise from these methods because of their inherent statistical nature and because it is generally not known how many antigen molecules are required to affect immunoisolation. Colocalization of proteins can also be assessed at an ultrastructural level of resolution using immunogold electron microscopy. However, fractionation and immunoisolation methods remain extremely useful both because they represent simple, easily accessible technologies and because they permit the analysis of multiple marker proteins in a single experiment.

Acknowledgments

We thank Amy Chang, Scott Emr, Todd Graham, Stephan te Heesen, Dan Klionsky, and Sean Munro for antibodies, Randy Schekman for the lyticase-producing strain, and Jennifer Blanchette, Tomoko Komiyama, E. J. Brace, and other members of the Fuller laboratory for critically reading the manuscript. This work was supported by NIH Grant GM50915 to R.S.F.

[21] Visualization and Purification of Yeast Peroxisomes

By RALF ERDMANN and STEPHEN J. GOULD

Introduction

Much of what we currently know about peroxisome morphology, chemical properties, metabolic roles, and biogenesis has come from yeast studies. These in turn have relied on technical developments throughout our field. This chapter presents some of the many protocols that can be used to detect and purify yeast peroxisomes, as well as for assessing peroxisomal matrix protein import in fixed and living cells. Although many yeast species have contributed greatly to our understanding of peroxisome biology, the protocols presented here are for *Saccharomyces cerevisiae*. Their application to other yeasts may require modification.

Basics of Peroxisome Biogenesis

Peroxisomes are single membrane-bound organelles that are present in virtually all eukaryotes.[1] Peroxisomes lack nucleic acids and all peroxisomal proteins are encoded in the nuclear genome. Peroxisomal proteins, both of the membrane and of the matrix, or lumen, are synthesized in the cytoplasm and are imported posttranslationally. Peroxisomal matrix enzymes contain di⁻... targeting signals that are sufficient to direct proteins into the peroxisome lumen. The type 1 peroxisomal targeting signal (PTS1) is a C-terminal tripeptide of the sequence serine-lysine-leucine or a conservative variant and is found on nearly all peroxisomal enzymes.[2] The PTS2 is a distinct type of targeting signal found on only a small number of enzymes (just one in *S. cerevisiae*, thiolase), is located near the N terminus,

[1] K. A. Sacksteder and S. J. Gould, *Annu. Rev. Genet.* **34**, 623 (2000).
[2] S. J. Gould, G. A. Keller, N. Hosken, J. Wilkinson, and S. Subramani, *J. Cell Biol.* **108**, 1657 (1989).

and has the sequence RLX_5HL, or a conservative variant.[3,4] A large number of PEX genes encode the peroxins that are required for assembling peroxisome membranes, import peroxisomal membrane proteins, and importing peroxisomal matrix proteins.[1,5] Additional information about peroxisome function and biogenesis can be obtained from the peroxisome website (www.peroxisome.org).

Variations in Size and Abundance of Peroxisomes and Expression of Peroxisomal Proteins

In all experiments with yeast peroxisomes it is important to consider that the abundance and the size of the organelle vary depending on the growth conditions. For example, yeast grown in high concentrations of glucose typically have just 1–3 small peroxisomes per cell whereas yeast that are grown in oleic acid medium may have >10 large peroxisomes per cell.[6–8] The increase in peroxisome size and abundance during growth on oleic acid reflects the essential role of yeast peroxisomes in fatty acid β-oxidation and is coordinated with the transcriptional activation of numerous peroxisomal enzymes and membrane proteins. A less pronounced increase in peroxisome abundance and size occurs when cells are grown on nonfermentable carbon sources[6–8] and when mitochondrial function is disrupted.[9] These environment-induced increases in peroxisome size and abundance are typically referred to as the peroxisome proliferation response. If an investigator wishes to couple the expression of a heterologous peroxisomal protein to the peroxisome proliferation process, we recommend expressing the gene of interest from a promoter of a fatty acid β-oxidation gene. Promoters of the acyl-CoA oxidase[10] (POX1), hydratase/dehydrogenase (FOX2), and thiolase[11] (POT1) genes are all capable of conferring oleate-induced expression on heterologous genes. For example, when the 500 bp upstream of these open reading frames (ORFs) were placed upstream of the luciferase cDNA on a centromeric plasmid, luciferase expression was induced >10 times during growth on oleic acid as compared to growth on ethanol (Table I). Repression effects during growth on glucose yield even lower expression from these promoters (S. J. Gould, unpublished observations, 2002).

---

[3] B. W. Swinkels, S. J. Gould, A. G. Bodnar, R. A. Rachubinski, and S. Subramani, *EMBO J.* **10**, 3244 (1991).
[4] S. Subramani, *Annu. Rev. Cell Biol.* **9**, 445 (1993).
[5] E. H. Hettema, W. Girzalsky, M. van Den Berg, R. Erdmann, and B. Distel, *EMBO J.* **19**, 223 (2000).
[6] M. Veenhuis, M. Mateblowski, W. H. Kunau, and W. Harder, *Yeast* **3**, 77 (1987).
[7] R. Erdmann and G. Blobel, *J. Cell Biol.* **128**, 509 (1995).
[8] P. Marshall, Y. Krimkevich, R. Lark, J. Dyer, M. Veenhuis, and J. M. Goodman, *J. Cell Biol.* **129**, 345 (1995).
[9] C. B. Epstein, J. A. Waddle, W. T. Hale, V. Dave, J. Thornton, T. L. Macatee, H. R. Garner, and R. A. Butow, *Mol. Biol. Cell* **12**, 297 (2001).
[10] T. Wang, Y. Luo, and G. M. Small, *J. Biol. Chem.* **269**, 24480 (1994).
[11] A. W. Einerhard, T. W. Voorn Brouwer, R. Erdmann, W. H. Kunau, and H. F. Tabak, *Eur. J. Biochem.* **200**, 113 (1991).

TABLE I
COMPARISON OF DIFFERENT YEAST PROMOTERS DURING CELL GROWTH
IN OLEIC ACID AND ETHANOL MEDIUM[a]

Promoter	Relative luciferase activity (light units/50 cells)	
	Ethanol	Oleic acid
None	0.2 ± 0	0.4 ± 0.2
CDC28	0.5 ± 0.2	0.6 ± 0.1
PGK1	3.4 ± 1.3	3.8 ± 0.6
POX1	0.5 ± 0.2	23 ± 3.4
FOX2	1.2 ± 0.6	19 ± 0.9
POT1	3.1 ± 0.7	21 ± 5.3

[a] In each plasmid, the luciferase cDNA is located between BamHI and NaeI sites of the Cen/Ars, HIS3 shuttle plasmid pRS313.[29] The promoters are inserted upstream of the luciferase cDNA, between SaII and BamHI sites. Luciferase assays of BY4733 cells[19] carrying each of these plasmids were performed using a rapid *in vivo* assay (see the protocol at the end of this article) following 16 hr growth in either SYOLT (0.67% yeast nitrogen base, 0.1% yeast extract, 0.2% oleic acid, and 0.02% Tween 40) or SYE (0.67% yeast nitrogen base, 0.1% yeast extract, 0.5% ethanol) medium. Results are the average ± the standard deviation from three independent trials. Each promoter consists of ~500 bp of sequence upstream of the start codon of the relevant open reading frame.

Visualization of Yeast Peroxisomes

The visualization of yeast peroxisomes in wild-type strains of *S. cerevisiae* can be accomplished by numerous techniques. These include *in vivo* microscopy of strains expressing fluorescent peroxisomal proteins, typically green fluorescent protein (GFP) and its spectral variants, as well as the analysis of fixed cells by fluorescence microscopy, immunofluorescence microscopy, and immunoelectron microscopy.

Detection of Peroxisomes and Peroxisomal Matrix Protein Import in Vivo

One limitation to studies of peroxisome dynamics and biogenesis is the lack of a membrane-permeable fluorescent dye that is specific for the peroxisomal compartment. The *in vivo* detection of peroxisomes in yeast requires the use of fluorescent peroxisomal proteins. Green fluorescent protein and its spectral variants can be targeted efficiently into the lumen of the peroxisome by addition of a peroxisomal targeting signal to the appropriate terminus of the protein. For example, a GFP–PTS1 fusion that contains the amino acids serine-lysine-leucine at the extreme C terminus of the protein is targeted to peroxisomes.[12] Similarly, addition of a

---

[12] J. E. Kalish, G. A. Keller, J. C. Morrell, S. J. Mihalik, B. Smith, J. M. Cregg, and S. J. Gould, *EMBO J.* **15**, 3275 (1996).

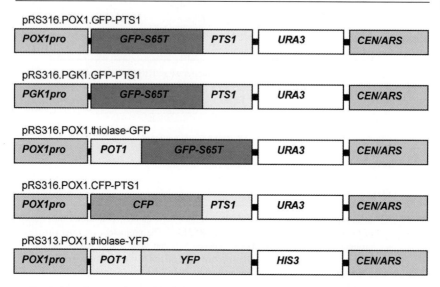

FIG. 1. Line diagram of plasmids designed to express peroxisomal forms of GFP and some of its color variants.

PTS2 to the N terminus of GFP and its variants will direct them to peroxisomes of wild-type strains.[12] Some fusion proteins in which GFP is appended to full-length yeast peroxisomal proteins may be targeted to peroxisomes with even greater apparent efficiency, such as GFP–LYS1 and GFP–CIT2 fusions.[13] However, the unpredictable nature of fusion protein folding and solubility makes it impossible to assume that all fusions between GFP and peroxisomal enzymes will be targeted to peroxisomes. Peroxisomal targeting of fusions between GFP and integral peroxisomal membrane proteins has also been reported,[12] and a PTS2–DsRed fusion protein has also been found to target properly to peroxisomes (R. Erdmann, unpublished observations, 2002). A few plasmids that express peroxisomal forms of GFP from yeast shuttle vectors (Fig. 1) are available on request from the authors.

To detect peroxisomes *in vivo,* the appropriate GFP expression vector is introduced into the strain of interest using standard yeast transformation techniques. Following selection for plasmid-containing strains, the cells can be visualized following selective growth of the strain in the appropriate medium.

1. Preculture the strain in 10 ml rich (YPD) or minimal (SC) medium at 30° to midlogarithmic phase. For oleic acid induction, harvest cells and grow for an

---

[13] M. T. Geraghty, D. Bassett, J. C. Morrell, G. J. Gatto, J. Bai, B. V. Geisbrecht, P. Hieter, and S. J. Gould, *Proc. Natl. Acad. Sci. U.S.A.* **96,** 2937 (1999).

additional 8–12 hr in either YPOLT (1% yeast extract, 2% Bacto-peptone, 0.2% oleic acid, 0.02% Tween 40) or YNO medium [0.67% yeast nitrogen base (Difco, Detroit, MI), 0.1% yeast extract, 0.1% oleic acid, 0.015% Tween 40].

2. Harvest cells by centrifugation at 4200g for 5 min at room temperature and wash twice with 10 ml of sterile water. Resuspend in 100 μl of sterile water.

3. Prepare poly (L-lysine)-coated cover glasses by spreading 100 μl of 1 mg/ml poly (L-lysine) over the entire surface of a cover glass, removing the excess, and allowing 10 min for the residual solution to dry. Wash twice with sterile water. Cells seem to adhere best if added shortly thereafter (within ~30 min).

4. Add a concentrated amount of cells to the cover glass. Allow 10 min for the cells to adhere. Wash once by dipping the cover glass in sterile water.

5. Remove excess fluid from the cover glass (e.g., by touching its edge to tissue paper) and place cells down over a droplet of 10 μl mounting solution [90% glycerol (v/v), 100 mM Tris–HCl, pH 8.5] on a glass slide. Anchor the cover glass to the slide using transparent nail polish or an equivalent adhesive/sealant.

6. Observe by fluorescence microscopy.

The distribution of peroxisomal GFPs can also be assessed in fixed cells using a modification of the same protocol. Simply incubate the cells in 5 volumes of 4% p-formaldehyde in phosphate-buffered saline (PBS) for 1 hr after step 1 and wash them twice in ~3 volumes of sterile water prior to adding them to the poly (L-lysine)-coated cover glasses.

Although GFP-labeled peroxisomes can be detected quite easily using a standard fluorescence microscope, superior images are typically obtained through deconvolution processing of image data or by using a confocal microscope for image acquisition. It should also be noted that fluorescent peroxisomal proteins that are targeted to peroxisomes via PTS1 or PTS2 targeting signals can be used to assess these two import pathways in mutant cells or in cells that have been exposed to potentially inhibitory compounds. This is evident from comparing the subcellular distribution of GFP–PTS1 in a wild-type strain (BY4733) as compared to a peroxisome biogenesis mutant, in this case a *pex12* disruption mutant of the same strain (BY4733, *pex12*Δ::*HIS3*) (Fig. 2). Assessing the integrity of the peroxisomal membrane protein import pathway solely through the use of GFP fusions to integral peroxisomal membrane proteins is not recommended, as a punctate distribution of such proteins may reflect mistargeting to other cellular organelles rather than correct targeting to peroxisomal membranes.[5] In fact, it is probably prudent to avoid using fusions between GFP and integral peroxisomal membrane proteins for the detection of peroxisome membranes even in wild-type cells. Several investigators have reported that overexpression of peroxisomal membrane proteins, particularly mutant membrane proteins such as fusion proteins, may have unpredictable effects on peroxisome morphology

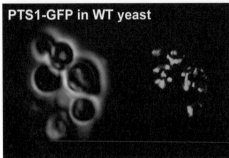 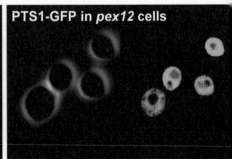

FIG. 2. Phase and fluorescence microscopy of GFP–PTS1 in wild-type [BY4733 (C. Baker-Brachmann, A. Davies, G. J. Cost, E. Caputo, J. Li, P. Hieter, and S. J. Gould, *Yeast* **14,** 115 (1998)] cells (left) and a *pex12Δ::HIS3* derivative of BY4733 (right).

and/or abundance and may even be misdirected to other organelles such as the endoplasmic reticulum.[14–16]

Detection of Peroxisomes by Immunofluorescence Microscopy

Peroxisomes can also be detected quite easily by immunofluorescence microscopy. Perhaps the most reliable reagent for peroxisome detection are antibodies to the peroxisomal β-oxidation enzyme thiolase.[17,18] Although thiolase expression is induced significantly during growth on oleic acid, it can also be detected quite easily during growth on glucose or galactose (unpublished observations, 2001), making it a useful marker for the immunodetection of peroxisomes under virtually all growth conditions. Our laboratories have generated antibodies to native or recombinant forms of *S. cerevisiae* thiolase and in virtually all cases the antibodies work well for immunodetection of peroxisomes (unpublished observations, 2001). Successful detection of peroxisomes by immunofluorescence microscopy has also been reported for epitope-tagged forms of peroxisomal proteins[5] and for proteins such as GFP that have been directed to peroxisomes via the addition of a PTS1 or PTS2 (S. South, E. Baumgart, and S. J. Gould, unpublished observations, 2001).

1. Preculture the strain in rich (YPD) or minimal (SD–0.3% dextrose) medium at 30° to midlogarithmic phase. For oleic acid induction, harvest cells and grow for an additional 8–12 hr in YNO medium containing 0.67% yeast nitrogen base, 0.1% yeast extract, 0.1% oleic acid, and 0.015% Tween 40.

[14] A. K. Stroobants, E. H. Hettema, M. van den Berg, and H. F. Tabak, *FEBS Lett.* **453,** 210 (1999).
[15] K. A. Sacksteder, J. M. Jones, S. T. South, X. Li, Y. Liu, and S. J. Gould, *J. Cell Biol.* **148,** 931 (2000).
[16] S. T. South, K. A. Sacksteder, X. Li, Y. Liu, and S. J. Gould, *J. Cell Biol.* **149,** 1345 (2000).
[17] R. Erdmann and W.-H. Kunau, *Yeast* **10,** 1173 (1994).
[18] I. van Der Leij, M. van Den Berg, R. Boot, M. Franse, B. Distel, and H. F. Tabak, *J. Cell Biol.* **119,** 153 (1992).

2. Harvest cells by centrifugation at 4200g for 5 min at room temperature and wash twice with 10 ml of sterile water. Resuspend in 100 µl of sterile water.

3. Induced cells (10 mg wet weight) are then incubated for 10 min in 100 mM Tris–HCl (pH 9.4) containing 10 mM dithiothreitol for 20 min at room temperature.

4. Harvest cells by centrifugation at 4200g for 5 min at room temperature and wash twice with 10 ml of sorbitol buffer [20 mM potassium phosphate (pH 7.4) 1.2 M sorbitol].

5. Harvest cells by centrifugation at 4200g for 5 min and resuspend in 200 µl sorbitol buffer containing 50 mg/ml Zymolyase 100T (ICN, Costa Mesa, CA). Incubate for 1 hr at 30° with occasional shaking. Cell wall digestion is monitored by phase-contrast microscopy. When about 80% of the cells burst upon the addition of distilled water, the reaction is considered complete.

6. Harvest cells by centrifugation in a Eppendorf microfuge (2000 rpm for 30 sec) and gently wash the cells three times with 1 ml sorbitol buffer.

7. Discard the final supernatant and gently resuspend the cells in 200 µl swelling buffer [20 mM MES (pH 6.0) 150 mM potassium acetate, 5 mM magnesium acetate, 750 mM sorbitol) and incubate for 20 min at room temperature.

8. Apply cells to poly (L-lysine)-coated cover glasses for 5 min at room temperature. Carefully remove excess liquid.

9. Fix and permeabilize cells by immersion in methanol for 5 min at room temperature, followed by immersion in acetone for 5 min at room temperature.

10. Absorb excess acetone by dabbing the edge of the cover glass on Whatman (Clifton, NJ) 3M paper (samples may be stored at $-20°$ at this point, although best results are obtained if the protocol is completed in 1 day).

11. Incubate the fixed cells with blocking solution (2% nonfat dry milk, 0,1% Tween 20 in 1× PBS [100 mM potassium phosphate (pH 7.4), 0.9% NaCl] for 15 min at room temperature. Repeat.

12. Incubate with primary antibody diluted into blocking buffer in a chamber humidified with blocking buffer for 2 hr at room temperature or overnight at 4°. Rabbit antithiolase serum may be used at a dilution of 1 : 3000. For double or triple labeling experiments, all primary antibodies should be mixed for a single incubation.

13. Wash six times with blocking buffer.

14. Incubate with fluorescently labeled secondary antibodies at 3 µg/ml each, again for 2 hr at room temperature or overnight at 4°.

15. Wash six times with blocking buffer followed by a final wash in 1% bovine serum albumin (BSA) in PBS.

16. Remove excess fluid from the cover glass (e.g., by touching its edge to tissue paper) and place cells down over a droplet of 10 µl mounting solution [90% glycerol (v/v) 100 mM Tris–HCl, pH 8.5] on a glass slide. Anchor the cover glass to the slide using transparent nail polish or an equivalent adhesive/sealant.

17. Observe by fluorescence microscopy.

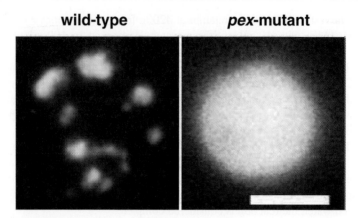

FIG. 3. Subcellular distribution of peroxisomal thiolase in wild-type (left) and *pex3* (right) cells, as determined by labeling the cells with antithiolase antibodies and fluorescein isothiocyanate-conjugated donkey anti-rabbit IgG. Each panel shows a single cell. Bar: 5 μm.

Antithiolase immunofluorescence can also be used to rapidly distinguish between strains that are capable of importing PTS2-targeted peroxisomal matrix proteins and those that are not, such as the *pex3* mutant (Fig. 3).

Detection of Peroxisomes by Immunoelectron Microscopy

The same properties that make antithiolase antibodies and combinations of antitag antibodies and tagged proteins so useful for immunofluorescence microscopy also make them useful for the immunoelectron microscopic detection of peroxisomes.

1. Grow cells in rich or selective medium at 30° to an OD_{600} of 1.0. Dilute the culture back to an OD_{600} of 0.2–0.3 in 20 ml of media and culture until the cells reach an OD_{600} of ~0.8.

2. Harvest the cells by centrifugation at 4200g for 5 min at room temperature and wash twice with 10 ml 100 mM Tris–HCl, pH 7.5, 50 mM EDTA, and 10 mM 2-mercaptoethanol.

3. Wash the cells once in 1 ml of 20 mM potassium phosphate, pH 7.5, 1.2 M sorbitol, and resuspend in 1 ml of 20 mM potassium phosphate, pH 7.5, 1.2 M sorbitol containing Zymolyase 20T (ICN Biomedicals) at a final concentration of 8 μg/ml. Incubate for 30 min at 30°. The conversion of cells to spheroplasts in this step is thought to improve the efficiency with which the cells can be embedded in the resin.

4. Gently harvest the spheroplasts by centrifugation at 2400g for 5 min at room temperature and resuspend in 1 ml of fixative (4% depolymerized formaldehyde, 0.05% glutaraldehyde in 20 mM potassium phosphate, pH 7.5, 1.2 M sorbitol). Incubate on ice for 1 hr.

5. Fixed cells are then dehydrated by washing three times with 1 ml of 70% (v/v) ethanol, three times with 1 ml of 80% ethanol, three times with 1 ml of 90% ethanol, and three times with 1 ml of 100% ethanol. Cells are incubated for 10 min in each wash, on ice, and are recovered by centrifugation for 5 min at 3000g at room temperature between each wash.

6. After the final wash, resuspend the cells in 50% ethanol, 50% Unicryl (Electron Microscopy Sciences, Fort Washington, PA) and incubate overnight at 4°.

7. Harvest the cells by centrifugation at 3000g for 5 min at room temperature, resuspend the cells in 100% Unicryl, and incubate for 24 hr at 4°.

8. Repeat step 7.

9. Harvest the cells by centrifugation at 3000g for 5 min at room temperature, resuspend the cells in 100% Unicryl, and polymerize the resin by incubating at 45° for 24 hr followed by incubation at 55° for an additional 24 hr.

10. Cut ultrathin sections using a diamond knife and capture the sections on Formvar-coated copper grids and process for immunolabeling.

11. Block the sections by placing the grids on a drop of 20 mM Tris–HCl, pH 7.5, 150 mM NaCl, and 4% (w/v) BSA for 1 hr at room temperature. Repeat.

12. Incubate the sections with primary antibody overnight at room temperature in a chamber humidified with TBSA (20 mM Tris–HCl, pH 7.5, 150 mM NaCl, 0.1% BSA). Dilute the primary antibody in TBSA and place a drop on each grid. A range of antibody dilutions need to be tested, typically around a range of 10 μg/ml for affinity-purified antibodies.

13. Wash the sections 12 times with 1 ml of TBSA.

14. Incubate the sections with 18 nm of conjugated gold donkey anti-rabbit antibodies (Jackson ImmunoResearch, West Grove, PA) for 1 hr in a humidified chamber.

15. Wash the sections eight times with TBSA and then four times with TBS (20 mM Tris–HCl, pH 7.5, 150 mM NaCl). Fix 10 min in 1% glutaraldehyde in TBS at room temperature. Wash the sections three times in TBS and then rinse thoroughly with distilled H_2O. Dry by gentle blotting on Whatman 3M paper.

16. To enhance contrast, incubate the sections for 30 sec in 2% (w/v) uranyl acetate. Rinse thoroughly with distilled H_2O and then incubate for 20 sec in 1% (w/v) lead citrate. Rinse thoroughly with distilled H_2O, dry by evaporation on Whatman 3M paper, and examine the grids in an electron microscope.

A representative image of peroxisome detection by this immunogold technique is shown in Fig. 4 for a variant of BY4733[19] that was grown in galactose medium, was expressing GFP–PTS1 from the PGK1 promoter, and was labeled with anti-GFP antibodies. It should be noted that other investigators may prefer to embed their cells in resins other than Unicryl, also with outstanding

[19] C. Baker-Brachmann, A. Davies, G. J. Cost, E. Caputo, J. Li, P. Hieter, and J. D. Boeke, *Yeast* **14**, 115 (1998).

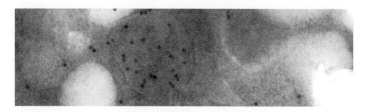

FIG. 4. Detection of peroxisomes by immunoelectron microscopy. A derivative of BY4733 cells expressing GFP–PTS1 from pRS316.PGK1.GFP-PTS1 was processed for immunoelectron microscopy using polyclonal rabbit antibodies to GFP (Abcam) and 18-nm gold-conjugated donkey anti-rabbit antibodies (Jackson ImmunoResearch). Height of image is 200 nm.

results.[20,21] Cryoimmunoelectron microscopy techniques have also been used for the immunodetection of peroxisomes and peroxisomal proteins in yeast,[18,22,23] providing the investigator with a wide range of options in visualizing peroxisomes by immunoelectron microscopy.

Purification of Peroxisomes

As with all subcellular organelles, peroxisomes can be purified to varying degrees from postnuclear supernatants of homogenized spheroplasts. It is not possible to purify peroxisomes by differential centrifugation alone, but the combination of differential centrifugation and Nycodenz or sucrose density gradient ultracentrifugation can yield highly purified peroxisomes. The following protocols have been used with BY4733 cells[19] and should therefore be applicable for virtually all mutants generated by the yeast gene disruption consortium.[24] The yield of peroxisomes varies considerably depending on the growth conditions of the cell, which may affect both the efficiency of cell homogenization and the abundance and size of peroxisomes. For routine purposes, cells are precultured in rich medium such as YPD and are induced to proliferate peroxisomes by incubation in oleic acid-containing medium prior to harvesting for organelle purification. To aid in identifying control proteins in these experiments, we have provided a list of known peroxisomal proteins of the yeast *S. cerevisiae* and their behavior in the following protocols (Table II).[6,21,27,28,30,31,32,33,34]

[20] A. C. Douma, M. Veenhuis, W. de Koning, and W. Harder, *Arch. Microbiol.* **143,** 237 (1985).
[21] J. Höhfeld, M. Veenhuis, and W. H. Kunau, *J. Cell Biol.* **114,** 1167 (1991).
[22] S. J. Gould, G. A. Keller, M. Schneider, S. H. Howell, L. J. Garrard, J. M. Goodman, B. Distel, H. Tabak, and S. Subramani, *EMBO J.* **9,** 85 (1990).
[23] Y. Elgersma, A. Vos, M. van den Berg, C. W. T. van Roermund, P. van der Sluijs, B. Distel, and H. Tabak, *J. Biol. Chem.* **271,** 26375 (1996).
[24] E. A. Winzeler, D. D. Shoemaker, A. Astromoff, H. Liang, K. Anderson, B. Andre, R. Bangham, R. Benito, J. D. Boeke, H. Bussey *et al., Science* **285,** 901 (1999).

TABLE II
PEROXISOMAL PROTEINS OF *S. cerevisiae* USED AS MARKERS

Gene	Protein	Function	Distribution	Usefulness as marker	Refs.
CTA1	Catalase	H_2O_2 elimination	Matrix	Good matrix marker. Best when used with anti-Ctal p antibodies. Catalase is the classic enzyme marker of peroxisomes in most species, but wild-type *S. cerevisiae* cells contain a cytosolic catalase (CTT1). Nevertheless, its the most commonly used enzyme marker.	6, 30
POX1	Acyl-CoA oxidase	First step in β-oxidation pathway	Matrix; some peripherally bound to inner face of membrane	Good matrix marker. Unique enzymatic marker of peroxisomes.	7
FOX2	Enoyl-CoA hydratase/ 3-hydroxyacyl-CoA dehydrogenase	Second and third steps in β-oxidation pathway	Matrix; some peripherally bound to inner face of membrane	Good matrix marker. Unique enzymatic marker of peroxisomes.	31
POT1	Thiolase (PTS2)	Fourth and final step in β-oxidation pathway	Matrix	Great matrix marker. Anti thiolase antibodies are commonly used for both IF and IEM microscopy, as well as for blotting, and it is a unique enzymatic marker.	17, 30
ECI1	Enoyl-CoA isomerase	Auxiliary enzyme of β-oxidation pathway	Matrix	Good matrix marker. Anti-Ecilp antibodies can be used in blotting experiments and it is a unique enzymatic marker.	27, 32
PEX4	Peroxisome-associated ubiquitin-conjugating enzyme	Matrix protein import	Peripheral outer peroxisomal membrane protein	Best peripheral membrane protein marker. Detect with antibodies.	33
PXA1	ABC transporter	Transporter of unknown substrates	Integral peroxisomal membrane protein	Good integral peroxisomal membrane protein marker. Detect with antibodies.	28, 34
PEX3	Required for peroxisome membrane synthesis	Unknown	Integral peroxisomal membrane protein	Good integral peroxisomal membrane protein marker. Detect with antibodies.	21

Enriching for Peroxisomes and Mitochondria by Differential Centrifugation

The purification of yeast peroxisomes first involves the generation of a postnuclear supernatant and the generation of an organelle pellet by differential centrifugation that is enriched for peroxisomes and mitochondria.

1. Inoculate 30 ml YPD (1% yeast extract, 2% Bacto-peptone, 2% glucose) with the strain of choice and grow overnight at 30°, with shaking.
2. Transfer enough of the culture to inoculate 500 ml YPD at an initial A_{600} of 0.1. Incubate at 30° with shaking until the culture reaches an A_{600} of 0.8.
3. Harvest the cells by centrifugation at 4200g for 5 min at room temperature and transfer to 1 liter YPOLT (1% yeast extract, 2% Bacto-peptone, 0.2% oleic acid, 0.02% Tween 40). Incubate at 30° with shaking for 14–16 hr.
4. Harvest 2500 A_{600} units of cells by centrifugation at 4200g for 5 min at room temperature and wash once with 50 ml sterile water.
5. Harvest the cells by centrifugation at 4200g for 5 min at room temperature, resuspend in 50 ml reducing buffer [50 mM potassium phosphate (pH 7.4), 1 mM EDTA, 10 mM 2-mecaptoethanol] and incubate at room temperature for 20 min.
6. Harvest the cells by centrifugation at 4200g for 5 min at room temperature, resuspend in 50 ml of spheroplast buffer [50 mM potassium phosphate (pH 7.4), 1.2 M sorbitol], add 0.1 unit of Zymolyase 20T per A_{600} unit of cells, and incubate at 30° for 30 min.
7. Harvest the spheroplasts by centrifugation at 3000g for 10 min and resuspend gently in 40 ml Dounce buffer at 4° [5 mM MES (pH 6.0), 0.6 M sorbitol, 1 mM KCl, 0.5 mM EDTA, 0.1% ethanol, 0.2 mM phenylmethylsulfonyl chloride (PMSF), 25 μg/ml aprotinin, 25 μg/ml leupeptin, 200 μg/ml NaF]. Lyse the spheroplasts in a 40-ml Dounce homogenizer by 10 passes with the tight-fitting pestle on ice.
8. Remove nuclei, unbroken cells, and other large debris by centrifugation at 3000g for 10 min at 4°, transfer the supernatant to a fresh tube, and repeat.
9. The resulting postnuclear supernatant is then spun at 25,000g for 30 min at 4° to generate an organelle pellet enriched for peroxisomes and mitochondria and a cytosolic/microsomal supernatant that is depleted of peroxisomes and mitochondria.

Notes. The supernatant contains 50–100 times the amount of protein in the pellet. Therefore, if one wishes to determine the relative amounts of a protein in these peroxisomal/mitochondrial and cytosolic/microsomal fractions, it is essential that the investigator assay equal *proportions* of the two fractions instead of assaying equal amounts of *protein* from the two fractions. It is also common to assay the supernatant and pellet fractions for catalase activity, an enzyme that is both peroxisomal and cytoplasmic, and succinate dehydrogenase (SDH), a mitochondrial enzyme. Due to the expression of both cytosolic (*CTT1*) and peroxisomal (*CTA1*) catalases in *S. cerevisiae*, the most catalase activity[25] one will detect in the pellet is ∼50%. In contrast, >95% of SDH activity[26] should be detected in the organelle pellet and greater levels of SDH activity in the supernatant indicate overhomogenization of the spheroplasts.

Purifying Peroxisomes by Nycodenz Density Gradient Ultracentrifugation

For purification of peroxisomes from mitochondria and other components of the 25,000g large organelle pellet, the organelle pellet is resuspended and spun through a density gradient by ultracentrifugation. Optimal separation is obtained using Nycodenz density gradients. Sucrose density gradients are not able to completely separate the peroxisomal and mitochondrial fractions but are also commonly used.[7,27]

1. The organelle pellet generated by centrifugation of the postnuclear supernatant at 25,000g is enriched for mature peroxisomes and mitochondria. To fractionate the components of this pellet, gently resuspend the organelle pellet in 1.5 ml gradient buffer at 4° [5 mM MES (pH 6.0), 10% Nycodenz, 1 mM KCl, 0.5 mM EDTA, 0.1% ethanol, 0.2 mM PMSF, 25 μg/ml aprotinin, 25 μg/ml leupeptin, 200 μg/ml NaF].

2. Apply the resuspended organelle pellet to the top of a 10-ml linear 15–35% (w/v) Nycodenz gradient, which lies on a 1.5-ml Maxidenz cushion in a 13-ml Beckman quick-seal tube for the vTi65.1 rotor, all at 4°.

3. Seal the tube and spin at 45,000g in a Beckman vTi65.1 rotor for 75 min at 4°.

4. Collect 750-μl fractions from the bottom of the tube. Assay every other fraction for catalase activity by the method of Peters *et al.*[25] and SDH activity by the method of Pennington.[26]

Notes: The peroxisomal and mitochondrial fractions should be clearly separated from one another, with peroxisomes near the bottom of the gradient (Fig. 5). Although it is common to detect some catalase at the top of the gradient, overly vigorous resuspension of the 25,000g pellet will rupture peroxisomes, resulting in significant catalase activity at the top of the gradient. Because the amount of protein in each fraction varies considerably, it is essential to assay equal proportions of each fraction rather than equal amounts of protein from each fraction.

[25] T. J. Peters, M. Muller, and C. de Duve, *J. Exp. Med.* **136**, 1117 (1972).
[26] R. J. Pennington, *Biochem. J.* **80**, 649 (1961).
[27] B. V. Geisbrecht, D. Zhu, K. Schulz, K. Nau, J. C. Morrell, M. Geraghty, H. Schulz, R. Erdmann, and S. J. Gould, *J. Biol. Chem.* **273**, 33184 (1998).

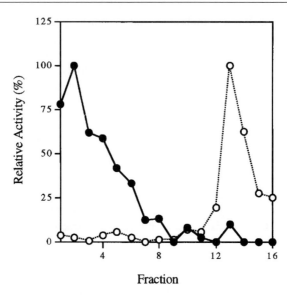

FIG. 5. Separation of peroxisomes from mitochondria by Nycodenz gradient centrifugation. A 25,000g organelle pellet was spun through a 15–35% Nycodenz gradient. Fractions are numbered from the bottom of the gradient. ●, catalase activity; ○, SDH activity.

Determining Subperoxisomal Distribution of Proteins

Once a protein is known to associate with peroxisomes, it is often desirable to assess its subperoxisomal distribution and/or the nature of its association with peroxisomes. Hypotonic lysis, salt extraction, and extraction by high pH are commonly used to determine whether a protein is a soluble matrix enzyme, a peripherally associated peroxisomal membrane protein, or an integral peroxisomal membrane protein. Due to the fragility and "leakiness" of purified peroxisomes, these experiments are typically performed on the 25,000g pellet fraction from a postnuclear supernatant. The following protocols are designed to be performed after resuspending the 25,000g pellet in 1.5 ml gradient buffer at 4° [5 mM MES (pH 6.0), 10% Nycodenz, 1 mM KCl, 0.5 mM EDTA, 0.1% ethanol, 0.2 mM PMSF, 25 μg/ml aprotinin, 25 μg/ml leupeptin, 200 μg/ml NaF].

Separating Soluble Peroxisomal Matrix Proteins from Peroxisomal Membrane Proteins by Hypotonic Lysis

1. Dilute the 25,000g pellet fraction in gradient buffer 1 : 25 in hypotonic lysis buffer at 4° [10 mM Tris–HCl (pH 8.5), 0.2 mM PMSF, 25 μg/ml aprotinin, 25 μg/ml leupeptin, 200 μg/ml NaF].
2. Homogenize the diluted sample in a Dounce homogenizer with 10 passes with the tight-fitting pestle on ice.
3. Incubate the sample for an additional 20 min at 4°.

4. Homogenize the diluted sample in a Dounce homogenizer with 10 passes with the tight-fitting pestle on ice.
5. Spin the resulting homogenate at 100,000g for 1 hr at 4° to pellet peroxisomal membranes.
6. Assay equal proportions of the resulting supernatant and pellet fractions for catalase activity and for the protein under consideration, either by enzyme assay or by immunoblot.

Salt Extraction of Peroxisomal Membranes. Once peroxisomal membranes have been generated by hypotonic lysis, they can be extracted with a high concentration of salt, which will strip many peripherally associated peroxisomal proteins from the membrane.

1. Resuspend the 100,000g membrane pellet in salt extraction buffer [1 M NaCl, 10 mM Tris–HCl (pH 8.5), 0.2 mM PMSF, 25 μg/ml aprotinin, 25 μg/ml leupeptin, 200 μg/ml NaF].
2. Homogenize the diluted sample in a Dounce homogenizer with 10 passes with the tight-fitting pestle on ice.
3. Incubate the sample for an additional 20 min at 4°.
4. Homogenize the diluted sample in a Dounce homogenizer with 10 passes with the tight-fitting pestle on ice.
5. Spin the resulting homogenate at 100,000g for 1 hr at 4° to pellet salt-stripped peroxisomal membranes.
6. Assay equal proportions of the resulting supernatant and pellet fractions for the protein under consideration, either by enzyme assay or by immunoblot.

High pH Extraction of Proteins from Peroxisomal Membranes. The standard method for determining whether a protein is peripherally associated with the peroxisome membrane or is an integral peroxisomal membrane protein is to treat the membranes with a high pH sodium carbonate solution. This treatment denatures the proteins without solubilizing the membranes, allowing the investigator to separate peripheral from integral membrane proteins by centrifugation.

1. Resuspend the 100,000g membrane pellet (generated from hypotonic lysis) in high pH extraction buffer [100 mM Na$_2$CO$_3$ (pH 11.5), 0.2 mM PMSF, 25 μg/ml aprotinin, 25 μg/ml leupeptin, 200 μg/ml NaF]. The protein concentration *must* be less than 0.8 mg/ml!
2. Homogenize the diluted sample in a Dounce homogenizer with 10 passes with the tight-fitting pestle on ice.
3. Incubate the sample for an additional 20 min at 4°.
4. Homogenize the diluted sample in a Dounce homogenizer with 10 passes with the tight-fitting pestle on ice.

5. Spin the resulting homogenate at 100,000g for 1 hr at 4° to pellet high pH-stripped peroxisomal membranes.
6. Assay equal proportions of the resulting supernatant and pellet fractions for the protein under consideration, either by enzyme assay or by immunoblot.

Notes. The extraction will quantitatively remove all nonintegrally inserted peroxisomal membrane proteins and may even remove some proportion of integral peroxisomal membrane proteins. Therefore, it may be necessary to compare the partitioning of the test protein with that of a known integral peroxisomal membrane protein, such as PXA1.[28]

In Vivo Luciferase Assay

Table I presents relative luciferase activities for strains expressing firefly luciferase from various promoters and under different growth conditions. These data were obtained using a novel *in vivo* assay for measuring luciferase activity, which is outlined here and may be more sensitive than standard *in vitro* luciferase assays.

1. Preculture the strain carrying the luciferase plasmid at 30° overnight.
2. Harvest cells by centrifugation at 4200g for 5 min and inoculate a 2-ml culture of the appropriate growth medium at an initiate A_{600} of 0.1. Grow until the culture reaches an A_{600} of 1.0.
3. Add 8 ml of sterile water, harvest cells by centrifugation at 4200g for 5 min, and wash twice with 10 ml of water (extensive washing is necessary for cells grown in oleic acid medium but may not be necessary for other applications).
4. Resuspend in 10 ml and determine cell density using a hemocytometer.
5. To a clean luminometer assay tube, add 100 µl of 10 mg/ml polymyxin B.
6. The timing of this step is critical: add 50 µl of the cell suspension, flick to mix the cells and the polymyxin B solution, and incubate for exactly 60 sec. Immediately after the 60-sec incubation, place the tube in a luminometer and initiate light detection by injecting 100 µl of 1 mM luciferin in PBS.

---

[28] N. Shani, P. A. Watkins, and D. Valle, *Proc. Natl. Acad. Sci. U.S.A.* **92**, 6012 (1995).
[29] R. S. Sikorski and P. Hieter, *Genetics* **122**, 19 (1989).
[30] R. Erdmann, F. F. Wiebel, A. Flessau, J. Rytka, A. Beyer, K. U. Fröhlich, and W. H. Kunau, *Cell* **64**, 499 (1991).
[31] J. K. Hiltunen, B. Wenzel, A. Beyer, R. Erdmann, A. Fossa, and W. H. Kunau, *J. Biol. Chem.* **267**, 6646 (1992).
[32] A. Gurvitz, A. M. Mursula, A. Firzinger, B. Hamilton, S. H. Kilpelainen, A. Hartig, H. Ruis, J. K. Hiltunen, and H. Rottensteiner, *J. Biol. Chem.* **273**, 31366 (1998).
[33] F. F. Wiebel and W.-H. Kunau, *Nature* **359**, 73 (1992).
[34] N. Shani and D. Valle, *Proc. Natl. Acad. Sci. U.S.A.* **93**, 11901 (1996).

7. Repeat the assay two more times for each culture and calculate the average and standard deviation.

Acknowledgments

Work in the Gould laboratory is supported by grants from the NIH (DK45787, DK59479, and HD10981). Studies in the Erdmann laboratory are supported by grants from the Deutsche Forschungsgemeinschaft (ER1782-3 and SFB449) and by the Fonds der Chemischen Industrie.

[22] Studying the Behavior of Mitochondria

By JODI NUNNARI, EDITH D. WONG, SHELLY MEEUSEN, and JENNIFER A. WAGNER

Introduction

Yeast mitochondria form dynamic tubular structures that are distributed uniformly at the cell cortex and contain an average of 50 to 100 copies of the mitochondrial genome (mtDNA) packaged into higher order nucleoid structures.[1] *Saccharomyces cerevisiae* is an ideal system for studying the mechanisms of both mitochondrial structure and mtDNA maintenance. Even after cells lose mtDNA, often as a secondary consequence of abnormal mitochondrial morphology, they can still be propagated and studied when a fermentable carbon source such as glucose is present. However, because of its role in various metabolic processes, the mitochondrial organelle, even in the absence of mtDNA, is an essential structure that cannot be created *de novo*. Therefore, a daughter cell survives only if a portion of the mitochondrion is inherited from the mother cell before cytokinesis.

Much insight into the mechanisms that govern the shape, distribution, and movement of mitochondria, as well as the behavior of mtDNA, has been gained through the use of genetic, cytological, and biochemical approaches.[2-4] This chapter describes some of the cytological and biochemical techniques that have been developed and used in our laboratory and by others in the field to study the behavior of this complex organelle.

---

[1] D. H. Willamson, *in* "Packaging and Recombination of Mitochondrial DNA in Vegetatively Growing Cells" (W. Bandlow, R. J. Schweyen, D. Y. Thomas, K. Wolf, and F. Kaudewitz, eds.), p. 99. Walter de Gruyter, Berlin, 1976.
[2] G. J. Hermann and J. M. Shaw, *Annu. Rev. Cell Dev. Biol.* **14**, 265 (1998).
[3] M. P. Yaffe, *Science* **283**, 1493 (1999).
[4] K. H. Berger and M. P. Yaffe, *Trends Microbiol.* **8**, 508 (2000).

Cytological Approaches

Visualization of Mitochondria

The behavior of mitochondria in live cells can be assessed through the use of membrane potential-dependent vital dyes or by mitochondrial-targeted forms of green fluorescent protein (GFP) variants or Dsred.[5,6] An advantage to using vital dyes is that mitochondria can be visualized relatively quickly and directly in cells without prior transformation of plasmids or the creation of chromosomal integrants. However, targeting of vital dyes to mitochondria requires a significant mitochondrial inner membrane potential.[7] Because of this, mitochondria cannot be visualized with vital dyes in cells grown using a fermentable carbon source, such as glucose, or in mutant cells that are respiratory incompetent, such as those lacking functional mtDNA. In contrast, efficient targeting of GFP to the various mitochondrial compartments occurs either under conditions of low membrane potential or by means independent of membrane potential (see later).[8] Thus, these mitochondrial-targeted versions of GFP are the tools of choice when examining mitochondrial structure in respiratory-incompetent cells or cells grown using glucose as the carbon source.

Vital Dyes. The two most commonly used vital dyes for the assessment of mitochondrial morphology are dimethylaminostyrylpyridiniummethyl iodine (DASPMI) (Sigma, St. Louis, MO) and MitoTracker CMXR (Molecular Probes, Eugene, OR). To visualize mitochondria with DASPMI, add 25 μg/ml DASPMI (from a 50-μg/ml stock solution prepared in water) to log-phase cells grown in YP glycerol (YPG) media.[9] Mount cells directly onto a microscope slide and visualize mitochondria using fluorescence microscopy with fluorescein isothiocyanate (FITC) filter sets.

To examine mitochondrial structure in cells using MitoTracker, add MitoTracker to a final concentration of 0.5 to 1 μM [from a 1 mM stock solution prepared in anhydrous dimethyl sulfoxide (DMSO)] to log-phase cells grown in YPG media. MitoTracker-labeled mitochondria can be visualized after short incubation times (2 to 5 min). Alternatively, after more extended incubation times (30 to 60 min), mitochondria become covalently labeled with MitoTracker and can be visualized after fixation of cells. This property makes MitoTracker a useful mitochondrial marker for double-labeling indirect immunofluorescence experiments and for monitoring mitochondrial fusion during mating (see later). To fix

---

[5] B. Westermann and W. Neupert, *Yeast* **16,** 1421 (2000).
[6] M. V. Matz, A. F. Fradkov, Y. A. Labas, A. P. Savitsky, A. G. Zaraisky, M. L. Markelov, and S. A. Lukyanov, *Nature Biotechnol.* **17,** 969 (1999).
[7] J. Llopis, J. M. McCaffery, A. Miyawaki, M. G. Farquhaar, and R. Y. Tsien, *Proc. Natl. Acad. Sci. U.S.A.* **95,** 6803 (1998).
[8] R. Rizzuto, M. Brini, F. De Giorgi, R. Rossi, R. Heim, F. Y. Tsien, and T. Pozzan, *Curr. Biol.* **6,** 183 (1996).
[9] S. J. McConnell, L. C. Stewart, A. Talin, and M. P. Yaffe, *J. Cell Biol.* **111,** 967 (1990).

TABLE I
MITOCHONDRIAL COMPARTMENT MARKERS USED FOR GFP FUSIONS

Mitochondrial compartment	Fusion protein	Refs.
Matrix	Matrix signal of CoxIV[a]	10
	Matrix signal of Su9[a]	5
Outer membrane	Tom70p	5
	Tom6p	18
	Fis1p	14
	Dnm1p	13
Inner membrane	Yta10p	18
Intermembrane space	Yme1p	32
Nucleoids	Abf2p	18
	Mgm101p	21

[a] Fusion proteins that have been constructed with GFP variants and/or DsRed [B. Westermann and W. Neupert, *Yeast* **16**, 1421 (2000); M. V. Matz, A. F. Fradkov, Y. A. Labas, A. P. Savitsky, A. G. Zaraisky, M. L. Markelov, and S. A. Lukyanov, *Nature Biotechnol.* **17**, 969 (1999); A. D. Mozdy, J. M. McCaffery, and J. M. Shaw, *J. Cell Biol.* **151**, 367 (2000); H. Sesaki and R. E. Jensen, *J. Cell Biol.* **152**, 1123 (2001)].

MitoTracker-labeled cells, wash cells several times with rich media to reduce nonspecific background labeling and incubate cells for 30 to 60 min in rich media with 3.7% (w/v) formaldehyde. MitoTracker-labeled mitochondria can be visualized by fluorescence microscopy using rhodamine or Texas Red filter sets.

Mitochondrial-Targeted GFP Fusions. Because of its versatility, GFP has been fused to known mitochondrial targeting signals and components and has been used to visualize various mitochondrial compartments, such as the matrix, intermembrane space, inner and outer membranes, and nucleoids. Table I summarizes some of the GFP fusions that have successfully been used to monitor mitochondrial behavior.

GFP fusions to matrix-targeting signals of either Su9 (amino acids 1 to 69 of *Neurospora crassa* F_0-ATPase subunit 9)[5] or Cox IV (amino acids of 1 to 21 of *S. cerevisiae* CoxIV)[10] are the most commonly used to study mitochondrial behavior in cells. The advantage of Su9–GFP as compared to CoxIV–GFP and vital dyes is that its targeting to the mitochondrial matrix is relatively insensitive to membrane potential, making it useful for mitochondrial visualization in respiratory-deficient cells.

Indirect Immunofluorescence. Standard immunofluorescence methods using antimitochondrial protein antibodies can also be used to assess mitochondrial structure in cells.[11] For this purpose, antibodies to porin, an outer membrane protein, are the best characterized and are available commercially (Molecular

[10] H. Sesaki and R. E. Jensen, *J. Cell Biol.* **147**, 699 (1999).
[11] J. R. Pringle, A. E. Adams, D. G. Drubin, and B. K. Haarer, *Methods Enzymol.* **194**, 565 (1991).

Probes). *Note.* Detection of antiporin antibodies (used at 5–20 μg/ml) by indirect immunofluorescence can be enhanced by the treatment of fixed and spheroplasted cells with 0.5% sodium dodecyl sulfate (SDS) for 2 min.

Vital Assay for Mitochondrial Fusion during Mating

An *in vivo* assay for mitochondrial content mixing (fusion) was developed by Nunnari *et al.*[11a] and has been optimized for use with W303 strains. This assay has been used to identify the specific functions of gene products involved in mitochondrial fusion, fission, and inner membrane remodeling.[12–17] In this assay, mitochondrial fusion is assessed by following the behavior of distinctively labeled haploid-derived mitochondrial proteins during mating. In the originally described assay (see later for details), mitochondria in haploid cells of opposite mating types are labeled with either mitochondrial matrix targeted-GFP or the covalent vital dye MitoTracker CMXR (Molecular Probes). It has been found that a mitochondrial matrix-targeted version of DsRed could be used instead of MitoTracker to successfully monitor mitochondrial fusion during mating.[14,17] Although the following assay primarily utilizes mitochondrial-targeted matrix markers, markers targeted to other mitochondrial compartments, such as the outer and inner membranes, have also been utilized in a similar manner to assess the diffusion behavior of various mitochondrial constituents following fusion during mating.[18]

Labeling Mitochondria with Mito-GFP. Grow cultures of haploid cells that have been transformed with a yeast expression vector containing a mitochondrial matrix-targeted form of GFP (mito-GFP) to early log phase. Harvest cells by centrifugation and wash into rich media at a concentration of 0.2 OD_{600}/ml. *Note.* To ensure that mito-GFP is produced only in the haploid parent before mating, expression can be placed under the regulation of the *GAL1/10* promoter and cells can be grown with 2% galactose and transferred to glucose prior to mating. However, we have found that mito-GFP expressed from strong constitutive promoters, such as *ADH*, is undetectable in mitochondria derived from the opposite mating type parent during the course of mating and assessment of content mixing and does not significantly interfere in the assay.

[11a] J. Nunnari, W. F. Marshall, A. Straight, A. Murray, J. W. Sedat, and P. Walter, *Mol. Biol. Cell* **8**, 1233 (1997).
[12] G. J. Hermann, J. W. Thatcher, J. P. Mills, K. G. Hales, M. T. Fuller, J. Nunnari, and J. M. Shaw, *J. Cell Biol.* **143**, 359 (1998).
[13] W. Bleazard, J. M. McCaffery, E. J. King, S. Bale, A. Mozdy, Q. Tieu, J. Nunnari, and J. M. Shaw, *Nature Cell Biol.* **1**, 298 (1999).
[14] A. D. Mozdy, J. M. McCaffery, and J. M. Shaw, *J. Cell Biol.* **151**, 367 (2000).
[15] Q. Tieu and J. Nunnari, *J. Cell Biol.* **151**, 353 (2000).
[16] E. D. Wong, J. A. Wagner, S. W. Gorsich, J. M. McCaffery, J. M. Shaw, and J. Nunnari, *J. Cell Biol.* **151**, 341 (2000).
[17] H. Sesaki and R. E. Jensen, *J. Cell Biol.* **152**, 1123 (2001).
[18] K. Okamoto, P. S. Perlman, and R. A. Butow, *J. Cell Biol.* **142**, 613 (1998).

Labeling Mitochondria with MitoTracker CMXR. Grow cultures of haploid cells of the opposite mating type of that used earlier in rich YP medium containing a nonfermentable carbon source, such as 2% (v/v) glycerol (YPG), to early log phase. Collect 10 OD_{600} by centrifugation and resuspend in 1 ml of YPG. To label mitochondria, add MitoTracker (from a 1 mM stock made in anhydrous DMSO) to a final concentration of 0.5 μM and incubate with shaking for 45 min. To remove noncovalently associated MitoTracker, wash cells repeatedly (10 to 15 times) by centrifugation and resuspension of cells in YP dextrose (YPD). After the final wash, resuspend cells in YPD to a final concentration of 0.2 OD_{600}/ml in YPD.

Mating Assays

1. Mix a 1-ml aliquot (OD_{600} = 0.2) of each labeled haploid strain by vortexing.
2. Collect the mixed cells by centrifugation and resuspend the cell pellet in 0.5 ml of YPD.
3. To facilitate mating of the haploid cells, concentrate cells by vacuum filtration onto a sterile nitrocellulose filter (25 mm, 0.8 μm, Metricel membrane filter, Gelman Sciences, Ann Arbor, MI). *Note.* Increasing or decreasing the concentration of cells will lower mating efficiency.
4. Place filter on a YPD plate with the cell side facing up and incubate plates at the desired temperature(s). *Note.* In the case of conditional mutant strains, filters can be cut into sections and several temperatures can be assessed simultaneously. At temperatures ranging from ambient to 37°, and depending on strain background, cells should mate efficiently between 1 and 3 hr.
5. To monitor mating, scrape a small number of cells from the filters using a sterile toothpick and examine using a light microscope. When large-budded zygotes are present in the mating mixture, recover the cells from the filters by placing them into a sterile microfuge tube containing 1 ml of YPD and vortexing. Remove and discard filter from the tube.
6. For microscopic analysis, zygotes can be visualized directly or fixed by adding formaldehyde to a final concentration of 3.7% and incubating at the mating temperature for 30 min. Remove media and/or fixative by centrifugation and resuspend cells in sterile phosphate-buffered saline (PBS).
7. Content mixing and mitochondrial fusion are assessed by examining and comparing the distribution of haploid-derived mitochondrial markers in unbudded and budded zygotes using fluorescence microscopy and the appropriate filter sets.

Visualization of Mitochondrial Nucleoids

The behavior of mitochondrial nucleoids can be studied by visualizing mtDNA directly in both live and fixed cells using the fluorescent dye, 4′,6-diamidino-2-phenylindole (DAPI), which binds selectively to mtDNA

FIG. 1. Visualization of nucleoid structures within the mitochondrial reticulum in yeast. Yeast cells expressing a GFP-tagged mitochondrial DNA-binding protein, Mgm101p (in yellow), were labeled with a mitochondrial-specific vital dye [in red; see S. Meeusen, Q. Tieu, E. Wong, E. Weiss, D. Schieltz, J. R. Yates, and J. Nunnari, *J. Cell Biol.* **145,** 291 (1999)].

in vivo.[19-21] Alternatively, mtDNA can be visualized directly in fixed cells following incorporation of the thymidine analog bromodeoxyuridine (BrdU), which can be detected by indirect immunofluorescence using anti-BrdU antibodies.[21] mtDNA can also be visualized indirectly through the use of GFP fusions to the mtDNA-binding proteins, Abf2p or Mgm101p (Fig. 1), or with antibodies directed against these proteins by indirect immunofluorescence of fixed cells[18,21] (Tables I and II). Methods for DAPI staining of mtDNA allow assessment of gross nucleoid morphology with little time expense. For example, this approach has been used successfully to assess mtDNA packaging, segregation, and partitioning in mutant cells.[18,21] Alternatively, the major advantage of the more time-consuming approach of BrdU labeling detection is that it can be utilized to directly assess mtDNA replication *in vivo*.[21]

DAPI for Detection of mtDNA. Grow cells to early log phase in either rich or minimal media. *Note.* Media containing nonfermentable carbon sources, such as glycerol, select for functional mtDNA. Resuspend approximately 1×10^6 cells in

---

[19] I. Miyakawa, H. Aoi, N. Sando, and T. Kuroiwa, *J. Cell Sci.* **66,** 21 (1984).
[20] L. Pon and G. Schatz, in "Biogenesis of Yeast Mitochondria" (J. Broach, J. Pringle, and E. Jones, eds.), p. 333. Cold Spring Harbor Laboratory Press, Cold Spring Harbor, NY, 1991.
[21] S. Meeusen, Q. Tieu, E. Wong, E. Weiss, D. Schieltz, J. R. Yates, and J. Nunnari, *J. Cell Biol.* **145,** 291 (1999).

TABLE II
USEFUL MARKER PROTEINS TO ASSESS MITOCHONDRIAL PURIFICATION
AND INTRAMITOCHONDRIAL LOCALIZATION

Protein	Location	Mature molecular mass (kDa)	Ref./Source
3-PGK	Cytosol	44	32; antibodies available from Molecular Probes
Por1p	OM	30	33; antibodies available from Molecular Probes
Fzo1p	OM	98	12, 33
Cytb$_2$p	IMS, soluble	45	34, 35
Tim23p	IMS, integral membrane protein	25	36, 37
Aac1p	IMS, integral membrane protein	34	38, 39
α-KDH	Matrix	90	40, 41
Abf2p	Nucleoid	20	21, 31
Mgm101p	Nucleoid	30	21

media containing 1 μg/ml DAPI (from a 10-mg/ml stock prepared in water) and incubate with shaking until nucleoids are visible (30 min to 4 hr depending on growth media and strain background) as small punctate structures at the cell periphery in the fluorescence microscope using DAPI/UV filter sets. Using this vital method, the detection of nucleoids precedes that of nuclear DNA because of the relatively A T-rich content of mtDNA. Alternatively, to examine nucleoid morphology more quickly, cells can be harvested and fixed in 70% ethanol containing 100 ng/ml DAPI for 2 to 3 min. After washing two to three times with 1 ml of PBS, mtDNA can be examined as described earlier; however, in fixed cells, nuclear DNA is also visualized.

Incorporation and Detection of Bromodeoxyuridine into mtDNA: Assessment of mtDNA Replication. S. cerevisiae lacks a thymidine kinase gene required to phosphorylate BrdU prior to its incorporation into DNA. Thus, to incorporate BrdU into mtDNA, the yeast strain of interest must possess a chromosomal or episomal copy of a thymidine kinase (TK) gene under regulation of a yeast promoter [see Nunnari et al.[11a] for strain construction using Herpes simplex virus-1 thymidine kinase (HSV-1 TK)].[22–24] Grow cells harboring the TK gene in rich media to early log phase. To minimize the incorporation of BrdU into nuclear DNA, mtDNA can be selectively labeled in *Mata* haploid cells arrested in G_1 by adding 10 μg/ml α factor and incubating at 30° for 1 hr. Arrest can be monitored by quantifying the percentage of unbudded cells in culture.

[22] J. Leff and T. R. Eccleshall, *J. Bacteriol.* **135,** 436 (1978).
[23] R. A. Sclafani and W. L. Fangman, *Genetics* **114,** 753 (1986).
[24] B. S. Dien and F. Srienc, *Biotechnol. Prog.* **7,** 291 (1991).

BrdU Labeling. To incorporate BrdU into mtDNA, the synthesis of deoxythymidylate is blocked by the addition of amethopterin and sulfanilamide to the media. Specifically, resuspend 1×10^7 cells in 2 ml of media lacking thymidine and containing 5 mg/ml sulfanilamide (added directly as solid) and 100 μg/ml amethopterin (from a 100× stock in DMSO). *Note.* To solubilize sulfanilamide, heat media to 65°. Add BrdU to a final concentration of 0.5 mg/ml to label cells and incubate cultures with shaking for up to 30 min at 30°.

Fixation and Processing of Cells for Indirect Immunofluorescence with α-BrdU Antibodies

1. To fix BrdU-labeled cells, harvest and wash cells into rich media containing 3.7% formaldehyde and incubate 2 hr with shaking at room temperature.
2. Wash cells three times in 1× PBS and twice in 0.1 M potassium phosphate, pH 7.4.
3. To digest the cell wall, resuspend cells in 1 ml of 0.1 M potassium phosphate, pH 7.4, with 100 μg/ml Zymolyase and incubate at 30° for 20 min.
4. Wash cells once with 0.1 M potassium phosphate and once with 1× PBS.
5. Resuspend cells in 250 μl 1× PBS and pipette 80 μl onto a poly (L-lysine)-coated slide well (Carlson Scientific, Inc., Peotone, IL).
6. To adhere cells to slides, incubate at 25° for 5–10 min in a humidified chamber.
7. Aspirate the remaining PBS from slide well and wash cells with 80 μl of PBS containing 0.1% Triton X-100 (PBS-T) for 30 min at 25°.
8. Aspirate PBS-T and treat cells with 80 μl of 0.6 N HCl for 5 min at 25°.
9. Aspirate HCl and neutralize the acid with 80 μl of 0.1 M sodium tetraborate, pH 8.5, for 5 min at 25°.
10. Aspirate sodium tetraborate and wash cells twice in 80 μl of 1× PBS, incubating for 5 min at 25° each time.
11. To block nonspecific binding of anti-BrdU antibody, incubate cells with 80 μl of PBS containing 5% bovine serum albumin (BSA) and 0.5% Tween 20 for 10 min at 25°.
12. To detect BrdU labeling, process cells using standard indirect immunofluorescence methods with monoclonal anti-BrdU antibodies (antibodies purchased from Becton Dickinson (Heidelberg, Germany) give the best results) at 1 : 200, followed by fluorescent-conjugated anti-mouse secondary antibodies.

Notes. The acid/base treatment in this protocol causes DNA strand breaks, which are required to render the BrdU epitope accessible to the antibody for detection. However, as a result, DAPI cannot be used to examine the total mtDNA/nucleoid population in cells. The fluorescent signal intensity of BrdU labeling can be correlated with the levels of mtDNA replication per nucleoid during the BrdU pulse. Cells lacking mtDNA (rho^0) are an excellent negative control for mtDNA labeling with BrdU.

Biochemical Approaches

Enrichment of Mitochondria

Cell fractions enriched with intact mitochondria can be isolated from *S. cerevisiae* by differential centrifugation of a homogenate of spheroplasted cells and can be further purified by equilibrium centrifugation.[25,26] These protocols are useful for assessing the cellular localization of proteins of interest and as a starting point for assessing the intramitochondrial localization of a given protein.

Spheroplast Formation and Cell Lysis. The following procedure has been adapted from Daum *et al.*[25] and has been optimized for isolating intact mitochondria from the W303 strain.

1. Grow 6 liters of cells overnight in rich media with 2% dextrose (YPD) to late log phase ($OD_{600} = 2$–3/ml).

2. Pellet cells at $1500g$ for 5 min at room temperature. Remove media by washing the cell pellet with distilled water.

3. To partially compromise the cell wall structure, resuspend cells in 100 mM Tris–HCl, pH 9.3, 50 mM 2-mercaptoethanol (2-ME) to 20 OD_{600}/ml and incubate with shaking at 30° for 10 min.

4. Pellet cells at $1500g$ for 5 min at room temperature. To remove Tris–HCl/2-ME and to osmotically stabilize cells during spheroplasting, wash cells with 100 ml of spheroplast buffer (SB): 1.2 M sorbitol, 20 mM potassium phosphate, pH 7.5, 5 mM magnesium chloride, and 10 mM 2-ME.

5. To digest cell walls, resuspend cells in SB to 50 OD_{600}/ml. Add yeast lytic enzyme (70,000 units/mg, ICN, Costa Mesa, CA) that has been resuspended in a small volume of 1.2 M sorbitol to a final concentration of 1.0 mg/ml and incubate with gentle shaking at 30°.

6. Monitor spheroplasting by examining cultures for the formation of phase-dark cells using a light microscope with phase optics. Harvest cells when spheroplast efficiency is 80–90%. Spheroplasting should be complete after 1 hr. *Note.* Spheroplasted cells should be handled gently to avoid prematurely lysing cells and releasing mitochondria.

7. From this step forward, ice-cold conditions are required.

8. Gently pellet cells at $1500g$ for 5 min at 4°. Gently wash spheroplasted cells twice with ice-cold SB to remove all yeast lytic enzyme.

9. On ice, resuspend cells at 50 OD/ml in ice-cold mitochondrial isolation buffer (MIB): 0.6 M sorbitol, 20 mM HEPES–KOH, pH 7.4, 0.5 mM phenylmethylsulfonyl fluoride (PMSF), and 3 mM benzamidine.

10. To release mitochondria, cool a manual Dounce on ice and rinse it with ice-cold MIB. Transfer the cells to the Dounce and homogenize with a *tight* pestle for

---

[25] G. Daum, P. C. Bohni, and G. Schatz, *J. Biol. Chem.* **257**, 13028 (1982).
[26] B. S. Glick and L. A. Pon, *Methods Enzymol.* **260**, 213 (1995).

20–50 strokes (Wheaton Scientific, Millville, NJ). Use phase-contrast microscopy to determine the completeness of cell lysis.

Enrichment of Mitochondria by Differential Centrifugation

1. Dilute homogenate with 1 volume of MIB. Centrifuge at 3000g for 5 min at 4° to pellet unlysed cells and nuclei.
2. Decant and save the supernatant fraction, which contains mitochondria, into a clean centrifuge tube. Rehomogenize the pellet in additional MIB and centrifuge again at 3000g for 5 min at 4°.
3. Combine the 3000g supernatant fractions, add additional 0.5 mM PMSF and 3 mM benzamidine to further protect against proteolysis, and centrifuge at 9500g for 10 min at 4°.
4. Carefully decant the 9500g supernatant.
5. Resuspend the pellet, which is enriched for mitochondria, in 30 ml of MIB. Centrifuge at 3000g for 5 min at 4° to remove any additional unlysed cells and nuclei. Decant the supernatant fraction containing mitochondria into a clean centrifuge tube.
6. Centrifuge supernatant fraction at 9000g for 10 min at 4°.
7. Resuspend the mitochondrial pellet in 30 ml ice-cold MIB and centrifuge at 9500g for 10 min at 4°.
8. Remove as much supernatant as possible without disturbing the mitochondrial-enriched pellet. Resuspend the pellet in a minimal amount of MIB (concentration should be approximately 20–30 mg protein/ml).
9. Crude mitochondria should be aliquoted and snap frozen in liquid nitrogen and stored at −80° for analysis or purified further by equilibrium centrifugation using an Optiprep gradient (see later).

Mitochondrial Purification by Equilibrium Centrifugation. To further purify crude mitochondria for analysis, equilibrium centrifugation using gradients made from sucrose, Nycodenz (Nycomed[26]), or Optiprep [iodixanol, 60% (w/v) Nycomed[21]] (Axis-Shield, Oslo, Norway) can be used. However, to easily maintain conditions isosmotic to mitochondria, an Optiprep gradient is ideal.

1. Prepare the Optiprep working solution (OWS): 50% iodixanol (Optiprep starts as a 60% solution), 7.43% (w/v) sorbitol, and 20 mM HEPES–KOH, pH 7.4. Density is 1.29 g/ml.
2. On ice, mix 1 ml of enriched mitochondria in MIB (10 mg protein/ml) with 5 ml of OWS.
3. Prepare the following density solutions using MIB, OWS, and the following formula:

Volume$_{MIB}$ = (1.29 − desired density)/(desired density − 1.03).
MIB density = 1.03 g/ml

For example, to prepare a solution of density of 1.1 g/ml: Volume$_{MIB}$ = (1.29 − 1.10)/(1.10 − 1.03) = 2.71 ml of MIB/1 ml OWS

Layer 1: density = 1.03   (MIB)
Layer 2: density = 1.10   (10.8 ml MIB, 4.0 ml OWS)
Layer 3: density = 1.16   (6.0 ml MIB, 6.0 ml OWS)
Layer 4: density = 1.29   (OWS)

4. Working in a 4° cold room, pipette the suspension of mitochondria and OWS into the bottom of a SW-28 (Beckman, Fullerton, CA) polycarbonate centrifuge tube (Nalgene, Rochester, NY). Carefully pipette 6 ml of each Optiprep solution from most dense (1.29 g/ml) to least dense (1.1 g/ml) on top of the mitochondrial/OWS suspension. Finish gradient by gently pipette 6–7 ml of MIB as the top layer.

5. Gently place the polycarbonate tubes in a SW-28 rotor. Centrifuge at 19,000 rpm for 3 hr at 4°, with the slowest acceleration and deceleration settings.

6. Remove the centrifuge tubes from the rotor and place on ice. In a 4° cold room, carefully remove the MIB layer and most of the 1.1-g/ml density layer, being careful not to disturb the 1.1/1.16-g/ml interface containing the mitochondria. Using a 1-ml pipette, harvest mitochondria from the 1.1/1.16-g/ml interface in as little volume as possible.

7. Aliquot floated mitochondria into microfuge tubes, quick freeze in liquid nitrogen, and store at −80° for further analysis.

Assessment of Mitochondrial Purification. The purification of mitochondria using the preceding procedures can be assessed by SDS–PAGE, followed by Western blotting with antibodies to proteins in different cellular compartments (Table II).

Determination of Submitochondrial Localization

To determine the intramitochondrial localization of a protein of interest, the sensitivity of the protein to exogenously added proteases is assessed in intact mitochondria prepared as described earlier and in mitoplasts, where the outer mitochondrial membrane is selectively ruptured.[26–28]

Preparing Mitoplasts by Hypoosmotic Shock. The outer mitochondrial membrane can be selectively disrupted by hypoosmotic shock by decreasing the molarity of sorbitol in MIB from between 0.05 and 0.5 M. This is accomplished by slowly adding ice-cold 20 mM HEPES, pH 7.4, to mitochondria resuspended at a concentration of 1 mg protein/ml with gentle mixing followed by an incubation of

---

[27] L. F. Sogo and M. P. Yaffe, *J. Cell Biol.* **126,** 1361 (1994).
[28] S. He and T. D. Fox, *Mol. Biol. Cell* **8,** 1449 (1997).

the mixture on ice for 15 min. The final concentration of sorbitol required to selectively disrupt the outer membrane depends greatly on strain background and growth conditions and therefore must be titrated to determine the correct conditions. Outer membrane disruption can be assessed by Western analysis to determine if the soluble intermembrane space marker, cytochrome b_2, is released into the supernate following centrifugation (Table II). Control mitochondrial samples diluted with isosmotic MIB instead of HEPES are required to ascertain whether the starting mitochondria are intact.

Treatment of Mitochondria with Exogenous Proteases. The two proteases used most commonly in mitochondrial protection assays are proteinase K (PK) and trypsin. Because the sensitivity and selectivity of a given protein to proteolytic cleavage are usually unknown, mitochondria should be treated with each protease separately in the range of 0 to 100 μg/ml. The proteases are incubated with mitochondria typically for 30 min on ice. Proteolysis is inhibited by the addition of 1 mM PMSF for PK and by the addition of 1 mM PMSF and 2.5 mg/ml soybean trypsin inhibitor for trypsin. To further ensure that exogenous proteases are inactivated, total protein is precipitated by the addition of trichloroacetic acid (TCA, final concentration 15%) and heated at 55° for 15 min.[29] To test whether a protein is intrinsically insensitive to proteolysis or protected by virtue of a membrane barrier, control reactions, where membranes are disrupted by detergent treatment (0.1% Triton X-100) or by sonication, should be performed.

Analysis of Protease Protection. Collect TCA precipitates of treated mitochondria by centrifugation at 13,000g for 20 min at 4°, solubilize pellets in SDS–PAGE sample buffer, and analyze samples by SDS–PAGE followed by Western blotting using antibodies directed against marker protein of various mitochondrial compartments (commonly used markers are listed in Table II). Intermembrane space or matrix markers susceptible to proteolysis in untreated mitochondria or mitochondria mock treated in isosmotic MIB buffer are indications that mitochondria in the starting preparation were not intact. This can occur, for example, if the osmolarity of the buffer is incorrect or if the membranes are sheared by rough treatment during the purification of mitochondria.

Enrichment of Mitochondrial Nucleoids

Mitochondrial nucleoids can also be enriched from purified mitochondria by equilibrium centrifugation and used for further analysis.[21,30,31] To enrich nucleoids,

---

[29] B. S. Glick, A. Brandt, K. Cunningham, S. Mueller, R. L. Hallberg, and G. Schatz, *Cell* **69,** 809 (1992).

[30] I. Miyakawa, N. Sando, S. Kawano, S. Nakamura, and T. Kuroiwa, *J. Cell Sci.* **88,** 431 (1987).

[31] S. M. Newman, O. Zelenaya-Troitskaya, P. S. Perlman, and R. A. Butow, *Nucleic Acids Res.* **24,** 386 (1996).

dilute purified mitochondria to a protein concentration of 2.5 mg/ml. Lyse mitochondria by slowly adding an equal volume of nucleoid extraction buffer [NXB: 20 mM HEPES, pH 7.4, 100 mM sucrose, 20 mM KCl, 1% Nonidet P-40 (NP-40), 1 mM spermidine, 100 ng/ml DAPI, 1 mM PMSF, 0.5 mM DTT) with gentle mixing. Homogenize the sample with a chilled, *loose* Dounce (Wheaton Scientific) and incubate on ice for 1 hr. Layer this mitochondrial lysate onto a 37.5/60/80% sucrose density step gradient. In these gradients, the 37.5% layer contains NXB and the 60% layer contains 20 mM HEPES, pH 7.4, 20 mM KCl, 1% Mega-8, 1 mM spermidine, 100 ng/ml DAPI, and 0.5 mM DTT. Centrifuge gradients at 100,000g for 90 min. Nucleoids band at the 37.5/60% interface. Collect this fraction and dilute twofold with NXB containing 2% NP-40 and rerun on another 37.5/60/80% step sucrose gradient. Assess nucleoid purification by the enrichment of DAPI stainable structures in isolated fractions, DNA content of fraction using the PicoGreen dye (Molecular Probes, Inc.), Southern blotting using mtDNA probes, and/or by analysis of nucleoid proteins such as Abf2p and Mgm101p in fractions by Western blotting (Table II).

Acknowledgment

We thank Eric Weiss for his development and optimization of the mitochondrial and nucleoid purification procedures.

[32] C. L. Campbell, N. Tanaka, K. H. White, and P. E. Thorsness, *Mol. Biol. Cell* **5,** 899 (1994).
[33] D. Rapaport, M. Brunner, W. Neupert, and B. Westermann, *J. Biol. Chem.* **273,** 20150 (1998).
[34] B. Guiard, *EMBO J.* **4,** 3265 (1985).
[35] S. P. Gygi, Y. Rochon, B. R. Franza, and R. Aebersold, *Mol. Cell. Biol.* **19,** 1720 (1999).
[36] J. Berthold, M. F. Bauer, H. C. Schneider, C. Klaus, K. Dietmeier, W. Neupert, and M. Brunner, *Cell* **81,** 1085 (1995).
[37] J. Blom, P. J. Dekker, and M. Meijer, *Eur. J. Biochem.* **232,** 309 (1995).
[38] G. S. Adrian, M. T. McCammon, D. L. Montgomery, and M. G. Douglas, *Mol. Cell. Biol.* **6,** 626 (1986).
[39] K. Dietmeier, V. Zara, A. Palmisano, F. Palmieri, W. Voos, J. Schlossmann, M. Moczko, G. Kispal, and N. Pfanner, *J. Biol. Chem.* **268,** 25958 (1993).
[40] B. Repetto and A. Tzagoloff, *Mol. Cell. Biol.* **11,** 3931 (1991).
[41] M. Perrot, F. Sagliocco, T. Mini, C. Monribot, U. Schneider, A. Shevchenko, M. Mann, P. Jeno, and H. Boucherie, *Electrophoresis* **200,** 2280 (1999).

[23] Isolation of Nuclear Envelope from *Saccharomyces cerevisiae*

By JULIA KIPPER, CATERINA STRAMBIO-DE-CASTILLIA, ADISETYANTARI SUPRAPTO, and MICHAEL P. ROUT

Introduction

The yeast *Saccharomyces cerevisiae* has been one of the systems of choice for the molecular biologist and the geneticist. This organism has become increasingly amenable to biochemical and cell biological techniques.[1-6] We have previously described a detailed method for the preparation of yeast nuclear envelopes (NEs) that was specifically designed for *Saccharomyces uvarum*.[7] This chapter presents modifications of published protocols[1,8] that are now utilized routinely in our laboratory to prepare nuclei and NEs from several wild-type and mutant *S. cerevisiae* strains. In particular, highly enriched yeast nuclei and NEs fractions can be prepared from *S. cerevisiae* strains that have been modified genetically to encode epitope-tagged versions of genes of interest. This method is highly reproducible in our laboratory and in other laboratories. The end of the chapter also describes a faster, cruder method for making nuclei that can be used for preparations in which the highest degree of enrichment is not crucial.

Strains, Materials, and Instrumentation

Pyrex 18-liter flasks, round magnetic stir bars with removable pivot rings, Nalgene autoclavable rubber tubing, an improved Neubauer hemacytometer, and glass gas dispersion tubes are obtained from Fisher Scientific (Pittsburgh, PA). Antifoam B, DNase I, mutanase (lysing enzymes), and protease inhibitor cocktail (PIC) are from Sigma (St. Louis, MO). Ten percent (v/v) Triton X-100 solution (Surfact-Amps X-100) is from Pierce (Rockford, IL). The refractometer (144974) is from Zeiss (Thornwood, NY). The Polytron (PCU 11 with PTA 10TS probe) is from Kinematica (Cincinnati, OH). Nycodenz (Accudenz) is from

---

[1] J. V. Kilmartin and J. Fogg, in "Microtubules and Microorganisms" (P. Cappucinelli and N. R. Morris, eds.), p. 157. Dekker, New York, 1982.
[2] M. P. Rout and J. V. Kilmartin, *J. Cell Biol.* **111**, 1913 (1990).
[3] C. Strambio-de-Castillia, G. Blobel, and M. P. Rout, *J. Cell Biol.* **131**, 19 (1995).
[4] A. Franzusoff, J. Rothblatt, and R. Schekman, *Methods Enzymol.* **194**, 662 (1991).
[5] M. P. Yaffe, *Methods Enzymol.* **194**, 627 (1991).
[6] P. D. Garcia, W. Hansen, and P. Walter, *Methods Enzymol.* **194**, 675 (1991).
[7] M. P. Rout and C. Strambio-de-Castillia, in "Cell Biology: A Laboratory Handbook" (J. E. Celis, ed.), Vol. 2, p. 143. Academic Press, London, 1998.
[8] T. H. Rozijn and G. J. M. Tonino, *Biochim. Biophys. Acta* **91**, 105 (1964).

Accurate Chemicals and Scientific (Westbury, NY). Zymolyase 20T is from Seikagaku America (Falmouth, MA). Glusulase is from NEN (Beverly, MA).

The preparation of highly enriched nuclei and nuclear envelopes is conducted routinely in our laboratory from several *S. cerevisiae* strains, including DF5,[9] W303,[10] and C13-ABYS86 (YW046[11]).

Solutions and Reagents

For nuclei and NE preparations, yeast cells are grown in Wickerham's medium. For an 18-liter preparation (large scale), dissolve 54 g malt extract (0.3%, w/v), 54 g yeast extract (0.3%, w/v), 90 g Bacto-peptone (0.5%, w/v) and 180 g glucose (1%, w/v) in 2 liters of warm tap water. Transfer the concentrated medium to a 18-liter flask and dilute it to 18 liters with tap water. Add 1 ml of antifoam and put a medium-size stir bar in the flask to facilitate good aeration of the medium during growth. Cover the flask with several layers of aluminum foil and autoclave for 90 min at 121°. For a 2-liter preparation (small scale), dissolve 6 g malt extract, 6 g yeast extract, 10 g Bacto-peptone, and 20 g glucose in 2 liters of warm tap water, add a drop of antifoam, and autoclave for 30 min at 121°. In the case of a large-scale preparation, the medium is prepared 2 days before the preparation; it is then cooled down to room temperature overnight and equilibrated to 30° for at least 12 hr before inoculation.

The materials described here are sufficient for a large-scale preparation. For a small-scale preparation, all amounts should be scaled down accordingly. All solutions are prepared 1 day in advance.

Distilled water: 5 liter, prechilled in a 4° cold room.

1.1 M sorbitol: Dissolve 400 g sorbitol in distilled water and bring to a total volume of 2 liter. Filter through a 0.45-μm filter and store at 4°.

Ficoll/sorbitol solution: Dissolve 30 g Ficoll-400 in ~250 ml of prewarmed 1.1 M sorbitol. It dissolves very slowly. Make up to 400 ml. Filter as just described. Store at 4°.

8% PVP solution: Dissolve 80 g polyvinylpyrrolidone-40 (PVP-40), 1.57 g KH_2PO_4, 1.46 g K_2HPO_4, and 1.52 mg $MgCl_2 \cdot 6H_2O$ in prewarmed distilled water. Dissolve the PVP first, as it dissolves slowly. Adjust the pH to 6.50 with H_3PO_4. Make up to 1 liter. Filter and store at 4°.

Solution P: Dissolve 2 mg pepstatin A and 90 mg phenylmethylsulfonyl fluoride (PMSF) in 5 ml of absolute ethanol. Store at −20°. Use appropriate care, as PMSF is highly toxic.

---

[9] D. Finley, E. Ozkaynak, and A. Varshavsky, *Cell* **48**, 1035 (1987).
[10] B. Thomas and R. Rothstein, *Cell* **56**, 619 (1989).
[11] W. Heinemeyer, J. A. Kleinschmidt, J. Saidowsky, C. Escher, and D. H. Wolf, *EMBO J.* **10**, 555 (1991).

Sucrose/PVP gradient solutions: These solutions can be stored indefinitely at −20°. It is convenient to prepare large volumes and store them in sterile 50-ml Falcon tubes for future use. Prepare the 2.3 M sucrose/PVP solution by prewarming 8% PVP to ~80°, weighing out 430 g sucrose, and adding 8% PVP solution in a beaker to a total weight of 680 g (approximately 1 liter). Stir vigorously on a hot stir plate; when the sucrose has dissolved, equilibrate to room temperature before measuring the refractive index (RI). Adjust the RI to 1.4540 (within 0.0003) by adding 8% PVP solution. For 2.01 M sucrose/PVP, mix 260 ml of 2.3 M sucrose/PVP with 35 ml of 8% PVP and adjust to a final RI of 1.4370. For 2.1 M sucrose/PVP, mix 270 ml of 2.3 M sucrose/PVP with 25 ml of 8% PVP and adjust to a final RI of 1.4420. Note that it is essential to prepare these solutions based on RI rather than molarity and to ensure complete mixture before each RI measurement.

BT buffer: Prepare a 0.1 M stock of Bis–Tris. Dissolve 104.5 g Bis–Tris in distilled water, adjust the pH to 6.5 with concentrated HCl solution, and make up to 500 ml. Filter as described earlier. Store at 4°. To prepare 1 liter of BT buffer, mix 100 ml of 0.1 M Bis–Tris–HCl, pH 6.5, stock with 100 μl of 1 M MgCl$_2$ and make up to 1 liter with distilled water. Filter sterilize.

Sucrose/BT gradient solutions: Prepare the 2.5 M sucrose/BT stock as described for the 2.3 M sucrose/PVP solution by dissolving 856 g sucrose in prewarmed BT buffer and making up to 1 liter with the same buffer. The RI should be 1.4533. Prepare 2.25 and 1.50 M sucrose/BT solutions by diluting the 2.5 M stock with BT buffer and adjusting the RI to 1.4414 and 1.4057, respectively.

Sucrose/Nycodenz solution: Warm 30 ml of distilled water to ~80° in a microwave oven. Add 20 g of Nycodenz and dissolve it by vigorous stirring on a heating plate. When the Nycodenz is dissolved, add 10 ml of 0.1 M Bis–Tris–HCl, pH 6.5, 10 μl of 1.0 M MgCl$_2$, 78.72 g of sucrose, and distilled water to just below 100 ml. After the sucrose has dissolved, let it cool down to room temperature and bring the volume to 100 ml with distilled water.

2% DNase I: Dissolve 5 mg of DNase I in 0.25 ml of 50 mM Tris–HCl, pH 6.8, 10 mM MgCl$_2$, 1.0 mM dithiothreitol (DTT), and 50% (v/v) glycerol. Store at −20°.

Growth of Cells

Large-Scale Preparations

1. Three days before the preparation, inoculate 5 ml of Wickerham's medium with the yeast strain of choice, starting from a fresh plate. Incubate with adequate aeration for 12 hr at 30°.

2. Two days before intending to perform the nuclei preparation, use a hemacytometer to count the cells in the 5-ml culture. Inoculate a 100-ml starter culture with enough cells to obtain a count of 1×10^7 cells/ml for a diploid strain and 2×10^7 cells/ml for a haploid strain the afternoon of the next day. Incubate the cells at 30° with appropriate aeration. The same day, insert a cotton plug at the end of a glass gas dispersion tube and connect to the same end to approximately 1 m of autoclavable rubber tubing. Wrap well into a package with aluminum foil and sterilize the whole assembly by autoclaving.

3. In the afternoon of the day before the preparation, count the cells in the 100-ml starter culture as just described. For most wild-type yeast strains, the growth of cells is optimal at 30°. In the case in which a 30° room is not available, cells can also grow at 25° or room temperature.

4. Inoculate an 18-liter flask with enough cells so that by the next morning the culture will have grown to a density of 1×10^7 cells/ml for diploid cells and 2×10^7 cells/ml for haploid cells. It is very important that the culture not be overgrown, as it will be difficult to remove the cell wall during spheroplasting. If the culture grows to a concentration above 2.5 and 5×10^7 cells/ml for diploid and haploid cells, respectively, it is not practical to precede with the preparation as the cells become difficult to spheroplast at higher concentrations (see later).

5. Place the flask on top of a stirring plate and stir vigorously for the duration of the incubation at either room temperature or at 30° (Fig. 1). Connect the bench air supply to a heating coil immersed in a water bath. It is paramount that the air be warmed before being bubbled through the culture, as the air bubbling through the flask will cool the culture down. The degree of heating necessary depends on each culture setup and must be determined experimentally, although we set the bath to 48°. Connect the outlet of the heating coil to the rubber tubing carrying the cotton-plugged glass gas dispersion tube. Use copious amounts of absolute ethanol to sterilize the top of the 18-liter flask and then insert the aeration assembly into the flask, making sure that the gas dispersion tip reaches the bottom of the flask but

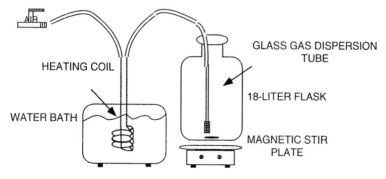

FIG. 1. Diagrammatic representation of the large-scale culture setup for yeast nuclei and nuclear envelopes preparations.

does not interfere with the magnetic stir bar. Open the air supply and regulate it so that air bubbles vigorously through the culture, but do not cause it to overflow.

6. The next morning, monitor the growth of the culture by counting cells with the hemacytometer.

Small-Scale Preparations

1. Prepare a 100-ml starter culture as described earlier for the large-scale preparation.
2. The afternoon of the day before the preparation, utilize this starter culture to inoculate a 2-liter flask as described previously.
3. Incubate the 2-liter flask in a shaker incubator at 30°. Shake vigorously overnight.
4. In the morning of the preparation, monitor the growth by counting the cells with the hemacytometer.

Preparation of Yeast Spheroplasts

Complete digestion of yeast cell walls is crucial to the successful isolation of nuclei. The following procedure applies to both large-scale and small-scale preparations.

1. Harvest yeast cells by centrifugation at $\sim$4000g_{av} for 5 min (e.g., Sorvall GS-3 rotor at 5000 rpm). Weigh an empty centrifuge bottle prior to harvesting in order to measure the weight of the resulting cell pellet. All centrifugation steps are performed at 4°. Keep the cell pellets on ice.

2. Resuspend cells in 100 ml ice-cold distilled water and pellet by centrifugation. For a small-scale preparation (e.g., 2-liter cultures), the cells can be resuspended in a minimal volume of water and transferred to a previously weighed 50-ml Falcon tube. Add enough water to bring the volume of the suspension to 40–50 ml and centrifuge at 2000g_{av} for 5 min (e.g., 3000 rpm in a Sorvall RT6000 centrifuge). Weigh the cell pellet, which is typically 25–35 g from a large-scale culture and 3–5 g from a small-scale culture.

3. Resuspend cells in 100 mM Tris, pH 9.4, 10 mM DTT. Use 100 ml per 20 g of cells and incubate in a 30° water bath for 10 min. Shake manually every 5 min or, if using a shaker, set the speed to 40 rpm.

4. Centrifuge the cell suspension at 2000g_{av} for 5 min and resuspend the resulting pellet in ice-cold distilled water (100–250 ml for large scale and 50 ml for small scale). Repeat this centrifugation step and then resuspend in ice-cold, filter-sterilized 1.1 M sorbitol (same volumes as used in step 3).

5. Centrifuge the cell suspension at 2000g_{av} for 5 min and resuspend the pellet in a minimal volume of 1.1 M sorbitol (a total volume of 100 ml for every 40 g of cells). Transfer this suspension to a flat-bottomed flask, such as a 500-ml Erlenmeyer glass flask.

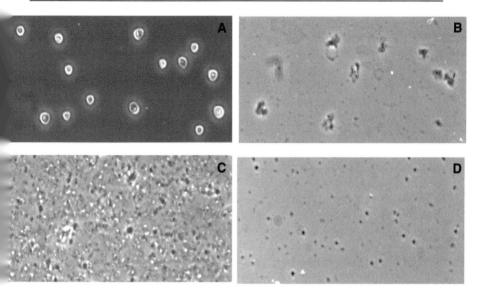

FIG. 2. Yeast spheroplasts and nuclei as observed by phase-contrast microscopy (40× objective). Spheroplasts diluted in 1.1 M sorbitol (A) and in water (B). (C) Spheroplasts lysed by Polytron homogenization as described in the text. (D) Isolated yeast nuclei (2.01/2.10/2.30). Bar: 10 μm.

6. To digest cell walls, add 10 ml glusulase, 3 ml of freshly made 10 mg/ml Zymolyase/mutanase cocktail (dissolve 10 mg of each enzyme powder in a total of 1 ml of 1.1 M sorbitol), and 0.5 ml of 1 M DTT for every 100 ml of cell suspension. Incubate the flask in a gently shaking 30° water bath (<60 rpm) for 3 hr. Be sure that the flask is not shaking too fast to prevent premature lysis of the cells. Swirl the suspension manually every 15–20 min to prevent clumping of the digesting cells. After 45 min, the suspension will separate into distinct layers due to aggregation and subsequent sedimentation of the cells. By 3 hr, the layers should disappear.

7. The best way to assess the extent of the digest is to check spheroplasts by light microscopy (Figs. 2A and 2B). Dilute 10 μl of cell suspension in 200 μl of 1.1 M sorbitol; allow to incubate at room temperature for 1–2 min. Spheroplasts should be spherical, phase bright, and not attached to buds. Dilute 10 μl of suspension in 200 μl distilled water, mix gently, and leave at room temperature for 1–2 min. The spheroplasts should be lysed almost completely, leaving small pieces of membranes and bits of cellular debris (the lysis is rarely as complete as that seen for *S. uvarum* strains[2]).

Isolation of Yeast Nuclei

The method utilized for the preparation of yeast nuclei is essentially as described by the Rout and Kilmartin[2] laboratory. During digestion of the cell wall, warm the 2.01, 2.10, and 2.30 M sucrose/PVP solutions for the fractionation gradient to room temperature. This will allow for easier dispensing of these very viscous solutions. Also, prepare the Ficoll/sorbitol cushions: for large-scale preparations, pour 50 ml of the Ficoll/sorbitol solution into a 250-ml centrifuge bottle suitable for a Beckman JS-4.2 rotor; for small-scale preparations, pour 15 ml of the Ficoll/sorbitol solution into a polypropylene HB-4 centrifuge tube (Sorvall). If the Polytron (for step 6, below) is not stored at 4°, move it to the cold room at the beginning of the day so it can cool.

Large-Scale Preparation

1. Add 100 ml ice-cold 1.1 M sorbitol to the spheroplast suspension and transfer to a centrifuge bottle. Centrifuge at $3000g_{av}$ for 5 min and decant or aspirate supernatant carefully.

2. Resuspend the resulting pellet in 100 ml of 1.1 M sorbitol by stirring gently with a pipette or glass rod.

3. Overlay the suspension onto a 50-ml Ficoll/sorbitol cushion, poured previously in a 250-ml centrifuge tube suitable for a Beckman JS4.2 rotor. To overlay, pipette the suspension slowly down the side of the tube until the cell suspension forms a layer on top of the cushion. Once the initial layer is formed, the suspension can be dispensed at a greater rate, taking care not to disrupt the interface between the cushion and the spheroplast suspension.

4. Centrifuge at $2000g_{max}$ (2700 rpm in a Beckman JS-4.2 rotor) for 25 min using the appropriate adapters for 250-ml bottles.

5. During the centrifugation step, prepare sucrose/PVP gradients for fractionation. Have at least 8 ml of each sucrose solution for each Beckman SW28 polypropylene tube. For a large-scale preparation, six SW28 gradients are typically required. However, be sure to adjust the number of gradients based on the weight of the cells (∼5–7.5 g cells per gradient). Add 1 : 100 dilution of solution P and 1 : 200 dilution of PIC to each sucrose/PVP solution and mix to homogeneity. Pipette 8 ml of the 2.3 M sucrose/PVP into the SW28 tube, followed by 8 ml of the 2.1 M sucrose/PVP, and finally 8 ml of the 2.01 M sucrose/PVP. The sucrose solutions should layer on top of each other with clear interfaces between them. Cover the tubes with Parafilm or foil and store in a rack at 4°.

6. After pouring the gradients, prepare the Polytron for the lysis procedure. Rinse the probe by running it in ice-cold distilled water in a 100-ml graduated cylinder three times, changing the water in between. Dry the probe with lint-free paper towels. If necessary, immerse the probe in ice to cool it completely.

7. At the conclusion of centrifugation, the spheroplasts will be at the bottom of the tube with the Ficoll/sorbitol cushion on top. This step removes the last traces of the digestive enzymes and removes small buds, which will not lyse and would thus contaminate the isolated nuclei fraction. Remove the Ficoll/sorbitol layer by

aspiration. Rinse the sides of the tube with ice-cold 1.1 M sorbitol and aspirate off the wash. Place the pellet on ice.

8. To minimize proteolysis, it is highly recommended that the spheroplasts be lysed immediately after removing the Ficoll/sorbitol cushion. Therefore, the lysis solution should be made in advance. For each large-scale preparation, prepare 150 ml of lysis solution: 150 ml 8% PVP, 375 μl 10% Triton X-100, 750 μl 1 M DTT, and 1.5 ml PIC.

9. This step should take place in the cold room. Add lysis solution to the spheroplast pellet, add 1.5 ml of solution P, and resuspend the pellet using the Polytron. To lyse the cells with the Polytron, use a circular/spiraling motion to move the probe along the side of the tube. For the first 20 sec, have the probe stay at the bottom to fully resuspend the spheroplast pellet. Afterward, move it upward in a spiraling motion to ensure that the probe makes contact throughout the liquid column. When the top is reached, spiral downward to the bottom. Maintain the motion for a total of 1 min. Keep the tube on ice for 10–20 sec and repeat at least two more times, but no more than a total of five runs. If the probe becomes the least warm to the touch, cool it again on ice. The speed setting for the Polytron varies with the instrument and with cell size, which is dependent on the strain background and whether the genome is diploid or haploid (the bigger the cells, the less shear force needed to lyse the plasma membrane, and thus the lower the speed setting). As a guideline, the DF5 strain background has smaller cells than W303 cells; DF5 haploid cells are homogenized at setting 6.5, whereas diploids are lysed at 5.5 or lower. In contrast, W303 haploid cells are lysed at 5.5, whereas diploids are lysed at 5 or lower.

10. Evaluate the extent of lysis by looking at 10 μl of the suspension under a phase-contrast light microscope (Fig. 2C). There should be no significant clumps of cells, a very small number of unbroken cells (less than 2%), shreds of membrane, intact vacuoles (large phase-bright spheres), and intact nuclei (small, gray spheres with a dark nucleolar crescent on one side). Nuclei can be seen clearly as small, dark gray balls with a 100× oil-immersion objective lens.

11. Pour half of the suspension into another 250-ml centrifuge tube. Underlay each tube with 75 ml of 0.6 M sucrose/PVP and mix thoroughly. Centrifuge at 10,000g_{av} (e.g., 8000 rpm in a Beckman JA-14 rotor) for 25 min.

12. Decant the supernatant into 50-ml centrifuge tubes; this is the crude cytosol and can be used for other experiments, such as the immunoaffinity purification of epitope-tagged karyopherins.[12,13] Store immediately at −80°.

13. Resuspend the remaining pellet (the crude nuclei fraction) in 60 ml of 2.1 M sucrose/PVP (for 25–45 g cells) using the Polytron, which is set to a low speed (usually setting 4–4.5) for 1 min. If necessary, adjust the RI to 1.4300 with

[12] J. D. Aitchison, G. Blobel, and M. P. Rout, *Science* **274**, 624 (1996).
[13] M. P. Rout, G. Blobel, and J. D. Aitchison, *Cell* **89**, 715 (1997).

PVP. This fraction will contain mainly nuclei, visible as gray fuzzy spheres under the microscope (see earlier discussion).

14. Overlay 10–12 ml of the suspension on each SW28 sucrose/PVP gradient. If necessary, overlay with 0.3 M sucrose/PVP to the top of the SW28 tube, which will prevent collapse of the tube during centrifugation. Centrifuge all gradients in a Beckman SW28 rotor at 28,000 rpm (103,000g_{av}), 4°, for 4 or 8 hr if preparing diploid or haploid nuclei, respectively.

15. To unload the gradients, use a 10-ml pipette to remove each layer, including the interface. It is often necessary to use the tip of the pipette to gently dislodge the interface from the sides of the tube by moving the tip with great care along the circumference of the tube at the interface. Start at the top, with the tip at the surface, and aspirate while moving the tip along the side of the tube. Keep the tip at the surface and do not disturb the lower layers during collection. The first fraction collected is labeled "S" and is 8–10 ml in volume. The second fraction is labeled "S/2.01," which includes the interface between the loaded sample and the 2.01 M sucrose/PVP layer. This interface is typically a thick, yellow-white band and the volume of this fraction is 8–10 ml as well. The first two fractions usually contain mitochondria, vesicles, and microsomes. The next *two* layers of the gradient are pooled together as one fraction (Nuclei), which is labeled "2.01/2.10/2.30" and is 15–20 ml in volume. The two interfaces should be less intense than the previous ones and be pearly white to beige in appearance. When viewed under the microscope, the "2.01/2.10/2.30" fraction should be considerably enriched for nuclei, with only occasional small contaminating particles and little to no unbroken cells present (Fig. 2D).

16. Store all fractions at $-80°$.

Small-Scale Preparation

Unless stated otherwise, all steps are as described in the large-scale preparation.

1. Add ice-cold 1.1 M sorbitol to the spheroplasts to a total volume of 50 ml and transfer to a 50-ml centrifuge tube.

2. Centrifuge at 2000g_{av} for 5 min (e.g., 3000 rpm for 5 min in a Sorvall RT6000 centrifuge). Carefully aspirate away the supernatant.

3. Resuspend the spheroplast pellet in a total volume of 20 ml 1.1 M sorbitol and overlay the suspension onto a 15-ml Ficoll/sorbitol in a polypropylene Sorvall HB-4 centrifuge tube (or equivalent).

4. Centrifuge at 8000 rpm (10,000g_{av}) for 15 min in a Sorvall HB-4 rotor and remove the Ficoll/sorbitol as described previously.

5. Add 20 ml of the lysis solution to the spheroplast pellet, followed by 200 μl of solution P. Resuspend the pellet as described earlier.

6. Underlay with 10 ml of 0.3 M sucrose/PVP (containing 1 : 100 dilution of solution P and 1 : 200 PIC) by adding the solution carefully to the lysed spheroplasts.

Centrifuge at 10,000 rpm ($16,000g_{av}$) for 20 min in a Sorvall HB-4 rotor and then gently decant off the supernatant. Take great care at this step, as the pellet is very loose.

7. Resuspend the pellet in 8 ml of 2.1 M sucrose/PVP containing a 1 : 100 dilution of solution P and 1 : 200 PIC using the Polytron as described earlier.

8. Overlay the suspension on the sucrose/PVP gradient and centrifuge at 28,000 rpm ($103,000g_{av}$) as described earlier. Unload the gradients as described previously.

Isolation of Yeast Nuclear Envelopes

Large-Scale Preparation

1. Measure the OD_{260} of the nuclear fraction after 1 in 100 dilution in an aqueous 1.0% SDS solution. One large-scale preparation results typically in a total of 500–1000 OD_{260}.

2. Adjust the nuclear fraction with 8% PVP solution (containing 0.01 volume of solution P) to a RI of 1.4340. This step usually takes ~0.2 volume of 8% PVP solution. Transfer the sample in centrifuge tubes suitable for Beckman Ty50.2Ti rotors (or equivalent), mix well by vortexing, and then overlay to the top of the tube with 8% PVP solution.

3. Centrifuge the nuclei suspension at 40,000 rpm ($193,000g_{max}$) in a Beckman Ty50.2Ti rotor for 1 hr at 4°. Aspirate the supernatant completely and wipe the sides clean. Keep nuclei pellets on ice if processed directly; they can be stored at $-80°$ if necessary.

4. The following procedure should be performed in the cold room one tube at a time. Make up the required amount of nuclear lysis solution 1 ml for every 100 OD_{260} (1.0 ml BT buffer, 1.0 μl 2.0% DNase I, 10 μl PIC, and 10 μl solution P) freshly each time and precool before use. *Immediately* resuspend the pellet by vortexing at the maximum setting. Note that it is extremely important that the solution swirls around the sides of the tube to develop enough shear force to lyse the nuclei efficiently; some foaming during this process is normal. After resuspension appears to be complete, continue to vortex for another ~2 min. Resuspension can take up to 10 min. Warm the tube to room temperature by holding it in your hands and then incubate for a further 5–10 min at room temperature. An incomplete DNA digestion (if the lysate has not been warmed up properly) can prevent NEs from floating up properly through the gradient during the next step.

5. Add one equal volume of the sucrose/Nycodenz solution to the nuclear lysate and mix thoroughly by vortexing. Note that the sucrose/Nycodenz solution for this and the subsequent step has to be at room temperature, as it will not mix properly with the NE suspension. Centrifuge at 6000 rpm ($4350g_{max}$) for 6 min in a Beckman Ty50.2Ti rotor to remove dead cells and cell wall debris. Transfer the supernatant to a fresh tube and add three additional volumes of sucrose/Nycodenz solution before mixing thoroughly. Measure the RI of the crude NE suspension

and, if necessary, adjust to at least 1.4430 with sucrose/Nycodenz solution. This step ensures that the NE suspension will underlay correctly in the subsequent flotation gradient.

6. Pour 9–12 ml of suspension in the appropriate number of Beckman SW28 centrifuge tubes. For a large-scale preparation, we usually use three gradient tubes. Overlay each sample with 9 ml of 2.25 M sucrose/BT solution and 9 ml of 1.50 M sucrose/BT solution. Fill the tube to the top with BT buffer and centrifuge at 28,000 rpm (103,000g_{av}) in a Beckman SW28 rotor for ∼24 hr at 4°.

7. Unload the gradient tubes from the top. A faint white band at the top of the tube contains a few vesicular remnants and should be removed completely (remove ∼6.0 ml per tube). NEs are found at the 1.50/2.25 M interface, appearing as a broad, white, and flocculent band (collect ∼12 ml per tube). The protein concentration of this fraction is typically ∼0.5–1.0 mg/ml, i.e., ∼15–30 mg total NE protein is obtained from a large-scale preparation. The next band is a dense, sharp yellowish/white band containing a few nuclear envelopes and dead cell remnants (collect ∼12.0 ml per tube; 2.25/S' fraction). The final ∼6.0 ml (S'fraction), including a dense brownish/white pellet, contains soluble and particulate matter mainly derived from chromatin and cell wall remnants. The NE-containing fraction can be stored at −80° indefinitely.

8. Successful NE preparation may be checked by light microscopy. Under a 100× phase-contrast oil-immersion objective, NEs appear as small "C"-shaped structures (retaining the original curvature of the NE).

Small-Scale Preparation

Unless otherwise stated, all steps are as described in the large-scale preparation.

1. To 6 ml of the nuclear fraction, add 1.5 ml of 8% PVP solution, 100 μl of solution P, and mix well by vortexing.
2. Load the nuclei suspension into Beckman Ty80Ti centrifuge tubes (or equivalent), overlay as before with 8% PVP, and pellet the nuclei for 1 hr at 50,000 rpm (235,200g_{av}) at 4°.
3. Remove the supernatant completely and resuspend the nuclear pellet in 0.5 ml lysis solution as described in step 4 of the large-scale preparation method.
4. After 10 min at room temperature, transfer the crude NE suspension into a fresh microcentrifuge tube and spin for 1 min in a VWR mini centrifuge at maximum speed (2000g_{max}).
5. Transfer the supernatant to a Beckman SW55 tube (or equivalent) and add 2 ml of sucrose/Nycodenz solution. Seal the top of the tube with Parafilm and vortex thoroughly.
6. Overlay the NE suspension with 1.5 ml of 2.25 M sucrose/BT solution followed by 1.2 ml of 1.50 M sucrose/BT solution. Top off with BT buffer. Both sucrose solutions contain solution P in a 1 : 200 dilution.

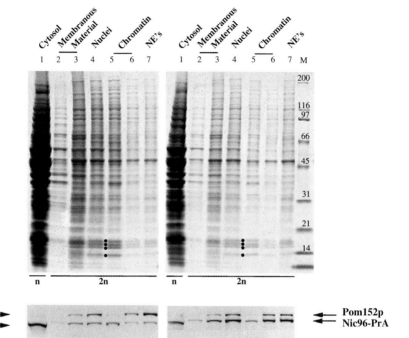

FIG. 3. Yeast cells expressing either the protein A (PrA)-tagged version of the karyopherin Kap95p or the nucleoporin Nic96p were subjected to subcellular fractionation as described in the text. Proteins present in each of the fractions were resolved by SDS–PAGE (5–20%). Gel samples were prepared as described in the text. Lane numbers on top of the gels correspond to the fraction numbers used in Table I. Numbers below the bars (loading equivalents) represent cell equivalents used to prepare each of the fractions. (A) Gels were stained with Coomassie Brilliant Blue R-250. Histones (dots) are indicated. (B) Immunoblots of gels identical to the ones presented in (A) were probed with rabbit anti-mouse IgG to detect the PrA-tagged protein and with MAb118C3[3] to detect the NPC marker Pom152p.

7. Spin the tubes in a Beckman SW55 rotor at 50,000 rpm ($175,000g_{av}$) for 24 hr at 4°.
8. Unload the gradient from the top, as described earlier.

Assessment of Yield and Enrichment of NE

The yield and enrichment of proteins of interest during the fractionation can be assessed by SDS–PAGE followed by Coomassie staining and immunoblot analysis (Fig. 3). Proteins that have been epitope tagged with the immunoglobulin G (IgG)-binding moiety of PrA from *Staphylococcus aureus* can be detected easily by rabbit affinity-purified antibody to mouse IgG (ICN/Cappel, Costa Mesa, CA[14]).

[14] M. P. Rout, J. D. Aitchison, A. Suprapto, K. Hjertaas, Y. Zhao, and B. T. Chait, *J. Cell Biol.* **148**, 635 (2000).

TABLE I
TYPICAL LOADING PROFILE FOR SDS–PAGE ANALYSIS OF FRACTIONS OBTAINED FROM
THE SMALL-SCALE YEAST NUCLEI AND NE ISOLATION PROCEDURE

Fraction No.	Fraction name	Total volume (ml)	Aliquot loaded on gel (μl)	Volume equivalent loaded on gel
1	Crude cytosol	30	15	1/2000
2	S	8	8	1/1000
3	S/2.01	8	8	1/1000
4	2.01/2.1/2.3 (nuclei)	16	16	1/1000
5	S'	1.5	3	1/500
6	2.25/S'	3	6	1/500
7	1.50/2.25 (NE)	3	6	1/500

To monitor the efficiency of the fractionation procedure, we commonly follow the marker nucleoporin Pom152p, an integral membrane protein in the NPC.[3,14]

Because the solutions used for fractionation often contain 8% PVP, we generally precipitate the proteins for SDS–PAGE using methanol. Commonly used protein precipitation protocols (such as TCA precipitation) do not work if the PVP concentration is higher than 1% because the polymer tends to coprecipitate with the proteins and interferes with the protein separation in SDS–PAGE.

1. To 100 μl of sample, add 900 μl of methanol and mix by vortexing.
2. Incubate for 1 hr at $-20°$.
3. Spin the sample in a microcentrifuge at maximum g ($20,800 g_{max}$) for 20 min at 4° and remove the supernatant carefully.

TABLE II
COMMONLY USED PROTEIN MARKER FOR IMMUNOBLOT ANALYSIS
OF YEAST NUCLEI AND NE FRACTIONATION PROCEDURES

Marker	Subcellular localization	Antibody name	Mw (kD)	Reference
Pom152p	NPC	MAb118C3	152	3
Nup1p	NPC	MAb414/M	114	15,16,17
Nup116p		Ab350	116	
Nsp1p			87	
p65^{NUP145N}			65	
Nup57p			57	
Nup49p			49	
Nop1p	Nucleolus	mAb D77	37	18
Sec61p	ER	Rabbit serum	70	19

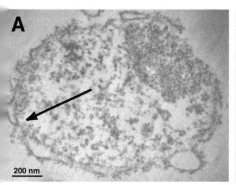
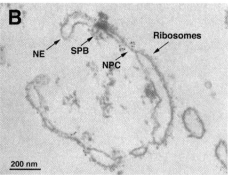

FIG. 4. (A) Electron micrograph of immunolabeled permeabilized nuclei (2.01/2.10/2.30) from yeast cells expressing a protein A-tagged nucleoporin (Nup1p-PrA). Colloidal gold is associated with an NPC in the lower left-hand corner. (B) Electron micrograph of an immunolabeled nuclear envelope (NE) (1.50/2.25) from yeast cells expressing Nup116p-PrA showing a clear "C"-shaped double-membraned structure with associated ribosomes, nuclear pore complexes (NPC), and spindle pole bodies (SPB). Colloidal gold is associated with NPCs.

4. Resuspend the protein pellet in 90% of methanol. Leave for 1 hr at $-20°$ and centrifuge the sample as before.

5. Remove the supernatant completely and resuspend the pellet in 50 μl solution A (0.5 M Tris–base, 5% SDS) if necessary by sonication. When the pellet is resuspended completely, add 50 μl solution B [75% (v/v) glycerol, 125 mM DTT, 0.05% (w/v) bromphenol blue], mix well, and heat for 10 min at 95°, or 65°, if the protein of interest appears as a smear after SDS–PAGE separation.

To accurately represent the distribution of the protein of interest throughout the fractionation procedure, it is necessary to load an aliquot of each fraction proportional in volume to the total volume of that fraction on the gel. Table I represents a typical loading profile for a small-scale preparation. To determine the quality of fractionation, a variety of subcellular markers can be used and are listed in Table II.[3,15–19]

A more extensive method of quality control from nuclei and NE preparations is to look at thin section EM of the enriched yeast nuclei and NE fractions (Fig. 4).

[15] L. I. Davis and G. Blobel, *Cell* **45**, 699 (1986).
[16] L. I. Davis and G. Blobel, *Proc. Natl. Acad. Sci. U.S.A.* **84**, 7552 (1987).
[17] L. I. Davis and G. R. Fink, *Cell* **61**, 965 (1900).
[18] J. P. Aris and G. Blobel, *J. Cell Biol.* **170**, 2059 (1988).
[19] C. J. Stirling, J. Rothblatt, M. Hosobuchi, R. Deshaies, and R. Schekman, *Mol. Biol. Cell.* **3**, 129 (1992).

Modifications

The following procedure can be adopted to prepare nuclei and NEs when simplicity, brevity, and yield are more important than a high degree of enrichment.

1. Perform the steps for isolating yeast nuclei as described earlier, up to the resuspension of the crude nuclear pellet (see "Isolation of Yeast Nuclei," large-scale preparation, steps 1–13).
2. Check the RI of the suspension and adjust it to 1.4380 with 2.3 M sucrose/PVP (usually requires 0.33 volume).
3. Transfer suspension to preweighed Beckman Ty50.2Ti tubes and overlay with 8% PVP to the top. Centrifuge at 40,000 rpm (120,000g_{max}) for 4 hr at 4°.
4. Remove supernatant and wipe the sides clean. If necessary, the nuclear pellet can be stored at −80° or processed directly.
5. Weigh the pellet (typically 0.3–1.0 mg), transfer a small weighed aliquot to a microcentrifuge tube, and measure the OD_{260} in 1% SDS. Calculate the total OD_{260} for the pellet.
6. Resuspend the pellet in the correct amount of nuclear lysis solution as described earlier. Proceed as usual.

[24] Studying Yeast Vacuoles

By ELIZABETH CONIBEAR and TOM H. STEVENS

Introduction

The vacuole of *Saccharomyces cerevisiae* is an acidic, degradative compartment comparable to the mammalian lysosome and the plant vacuole. In addition to its role in protein turnover and nutrient recycling, it has important functions in osmoregulation and ion homeostasis (e.g., Ca, Fe), as well as the storage of amino acids and inorganic phosphate.[1]

Despite these diverse roles, most functions of the yeast vacuole are dispensable for life when yeast are grown in defined media under laboratory conditions, making the study of vacuolar biogenesis amenable to genetic studies. Regulated transport to the vacuole is important in such processes as the clearance of mating pheromone receptors from the cell surface, the redistribution of amino acid permeases during

---

[1] E. W. Jones, G. C. Webb, and M. A. Hiller, "Molecular Biology of the Yeast *Saccharomyces cerevisiae*," Vol. III, p. 363. Cold Spring Harbor Laboratory Press, Cold Spring Harbor, NY, 1997.

changes in nutrient availability, and starvation-induced autophagy. Newly synthesized vacuolar hydrolases and other enzymes, such as the V-ATPase responsible for vacuolar acidification, must be constitutively targeted to the vacuole. Consequently, the yeast vacuole has become an important model system for investigating the sorting of proteins from the Golgi, cell surface, or cytosol and for other studies ranging from vacuolar inheritance to homotypic vacuolar fusion.[2–5]

The chapter is organized in four parts. We first describe techniques for visualizing the vacuole in living and in fixed cells. Because protein targeting to the vacuole is central to its function (and a major focus of our laboratory), methods are provided in the next section for assessing the fidelity of protein sorting using soluble and membrane-bound hydrolases as marker proteins. The third section outlines techniques for determining if an uncharacterized protein is associated with and/or degraded in the vacuole. The use of stage-specific transport mutants to identify the route by which a particular protein reaches the vacuole is also examined. Finally, we describe methods for purifying vacuoles based on their low buoyant density in order to carry out biochemical studies, such as the assembly and activity of the V-ATPase.

Microscopic Visualization of Vacuoles

Three different approaches for visualizing the yeast vacuole are presented: staining with vital dyes, using green fluorescent protein (GFP)-tagged vacuolar marker proteins, and immunofluorescence microscopy of formaldehyde-fixed cells. In choosing a particular technique, it is important to be aware of its advantages and limitations. For example, changes in the size, shape, and number of vacuoles over time can best be appreciated by examining live cells. Vital staining is quick and simple to carry out and reflects the status of the vacuole as a whole. Fixation artifacts are avoided, but vital dyes themselves can introduce artifacts, as discussed later. Using a GFP-tagged marker protein allows the vacuole to be visualized without further experimental manipulation once the appropriate strain is constructed. Indirect immunofluorescence microscopy provides a highly sensitive and specific way to examine the localization of an endogenous marker protein, even though it is laborious and time-consuming. In practice, it is often desirable to use a combination of one or more of these techniques.

Staining with Vital Dyes

The vacuole is a dynamic organelle, undergoing morphological changes in response to osmotic stress, changing nutrient conditions and during the cell cycle

---

[2] E. Conibear and T. H. Stevens, *Biochim. Biophys. Acta* **1404**, 211 (1998).
[3] N. J. Bryant and T. H. Stevens, *Microbiol. Mol. Biol. Rev.* **62**, 230 (1998).
[4] D. J. Klionsky and S. D. Emr, *Science* **290**, 1717 (2000).
[5] W. Wickner and A. Haas, *Annu. Rev. Biochem.* **69**, 247 (2000).

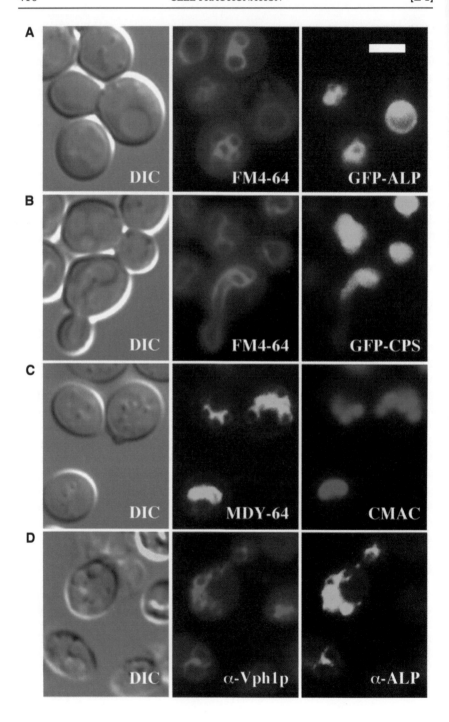

when it is actively partitioned into the forming bud. Wild-type cells typically contain one to five vacuoles that can be observed in the light microscope using suitable optics (e.g., differential interference microscopy; Fig. 1). Low levels of glucose and a number of other environmental stresses cause the rapid fusion of vacuoles into a single large, round structure, whereas the formation of segregation structures during mitosis causes the vacuole to appear elongated and fragmented. Specific morphological defects are associated with different classes of vacuolar protein-sorting mutants,[6] and screening for mutants with aberrant vacuolar morphology has led to the isolation of components of the vacuolar fusion machinery.[7]

A number of fluorescent vital dyes have been developed that accumulate either in the lumen or in the limiting membrane of the vacuole. In general, these staining procedures are simple and quick to carry out and many are compatible with the dual visualization of GFP-tagged proteins or other fluorescent markers.

General Considerations for Viewing Live Cells

The sensitivity of vacuolar morphology to changes in the environment means that care must be taken in the growth, handling, staining, and viewing of live cells. Certain fluorescent compounds that label the vacuole may themselves influence vacuolar morphology or function (e.g., quinacrine and the red fluorescent pigment that accumulates in the vacuole of *ade2* mutants[8]). It is advisable to supplement the growth medium of *ade2* strains with 40 mg/liter adenine sulfate to suppress the accumulation of this endogenous fluorophore. To maintain both vacuolar morphology and integrity of the V-ATPase,[9] cells should be grown to early log phase in glucose-containing medium and harvested by centrifugation at low speed or by filtration.

Cells can be immobilized for viewing and photography on a slide that has been coated with 1 mg/ml concanavalin A (Con A) and air-dried. Examine live cells soon after staining (within 30 min), as anaerobic conditions and nutrient depletion develop quickly on the slide. Cell viability can be extended by mounting in a thin

---

[6] C. K. Raymond, I. Howald-Stevenson, C. A. Vater, and T. H. Stevens, *Mol. Biol. Cell* **3,** 1389 (1992).
[7] Y. Wada, Y. Ohsumi, and Y. Anraku, *J. Biol. Chem.* **267,** 18665 (1992).
[8] J. R. Pringle, R. A. Preston, A. E. Adams, T. Stearns, D. G. Drubin, B. K. Haarer, and E. W. Jones, *Methods Cell Biol.* **31,** 357 (1989).
[9] P. M. Kane, *J. Biol. Chem.* **270,** 17025 (1995).

FIG. 1. Different ways to look at the vacuole. (A) GFP–ALP is found on vacuolar membranes that label with FM4-64 (B) GFP–CPS is found within the lumen of FM4-64-labeled membranes. (C) Double labeling of vacuoles with the lumenal vital dye CMAC and the vacuolar membrane vital dye MDY-64. (D) Colocalization of Vph1p and ALP in fixed cells by double-label immunofluorescence microscopy. Because GFP–ALP (A), GFP–CPS (B), and ALP (D) are expressed from plasmids, staining may not be present in every cell. Note that vacuoles look like craters by differential interference contrast microscopy (DIC). Bar: 3 μm.

layer of 25% (w/v) gelatin,[10] using an agarose pad,[11] or by the "hanging agar block" method.[12]

Labeling Vacuolar Membrane

FM4-64 (Molecular Probes, Eugene, OR) is a lipophilic dye that exhibits a long wavelength red fluorescence only when bound to lipid. It is particularly useful for colocalization studies with GFP-tagged proteins (Figs. 1A and 1B). Because it binds to the plasma membrane and follows the endocytic pathway to reach the vacuole, its uptake is energy and temperature dependent.[13] Under conditions that slow the rate of endocytosis (e.g., certain mutations, or low temperature) or after short chase times, this dye will label the plasma membrane and endocytic structures in addition to the vacuole. FM4-64 can also be used to follow the rate of endocytosis to the vacuole and to quantify membrane recycling from intracellular compartments back to the plasma membrane.[14,15]

PROTOCOL

1. Grow cells to early log phase. Pellet cells at low speed (3500g for 30 sec) and resuspend in YEPD at a density of 2–4 OD$_{600}$ units/ml.
2. Add FM4-64 to a final concentration of 40 μM [from a 16 mM stock in dimethyl sulfoxide (DMSO)] and incubate for 15 min at 30°.
3. Resuspend cells in fresh YEPD and incubate for 30–60 min at 30°. Examine by fluorescence microscopy using Texas Red filters.

MDY-64 (Molecular Probes) is a bright, green lipophilic dye that preferentially labels the vacuole[16] (Fig. 1C). Longer incubations with the dye result in intense labeling and the staining of other cellular membranes.

PROTOCOL

1. Resuspend cells at 2–4 OD$_{600}$ units/ml in labeling buffer (10 mM HEPES, pH 7.4, 5% glucose).
2. Add MDY-64 to a final concentration of 10 μM (from a 10 mM stock solution in DMSO) and incubate for 3–5 min at room temperature.
3. Resuspend cells in fresh labeling buffer. Examine immediately by fluorescence microscopy using fluorescein filters.

[10] S. L. Shaw, E. Yeh, K. Bloom, and E. D. Salmon, *Curr. Biol.* **7,** 701 (1997).
[11] J. A. Waddle, T. S. Karpova, R. H. Waterston, and J. A. Cooper, *J. Cell Biol.* **132,** 861 (1996).
[12] S. D. Kohlwein, *Microsc. Res. Tech.* **51,** 511 (2000).
[13] T. A. Vida and S. D. Emr, *J. Cell Biol.* **128,** 779 (1995).
[14] A. Wiederkehr, S. Avaro, C. Prescianotto-Baschong, R. Haguenauer-Tsapis, and H. Riezman, *J. Cell Biol.* **149,** 397 (2000).
[15] T. Vida and B. Wendland, *Methods Enzymol.* **351,** [35], 2002 (this volume).
[16] R. P. Haugland, "Handbook of Fluorescent Probes and Research Chemicals," 6th Ed. Molecular Probes, Eugene, OR, 1996.

Labeling Vacuole Lumen

The blue fluorescent dye CMAC (7-amino-4-chloromethylcoumarin; Molecular Probes) accumulates in the lumen of the vacuole and is compatible with double staining with MDY-64 (Fig. 1C), as well as the colocalization of GFP-tagged proteins.[16,17] Two derivatives, CMAC-Arg and CMAC-Ala-Pro (Molecular Probes), are nonfluorescent until cleaved by vacuolar peptidases. CMAC dyes are reported to be aldehyde fixable.[16]

PROTOCOL

1. Resuspend cells at 2–4 OD_{600} units/ml in labeling buffer (10 mM HEPES, pH 7.4, 5% glucose).
2. Add CMAC to a final concentration of 100 μM (from a 10 mM stock solution in DMSO) and incubate for 15–30 min at room temperature.
3. Examine by fluorescence microscopy using UV filters.

CDCFDA (carboxydichlorofluorescein diacetate; Molecular Probes) is nonfluorescent until both acetate groups are hydrolyzed by esterases. Its accumulation in the vacuole is not affected by mutations that disrupt vacuolar acidification or the activation or accumulation of vacuolar hydrolases. In fact, it can stain the vacuoles of certain *vps* mutants very brightly. The related dye 6-CFDA has been used to measure the pH of the yeast vacuole by fluorescence ratio microscopy using the difference in emission intensity at excitation wavelengths of 450 and 495 nm.[8]

PROTOCOL

1. Resuspend cells at 2–4 OD_{600} units/ml in 50 mM sodium citrate, pH 5.0, plus 2% glucose.
2. Add CDCFDA to a final concentration of 10 μM (from a 10 mM stock solution in DMSO) and incubate for 15–30 min at room temperature.
3. Examine by fluorescence microscopy using fluorescein filters.

Staining with quinacrine is used primarily to assess the acidity of the vacuole. Because this weak base will accumulate until the pH of the vacuole has been neutralized, staining intensity provides an indication of intralumenal pH. Using a modified procedure, the vacuolar accumulation of quinacrine can be quantified in a spectrophotometer.[18] Quinacrine is not recommended for morphological studies because it causes the multiple vacuoles seen in wild-type cells to coalesce. Like other lysosomotropic weak bases, it affects protein-sorting pathways in mammalian cells and may also do so in yeast.[8]

---

[17] C. J. Stefan and K. J. Blumer, *J. Biol. Chem.* **274**, 1835 (1999).
[18] J. H. Seol, A. Shevchenko, A. Shevchenko, and R. J. Deshaies, *Nature Cell Biol.* **3**, 384 (2001).

PROTOCOL

1. Resuspend cells at 2–4 OD_{600} units/ml in YEPD buffered to pH 7.6 with 50 mM Na_2HPO_4 containing 200 μM quinacrine.
2. Incubate for 5–10 min at room temperature.
3. Wash once in minimal medium buffered to pH 7.6 with 50 mM Na_2HPO_4 and resuspend cells in the same solution (buffered YEPD can be used, although it will inhibit the immobilization of cells on Con A-coated slides).
4. Examine within 10 min by fluorescence microscopy using fluorescein filters.

Vacuolar Localization of Green Fluorescent Protein Fusions

The use of suitable marker proteins tagged with GFP from *Aequorea victoria* is an effective way to visualize the vacuole in living yeast cells.[19] The expression of fully functional tagged forms of a protein at endogenous levels should not perturb organelle morphology or function. Distribution of a GFP-tagged protein can be compared with other markers using different color variants of GFP (e.g., CFP and YFP[20]), labeling the cells with compatible vacuolar dyes, such as FM4-64 or CMAC (see earlier discussion), or fixing and processing the cells for double-label immunofluorescence microscopy (see later). The fluorescence of GFP-tagged proteins is often retained after fixation and, if necessary, the signal can be enhanced using commercially available anti-GFP antibodies and fluorescein isothiocyanate (FITC)-conjugated secondary antibodies. Mild fixation can also be helpful in immobilizing dynamic intracellular structures for photography.[21]

There are some drawbacks to the use of GFP-tagged proteins. The long folding times of GFP (30 min–2 hr depending on the variant used) mean that newly synthesized material cannot be detected. Furthermore, if the GFP is transported to the vacuolar lumen before folding is complete, it is degraded and can only been seen in vacuolar protease-deficient (*pep4*) mutant strains. Conversely, fully folded GFP is resistant to vacuolar proteases and can accumulate in the vacuole after being cleaved from the fusion protein. Because the signal from GFP is not always very strong, autofluorescence can be a significant problem. Dead cells fluoresce strongly, and culture media (particularly YEPD) contain green fluorescent compounds that can accumulate within the vacuole. To reduce autofluorescence, grow and mount the cells in a low fluorescence synthetic medium,[11] use cells in early log phase, and view using band-pass filters that are optimized for GFP (Chroma Technology Corp., Brattleboro, VT).

GFP-Tagged Vacuolar Marker Proteins

A number of different GFP-tagged vacuolar proteins have been reported. GFP-tagged alkaline phosphatase (ALP) is a suitable marker for most applications

[19] C. G. Burd, *Methods Enzymol.* **327**, 61 (2000).
[20] K. Tatchell and L. C. Robinson, *Methods Enzymol.* **351**, [39], 2002 (this volume).
[21] O. W. Rossanese, C. A. Reinke, B. J. Bevis, A. T. Hammond, I. B. Sears, J. O'Connor, and B. S. Glick, *J. Cell Biol.* **153**, 47 (2001).

TABLE I
GFP FUSION PROTEINS THAT LOCALIZE TO THE VACUOLE

Marker	Vacuolar staining	Pathway to vacuole	PEP4-dependent cleavage of GFP tag	Ref.
Ste3-GFP	Lumenal	Endocytic	Yes	a
Ste2-GFP	Lumenal	Endocytic	Yes	b
Vph1-GFP	Membrane	CPY	No	a
GFP-DPAP-B	Membrane	CPY	No	c
GFP-CPS	Lumenal	CPY	Yes	c
GFP-Nyv1	Membrane	ALP	No	d
GFP-Vam3	Membrane	ALP	No	d
GFP-ALP	Membrane	ALP	No	e
Vma2-GFP	Peripheral membrane	From cytosol	No	f

[a] J. L. Urbanowski and R. C. Piper, *J. Biol. Chem.* **274**, 38061 (1999).
[b] C. J. Stefan and K. J. Blumer, *J. Biol. Chem.* **274**, 1835 (1999).
[c] G. Odorizzi, M. Babst, and S. D. Emr, *Cell* **95**, 847 (1998).
[d] F. Reggiori, M. W. Black, and H. R. Pelham, *Mol. Biol. Cell* **11**, 3737 (2000).
[e] C. R. Cowles, G. Odorizzi, G. S. Payne, and S. D. Emr, *Cell* **91**, 109 (1997).
[f] J. H. Seol, A. Shevchenko, A. Shevchenko, and R. J. Deshaies, *Nature Cell Biol.* **3**, 384 (2001).

because it labels the vacuolar membrane brightly and specifically.[22] Other markers follow different pathways to the vacuole and have different distributions within the vacuole (Table I, Fig. 2).

A subset of proteins is sorted within a prevacuolar endosomal compartment (the PVC) into vesicles that bud from the limiting membrane of this compartment into the lumen (Fig. 2).[23] Subsequent fusion with the vacuole delivers these vesicles into the proteolytic environment of the vacuolar lumen, where both the vesicles and their contents are degraded. This pathway is often used by proteins that are destined for degradation (e.g., Ste2p[17]), as well as by newly synthesized carboxypeptidase S (CPS).[23] In contrast, a number of other proteins, such as ALP[22] and dipeptidylaminopeptidase B (DPAP B),[23] are restricted to the limiting vacuolar membrane. Because the contents of the vacuole can sometimes be difficult to preserve by fixation techniques used commonly for immunofluorescence microscopy, GFP tagging is the most straightforward way to look at proteins that are sorted into intralumenal vacuolar vesicles.

GFP Tagging of Uncharacterized Proteins

When using epitope tagging to investigate the localization of an uncharacterized protein, it is always important to check that the tagged protein can complement all of the defects of a null mutant strain. This is still no guarantee that the modified protein will be localized correctly. In the case of membrane proteins that are

[22] C. R. Cowles, G. Odorizzi, G. S. Payne, and S. D. Emr, *Cell* **91**, 109 (1997).
[23] G. Odorizzi, M. Babst, and S. D. Emr, *Cell* **95**, 847 (1998).

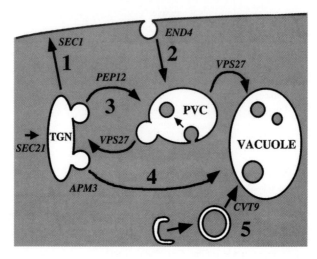

FIG. 2. Pathways to the yeast vacuole. A subset of the membrane-associated proteins that follow the secretory pathway (1) to the cell surface are internalized and reach the vacuole by way of the endocytic pathway (2). In contrast, newly synthesized vacuolar proteins that enter the secretory pathway are diverted from the bulk of the secretory traffic at the last compartment of the Golgi complex (the TGN). These hydrolases are sorted into either the CPY pathway (3), which transports cargo to the vacuole via the prevacuolar compartment (PVC), or the ALP pathway (4), which bypasses the PVC. Finally, cytosolic proteins can reach the vacuole directly through the cytoplasm-to-vacuole pathway (5). Selected genes that regulate transport along these pathways are indicated (see text for details).

transported to the vacuole, there are some additional considerations in designing tagged proteins.

1. Put the epitope tag on the cytosolic side. Lumenal tags will be subjected to proteolysis in the vacuole and, in some cases, the folding or fluorescence of a particular tag may be affected by the different environments encountered during transit through the secretory pathway. Although folded GFP is stable, the rate of vacuolar delivery for many proteins is likely to be faster than that of GFP folding. Even a tag on the cytosolic side can be degraded if the protein is internalized into intralumenal vesicles. It is helpful to check for *PEP4*-dependent cleavage of the tag by Western blotting.

2. Express the tagged protein at endogenous levels. Expression of the tagged protein from the chromosomal locus under control of its own promoter gives the most consistent and reliable staining. Increased expression of membrane proteins can saturate the sorting machinery, causing mislocalization to the vacuole. High levels of expression can also result in misfolded proteins that may be retained in the ER or sent to the vacuole for degradation. Proteins that are difficult to detect at single copy can often be functionally tagged with multiple copies of GFP.[21]

Although many of the vacuolar markers listed in Table I are expressed from high-copy plasmids, their distribution has been shown to correspond to that of the endogenous protein.

Visualizing Vacuole by Immunofluorescence Microscopy

Immunofluorescence microscopy allows the sensitive detection of endogenous proteins. Using polylysine-coated multiwell slides, many strains can be stained in parallel with small amounts of antibody and stored for weeks or months. To preserve vacuolar structure, we have modified standard methods[8] to include an extended, two-step fixation procedure prior to removal of the cell wall, followed by treatment with sodium dodecyl sulfate (SDS) to expose antigens.

Choice of Antibodies

To obtain the strongest signal in the widest range of fixation conditions, it is best to use polyclonal antibodies that have been both affinity purified[8] and cross-absorbed.[24] Rabbit sera typically contain a number of cross-reacting antibodies, particularly to yeast cell wall components, which are not always removed by affinity purification procedures. When studying the localization of nonessential proteins, a deletion strain lacking the antigen of interest provides an important control for specificity. We routinely cross-absorb small amounts of working-strength antiserum against an aliquot of the deletion strain that has been fixed and processed for immunofluorescence microscopy, as described in the following protocol. This is a simple and effective way to remove contaminating antibodies even from crude serum.

The appropriate working concentration for each new antibody (primary or secondary) must be determined empirically. Antibodies that are used at 1 : 1000 for Western blotting are typically tested at dilutions ranging from 1 : 100 to 1 : 1000 on wild-type and deletion strains to determine the best signal-to-noise ratio. Overnight incubations at 4° can enhance binding greatly for some antibodies, but increase the background for others. Fixation conditions can also be important for optimal signal. Glutaraldehyde fixation and methanol and/or acetone treatments have all been used to improve labeling with particular antibodies.[21]

Polyclonal antibodies raised against subunits of the abundant vacuolar H-ATPase are particularly effective for visualizing the vacuolar membrane. Commercially available monoclonal antibodies to alkaline phosphatase (MAb 1D3-A10; Molecular Probes) and to the 60-kDa subunit of the V-ATPase (1D11-B2; Molecular Probes) are also widely used.

The immunofluorescence microscopy of epitope-tagged proteins has many advantages. A wide range of good polyclonal and monoclonal antibodies to a

[24] C. J. Roberts, C. K. Raymond, C. T. Yamashiro, and T. H. Stevens, *Methods Enzymol.* **194,** 644 (1991).

variety of epitopes is available commercially, it is easy to control for (and cross-absorb) nonspecific staining using untagged strains, and dilutions and staining conditions generally do not have to be optimized for each protein. In addition, the signal can be enhanced by increasing the number of epitope tags rather than the copy number of the protein; e.g., constructs are available for adding 3, 9, or 13 copies of the *myc* epitope tag to yeast proteins.[25,26] Points to consider when designing epitope-tagged versions of membrane proteins that may localize to the vacuole were described earlier. We have been successful in detecting *myc*, hemagglutinin (HA), and GFP tags with commercially available rabbit and mouse antibodies.

Fluorescein and Texas Red-conjugated secondary antibodies are used commonly in double-label immunofluorescence experiments. However, new fluorophores that are brighter and less sensitive to photobleaching, such as Alexa 488 and Alexa 594 (Molecular Probes), enhance the detection of weak signals. Antibody "sandwiching," which involves alternating as many as four layers of fluorochrome-conjugated secondary antibodies, has been used successfully in a number of laboratories for signal amplification,[8] but in our hands, the background staining is also amplified such that the overall signal-to-noise ratio decreases. We find that an incubation with biotin-conjugated anti-mouse or anti-rabbit antibody followed by FITC–streptavidin (Jackson ImmunoResearch Laboratory, Inc., West Grove, PA) gives staining that is slightly more intense compared to a single layer of Alexa 488-conjugated antibody (note that Alexa–streptavidin conjugates give a high background in our hands). For double-label immunofluorescence experiments, it is essential to use secondary antibodies that have been cross-absorbed against IgG from the relevant species and to include the appropriate controls to detect any cross-reactivity. Omit each of the primary antibodies in turn to assess the background contributed by the secondary antibodies. If possible, a strain lacking the antigen of interest (e.g., a knockout strain or one not expressing the epitope-tagged version of the protein) should be included as a control in every immunofluorescence experiment.

SOLUTIONS AND MATERIALS REQUIRED

FIX: Heat 50 ml water. Add 375 μl of 6N NaOH and 2 g of paraformaldehyde and stir until it dissolves. Take care not to breathe the paraformaldehyde powder while weighing and prepare the solution in a fume hood. Cool to room temperature and add 0.68 g of KH_2PO_4 to neutralize the solution to a final pH of 6.5. Make fresh before use.

[25] M. Knop, K. Siegers, G. Pereira, W. Zachariae, B. Winsor, K. Nasmyth, and E. Schiebel, *Yeast* **15**, 963 (1999).
[26] M. S. Longtine, A. McKenzie, D. J. Demarini, N. G. Shah, A. Wach, A. Brachat, P. Philippsen, and J. R. Pringle, *Yeast* **14**, 953 (1998).

TEB: 200 mM Tris, pH 8.0, 20 mM EDTA, 1% 2-mercaptoethanol (make fresh)

SPM: 1.2 M sorbitol, 50 mM KPO$_4$, pH 7.3, 1 mM MgCl$_2$ containing 150 μg/ml Zymolyase (add fresh). Zymolyase-100T (Seikagaku America, Inc., Falmouth, MA) can be stored as a 15-mg/ml stock solution in 50 mM KPO$_4$, pH 7.3, 50% (v/v) glycerol that is stable for months at 4°. If Zymolyase is added to the SPM in powdered form, allow at least 20 min for it to dissolve before use.

PBS–BSA: Phosphate-buffered saline (PBS), 5 mg/ml bovine serum albumin (BSA), 10 mM NaN$_3$. Store at 4°.

Mounting medium: Dissolve 10 mg of p-phenylenediamine in 1 ml of PBS, pH 9. Add 9 ml of 100% glycerol and, if desired, 0.2 μl of 4',6'-diamidino-2-phenylindole (DAPI) (from a 10-mg/ml stock in water) to stain nuclei. Store in aliquots, protected from light, at $-70°$. Discard when it turns brown.

Preparation of slides and humid chamber: Add 20 μl of a 1-mg/ml polylysine solution to each well of an 8-well multiwell slide (ICN Pharmaceuticals, Inc., Costa Mesa, CA). Incubate for 1 min, rinse thoroughly in distilled water (we hold the slide under running water), and let dry. Place the slide on a damp paper towel in the bottom of a large petri dish or other shallow container and cover with a light-proof lid.

PROTOCOL

1. Grow yeast strains in 10 ml of YEPD to a density of 2×10^7 cells/ml. If the strain contains a plasmid, grow overnight in selective media, transfer to YEPD, and allow to double twice before fixing. If possible, prepare a strain lacking the antigen of interest as a negative control and for cross-absorbing antisera (see later).

2. Add 1.2 ml of 37% formaldehyde to each culture and shake at 30° for 30 min.

3. Pellet the cells and resuspend in 2 ml of FIX. Incubate at room temperature or at 30° overnight with gentle shaking.

4. Harvest cells, resuspend in 1 ml of TEB, and transfer to a microfuge tube. Incubate for 10 min at 30°.

5. Pellet cells and resuspend in 1 ml of SPM. Incubate for 30–45 min at 30°.

6. Pellet spheroplasted cells for 30 sec at low speed (3000g), wash once with 1.2 M sorbitol, and resuspend in 500 μl of 1.2 M sorbitol. Note that cells are very fragile after spheroplasting and must be handled carefully in all subsequent centrifugation and resuspension steps.

7. Add an equal volume (500 μl) of 10% SDS in 1.2 M sorbitol and incubate for 5 min at room temperature. The concentration of SDS may have to be optimized for a particular antibody (see later).

8. Wash cells twice in 1 ml of 1.2 M sorbitol and resuspend the final pellet in 1 ml of 1.2 M sorbitol.

9. Place 25–40 µl of cells in each well of a multiwell, polylysine-coated slide in a humid chamber.

10. Allow cells to settle for 15 min and then wash twice with 20 µl of PBS-BSA solution. A repeating pipetter and an aspirator are very useful in performing many washes quickly. To prevent drying, make sure that cells are covered with buffer immediately after aspiration in this and all subsequent steps.

11. Add 15 µl of 5% horse serum in PBS-BSA to block nonspecific binding and incubate for 30 min to 1 hr.

12. Meanwhile, prepare primary antibody solutions. If a control strain lacking the antigen of interest has been prepared, pellet approximately 200 µl of the cell suspension from step 8 and resuspend in the diluted antibody solution. Incubate for 1 hr at room temperature with constant mixing, pellet cells at high speed, and use the supernatant for staining cells.

13. Aspirate blocking solution and add 10 µl of primary antibody to each well. Incubate for 1 hr at room temperature.

14. Wash six times with PBS-BSA (over a period of at least 5 min).

15. Add 10 µl of secondary (fluorochrome-conjugated) antibody. Incubate for 1 hr.

16. Wash nine times with PBS-BSA.

17. Add a small drop of mounting medium to each well. Carefully lower a coverslip (24 × 60 mm) over cells to avoid trapping air bubbles and fix in place with nail polish. The slide may be viewed immediately or stored at −20° for 1–2 months. However, there may be significant loss of signal after as little as 2 weeks of storage.

Comments

Spheroplasting is most efficient in cells grown to early log phase in rich medium. Treating cells with reducing agents at high pH helps loosen mannans in the cell wall and enhance spheroplasting (step 5). We find that Zymolyase is appropriate for most applications; however, for cell surface antigens, oxalyticase (Enzogenetics, Corvallis, OR) is less likely to have contaminating proteases. A mixture of glusulase (25 µl/ml; Du Pont NEN, Boston, MA) and Zymolyase may be needed to digest the thick cell wall of stationary-phase cells.

The SDS treatment serves a number of purposes. It permeabilizes the cells, denatures and exposes the antigens, and enhances the binding of cells to polylysine-covered slides. The optimal concentration of SDS must be determined empirically for each antibody. For example, treatment with 5% SDS for 5 min is critical for staining with the anti-ALP MAb 1D3 A10, whereas the Vma2p MAb 1D11-B2 works best after treatment with 2% SDS for less than 1 min. Other antibodies, such as the anti-HA MAb HA.11 and the anti-myc MAb 9E10 (Covance Research Products, Inc., Cumberland, VA), work over a wide range of SDS concentrations.

In our hands, ProLong (Molecular Probes) mounting medium is more effective at reducing photobleaching than the one described here, although it is less convenient to use. It is recommended that a fresh aliquot of the mounting medium be prepared according to the manufacturer's directions. However, unused ProLong can be stored at $-70°$ for several weeks and reused as long as it has been protected from light and is not discolored. The slide must be left at $4°$ overnight or at room temperature for several hours before viewing to allow the mounting medium to harden. Slides can be stored at $4°$ for months or even years with little deterioration of the signal.

Techniques for Studying Protein Sorting

Newly synthesized proteins can be transported to the vacuole by a number of different pathways (Fig. 2; reviewed in Refs. 2 and 3). The soluble vacuolar hydrolase carboxypeptidase Y (CPY) is recognized at the late Golgi compartment by its receptor and is transported first to an endosomal compartment (i.e., the prevacuolar compartment) before reaching the vacuole. In contrast, newly synthesized ALP is sorted into a distinct set of transport vesicles at the late Golgi and is transported to the vacuole by an alternative pathway that bypasses the PVC. Other proteins, such as pheromone receptors Ste2p and Ste3p, follow the secretory pathway to the surface where they undergo endocytosis and subsequent turnover in the vacuole.

Detecting Secreted Proteins by Colony Immunoblotting

If newly synthesized, soluble vacuolar hydrolases are not sorted correctly in the late Golgi, they are mislocalized into the secretory pathway and released into the extracellular medium. The missorting of vacuolar hydrolases has been the basis of a number of large-scale mutant screens.[2] Although these hydrolases are usually secreted in an inactive, late Golgi-modified form, a fraction of the secreted enzyme can become activated in the medium and detected by activity assays.[24,27] Alternatively, sorting of an invertase fusion construct containing vacuolar targeting signals can be measured in *suc2* strains using a growth or colormetric assay.[28]

We prefer to use a colony immunoblotting assay to detect the missorting of proteins such as CPY, PrA, PrB, and pro-alpha (α) factor. In this assay, a nitrocellulose filter is placed over growing yeast colonies to capture any secreted proteins. After washing off adherent yeast cells, proteins retained on the filter can be visualized using a standard Western-blotting protocol. Colony overlay assays have been widely used to isolate vacuolar protein-sorting (*vps*) mutants that secrete CPY and

---

[27] E. W. Jones, *Methods Enzymol.* **194**, 428 (1991).
[28] V. A. Bankaitis, L. M. Johnson, and S. D. Emr, *Proc. Natl. Acad. Sci. U.S.A.* **83**, 9075 (1986).

to characterize the CPY sorting of new mutants using the commercially available antibody 10A5-B5 (Molecular Probes).

Protocol

1. Replica plate freshly grown yeast patches or colonies onto a new plate and overlay with a dry 85-mm nitrocellulose filter (0.45 μm; Millipore Corp., Bedford, MA). The filter will absorb moisture from the plate and adhere closely to the yeast. We find that minimal medium gives better discrimination than rich plates.

2. Incubate the plates at 30° for 12–18 hr. Longer periods of growth result in higher backgrounds and distortion due to the trapping of gases under the filter. Remove filter and rinse off adherent yeast cells under a stream of distilled water.

3. Incubate filter for 15–30 min in 5% nonfat milk in TTBS (20 mM Tris, pH 7.5, 0.5 M NaCl, 0.1% Tween 20) to block nonspecific binding.

4. Incubate filter at room temperature for 2 hr or more in the appropriate primary antibody diluted in TTBS+5% milk (e.g., anti-CPY MAb 10A5-B5 at 0.5 μg/ml). For library screening, as many as 50 filters can be stained simultaneously, stacked on top of each other in a beaker, with continuous agitation.

5. Wash three times, for a total of 15 min, in TTBS.

6. Incubate for 1 hr with a suitable secondary antibody (e.g., HRP- or AP-conjugated antimouse antibody). Wash as before and detect as appropriate for the particular secondary antibody used.

Comments

It is important to include appropriate controls, such as a wild-type strain and a *vps* mutant, on each plate. To ensure that CPY retained on the filter is secreted and not the result of cell lysis, a similar filter overlay assay can be carried out in parallel, staining instead for the presence of a soluble cytosolic protein, such as phosphoglycerate kinase (PGK).

Measuring Kinetics of Vacuolar Transport by Pulse-Chase Immunoprecipitation

Vacuolar hydrolases are generally synthesized in an inactive pro form and are cleaved in the vacuole to their mature, active form, a process that requires the *PEP4* gene product, proteinase A. The difference in mobility on SDS–PAGE gels between pro and mature forms provides a convenient way to assess transport to the vacuole. In addition, the turnover of membrane proteins in the vacuole can be followed as the *PEP4*-dependent loss of protein over time. Soluble hydrolases that are not targeted correctly to the vacuole are secreted from the cell; therefore, protein is immunoprecipitated from both the cells and the extracellular medium to assess the sorting accuracy. Two different immunoprecipitation protocols are given here to assess the transport of either soluble or membrane-associated proteins to the vacuole.

Immunoprecipitation of CPY and Other Soluble Hydrolases

Measuring the amount of cell-associated CPY by the Western blotting of whole cell extracts is not a very sensitive way to assess vacuolar protein sorting and has led to some confusion in the literature. The overlay assay described earlier is simple and relatively foolproof, but it is not quantitative and does not distinguish between Golgi and vacuolar forms of the enzyme. Pulse-chase immunoprecipitation is a quantitative method of monitoring transport through different stages of the secretory pathway to the vacuole, at least for those enzymes that, like CPY, are glycosylated or proteolytically processed during transport, resulting in gel mobility shifts.

SOLUTIONS

Labeling medium: minimal medium lacking methionine, 50 mM KPO$_4$, pH 5.7, 2 mg/ml BSA. Buffering is essential for the efficient recovery of CPY from the extracellular medium, and BSA allows the cells to be recovered in a tight pellet.

10× IP buffer: 0.9 M Tris–HCl, pH 8.0, 1% Triton X-100, 20 mM EDTA

10× IP + SDS buffer: 0.9 M Tris–HCl, pH 8.0, 1% SDS, 1% Triton X-100, 20 mM EDTA

Spheroplast mix: 50 mM Tris–HCl, pH 7.4, 1.4 M sorbitol, 2 mM MgCl$_2$, 10 mM NaN$_3$. Add before use: 3 μl/ml 2-mercaptoethanol and 11.2 μl/ml oxalyticase (Enzogenetics).

1× IP wash buffer: 10 mM Tris–HCl, pH 8.0, 0.1% SDS, 0.1% Triton X-100, 2 mM EDTA

100× PI: 50 mM phenylmethylsulfonyl fluoride (PMSF), 0.1 mM leupeptin, 0.1 mM pepstatin

1% SDS 8 M urea: Make fresh or store at room temperature and check before use that the pH (which will increase during storage due to breakdown of the urea) does not exceed pH 8.

PROTOCOL FOR SOLUBLE HYDROLASES

1. Grow yeast cells in minimal medium supplemented with appropriate nutrients but lacking methionine to a density of 0.4–1.2 OD$_{600}$ units/ml. Pellet 1 OD$_{600}$ unit (2 × 10^7 cells) for each strain to be tested and resuspend in 1 ml of labeling medium. Let equilibrate in a 30° water bath, with shaking, for 10–15 min.

2. Add 20 μl of ^{35}S-Express label (200 μCi; NEN Life Sciences Products, Inc., Boston, MA) for every 1 OD$_{600}$ unit of cells. Incubate for 10 min at 30°.

3. Initiate chase by adding 100 μl of a solution containing 5 mg/ml each of unlabeled cysteine and methionine. Mix and immediately remove 0.5 ml to a microfuge tube on ice containing 5 μl of 1 M NaN$_3$ on ice ($t = 0$ time point).

4. Incubate in a 30° shaking water bath for 60 min.

5. Harvest remaining 0.5 ml of cells and transfer to a microfuge tube containing 5 μl of 1 M NaN$_3$ on ice ($t = 60$ time point).

6. Pellet cells at maximum speed for 20 sec and transfer supernatants to fresh tubes ("extracellular" samples). Keep the cell pellets ("intracellular" samples) on ice.

7. To the supernatants from step 6, add 100 μl of 10× IP buffer. Heat at 100° for 5 min and cool on ice. Add 350 μl of water and 50 μl of IgGSorb (The Enzyme Center, Malden, MA). Incubate at least 15 min at 4° to preabsorb proteins that bind nonspecifically to the IgGSorb (store on ice until step 10).

8. Resuspend the cell pellets from step 6 in 150 μl of spheroplast mix. Incubate for 30 min at 30°.

9. After spheroplasting, lyse the cells by adding 50 μl of 2% SDS and incubate for 5 min at 100° to denature proteins. Cool on ice.

10. Add 100 μl of 10× IP buffer, 650 μl water, and 50 μl IgGSorb. Incubate at least 15 min at 4°. (The cells are not pelleted from the spheroplast medium to prevent any CPY trapped in the periplasmic space from being discarded.)

11. Remove IgGSorb by spinning for 5 min at 15,000g and transfer supernatants to fresh microfuge tubes that contain the primary antibody (e.g., 1 μl of an anti-CPY antiserum). Incubate at least 1 hr at 4°.

12. Add 50 μl of IgGSorb and incubate 1 hr at 4° with periodic mixing. The incubation can be extended to several hours, but overnight incubations with crude antiserum can lead to protein degradation.

13. Centrifuge at 11,000 rpm for 60 sec and resuspend in 1 ml of 1× IP + SDS buffer by vortexing vigorously. We shake on a vortex mixer for an additional 5 min to allow the dissociation of nonspecifically bound material during the wash step and to break up the pellet. Repeat the wash once.

14. Resuspend the final pellet in 30 μl of 2× sample buffer and heat at 100° for 5 min. Spin at 15,000g for 2 min and load the supernatant on an 8% SDS–PAGE gel.

15. Fix and dry the gel, and expose to X-ray film (Hyperfilm MP, Amersham Pharmacia Biotech, Piscataway, NJ) for 18–24 hr.

To assess the fidelity of CPY sorting, compare the proportion of CPY immunoprecipitated from the extracellular fraction to the total CPY (intracellular + extracellular) at the 60-min time point.

Comments

Carrying out immunoprecipitations in parallel on both the wild-type parent strain and a well-characterized *vps* mutant strain provides a positive and a negative control. Some laboratories report difficulty in quantitatively recovering extracellular CPY by immunoprecipitation, perhaps as a result of degradation or precipitation of CPY from the medium. If CPY is being lost during the labeling procedure, the total amount of CPY recovered at the 60-min time point will be reduced compared to the 0-min time point.

PROTOCOL FOR MEMBRANE-BOUND PROTEINS

Metabolic labeling and harvesting of cells in 0.5-ml aliquots for each time point of the chase are carried out essentially as described for soluble hydrolases. The length of the chase period required depends on the protein being studied (e.g., 30 min for ALP and 2–3 hr for DPAP A).

1. Grow yeast cells in minimal medium supplemented with appropriate nutrients but lacking methionine to a density of 0.4–1.2 OD_{600} units/ml. Pellet 0.5 OD_{600} unit (1×10^7 cells) for each time point to be tested and resuspend in 0.5 ml of labeling medium. *Note.* For metabolic labeling of ALP, omit potassium phosphate from the labeling medium. Let equilibrate in a 30° water bath, with shaking, for 10–15 min.

2. Add 10 μl of ^{35}S-Express label (100 μCi) for every 0.5 OD_{600} of cells. Incubate for 10 min at 30°.

3. Initiate chase by adding 1/10 volume of a solution containing 5 mg/ml each of unlabeled cysteine and methionine. Mix and immediately remove 0.5 ml to a microfuge tube containing 5 μl of 1 M NaN$_3$ on ice ($t = 0$ time point).

4. Incubate in a 30° shaking water bath and remove 0.5-ml aliquots of cells to microfuge tubes containing 5 μl of 1 M NaN$_3$ on ice at the desired intervals. Cells can be kept on ice for extended periods (>3 hr) until all aliquots have been harvested.

5. Pellet cells at full speed for 20 sec, discard the supernatant, and resuspend in 150 μl of spheroplast mix. Incubate at 30° for 30–45 min.

6. Pellet cells and discard supernatant. Resuspend cells in 50 μl of 1% SDS/8 M urea and heat at 95° for 5 min. Vortex and then cool on ice.

7. Add 100 μl of 10× IP buffer, 790 μl of water, 10 μl of 100× PI, and 50 μl of IgGSorb. Incubate on ice for 15 min with occasional mixing.

8. Proceed with steps 11–15 described in the protocol for soluble hydrolases.

Comments

Weak signals can be enhanced by fluorography: after fixing the gel, wash in water and incubate for 15 min in 1 M salicylate before drying. The signal can also be improved by increasing the amount of ^{35}S-Express label from 10 to 15 μl or by harvesting more cells for each time point. Each antiserum should be titered to optimize the signal-to-noise ratio. We typically use 0.5–5 μl of crude antiserum for each 0.5OD_{600} unit of cells. High backgrounds can be reduced by more stringent washing (e.g., in high salt or different detergent combinations) or by using protein A-Sepharose instead of IgGSorb. Blocking agents other than BSA can also be tried, the most effective being an unlabeled extract from a yeast strain lacking the antigen of interest.

It may be more convenient to carry out the procedure over more than 1 day, especially when chase times are long. Although incubation with the primary antibody can be left to proceed overnight, in many cases this leads to increased protein degradation. Instead, the cell pellets can be frozen after spheroplasting (step 6 of the second protocol) or after resuspending the washed IgG sorb pellets in sample buffer (step 14). Alternatively, the samples can be separated on the SDS–PAGE gel overnight at low current. This has the added advantage of increasing the resolution of high molecular weight proteins, although very low molecular weight proteins will tend to diffuse.

Studying Vacuolar Association and Transport of Uncharacterized Protein

The protocols described so far outline a number of techniques that are useful for studying the localization and transport of known vacuolar marker proteins. However, these techniques are equally useful in the analysis of uncharacterized proteins that are suspected to be associated with the vacuole. Membrane proteins that are expressed at high levels are often transported to the vacuole, even though they may function elsewhere. Misfolded proteins, whether they are soluble or membrane bound, may also be targeted to the vacuole for degradation.

Proteins that are degraded in the vacuole are typically stabilized in a *pep4* strain, which lacks active vacuolar proteases. Because *PEP4*-dependent proteolysis can also occur in the protease-active prevacuolar compartment that accumulates in certain mutants,[6] it is often necessary to demonstrate localization to vacuolar membranes by other means. Distribution of a protein can be compared with that of known vacuole markers by double-label immunofluorescence microscopy as described previously. Vacuoles pellet at low speed (13,000g) and float in density gradients, making subcellular fractionation another useful way to demonstrate association with the vacuole.[29]

Using Transport Mutants to Dissect Trafficking Pathways

Once association with the vacuole has been established, it is often important to identify the route by which the protein of interest is transported to the vacuole. As discussed earlier, proteins can reach the vacuole in a number of different ways. Because each pathway has been subjected to a thorough genetic analysis, a few well-defined mutants can be used to distinguish the trafficking pathway taken by a given protein (Fig. 2). For this approach it is important to have a way of monitoring arrival at the vacuole, typically by following either the *PEP4*-dependent

---

[29] G. Sipos and R. S. Fuller, *Methods Enzymol.* **351**, [20], 2002 (this volume).

proteolytic processing or the vacuolar localization of the protein of interest. Transport of the uncharacterized protein can then be evaluated in mutant strains in which specific trafficking pathways are blocked and compared to the transport of well-characterized marker proteins, as outlined later.

Cytoplasm-to-Vacuole Targeting

Proteins that are targeted to the vacuole directly from the cytoplasm (e.g., API, Ams1p) do not transit the early stages of the secretory pathway. Therefore, their transport is not affected by the ER-to-Golgi mutants *sec12* or *sec23* (unlike CPY and ALP).[30] However, the transport of API is blocked by cytoplasm-to-vacuole targeting (*cvt*) mutants, which show considerable genetic overlap with autophagy mutants (*aut* and *apg*; e.g., *aut3/cvt10/apg1*[31]).

Secretion Followed by Endocytosis

Membrane proteins that follow the secretory pathway to the cell surface can be internalized, reaching the vacuole by way of the endocytic pathway (e.g., Ste2p, Ste3p). Their transport will be blocked by mutations in components of the late secretory pathway that direct fusion with the plasma membrane, such as *sec1-ts* and *sec4-ts*, and will be trapped in secretory vesicles that accumulate near the bud. Blocking endocytosis using *end4-ts* or *end3-ts* mutations[32] will also prevent the *PEP4*-dependent degradation of these proteins, causing them to be trapped on the cell surface.[33]

Transport via PVC: CPY and Endocytic Pathways

Class E *vps* mutations, which prevent the invagination of vesicles into the lumen of the PVC, also prevent recycling out of this compartment. In these mutant strains, an enlarged, aberrant form of the PVC containing active proteases accumulates next to the vacuole. Consequently, in *vps27 pep4* strains, endocytosed proteins, as well as those that follow the CPY pathway, are found in one or two perivacuolar structures that label brightly with FM4-64 or with antibodies to the t-SNARE Pep12p.[13,34] Proteins that are directed into the CPY pathway at the late Golgi compartment reach the PVC without passing through the cell surface (e.g., DPAP A, Vps10p, CPS). Therefore, they can be distinguished from endocytosed proteins because their transport is independent of late secretory (*SEC1*, *SEC4*) and early endocytic (*END3*, *END4*) genes.

---

[30] D. J. Klionsky, R. Cueva, and D. S. Yaver, *J. Cell Biol.* **119**, 287 (1992).
[31] T. M. Harding, A. Hefner-Gravink, M. Thumm, and D. J. Klionsky, *J. Biol. Chem.* **271**, 17621 (1996).
[32] S. Raths, J. Rohrer, F. Crausaz, and H. Riezman, *J. Cell Biol.* **120**, 55 (1993).
[33] L. Hicke, B. Zanolari, M. Pypaert, J. Rohrer, and H. Riezman, *Mol. Biol. Cell* **8**, 13 (1997).
[34] K. Bowers, B. P. Levi, F. I. Patel, and T. H. Stevens, *Mol. Biol. Cell* **11**, 4277 (2000).

Proteins that become trapped in the PVC in *vps27* mutants will still be subject to *PEP4*-dependent turnover. Because the fusion of transport vesicles with the PVC requires the t-SNARE Pep12p, *pep12* mutations can be used to prevent the *PEP4*-dependent proteolysis of proteins that reach the vacuole by way of the PVC. These proteins accumulate in transport vesicles in *pep12* strains, giving rise to a hazy, cytoplasmic-staining pattern.

Transport via ALP (AP3) Pathway

If a protein transits the early secretory pathway, but reaches the vacuole independent of *SEC1*, *END4*, *VPS27*, and *PEP12* function, then it is likely to follow the ALP pathway. ALP enters a distinct class of AP3-dependent vesicles at the late Golgi and is transported to the vacuole without passing through the PVC.[22,35] Mutation of the subunits of the AP3 coat complex (e.g., *apm3*) blocks the ALP pathway and causes the cargo to be rerouted into the *PEP12*-dependent CPY pathway.[22]

Biochemical Methods for Studying Yeast Vacuoles

The purification of vacuolar membranes on density gradients is an effective way to assess the vacuolar association of a particular protein. It is also a useful first step in the purification of vacuolar membrane proteins for subsequent biochemical analysis. The V-ATPase is one of the most abundant and widely studied vacuolar membrane proteins. Therefore, we present a protocol for the preparation of purified vacuolar membranes suitable for the subsequent biochemical characterization of the vacuolar ATPase, as well as methods for determining V-ATPase activity.

Purification of Vacuoles by Flotation

Due to their low buoyant density, vacuoles can be purified by density gradient centrifugation. We have modified the method of Kakinuma *et al.*[36] to isolate vacuoles from osmotically lysed spheroplasts by flotation on discontinuous Ficoll gradients. Two sequential flotation steps are required for optimal purity but may reduce the yield; vacuoles collected from the first Ficoll gradient may be sufficiently pure for some applications. Because vacuoles contain a number of hydrolases and are lysed easily, it is important to carry out the procedure as quickly as possible and to keep the samples, buffers, and rotors cooled throughout.

SOLUTIONS

Spheroplast buffer: 1.2 M sorbitol, 25 mM MOPS/25 mM MES, pH 7.4, 5 mM MgCl$_2$
100× PI: 50 mM PMSF, 0.1 mM leupeptin, 0.1 mM pepstatin. Add a 1 : 50

[35] R. C. Piper, N. J. Bryant, and T. H. Stevens, *J. Cell Biol.* **138**, 531 (1997).
[36] Y. Kakinuma, Y. Ohsumi, and Y. Anraku, *J. Biol. Chem.* **256**, 10859 (1981).

dilution of this protease inhibitor cocktail to buffers A, B, and C immediately before use.

Buffer A: 10 mM MOPS/10 mM MES, pH 6.9, 0.1 mM MgCl$_2$, 12% Ficoll 400. Filter sterilize.

Buffer B: 10 mM MOPS/10 mM MES, pH 6.9, 0.1 mM MgCl$_2$, 8% Ficoll 400. Filter sterilize.

Buffer C: 10 mM MOPS/10 mM MES, pH 6.9, 5 mM MgCl$_2$, 25 mM KCl

Note: If purified vacuolar membranes are to be used for the determination of V-ATPase activity by the plate assay, it is important to use highly purified water when preparing buffers to reduce contamination by inorganic phosphate (we use Nanopure water with a resistivity of 16 MΩ). The pH of the buffers can be adjusted using concentrated Tris base.

PROTOCOL

1. Harvest 4000 OD$_{600}$ units of yeast, grown in YEPD at 30°, to midlog phase (2–4 OD$_{600}$ units/ml) by centrifugation at 4000g for 5 min. Wash once with water at room temperature.

2. Resuspend pellet in 100 ml of 50 mM Tris, pH 9.5, + 10 mM DTT and incubate for 15–30 min at 30°.

3. Spin and resuspend cells in 100 ml of spheroplast buffer. Add 1 ml of a Zymolyase-100T stock solution (4 mg/ml in 50 mM Tris, pH 7.4, 50% glycerol). Incubate with gentle mixing at 30° for 60–90 min. If spheroplasting is complete, an aliquot of the cells diluted 10-fold into buffer A should be noticeably less turbid than a similar aliquot diluted into 1.2 M sorbitol.

4. Collect spheroplasts by centrifugation at 1000g for 5 min. Gently wash spheroplasts once with 150 ml of spheroplast buffer and once with 150 ml of spheroplast buffer containing 1% glucose to maintain the association between V$_1$ and V$_0$ subunits of the V-ATPase.

5. Resuspend the washed spheroplast pellet in 25 ml ice-cold buffer A, homogenize 10–15 strokes with a Dounce homogenizer, and transfer into a polyallomer tube. Overlay with approximately 13 ml of cold buffer A to fill the tube and centrifuge at 60,000g for 35 min at 4° in prechilled Beckman SW28 rotor.

6. Collect the floating vacuole layer (which looks like a white wafer) with a spoon-shaped spatula dipped in buffer A. Homogenize in 6 ml buffer A and transfer to a polyallomer tube. Overlay with 6 ml buffer B and centrifuge at 60,000g for 35 min at 4° in a prechilled Beckman SW 41 Ti rotor.

7. Collect the vacuole layer (using the spatula dipped in buffer C), resuspend in a small volume (0.4 ml) of 2× buffer C, and add 2 volumes of 1× buffer C. This step fragments vacuoles into vacuolar vesicles of more uniform size, with some loss of lumenal content.

8. To wash vacuolar vesicles, dilute membranes into 10 ml cold TE (10 mM Tris, pH 7.4, 1 mM EDTA) containing 10% (v/v) glycerol and protease inhibitors, homogenize gently, and spin at 37,000g for 30 min at 4°. Repeat wash twice.

9. Resuspend final pellet in TE +10% glycerol.

Comments

The protein concentration of the final preparation is determined using a modified Lowry assay after adding 2% SDS to solubilize the vacuoles.[37] Alternately, proteins are precipitated by adding 950 μl of a 1 : 1 mixture of ice-cold ethanol and acetone to 50 μl of the vacuolar vesicles and incubating for 30 min on ice. The precipitated protein is recovered by centrifugation for 10 min at 15,000g, solubilized in 50 μl of 10% SDS, and quantified using a BCA protein assay (Pierce, Rockford, IL).

We typically adjust the protein concentration of the final preparation to 1 mg/ml if the vacuolar vesicles are to be used in activity assays (see later). If it is not possible to perform the assays the same day, the purified membranes can be snap frozen in aliquots and stored at −80. However, freeze/thaw will cause some loss of V-ATPase integrity and activity. The thawed membranes should be washed once in TE + glycerol immediately before use to remove free V_1 subunits.

Measurement of Vacuolar H^+-ATPase Activity

Method 1: Coupled Spectrophotomeric Assay

This is a sensitive assay for the determination of V-ATPase activity using a coupled enzyme–ATP regeneration system.[24,38] To determine the extent of contamination by mitochondrial and PM ATPases, we measure azide- and vanadate-sensitive ATPase activity, respectively, in the starting material and the purified vacuoles (typically less than 0.5% of the total activity).[24] Conversely, performing a parallel reaction in the presence of the V-ATPase inhibitor concanamycin A (100–300 nM) permits determination of the V-ATPase-independent ATPase activity.

Assay buffer: 25 mM HEPES, pH 7.0, 25 mM KCl, 5 mM MgCl$_2$, 2 mM phosphoenolpyruvate, 2 mM ATP, 0.5 mM NADH. Adjust pH to 7.0 with KOH and add 30 units of L-lactate dehydrogenase and 30 units of pyruvate kinase.

PROTOCOL. Add 5-100 μl of vacuolar vesicles directly to a cuvette containing 1 ml of the assay solution. Immediately observe the change in absorbance at 340 nm using the time drive mode on the spectrophotometer.

[37] M. A. Markwell, S. M. Haas, L. L. Bieber, and N. E. Tolbert, *Anal. Biochem.* **87,** 206 (1978).
[38] H. R. Lotscher, C. deJong, and R. A. Capaldi, *Biochemistry* **23,** 4128 (1984).

Absorbance readings are linear up to an A_{340} value of 3.0. The molar extinction coefficient for NADH (ε) is 6.22 mM^{-1}cm^{-1} and depletion of NADH is directly correlated to ATP hydrolysis. Specific activity corresponds to micromoles ATP hydrolyzed per minute per milligram protein.

Comments

Activity of the ATP regeneration system is critical to the success of this assay. Lactate dehydrogenase and pyruvate kinase purchased as a suspension must first be centrifuged to remove ammonium sulfate before being resuspended in the assay buffer. The assay mixture can be frozen in batches and stored at $-80°$ for 4–6 weeks with minimal loss of activity.

Method 2: Plate Reader Assay for Inorganic Phosphate Release

ATPase activity can be determined by incubating vacuolar membranes with Mg-ATP and measuring the time-dependent release of inorganic phosphate in a 96-well plate format.[39] This assay is less sensitive than that described earlier, but does not rely on the activity of a coupled enzyme–ATP regeneration system. Because ATP depletion will affect the linearity of the assay over time, phosphate release is measured for 30 min and the activity is determined over the linear part of the curve. Phosphate contamination from water or residual detergent on laboratory glassware can be a significant problem. Therefore, all solutions should be prepared using disposable plastic labware and highly purified water (i.e., resistivity of 16 MΩ).

SOLUTIONS

Buffer 1 : 50 mM MES/Tris, pH 6.9, 5 mM MgSO$_4$, 0.01% NaN$_3$
Phosphate reagent: 1% (w/v) SDS, 1% (w/v) ammonium molybdate, 4% (w/v) sulfuric acid
Ascorbate solution: 10% ascorbic acid in water (make fresh)

PROTOCOL

1. Prepare the stop solution by mixing the phosphate reagent and the ascorbate solution in a 50 : 1 ratio. Dispense 80 μl into each well of a 96-well microtiter plate.

2. Set up 500-μl reactions containing vacuolar membranes as follows: 50–100 μg vacuolar membranes (50 μl of a 1-mg/ml preparation), 0.5 μl NaN$_3$ (from a 10% stock), and 449.5 μl buffer 1.

3. If desired, add inhibitors and preincubate for 10 min. Contaminating mitochondrial F-ATPases are inhibited by 2 mM azide, whereas 0.1 mM sodium

[39] M. A. Harrison, P. C. Jones, Y. I. Kim, M. E. Finbow, and J. B. Findlay, *Eur. J. Biochem.* **221**, 111 (1994).

vanadate inhibits plasma membrane ATPases. Conversely, 300 nM concanamycin A can be added to determine the extent of V-ATPase-independent ATPase activity.

4. Start the reaction by adding Mg-ATP to a final concentration of 5 mM (5 μl of a 0.5 M stock).

5. Mix and immediately remove a 20- to 50-μl aliquot to the 96-well plate to give the 0-min time point. Incubate the rest of the reaction at room temperature and take aliquots at desired time points over a 0- to 30-min range, adding them to the wells of the 96-well plate.

6. Leave reactions at room temperature for 10–15 min and measure absorbance at 630 nm.

Prepare a standard curve using KH_2PO_4 (0–100 μM: adjust depending on activity range desired). ATPase activity is expressed as μM P_i/mg protein/min.

We routinely carry out three additional reactions in parallel to control for the presence of inorganic phosphate or contaminating ATPases. In the first, we omit vacuolar membranes; in the second, we omit ATP; and in the third, we include 300 nM concanamycin A to measure V-ATPase independent activity.

Conclusion

This chapter described methods for studying different aspects of vacuole biogenesis and function. The development of new vacuole-specific dyes and vacuolar marker proteins (particularly those tagged with GFP) has expanded the range of techniques available for visualizing the yeast vacuole. The ability to monitor the transit of particular proteins to the vacuole from the Golgi, plasma membrane, or cytosol has been instrumental in characterizing the vesicle transport machinery. By exploiting mutants blocked in specific transport steps, the intracellular itinerary of still uncharacterized membrane proteins can be elucidated. Finally, the ease with which vacuoles can be isolated based on their low buoyant density has been important for biochemical studies of vacuolar enzymes, such as the V-ATPase.

Acknowledgments

We thank Kate Bowers for critical reading of the manuscript and Laurie Graham, Andrew Flannery, Ben Powell, and members of the Capaldi laboratory for their contributions to the vacuole purification and ATPase activity assay protocols.

[25] Purification of Yeast Actin and Actin-Associated Proteins

By BRUCE L. GOODE

Introduction

Saccharyomyces cerevisiae is an ideal model organism in which to study actin functions because actin and actin-associated proteins (AAPs) are highly conserved between yeast and other eukaryotes and because of the opportunity to combine genetic and biochemical approaches. Specific mutations in actin and actin-associated proteins are introduced rapidly in yeast, allowing parallel analysis of their resulting phenotypes *in vivo* and changes in biochemical activities. The goal of this chapter is to provide updated methods for isolating yeast actin (first section) and actin-associated proteins (second section), that may facilitate biochemical analyses of the yeast actin cytoskeleton.

The first section is a detailed outline of the DNase I affinity method for purifying yeast actin originally described by Kron *et al.*[1] Minor modifications are included to save time in the preparation, provide added versatility, and remove contaminating cofilin/ADF (actin-depolymerizing factor), which can strongly influence actin kinetic studies. The second section presents a new method we have developed for isolating actin and actin-associated proteins from yeast. This method is based on the reconstitution of actin assembly in soluble yeast extracts and allows a one-step isolation of a semi-intact actin cytoskeleton. Using additional chromatography steps, various components can be isolated, providing the opportunity to study the activities of native actin-associated proteins from wild-type and mutant strains.

Purification of Yeast Actin Using DNase I Affinity

Some of the most common uses for purified yeast actin are (i) testing the ability of a protein to cosediment with filamentous actin,[2–6] (ii) testing the effects

---

[1] S. J. Kron, D. G. Drubin, D. Botstein, and J. A. Spudich, *Proc. Natl. Acad. Sci. U.S.A.* **89**, 4466 (1992).
[2] D. A. Holtzman, K. F. Wertman, and D. G. Drubin, *J. Cell Biol.* **126**, 423 (1994).
[3] J. E. Honts, T. S. Sandrock, S. M. Brower, J. L. O'Dell, and A. E. Adams, *J. Cell Biol.* **126**, 413 (1994).
[4] B. L. Goode, D. G. Drubin, and P. Lappalainen, *J. Cell Biol.* **142**, 723 (1998).
[5] B. L. Goode, J. J. Wong, A.-C. Butty, M. Peter, A. McCormack, J. R. Yates, D. G. Drubin, and G. Barnes, *J. Cell Biol.* **144**, 83 (1999).
[6] D. Winter, T. Lechler, and R. Li, *Curr. Biol.* **9**, 501 (1999).

of an actin-binding protein on actin assembly and/or actin disassembly kinetics,[4-8] and (iii) determining the structures and/or activities of mutant actins.[9,10] For such analyses, it is desirable to have milligram quantities of yeast actin on hand. The preparation described here yields approximately 5 mg of yeast actin.

For some biochemical assays (e.g., actin-associated proteins with different affinities for yeast and vertebrate actin), it is desirable to use 100% yeast actin. Similar to mammalian actin, yeast actin can be modified with a pyrene label for use in fluorometric assays and a rhodamine label for use in microscopic assays. For other applications, yeast actin can be spiked with 5% pyrene-labeled or rhodamine-labeled rabbit muscle actin (Cytoskeleton Inc., Denver, CO).

Growing Yeast for Actin Preparations

The DNase I affinity procedure described by Kron et al.[1] remains the method of choice for purifying yeast actin. Included here are minor modifications in the protocol to help save time in preparation and to minimize cofilin contamination. Using the DNase I affinity strategy, >90% of the cellular actin in yeast is isolated in a 1- to 2-day preparation, and a typical yield is 5 mg actin from 200 g packed cells.

Because these preparations require such large quantities of yeast, a commercial source of yeast such as Red Star Yeast, Inc. (Emeryville, CA) can be used. Companies often will donate 10–20 pounds of yeast at no cost to researchers. If such a commercial source is used, it is important to obtain the yeast on the same day cells are harvested from the fermenters. The yeast are prepressed into semidry "bricks," which, importantly, are not bone dry (i.e., do not use freeze-dried yeast). However, if a specific mutant strain must be grown and/or a commercial source is not available, yeast can be grown in the laboratory to saturation in liquid YPD culture. A saturated culture yields about 10–15 g packed cells per liter.

Regardless of the source, yeast cells should be rinsed and pelleted and then washed once with 3–5 volumes of cold H_2O. Using a large metal spatula, resuspend the cells with 0.2–0.3 volumes of cold H_2O to achieve a thick soupy consistency. Freeze by slowly pouring a thin stream of cells into a large liquid nitrogen bath, breaking up the frozen chunks with a metal spatula. Pour off the excess liquid nitrogen and allow residual nitrogen to evaporate completely. Transfer the frozen yeast to plastic containers with screw cap lids and store at $-80°$. Frozen yeast samples can be kept for many years.

---

[7] B. L. Goode, A. A. Rodal, G. Barnes, and D. G. Drubin, *J. Cell Biol.* **153**, 627 (2001).
[8] X. Chen and P. A. Rubenstein, *J. Biol. Chem.* **270**, 11406 (1995).
[9] L. D. Belmont, A. Orlova, D. G. Drubin, and E. H. Egelman, *Proc. Natl. Acad. Sci. U.S.A.* **96**, 29 (1999).
[10] A. Orlova, X. Chen, P. A. Rubenstein, and E. H. Egelman, *J. Mol. Biol.* **271**, 235 (1997).

Lysing Yeast Cells

Yeast cells have a thick wall that is resistant to lysis by osmotic shock and normal mechanical perturbation. The cell wall can be digested enzymatically, but this is not an economical strategy for large-scale cell lysis. Instead, many laboratories use bead beating[1] or a French press[6] for large-scale lysis. However, a third method, originally developed in the laboratory of Dr. Peter Sorger (Massachusetts Institute of Technology), may be the most convenient because it allows lysed cells to be stored at $-80°$ indefinitely before their use in biochemical experiments. This method uses liquid nitrogen and the mechanical shearing of a stainless-steel Waring blender to lyse yeast cells.[11] The following paragraph describes our exact adaptation of the method.

In a cold room, add 100–150 g frozen yeast to a 1-liter stainless-steel Waring blender (available from Fisher Scientific, Pittsburgh, PA). Make sure that the blender is *completely* dry before starting the lysis, as the blender rotor may freeze when liquid nitrogen is added. Also note that the range in volume of yeast for efficient lysis is 40–200 g for a 1-liter blender. Pour in liquid nitrogen just over the top of the yeast, cover the chamber with the *vented* rubber cap, and use freezer gloves to hold down the cap. Blend on low speed for 1–2 sec and then switch to high speed. If the pressure is too strong, stop the blender and allow some of the nitrogen to evaporate before continuing. Blend until you hear a distinct change in the sound of the grinding, which signifies that the nitrogen is running out. Turn off the blender, remove the chamber from the base and tap it a few times to get the powder off the sides. Add more nitrogen as described earlier and repeat the blending cycle three more times. After the fourth round of blending, allow all excess nitrogen to evaporate, transfer the frozen yeast lysate to plastic containers, and store at $-80°$. These samples can be kept for years.

Constructing DNase I Affinity Column

When constructing a DNase I affinity column, use 40–50 mg DNase I per 1 mg actin isolated. For the 5-mg actin preparation described here, combine 10 ml Affi-Gel 10 and 200 mg DNase I to generate a column containing 20 mg/ml DNase I.

1. Resuspend 200 mg purified DNase I in 10 ml coupling buffer [100 mM HEPES–KOH (pH 7.2), 80 mM CaCl$_2$, 1 mM phenylmethylsulfonyl fluoride (PMSF)]. We have found that it is unnecessary to perform the overnight dialysis of DNase I suggested by Kron and co-workers.[1]

2. The most critical step in the column preparation is washing the resin rapidly. In doing so, avoid exposure of the resin to air and minimize its time of exposure to

---

[11] P. K. Sorger and H. R. B. Pelham, *EMBO J.* **6**, 3035 (1987).

the aqueous solution before addition to DNase I. The Affi-Gel 10 resin (Bio-Rad, Hercules, CA) is shipped and stored in ethanol, where it remains inactive. The resin is activated when exchanged into an aqueous environment and/or by exposure to air. To facilitate rapid washing of the resin, we use a 20- to 40-ml ground glass filter and apply very low vacuum to perform three quick washes with 20 ml H_2O, followed by two quick washes with 20 ml coupling buffer. While the resin is still wet from the last wash, immediately scoop it into a tube on ice containing 10 ml DNase I (20 mg/ml) in coupling buffer.

3. Rotate the reaction at 4° for 4 hr to overnight to allow coupling.

4. Pour the resin into a 2.5 × 10-cm Econo-Column (Bio-Rad). Wash the column at 2 ml/min with 50 ml 0.1 M Tris–HCl (pH 7.5) to inactivate all remaining cross-linking groups. Then, wash with 100 ml G buffer [10 mM Tris–HCl (pH 7.5), 0.5 mM ATP, 0.2 mM dithiothreitol (DTT), 0.2 mM $CaCl_2$], followed by 50 ml G buffer +0.2 M ammonium chloride, and finally 50 ml G buffer. The column is now ready for use in the actin preparation (later). After a preparation, the column should be washed extensively with G buffer to remove all formamide and stored in G buffer plus sodium azide. If stored properly, DNase I columns can be used several times over a 1- to 3-month period. Other laboratories suggest that the columns remain good for several years.

Purification of Yeast Actin

1. Weigh 200 g frozen yeast lysate in a 1000-ml glass beaker. Thaw the lysate by adding 200 ml room temperature G buffer and stirring constantly with a spatula. To accelerate thawing, place the beaker containing the lysate into a shallow bath of warm water and stir constantly. While stirring, and as the lysate is thawing, add 1 ml PMSF (200 mM) and 200 μl 1000× aqueous protease inhibitor cocktail (0.5 mg/ml each of antipain, leupeptin, pepstatin A, chymostatin, and aprotinin).

2. When the lysate is thawed completely (no visible chunks), but still cold, transfer to centrifuge tubes for a GSA rotor (Sorvall). Centrifuge at 12,000 rpm, 4°, 20 min. Pour the resulting low-speed supernatant (LSS) through eight layers of cheesecloth wedged in a plastic funnel over a 250-ml graduated cylinder on ice. Leave behind the murky LSS at the bottom of the tube (last 10 ml).

3. Using a pipette, transfer 26 ml LSS to each of 10 tubes (polycarbonate with metal cap assemblies). Centrifuge in a Ti60 rotor (Beckman), 70–80 min, 4°, 60,000 rpm. Alternatively, spin in a Ti45 rotor, 2 hr, 45,000 rpm.

4. Harvest the high-speed supernatant (HSS) with extreme care using a 10-ml pipette (an electric pipet-aid with speed control is ideal for this step). Harvest only the clearest HSS, minimizing uptake of the upper lipid phase and absolutely avoiding the lower debris. Filter the HSS through cheesecloth into a graduated cylinder on ice.

5. Load the HSS onto the DNase I affinity column at a rate of about 2 ml/min. Then, wash the column with 20 ml G buffer + 10% deionized formamide, 20 ml G buffer + 0.2 M ammonium chloride, and finally 20 ml G buffer alone.

6. Elute the actin with 20 ml G buffer + 50% (v/v) deionized formamide. It is important to use either freshly deionized formamide or a freshly thawed aliquot stored at $-80°$. To deionize formamide, add approximately 20 g AG-501 resin (Bio-Rad) to 500 ml of formamide. Stir the mixture very gently (without crushing the resin) for 1 hr at room temperature and filter to remove resin. Aliquot and store at $-80°$ immediately.

7. The actin eluate should be exchanged into G buffer, as prolonged exposure to formamide can denature actin. To accomplish this, Kron and co-workers[1] applied the actin eluate directly to a DEAE column, washed the column, and eluted the actin with a 100–400 mM KCl gradient in G buffer. Instead, we substitute a Mono Q column (Amersham/Pharmacia, Piscataway, NJ) because of the superior resolving power over DEAE. This added resolution is important for removing contaminant cofilin (see later). Load the actin eluate from the DNase I column onto the Mono Q column at 0.5 ml/min. Wash with 5 column volumes (5 ml) G buffer and elute the actin using a 20-ml linear KCl gradient (100–300 mM) in G buffer, collecting 0.5-ml fractions. Analyze 10 μl of each column fraction by SDS–PAGE and Coomassie staining. Actin elutes at approximately 250 mM KCl.

8. Pool the peak actin-containing fractions (usually 2 ml total volume) and dialyze overnight against 1 liter of G buffer. Harvest only the actin peak because cofilin begins to elute from the Mono Q column shortly after actin.

9. This step is optional. In many cases, it is important to demonstrate that the actin isolated is assembly competent. This step is especially recommended if purifying a yeast actin mutant for the first time, as some yeast actin mutants may denature on exposure to 50% (v/v) formamide. To initiate polymerization of the actin, add 1/19th volume of 20× initiation mix (1 M KCl, 40 mM MgCl$_2$, 10 mM ATP). Incubate for 60 min at 25° and then pellet the actin in a TLA100.3 rotor, 90,000 rpm, 20 min, 25°. Resuspend the actin pellet in G buffer to a convenient stock concentration (1–2 mg/ml, or approximately 50–100 μM). Dounce the resuspended material thoroughly (omission of this step can reduce the yield drastically). Dialyze the actin overnight versus 1 liter G buffer, 4°, and proceed to step 10.

10. Concentrate the pooled/dialyzed actin fractions from the Mono Q column to 50–100 μM using Microcon-10 devices (Amicon, Danvers, MA). Clear any denatured actin in the sample by ultracentrifugation in a TLA100 or TLA100.3 rotor (Beckman, Fullerton, CA), 90,000 rpm, 20 min, 4°. Snap freeze 20- to 50-μl aliquots in liquid nitrogen and store at $-80°$.

11. If cofilin contamination is not a concern, the ion-exchange column (step 7) can be bypassed. In this case, use a Centriprep-10 device (Amicon) to concentrate the 20- to 40-ml actin eluate from the DNase I column. Concentrate to 1 ml and

exchange into G buffer on a disposable NAP-10 column (Amersham/Pharmacia) preequilibrated in G buffer. Proceed to step 10. Actin isolated in this manner may contain as high as a 1 : 50 molar stoichiometry of cofilin : actin contamination.

Optional Steps for Removing Cofilin/ADF Contamination

Using the method of Kron et al.,[1] yeast actin is isolated in a purified form, but contains substoichiometric levels (1 : 50 to 1 : 200) of the major actin filament-depolymerizing factor in cells, cofilin/ADF. Cofilin contamination can be seen on overloaded Coomassie-stained gels (50 μg actin in a single lane), appearing as an 18-kDa band recognized by a yeast cofilin antibody (A. Rodal and B. Goode, unpublished data, 2001). Such small amounts of cofilin generally have little effect on actin filament cosedimentation assays, but can strongly influence actin kinetic studies. Even low stoichiometric quantities of cofilin can significantly reduce the lag phase for actin assembly and increase the rate of actin filament disassembly and turnover. To remove cofilin contamination, we substitute a Mono Q column for the DEAE column, as described earlier, which resolves actin and cofilin.

Isolation of Actin-Associated Proteins from Yeast Extracts

Technical barriers have prevented the analysis of actin assembly in membrane-free soluble extracts of *S. cerevisiae*. The yeast actin cytoskeleton disassembles rapidly on cell lysis, and past efforts to restimulate actin assembly in lysates have been unsuccessful. We have established a method for the reconstitution of actin assembly in soluble yeast extracts. A detailed outline of this method is described. A comprehensive study based on this procedure will be reported elsewhere (manuscript in preparation).

In brief, yeast cells are lysed, and a clarified extract is prepared in a low salt buffer. This soluble lysate contains the vast majority of the cellular actin in a monomeric form, along with a free pool of soluble actin filament-associated proteins. Actin assembly is triggered by the addition of 5 mM magnesium chloride and 10% glycerol and is then incubated on ice for 3 hr to overnight. The actin structures formed in the extracts are isolated by centrifugation, washed, and fractionated into individual components by ion-exchange chromatography.

Using mass spectrometry analyses and immunoblotting, we have identified all of the major bands on SDS–PAGE gels of the column fractions (Fig. 1). These include many of the known yeast actin-binding proteins (e.g., Aip1p, Abp1p, Abp140p, Arp2/3 complex, Cap1p, Cap2p, Cof1p, Crn1p, Sac6p/fimbrin, Sla2p, Srv2p/CAP, Twf1p; the functions of these factors are reviewed elsewhere[12]). In addition, we have identified several new putative actin-associated factors. This

[12] D. Pruyne and A. Bretscher, *J. Cell Sci.* **113**, 571 (2000).

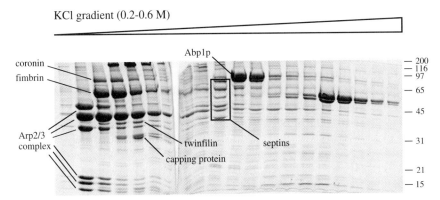

FIG. 1. Fractionation of a yeast actin-associated protein (AAP) mixture by anion-exchange chromatography. The actin filament structures isolated from yeast extracts were stripped of AAPs by exposure to 0.5 M KCl (see method in text). Next, the AAP mixture was exchanged into HEK buffer [20 mM HEPES–KOH (7.5), 1 mM EDTA, 50 mM KCl] and applied to a Mono Q 5/5 column (Amersham/Pharmacia). Bound proteins were eluted from the column using a 20-ml linear gradient of KCl (0.1–0.6 M) in HEK buffer. Column fractions (500 ml) were collected, and 10 μl of each column fraction was analyzed on a 12% SDS–PAGE gel stained with Coomassie blue. The identities of almost every protein visible in gels like this have been determined by a comprehensive tandem mass spectrometry approach (manuscript in preparation). Some of the more pertinent AAPs are labeled; the identities of these proteins were confirmed by immunoblotting.

procedure opens the door to comparing actin assembly in wild-type and mutant yeast extracts and to isolating native wild-type and mutant yeast actin-associated proteins to define their activities.

Strain Requirements and Cell Lysis

A normal large-scale preparation requires about 100 g of pelleted yeast cells grown to either log or stationary phase. Importantly, 100 g of pelleted cells resuspended, frozen, and lysed yields ~130 g frozen yeast lysate powder. Although we optimized this procedure using a w303 derivative strain, BGY12 (*ura3, his3, leu2, trp1, ade2*), we have found that the method is equally successful using other strains, including s288c. Yeast cells used in the procedure are washed, frozen, and lysed as described in the first section.

1. Thaw the lysate. Weigh 130 g frozen yeast lysate powder (see the first section) in a 500-ml beaker at room temperature. Immediately add 130 ml HEK buffer [20 mM HEPES–KOH (7.5), 1 mM EDTA, 50 mM KCl], thaw, and add protease inhibitors as described in step 1 of the first section.

2. Generate a low-speed supernatant. Centrifuge the lysate in a GSA rotor (Sorvall), 12,000 rpm, 4°, 20 min. Pour the supernatant through eight layers of

cheesecloth wedged in a plastic funnel over a 250-ml graduated cylinder on ice. Leave behind the murky supernatant at the bottom of the tubes (last 10 ml). The total volume recovered should be ~190 ml.

3. Add glycerol. Add a stir bar and transfer the graduated cylinder to a stir plate in the cold room. While mixing, add a thin stream of 100% glycerol to a final concentration of 10% for a new total volume of ~210 ml. If necessary, stir with a spatula to assist mixing the LSS and glycerol.

4. Generate a high-speed supernatant. Using a disposable pipette, transfer 26 ml of LSS/10% glycerol to each of 10 ultracentrifuge tubes (polycarbonate with metal cap assemblies), for a Ti60 rotor (Beckman). Centrifuge 70–80 min, 4°, 60,000 rpm. Carefully harvest the HSS as described in step 4 of the first section. The total volume recovered should be ~150 ml.

5. Add magnesium to induce actin assembly. To trigger actin assembly, add 5 mM MgCl$_2$ (from a 1 M stock) and 0.2 mM ATP (from a 100 mM stock). Stir well to mix. Incubate the mixture (without stirring) for 2 hr (or up to overnight) in the cold room to allow the assembly of actin. The rate of actin assembly will vary depending on the protein concentration of the extract: the higher the protein concentration, the faster the actin assembly. Thus, efficient actin assembly requires efficient cell lysis. If yield is low and/or the minimum incubation time (1 hr) is desired, generate a more concentrated cell extract. This is accomplished by altering step 1: in step 1, resuspend the 130-g frozen cell lysate in 14 ml of 5× HEK buffer and then proceed.

6. Isolate actin structures by centrifugation. Transfer (by pouring) ~30 ml of lysate into each centrifuge tube (open-mouth, polypropylene thick-walled tubes with ~50 ml capacity). Centrifuge in a SA600 rotor (Sorval), 4°, 17,000 rpm, 90 min to pellet the actin. Although these g forces are not sufficient to pellet purified actin filaments, they will precipitate the actin structures from cell extracts. This is because the actin filaments in extracts are bundled and organized by actin-associated proteins (e.g., Sac6p, Crn1p, Abp140p, EF-1α).

7. Wash actin structures. Decant the supernatants and invert the tubes on paper towels for 1–2 min. Using a P1000 Pipetman and a single cutoff tip (terminal 5 mm removed with a razor blade), resuspend all 10 pellets in a *total volume* of 10 ml HEK buffer plus 5% glycerol and 5 mM MgCl$_2$. Transfer approximately 2.5 ml of material to each of four polyallomer tubes for a tabletop ultracentrifuge TLA100.3 rotor (Beckman). It is not necessary to homogenize this material. Spin 70,000 rpm, 4°, 15 min to wash the actin structures. Keeping the volume low at this step is critical for obtaining a high yield because actin will disassemble rapidly to its critical concentration (~0.2 μM) in the wash buffer. Thus, losing ~10 ml (wash volume) of 0.2 μM actin.

8. Rinse the pelleted actin structures. Remove supernatants by aspiration and very gently rinse each pellet with 1 ml HEK buffer +5% glycerol +5 mM MgCl$_2$ +0.2 mM ATP.

9. Salt strip the actin-associated proteins. Resuspend the pellets in a total of 2 ml HEK buffer +5% glycerol +5 mM MgCl$_2$ +0.2 mM ATP +0.5 mM KCl. The high salt strips the actin-associated proteins, leaving the actin filaments intact (although a small amount of polymer disassembles in the presence of 0.5 M KCl). Using a P200 Pipetman and a cutoff tip, resuspend each of the four pellets with 0.5 ml buffer and transfer the material to a 10-ml Wheaton (Millville, NJ) Dounce homogenizer on ice. Rinse the tubes with ∼0.4 ml buffer to recover residual pellet material and combine in Dounce. Using a tight-fitting pestle, homogenize the material with ∼20 strokes, minimizing aeration of the sample.

10. Centrifuge to remove actin filaments. With a P1000 and cutoff tip, transfer the homogenized material to a TLA100.3 rotor tube on ice. Rinse the Dounce with another 0.5 ml at high salt buffer and combine for a total volume of ∼3 ml. Centrifuge for 20 min, 80,000 rpm, 4° to pellet actin filaments. Remove the supernatant (∼2.5 ml) using a P1000, avoiding the upper lipid phase. This supernatant is the AAP mixture.

11. Lower the salt concentration of the AAP mixture. Transfer the AAP mixture to a 50-ml tube on ice. Reduce the salt concentration to 50–80 mM KCl by diluting the mixture with 8 volumes of low salt buffer [20 mM HEPES (7.5), 1 mM EGTA, 5% glycerol]. The total volume after this step is ∼23 ml. Alternatively, the salt concentration can be reduced by dialysis versus HEK buffer; however, this requires more time.

12. Fractionate the AAP mixture by anion-exchange chromatography. Using a 50-ml superloop, load the AAP onto an FPLC Mono Q [(5/5) = 5 mm × 5 cm] column (Amersham/Pharmacia) at ∼0.8 ml/min. Wash the column with 10 ml HEK buffer. Elute bound AAPs with a 20-ml linear KCl gradient (0.1–0.6 M) in HEK buffer at a rate of 0.5 ml/min, collecting 0.5-ml fractions.

13. Further purification of individual AAPs. Run 10 μl of each column fraction on a 12% SDS–PAGE gel and stain with Coomassie blue. This should resolve clearly bands of the Arp2/3 complex, coronin, Sac6p/fimbrin, the Abp1p doublet, capping protein doublet (Cap1/2p), and twinfilin. Individual components of the AAP can be purified further by pooling fractions of interest, reducing their volume using a Microcon-10 concentration device (Amicon), and fractionating on a Superdex 12 gel-filtration column (or other columns, such as Mono S and HAP).

[26] Identifying Functional Interactions with Molecular Chaperones

By JILL L. JOHNSON and ELIZABETH A. CRAIG

All organisms have proteins that function in the maturation of other proteins, including their folding and translocation across membranes. *Saccharomyces cerevisiae* is no exception, as it has molecular chaperones in all major cellular compartments. Some are induced by stresses such as increased temperatures, while expression and function of others are constitutive. Many investigators studying diverse cellular processes encounter molecular chaperones in the course of their work. The challenge facing these scientists is whether the interactions observed are biologically meaningful or if they are artifacts caused by the propensity of molecular chaperones to interact with hydrophobic stretches of proteins, particularly of partially unfolded proteins. This is not an easy question for anyone to answer. However, this chapter describes tools and approaches that can be used to address the relevance of such interactions.

Heat Shock (Stress)-Inducible Genes

The highly conserved molecular chaperones called heat shock proteins (Hsps) are divided into classes. As this classification was originally done according to their migration in SDS–PAGE gels, the major groupings are called (somewhat inaccurately), Hsp104, Hsp90, Hsp70, Hsp60, and "small Hsps" (Hsp26). Although some of these proteins are encoded by single genes, others form very complex multigene families. For example, the cytosolic chaperones Hsp104 and Hsp26 are encoded by single genes. However, a mitochondrial protein related to Hsp104, called Hsp78, also exists in yeast. Hsp90 is a collective name for two closely related cytosolic proteins: the heat-inducible Hsp82 and the constitutively expressed Hsc82. The name Hsp70 encompasses 14 different proteins, including two members of the Hsp110 subfamily, Sse1/2. As an additional layer of complexity, many of these chaperones work with cochaperones, e.g., Hsp70s function with Hsp40s (DnaJ proteins). In addition, a number of heat-inducible genes of *S. cerevisiae* are not obviously related to the ubiquitous heat shock proteins that have been highly conserved, some of which may be molecular chaperones as well.

More detailed information about the individual yeast chaperone families can be found in the volume "Guidebook to Molecular Chaperones and Protein-Folding Catalysts," edited by Gething,[1] as well as reviews that have references to many

[1] M.-J. Gething, "Guidebook to Molecular Chaperones and Protein-Folding Catalysts." Oxford Univ. Press, New York, 1997.

relevant papers.[2-7] In addition, the *Saccharomyces* Genome Database[8] and the Proteome YPD database[9] are excellent references for finding information about gene expression and known protein–protein interactions of all yeast proteins. However, this chapter focuses on Hsp70s and Hsp90s, the general chaperones most commonly found when studying other cellular processes.

Test for Induction of Heat Shock (Stress) Response

Some investigators actively study the heat shock response and its physiological consequences. Probably many more do not realize that the heat shock/stress response is active under their experimental conditions. For example, conversion of yeast cells to spheroplasts often results in the induction of heat shock genes.[10] Two types of transcription factors function to regulate the response of *S. cerevisiae* to stress: the "heat shock factor," Hsf1, and "general stress response factors," Msn2 and Msn4. Hsf1 binds to promoter elements termed HSEs, whereas Msn2/4 bind to STRE elements (reviewed in Estruch[11]). Induction of stress genes by these transcription factors may occur due to stresses other than an increase in temperature. Induction by Hsf1 is triggered by other factors, such as accumulation of abnormally folded proteins, particularly in the cytosol/nucleus, high concentrations of ethanol, and perhaps oxidative stress. Msn2/4 are activated not only by heat shock and ethanol, but are also under negative control by protein kinase A. Thus, as cyclic AMP levels fall due to nutrient limitation, particularly as cells approach stationary phase, genes containing STREs in their promoters are activated. In addition, such genes are responsive to osmotic stress via the *HOG1* pathway. Several approaches can be used to determine whether particular conditions induce the expression of stress-inducible genes: promoter fusions to indirectly measure expression from a stress-inducible promoter or RNA (Northern) blots or immunoblots to measure the expression of a particular mRNA or protein.

---

[2] K. Morano, P. Liu, and D. Thiele, *Curr. Opin. Microbiol.* **1**, 197 (1998).
[3] B. Bukau and A. L. Horwich, *Cell* **92**, 351 (1998).
[4] M. Leroux and F. Hartl, *Curr. Biol.* **10**, 260 (2000).
[5] E. Craig, W. Yan, and P. James, in "Molecular Chaperones and Folding Catalysts: Regulation, Cellular Function and Mechanisms" (B. Bukau, ed.), p. 139. Harwood Academic, Amsterdam, 1999.
[6] L. H. Pearl and C. Prodromou, *Curr. Opin. Struct. Biol.* **10**, 46 (2000).
[7] M. P. Mayer and B. Bukau, *Curr. Biol.* **9**, R322 (1999).
[8] J. M. Cherry, C. Ball, K. Dolinski, S. Dwight, M. Harris, J. C. Matese, G. Sherlock, G. Binkley, H. Jin, S. Weng, and D. Botstein, http://genome-www.stanford.edu/Saccharomyces/.
[9] M. C. Costanzo, M. E. Crawford, J. E. Hirschman, J. E. Kranz, P. Olsen, L. S. Robertson, M. S. Skrzypek, B. R. Braun, K. L. Hopkins, P. Kondu, C. Lengieza, J. E. Lew-Smith, M. Tillberg, and J. I. Garrels, *Nucleic Acid. Res.* **29**, 75 (2001).
[10] C. Adams and D. Gross, *Mol. Cell. Biol.* **173**, 7429 (1991).
[11] F. Estruch, *FEMS Microbiol. Rev.* **24**, 469 (2000).

Promoter Fusions

The use of promoter fusions has the advantage of ease of measurements, as fusions to *lacZ*, encoding the enzyme β-galactosidase, whose activity is measured very easily, are commonly employed.[12] However, each strain must be transformed with the plasmid carrying the gene fusion and selective pressure must be applied to maintain the plasmids containing the constructs, unless they are integrated into the chromosome. The fusion we find particularly useful is a translational fusion between the promoter of *SSA4* encoding a cytosolic Hsp70 and *lacZ* (pWB213, a centromeric plasmid carrying the *TRP1* gene[13]). *SSA4* is the archetypal heat shock gene. It is expressed at extremely low levels under optimal growth conditions and is induced greater than 100-fold under stress conditions. *SSA4* is a sensitive indicator of cell stress, as we have seen subtle stress conditions where *SSA4* shows substantial induction and other heat-inducible genes do not. However, it should be remembered that Ssa4 has HSEs in its promoter, but not STRE elements, and is thus regulated by Hsf1, but not Msn2/4.

CTT1, encoding catalase T, is regulated by Msn2/4. Analysis of fusions to *lacZ* integrated into the chromosome at the URA3 locus (pTB3 and derivatives) have been described (e.g., in Wieser *et al.*[14]). Fusions of 390 bp of the *CTT1* promoter to *lacZ* show a 35-fold induction after a heat shock from 23 to 37°. It should be kept in mind that the *CTT1* promoter contains promoter elements in addition to STREs, including a binding site for the heme-regulated transcription factor Hap1. However, a segment of the promoter–325 to 382 acts as a UAS in the context of the *LEU2* promoter; this fusion (AW2X) is 15-fold inducible by heat shock.

Two stress-inducible genes, *HSP104* and the Hsp70 gene, *SSA3*, appear to be regulated by both systems, although in the case of *SSA3* this regulation is complex.[11] *lacZ* fusions have been used to measure expression from the promoters of both these genes.[15,16] The *HSP104 : lacZ* fusion showed a 8- to 10-fold induction on a heat shock of cells grown on glucose-based media. This fusion was constructed in the *URA3*-based integrative *lacZ* fusion vector YIP358R.[17] The *SSA3* promoter fusion, carried on a centromeric vector harboring the *TRP1* gene, has been characterized more thoroughly.[18] A wild-type strain containing a *SSA3 : lacZ* translational fusion (pWB204Δ-583) and growing logarithmically on glucose-based media has approximately 3.5 units of β-galactosidase activity. On heat shock the activity

---

[12] C. M. Nicolet and E. A. Craig, *Methods Enzymol.* **194,** 710 (1991).
[13] W. R. Boorstein and E. A. Craig, *J. Biol. Chem.* **265,** 18912 (1990).
[14] R. Wieser, G. Adam, A. Wagner, C. Schuller, G. Marchler, H. Ruis, Z. Krawiec, and T. Bilinski, *J. Biol. Chem.* **266,** 12406 (1991).
[15] B. Hazell, H. Nevalainen, and P. Attfield, *FEBS Lett.* **377,** 457 (1995).
[16] W. Boorstein and E. A. Craig, *Mol. Cell. Biol.* **10,** 3262 (1990).
[17] A. Meyers, A. Tzagoloff, D. Kinney, and C. Lusty, *Gene* **299,** 299 (1986).
[18] W. R. Boorstein and E. A. Craig, *EMBO J.* **9,** 2543 (1990).

increases to 61 units; continued growth until the diauxic shift is reached results in an increase to 290 units. Therefore, this promoter fusion provides an indication of both "heat shock" induction and glucose depletion.

RNA (Northern) and Immunoblots

The induction of heat shock genes can be monitored by direct detection of either the mRNA or the protein using standard methods. However, several things should be kept in mind when choosing tools for this analysis. Some of the heat-inducible proteins mentioned in the previous section are closely related to proteins that are constitutively expressed. Simple hybridization experiments will not distinguish between expression of these genes. The Hsp70 *SSA* family is particularly problematic in this regard, as *SSA2* and *SSA1* are expressed at high levels normally, whereas *SSA3* and *SSA4* are not. *HSP104* is a unique heat-inducible gene and is therefore more useful in simple hybridization experiments. However, it should be noted that expression of *HSP104* increases as cells approach stationary phase and is normally higher in cells grown on carbon sources other than glucose. HSP26 is also a useful probe in hybridization experiments. Like *HSP104*, *HSP26* is a unique gene that is expressed at high levels on approach to stationary phase.[19]

Detection of the presence of heat shock proteins directly is the most accurate reflection of induction of the response because it measures the level of the proteins themselves. However, antibodies that react with most yeast heat shock proteins are not available commercially. An exception is Hsp104. Antibodies can be purchased from Affinity BioReagents, Inc. (Golden, CO) or StressGen Biotechnologies Corp. (Victoria, BC, Canada).

Importance of Interaction with Chaperone

Researchers most commonly come upon molecular chaperones in their work through either genetic or biochemical interactions. The challenge is to determine whether these interactions are biologically important. Typically, a biochemical interaction is found because of coimmunoprecipitation of a chaperone with the protein of interest. Genetic interactions can be found in synthetic lethal, multicopy suppressor or intragenic suppressor screens or selections. In our experience, the most productive way to test the importance of an interaction is to play genetics off of biochemistry and vice versa. If a biochemical interaction is found, use genetics to test its biological importance. If a genetic interaction is found, test for a direct biochemical interaction as well. Focusing on Hsp70 and Hsp90 interactions, we discuss ways to go about testing the nature of these interactions.

[19] R. Susek and S. Lindquist, *Mol. Cell. Biol.* **10,** 6362 (1990).

Hsp70 Interactions

Hsp70s, found in several cellular compartments, are involved in many physiological functions, including protein folding, translocation of proteins into organelles, and assembly and disassembly of protein complexes. Hsp70s are often restricted to a particular cellular compartment, some to mitochondria, the endoplasmic reticulum (ER) lumen, or the cytosol and/or nucleus. All Hsp70s function with cochaperones called J-type chaperones (Hsp40s/DnaJs). At least 16 genes of the *S. cerevisiae* genome encode J-type chaperones. Some of these are experimentally well defined. In others cases it is only known that the encoded protein contains a signature "J" domain with the highly conserved HPD (histidine, proline, aspartic acid) motif. Fourteen members of the Hsp70 chaperone family are encoded in the *S. cerevisiae* genome. Therefore, it is impossible to comprehensively deal with individual Hsp70s in this chapter. Rather, we discuss methods that can be applied to the analysis of many different Hsp70s.

Hsp70 : Hsp40 Interactions. Hsp70s have an amino-terminal ATPase domain and a carboxy-terminal substrate-binding region. The interaction of Hsp70s with unfolded or partially unfolded polypeptides is regulated by ATP.[3] Hsp40s bind the ATPase domain and, by stimulating ATP hydrolysis, promote the ADP-bound form of Hsp70, which has a higher affinity for unfolded protein substrates. All experiments to date indicate that Hsp70s always function together with cochaperones of the J class. Table I lists the Hsp70 : Hsp40 partnerships that are suggested by published data. Some of these partnerships, such as the Ssa Hsp70s and the J-type Ydj1 and Sis1 cochaperones of the cytosol, are well defined by both genetic and biochemical experiments. Others, such as the Ssb Hsp70s and the J-cochaperone Zuo1, are less well defined, based on genetic and *in vivo* colocalization studies. In individual cases the literature must be evaluated to determine the experimental foundation of the classification.

Analysis of Direct Interaction between Hsp70 and Protein of Interest. Coimmunoprecipitation of a molecular chaperone with the protein of interest leads to the question of the nature of the interaction and whether this interaction is important *in vivo*. Because of the propensity of Hsp70s to bind to exposed hydrophobic sequences in proteins, the first question is whether the interaction is occurring inside the cell or isolated organelle or whether the interaction occurs in the cell extract after lysis. This problem can be addressed by the addition of excess Hsp70 protein when lysing the cell/organelle, as has been done in the analysis of the mitochondrial Hsp70 Ssc1.[20,21] If radiolabeled cells are being used, addition of an excess of unlabeled chaperone can be added prior to lysis. If binding

---

[20] J. Rassow, A. Maarse, E. Krainer, M. Kubrich, H. Muller, M. Meijer, E. Craig, and N. Pfanner, *J. Cell Biol.* **127**, 1547 (1994).

[21] N. G. Kronidou, W. Opplinger, L. Bolliger, K. Hannavy, B. Glick, G. Schatz, and M. Horst, *Proc. Natl. Acad. Sci. U.S.A.* **91**, 12818 (1994).

TABLE I
Hsp70/Hsp40 Chaperone Pairs

Site	Hsp70	Hsp40	Ref.
Cytoplasm	Ssa1-4	Ydj1^m	26
		Sis1	a
	Ssb1/2	Sis1	b
		Zuo1	c
	Pdr13	Zuo1	d
	Sse1/2	?	e
Endoplasmic reticulum	Kar2	Sec63^m	f
		Scj1	g
		Jem1	h
	Lhs1 (Cer1)	Scj1	i
Mitochondria	Ssc1	Mdj1^m	j
	Ssq1	Jac1	k
	Ecm10	?	l

^{a} Z. Lu and D. M. Cyr, *J. Biol. Chem.* **273**, 27824 (1998).
^{b} M. Ohba, *FEBS Lett.* **409**, 307 (1997).
^{c} W. Yan, B. Schilke, C. Pfund, W. Walter, S. Kim, and E. A. Craig, *EMBO J.* **17**, 4809 (1998).
^{d} T. Michimoto, T. Aoki, A. Toh-e, and Y. Kikuchi, *Gene* **257**, 131 (2000).
^{e} H. Mukai, T. Kuno, H. Tanaka, D. Hirata, T. Miyakawa, and C. Tanaka, *Gene* **132**, 57 (1993).
^{f} J. L. Brodsky and R. Schekman, *J. Cell Biol.* **123**, 1355 (1993).
^{g} G. Schlenstedt, S. Harris, B. Risse, R. Lill, and P. A. Silver, *J. Cell Biol.* **129**, 979 (1995).
^{h} V. Brizzio, W. Khalfan, D. Huddler, C. T. Beh, S. S. Andersen, M. Latterich, and M. D. Rose, *Mol. Biol. Cell* **10**, 609 (1999).
^{i} T. G. Hamilton and G. C. Flynn, *J. Biol. Chem.* **271**, 30610 (1996).
^{j} B. Wetermann, B. Gaume, J. M. Herrmann, W. Neupert, and E. Schwarz, *Mol. Cell. Biol.* **16**, 7063 (1996).
^{k} C. Voisine, Y. C. Cheng, M. Ohlson, B. Schilke, K. Hoff, H. Beinert, J. Marszalek, and E. Craig, *Proc. Natl. Acad. Sci. U.S.A.* **98**, 1483 (2001).
^{l} F. Baumann, I. Milisav, W. Neupert, and J. M. Herrmann, *FEBS Lett.* **487**, 307 (2000).
^{m} Most thoroughly established Hsp70–Hsp40 interactions.

of the chaperone occurs during or after lysis, the amount of radiolabeled chaperone pulled down in the experiment will be decreased due to competition for binding with the unlabeled chaperone. However, if the interaction is occurring in the cell and is stable, addition of an exogenous chaperone should not affect the amount of radiolabeled chaperone coimmunoprecipitating with the protein of interest. Variations on this theme can also be used. For example, His-tagged chaperone can be added to cells on lysis, prior to immunoprecipitation. Assuming the His-tagged and untagged protein migrate differently in SDS–PAGE, the amount of binding by the native cellular chaperone in the presence or absence of an exogenously added His-tagged chaperone can be determined, without resorting to radiolabeling.

Interactions with Hsp70s can be divided into two categories: Hsp70 : substrate interactions and "typical" protein : protein interactions. Usually, a protein binding as a substrate (i.e., binding in the peptide-binding cleft) will be released on incubation with ATP, as binding of ATP increases the off rate of substrates dramatically (e.g., see Zhang et al. [22]). Obviously, proteins may interact at sites other than the peptide-binding cleft. Such interactions may be independent of nucleotide, as is the interaction between the mammalian protein Hop and the extreme C terminus of Hsp70.[23] However, one such interaction that is disrupted by ATP is the interaction of the ATPase domain of the mitochondrial Hsp70 Ssc1 with the nucleotide exchange factor Mge1.[24] In addition, initiation of interactions between Hsp70s and J-type chaperones is dependent on ATP.[25]

Finding the Meaning of Genetic Interactions with Hsp70s

In the case of genetic interactions, be it a synthetic lethal interaction of mutations or a suppression of mutant phenotypes, the question is whether there is direct involvement of the chaperone in the process being studied. Alternatively, the effect could very well be indirect, as by their nature increases or decreases in the activity of chaperones often have very pleiotropic effects. Strains carrying mutants in a number of HSP70 genes are available that allow testing to determine if a decrease in chaperone activity has an affect on the pathway or physiological process of interest. Table II lists some of the mutants that may be of use in such studies. Analysis of mutant *SSA* strains is complicated because Ssa proteins are encoded by four genes, at least one of which must be present for viability. Strains containing multiple mutations in genes encoding Ssa are available and have been used to demonstrate a role for Ssa in protein import into mitochondria, the ER, and vacuolar vesicles.[26,27]

Hsp90 Interactions

Hsp90 is an essential, cytosolic chaperone accounting for 1–2% of all cytosolic proteins.[28] In *S. cerevisiae,* Hsp90 is encoded by two genes: the constitutively expressed *HSC82* and the heat-inducible *HSP82*. In mammalian cells, Hsp90 is critical for the activity of a number of signal-transducing proteins, such as steroid receptors, oncogenic tyrosine kinases, and additional diverse proteins involved in

---

[22] S. Zhang, C. J. Williams, K. Hagan, and S. W. Peltz, *Mol. Cell. Biol.* **19,** 7568 (1999).
[23] J. Demand, J. Luders, and J. Hohfeld, *Mol. Cell. Biol.* **18,** 2023 (1998).
[24] B. Miao, J. E. Davis, and E. A. Craig, *J. Mol. Biol.* **265,** 541 (1997).
[25] A. K. Corsi and R. Schekman, *J. Cell Biol.* **137,** 1483 (1997).
[26] J. Becker, W. Walter, W. Yan, and E. A. Craig, *Mol. Cell. Biol.* **16,** 4378 (1996).
[27] C. R. Brown, J. A. McCann, and H. L. Chiang, *J. Cell Biol.* **150,** 65 (2000).
[28] K. A. Borkovich, F. W. Farrelly, D. B. Finkelstein, J. Taulien, and S. Lindquist, *Mol. Cell. Biol.* **9,** 3919 (1989).

TABLE II
Hsp70 MUTANT STRAINS

Site	Hsp70	Available mutants	Growth defects	Ref.
Cytoplasm	Ssa1–4	ssa1ssa2	Temperature sensitive	a
		ssa1[45]ssa2ssa3ssa4	Temperature sensitive	26
	Ssb1/2	Null	Cold sensitive	b
	Pdr13	Null	Cold sensitive	c
	Sse1	Null	Temperature sensitive	d
	Sse2	Null	None	d
Endoplasmic reticulum	Kar2	Conditional alleles	Temperature sensitive	e, f
	Lhs1	Null	Slightly cold sensitive	g–i
Mitochondria	Ssc1	Conditional alleles	Temperature sensitive	j,k
	Ssq1	Null	Cold sensitive	l
	Ecm10	Null	No reported defects	m

[a] E. A. Craig and K. Jacobsen, *Cell* **38**, 841 (1984).
[b] R. J. Nelson, T. Ziegelhoffer, C. Nicolet, M. Werner-Washburne, and E. A. Craig, *Cell* **71**, 97 (1992).
[c] T. C. Hallstrom, D. J. Katzmann, R. J. Torres, W. J. Sharp, and W. S. Moye-Rowley, *Mol. Cell. Biol.* **18**, 1147 (1998).
[d] H. Mukai, T. Kuno, H. Tanaka, D. Hirata, T. Miyakawa, and C. Tanaka, *Gene* **132**, 57 (1993).
[e] J. Polaina and J. Conde, *Mol. Gen. Genet.* **186**, 253 (1982).
[f] J. P. Vogel, L. M. Misra, and M. D. Rose, *J. Cell Biol.* **110**, 1885 (1990).
[g] R. A. Craven, M. Egerton, and C. J. Stirling, *EMBO J.* **15**, 2640 (1996).
[h] T. G. Hamilton and G. C. Flynn, *J. Biol. Chem.* **271**, 30610 (1996).
[i] B. K. Baxter, P. James, T. Evans, and E. A. Craig, *Mol. Cell. Biol.* **16**, 6444 (1996).
[j] P. J. Kang, J. Ostermann, J. Shilling, W. Neupert, E. A. Craig, and N. Pfanner, *Nature* **348**, 137 (1990).
[k] B. D. Gambill, W. Voos, P. J. Kang, B. Miao, T. Langer, E. A. Craig, and N. Pfanner, *J. Cell Biol.* **123**, 109 (1993).
[l] B. Schilke, J. Forster, J. Davis, P. James, W. Walter, S. Laloraya, J. Johnson, B. Miao, and E. Craig, *J. Cell Biol.* **134**, 603 (1996).
[m] F. Baumann, I. Milisav, W. Neupert, and J. M. Herrmann, *FEBS Lett.* **487**, 307 (2000).

signaling pathways and the cell cycle control.[7,29] Unlike Hsp70, which is believed to play a general role in chaperoning a wide range of cellular proteins, Hsp90 is probably not required for the bulk folding of cytosolic proteins,[30] rather it appears to be involved in the maturation of a diverse subset of proteins. In *S. cerevisiae*, Hsp90 interacts with heterologous substrates expressed in yeast and native substrates such as the MEK kinase Ste11,[31] Gcn2, a member of the eIF-2α kinase family,[32] the

[29] W. B. Pratt and D. O. Toft, *Endocr. Rev.* **18**, 306 (1997).
[30] D. F. Nathan, M. H. Vos, and S. Lindquist, *Proc. Natl. Acad. Sci. U.S.A.* **94**, 12949 (1997).
[31] J. F. Louvion, T. Abbas-Terki, and D. Picard, *Mol. Biol. Cell* **9**, 3071 (1998).
[32] O. Donze and D. Picard, *Mol. Cell. Biol.* **19**, 8422 (1999).

heme-responsive transcription factor Hap1,[33] the yeast heat shock factor Hsf1,[34] and Cna2, the catalytic subunit of calcineurin.[35]

In yeast and mammalian systems, Hsp90 is found in a conserved complex with a number of cochaperones.[29,36] In yeast, known Hsp90 cochaperones are Sti1, Sba1, Ydj1, Cdc37, Cns1, Cpr7, and Sse1.[37] Thus the appearance of Hsc82/Hsp82 or any of the cochaperones in a genetic or two-hybrid screen may indicate an interaction between the protein of interest and the Hsp90 complex. The following section outlines some approaches used to determine the significance of this interaction. As discussed earlier, a combination of genetic and biochemical approaches is most convincing, so if the interaction was first found genetically, determine a biochemical interaction, and viceversa. The three main approaches to determining a functional relationship between Hsp90 and a protein of interest are examining a direct interaction between Hsp90 and the protein of interest by coimmunoprecipitation or pulldown assays, determining the effect of pharmocological inhibition of Hsp90 function on the activity of the protein of interest, and assaying defects in substrate activity in the presence of mutations in Hsp90 or cochaperones.

Analysis of Direct Interaction between Hsp90 and Protein of Interest. Much of what is known about the interaction of Hsp90 with substrate proteins comes from the analysis of vertebrate steroid receptor complexes.[29] Hsp90 and cochaperones form stable complexes with receptors in the absence of hormone, when receptors are inactive as transcription factors. On hormone binding, Hsp90 and cochaperones dissociate, resulting in a transcriptionally active receptor. Hsp90 association is required for repression of receptor activity in the absence of hormone, as well as maintenance of the high-affinity hormone-binding state of the receptor. Hsp90 also has two roles in the maturation of Ste11, as Hsp90 is required for both repression of the pheromone pathway in the absence of pheromone and pheromone induction.[31]

Coimmunoprecipitation of Hsp90 with a protein of interest may be used to determine an Hsp90 interaction. Hsp90 interactions with substrate proteins are frequently stabilized in the presence of molybdate, albeit through an unknown mechanism.[29] Thus, the addition of 10 mM sodium molybdate to yeast lysis buffers may help stabilize the interaction between Hsp90 and the protein of interest. Generally, coprecipitation of Hsp90 with antibodies against a substrate protein is not difficult. However, because Hsp90 is very abundant, the level of any given substrate protein is likely much lower than the Hsp90 level, which may make detection of a coimmunoprecipitating substrate protein difficult. In some cases, overexpression of the substrate protein has been necessary for detection.[31] An antibody to

---

[33] L. Zhang, A. Hach, and C. Wang, *Mol. Cell. Biol.* **18**, 3819 (1998).
[34] A. A. Duina, H. M. Kalton, and R. F. Gaber, *J. Biol. Chem.* **273**, 18974 (1998).
[35] J. Imai and I. Yahara, *Mol. Cell. Biol.* **20**, 9262 (2000).
[36] H. C. Chang and S. Lindquist, *J. Biol. Chem.* **269**, 24983 (1994).
[37] A. J. Caplan, *Trends Cell Biol.* **9**, 262 (1999).

yeast hsc82/hsp82 is not available commerically. However, investigators have successfully coimmunoprecipitated Hsp90–substrate complexes using Hsc82/Hsp82 antibodies[35] or Flag-tagged Hsc82.[31]

Pharmacological Inhibition of Hsp90 Interaction. Once a direct interaction between Hsp90 and the protein of interest has been observed, the challenge is to determine whether the interaction is specific and has functional consequences. Hsp90 has ATPase activity that is essential for its *in vivo* function. However, the interaction between Hsp90 and substrates is complex, and the interaction may not be monitored by binding in the presence or absence of ATP, as for Hsp70.[6,7] However, Hsp90 may be inhibited *in vitro* and *in vivo* with the ansamycin antibiotics geldanamycin (GA) and macbecin I. These drugs are available through the Developmental Therapeutics Program at the National Cancer Institute (Bethesda, MD). GA has been found to specifically decrease the activity of a number of mammalian Hsp90–substrate proteins, such as steroid receptors and oncogenic tyrosine kinases. The related drug, macbecin I, is more effective than GA at inhibition of *in vivo* Hsp90 activity in *S. cerevisiae*.[38] As discussed later, these drugs may be used to establish the functional importance of the interaction between substrate proteins and Hsp90 *in vitro* and *in vivo*.

Conservation of Hsp90 function between mammalian and yeast systems allows functional analysis of heterologous substrates expressed in yeast Hsp90, as well as yeast proteins expressed in rabbit reticulocyte lysates. GA is particularly useful when the protein of interest has an activity that may be assayed after transcription and translation in reticulocyte lysate. For example, the synthesis of Gcn2 in reticulocyte lysates in the presence of GA resulted in a dramatic reduction in *in vitro* kinase activity, helping to demonstrate a role for Hsp90 in kinase maturation. In addition, GA treatment of reticulocyte lysate resulted in an increased Hsp90–Gcn2 interaction, possibly by inhibiting the release of Hsp90 required for kinase activity.[32] This result, which contrasts with studies in which GA treatment decreases Hsp90 interaction with steroid receptors,[29] is likely an indication of the many ways in which Hsp90 association may affect substrate activity.

Geldanamycin and the related drug macbecin I may also be used to inhibit Hsp90 activity *in vivo*. The role of Hsp90 in substrate maturation may then be assayed by the treatment of cells with macbesin I prior to coimmunoprecipitation or enzymatic assay. A caveat of using these drugs to inhibit Hsp90 association is that the level of the substrate protein is frequently decreased in the absence of Hsp90 interaction, presumably due to the intrinsic instability of many Hsp90 substrates in the absence of Hsp90 or their respective ligands. For example, the *in vivo* levels of Gcn2 or Cna2 decreased on treatment with macbesin I or GA.[32,35]

Mutations in Hsp90 and Cochaperones. Mutations in Hsp90 and cochaperones have been found to specifically affect the activity of heterologous steroid receptors

---

[38] S. P. Bohen, *Mol. Cell. Biol.* **18**, 3330 (1998).

TABLE III
Hsp90 AND COCHAPERONE MUTANT STRAINS

Protein	Available mutants	Growth defects	Ref.
Hsc82/Hsp82	Reduced level	Temperature sensitive	39
	Reduced activity	Temperature sensitive	40[a,b]
	Conditional alleles	Temperature sensitive	40
Sba1	Null	None	38[c]
Sti1	Null	Temperature sensitive	d
Sse1	Null	Temperature sensitive	e, f
Cpr7	Null	Slow growth	g
Cdc37	Conditional alleles	Temperature sensitive	h–j
Ydj1	Null, conditional alleles	Temperature sensitive	k,l

[a] Y. Kimura, S. Matsumoto, and I. Yahara, *Mol. Gen. Genet.* **242**, 517 (1994).
[b] S. P. Bohen and K. R. Yamamoto, *Proc. Natl. Acad. Sci. U.S.A.* **90**, 11424 (1993).
[c] Y. Fang, A. E. Fliss, J. Rao, and A. J. Caplan, *Mol. Cell. Biol.* **18**, 3727 (1998).
[d] C. Nicolet and E. Craig, *Mol. Cell. Biol.* **9**, 3638 (1989).
[e] X. D. Liu, K. A. Morano, and D. J. Thiele, *J. Biol. Chem.* **274**, 26654 (1999).
[f] H. Mukai, T. Kuno, H. Tanaka, D. Hirata, T. Miyakawa, and C. Tanaka, *Gene* **132**, 57 (1993).
[g] A. A. Duina, J. A. Marsh, and R. F. Gaber, *Yeast* **12**, 943 (1996).
[h] B. Dey, J. J. Lightbody, and F. Boschelli, *Mol. Biol. Cell.* **7**, 1405 (1996).
[i] M. R. Gerber, A. Farrell, R. J. Deshaies, I. Herskowitz, and D. O. Morgan, *Proc. Natl. Acad. Sci. U.S.A.* **92**, 4651 (1995).
[j] S. I. Reed, *Genetics* **95**, 561 (1980).
[k] J. L. Johnson and E. A. Craig, *Mol. Cell. Biol.* **20**, 3027 (2000).
[l] B. Dey, A. J. Caplan, and F. Boschelli, *Mol. Biol. Cell* **7**, 91 (1996).

and v-src expressed in yeast. Now that the effect of these mutations on known Hsp90 substrates has been demonstrated, they become valuable tools in helping identify novel Hsp90 substrates. Table III lists some of the mutant strains that have been shown to affect the activity of Hsp90 substrates.

Assays of Substrate Activity in Hsp90 Mutant Cells. Hsp90 is essential in yeast, and the available *hsc82 hsp82* mutant yeast strains exhibit either reduced Hsp90 activity at all temperatures or temperature-sensitive activity. The most commonly used mutant Hsp90 yeast strain, GRS4, contains chromosomal deletions of *hsc82* and *hsp82* in combination with a plasmid-borne copy of *HSP82* under the *GAL1* promoter. Wild-type levels of Hsp90 are produced when cells are grown in the presence of galactose. However, in the presence of glucose, the leaky promoter results in expression of 5–10% of the wild-type level of Hsp90.[39] This strain was used to demonstrate a role for Hsp90 in Hap1 activation, as Hap1 activity in the presence of glucose was sharply reduced relative to that observed in the

[39] D. Picard, B. Khursheed, M. Garabedian, M. Fortin, S. Lindquist, and K. Yamamoto, *Nature* **348**, 166 (1990).

presence of galactose.[33] Although the level of Hap1 was unaffected, the levels of Cna2 were dramatically decreased in cells grown in glucose,[35] indicating that the stability of some Hsp90 substrates may be affected by decreased levels of Hsp90.

Hsp82 point mutants have been isolated in three independent genetic screens (see Table III). These mutants were generated in *hsc82 hsp82* disruption strains containing mutagenized *HSP82*. Most of these mutants exhibit reduced Hsp90 activity at 25°. Thus, the activity of Hsp90 substrates, such as GCN2, is decreased in strains expressing these mutants. Another tool for analyzing the role of Hsp90 is use of the conditional temperature-sensitive allele G170D.[40] While strains expressing this mutant behave like wild type at 25°, the protein rapidly becomes inactive at 37° (within 90 min), resulting in loss of Hsp90 function. This mutant was used to show that loss of Hsp90 function has a dramatic effect on the repression of Hsf1.[34]

Evidence that Hsp90 plays a role in the maturation of a protein of interest is further supported when Hsp90 cochaperones are also shown to be required. The observation that Gcn2 activity decreased in strains containing mutations in *STI1*, *CDC37*, and *SBA1* was used to help establish a role for Hsp90 in kinase maturation.[32] Table III lists many of the available mutations in Hsp90 cochaperones. The Hsp90 cochaperones Sba1(p23), Sti1 (Hop), Cpr7 (Cyp-40), Ydj1 (Hsp40), and Sse1 (a member of the Hsp110 family) are encoded by single nonessential genes. Cdc37, which may be associated with only a subset of Hsp90 substrates,[41] is encoded by an essential gene. Although it may be possible that cochaperones have cellular functions independent of Hsp90 function, the involvement of Hsp90 cochaperones in a particular cellular process provides valuable evidence that Hsp90 is also involved in that process.

---

[40] D. F. Nathan and S. Lindquist, *Mol. Cell. Biol.* **15,** 3917 (1995).
[41] J. Rao, P. Lee, S. Benzeno, C. Cardozo, J. Albertus, D. M. Robins, and A. J. Caplan, *J. Biol. Chem.* **276,** 5814 (2001).

Section IV
Cell Biology

[27] Synchronization Procedures

By ANGELIKA AMON

Introduction

Budding yeast is an ideal model system to study the regulatory mechanisms governing cell cycle progression. The ability to generate cell populations progressing through the cell cycle in a synchronous manner (synchronous cultures) has contributed significantly to the rapid progress in our understanding of cell cycle control in this organism.

Cell populations progressing through the cell cycle in a synchronous manner are generated by arresting cells at a certain stage of the cell cycle (block) followed by removing the cell cycle block, allowing cells to resume cell division (release). Time points are taken at certain times after release, representing different cell cycle stages. Several methods have been developed to generate synchronous cell populations.[1] It is, therefore, important to first decide which type of block–release experiment is most suited to address the question of interest. As the degree of synchrony in a culture declines as cells progress through the cell cycle, it is best to arrest cells in a cell cycle stage close to the window of the cell cycle to be analyzed. A second important consideration is the degree of synchrony obtained with the different procedures, as some types of block–release experiments generate cell cultures with a higher degree of synchrony than others. Finally, it is important to bear in mind that cell cycle arrests imposed by a compound or by inactivating/ depleting a protein critical for a particular cell cycle transition can lead to artifacts. Some, if not all, compounds used to arrest cells induce stress responses or developmental programs, causing major physiological changes in the cell.[2-4] It is thus best to examine a parameter of interest using more than one synchronization method.

Cell cycle arrests can be imposed by compounds and released from the block by removal thereof.[1,5-7] Alternatively, cells can be arrested at a certain cell cycle stage by depleting or inactivating a gene product critical for a certain cell cycle

---

[1] B. Futcher, *Methods Cell Sci.* **1**, 79 (1999).
[2] R. J. Cho, M. J. Campbell, E. A. Winzeler, L. Steinmetz, A. Conway, L. Wodicka, T. G. Wolfsberg, A. E. Gabrielian, D. Landsman, D. J. Lockhart, and R. W. Davis, *Mol. Cell* **2**, 65 (1998).
[3] P. T. Spellman, G. Sherlock, M. Q. Zhang, V. R. Iyer, K. Anders, M. B. Eisen, P. O. Brown, D. Botstein, and B. Futcher, *Mol. Biol. Cell* **9**, 3273 (1998).
[4] J. J. Wyrick, F. C. P. Holstege, E. G. Jennings, H. C. Causton, D. Shore, M. Grunstein, E. S. Lander, and R. A. Young, *Nature* **402**, 418 (1999).
[5] L. L. Breeden, *Methods Enzymol.* **283** (1997).
[6] C. Price, K. Nasmyth, and T. Schuster, *J. Mol. Biol.* **218**, 543 (1991).
[7] O. Cohen-Fix, J.-M. Peters, M. W. Kirschner, and D. Koshland, *Genes Dev.* **15**, 3081 (1996).

transition and released by reactivation or resynthesis of this gene product.[8–11] Finally, centrifugal elutriation can be used to obtain cells of uniform cell size.[12–15] Because cell size correlates with cell cycle stage, cells obtained by centrifugal elutriation are synchronized with respect to their position in the cell cycle. This chapter will point out advantages and disadvantages of the various methods and indicate the degree of synchrony obtained with the various block–release methods.

Synchronization Procedures Using Compounds

α-Factor Block–Release Method

In the α-factor block–release method, cells are synchronized in the G_1 stage of the cell cycle by the addition of α-factor pheromone and released from the block by removal thereof. Strains wild-type or deleted for the *BAR1* protease gene can be employed. Bar1 degrades α factor and cells lacking Bar1 are supersensitive to pheromone.[16] The advantage of using *BAR1* cells in synchronous cultures is that cells enter the cell cycle readily on removal of pheromone. However, the G_1 arrest imposed by α factor is transient as the protease degrades the pheromone in the medium. Thus, when using *BAR1* strains it is critical that the cell density in the culture be low when pheromone is added. The advantage of using *bar1* cells is that cells can be grown to higher density prior to arrest and less pheromone is needed to arrest cells. However, *bar1* cells do not enter the cell cycle as readily as *BAR1* cells on removal of pheromone, as even small amounts of pheromone not eliminated by the washing procedure prevent cell cycle progression. Release from the pheromone block can be improved by adding protease type XIV, which degrades α factor, at the time of release. Generally, we use *BAR1* strains, as the synchrony is higher on release from the G_1 block, but when large amounts of cells are needed, it is advisable to use *bar1* cells.

Reagents and Equipment

α-Factor pheromone [soluble in dimethyl sulfoxide (DMSO), ethanol, or methanol and stored at $-20°$] can be obtained from various vendors. However, α factor is expensive and the quality is often poor. We, therefore, have the peptide

[8] L. J. Oehlen and F. R. Cross, *Genes Dev.* **8,** 1058 (1994).
[9] A. Amon, S. Irniger, and K. Nasmyth, *Cell* **77,** 1037 (1994).
[10] M. P. Cosma, T. Tanaka, and K. Nasmyth, *Cell* **97,** 299 (1999).
[11] F. Uhlmann, D. Wernic, M. A. Poupart, E. V. Koonin, and K. Nasmyth, *Cell* **103,** 375 (2000).
[12] L. H. Johnston and A. L. Johnson, *Methods Enzymol.* **283,** 342 (1997).
[13] L. H. Johnston, J. H. White, A. L. Johnson, G. Lucchini, and P. Plevani, *Mol. Gen. Genet.* **221,** 44 (1990).
[14] L. J. Oehlen, J. D. McKinney, and F. R. Cross, *Mol. Cell. Biol.* **16,** 2830 (1996).
[15] E. Schwob and K. Nasmyth, *Genes Dev.* **7,** 1160 (1993).
[16] V. L. MacKay, S. K. Welch, M. Y. Insley, T. R. Manney, J. Holly, G. C. Saari, and M. L. Parker, *Proc. Natl. Acad. Sci. U.S.A.* **85,** 55 (1988).

synthesized and purified. The protein sequence is as follows:

NH_3 - Trp - His - Trp - Leu - Gln - Leu - Lys - Pro - Gly - Gln - Pro - Met - Tyr - COO

Protease to degrade α factor (necessary when using *bar1* cells): Protease type XIV (Sigma, St. Louis, MO)
Filtration apparatus (Millipore, Bedford, MA)
Circular nitrocellulose filters
Microscope

Procedure

1. Exponentially growing *BAR1* cells are diluted to OD_{600} of 0.2 and α factor is added [3 μg/ml (1.8 μM)]. For *bar1* cells, 0.5 μg/ml (0.3 μM) α factor is sufficient to arrest cells. Sixty minutes after pheromone addition, cells are examined to determine how many cells are arrested in G_1 phase (cells are unbudded and form a mating projection). This analysis should be repeated every 15 min thereafter. At 30° in YEPD, cells are usually arrested in G_1 phase 90–120 min after pheromone addition.

2. When more than 90% of cells are arrested in G_1, the medium is removed by filtration using a nitrocellulose filter to retain cells (Fig. 1). Cells are then washed

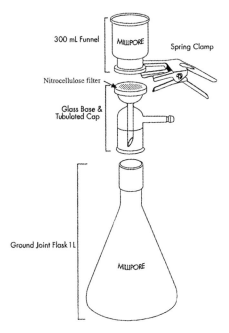

FIG. 1. Filter apparatus used to filter and wash cells (courtesy of Millipore Corp.).

with 4 culture volumes of media. It is also possible to wash cells by centrifugation, but washing by filtration is much faster.

3. The nitrocellulose filter is then placed in a flask with fresh medium (for *bar1* cells, this medium should contain 0.1 mg/ml protease type XIV) and shaken vigorously to detach cells from the filter.

4. Time points are taken at the desired times after release. The synchrony of the release and cell cycle progression are monitored under the microscope. In a good release, more than 75% of cells should have small buds 45 min after transfer into fresh medium. It is noteworthy that cells are clumpy initially and it is important to shake cultures well before removing a sample.

Evaluation of Cell Cycle Progression and Cell Synchrony. It is important to take samples at the various times after release to evaluate cell cycle progression and synchrony of the culture. Several methods are available to assess cell cycle progression and synchrony of the culture:

BUDDING INDEX. In this method, the percentage of unbudded (G_1) and budded (other cell cycle stages) is determined.

DNA CONTENT ANALYSIS. Using flow cytometry, the proportion of cells with unreplicated and replicated DNA is analyzed.[17]

MITOTIC SPINDLE FORMATION. Mitotic spindles are visualized by indirect *in situ* immunofluorescence using anti-tubulin antibodies.[18]

We usually analyze the status of the mitotic spindle by indirect *in situ* immunofluorescence to determine when cells enter mitosis. At the same time, we analyze the DNA content of cells by flow cytometry to determine when cells enter and complete S phase.

Comments. α-Factor block–release and elutriation, where small G_1 cells are isolated from a sedimentation gradient, are the most commonly used methods to generate synchronous cell cultures. These two methods are also the methods of choice to obtain a first clue as to whether a parameter changes during the cell cycle. The α-factor block–release method creates highly synchronous cell populations. In a good release, 70% or more cells are found to enter the various cell cycle stages simultaneously. However, α factor induces not only morphological changes (mating projection formation), but also significantly alters the transcriptional program normally occurring during G_1.[2,3] Thus, it is important to bear in mind that pheromone treatment could affect the parameter of interest, particularly during G_1 arrest.

---

[17] A. B. Futcher, "The Cell Cycle: A Practical Approach," p. 69. IRL Press at Oxford Univ. Press, 1993.

[18] J. V. Kilmartin and A. E. Adams, *J. Cell Biol.* **98,** 922 (1984).

Hydroxyurea and Nocodazole Block–Release Method

Hydroxyurea arrests cells in early S phase by blocking ribonucleotide reductase.[19] Nocodazole depolymerizes microtubules, leading to cell cycle arrest in metaphase.[20] Removal of the drugs will cause cells to release from the block.

Reagents and Equipment

Hydroxyurea (Sigma) or nocodazole (Sigma, 1.5 mg/ml stock dissolved in DMSO, stored at −80°)
DMSO (for nocodazole block–release only)
Filtration apparatus (Millipore)
Circular nitrocellulose filters
Microscope

Method

1. Exponentially growing cells are diluted to OD_{600} of 0.4–0.5, and 10 mg/ml (130 mM) of hydroxyurea is added to cultures in crystalline form. To arrest cells with nocodazole, 15 μg/ml nocodazole (50 μM, dissolved in DMSO) is added to cultures. Sixty minutes after addition of the drug, cells are examined under the microscope to assess the percentage of arrested cells (arrested cells appear dumbbell shaped). This is repeated every 15 min thereafter. At 30° in YEPD, cells are usually arrested 90–120 min after addition of the drug.

2. When more than 90% of cells are arrested, the medium is removed by filtration or centrifugation. Cells are then washed with 4 culture volumes of media either by filtration or by centrifugation.

Note: When releasing cells from a nocodazole arrest, it is critical that the medium used for washing cells contains 1% DMSO.

3. The nitrocellulose filter containing cells is then placed in a flask with fresh medium (for nocodazole block–release the medium should contain 1% DMSO) and shaken vigorously to detach cells from the filter.

4. Time points are taken at the desired times after release. The synchrony of the release and cell cycle progression are monitored as described for α-factor block–release.

Comments. The synchrony of cultures after release obtained with these methods is not nearly as good as that achieved with an α-factor block–release experiment or elutriation.

[19] S. J. Elledge, Z. Zhou, J. B. Allen, and T. A. Navas, *Bioessays* **15,** 333 (1993).
[20] C. W. Jacobs, A. E. Adams, P. J. Szaniszlo, and J. R. Pringle, *J. Cell Biol.* **107,** 1409 (1988).

Synchronization of Cells by Inactivation/Reactivation of CDC Genes

Block–Release Experiments Using Temperature-Sensitive cdc Mutants

Whether a temperature-sensitive *cdc* (cell division cycle) mutant can be used to generate cell populations progressing through the cell cycle in a synchronous manner largely depends on (1) how uniformly and promptly a *cdc* mutant arrests at the restrictive temperature and (2) how quickly cells recover from the cell cycle block when returned to the permissive temperature. Whether a *cdc* mutant is suitable for a block–release experiment cannot be predicted, but needs to be determined experimentally. *cdc28-13* mutants, which arrest in G_1, and *cdc15-2* mutants, which arrest in telophase, have been used successfully to create synchronous cell populations.[2,21–23] While there are alternatives for blocking cells in G_1 (α factor, elutriation), the *cdc15-2* mutation stands alone in arresting cells late in the cell cycle, allowing for a highly synchronous release from the block on transfer to the permissive temperature. The *cdc15-2* mutant is particularly useful when a detailed analysis of exit from mitosis and entry into and progression through G_1 is needed.[21–23]

Reagents and Equipment

Shaking water bath (New Brunswick, Edison, NJ, or Lab Line, Melrose Park, IL)
Filtration apparatus (Millipore)
Circular nitrocellulose filters
Microscope

Procedure

1. Cells are grown to exponential phase at the permissive temperature (for many *cdc* mutants 23°) and transferred (by filtration or centrifugation) into medium prewarmed to the restrictive temperature (for many *cdc* mutants 37°) at a density of $OD_{600} = 0.3$–0.4. It is better to transfer cells into prewarmed medium rather than transferring the entire flask to the restrictive temperature, as it takes time for the medium to heat up, leading to a longer than necessary time for the arrest to occur. We also perform our temperature shift experiments in shaking water baths (New Brunswick or Lab Line), as these types of incubators ensure a more constant temperature environment. Cell cycle arrest is followed by microscopic examination. At 37° in YEPD medium, cells are usually arrested after 90–120 min.

[21] I. Fitch, C. Dahmann, U. Surana, A. Amon, K. Nasmyth, L. Goetsch, B. Byers, and B. Futcher, *Mol. Biol. Cell.* **3** (1992).
[22] U. Surana, A. Amon, C. Dowzer, J. McGrew, B. Byers, and K. Nasmyth, *EMBO J.* **12**, 1969 (1993).
[23] B. Kovacech, K. Nasmyth, and T. Schuster, *Mol. Cell. Biol.* **16**, 3264 (1996).

2. When more than 90% of cells are arrested, the medium is removed by filtration or centrifugation.

3. The nitrocellulose filter containing the cells is placed into a flask with fresh medium prewarmed to the permissive temperature and shaken vigorously to detach cells from the filter.

4. Time points are taken at the desired times after release, and synchrony of the release and cell cycle progression are monitored as described for α-factor block–release.

Comments. When a *cdc* mutant is suitable for a block–release experiment (cells arrest well at the restrictive temperature and release well from the block after transfer to the permissive temperature), the synchrony of cultures after release is usually very good. For example, in a *cdc15-2* block–release experiment, 90% of cells disassemble their mitotic spindles within 15 min.

Synchronization of Cells by Depletion/Resynthesis of Unstable Cdc Proteins

In this type of synchronization procedure, a cell cycle arrest is imposed by depleting cells of a protein critical for a particular cell cycle transition. Release from the block is brought about by inducing synthesis of this particular gene product. For this method, only cell cycle proteins that are unstable, at least during part of the cell cycle, are suitable. As for temperature-sensitive *cdc* mutants, whether a Cdc protein is employable for such a synchronization procedure cannot be predicted, but needs to be determined experimentally.

Once an unstable Cdc protein has been identified, its encoding gene is cloned under the control of an inducible promoter, either the galactose-inducible *GAL1-10* promoter or the methionine-repressible *MET3* promoter. The *GAL1-10* promoter is a very strong promoter. If high levels of a Cdc gene product lead to defects in cell cycle progression, a weaker promoter, such as the *MET3* promoter, should be employed. Strains are then engineered that contain the *MET3–CDC* or *GAL1-10– CDC* fusion as the sole source of this *CDC* gene.

Reagents and Equipment

Filtration apparatus (Millipore)
Circular nitrocellulose filters
Microscope

Procedure for Use of GAL1-10 Promoter

1. Cells carrying the *CDC* gene to be depleted under the *GAL1-10* promoter are grown to exponential phase to a density of $OD_{600} = 0.3$–0.4 in YEP medium

containing 2% raffinose and 2% galactose. Cells are then transferred to YEP containing 2% raffinose (YEP-Raf) either by filtration or by centrifugation. The arrest is monitored microscopically. Cells are usually arrested 120–180 min after transfer into YEP-Raf medium.

2. When more than 90% of cells are arrested, galactose (2%) is added. The synchrony of the release and cell cycle progression are monitored as described for α-factor–block release.

Procedure for Use of MET3 Promoter

1. Cells carrying the *CDC* gene to be depleted under the control of the *MET3* promoter are grown to exponential phase to a density of $OD_{600} = 0.3$–0.4 in synthetic medium lacking methionine. Cells are then transferred to YEPD containing 2 mM methionine. Methionine (2 mM) is readded after 60 min to ensure that sufficient methionine is present in the medium to complete and maintain cell cycle arrest. Cells are usually arrested 120–180 min after transfer to YEPD + methionine medium.

2. When more than 90% of cells are arrested, the medium is removed by filtration or centrifugation and cells are washed with 4 volumes of synthetic medium lacking methionine.

3. The nitrocellulose filter containing cells is then placed in a flask with fresh synthetic medium lacking methionine and shaken vigorously to detach cells from the filter.

4. Synchrony of the release and cell cycle progression are monitored as described for α-factor–block release.

Comments. Depletion of a *CDC* gene is mostly employed to determine the null phenotype for a particular *CDC* gene, but several *CDC* genes have been shown to be very useful for creating highly synchronous cell populations. These genes include the G_1 *CLN* cyclins, depletion of which causes cells to arrest in G_1, and *CDC20*, depletion of which causes cells to arrest in metaphase.[8–11] Like nocodazole, depletion of *CDC20* causes cells to arrest in metaphase. Cell cycle progression on resynthesis of *CDC20* is much more synchronous than that on removal of nocodazole and is, thus, the method of choice when cultures are needed that progress through mitosis and enter G_1 in a highly synchronous manner.

Isolation of G_1 Cells by Centrifugal Elutriation

During elutriation, small G_1 cells are isolated from a sedimentation gradient. Cells are placed in a chamber within a spinning rotor. Due to centrifugal forces, the cells sediment toward the bottom of the chamber. Liquid flowing through the chamber from bottom to top counteracts the centrifugal force and pushes smaller cells toward the chamber exit, whereas larger cells migrate toward the area

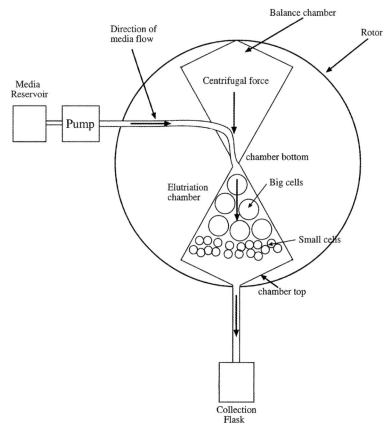

FIG. 2. Schematic illustration of the elutriation system (not to scale).

of highest centrifugal force (bottom of the chamber, Fig. 2). By increasing the flow rate of the liquid flowing through the chamber, cells are pushed toward the chamber exit. The smallest cells (G_1 daughter cells) exit the chamber first and are collected.

Reagents and Equipment

Beckman J6 centrifuge
Beckman JE-5.0 rotor
40-ml chamber
Peristaltic pump
Microscope
8–10 liters of YEP
5 liters sterile deionized water

Procedure

1. One to 2 liters of cells are grown to late logarithmic phase ($OD_{600} = 2$–3) in YEP containing either 2% raffinose or 2% sucrose. The use of raffinose or sucrose as a carbon source increases the percentage of G_1 cells in the culture.

2. Before harvesting the cells, the equipment is set up. The elutriation chamber and tubing are filled with YEP medium chilled to 4°. (We usually do not add any carbon source to YEP as it enhances growth of bacteria and fungi in the tubing when the elutriator is not in use.) All media used during elutriation need to be kept cold (we usually keep them in an ice bucket) to prevent elutriated cells from progressing through the cell cycle. It is very important to eliminate all air bubbles within the system prior to filling the chamber with cells, as bubbles passing through the system during elutriation will disrupt the sedimentation gradient in the chamber. Initially, while filling the system, some air bubbles will remain trapped in the chamber. These will escape once the centrifuge is spinning.

3. When the system is filled with liquid, the centrifuge is turned on. The speed is set at 4000 rpm and the temperature at 4°. The peristaltic pump is set at 20 ml/min. (The pump setting needed to allow for 20 ml of liquid to be pumped through the system per minute needs to be determined empirically).

4. Once the centrifuge reaches 4000 rpm, the strobe light is adjusted so that the elutriation chamber is visible through the porthole on the centrifuge lid.

5. To load cells into the elutriation chamber, cells are collected either by centrifugation or by filtration and resuspended in 30–50 ml YEP medium. It is important that cells are resuspended in the same type of medium used for elutriation, as media differences will interfere with the establishment of the sedimentation gradient in the chamber. Cells are sonicated briefly to disrupt cell clumps.

6. To fill the chamber with cells, the loading tube (situated in the media reservoir; Fig. 2) is transferred into the flask containing the cells. Cells are loaded until the elutriation chamber is filled up to 80%. The degree to which the elutriation chamber is filled with cells is monitored through the porthole on the centrifuge lid. When loading of cells is complete, the loading tubing is transferred back into the flask containing YEP medium. This needs to be done quickly to minimize the amount of air bubbles entering the system.

7. The elutriator is then run for 15 min at 4000 rpm with the pump speed set at 20 ml/min to establish the sedimentation gradient in the chamber.

8. To elutriate G_1 cells, the pump speed is increased slowly (one increment on the pump dial). After a few minutes the liquid exiting the elutriation chamber is examined under the microscope to determine whether small G_1 cells are leaving the chamber. The routine is as follows. Increase pump speed one increment on the pump dial and wait for 1 min. Then, determine whether cells are in the outflow by analyzing a sample under the microscope. If no cells leave the elutriation chamber, increase pump speed and continue to do so until cells start appearing in the outflow.

9. When small G_1 cells start to leave the elutriation chamber, 250-ml fractions are collected into centrifuge tubes. When the cell number in the outflow declines, the pump speed is increased. The fewer times the pump speed is increased during the collection of cells, the tighter the size distribution in the culture and the higher the synchrony of the cell population will be.

10. Collected cells are centrifuged and resuspended in 2–5 ml of medium, transferred into smaller tubes, and kept on ice until the completion of the elutriation run.

11. At the end, fractions are pooled and inoculated in fresh medium. Elutriated cells take a long time (compared to α-factor-arrested cells) to enter the cell cycle. This is likely due to the fact that elutriated cells are extremely small in size (α-factor-arrested cells are much larger) and, thus, it takes time for cells to grow to reach the critical cell size to enter the cell cycle. It is, therefore, not necessary to take any time points prior to 60 min after inoculation into fresh medium.

12. After the desired amount of small G_1 cells is elutriated, the elutriation chamber has to be cleared from the remaining cells. The centrifuge speed is lowered to 2000 rpm and the pump speed is increased slowly. Do not turn off the centrifuge and pump as this will cause clogging of the instrument. When most of the cells have been pumped out of the elutriation chamber, the centrifuge is turned off and large amounts of sterile water (5–10 liters) are pumped through the system. When the system is free from cells and medium, the tubing and chamber are emptied and kept dry until further use.

Comments. As cells are not treated with any compounds or genetically manipulated to bring about cell cycle arrest, G_1 cells obtained by centrifugal elutriation are probably as close to "true G_1 cells" as experimentally feasible. Moreover, cultures obtained by centrifugal elutriation progress through the cell cycle in a highly synchronous manner (comparable to that obtained in the α-factor block–release procedure). Disadvantages of this method are that it is very time-consuming (one elutriation run takes about 3 hr), requires a lot of medium and special equipment, and the yield of G_1 cells is low compared to pheromone arrest.

[28] Separation of Mother and Daughter Cells

By PETER U. PARK, MITCH MCVEY, and LEONARD GUARENTE

Introduction

The budding yeast *Saccharomyces cerevisiae* divides asymmetrically. In vegetative growth, yeast cells reproduce by budding, and the position where the bud forms ultimately determines the plane of cell division.[1] A bud emerges during the late G_1 stage of the cell cycle and continues to grow, first at the tip and then throughout the bud, until late nuclear division and cytokinesis. The fact that budding yeast undergo polarized cell growth and directional cell division has been useful in understanding fundamental processes that are essential for the development of higher eukaryotes, such as early embryogenesis in *Caenorhabditis elegans* and neurogenesis in *Drosophila*.[2,3] Furthermore, asymmetric division plays a role in mating type switching and some aspects of cell cycle regulation.[4] Additionally, this asymmetry has been very useful in studying the aging process in yeast.

A key requirement for tracing aging in an organism is the ability to distinguish and follow individuals over time. Asymmetric cell division in yeast makes it possible to follow the fate of an individual cell throughout many cell divisions. The mother cell buds to give rise to a daughter cell that is clearly smaller in size. By micromanipulating away the daughters from a mother, Mortimer and Johnston[5] found that mother cells divide a relatively fixed number of times before stopping, and the probability of stopping increases exponentially as the number of prior divisions increases. These experiments showed that the replicative life span of yeast cells fits the mathematical definition of aging also seen in humans.[6,7] The replicative life span in yeast depends on the number of progressions through the cell cycle, as opposed to the chronological life span, which is measured in stationary-phase cells under conditions of nutrient starvation.

Tracing the fates of individual yeast cells over time by micromanipulation allows for not only the determination of life span, but also the characterization of a number of phenotypic changes that occur during the aging process. These changes include bud scar accumulation, enlargement of cell size, slowing of cell cycle, and

---

[1] J. Chant, *Annu. Rev. Cell Dev. Biol.* **15,** 365 (1999).
[2] A. A. Hyman and J. G. White, *J. Cell Biol.* **105,** 2123 (1987).
[3] R. Kraut, W. Chia, L. Y. Jan, Y. N. Jan, and J. A. Knoblich, *Nature* **383,** 50 (1996).
[4] J. E. Haber, *Annu. Rev. Genet.* **32,** 561 (1998).
[5] R. K. Mortimer and J. R. Johnston, *Nature* **183,** 1751 (1959).
[6] B. Gompertz, *Philos. Trans. R. Soc.* **115,** 513 (1825).
[7] C. Finch, "Longevity, Senescence, and the Genome." University of Chicago Press, Chicago, 1990.

loss of fertility.[5,8,9] However, although micromanipulation allows for the visual characterization of changes associated with aging, it does not allow for the molecular and biochemical analyses of those changes. Such analyses require an isolation of a large number of aged cells. Several methods have been developed that allow for this large-scale isolation. Using these methods, a number of additional changes in old cells have been characterized. These include changes in the expression of *LAG1* and *LAG2*, nucleolar fragmentation, loss of silencing, movement of the Sir complex from telomeres to nucleolus, and rDNA circle accumulation.[10–15]

This section describes the detailed procedures for separation and isolation of mothers and daughters. These protocols have been used by investigators studying aging, bud site selection, and other aspects of asymmetric cell division. The first part of this chapter describes the procedures for performing life span analysis by micromanipulation, and the second part describes steps for the large-scale collection of old cells.

Isolation of Mothers from Daughters by Micromanipulation

The most accurate method for isolating mother cells of specific age involves following each mother cell through multiple divisions. Practically, this is accomplished by isolating virgin mother cells on an agar plate and separating daughter cells from each mother by micromanipulation. For life span analysis, this process is continued until the mother cell ceases division and eventually lyses. This technique is of greatest use in experiments where knowledge of the exact age of the mother cells is required. However, the small number of old cells obtained and their isolation on an agar plate preclude their use in either immunofluorescence or biochemical experiments.

Life Span Analysis

Prior to beginning a life span, it is essential to ensure that the population of yeast cells is free of petites (which have a longer life span in some strains)[16] and is growing robustly. Then, the life span begins with cells that have not divided previously. From these virgin mother cells, buds are removed as they form and are discarded. If desired, the daughters can be monitored through additional cell divisions, as in the case of pedigree analysis.[8,15]

[8] B. K. Kennedy, N. R. Austriaco, Jr., and L. Guarente, *J. Cell Biol.* **127**, 1985 (1994).
[9] D. Sinclair, K. Mills, and L. Guarente, *Annu. Rev. Microbiol.* **52**, 533 (1998).
[10] P.-A. Defossez, P. U. Park, and L. Guarente, *Curr. Opin. Microbiol.* **1**, 707 (1998).
[11] N. K. Egilmez, J. B. Chen, and S. M. Jazwinski, *J. Biol. Chem.* **264**, 14312 (1989).
[12] T. Smeal, J. Claus, B. Kennedy, F. Cole, and L. Guarente, *Cell* **84**, 633 (1996).
[13] B. K. Kennedy, M. Gotta, D. A. Sinclair, K. Mills, D. S. McNabb, M. Murphy, S. M. Park, T. Laroche, S. M. Gasser, and L. Guarente, *Cell* **89**, 381 (1997).
[14] D. A. Sinclair, K. Mills, and L. Guarente, *Science* **277**, 1313 (1997).
[15] D. A. Sinclair and L. Guarente, *Cell* **91**, 1033 (1997).
[16] P. A. Kirchman, S. Kim, C. Y. Lai, and S. M. Jazwinski, *Genetics* **152**, 179 (1999).

Procedure

1. Start with a single colony of the yeast strain whose life span is to be determined. Streak this colony onto a plate containing glycerol as the sole carbon source (YPG plate) in order to remove any petite cells from the population.[16]

2. From the YPG plate, pick a single colony and patch this onto rich medium (YPD plate). Allow the patch to grow at 30° for 1 day. From the overnight patch, streak cells to a fresh YPD plate and grow for 1 more day at 30°.

3. Transfer several thousand cells (a small dab on the end of a toothpick) to the side of a fresh plate and allow to divide for several hours at 30°.

4. Using a fiber-optic glass needle attached to a micromanipulator (see section on tetrad dissection), arrange at least 40 cells in a gridded pattern in the middle of the plate. There should be about 50 µm between each cell to provide enough room to separate mothers from daughters without disturbing neighboring cells. It is often useful to poke holes in the agar, with the fiber optic needle, between every 10 mother cells, thereby providing convenient reference points on the plate (Fig. 1). *Note.* Micromanipulation is generally performed at room temperature (20–25°). All incubations are performed at 30°.

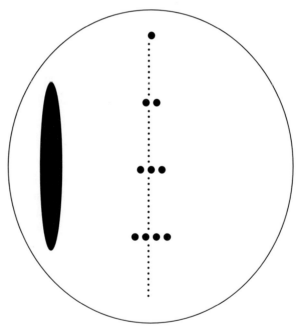

FIG. 1. Schematic of a life span plate. Cells are initially patched onto the plate (oval, left) and allowed to divide for several hours. Mother cells (small circles) are isolated in the middle of the plate in groups of 10 and their positions are marked by holes in the agar (large circles). After incubation, daughters are separated from mothers and moved to the side of the plate (oval).

5. Allow the isolated cells to divide at least once. Generally, this will take 1–2 hr, depending on the strain.

6. Separate the mother cell from the daughter cell (the mother is the larger of the two). Keep the daughter cell and drag the mother cell back to the patch of cells on the side of the plate. This daughter cell is now a virgin mother cell.

7. Incubate the virgin mother cells at 30° for one cell division. At the beginning of the life span, the cells will divide once approximately every 90 min at 30°. Therefore, it is extremely important not to incubate the plates for too long so that the daughters do not divide too many times and obscure the mother.

8. As daughter cells emerge from the mother, separate them by gently placing the needle on the plate just to the side of them and tapping the base of the microscope gently.

9. Keep the mother cell and remove each daughter cell to the side of the plate. Each bud that is removed from the mother cell (not from other daughter cells) represents one generation for the mother cell that produced it. After each micromanipulation, record the number of times that each mother cell has divided, using a gridded data sheet.

10. The plates can be incubated overnight at temperatures lower than room temperature (4–10°). Generally, mother cells will bud once during a 12-hr period at 10° and growth ceases at 4°. Incubation at a lower temperature does not significantly affect life span.[8]

11. Continue this procedure until all mother cells cease to divide and lyse. As cells get older, the generation time slows down markedly. Also, cells will often fail to divide for several hours and then resume division. The reasons for this are unknown. Therefore, it is important to follow each mother cell until it lyses.

12. A mortality curve can be constructed that displays the life span of all of the mother cells in the population by graphing the percentage of cells still dividing at each generation point.[5]

Useful Hints

1. Be extremely careful not to contaminate the life span plate with bacteria or other fungi. Any contamination will quickly overtake the plate and ruin the experiment. Contamination, if caught early enough, can be removed from the agar by sterilizing the round end of a glass pipette and pressing the pipette into the agar, thereby removing the contamination in a round agar plug.

2. To keep the life span plate from drying out and killing the cells, use thick plates and keep them sealed with Parafilm during the 30° incubations.

3. Be careful not to touch the needle into large colonies, as cells can stick to the sides of the needle. During subsequent manipulations, these cells may be transferred accidentally to the part of the plate where the mother cells are dividing. These extra cells will divide and overtake the mother cells rapidly.

4. It is possible to determine the life spans of up to six strains (with 40 cells each) at one time. This is accomplished by performing micromanipulation on one strain while the others are incubating at 30°.

5. When directly comparing the life spans of several strains, it is advisable to only compare mortality curves generated in a single experiment using plates poured at the same time. The life spans of strains have been found to vary slightly using different batches of plates.

6. Life spans can be performed on synthetic complete medium, although we have found that this can affect the life span of certain strains negatively. In addition, it has been shown that altering the nutrient content of the plates can both extend and decrease life span.[17,18]

Visual Identification of Mother Cells

At the beginning and the end of a life span, it can be difficult to distinguish mothers from daughters. At most points in the life span, daughter cells are smaller than the mothers that produced them. In addition, mother cells will generally bud a second time before their daughter cells form their first bud. Therefore, if you are unsure which of two cells is a mother cell, you can wait until one of them begins to bud. This cell is usually the mother cell. However, this pattern breaks down as mother cells approach their maximum life span because older cells have a longer generation time than younger cells.[8]

In addition, very old mother cells often produce identically sized daughters during their last few divisions.[8] Therefore, at the end of a life span, it is generally advisable to keep both mother and daughter cells if they are of similar sizes. The cell that stops dividing first can be assumed to be the mother cell.

Finally, we have found it useful after each micromanipulation to make a quick drawing of the positions of the mother cells and buds that cannot be removed. We include these drawings on the same data sheet used to record the number of divisions. This practice aids in the identification of the original mother cell after subsequent incubation.

Large-Scale Separation of Mother and Daughter Cells

Although micromanipulation of yeast cells used for life span determination allows isolation of a small number of old mother cells, molecular and biochemical analyses of old mother cells require a large-scale isolation of aged cells. The separation of large quantities of old cells is a difficult task because the proportion of old cells in a random population of exponentially growing cells is miniscule. In a given population, one-half of the cells are virgin daughter cells, one-quarter are

[17] J. C. Jiang, E. Jaruga, M. V. Repnevskaya, and S. M. Jazwinski, *FASEB J.* **14,** 2135 (2000).
[18] S. J. Lin, P. A. Defossez, and L. Guarente, *Science* **289,** 2126 (2000).

cells that have divided once, one-eighth are two divisions old, one-sixteenth are three divisions old, etc. Therefore, exponentially growing cells can be considered as young cells, as more than 87% of exponentially growing cells have divided twice or less. In fact, there are only about 100 twenty-generation-old mother cells in 10^8 cells. Several techniques have been developed for this large-scale separation of old mothers and daughters.

One method for effective isolation of virgin daughter cells from mother cells, but not for recovery of old mothers, is called a "baby machine."[19,20] Mother cells are attached to a membrane and allowed to divide. Daughter cells from these attached cells are eluted continuously by washing the membrane. Two other methods take advantage of the difference in size between mother and daughter cells. One method relies on centrifugal elutriation to continuously separate daughters from mothers.[21] After mother cells are grown in the chamber of elutriation rotor, cells are eluted by centrifugation. Eluted mother cells are kept while daughter cells are discarded. The process can be repeated many times to enrich for old cells. Another technique uses sucrose gradient centrifugation.[22,23] First, virgin daughters and mothers are separated into two distinct bands on a 10–30% sucrose gradient by centrifugation. Daughter cells are recovered and synchronized with mating pheromone. Then, daughters are allowed to grow for several generations and are again separated by centrifugation. This time, mother cells are collected, synchronized, and grown before another round of sucrose gradient separation. A relatively pure population (90%) of old cells can be isolated after repeating the technique many rounds.

Another method for enriching old mother cells comes from the observation that the cell surface of emerging daughter cells is synthesized *de novo* at the budding site.[24] The previously synthesized components of the mother cell surface do not contribute to the daughter cell surface. Therefore, if the mother cells are tagged at the surface with a label, then the label will stay with the mother cells and can be used to distinguish them from the daughter cells after many divisions. Subsequent recovery of labeled cells gives a pure population of cells that have divided a predetermined number of times. One method that has been used involves a biotin–avidin labeling system. Proteins on the cell surfaces are first conjugated with biotin and then cells are grown for a desired number of generations. The biotin-labeled cells can be recovered by the addition of either fluorochrome-conjugated avidin or streptavidin-coated magnetic beads. Fluorescence-activated cell sorting (FACS) can be performed to isolated biotin-labeled old cells if fluorochrome-conjugated

[19] C. E. Helmstetter, *New Biol.* **3**, 1089 (1991).
[20] A. Grzelak, J. Skierski, and G. Bartosz, *FEBS Lett.* **492**, 123 (2001).
[21] C. L. Woldringh, K. Fluiter, and P. G. Huls, *Yeast* **11**, 361 (1995).
[22] N. K. Egilmez, J. B. Chen, and S. M. Jazwinski, *J. Gerontol.* **45**, B9 (1990).
[23] N. K. Egilmez and S. M. Jazwinski, *J. Bacteriol.* **171**, 37 (1989).
[24] C. E. Ballou, "The Molecular Biology of the Yeast *Saccharomyces*." Cold Spring Harbor Laboratory Press, Cold Spring Harbor, NY, 1982.

avidin is used. This technique allows isolation of biotin-labeled old cells with >99% purity, but in relatively small quantities (10^4 cells).[12] Magnetic sorting with streptavidin-coated paramagnetic iron beads allows procurement of more than 10^8 cells. The magnetic beads are incubated with the bulk culture containing biotin-labeled, old cells.[12,15] By placing the culture near a magnet, the old cells coated with beads are separated from the unlabeled, young cells (Fig. 2). Magnetic

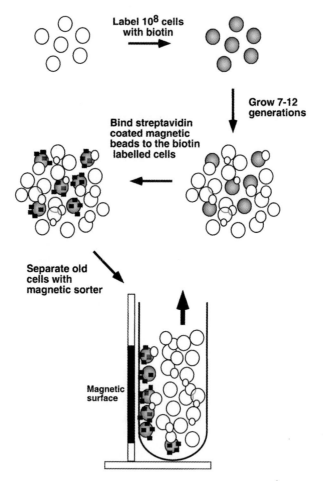

FIG. 2. Biotin–streptavidin magnetic sorting of old cells. The surface of 10^8 cells is labeled with biotin. Labeled cells are grown for 7–12 generations. The biotin label stays with the original mother cells because the cell surface of emerging daughter cells is synthesized *de novo* at the budding site. The culture now contains the original cells, now old, and a large number of young cells. Labeled, old cells are then bound to streptavidin-coated magnetic beads. Old cells are separated from young cells using a magnetic sorter.

sorting is the preferred technique for the isolation of old cells in our laboratory because it allows for separation of a high-yield, pure population of old cells with a relatively small number of manipulation steps.

Biotin–Streptavidin Magnetic-Sorting Procedure

1. Grow 5 ml of overnight culture in YPD.
2. Dilute the culture in 50 ml of YPD and grow the cells at least 5 hr to an OD_{600} of 0.7–1.0.
3. Harvest and resuspend the cell pellet with 1 ml of sterile phosphate-buffered saline (PBS). Transfer the cells into a 1.5-ml Eppendorf tube and determine the cell density using a hemocytometer. Transfer 10^8 cells into a new tube and wash with 1 ml of PBS. Resuspend the cells in 0.4 ml of PBS.

Labeling Cells

4. Add 8 mg of sulfo-NHS-LC-biotin (Pierce, Rockford, IL), dissolved in 0.3 ml of PBS just before use, to 10^8 cells. Sulfo-NHS-LC-biotin is very sensitive to moisture. Prewarm the bottle to room temperature before opening the bottle. Batches of sulfo-NHS-LC-biotin were found to vary. The quality of sulfo-NHS-LC-biotin is the most critical element determining the success of this procedure. Make sure sulfo-NHS-LC-biotin, when opened for the first time, is a fine powder and is not contaminated with moisture. If it is contaminated, try to obtain a different batch of the powder from the supplier. Close the bottle containing sulfo-NHS-LC-biotin immediately after use and wrap the bottle in Parafilm. Store the powder in a $-20°$ freezer.
5. Incubate the cells at room temperature for 15 min with gentle shaking.
6. Wash the cells three times with 1 ml of PBS to remove excess sulfo-NHS-LC-biotin. The cells should be pelleted gently with speed <6000 rpm for 2 min at room temperature for each wash.
7. Resuspend the cells in 1 ml of YPD.
8. Add 10^8 cells per liter of YPD and grow the cells at $30°$ with shaking. The cells are usually grown for about 11–13 hr. It is important that the cells are not overgrown to OD_{600} exceeding 1.0 because we have noticed that there is a poor recovery rate of old cells if OD_{600} exceeds 1.0. The YPD used for this procedure contains 2.5% glucose and is made freshly just before use.

Sorting Old Cells

9. Harvest the cells by centrifugation at 5000 rpm for 10 min at $4°$. Remove the medium thoroughly.
10. Resuspend the cells into 30–50 ml of prechilled ($4°$) PBS. From this point, cells must be kept cold to prevent the formation of new daughters.

11. Wash twice 0.3 ml of streptavidin-coated magnetic beads (PerSeptive Biosystems, Framingham, MA) in 1 ml cold PBS. About 0.3 ml of magnetic beads is needed per 10^8 cells.

12. Add the beads to the cells and incubate on ice for 2–3 hr with occasional swirling to bind the cells to the beads.

13. From this point, perform the sorting steps at 4°. Add the cells bound to beads into several test tubes and place the test tubes in a magnetic sorter (rack) (PerSeptive Biosystems). Add the appropriate amount of cells into each tube to level with the top of the magnet. Ensure that the test tubes are firmly touching the side surface of the rack. Incubate for 20 min to allow the beads and the cells bound to the beads to move toward the magnetic surface.

14. Gently remove the unlabeled young cells with a 10-ml pipette without disturbing the strip of cells and beads magnetized to the surface of the test tube. Remove the cells by slowly pipetting from the top of the tube and descending. Leave about 1 ml of cells at the bottom of the test tube because some portion of old cells could have settled at the bottom of the tube.

15. Add prechilled (4°) YPD to level with the top of the magnet and incubate for 15 min.

16. Repeat steps 14 and 15.

17. Remove unbound cells and add more prechilled YPD as before, but this time, remove the test tubes from the magnetic sorter and resuspend the bound cells gently by flicking the tubes. Place the tubes back to the sorter and incubate for 15 min. This further separates young and old cells.

18. Repeat step 17 seven to eight more times.

19. Pool the remaining cells in 1.5 ml of cold YPD and check the yield using a hemocytometer.

Counting Bud Scars to Determine Age

20. Dissolve 10 mg of Calcofluor white M2R (fluorescent brightener 28, Sigma) in 1 ml PBS. Centrifuge at the maximum speed and use the top 0.9 ml of supernatant to remove undissolved crystals, which can interfere with bud scar counting.

21. Wash 10^6 cells in 1 ml PBS and spin down the cells at 6000 rpm for 1 min. Resuspend the cells with 0.4 ml of prepared Calcofluor solution and incubate for 5 min.

22. Add 1 ml PBS and spin again at 6000 rpm for 1 min. Wash the cells once with 1 ml of PBS. Resuspend the cells in 20 μl. Place the cells on a microscopic slide with a coverslip and count the number of bud scars using UV fluorescence. To determine the total number of bud scars, move the microscopic field slightly up and down. Bud scars consist of chitin rings that are deposited on the cell surface of the mother cell and indicate previous sites of cell division, thus indicating age. Old cells from a single sort will have an average bud scar count of 7 to 12, depending on the strain.

Second and Third Sorting

23. To obtain even older cells, second and third sortings must be performed. The remaining cells from the first sort are inoculated at 10^8 cells per liter of YPD and are grown for 11–13 hr. The sorting procedure is then repeated. Because the cells from the first sort are already biotinylated, it is not necessary to biotinylate the cells again. For second and third sorts, it is critical that all the procedures in the previous sort are carried out in a sterile condition. Bacterial contamination can ruin the separation of old cells. Old cells from two sorts will have an average bud scar count of 14–24.

[29] Assays of Cell and Nuclear Fusion

By ALISON E. GAMMIE *and* MARK D. ROSE

Introduction

The pathway of yeast mating forms a microcosm of the cell biological universe, encompassing fundamental problems in signaling, transcriptional regulation, polarization, motility, and membrane fusion. Each step in the pathway has been defined and dissected by the isolation and analysis of numerous mutations that cause specific defects in the efficiency of mating. Certain profoundly mating defective mutants may reduce the efficiency of mating by more than five orders of magnitude. Such mutants are generally called sterile or *ste* and are caused by defects in the initial pheromone signaling or the subsequent response pathways. Analysis of *ste* mutants has led to a deep understanding of the processing, export, and reception of the pheromones, as well as signal transduction through the MAP kinase cascade, ultimately leading to the activation of genes required for cell and nuclear fusion. Several excellent reviews of the pheromone response pathway and methods for assaying defects in the mating pathway have been published.[1–5]

Although less well understood, mutations that cause defects in the mechanism of mating per se (e.g., cell and nuclear fusion) provide insight into an even broader spectrum of cell biological issues. Cell fusion mutations (many, but not all, of which are called *fus*) provide information about cell–cell signaling, cell

[1] G. F. Sprague, Jr., *Methods Enzymol.* **194**, 77 (1991).
[2] G. F. Sprague and J. W. Thorner, in "Pheromone Response and Signal Transduction during the Mating Process of *Saccharomyces cerevisiae*" (E. W. Jones, J. R. Pringle, and J. R. Broach, eds.), p. 657. Cold Spring Harbor Laboratory Press, Cold Spring Harbor, NY, 1992.
[3] J. Kurjan, *Annu. Rev. Genet.* **27**, 147 (1993).
[4] E. Leberer, D. Y. Thomas, and M. Whiteway, *Curr. Opin. Genet. Dev.* **7**, 59 (1997).
[5] E. A. Elion, *Curr. Opin. Microbiol.* **3**, 573 (2000).

polarization, and its underlying cytoskeletal basis and membrane fusion. Nuclear fusion mutations (many of which are called *kar* for karyogamy defective) provide information about nuclear envelope fusion and microtubule-based mechanisms of nuclear orientation and movement. In contrast to sterile mutants, cell and nuclear fusion mutations are more subtle in their effects. For cell fusion mutants, genetic or physiological redundancy often results in only slight defects in the efficiency of mating (perhaps reduced by 50% or less) or no defect except when mated to specific mutants. Similarly, because of redundancy or diffusion in the zygote, most nuclear fusion mutants are only partially penetrant or invisible unless mated to an appropriate (usually identical) mutant. Accordingly, specialized methods of analysis must be employed to detect the effects of these mutations. Several papers have described the isolation and characterization of cell and nuclear fusion mutants using a variety of techniques.[6-14] The overall pathway of cell and nuclear fusion has been reviewed elsewhere.[15,16]

In general, there are two basic ways to detect defects in cell and nuclear fusion. The first approach uses sensitive methods of detecting either the reduced efficiency of diploid formation or, in the case of nuclear fusion, the increased production of unique progeny cells called cytoductants. Cytoductants contain a haploid nucleus from one parent but the cytoplasm from both parents. While these methods give precise measures of the overall efficiency of the mating pathway, they provide little or no detail about the specific stage of the defect. The second approach uses microscopic methods to detect the distinct zygotes in which cell or nuclear fusion has failed. It has the virtue of providing information about the nature of the mating defect, but, because of the relatively fewer numbers of cells examined, may provide quantitatively less precise data.

Critical Parameters of Mating: Growth Phase, Density, and Timing

Because cell and nuclear fusion mutants may have subtle defects, the mutant phenotype may remain undetected if care is not taken to optimize the conditions

---

[6] J. Chenevert, N. Valtz, and I. Herskowitz, *Genetics* **136,** 1287 (1994).
[7] J. Conde and G. R. Fink, *Proc. Natl. Acad. Sci. U.S.A.* **73,** 3651 (1976).
[8] R. Dorer, C. Boone, T. Kimbrough, J. Kim, and L. H. Hartwell, *Genetics* **146,** 39 (1997).
[9] S. Erdman, L. Lin, M. Malczynski, and M. Snyder, *J. Cell Biol.* **140,** 461 (1998).
[10] L. J. Kurihara, C. T. Beh, M. Latterich, R. Schekman, and M. D. Rose, *J Cell Biol.* **126,** 911 (1994).
[11] V. Berlin, J. A. Brill, J. Trueheart, J. D. Boeke, and G. R. Fink, *Methods Enzymol.* **194,** 774 (1991).
[12] L. Elia and L. Marsh, *J. Cell Biol.* **135,** 741 (1996).
[13] J. Philips and I. Herskowitz, *J. Cell Biol.* **143,** 375 (1998).
[14] G. McCaffrey, F. J. Clay, K. Kelsay, and G. F. Sprague, Jr., *Mol. Cell. Biol.* **7,** 2680 (1987).
[15] L. Marsh and M. D. Rose, in "The Pathway of Cell and Nuclear Fusion during Mating in *S. cerevisiae*" (J. R. Pringle, J. R. Broach, and E. W. Jones, eds.), p. 827. Cold Spring Harbor Laboratory Press, Cold Spring Harbor, NY, 1997.
[16] M. D. Rose, *Annu. Rev. Cell Dev. Biol.* **12,** 663 (1996).

of the mating reaction. Dutcher and Hartwell[17] carefully explored the parameters of mating and identified three that are critical for efficient mating: growth phase, cell ratio/density, and time. The most efficient matings were those in which freshly grown cells in the early logarithmic phase of growth were combined on the surface of a filter at a 1 : 1 ratio and at a defined cell density. The overall efficiency of zygote formation could be diminished greatly when cells were too close to the stationary phase, too sparse or dense on the filter, or at an uneven ratio. The duration of the mating reactions should be determined empirically to maximize the mutant phenotype. Many mating defective mutants do not have a complete block in the process; thus, given sufficient time to mate, many diploids will be formed and the defect can be overlooked. Finally, many mutants show reduced rates of zygote formation as well as specific blocks in mating. From an experiment showing inefficient mating, it may be difficult to microscopically identify enough zygotes for statistically significant data, underscoring the importance of optimizing the mating reaction.

Quantitative and Semiquantitative Plate Mating Assays

Filter Matings to Assay Cytoductants

Karyogamy mutants (*kar*) complete cell fusion, but fail to fuse the haploid nuclei (see reviews[15,16]). For this reason, matings with *kar* mutants result in the formation of aberrant zygotes, which may contain two or more unfused haploid nuclei (heterokaryons) and aberrant progeny that contain a haploid nucleus from one parent and the cytoplasm from both (cytoductants). Using appropriate nuclear and cytoplasmic genetic markers, cytoductants are distinguishable from the parental strains and true diploids in a mating mixture. For example, when a ρ^0 (lacking mitochondrial genome) strain, containing a recessive nuclear allele for cycloheximide resistance (cyh^r), is crossed to a ρ^+ (wild-type mitochondrial genome) strain, containing the wild-type dominant nuclear allele for cycloheximide sensitivity (CYH^s), neither the resulting diploids (CYH^s/cyh^r) nor the haploid parents (ρ^0 and CYH^s) can grow on glycerol medium containing cycloheximide. However, haploid cytoductants that contain the cyh^r nucleus and ρ^+ mitochondria can grow on glycerol medium containing cycloheximide. Cytoduction assays are extremely sensitive and specific for karyogamy defects, but require that specific strains be constructed to allow the selection for the cytoductants. In principle, any recessive drug resistance may be used (e.g., canavanine resistance) singly or in combination. The quantitative mating assays for cytoductants detailed here are based on those described by Dutcher and Hartwell.[17]

---

[17] S. K. Dutcher and L. H. Hartwell, *Cell* **33**, 203 (1983).

Preparation of ρ^0 Strains[18,19]

1. ρ^0 mutants may be obtained by growing cultures in YEPD + 10 μg/ml ethidium bromide wrapped in foil at 30° overnight. Streak the cultures to form single colonies on YEPD plates and incubate at 30°. Test single colonies by patching onto YEP glycerol and YEPD and incubate at 30°. Colonies that fail to grow on YEP glycerol are analyzed further to assess whether they are ρ^- (loss of mitochondrial protein synthesis) or ρ^0 (complete loss of mitochondrial DNA).

2. Strains lacking mitochondrial DNA may be identified easily using DAPI (4′,6′-diamidino-2-phenylindole) staining of cultures with wild-type and ρ^0 controls. To fix cells prior to staining, centrifuge 1 ml of an exponentially growing culture. Unless stated otherwise, all centrifugations are at 2000g for 5–10 min at room temperature. Resuspend the cells in 1 ml of phosphate-buffered saline (PBS, 0.04 M K$_2$HPO$_4$, 0.01 M KH$_2$PO$_4$, 0.15 M NaCl, pH 7.4). Centrifuge the cells and resuspend them in methanol: acetic acid (3:1, v/v) for 30–60 min on ice. Centrifuge and wash the cells several times in PBS.

3. Stain the cells by adding DAPI from a 10-mg/ml stock (stored at −20°) to a final concentration of 1 μg/ml in PBS. Stain for 5 min and wash twice with PBS. Resuspend the final pellet in 100 μl of PBS. Visualize the DNA with ultraviolet fluorescence microscopy using DAPI filter sets. ρ^0 and ρ^- cells can be distinguished from ρ^+ cells by the absence of mitochondrial DNA staining. Wild-type yeast cells will show both nuclear and mitochondrial DNA staining.

4. As a final confirmation, ρ^0 cells can be distinguished from ρ^- cells by mating the strains in question to wild-type ρ^+ cells. Matings between ρ^0 and ρ^+ give rise to only ρ^+ haploid progeny (growth on YEP glycerol), whereas matings between ρ^- and ρ^+ result in the formation of some ρ^- haploid progeny (failure to grow on YEP glycerol).

Isolation of cyh^r Strains[18]

Low-level cycloheximide resistance can be conferred by the loss of function of a number of different genes; however, resistance to high levels of cycloheximide is a consequence of rare mutations at the *CYH2* locus. cyh^r strains can be selected for by plating 200 μl of a saturated culture on YEPD plus cycloheximide (10 μg/ml). A 10-mg/ml stock solution of cycloheximide may be filter sterilized and stored at 4° (add to plates after autoclaving). Confirm the resistance phenotype by patching single colonies onto YEPD and YEPD plus cycloheximide (10 μg/ml).

---

[18] M. D. Rose, F. Winston, and P. Hieter, "Methods in Yeast Genetics: A Laboratory Course Manual." Cold Spring Harbor Laboratory Press, Cold Spring Harbor, NY, 1990.

[19] T. D. Fox, L. S. Folley, J. J. Mulero, T. W. McMullin, P. E. Thorsness, L. O. Hedin, and M. C. Costanzo, *Methods Enzymol.* **194**, 149 (1991).

Preparing Mating Mixtures from Liquid Cultures

1. Inoculate 5 ml of media (usually YEPD, unless selecting for plasmids) with each of the *MATα* and *MAT*a cells to be mated to prepare fresh overnight cultures. To have the strains in the early logarithmic phase of growth at a convenient time, approximately 16 hr prior to the experiment, dilute the overnight cultures into fresh media using a range of dilutions (e.g., for YEPD: 5×10^{-4}, 10^{-4}, and 5×10^{-3}; for synthetic media: 5×10^{-3}, 10^{-3}, and 5×10^{-2}). If the strains have a growth defect, the proper dilution should be determined empirically.

2. To initiate the mating, mix together equal quantities ($3-5 \times 10^6$ cells) of exponentially growing cells from each parent (optimal mating was achieved for cultures at a density between 5×10^6/ml to 1.5×10^7/ml). Concentrate the mating mixture on a 0.45-μm pore size nitrocellulose filter disk using a vacuum-filtration system (Millipore, Bedford, MA). The support base for the filter is inserted into the mouth of a side-arm flask attached to a vacuum line. The base and funnel may be sterilized by autoclaving, or with 70% (v/v) ethanol. Using sterilized forceps, place a filter disk on the base. Attach the funnel if large volumes are required. Concentrate the mating mixture onto the filter by applying the cells with the vacuum turned on. Use sterile forceps to remove the filter and place on a prewarmed YEPD plate with the cell-side up. Incubate the mating mixtures for 2 to 4 hr at 30° (optimal length of time must be determined). Note also that many strains are inherently temperature sensitive for mating and that the frequency of zygotes will be reduced drastically at 35° and above.

3. Lift the filter and rinse the mating mixture into a small culture tube with sterile water. Place the samples in a bath sonicator at low power for 3 min to disrupt mating aggregates. Prepare 10-fold serial dilutions (10^{-1}, 10^{-2}, 10^{-3}, and 10^{-4}) using sterile water.

4. Cytoductants in the mating mixtures can range from 0.01% to greater than 90% of the zygotes. If the mating defect has not been characterized previously, plate 100-μl samples from the undiluted suspension and each of the dilutions onto selective synthetic medium (to determine diploid formation) and YEP glycerol plus cycloheximide (3 μg/ml) (to determine cytoductants). Determine the viable cell count by plating 100 μl from the 10^{-3} and 10^{-4} dilutions onto YEPD plates.

5. Two types of colonies typically may grow on YEP glycerol + cycloheximide plates. The larger colonies are the $\rho^+ cyh^R$ cytoductants and should be scored. There may also be very small colonies, which are petites that grow very slowly and should not be scored. Count the appropriate colonies on all of the plates.

6. The efficiency of mating is determined by the percentage of diploids formed versus total viable cells. The frequency of cytoduction formation is calculated as the percentage of $\rho^+ cyh^r$ colonies versus total viable cells. The frequency of defective nuclear fusion is expressed as the number of cytoductants divided by the number of diploids formed,[18] termed the C/D ratio. Wild-type cells have a

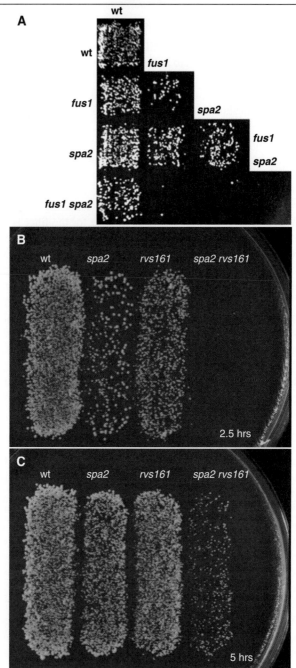

typical C/D ratio of 0.001, and karyogamy mutants have a C/D ratio between 0.1 and 1000.

Semiquantitative Mating Assays

In typical petri plate mating assays to determine the mating types, strains are allowed to mate overnight to maximize the yield of diploids before being replica plated to appropriate media to select for diploids. Under these conditions, even poor maters give luxuriant growth, and cell and nuclear fusion mutants appear to mate like a wild-type strain. However, even subtle mating defects can be observed in semiquantitative "limited" plate-mating assays, which have been optimized to maximize the difference between mutant and wild-type strains (Fig. 1). The major difference is that limited matings are allowed to proceed for a very short time and measures are taken to reduce the inoculum size during the replica printing to diploid selective media. Although less rigorous than the quantitative assay described earlier, the limited plate-mating assay allows multiple strains (including both mutant and wild-type controls) to be assayed on the same plate and very little time is required for its execution.

The semiquantitative plate mating method can detect mutants that display at least a 50% mating defect by a quantitative assay. For more subtle defects, microscopic analysis of mating mixtures is advised (see later). Alternatively, the use of enfeebled mating partners can allow for the observation of even more subtle mating defects. For example, when mated to a *fus1 fus2* double mutant, many cell fusion mutants will show defects that are significantly more severe than

---

FIG. 1. Semiquantitative plate-mating assays. (A) The plate shows the growth of diploid colonies formed from 10 different crosses in a limited plate-mating assay. Various *MATa* and *MATα* strains were patched as wide stripes onto YEPD plates. After overnight growth, they were then crossed by replica printing them together at a 90° angle onto a YEPD plate and allowing the cells to mate for 2.5 hr. The mating plate was then replica printed to minimal medium to select for diploids. The relevant genotypes of the parent strains in the patches are indicated on the left of the horizontal stripes and the top of the vertical stripes. Wild-type (wt), *fus1* and *spa2* single mutants, and *fus1 spa2* double mutants were used in this analysis. Note that each parent contains a copy of both genes; taken together, there are a total of four genes controlling the phenotype of the zygote. In this example, mutations in one or two of the four genes cause only a weak defect in mating, whereas mutations in three or four genes cause strong defects (bottom row). (B) The plate shows the growth of diploid colonies formed from 4 different crosses in a limited mating. Patches of the strains to be tested and a lawn of a tester strain were replica printed together onto YEPD and allowed to mate for 2.5 hr. The plate was then replica printed to minimal medium to select for diploids. The lawn was an *rvs161 spa2* double mutant, and the patches were a wild-type control (wt), a *spa2* single mutant, a *rvs161* single mutant, and a *spa2 rvs161* double mutant. (C) A duplicate of the plate mating shown in B was allowed to mate for 5 hr. Note that increasing the time of mating reduces the ability to discriminate between mutant and wild-type matings.

those observed in matings to wild type or in self-matings.[6,20] Use of the *fus1 fus2* double mutant underscores another important property of semiquantitative mating assays. Single mutants often show very little or no defect in mating assays, even when both parents are mutant. However, by crossing double mutant parents together, a more severe defect is commonly observed (cf. Fig. 1A) than with either single mutant.[20,21]

1. Patch equal amounts of freshly grown cells of one mating type onto distinct regions of a plate (usually YEPD unless selecting for plasmids). Maximizing the area for the patches often yields more reproducible results. Include both a wild-type and a mating-defective control strain on each plate. Prepare the mating partner lawn by spreading 100–200 μl of a freshly grown, saturated culture onto a plate. Care should be taken to ensure that the patches and lawns are spread evenly. Allow the patches and lawns to grow overnight at 30°.

2. If the patches and lawn have grown evenly, proceed with the mating. Mate the strains by replica printing both together onto a prewarmed YEPD plate. For certain mutant phenotypes, the concentration of cells on the mating plate is crucial. It is advised to try different replica-printing techniques to dilute the lawns and patches to optimize the scoring of the mating defect. Successful strategies include using the second replica print rather than the first, or performing serial replica prints.

3. Allow the strains to mate for 2 to 4 hr (determined empirically for each mutant phenotype) at 30°. Shorter mating times may be required to observe subtler mating defects. For example, simply increasing the mating time from 2.5 to 5 hr makes some single mutants indistinguishable from the wild type (Figs. 1B and 1C). Replica print the mating plate to synthetic medium to select for diploids. Serial replica printing may be used to reduce the inoculum.

4. The frequency of diploid formation may be assessed qualitatively using wild-type and mutant mating controls as references. The plates should be incubated between 16 and 24 hr and scored at intervals. Allowing the selection plates to grow for a more extended period of time (greater than 24 hr) can result in a failure to observe the mutant phenotype, as patches from the mutant mating will eventually grow to the same extent as the wild-type mating patch.

Micromanipulation of Zygotes

A useful technique for analyzing mating-defective mutants is to micromanipulate zygotes from a limited mating mixture using the same setup as that used for tetrad dissection. This method, termed zygote pulling, allows for subsequent measurement of cell viability and diploid formation from individual mating events.

---

[20] J. Trueheart, J. D. Boeke, and G. R. Fink, *Mol. Cell. Biol.* **7**, 2316 (1987).
[21] A. E. Gammie, V. Brizzio, and M. D. Rose, *Mol. Biol. Cell* **9**, 1395 (1998).

1. The mating mixtures can be prepared as described earlier for the quantitative cytoductant assays. Alternatively, limited matings can be prepared by simply mixing equal amounts of each parent on the surface of a YEPD plate and allowing the mating to proceed for 2 to 4 hr. A wild-type mating should also be set up to control for the loss of viability as a consequence of manipulation.

2. After the limited matings, rinse the mating mixture off the 0.45-μm filter, or scrape the mixtures off the plate into 1 ml of sterile water and pellet the cells. Resuspend the cells at $\sim 10^6$ cells/ml with sterile water. Transfer 50 to 100 μl to the edge of a YEPD plate. Spread the cells in a single line across the edge of the plate.

3. Place the plate on the stage of a dissecting microscope. The magnification of most dissection microscopes provides limited resolution so that early zygotes may be difficult to distinguish from large-budded cells. Therefore, it is important to select zygotes with a large bud. Large-budded zygotes have a distinctive propeller shape. Identify a zygote and move it to an isolated location on the dissection plate. Because zygotes seem to be more sensitive to manipulation and drying than spores, it is advised to drag the zygote along the surface of the agar to a distinct region of the plate.

4. Allow the cells to grow for 2 days at 30°. Score the viability as compared to zygotes pulled from a wild-type mating. Replica print to media that will distinguish between diploid and haploid cells.

After micromanipulation, wild-type zygotes give rise to colonies that are uniformly diploid. In contrast, zygotes in which cell or nuclear fusion has failed will give rise to mixed colonies or colonies with one predominant haploid genotype. To ensure that a mating has occurred, one parent may be cytoplasmically marked (e.g., ρ^0) and the transfer of cytoplasm can be scored (see earlier discussion for cytoductants).

Microscopic Methods

Microscopic methods have the advantage of providing information about the specific stage of mating that is defective. In a practical sense, microscopic methods also do not require elaborate genetic markers to be built into the strains before examining their mating defects. Successful microscopic analysis of mating mixtures depends on the accurate identification of a zygote. To the novice, a zygote can be confused easily with a large-budded cell, or with two closely associated cells. Zygotes may be distinguished from mitotic cells by their distinct cell/cell contact region. Zygotes have a smooth, curved junction between the parent cells (resembling a peanut shell with two nuts), whereas budded cells have a sharp constriction at the mother/bud neck (resembling two closely associated spheres). Representative zygotes observed by light microscopic methods are shown in Figs. 2–4.

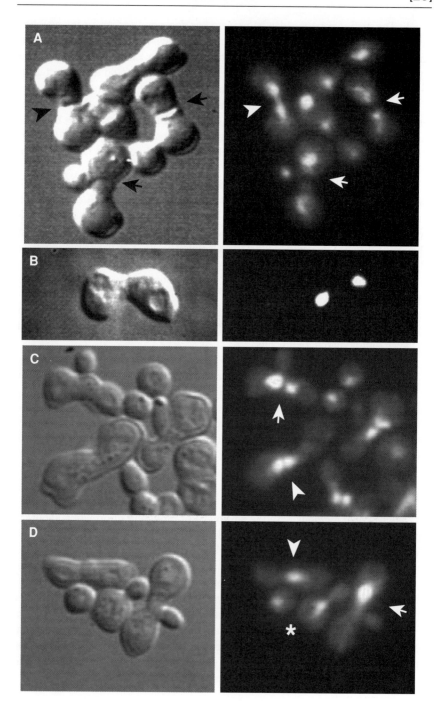

Detection of Cell Fusion and Nuclear Fusion Defects by Fluorescence Microscopy

The easiest microscopic analysis of karyogamy and cell fusion phenotypes utilizes differential interference contrast (DIC) or phase-contrast optics to assess zygote morphology and DAPI fluorescence to evaluate the position of the nuclear DNA (Fig. 2). For detailed visual observation of the zygote, 100× objectives are necessary and DIC optics are preferable, as the bright halo surrounding cells in phase-contrast optics may obscure details of the cell fusion zone.

1. Mating mixtures should be prepared as described earlier for quantitative cytoductant assays. At the appropriate times, rinse the mating mixtures from the 0.45-μm filter with 1 ml of ice-cold PBS into a microcentrifuge tube. Typical matings are allowed to proceed for 2 to 4 hr at 30°, depending on the phenotype and mutant being assayed. For example, karyogamy defects are best scored using shorter times of mating (see "Scoring Karyogamy Mutants") Pellet the cells briefly in a microcentrifuge and resuspend the cells in cold PBS. Repeat the wash twice.

2. Centrifuge the mating mixtures and resuspend in methanol : acetic acid (3 : 1) to fix the cells. Fix the cells for 60 min on ice. Pellet and remove as much of the fix as possible. Resuspend the cells in PBS and incubate on ice for at least 30 min to allow the cells to rehydrate. Pellet and wash the cells twice in PBS

3. Stain the mating mixtures by adding DAPI from a 10-mg/ml stock (stored at −20°) to a final concentration of 1 μg/ml in PBS. Stain for 5 min and wash twice with PBS.

4. Resuspend the final pellets in 100 μl of PBS. The mating mixtures may be kept at 4° for up to 1 week before examination. However, both fluorescence and cellular morphology will deteriorate over time.

5. Sonicate the samples at low power for approximately 3 min to break up mating aggregates before scoring. A bath sonicator is useful for this purpose.

FIG. 2. Phenotypes of cell and nuclear fusion mutant zygotes detected by fluorescence microscopy. Mutants were examined for zygote morphology using differential interference contrast optics (left) and for nuclear positioning using the DNA-specific dye DAPI and fluorescence microscopy (right). (A) Cell fusion-defective zygotes from an *rvs161* × *rvs161* mating. Note the presence of septa separating cell bodies in the zygotes and the well-separated nuclear masses. Arrows indicate zygotes with complete septa; the arrowhead indicates a zygote with a partial septum. (B) A class I karyogamy-defective zygote from a *kar4* × *kar4* mating. Note the absence of the septum and widely separated nuclei. (C) Class II karyogamy-defective mutant zygotes from a *kar5* × *kar5* mating. Note the absence of septa and closely apposed nuclei. The arrow indicates a budded zygote; the arrowhead indicates an unbudded zygote. (D) Examples of wild-type zygotes. Note the absence of septa and single nuclei in both budded (arrow) and unbudded (arrowhead) zygotes. Note also the distinct difference in zygote morphology compared to the budded cell (asterisk).

6. Prepare single, wet-mount slides. Place 3 μl of the cell suspension on a standard 1 × 3-in. glass microscope slide. Place the edge of a coverslip (#1 thickness, 18 mm^2) into the sample and slowly lower the coverslip at an angle such that air bubbles are expelled at the far edge. Excess liquid can be removed by patting the slide with a lint-free absorbent tissue. If the cells and zygotes are not well separated, dilute the sample, repeat step 5, and remake the slide.

7. Examine the slides by differential interference or phase contrast to assess zygote morphology and by ultraviolet fluorescence microscopy using a DAPI filter set to examine the position of the nuclear masses.

Scoring Karyogamy Mutants

Wild-type zygotes with small buds invariably contain a single nucleus (Fig. 2D), whereas karyogamy-defective zygotes with buds contain two or more nuclei.[10] Class I karyogamy mutant zygotes have widely separated nuclei (2.3 to 3.6 μm apart) because of defects in microtubule function required for nuclear congression (Fig. 2B). Class II karyogamy mutant zygotes contain two or more closely associated unfused nuclei (less than 0.5μm apart) because of defects in nuclear envelope fusion (Fig. 2C). The molecular basis for class I and class II defects has been reviewed.[15,16] Note that as class I mutant zygotes resume the cell cycle, unfused nuclei may be pulled together toward the bud and become closely associated. As a consequence, some zygotes would be placed incorrectly in the class II category and cause a large variance for the distance measurements. Therefore, only zygotes with no buds or small buds should be scored for an accurate characterization of the nuclear fusion defect. A minimum of 100 or more zygotes should be scored, depending on the subtlety of the defect.

Scoring Cell Fusion Defects

Cell fusion mutants (*fus*) can be distinguished by the presence of a septum between the mating partners (Figs. 2A, 3, and 4). Typically, three classes of zygotes are present in a mating mixture from Fus$^-$ mutant parents. Some Fus$^-$ zygotes display a complete block in cell fusion. Zygotes in this class contain a complete septum between the partners and two unfused nuclei on either side of the septum (Figs. 2A, 3C, and 4). Partially defective Fus$^-$ zygotes have a pronounced septum (or septum remnants) and a single nucleus. For this class, cytoplasmic fusion has occurred, but was apparently less efficient than normal (Figs. 3B and 4). These zygotes will form viable diploids and thus their defect would be missed if using a quantitative mating assay selecting for diploid growth. The final class of zygotes in a Fus$^-$ mating mixture appears to have completed cell and nuclear fusion with wild-type proficiency (Figs. 2D, 3A, and 4). Because all three zygote types may be present in a given mutant mating, it is critical that a large number of zygotes are scored (>200) to ensure statistical significance.

Microscopic Assays for Cell and Nuclear Fusion Phenotypes Using Green Fluorescent Protein

More rigorous assessments of cell and nuclear fusion utilize fluorescent tracers to report on the state of the relevant cell structure. One approach is to express green fluorescent protein (GFP), or one of its variants, in one of the parents and look for the transfer of GFP into the other cell body in the zygote as an assay for cell fusion[21,22] (Fig. 3). Indeed, the analysis of cytoplasmic mixing using GFP is much more sensitive and specific for cell fusion than the DAPI/DIC analysis described earlier. It seems that even very small sites of plasma membrane fusion lead to cytoplasmic mixing, which can be detected by GFP transfer, although these may not be large enough to allow nuclear fusion and diploid formation. For example, the DAPI/DIC analysis of zygote and nuclear morphology can lead to an overestimation of the severity of the cell fusion phenotype for mutants such as *fus2*.[21,23] Because of its sensitivity, this assay is very useful for phenotypically distinguishing among different mutants with strong defects.[21]

At least three different GFP constructs have been used in the mating assays. A cytoplasmic P_{GAL10}–GFP (T. Stearns, Stanford University) construct has been used with success[21]; however, the assay requires that GFP expression be induced using galactose prior to mating. A *RAS2–GFP* fusion construct (J. Whistler and J. Rine, University of California at Berkeley) has also been employed.[22] Ras2p–GFP mainly localizes to the cell membrane and may be slower to diffuse into the partner after fusion. Finally an endoplasmic reticulum (ER) resident GFP has been used to label the nuclear envelope to determine whether nuclear fusion has occurred (Fig. 3D; N. Erdeniz and M.D. Rose, unpublished observations, 2001).

1. The mating mixtures should be prepared as described earlier for the quantitative cytoductant assays. For cell fusion assays, only one partner should express GFP. Using constitutively expressed GFP will simplify the procedure. For nuclear envelope fusion assays, GFP is best used simply as a morphological marker; *de novo* synthesis after cell fusion complicates its use as a tracer for the kinetics of ER lumenal mixing.

2. After limited matings, rinse the mating mixtures off the 0.45-μm filter with 1 ml of ice-cold PBS into a microcentrifuge tube. Pellet the cells briefly in a microcentrifuge. Resuspend the cells in cold PBS. Repeat the wash twice.

3. Fix the mating mixtures briefly (5–15 min) in 4% formaldehyde to arrest mating without destroying GFP fluorescence. Wash several times in 1 ml of PBS.

---

[22] J. Philips and I. Herskowitz, *J. Cell Biol.* **138**, 961 (1997).
[23] E. A. Elion, J. Trueheart, and G. R. Fink, *J. Cell Biol.* **130**, 1283 (1995).

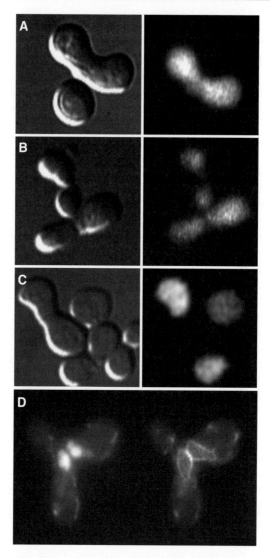

FIG. 3. Cell and nuclear fusion phenotypes scored using green fluorescent protein (GFP). For the top three panels, strains were examined for zygote morphology using differential contrast optics (left panels A through C) and for cytoplasmic mixing using soluble GFP and fluorescence microscopy (right panels A through C). (A) A wild-type zygote with cytoplasmic mixing. Note the absence of a septum. (B) Partially defective zygotes with visible septa between the mating partners (left) in which cytoplasmic mixing has occurred (right). (C) Completely cell fusion-defective zygotes with visible septa between mating partners (left) and in which no cytoplasmic mixing has occurred (right). (D) A class II nuclear fusion-defective zygote ($kar5 \times kar5$) showing the close apposition of nuclei using the DNA-specific dye DAPI and fluorescence microscopy (left) and the presence of a nuclear envelope (marked with lumenal GFP in the endoplasmic reticulum) separating the two nuclei (right).

4. Stain the mating mixtures by adding DAPI from a 10-mg/ml stock (stored at $-20°$) to a final concentration of 1 μg/ml in PBS. Stain for 5 min and wash twice with PBS.

5. Prepare wet-mount slides. If necessary, perform dilutions to ensure that the cells and zygotes are well separated. Light sonication may be used to separate the cells.

6. Score the defect by identifying a zygote using DIC or phase-contrast optics before observing the fluorescence of the zygote. Switch to epifluorescence illumination using FITC or GFP filter sets. For cell fusion assays, mutant zygotes will have bright fluorescence in one-half of the zygote (Fig. 3C). For nuclear fusion assays using an ER resident GFP, a bright rim of nuclear envelope will surround the DAPI-stained nuclei. For class II karyogamy mutants, DAPI-stained nuclei will be closely apposed but separated by a bright line of nuclear envelope (Fig. 3D).

Assays for Cell Wall Removal and Plasma Membrane Fusion Using Lipophilic Styryl Dye

An adjunct to DIC microscopy for studies of cell fusion is to stain cells with a fluorescent dye that specifically labels the membranes (Fig. 4). Certain lipophilic styryl dyes, such as FM 4-64, *N*-(3-triethylammoniumpropyl)-4-(*p*-diethylaminophenylhexatrienyl)pyridinium dibromide (Molecular Probes, Eugene, OR), exhibit increased fluorescence in a lipid environment. Originally used to examine endocytosis in yeast,[24] FM 4-64 has also been used to visualize plasma membranes during cell fusion.[25] Because it looks only for the presence of a residual plasma membrane at the cell fusion zone, FM 4-64 staining is more sensitive to inefficient fusion than the cytoplasmic GFP transfer assay. Thus, it may be particularly useful for examining mutants that exhibit a high degree of partial cell fusion. FM 4-64 also has the advantage of not requiring additional strain construction. Note that FM 4-64 initially stains the plasma membrane, but over time it will be internalized and ultimately accumulate in the vacuole.[24]

1. Prepare mating mixtures as described previously for the quantitative cytoductant assays.

2. Rinse the mating mixtures off the 0.45-μm filter with 1 ml of ice-cold PBS into a microcentrifuge tube. Pellet the cells briefly in a microcentrifuge. Resuspend the cells in cold PBS. Repeat the wash twice. Resuspend the mating mixtures in ice-cold H_2O. Place the samples on ice.

3. Add FM 4-64 (Molecular Probes) to a final concentration of 33 μM just before fluorescence microscopic examination. The membranes are visualized using

[24] T. A. Vida and S. D. Emr, *J. Cell Biol.* **128,** 779 (1995).
[25] V. Brizzio, A. E. Gammie, G. Nijbroek, S. Michaelis, and M. D. Rose, *J. Cell Biol.* **135,** 1727 (1996).

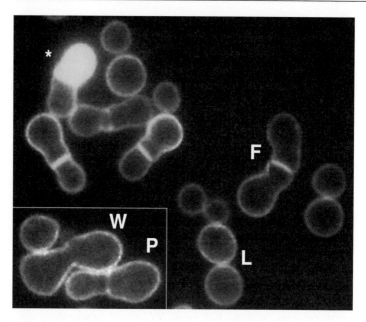

FIG. 4. Visualization of plasma membranes in cell fusion-defective zygotes using the lipophilic styryl dye FM 4-64. Mating mixtures from a *spa2* × *spa2* cross (larger image) or from a *MATα* wild-type × *MATa axl1* cross (insert) were stained with FM 4-64 and visualized using rhodamine filter sets and fluorescence microscopy. A phenotypically wild-type zygote (W), a zygote showing a partial block in cell fusion (P), and a zygote showing a complete block in cell fusion (F) are indicated. A large-budded cell (L) is marked to illustrate the difference in cell shape compared to a zygote. The asterisk (*) indicates a zygote in which one-half shows intense fluorescence, a phenomenon sometimes observed with FM 4-64 staining.

ultraviolet fluorescence microscopy and rhodamine filter sets. As the cells warm on the slide, the dye will be internalized, thus the samples should be photographed quickly for scoring.

4. The presence of unfused plasma membranes at the zone of cell fusion is observed by noting enhanced fluorescence at the region of cell–cell contact.

Immunofluorescence of Mating Mixtures

Simple microscopic examination (e.g., methanol–acetic acid fixation and DIC/DAPI visualization) is a rapid method for determining if cell or nuclear fusion has occurred, but does not provide much information about the specific cellular structures affected. To identify the underlying defect/structures responsible for the failure in cell or nuclear fusion, immunofluorescent and fluorescent-tagging methods must be employed. This is particularly true for studies of nuclear congression where a variety of very different microtubule defects can result in similar class I karyogamy phenotypes.[10] The following immunofluorescence procedure is based on previously detailed methods.[18,26-28] Note that other fixation methods may yield different or better results depending on the particular antigen/structure (e.g., membrane proteins[29] or the spindle pole body[30]). It is important to try multiple methods and conditions to optimize both the signal and the quality of preservation.

1. The mating mixtures should be prepared as described earlier for the quantitative cytoductant assays. If possible, a mating with parents lacking the protein/epitope under investigation should be included.

2. After the limited matings, rinse the mating mixtures off the 0.45-μm filter with 5 ml of 0.1 M potassium phosphate (pH 6.5) into a culture tube.

3. Add 0.6 ml of 37% formaldehyde. Incubate at room temperature with rotation for 5 min to 2 hr. Note that certain epitopes are sensitive to formaldehyde fixation (e.g., the hemagglutinin epitope tag) and that some structures may become inaccessible to antibodies with extended fixation (e.g., the spindle pole body). However, some structures require longer fixations for a good preservation of structure (e.g., microtubules). For best results, it is advised to try a spectrum of fixation times ranging from 5 min to 2 hr.

4. Pellet the cells by centrifugation for 2 min at 2000 rpm. Wash the cells twice in 5 ml of 0.1 M potassium phosphate, pH 6.5. Wash the cells once in 5 ml of phosphate-buffered sorbitol (1.2 M sorbitol, 0.1 M potassium phosphate, pH 6.5). The cells may be stored at 4° at this stage for several days.

5. Pellet the cells as in step 4. Resuspend in 1 ml of phosphate-buffered sorbitol. Add 5 μl of 2-mercaptoethanol and 30 μl of a 10-mg/ml solution of Zymolyase 100,000. Digest the cells for 30 to 90 min at 30° with rotation. Beginning at 30 min after Zymolyase treatment, examine the cells by phase-contrast microscopy. Adequately digested cells should be dark, translucent gray. Bright (refractile) cells are insufficiently digested. "Ghost" cells (pale gray cells with little internal structure) have been overdigested.

6. Collect the cells by centrifugation at 2000 rpm for 2 min. Wash once with 5 ml of phosphate-buffered sorbitol. Resuspend in 1 ml of phosphate-buffered sorbitol. The cells may be stored at 4° at this stage for several days.

7. Place two tightly sealed Coplin jars, one with 100% methanol and the other with 100% acetone, at −20°. Allow the solutions to reach −20°.

8. Thaw a 1% (w/v) polylysine (molecular weight >300,000) stock from −20° storage. Centrifuge the stock for ∼10 min in a microfuge. Dilute the stock

[26] A. E. Adams and J. R. Pringle, *J. Cell Biol.* **98**, 934 (1984).
[27] J. V. Kilmartin and A. E. Adams, *J. Cell Biol.* **98**, 922 (1984).
[28] J. R. Pringle, A. E. Adams, D. G. Drubin, and B. K. Haarer, *Methods Enzymol.* **194**, 565 (1991).
[29] C. J. Roberts, C. K. Raymond, C. T. Yamashiro, and T. H. Stevens, *Methods Enzymol.* **194**, 644 (1991).
[30] M. P. Rout and J. V. Kilmartin, *J. Cell Biol.* **111**, 1913 (1990).

to 0.1%. Place 15 µl of the 0.1% polylysine solution onto each of the wells of a Teflon-masked slide. Incubate for 1 to 5 min at room temperature. Aspirate the excess and wash three times with 1 drop of sterile distilled water.

9. Place 15 µl of the fixed cells in each of the wells of the slide. Allow the cells to settle for a few minutes. Aspirate the liquid and wash with 1 drop of BSA–PBS (10 mg/ml bovine serum albumin, 0.1% NaN_3 in PBS) three times. Rest the chilled methanol and acetone jars in dry ice.

10. Aspirate the excess solution from each well and allow the cells to air dry for approximately 2 min. Plunge the slide into the −20° methanol and incubate for 6 min. Remove the slide and submerge it in the −20° acetone jar for 30 sec.

11. Remove the slide and allow the acetone to evaporate at room temperature.

12. Add 1 drop of BSA–PBS. Incubate at room temperature for at least 5 min. Aspirate excess fluid and proceed with the antibody binding.

13. Centrifuge the primary antibody stock for ∼10 min at 12,000 rpm in a microcentrifuge at room temperature. Dilute the primary antibody in BSA–PBS. It is important to use affinity-purified polyclonal antibodies to prevent background staining. A range of antibody dilutions should be used the first time. The range of dilution is epitope and antibody specific. It is worthwhile to set up a broad range to find the optimal conditions.

14. Add 15 µl of diluted antibody per well. Include at least one control well with BSA–PBS with no primary antibody added. Incubate for 45 min or up to overnight at room temperature or at 37° in a sealed moist chamber to prevent evaporation. Typically, 2 hr at room temperature is sufficient, but some epitope/antibody combinations may require prolonged incubation. In this and all subsequent steps, do not allow the slides to dry out until instructed.

15. Aspirate excess solution and add 1 drop of BSA–PBS. Incubate for 5 min and repeat the process at least four times.

16. Centrifuge the fluorescently labeled secondary antibody stock in a microcentrifuge as in step 13. Dilute the secondary antibody in BSA–PBS. The proper dilution for the secondary antibody should be determined empirically. Add 15 µl of diluted secondary antibody to each well, including the no primary control well. A well in which no secondary antibody has been added will allow for the determination of background fluorescence in the cells. Incubate for 1 to 2 hr at room temperature in a sealed moist chamber in the dark.

17. Wash the cells as described previously at least four times.

18. If staining simultaneously with DAPI, wash the cells twice more with PBS without BSA. Aspirate excess PBS and add 1 drop of freshly diluted DAPI (to 1 µg/ml in PBS). Incubate for 5 min at room temperature. Wash twice with PBS.

19. Aspirate all of the supernatant and allow the slide to air-dry.

20. Place a drop of mounting medium at one end of the slide (to prepare mounting medium, dissolve 100 mg of p-phenylenediamine in 10 ml of PBS, adjust the pH to above 8 with 0.5 M sodium carbonate buffer, pH 9.0, adjust the volume to 100 ml with glycerol, and store at −20° in the dark). Carefully position a

50 × 20-mm coverslip onto the slide so that all of the air is expelled. To accomplish this, place the coverslip down at an angle and allow the mounting medium to come in contact with the short edge and run under the coverslip by capillary action as the angle between the coverslip and the slide is decreased. Blot excess mounting medium from the slide by laying the slide face down on a paper towel.

21. Seal the edge of the slide with a thin layer of clear nail polish.
22. After the nail polish has hardened, clean the slide under running water. Dry and polish with a Kimwipe. Slides can be stored for an extended period of time in the dark at −20°.

Electron Microscopy: Fixation and Embedment of Zygotes

Glutaraldehyde/Permanganate Fixation of Membranes with Sodium Periodate Oxidation of Cell Wall. To improve cell embedment, many published techniques for electron microscopy of yeast cells tend to destroy the ultrastructure of the cell wall. However, investigation of cell fusion requires examination of zygotes where cell wall integrity has been preserved. An effective method that gives good embedment, with minimal effects on the cell wall and good membrane contrast, uses a combination of sodium periodate treatment with permanganate/uranyl acetate staining (Fig. 5). This method works well for S288C-derived strains; however, it may not work for all strains.

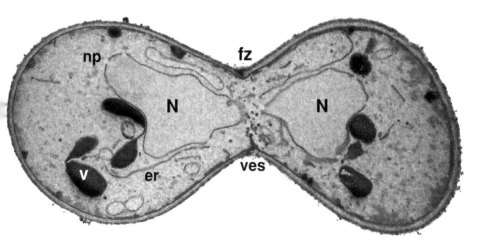

FIG. 5. Electron micrograph of a wild-type zygote. The zygote was fixed and stained using gluteraldehyde and permanganate as described in the text with sodium periodate oxidation of the cell wall. The micrograph is of a wild-type zygote after cell fusion but prior to nuclear fusion. Nuclei (N) are outlined by darkly staining nuclear envelope with nuclear pores (np) appearing as light gaps. It is also likely that the spindle pole body appears as a gap using this procedure. The long membranous structures are peripheral endoplasmic reticulum (er). Balloon-shaped structures with dark interiors are vacuoles (v). Note the presence of numerous darkly stained vesicles (ves) at the zone of cell fusion (fz). Also present in this micrograph are mitochondria (not labeled).

1. To produce sufficient numbers of cells for electron microscopy, the limited filter-mating protocol should be scaled up. Set up overnight cultures of the parents to be mated. Approximately 16 hr before the mating, dilute *MAT*α and *MAT*a overnight cultures (10^{-4}) into 20 ml of YEPD media. Grow the cultures overnight at 30° with aeration.

2. If the strains are both at early exponential phase, proceed with the mating. Mix together the 20 ml from the *MAT*a and *MAT*α strains in a 50-ml tube and centrifuge to pellet the cells. Pour off most of the liquid, leaving about 1–5 ml. Resuspend the mating mixture in the residual liquid. Have a filtration–aspirator set up and sterilized (see earlier discussion for quantitative mating). If the efficiency of zygote formation is not affected in the mutant mating, then 3–5 filters can be used for the mating. If mating involves mutants that display a reduced efficiency of zygote formation, it is best to spread the mating mix over 5–10 filters. Dispense and filter the mating mixtures over the appropriate number of 0.45-μm filters and place the disks cell-side up on prewarmed YEPD plates. Mate the strains for 2 to 4 hr at 30°.

3. Wash the mating mixture into a microcentrifuge tube with 1 ml of FIX [40 mM potassium phosphate, pH 7.4; 1 mM CaCl$_2$; 1 mM MgCl$_2$; 0.2 M sorbitol; 2% (v/v) fresh gluteraldehyde (Polysciences, Warrington, PA)]. Centrifuge cells and resuspend in 500 μl of fresh FIX. The total length of time in FIX should be 30 min (including the initial washing into the tubes) at room temperature. After the fixation, wash three times with 50 mM potassium phosphate, pH 7.4.

4. Resuspend cells in 1 ml of fresh 4% (w/v) potassium permanganate that has been filtered (Whatman, Clifton, NJ, #1 filter) to remove undissolved crystals. Mix the cells in the permanganate at 4° for 2 to 6 hr. Wash the cells at least four times with distilled H$_2$O until the supernatant is clear (i.e., no longer purple).

5. Resuspend the cells in 1 ml of 1% (w/v) sodium periodate (Sigma, St. Louis, MO). Centrifuge the cells and resuspend in fresh 1% (w/v) sodium periodate. Incubate for a total of 15 min. Treatment of the cells with 0.5 to 1.0% (w/v) periodate does not alter cell wall appearance but significantly improves infiltration of the resin. Centrifuge and wash once with 50 mM potassium phosphate, pH 7.4.

6. Resuspend the pellet in 1 ml of 50 mM ammonium phosphate. Centrifuge and resuspend in 1 ml of 50 mM ammonium phosphate. Incubate for a total of 15 min. Wash twice with distilled H$_2$O.

7. Resuspend the pellet in 2% uranyl acetate that has been filtered through a 0.45-μm syringe filter. Incubate at 4° overnight with mixing. Contact the local environmental health and safety office for the appropriate disposal of uranyl acetate waste. Centrifuge the cells and do the following dehydration washes with ethanol solutions: 50% ethanol, 5 min, two times; 70% ethanol, 5 min, two times; 95% ethanol, 5 min, one time; and 100% ethanol, 5 min, three times. Use a freshly opened bottle of absolute ethanol for the final dehydration wash.

8. Resuspend the final pellet in a 50:50 of 100% ethanol:LR White resin (Polysciences). Work in a ventilated hood and wear gloves when using

unpolymerized resin. Incubate several hours with gentle rotation. A slow roller drum works well for this. Centrifuge at 14,000 rpm for 1 min and resuspend in 100% LR White resin. Incubate overnight with gentle rotation.

9. Centrifuge and resuspend in 100% LR White resin. Centrifuge at 14,000 rpm for 1 min and resuspend in ~200–500 µl of 100% LR White and transfer to prebaked (overnight at 60°) Beem-embedding molds. Check that there are no tiny bubbles in the bottom of the tube. Prepare labels that are printed in small font with a laser printer or written in pencil. Do not use pen or ink jet-printed labels! Place the labels in the tubes and fill the tubes to the top with fresh resin. Place the samples under vacuum for about 30 min to remove air bubbles. Incubate overnight in a 60° vacuum oven (preflushed with nitrogen) for about 24 hr. Oxygen inhibits the polymerization, thus removal of air bubbles in the sample and flushing the oven are critically important.

10. Cut 70- to 90-nm sections. Stain the sections for 5 min with lead using Reynold's lead citrate.[31] To prepare the lead citrate solution, combine 1.33 g lead nitrate, 1.76 g sodium citrate, and 30 ml distilled water. Mix well for 1 min and allow the solution to stand for 30 min with occasional shaking. Add 8 ml of 1 N NaOH and mix. Dilute to 50 ml with distilled water. The final pH should be pH 12. Store up to 6 months.

11. Rinse the grids with distilled water. Air-dry and examine by transmission electron microscopy. Using this protocol, membranes appear dark, and proteinaceous structures are light. Thus nuclear pores appear as gaps in the nuclear envelope. One notable feature of the cell fusion zone is the high concentration of vesicular structures. The cell wall is well preserved and several layers are evident.

OsO_4/Uranyl Acetate Staining for Spindle Pole Bodies and Microtubules

The permanganate/uranyl acetate method does not allow for the visualization of important cellular structures such as microtubules and spindle pole bodies. Byers and Goetsch[32,33] have developed an OsO_4/uranyl acetate-staining method, which allows for the detection of these and other key components in cell and nuclear fusion.

1. Follow steps 1 and 2 from the permanganate/uranyl acetate method described previously.

2. Wash the mating mixture into a culture tube with 10 ml of PEM (0.1 M PIPES, pH 6.9, 2 mM EGTA, 1 mM $MgCl_2$). Pellet the cells and resuspend in 8 ml of PEM + 2 ml 10% glutaraldehyde (Polysciences). Fix for 2 hr at 23° with rotation.

---

[31] E. S. Reynolds, *J. Cell Biol.* **17**, 208 (1963).
[32] B. Byers and L. Goetsch, *J. Bacteriol.* **124**, 511 (1975).
[33] B. Byers and L. Goetsch, *Methods Enzymol.* **194**, 602 (1991).

3. Pellet the mating mixtures and wash with 5 ml of PEM. (The cells may be kept overnight at 4° at this stage.)

4. Pellet cells and resuspend in 5 ml of PEM + 0.2% (w/v) tannic acid. Incubate for 20 min at 23°.

5. Pellet the cells and wash with 5 ml of 0.1 M potassium phosphate, pH 7.5. Pellet cells and resuspend in 2.25 ml of potassium phosphate. Add 0.25 ml of 2 mg/ml Zymolyase 100T (ICN Biomedicals, Inc., Aurora, OH) and incubate for 45 min at 30° with rotation.

6. Pellet the cells and wash with 5 ml of 0.1 M sodium cacodylate, pH 6.8. Cells can be kept at 4° overnight at this step. Centrifuge at 2000g for 5–10 min at room temperature and resuspend in 1.5 ml of 0.1 M sodium cacodylate, pH 6.8.

7. Transfer cells to a microcentrifuge tube. Pellet the cells and resuspend in 2% OsO_4, 0.1 M sodium cacodylate, pH 6.8. Incubate for 1 hr at room temperature with rotation. *Note.* Do not add osmium to conical tubes because the cells stick to the walls. Pellet cells and wash with distilled water three times until the supernatant clears.

8. Resuspend in 0.5% uranyl acetate (from 1% stock). Incubate at room temperature for 1 hr and overnight at 4° in the dark with rotation.

9. Wash with distilled water three times. Complete the dehydration protocol as follows: 50% ethanol, 3 min; 70% ethanol, 3 min; 90% ethanol, 5 min; and 100% ethanol, 10 min five times. For the last wash, use a freshly opened bottle of 100% ethanol.

10. Resuspend the final pellet in 70:30 ethanol:Spurr low-viscosity embedding resin (Polysciences). Work in a ventilated hood and wear gloves when using unpolymerized resin. Incubate several hours with gentle rotation. Centrifuge and resuspend in 50:50 ethanol:Spurr resin. Incubate overnight at room temperature with gentle rotation.

11. Centrifuge and resuspend in 100% Spurr resin. Change the 100% Spurr resin three to four times at 1-hr intervals. Centrifuge and resuspend in 200 to 500 μl of 100% Spurr resin and transfer to prebaked (overnight at 60°) Beem-embedding molds (Polysciences). Follow steps 10 through 12 from the permanganate/uranyl acetate method. In this protocol, nuclear pores and the spindle pole body appear as darkly stained structures embedded in much lighter membranes. The spindle pole body is distinguished by radiating microtubules on the nuclear face as well as additional morphological features that may be observed in favorable sections.[32,33]

Acknowledgments

We thank all members of the Rose Laboratory, past and present, for their participation in perfecting these procedures. We thank Naz Erdeniz for helpful comments on this manuscript.

[30] Analysis of Prion Factors in Yeast

By YURY O. CHERNOFF, SUSAN M. UPTAIN, and SUSAN L. LINDQUIST

Introduction

Prions are unique proteins that can adopt two or more distinctly different conformational states *in vivo*.[1] In yeast and other fungi, prions act as heritable protein-based genetic elements because they can stimulate conversion of the normal form of the protein to the prion form. This conformational switch is often accompanied by a change of phenotype that is propagated from generation to generation as the prion protein is transferred from mother to daughter cell, continuing the cycle of conformational conversion. For example, the yeast prion [PSI^+] results from a self-perpetuating conformational change in Sup35, a component of the translational termination factor, that partially inactivates Sup35 and leads to reduced translational termination fidelity.[2-4] Another yeast prion, [$URE3$], a partially inactive isoform of Ure2 protein, enables yeast growing on rich nitrogen sources to import and utilize poor nitrogen sources, thereby eliminating nitrogen catabolite repression.[2] Thus, prions can cause biologically important phenotypic changes without any underlying change in the nucleic acid sequence.[4]

Although they affect distinct biological processes, yeast prions and their underlying determinants share many unusual genetic, cell biological, and biochemical properties. These distinguishing characteristics form the basis of diagnostic tests to identify, study, manipulate, and utilize prions. Many of these tests detect changes in a phenotype. The [PSI^+] prion causes nonsense suppression, which can be monitored easily by growth when the strain has an auxotrophic marker interrupted by a [PSI^+]-suppressible stop codon. The prion [$URE3$] enables *ura2* cells, which are normally unable to grow on media lacking uracil, to import ureidosuccinic acid (USA) and make uracil.[2]

A hallmark of yeast prions is that they show non-Mendelian patterns of inheritance during meiosis.[5-7] This and other unusual genetic properties of prions initially baffled yeast geneticists but are now indispensable criteria to identify and study them. Because all known yeast prions are propagated cytoplasmically,[2] the

---

[1] S. B. Prusiner, *Proc. Natl. Acad. Sci. U.S.A.* **95**, 13363 (1998).
[2] R. B. Wickner and Y. O. Chernoff, in "Prion Biology and Diseases" (S. B. Prusiner, ed.), p. 229. Cold Spring Harbor Laboratory Press, Cold Spring Harbor, NY, 1999.
[3] T. R. Serio and S. L. Lindquist, *Annu. Rev. Cell Dev. Biol.* **15**, 661 (1999).
[4] Y. O. Chernoff, *Mutat. Res.* **488**, 39 (2001).
[5] B. Cox, *Heredity* **20**, 505 (1965).
[6] F. Lacroute, *J. Bacteriol.* **206**, 519 (1971).
[7] R. B. Wickner, *J. Science* **264**, 566 (1994).

transmission of an altered phenotype from one cell to another by cytoduction is an important tool to characterize and manipulate yeast prions. Other genetic determinants are transmitted cytoplasmically, including mitochondrial DNA and viruses; thus, transmission by cytoduction is consistent with, but does not by itself prove, the presence of a prion. Another unusual property of yeast prions is that transient overproduction of the protein determinant of a prion dramatically increases the frequency at which the prion is formed.[7-9] Yeast can be cured of their prions by growing the cells for several generations in the presence of low concentrations of guanidinium hydrochloride (GuHCl)[10] or by transiently changing the expression level of molecular chaperones such as Hsp104.[11] That is, once the conformational state is established, it is self-perpetuating, and once it is lost, it has a low probability of reforming.

Many assays exploit the physical differences between normal and prion conformational states. Sup35 in the prion state is insoluble and easily sedimentable, whereas the normal form is soluble.[12,13] The prion forms of Sup35 and Ure2 are more protease resistant than the normal forms.[13,14] Fusions of the prion determinants to green fluorescent protein (GFP) enable detection of the prion in living yeast cells.[12,15] The insoluble prion form appears as bright, tight green fluorescent foci, whereas the soluble form fluoresces diffusely green. There are also assays that model the prion conversion process *in vitro*. Soluble prion determinants can form amyloid fibers *in vitro*.[16-19] Fiber formation can be stimulated greatly by adding preformed amyloid, mimicking the ability of previously converted prion determinants from a mother cell to perpetuate the conversion process when it is passed to the daughter cell. Sup35 prion aggregates are capable of self-seeded propagation in cell-free extracts,[16,20] and the efficiency of conformational conversion can be monitored quantitatively *in vitro*.

[8] Y. O. Chernoff, I. L. Derkach, and S. G. Inge-Vechtomov, *Curr. Genet.* **24,** 268 (1993).
[9] I. L. Derkatch, Y. O. Chernoff, V. V. Kushnirov, S. G. Inge-Vechtomov, and S. W. Liebman, *Genetics* **144,** 1375 (1996).
[10] M. F. Tuite, C. R. Mundy, and B. S. Cox, *Genetics* **98,** 691 (1981).
[11] Y. O. Chernoff, S. L. Lindquist, B. Ono, S. G. Inge-Vechtomov, and S. W. Liebman, *Science* **268,** 880 (1995).
[12] M. M. Patino, J. J. Liu, J. R. Glover, and S. Lindquist, *Science* **273,** 622 (1996).
[13] S. V. Paushkin, V. V. Kushnirov, V. N. Smirnov, and M. D. Ter-Avanesyan, *EMBO J.* **15,** 3127 (1996).
[14] D. C. Masison and R. B. Wickner, *Science* **270,** 93 (1995).
[15] H. K. Edskes, V. T. Gray, and R. B. Wickner, *Proc. Natl. Acad. Sci. U.S.A.* **96,** 1498 (1999).
[16] J. R. Glover, A. S. Kowal, E. C. Schirmer, M. M. Patino, J. J. Liu, and S. Lindquist, *Cell* **89,** 811 (1997).
[17] K. L. Taylor, N. Cheng, R. W. Williams, A. C. Steven, and R. B. Wickner, *Science* **283,** 1339 (1999).
[18] C. Thual, A. A. Komar, L. Bousset, E. Fernandez-Bellot, C. Cullin, and R. Melki, *J. Biol. Chem.* **274,** 13666 (1999).
[19] M. Schlumpberger, H. Wille, M. A. Baldwin, D. A. Butler, I. Herskowitz, and S. B. Prusiner, *Protein Sci.* **9,** 440 (2000).
[20] S. V. Paushkin, V. V. Kushnirov, V. N. Smirnov, and M. D. Ter-Avanesyan, *Science* **277,** 381 (1997).

The purpose of this chapter is to review the principal techniques used for genetic, cell biological, and biochemical characterization of yeast prions. Our major focus will be on [PSI⁺]; however, [URE3] and composite prions will be discussed for comparison. Although many of the protocols described here are in their details specific to a single yeast prion, it is hoped that by analogy, they will help guide the design of new methods to characterize as yet undiscovered prions. We begin with methods to detect yeast prions based on changes in phenotype.

Protocols

Phenotypic Detection of Yeast Prions

Because the [PSI⁺] state partially inactivates the translational termination factor, Sup35, the most convenient phenotypic assay for [PSI⁺] is translational readthrough, or nonsense suppression (Figs 1A and 1B). [PSI⁺] is a relatively weak translational suppressor. Usually, products of [PSI⁺]-mediated translational readthrough accumulate to only a few percent of wild-type protein levels.[21,22] Thus, it is not surprising that most nonsense alleles in auxotrophic markers are not sufficiently suppressed by [PSI⁺] to support easily detectable growth on selective medium. There are some nonsense alleles, such as *ade1-14* (UGA),[11,23] *lys2-187* (UGA),[24,25] *met8-1* (UAG),[26] and *trp5-48* (UAA),[26] that can be suppressed by [PSI⁺] to such an extent that growth on the corresponding selective media is seen in reasonable periods of time. Among those, probably the most useful one is the *ade1-14* allele,[23] where the UGA stop codon has been substituted for the UGG (Trp) codon at position 244 of the *ADE1* gene[27] (L. Osherovich and J. Weissman, personal communication, 2000). In [psi⁻] cells, the absence of functional Ade1 protein prevents growth on –Ade medium (Fig. 1C) and causes the accumulation of a red pigment on rich medium (Fig. 1D). This provides a convenient color assay for the prion: [PSI⁺] strains grow on –Ade medium and form pink or white colonies on rich medium (YPD). Because the red pigment accumulates more readily in nondividing cells, color differences are more apparent in older colonies (grown 3 to 4 days) and can be exacerbated by incubating plates 1 to 2 days in a refrigerator. In some strains, [PSI⁺] colonies can be differentiated from [psi⁻] colonies by color on synthetic medium; however, color development is less reproducible on synthetic medium than on YPD.

---

[21] S. W. Liebman and F. Sherman, *J. Bacteriol.* **139**, 1068 (1979).
[22] G. P. Newnam, R. D. Wegrzyn, S. L. Lindquist, and Y. O. Chernoff, *Mol. Cell. Biol.* **19**, 1325 (1999).
[23] S. G. Inge-Vechtomov, O. N. Tikhodeev, and T. S. Karpova, *Genetika* **24**, 1159 (1988).
[24] B. B. Chattoo, E. Palmer, B.-I. Ono, and F. Sherman, *Genetics* **93**, 67 (1979).
[25] Y. O. Chernoff, S. G. Inge-Vechtomov, I. L. Derkach, M. V. Ptyushkina, O. V. Tarunina, A. R. Dagkesamanskaya, and M. D. Ter-Avanesyan, *Yeast* **8**, 489 (1992).
[26] B.-I. Ono, Y. Ishino-Arao, M. Tanaka, U. Awano, and S. Shinoda, *Genetics* **114**, 363 (1986).
[27] T. Nakayashiki, K. Ebihara, H. Bannai, and Y. Nakamura, *Mol. Cell* **7**, 1121 (2001).

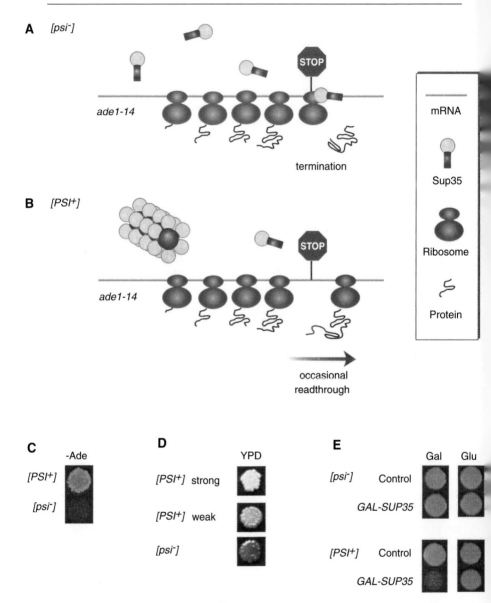

FIG. 1. Diagram of the effect of [PSI^+] on Sup35 and translation termination and examples of phenotypic assays for [PSI^+]. (A) In [psi^-] cells, a complex of Sup35 (see legend at right) and Sup45 (not shown) binds ribosomes at stop codons to mediate translational termination. Often [PSI^+] and [psi^-] are studied in strains carrying the ade1-14 allele, which has a nonsense mutation within the open reading frame, because it provides a convenient phenotypic assay to distinguish the two cell types. In [psi^-] ade1-14 strains, a truncated, inactive Ade1 protein is produced, preventing yeast from

Because the translational readthrough caused by [PSI^+] can be rather weak, it is easier in some cases to detect [PSI^+] through allosuppression, an augmentation of the efficiency of suppression caused by a nonsense suppressor mutation when [PSI^+] is present. A commonly used assay for [PSI^+] employs the serine tRNA suppressor $SUQ5$ ($SUP16$) and the $ade2$-1 (UAA) allele. $SUQ5$ suppresses $ade2$-1 only in the presence of [PSI^+].[5,28] As in $ade1$-14 strains, [PSI^+] and [psi^-] cells can be differentiated by color on YPD medium in strains with both the $ade2$-1 allele and $SUQ5$.[5,28] Note that $SUQ5$ can efficiently suppress other UAA alleles, even in a [psi^-] cell; i.e., allosuppression may cause significant, undesirable, and undetectable secondary effects and unusual genetic segregation patterns.

To detect [PSI^+] by growth, strains with a [PSI^+]-suppressible nonsense marker (e.g., $ade1$-14) should be either velveteen replica plated or suspended in water and then spotted by pipette onto appropriate selective medium (in this case lacking adenine). We do not recommend simply patching or streaking cells on selective medium: all [PSI^+]-suppressible markers exhibit high frequencies of spontaneous reversion due to suppressor mutations, which might complicate comparisons if the number of cells plated is not normalized. Depending on the nonsense-suppressible allele, genotypic background, and isolate of [PSI^+], it might take from 1–2 days to 2–3 weeks of incubation to detect growth. Including a control [psi^-] strain of the same genotype on the same plate for comparison is essential. A [psi^-] isolate can be obtained by guanidinium HCl (GuHCl) treatment or by overproduction of Hsp104, as described later. Background growth of the whole spot, which is indicative of [PSI^+], should be carefully distinguished from strong growth of rare secondary colonies, which appear in [PSI^+] and [psi^-] isolates due to reversion of the stop codon in the auxotrophic marker. [PSI^+]-mediated suppression is usually higher at 20–25° than 30° and less at 37°.

[28] B. S. Cox, M. F. Tuite, and C. S. McLaughlin, Yeast **4**, 159 (1988).

---

growing on synthetic media lacking adenine (-Ade) and causing the cells to accumulate a red-pigmented by-product on rich media (YPD). Sup35 is composed of two regions; a prion-forming domain (red rectangle) and a domain that is required for translational termination (yellow sphere). (B) In [PSI^+] cells, prion-forming domains of the majority of Sup35 adopt the prion conformation and self-assemble into an aggregated, possibly amyloidic structure (depicted as red cylinder). This conformational change impairs the ability of Sup35 to participate in translational termination; consequently, stop codons are read through occasionally. In $ade1$-14 strains, translational readthough produces enough full-length Ade1 protein that cells can synthesize sufficient adenine to grow on –Ade media and to prevent accumulation of the red-pigmented by-product. (C) The [PSI^+] $ade1$-14 (UGA) strain grows on -Ade medium, whereas an isogenic [psi^-] strain does not. (D) Suppression of $ade1$-14 by [PSI^+] decreases the accumulation of a red pigment on YPD medium. Three isogenic derivatives of the strain 74-D694 are shown. This approach can be used to differentiate the strong [PSI^+] isolates, which are whiter on YPD, from the weak [PSI^+] isolates, which are pink. (E) [PSI^+]-dependent inhibition of growth by overproduced Sup35 protein. Overproduction of the GAL–$SUP35$ construct is induced on galactose (Gal) but not on glucose (Glu) medium.

For quantitative assessment of [PSI^+]-mediated suppression, one can use a series of test constructs that contain a portion of the yeast 3-phosphoglycerate kinase gene (*PGK*) fused in frame to the *Escherichia coli lacZ* gene[29,30] either without or with an intervening UAA, UAG, or UGA stop codon. When using constructs with stop codons, active β-galactosidase is synthesized only when there is a termination defect (nonsense suppression). These constructs exist in both multicopy[29] and single-copy versions[30]; however, we recommend using single-copy constructs for more precise measurements, as multicopy constructs without a stop codon can inhibit cell growth.

Method for Quantitative Analysis of [PSI^+] Suppression

1. Transform each yeast strain individually with the control and the three nonsense codon-containing *PGK–lacZ* constructs.

2. Grow individual fresh transformants in liquid synthetic medium selective for the plasmid to the middle of exponential phase. A 5-ml culture is sufficient if chemiluminescent detection is used, but 50-ml cultures should be used if colorimetric detection is used (see step 5).

3. Sediment the cells and resuspend them in 200–400 μl of breaking buffer [0.1 M Tris–HCl, pH 8.0, 1 mM dithiothreitol (DTT), 20%(v/v) glycerol, or per kit instructions] and lyse cells by vortexing with 1 volume glass beads (keep culture on ice when not vortexing). Cell extracts can be stored at −70° until use, but they should not be frozen and thawed more than once.

4. Sediment the unbroken cells, cellular debris, and glass beads at 4°, 2000g, 3 to 5 min.

5. Determine β-galactosidase activity in each supernatant fraction using either a colorimetric assay with *o*-nitrophenyl-β-D-galactopyranoside (ONPG, Sigma, St. Louis, Mo)[31] or a chemiluminescent assay (Applied Biosystems, Foster City, CA).[32] When working with single-copy constructs, the more sensitive chemiluminescent assay is preferred. It is imperative that chemiluminscent assays be performed within the linear range of the luminometer. Usually, 1–2 μl (control construct) or 10–20 μl (nonsense codon-containing construct) of the supernatant fraction is sufficient for a chemiluminescent reaction. Reactions should be performed in triplicate. To determine the actual amount of enzyme present in the samples, use a standard curve prepared with commercially available β-galactosidase (Sigma). Normalize data using protein concentrations determined by the Bradford or Lowry assays.

---

[29] M. Firoozan, C. M. Grant, J. A. Duarte, and M. F. Tuite, *Yeast* **7,** 173 (1991).
[30] I. Stansfield, Akhmaloka, and M. F. Tuite, *Curr. Genet.* **27,** 417 (1995).
[31] D. Burke, D. Dawson, and T. Stearns, "Methods in Yeast Genetics: A Cold Spring Harbor Laboratory Course Manual." Cold Spring Harbor Laboratory Press, Cold Spring Harbor, NY, 2000.
[32] V. K. Jain and I. T. Magrath, *Anal. Biochem.* **199,** 119 (1991).

6. Calculate the ratio between the normalized activities of β-galactosidase in the two types of lysates. This gives an efficiency of nonsense suppression relative to wild-type levels of the Pgk–lacZ protein.

For UGA-containing constructs, background suppression in [psi^-] cells is usually not higher than 0.1 to 0.2%, whereas [PSI^+] causes a 5- to 50-fold increase depending on both genotypic background and [PSI^+] isolate.[33] Because the efficiency of nonsense suppression is context dependent,[34] the range of nonsense suppression may vary with other such test constructs.

Detection of [PSI$^+$] by Incompatibility with Overproduced Sup35. Another phenotypic assay used to identify [PSI^+] is differential growth of [PSI^+] and [psi^-] cells transformed with Sup35-overexpressing plasmids. Overproduced Sup35 inhibits the growth of [PSI^+] cells, but not [psi^-] cells,[25] via a poorly characterized mechanism. One possibility is that increased aggregation caused by Sup35 overexpression depletes the level of functional Sup35, causing death. An advantage of this assay is that nonsense-suppressible markers are not required. To avoid growth inhibition of initial transformants, use a construct that expresses *SUP35* controlled by a regulated promoter such as *GAL*. Transformants, bearing the *GAL–SUP35* plasmid or matching vector plasmid, should be selected on glucose (Glu) medium where the promoter is turned off. Resulting colonies should be resuspended in water, and equal amounts of cells should be spotted on both Glu and galactose (Gal) synthetic medium selective for the plasmid. Growth inhibition on Gal versus Glu medium by the *GAL–SUP35* construct is indicative of the presence of [PSI^+] (Fig. 1E). The extent of growth inhibition can vary with both the genetic background and the [PSI^+] isolate. Moreover, high levels of Sup35 overproduction will become toxic for [psi^-] strains once they convert to [PSI^+] in response to Sup35 overproduction (see later). Therefore, if growth inhibition is observed, check whether it can be eliminated or partly relieved by the [PSI^+] curing agents, such as an excess of Hsp104 (see later).

Phenotypic Analysis of [URE3]. [*URE3*] is the prion form of the Ure2 protein, whereas [*ure-o*] is the nonprion form. Normally, Ure2 negatively regulates the positive transcriptional regulator Gln3 at the posttranslational level (Figs. 2A and 2B). A mutation in *ure2* or the conversion of Ure2 to the [*URE3*] prion form results in Gln3 activation, which leads to transcription of genes that are normally repressed in the presence of rich nitrogen sources. Among these is *DAL5*, which encodes the transporter of the poor nitrogen source, allantoate. Expression of Dal5 enables *ure2* and [*URE3*] yeast-cells to import ureidosuccinic acid (USA), an intermediate metabolite of uracil biosynthesis that is structurally similar to allantoate.[2] [*URE3*] can be detected in yeast cells deficient in aspartate transcarbamylase, an enzyme

---

[33] Y. O. Chernoff, G. P. Newnam, J. Kumar, K. Allen, and A. D. Zink, *Mol. Cell. Biol.* **19**, 8103 (1999).
[34] B. Bonetti, L. Fu, J. Moon, and D. M. Bedwell, *J. Mol. Biol.* **251**, 334 (1995).

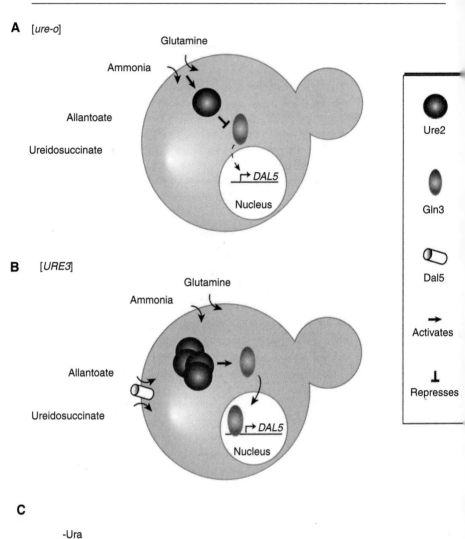

FIG. 2. Diagram of the effect of [URE3] on Ure2 and uptake of ureidosuccinate and a phenotypic assay for [URE3]. (A) In [ure-o] cells, the uptake of poor nitrogen sources, such as ureidosuccinate and allantoate, is repressed in the presence of good nitrogen sources, such as glutamine and ammonia. The presence of good nitrogen sources is relayed through Ure2, which blocks the action of the transcription factor, Gln3. Without transcriptional activation, the allantoate transporter, Dal5, is not

catalyzing USA biosynthesis in the cell and coded by the gene *URA2*, as [*URE3*] *ura2* cells can grow on the medium lacking uracil but containing 100 µg/ml of sodium ureidosuccinate (-Ura + USA) (Fig. 2C). [*URE3*] can also be detected in *URA2* strains. [*URE3*] and *ure2* cells excrete uracil in the presence of excess USA; thus, Ura⁻ cells surrounding [*URE3*] cells can grow on media lacking uracil. To test for [*URE3*], the strains under question are patched on a freshly prepared lawn (10^6–10^7 cells per plate) of *ura2/ura2* α/**a** diploid (or a haploid heterothallic *ura2* strain of the same mating type) on -Ura + USA medium. A halo will appear surrounding the [*URE3*] patches after 2 days.

Phenotypic Analysis of Composite Prions. The ability of Sup35 and Ure2 to adopt a prion conformation is conferred by a region of unusual amino acid composition at their N termini. These prion-forming domains (PFDs), containing an unusually high number of glutamine(Q) and/or asparagine(N) residues, are essential for prion propagation.[14,35] Overproduction of these N-terminal regions can induce prion formation more efficiently than overproduction of the entire protein.[9,14] These regions of Sup35 and Ure2 are modular and transferable, as fusion of their PFDs to other proteins, such as GFP or the glucocorticoid receptor, can confer prion-like properties to the fusion protein.[12,15,36]

In many cases, potential prion proteins may have no detectable phenotype, making their characterization difficult. One solution is to construct a composite prion by replacing the PFD of one prion protein, such as Sup35, with the putative PFD of a prion candidate. Then the effect of the putative prion domain on the chimeric protein can be analyzed using established methods. In this way, composite prions containing the PFDs of Sup35 homologues of evolutionarily distant yeasts[37–39] or the putative PFDs of the yeast proteins Rnq1[40] and New1[38] were studied.

Perhaps the quickest approach involves ectopic expression of the putative composite prion. This method works particularly well with fusions to the C terminus of

---

[35] M. D. Ter-Avanesyan, A. R. Dagkesamanskaya, V. V. Kushnirov, and V. N. Smirnov, *Genetics* **137**, 671 (1994).
[36] L. Li and S. Lindquist, *Science* **287**, 661 (2000).
[37] Y. O. Chernoff, A. P. Galkin, E. Lewitin, T. A. Chernova, G. P. Newnam, and S. M. Belenkiy, *Mol. Microbiol.* **35**, 865 (2000).
[38] A. Santoso, P. Chien, L. Z. Osherovich, and J. S. Weissman, *Cell* **100**, 277 (2000).
[39] V. V. Kushnirov, N. V. Kochneva-Pervukhova, M. B. Chechenova, N. S. Frolova, and M. D. Ter-Avanesyan, *EMBO J.* **19**, 324 (2000).
[40] N. Sondheimer and S. Lindquist, *Mol. Cell.* **5**, 163 (2000).

---

produced. (B) In [*URE3*] cells, conversion of Ure2 into the prion conformation interferes with the repression of Gln3. Thus, even in the presence of preferable nitrogen sources, Gln3 activates transcription of the allantoate importer, Dal5. Because ureidosuccinate is structurally similar to allantoate, it enters the yeast cell via the Dal5 importer. (C) Detection of [*URE3*] by growth on -Ura medium containing ureidosuccinic acid (USA).

Sup35, the region of the protein that is involved in translational termination.[41–43] Usually, the expression of *SUP35* derivatives lacking the PFD restores efficient translation termination in [*PSI*⁺] cells, a phenomenon known as an antisuppression. Thus, a [*PSI*⁺] *ade1-14* yeast strain that is normally white and Ade⁺ would become red and Ade⁻ in the presence of such a plasmid. Transformation of a plasmid bearing a chimeric gene, in which *SUP35PFD* is substituted for another protein domain X that does not behave like a prion, would mask the [*PSI*⁺] phenotype in the same way unless this substitution impairs normal Sup35 function. If domain X adopts a prion conformation, antisuppressor activity will be lost and the colonies will be white and Ade⁺.

An alternative tactic is to fuse a PFD to a reporter protein. A good choice for a reporter would be a protein that must be targeted to a particular location to perform its function, such as a transcription factor. For example, the rat glucocorticoid receptor (GR) behaves like a prion once it is fused to the Sup35PFD and introduced into a [*PSI*⁺] cell.[36] In a specifically engineered yeast strain, GR induces the expression of *lacZ* that is controlled by a GR-regulated promoter. Normally, this results in blue yeast colonies on medium containing X-Gal. When Sup35PFD-GR converts to the prion state, GR is inactivated and the colonies appear white on X-Gal medium. However, not every fusion will form a prion with a detectable phenotype. Enzymes that catalyze reactions utilizing easily diffusible substrates may remain active in the aggregated form, as is the case when the PFD of Ure2 is fused to β-galactosidase.[14]

Genetic Analysis of Prions

Non-Mendelian Inheritance of Yeast Prions: Mating, Tetrad Analysis, and Cytoduction. All known yeast prions are dominant.[5,6] This makes it possible to uncover the presence of a prion in a yeast strain lacking an appropriate phenotypic assay system by mating the prion candidate to another strain that does. Consider, for example, testing for [*PSI*⁺] in a strain lacking a [*PSI*⁺]-suppressible auxotrophic marker but containing a stable mutation in a gene such as *ADE1*, *ADE2*, or *LYS2*. In such strains, [*PSI*⁺] can be detected by mating [*PSI*⁺] cells to a [*psi*⁻] strain bearing a [*PSI*⁺]-suppressible allele in the corresponding gene. Because [*PSI*⁺] is dominant, it will suppress the nonsense allele in the heteroallelic diploid. An isogenic [*psi*⁻] diploid is needed as a control to distinguish between [*PSI*⁺]-mediated suppression and papillation due to spontaneous heteroallelic recombination.

[41] L. Frolova, X. Le Goff, H. H. Rasmussen, S. Cheperegin, G. Drugeon, M. Kress, I. Arman, A. L. Haenni, J. E. Celis, M. Philippe, J. Justesen, and L. Kisselev, *Nature* **372**, 701 (1994).

[42] G. Zhouravleva, L. Frolova, X. Le Goff, R. Le Guellec, S. Inge-Vechtomov, L. Kisselev, and M. Philippe, *EMBO J.* **14**, 4065 (1995).

[43] I. Stansfield, K. M. Jones, V. V. Kushnirov, A. R. Dagkesamanskaya, A. I. Poznyakovski, S. V. Paushkin, C. R. Nierras, B. S. Cox, M. D. Ter-Avanesyan, and M. F. Tuite, *EMBO J.* **14**, 4365 (1995).

All known yeast prions are non-Mendelian genetic elements.[5–7] In most cases, when a $[PSI^+]$ strain is mated to a $[psi^-]$ strain, $[PSI^+]$ segregates 4 : 0 ($[PSI^+]$: $[psi\text{-}]$), in strong contrast to the 2 : 2 Mendelian segregation of nuclear encoded traits. Some yeast prions show other types of unusual segregation ratios. For example, some proportion of $[URE3]$ derivatives lose the prion during meiosis, leading to all types of ratios from 4 $[URE3]$: 0 [$ure\text{-}o$] to even 0 $[URE3]$: 4 $[ure\text{-}o]$.[6] Furthermore, phenotypes used to score for the prion, such as nonsense suppression in case of $[PSI^+]$ or growth on -Ura+USA medium in the case of $[URE3]$ $ura2$ cells, can be affected by the genetic variation among the meiotic progeny. Thus, whenever possible, isogenic strains lacking the prion should be used in matings.

Meiotic progeny of crosses between prion-containing and normal cells can exhibit a normal Mendelian ratio of 2 : 2 (prion : nonprion) if one parent contains a mutation that prevents prion propagation. Such was the case of a strain where the PFD of Sup35 was deleted. That strain behaves as a recessive "$[PSI]$ no more" (pnm) strain, as the prion cannot be maintained.[35] Some point mutations in the PFD of Sup35 can behave as either dominant PNM or recessive pnm.[44,45] The severity of the $[PSI^+]$ propagation defect caused by these point mutations, however, frequently depends on both $[PSI^+]$ isolate and genetic background.[46]

Cytoduction is a useful test for the non-Mendelian inheritance of yeast prions. At a low frequency during mating, nuclei fail to fuse, giving rise to haploid cytoductants that contain the cytoplasmic contents of both parents, but the nuclear content of only one parent. Mutants, such as $kar1$, have been identified that increase the frequency of cytoduction greatly,[47] making this a feasible technique to transfer yeast prions. The strain to be tested for a prion is called the donor strain, whereas the strain without the prion is called the recipient. In addition to using a $kar1$ strain as either one of the parents, nuclear and cytoplasmic markers are needed. The recipient strain can be converted into a mitochondrially defective rho^- or rho^0 derivative by treatment with medium containing ethidium bromide.[48] Cytoplasmic mixing is detected by the transfer of mitochondria from the rho^+ donor to the rho^- recipient. The recipient strain is also frequently marked by a recessive nuclear selectable marker, such as canavanine resistance ($can1$) or cycloheximide resistance (cyh^R). (Such a counterselectable antibiotic resistance marker is further designated as Ant^R). This allows for counterselection of rarely occurring diploid cells.

1. Mix on a YPD plate (or appropriate selective media) a Rho^+ Ant^S donor strain bearing the suspected prion and a Rho^- Ant^R recipient strain of the opposite

---

[44] S. M. Doel, S. J. McCready, C. R. Nierras, and B. S. Cox, *Genetics* **137,** 659 (1994).
[45] A. H. DePace, A. Santoso, P. Hillner, and J. S. Weissman, *Cell* **93,** 1241 (1998).
[46] I. L. Derkatch, M. E. Bradley, P. Zhou, and S. W. Liebman, *Curr. Genet.* **35,** 59 (1999).
[47] J. Conde and G. R. Fink, *Proc. Natl. Acad. Sci. U.S.A.* **73,** 3651 (1976).
[48] T. D. Fox, L. S. Folley, J. J. Mulero, T. W. McMullin, P. E. Thorsness, L. O. Hedin, and M. C. Costanzo, *Methods Enzymol.* **194,** 149 (1991).

mating type, not containing a prion. Incubate them for about 1 day. (One of the strains should contain a *kar1* mutation.)

2. Velveteen replica plate the cells onto two types of media: (A) synthetic (S) medium containing all the compounds needed for the recipient strain, as well as ethanol and glycerol (EG) instead of glucose, and the antibiotic (Ant) to counterselect against rare diploids and donor growth (SEG+Ant) and (B) SEG+Ant medium to test for a prion. For [*PSI*$^+$] *ade1-14* strains, use SEG−Ade+Ant media or in the case of [*URE3*], use SEG−Ura+USA+Ant media.

Growth on the first medium indicates successful cytoduction, whereas growth on the second medium confirms prion transfer. Because EG or Ant media can interfere with prion testing, the cytoductants selected in the first medium can also be subsequently checked on an appropriate medium. It should be noted that nuclear chromosomes can be transmitted by cytoduction, albeit with low efficiency[49]; therefore, low cytoduction efficiency does not necessarily prove cytoplasmic inheritance. Putative cytoductants should be checked for the presence of all markers. Prions are usually transferred with high efficiency so essentially all cytoductants should contain a prion.

Prion Curing by Chemical Agents: Guanidinium Hydrochloride and Latrunculin A. Yeast prions can be cured by a variety of environmental agents. These include conventional mutagens, such as EMS and UV light, as well as agents not affecting the frequency of gene mutations, such as dimethyl sulfoxide (DMSO), methanol, and high concentrations of KCl and glycerol.[28] It is likely that the mechanism by which EMS and UV light cure yeast prions is by increasing the expression of the molecular chaperone, Hsp104 (see later),[11,50] rather than by mutation of the prion determinant. Here, we consider two nonmutagenic prion-curing agents, the protein denaturant, guanidinium-hydrochloride (GuHCl), and the anticytoskeletal agent, latrunculin A (Lat-A), which differ from each other by phenomenology and mechanism of action.

PRION CURING BY GuHCl. GuHCl is probably the most frequently used prion-curing agent.[10] It cures all the forms of [*PSI*$^+$],[4,9,28] as well as [*URE3*][7] and newly identified candidate prion [*PIN*$^+$].[51] Whether a phenotype is cured by GuHCl is increasingly being applied as an operational tool that suggests, but does not prove, a prion-based mechanism. At high concentrations, GuHCl causes protein denaturation; however, the millimolar concentrations used to cure prions are clearly too low to denature prion proteins. Instead, GuHCl blocks [*PSI*$^+$] proliferation, apparently

---

[49] S. K. Dutcher, *Mol. Cell. Biol.* **1,** 245 (1981).
[50] S. Lindquist, M. M. Patino, Y. O. Chernoff, A. S. Kowal, M. A. Singer, S. W. Liebman, K. H. Lee, and T. Blake, *Cold Spring Harb. Symp. Quant. Biol.* **60,** 451 (1995).
[51] I. L. Derkatch, M. E. Bradley, P. Zhou, Y. O. Chernoff, and S. W. Liebman, *Genetics* **147,** 507 (1997).

without affecting preexisting prion aggregates.[52] Thus, GuHCl cures [PSI^+] only in proliferating cells due to the dilution of prion particles upon cellular division. There is a lag phase of several generations before cells are cured, apparently due to the large number of preexisting prion particles in the yeast cell.[52]

Generally, GuHCl-mediated curing of [PSI^+], [PIN^+], or [$URE3$] is performed by patching a small amount of a culture on solid YPD medium containing 5 mM GuHCl (YPD+Gu). Although the effectiveness of GuHCl varies among yeast strains and type of prion, 5 mM GuHCl cures most strains. Higher concentrations can be used, but with caution because GuHCl can affect yeast growth adversely. Incubate yeast cells until the whole patch grows (usually 1–2 days) and then velveteen replica plate them onto the fresh YPD+Gu plate. Alternatively, take a small amount of cells from the patch and make a new patch on the same or new YPD+Gu plate. Let the patch grow again and repeat this procedure once more. Three consecutive passages on YPD (20 or more cell generations) are usually sufficient to cure almost all yeast cells. Still, a small fraction of prion-containing cells can remain; therefore, cells from the third passage should be restreaked out on YPD to get single colonies, and these colonies should be checked for the presence of the prion using an appropriate assay.

To measure prion curing by GuHCl quantitatively, yeast cultures should be inoculated to between 10^5 and 10^6 cells/ml and grown in liquid YPD+Gu medium (see Fig. 3A). Monitor growth of the culture by checking optical density (OD) or counting cells after certain periods of time. Take aliquots, dilute appropriately, and spread between 100 and 1000 cells onto media appropriate for the prion test method being used. Determine the number of prion- and nonprion-containing cells per generation. The number of generations (G) for the time period t can be calculated according to the following formula: $G = \log_2(C_t/C_0)$, where C_t is the concentration of viable cells at time t and C_0 is the concentration of viable cells at the starting point. In the case of [PSI^+], [psi^-] cells are not usually observed until the fifth or sixth cell division in the presence of GuHCl. Mosaic colonies containing both [PSI^+] and [psi^-] cells are observed frequently. For example, such mosaics appear as white colonies with red sectors in $ade1$-14 strains. These mosaics probably result from cells maintaining only one or very few prion seeds and losing them in the first few divisions after plating. Once a culture reaches the concentration of 10^8 cells/ml or higher, dilute it down to 10^5–10^6 cells/ml in fresh YPD+Gu medium and continue monitoring and testing for the prion. GuHCl also cures prions in synthetic medium; however, cultures grown in synthetic medium are usually more resistant to GuHCl compared to those grown in YPD medium (Y. Chernoff, G. Newnam, and S. Belenkiy, unpublished data, 1999).

[52] S. S. Eaglestone, L. W. Ruddock, B. S. Cox, and M. F. Tuite, *Proc. Natl. Acad. Sci. U.S.A.* **97**, 240 (2000).

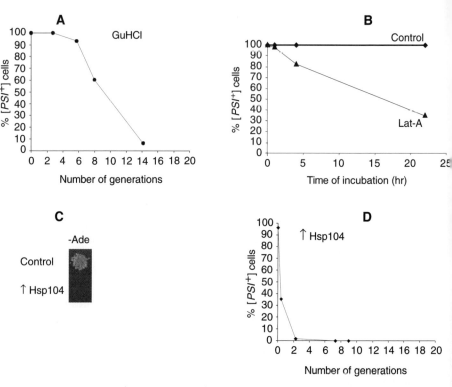

FIG. 3. Curing of [PSI^+]. (A) Curing [PSI^+] strains by GuHCl treatment requires cellular division. A weak [PSI^+] isolate of 74-D694 was incubated in YPD medium containing 5 mM GuHCl. The culture was kept in a proliferating state by diluting it periodically with fresh YPD medium. After various periods of time, cells were washed and plated onto YPD medium. Numbers of generations were calculated based on the concentrations of colony-forming units at various times. Percentages of [PSI^+] colonies are shown. The long lag period indicates that GuHCl blocks [PSI^+] proliferation, but does not affect preexisting [PSI^+] units, which must be diluted by several cell divisions before they are lost. (B) [PSI^+] curing by latrunculin A does not require cellular growth. Two exponential phase cultures of the same [PSI^+] isolate were treated with or without 200 μM of the antiactin drug latrunculin A, which blocks growth. Aliquots were taken at various time points, and cells were washed and plated onto YPD medium. Percentages of [PSI^+] colonies are shown. (C) [PSI^+] curing by constitutively expressed Hsp104. The yeast [PSI^+] $ade1$-14 strain was transformed with the plasmid bearing $HSP104$ under the strong constitutive GPD promoter (↑Hsp104) or matching control plasmid. Transformants were grown on media selective for the plasmid and velveteen replica plated onto -Ade medium nonselective for the plasmid. Complete absence of growth on -Ade in the ↑Hsp104 culture confirms that [PSI^+] was completely cured by constitutively overproduced Hsp104 and was not recovered even in cells that lost the plasmid. (D) [PSI^+] curing by inducing Hsp104 expression. The same [PSI^+] isolate of 74-D694 as in A and B was transformed with the GAL–$HSP104$ plasmid. Expression of the GAL–$HSP104$ construct was induced by shifting to the galactose/raffinose medium as described in the text. Aliquots were taken at various time points after induction and plated onto synthetic glucose medium where GAL–$HSP104$ expression is turned off. Numbers of generations were calculated based on the concentrations of colony-forming units at various times. Percentages of [PSI^+] colonies are shown. In contrast to [PSI^+] curing by GuHCl, [PSI^+] curing by excess Hsp104 does not exhibit any lag period.

[*PSI*⁺] CURING BY LATRUNCULIN A. Unlike GuHCl, curing of [*PSI*⁺] by latrunculin A (Lat-A) does not rely on cell division.[53] Lat-A is a marine toxin isolated from the spongi *Latrunculia magnifica*. It disrupts the actin cytoskeleton by sequestering monomeric actin.[54] Although the cytoskeleton is disrupted after several minutes of Lat-A treatment, it takes several hours to observe significant loss of [*PSI*⁺]. The molecular mechanism by which Lat-A cures [*PSI*⁺] is unknown. In contrast to the antiactin effect of Lat-A, [*PSI*⁺] curing by Lat-A requires continuous protein synthesis in the presence of Lat-A.[53] Yeast cells do not divide in the presence of Lat-A, suggesting that this drug cures yeast cells of [*PSI*⁺] by disrupting prion aggregates or by interfering with conformational conversion. Lat-A does not cure the candidate prion [*PIN*⁺].[53] Whether Lat-A affects other prions is unknown.

To cure yeast cells of [*PSI*⁺] by Lat-A, prepare a 10 mM stock solution of Lat-A in DMSO. Although DMSO exhibits a slight [*PSI*⁺] curing effect,[28] low concentrations of DMSO used in these experiments usually do not affect [*PSI*⁺] maintenance.[53] Add an equal volume of DMSO without Lat-A to another culture, which serves as a control. [*psi*⁻] cells appear at Lat-A concentrations as low as 40 μM; however, curing was most effective at higher concentrations (200–500 μM). Add the appropriate amount of the drug to an exponentially growing yeast culture in liquid YPD medium (OD$_{600}$ = 0.5–1.0). After desired periods of time, wash cells three times in liquid medium and plate onto solid medium appropriate to score [*PSI*⁺] and [*psi*⁻] colonies. The efficiency of curing depends on the [*PSI*⁺] isolate. About 10–15% of a weak [*PSI*⁺] isolate OT55 converts to [*psi*⁻] after 4 hr of incubation with 200 μM Lat-A and more than 60% convert to [*psi*⁻] after overnight incubation with 200 μM Lat-A (Fig. 3B[53]). Overnight incubation with 500 μM Lat-A cured weak isolates of [*PSI*⁺] completely (R. Wegrzyn and Y. Chernoff, unpublished data, 2000). Strong [*PSI*⁺] isolates are much more resistant to Lat-A action. [*PSI*⁺]/[*psi*⁻] mosaics usually constitute a relatively small fraction of colonies. An important caveat is that overnight incubation with Lat-A causes a significant cell death (more than 99% in the case of 200 μM Lat-A), which should be taken into consideration once dilutions are made.

Prion Curing by Chaperone Manipulations. The yeast chaperone Hsp104 has a unique relationship with [*PSI*⁺]: an intermediate level of Hsp104 is required for prion propagation.[11] Inactivation of Hsp104 cures all [*PSI*⁺] isolates efficiently,[11] whereas the efficacy of curing by Hsp104 overproduction depends on the level of Hsp104 expression, genetic background, and [*PSI*⁺] isolate.[9,13,22] Often, transient overproduction of Hsp104 is sufficient to cure [*PSI*⁺]. High overexpression of Hsp104 usually cures [*PSI*⁺] completely. Low levels of excess Hsp104 are inefficient in curing yeast cells of [*PSI*⁺], but inhibit [*PSI*⁺]-mediated nonsense

---

[53] P. A. Bailleul-Winslett, G. P. Newnam, R. D. Wegrzyn, and Y. O. Chernoff, *Gene. Expr.* **9**, 145 (2000).
[54] M. Coue, S. L. Brenner, I. Spector, and E. D. Korn, *FEBS Lett.* **213**, 316 (1987).

suppression,[11] possibly by partially solubilizing the prion form of Sup35[13] and/or releasing some other translational factors sequestered by the prion.[55] Inactivation of Hsp104 also cures yeast cells of the prion [PIN^+][51] and, at least in some genetic backgrounds, [$URE3$],[56] whereas overproduction of Hsp104 has no effect on these prions. Some other chaperones, such as the yeast Hsp70 homologues Ssa[22,57] and Ssb[33] and the yeast Hsp40 homologs Ydj1[56,58] (K. Allen and Y. Chernoff, unpublished data, 1999) and Sis1[59] (S. M. Uptain and S. Lindquist, unpublished data, 2000), also influence the propagation of yeast prions, although the effects of these chaperones are usually not as strong as the effects of Hsp104. Quite remarkably, Hsp70 and Hsp40 proteins also cooperate with Hsp104 in protecting yeast cells from heat-damaged protein aggregates.[60]

Here, we describe four methods to cure a [PSI^+] strain by manipulating Hsp104 levels. One method involves deletion of the $HSP104$ gene. The others require overproduction of the chaperone using constitutive or inducible overexpression constructs. In some cases, it may be necessary to quantitatively examine curing; thus, we include a method for this using an inducible expression system. To simplify detection of [psi^-] cells, all the methods described use an $ade1$-14 [PSI^+] strain.

CURING [PSI^+] BY CONSTITUTIVE HSP104 OVERPRODUCTION (FIG. 3C). Transform a [PSI^+] $ade1$-14 yeast strain with a plasmid encoding $HSP104$ controlled by its own promoter or by a strong constitutive promoter such as the glyceraldehyde-3-phosphate dehydrogenase promoter (GPD). The same plasmid without the $HSP104$ gene should be used as a control. Patch transformants on synthetic medium selective for the plasmid (e.g., -Leu if the $LEU2$ marker is used), and velveteen replica plate them onto three types of media: (1) YPD; (2) synthetic medium selective for plasmid and lacking adenine (e.g., -Leu-Ade); and (3) the same medium not selective for the plasmid (e.g., -Ade). No growth on -Leu-Ade plates confirms that excess Hsp104 inhibited [PSI^+]-mediated suppression. If [PSI^+] is completely cured, there will be no growth on -Ade, and dark red patches will be observed on YPD. If curing of [PSI^+] is incomplete, as is often the case, residual growth will be detected on -Ade due to recovery of [PSI^+] in the cells that have lost the plasmid, and mixed pink/red patches will be observed on YPD. To isolate [psi^-] derivatives, take red cells from the YPD plate and streak them on a fresh YPD plate to get single colonies. Then, check those colonies to confirm the absence of [PSI^+] by color on YPD and an inability to grow on -Ade medium. It is also

---

[55] S. V. Paushkin, V. V. Kushnirov, V. N. Smirnov, and M. D. Ter-Avanesyan, *Mol. Cell. Biol.* **17**, 2798 (1997).

[56] H. Moriyama, H. K. Edskes, and R. B. Wickner, *Mol. Cell. Biol.* **20**, 8916 (2000).

[57] G. Jung, G. Jones, R. D. Wegrzyn, and D. C. Masison, *Genetics* **156**, 559 (2000).

[58] V. V. Kushnirov, D. S. Kryndushkin, M. Boguta, V. N. Smirnov, and M. D. Ter-Avanesyan, *Curr. Biol.* **10**, 1443 (2000).

[59] N. Sondheimer, N. Lopez, E. A. Craig, and S. Lindquist, *EMBO J.* **20**, 2435 (2001).

[60] J. R. Glover and S. Lindquist, *Cell* **94**, 73 (1998).

necessary to screen for the loss of the Hsp104-encoding plasmid by plating [psi^-] cells on media selective for the plasmid (-Leu medium).

[PSI^+] CURING BY INDUCIBLE Hsp104 OVERPRODUCTION: A QUALITATIVE ASSAY. For these experiments, use plasmids encoding Hsp104 controlled by a tightly regulated, inducible promoter, such as *GAL*. Plasmids encoding *HSP104* regulated by the copper-inducible *CUP1* promoter can cure [PSI^+] efficiently; however, high background levels of *HSP104* expression in the absence of copper can cure some cells of [PSI^+]. Another advantage of using a *GAL::HSP104* construct is that expression can also be induced on glucose medium containing β-estradiol when a second plasmid coding for the chimeric Gal4-VP16-Er transcriptional activator is cotransformed. This renders the *GAL* promoter β-estradiol inducible.[61] We strongly suggest using centromeric plasmids bearing *GAL4-VP-ER* under a low-level constitutive promoter,[62] as high levels of Gal4-VP16-Er can activate the *GAL* promoter in the absence of β-estradiol.

Select the [PSI^+] *ade-14* transformants bearing *GAL–HSP104* on glucose synthetic medium selective for the corresponding plasmid (e.g., -Ura in case of *URA3* marker) and velveteen replica plate them onto the same medium with galactose (Gal-Ura) instead of glucose to induce the *GAL* promoter. After 2 to 4 days of incubation, velveteen replica plate transformants onto the following media: (1) the same medium lacking adenine to score for inhibition of [PSI^+]-mediated suppression); (2) synthetic glucose medium, selective for the plasmid and lacking adenine, to score for [PSI^+] loss (note that selection for the plasmid is to assure that only cells that retain the *GAL–HSP104* plasmid throughout the galactose induction period are being assayed) and; (3) YPD for the color assay for [PSI^+]. The efficiency of curing depends on both the genotypic background and the [PSI^+] isolate. If curing is inefficient after one passage on Gal medium, try two or three consecutive passages on Gal medium. To obtain [psi^-] derivatives, cells from the initial Gal-Ura plate are restreaked onto YPD to obtain single colonies. Screen the colonies by color and growth on -Ade.

[PSI^+] CURING BY INDUCIBLE Hsp104 OVERPRODUCTION: A QUANTITATIVE ASSAY (FIG. 3D). To quantify [PSI^+] curing by overproduced Hsp104, grow transformants containing the *GAL–HSP104* plasmid in synthetic liquid glucose medium that is selective for the plasmid for 1 day with shaking. Wash the cells twice and use them to inoculate synthetic medium selective for the plasmid and containing 2% raffinose and 2% glucose (Gal + Raf) instead of glucose at the starting concentration 10^5 cells/ml. (In our hands, liquid Gal + Raf medium works more reproducibly with a broader range of strains than Gal medium.) Incubate with intense shaking (200–250 rpm). Monitor growth by measuring OD or by counting cells. Take aliquots after certain periods of time and plate appropriate dilutions

---

[61] J. F. Louvion, B. Havaux-Copf, and D. Picard, *Gene* **131,** 129 (1993).
[62] C. Y. Gao and J. L. Pinkham, *Biotechniques* **29,** 1226 (2000).

onto synthetic plasmid-selective glucose medium. Count colonies and velveteen replica plate them onto YPD and -Ade media to score for the presence of [PSI^+] as described previously. Calculate the number of [psi^-] clones out of the total cells per generation. The number of generations (G) for each time t can be calculated according to the following formula: $G = \log_2(C_t/C_0)$, where C_t is the concentration of viable plasmid-containing cells at time point t (obtained from numbers of colonies grown on selective medium) and C_0 is the concentration of viable plasmid-containing cells at the starting point. In contrast to the GuHCl-induced curing described earlier, [PSI^+] curing by overproduced Hsp104 begins in the very first generation after induction and does not exhibit a lag period.[63,64] Mosaic [PSI^+]/[psi^-] colonies are sometimes observed, but usually constitute a minor fraction of cured cells. We count each mosaic colony as $\frac{1}{2}$ [PSI^+] and $\frac{1}{2}$ [psi^-].

[PSI^+] CURING BY hsp104 DELETION. To assess [PSI^+] curing by hsp104 deletion, one can either disrupt the HSP104 gene in a [PSI^+] strain directly using standard gene replacement techniques[65] or mate a [PSI^+] HSP104$^+$ strain to a [psi^-] hsp104Δ strain. When the resulting diploids are sporulated and the tetrads dissected, all hsp104Δ spore clones will be [psi^-]. Pedigree analysis indicates that [PSI^+] is not lost immediately after HSP104 inactivation: [PSI^+] clones can be recovered by crossing the hsp104Δ meiotic progeny to a [psi^-] HSP104$^+$ cell within the first few cell divisions after meiosis.[64] Our assumption is that [PSI^+] is maintained by preexisting Hsp104 protein within the initial meiotic segregants. Hsp104 is very stable and does not degrade immediately after inactivation of the HSP104 gene. Once a hsp104Δ colony is visible on a plate, however, all cells in that colony are [psi^-].

Prion Induction de Novo by Overproduction of Protein Determinant. [PSI^+] and [URE3] can be induced by transient overproduction of Sup35[8] or Ure2,[7] respectively. In both cases, overproduction of the prion-forming domain alone was sufficient for prion induction; moreover, overproducing either PFD induced prion formation even more efficiently than overproducing the complete protein.[9,14] Other PFDs, including heterologous Sup35,[37–39] Rnq1,[40] and *Podospora* Het-s,[66] can induce their respective prions upon transient overproduction. Thus, testing for prion induction by overproduction of the putative determinant is an effective tool for identifying the prion elements in yeast.[2]

Unfortunately, use of this test with [PSI^+] is complicated for three reasons. First, overproduction of Sup35 or the Sup35PFD can cause nonsense suppression

---

[63] Y. O. Chernoff, S. W. Liebman, M. M. Patino, and S. L. Lindquist, *Trends Microbiol.* **3**, 369 (1995).
[64] R. D. Wegrzyn, K. Bapat, G. P. Newnam, A. D. Zink, and Y. O. Chernoff, *Mol. Cell. Biol.* **21**, 4656 (2001).
[65] R. Rothstein, *Methods Enzymol.* **194**, 281 (1991).
[66] V. Coustou, C. Deleu, S. Saupe, and J. Begueret, *Proc. Natl. Acad. Sci. U.S.A.* **94**, 9773 (1997).

without converting the cells to $[PSI^+]$,[8,25,67] possibly due to aggregation of Sup35 and/or sequestration of other translation proteins by the Sup35 aggregates. Moreover, some derivatives of Sup35 cause nonsense suppression but are not able to induce stably propagated $[PSI^+]$ when overproduced.[9] To test whether $[PSI^+]$ was induced, cells should be checked for nonsense suppression in the absence of Sup35 overproduction.

Second, most of the Sup35-derived constructs require the presence of another prion-like non-Mendelian element, $[PIN^+]$,[51] to induce nonsense suppression and efficient $[PSI^+]$ formation. The presence of $[PIN^+]$ can be tested by either transforming the strain in question with a multicopy plasmid encoding *SUP35* or mating it to a $[PIN^+]$ partner containing this plasmid. In both cases, the multicopy *SUP35* plasmid will cause nonsense suppression in a $[PIN^+]$ but not in a $[pin^-]$ background. A $[pin^-]$ derivative of the same strain should be used as a control and can be obtained easily by passaging the cells three or more times on YPD+5 mM GuHCl (see earlier discussion), which efficiently cures yeast cells of $[PIN^+]$.[51] The $[PIN^+]$ requirement for $[PSI^+]$ induction can be overcome by overproducing specific Sup35PFD derivatives.[51,68]

A third complication is that genotypic factors other than $[PIN^+]$ can also modulate the efficiency of $[PSI^+]$ induction. For example, deletion of both genes coding for the Hsp70 chaperone Ssb, *ssb1*Δ *ssb2*Δ, significantly increases (about 10-fold) both the spontaneous appearance of $[PSI^+]$ and the induction of $[PSI^+]$ by overproduced Sup35 or Sup35PFD.[33] In contrast, deletion of the gene encoding the cytoskeletal assembly protein *SLA1* exhibits an opposite effect.[69] Other unidentified weak prion modulators are probably present in yeast, and these may cause considerable variation in the efficiency of $[PSI^+]$ formation.

To induce $[PSI^+]$ by constitutive overproduction of Sup35, transform a $[psi^- PIN^+]$ *ade1-14* strain with the multicopy plasmid encoding *SUP35* or *SUP35PFD* regulated by its own promoter. Alternatively, use a single-copy plasmid encoding *SUP35* or *SUP35PFD* regulated by a strong constitutive promoter such as *GPD*. A matching plasmid without *SUP35* serves as a control. Once transformants are selected, velveteen replica plate them on YPD medium and –Ade medium. If nonsense suppression occurs, colonies will be white or light pink on YPD and grow on –Ade. Next, restreak or plate serial dilutions of the white or pink cells onto YPD to get single colonies. Identify colonies that have lost the *SUP35* plasmid by velveteen replica plating onto medium lacking the appropriate nutrient. (*URA3* plasmids can be counterselected by using 5-FOA medium.) Check these colonies for the presence of $[PSI^+]$ on –Ade and YPD as described earlier. Although the frequency

---

[67] M. D. Ter-Avanesyan, V. V. Kushnirov, A. R. Dagkesamanskaya, S. A. Didichenko, Y. O. Chernoff, S. G. Inge-Vechtomov, and V. N. Smirnov, *Mol. Microbiol.* **7**, 683 (1993).
[68] I. L. Derkatch, M. E. Bradley, S. V. Masse, S. P. Zadorsky, G. V. Polozkov, S. G. Inge-Vechtomov, and S. W. Liebman, *EMBO J.* **19**, 1942 (2000).
[69] P. A. Bailleul, G. P. Newnam, J. N. Steenbergen, and Y. O. Chernoff, *Genetics* **153**, 81 (1999).

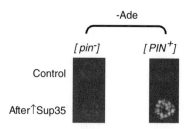

FIG. 4. Formation of [PSI^+] de novo. The isogenic [pin^- psi^-] and [PIN^+ psi^-] strains, bearing ade1-14, were transformed with the GAL–SUP35 construct (↑Sup35) or matching control plasmid. Transformants were grown on galactose medium, where GAL–SUP35 is induced, and velveteen replica plated onto -Ade/glucose medium, where GAL–SUP35 is turned off. Papillation results from [PIN^+]-dependent induction of [PSI^+] by overproduced Sup35.

of [PSI^+] induction varies greatly with both construct and genetic background, usually 1 to 30% of the colonies will contain [PSI^+]. These numbers, however, do not represent a reliable estimate of the actual efficiency of [PSI^+] formation, as an accumulation of excess Sup35 or Sup35PFD will inhibit the growth of [PSI^+] cells.

To avoid problems with growth inhibition, use constructs encoding SUP35 regulated by an inducible promoter (such as GAL) (Fig. 4). Transform cells with either GAL–SUP35 or GAL–SUP35PFD constructs or a matching control plasmid lacking Sup35 and plate the cells on glucose synthetic medium selective for the plasmid. Next, velveteen replica plate them onto two types of plates: glucose or galactose synthetic media selective for the plasmid. After 3 to 4 days of incubation, velveteen replica plate the cells from each plate onto glucose –Ade medium. [PSI] induction will cause heterogeneous growth or papillation on –Ade medium following GAL induction but not following growth on glucose after 7–10 days. The efficiency of [PSI^+] induction is usually higher if Gal plates are incubated at 25° rather than 30°. Induction can also be increased greatly by incubating Gal plates in a refrigerator for 3–4 weeks before velveteen replica plating them onto –Ade.

For a more accurate measurement of [PSI^+] induction frequency, grow transformants bearing the GAL–SUP35 or GAL–SUP35PFD plasmid or the matching control plasmid for 1 to 2 days in synthetic glucose medium selective for the plasmid. Wash the cells twice with water and inoculate into Gal + Raf medium selective for the plasmid at the starting concentration of 10^5 cells/ml. Take aliquots before induction and after various periods of incubation in Gal + Raf medium, and plate appropriate dilutions of cells from each time point, about 100 cells per plate, onto glucose medium selective for the plasmid. After 3–4 days of growth, velveteen replica plate the cells onto YPD and –Ade media to identify [PSI^+] by colony color and growth as described previously. [PSI^+] induction can be detected as early as 1 day after incubation in Gal + Raf medium; however, the highest levels of induction are achieved after 4–5 days.[33,37] It is very important to use fresh colony-purified cultures for quantitative experiments, as [psi^- PIN^+] strains will accumulate [PSI^+] colonies spontaneously during storage in the refrigerator.[68,69b]

For Gal$^-$ strains or when variable levels of *GAL–SUP35* are to be tested, the dual-plasmid system employing the Gal4-VP16-Er transcriptional activator and β-estradiol is helpful[9] (see earlier discussion for [PSI^+] curing with *GAL–HSP104*). Constructs encoding *SUP35* or its derivatives regulated by the *CUP1* promoter can also induce [PSI^+] efficiently in the presence of 30–50 μM CuSO$_4$. Background levels of *CUP1* expression are usually not sufficient to cause detectable induction of [PSI^+].

Analysis of Prion Aggregation in Vivo

Differential Centrifugation Analysis of Yeast Extracts. The prion isoform of Sup35 protein (Sup35^{PSI+}) is insoluble and can be distinguished from the normal form of Sup35 (Sup35^{psi-}) *in vitro* by differential centrifugation of crude cell lysates. Typically, the majority of Sup35 from [PSI^+] lysates partitions to the pellet fraction, whereas most of the Sup35 from [psi^-] lysates partitions to the soluble fraction.[12] Although [PSI^+] cells contain some soluble Sup35 protein, the ratio between soluble and insoluble Sup35 is altered drastically in [PSI^+] cells compared to [psi^-] cells (Fig. 5A). The fractionation of other proteins, such as the ribosomal protein L3 or molecular chaperone Hsp70, is the same in [PSI^+] and [psi^-] cells.[12] Moreover, it is becoming increasingly clear that the amount of soluble Sup35 can vary between different kinds of [PSI^+] isolates.[70,71,71b] Thus, determination of the amount of soluble Sup35 not only provides a means to distinguish between [PSI^+] and [psi^-] cells, but can also be used quantitatively to discriminate between [PSI^+] variants that may be phenotypically identical.[71b]

Yeast crude cell lysates can be prepared by either disintegration using glass beads or enzymatic digestion of the cell wall. Following lysis, the crude lysate can be sedimented directly, as described later, or can be layered on a sucrose cushion or gradient with similar results.[13] The partitioning of Sup35 to the pellet or supernatant fractions can be detected qualitatively by separating proteins from equal volumes of each fraction by SDS–PAGE followed by immunoblot analysis. Alternatively, proteins from the supernatant fractions can be serially diluted and applied

---

[69b] A. S. Borchsenius, R. D. Wegrzyn, G. P. Newnam, S. G. Inge-Vechtomov, and Y. O. Chernoff, *EMBO J.* **20,** 6683 (2001).

[70] P. Zhou, I. L. Derkatch, S. M. Uptain, M. M. Patino, S. Lindquist, and S. W. Liebman, *EMBO J.* **18,** 1182 (1999).

[71] N. V. Kochneva-Pervukhova, M. B. Chechenova, I. A. Valouev, V. V. Kushnirov, V. N. Smirnov, and M. D. Ter-Avanesyan, *Yeast* **18,** 489 (2001).

[71b] S. M. Uptain, G. J. Sawicki, B. Caughey, and S. Lindquist, *EMBO J.* **20,** 6236 (2001).

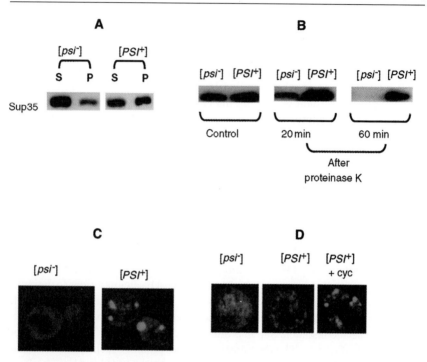

FIG. 5. Detection of the Sup35^{PSI+} protein *in vivo*. (A) Detection of Sup35^{PSI+} by differential centrifugation. Isogenic [*psi*⁻] and [*PSI*⁺] cultures were lysed with glass beads as described in the text. Cell extracts were fractionated by centrifugation at 8000g. The Sup35 protein was detected by SDS–PAGE and Western blotting with Sup35-specific antibodies. Most of the Sup35 protein from [*PSI*⁺] cells was partitioned to the supernatant (S) fraction, whereas the majority of Sup35 from a [*PSI*⁺] strain was partitioned to the pellet (P) fraction. (B) Detection of Sup35^{PSI+} by proteinase K resistance. Cell extracts of the isogenic [*psi*⁻] and [*PSI*⁺] derivatives of D1142-1A, prepared as described in the text, were treated with 1 μg/ml proteinase K for various periods of time. At 20 and 60 min, the reactions were terminated, and equal amounts of lysates were analyzed by SDS–PAGE and immunoblot analysis using Sup35-specific antibodies. Although a significant portion of the Sup35^{psi-} was digested after 20 min of incubation, most of the Sup35^{PSI+} remained intact, even after 60 min of incubation. (C) Aggregation of overproduced Sup35PFD–sGFP in [*PSI*⁺] cells. Isogenic [*psi*⁻] and [*PSI*⁺] strains were transformed with the *CUP1-SUP35PFD-sGFP* plasmid. Expression was induced by adding 50 μM CuSO₄, leading to the formation of large fluorescent clumps of the fusion proteins in [*PSI*⁺] but not in [*psi*⁻] cells. (D) Aggregation of Sup35PFD–sGFP expressed under the endogenous *SUP35* promoter in [*PSI*⁺] cells. The *SUP35PFD–sGFP* construct under the endogenous *SUP35* promoter caused formation of numerous small fluorescent clumps in [*PSI*⁺] but not in isogenic [*psi*⁻] cells. The size of the clumps was increased by the addition of 100 μg/ml cycloheximide at 15 min before detection.

directly to a membrane using a dot-blot apparatus for a more quantitative analysis. As with all fractionations, introduction of either cushions or gradients diminishes cross-contamination between fractions. A potential disadvantage of the centrifugation step included in the sucrose cushion method to remove cellular debris and unbroken cells is that some of the Sup35 aggregates may be removed from [*PSI*⁺]

lysates. The degree of purity and yield required for subsequent analysis should, therefore, dictate the method employed. Due to the great difficulty in working with aggregation-prone proteins, considerable time should be invested to ensure reproducibility with known samples before analyzing the behavior of unknowns.

Lysis with Glass Beads

1. Grow yeast cultures to midlog phase ($\sim$1-5 × 10^7 cells/ml) in complete (YPD) or $\sim$2-4 × 10^6 cells/ml in synthetic (SD) medium at 30° with constant shaking (250 rpm). It is important to keep the compared cultures at the same OD. The [PSI^+] and [psi^-] states can be differentiated using stationary-phase cultures. However, these cells are much more difficult to lyse and contain much higher levels of cellular proteases than exponential-phase cells. Fifteen minutes prior to collection, add cycloheximide to 200 μg/ml to stabilize polysomes and allow newly synthesized proteins to achieve their characteristic conformations.

2. Cool the cells on ice for 15 min and then harvest by low-speed centrifugation (2000g, 5 min, 4°). Discard the supernatant and wash the cell pellet once with an equal volume of cold water containing 200 μg/ml cycloheximide.

3. Wash the pellet once with an equal volume of lysis buffer [50 mM Tris–HCl, pH 7.5, 5 mM MgCl$_2$, 10 mM KCl, 0.1 mM EDTA, pH 8.0, 1 mM DTT, 100 μg/ml cycloheximide, 1 mM benzamidine, 2 mM phenylmethylsulfonyl fluoride (PMSF), 10 μg/ml leupeptin, 2 μg/ml pepstatin A, 100 μg/ml ribonuclease A].

4. Transfer cells to a 1.5-ml microcentrifuge tube and pellet at 2000g, 5 min, 4°. If desired, the cells may be flash frozen and stored at $-80°$ at this point.

5. Resuspend pellet in lysis buffer at a concentration of $\sim$3 × 10^6 cells/μl. Add an equal volume of 425- to 600-μm acid-washed glass beads (Sigma).

6. Homogenize cells in a Mini-BeadBeater 8 (Biospec Products, Bartlesville, OK) at 4°, vortexing 1 min followed by incubating 2 min on ice. Repeat this for at least four cycles or as many times as needed to achieve sufficient lysis. The extent of lysis can be determined by light microscopy.

7. Carefully remove supernatant and wash glass beads twice with one-half the volume of lysis buffer originally used. Combine washes with lysate from step 7.

8. Sediment the lysate at 2500g for 10 min at 4° to pellet unbroken cells and cellular debris. Remove the supernatant fraction to a new tube without disturbing the pellet.

9. Determine the protein concentration of the lysate. Typical yields are 5–15 mg/ml using the Bio-Rad (Hercules, CA) protein assay reagent with BSA as a standard. Lysates may be flash frozen and stored at $-80°$ at this point.

Lysis by Spheroplasting

1. Culture growth conditions are the same as just described. Resuspend pellets at 20 OD$_{600}$/ml in 0.1 M Tris–SO$_4$, pH 9.4, 10 mM DTT. Incubate at room temperature for 10 min. Pellet cells in a microcentrifuge for 10 sec at room temperature.

2. Wash pellet with 1 ml of 1.2 M sorbitol.

3. Resuspend pellet at 50 OD_{600}/ml in spheroplasting buffer [1.2 M sorbitol, 10 mM potassium phosphate, pH 7.0, 125–250 μg/ml Zymolyase 100T (Seikagaku America, Inc., Falmouth, MA)] that has been preincubated at 30°. (The amount of Zymolyase 100T can vary with the lot of enzyme and with cell strain; therefore, it should be titrated beforehand.) Incubate cells at 30° for approximately 5–10 min or until more than 90% of cells are spheroplasts. [Monitor spheroplast formation by diluting cells 1 : 10 into both 1.2 M sorbitol and 0.1% (w/v) SDS. Spheroplasts will lyse in the latter.]

4. Collect cells by centrifugation at 1500g for 5 min at 4°. Sedimentation at higher speeds can result in premature lysis of fragile cells. Gently wash pellet with 1 ml of ice-cold 1.2 M sorbitol, 10 mM potassium phosphate, pH 7.0, using a wide-bore 1-ml pipette tip.

5. Collect cells by centrifugation as in step 4. Resuspend pellet in ice-cold lysis buffer [50 mM Tris–HCl, pH 7.2, 10 mM KCl, 100 mM EDTA, pH 8.0, 1 mM DTT, 0.2% (w/v) SDS, 1% (v/v) Triton X-100, 1 mM benzamidine, 2 mM PMSF, 10 μg/ml leupeptin, and 20 μg/ml pepstatin] at 50 ODs/ml. Incubate on ice for 5 min, and check for complete lysis by microscopy. If unlysed cells remain, they should be removed before continuing. *Note:* Cell lysates obtained by either approach can be fractionated by either low-speed, for a quick qualitative assessment, or high-speed centrifugation.

Low-Speed Centrifugation

1. Retain a portion of the lysate for analysis by electrophoresis and sediment the remaining lysate by centrifugation at 6000–12,000g for 10 min at 4°. (In our strains, we observed best differentiation at 8000g).

2. Remove supernatant fraction to a tube containing an appropriate volume of 6× sample buffer [350 mM Tris–HCl, pH 6.8, 30% (v/v) glycerol, 10% (w/v) SDS, 600 mM DTT, 0.12% (w/v) bromphenol blue] to give a 1× concentration.

3. Resuspend the pellet in the same volume of lysis buffer used in step 1 of this section. Add an appropriate volume of 6× sample buffer to give a 1× concentration.

4. Boil the total, supernatant, and pellet fractions in a water bath for 10 min.

5. Load equal volumes of each fraction in separate lanes of a 10% SDS–PAGE gel, run the gel at 25 mA/gel, electrotransfer the proteins to Immobilon-P (Millipore, Bedford, MA), and analyze by immunoblotting. A total of 10–40 μg of protein/lane yields a sufficient Sup35 signal for detection with our antiserum.[12]

High-Speed Centrifugation

1. Remove 200 μl of lysate to a chilled polycarbonate Beckman tube.

2. Centrifugate at 100,000g (48,000 rpm in a TLA-100 rotor, Beckman) for 15 min at 4° in a prechilled rotor.

3. Remove supernatant fraction to an Eppendorf tube containing the appropriate amount of the 6× sample buffer.

4. Resuspend pellet in 200 µl of lysis buffer and remove to an Eppendorf tube containing the appropriate amount of 6× sample buffer. (The pellet contains an easily resuspended outer translucent layer and a dense white layer that is difficult, if not impossible, to resuspend. Move the white pellet with the pipette tip directly to the sample buffer without resuspending it.)

5. Remove 60 µl of the unspun lysate to an Eppendorf tube containing the appropriate amount of sample buffer.

6. Boil all samples 20 min and run on SDS–PAGE as described previously.

SEMIQUANTITATIVE DOT-BLOT ANALYSIS

1. Normalize the total protein present in the supernatant fractions to between 0.1 and 1 mg/ml. Typically, we use the uppermost fractions of sucrose cushions that were subjected to $100,000g^{70}$; however, supernatant fractions obtained by either low- or high-speed sedimentation of the crude lysate can also be used.

2. Serially dilute the proteins in two- or four-fold increments.

3. If a protein standard is desired, dilute purified Sup35 or Sup35PFD starting at 50 µg/ml (see later for purification procedure).

4. Apply an equal volume of each dilution to a PVDF membrane using a dot-blot apparatus (V&P Scientific, Inc., San Diego, CA, or equivalent). The total amount of the supernatant fraction applied ranges between 100 and 0.78 µg, whereas the amount of purified Sup35PFD ranges from 50 to 0.2 ng.

5. Detect the distribution of Sup35 by immunoblotting with antisera specific for Sup35. Assess the total amount of protein applied to the membrane using the nonspecific protein stain Ponceau-S.

Proteinase Resistance. For mammalian prions, the most definitive assay to detect prion conformation *in vitro* is higher proteinase K resistance of the prion form compared to the nonprion form. This assay is very robust and the prion protein is several orders of magnitude more stable than the normal form.[72] Moreover, digestion of the prion conformation with proteinase K produces a diagnostic and persistent 27- to 30-kDa fragment. Similarly, the prion forms of Sup35 and Ure2 are resistant to proteolytic degradation compared to the nonprion forms.[13,14] This provides an experimental assay for the identification of prion isoforms; however, this assay is used much less often to study Sup35^{PSI+} for two reasons. First, the difference in the rate that the two isoforms are digested is not as great as for other prions. Second, digestion does not produce any persistently protected proteolytic fragments. Unlike the prion form of Ure2,[14] Sup35^{PSI+} digestion produces only poorly resolved, weak bands of partially digested products. As a result, the

[72] S. A. Priola and B. Chesebro, *J. Biol. Chem.* **273**, 11980 (1998).

difference in proteinase resistance is best detected by the amount of full-length protein remaining (Fig. 5B).

The simplest approach used to identify proteinase-resistant aggregates of Sup35^{PSI+} involves lysis of yeast cells with glass beads in 0.1 M Tris–HCl, pH 7.5–8.0, 1 mM DTT, 1 mM EDTA buffer lacking proteinase inhibitors and treating the extract with proteinase K.[13] The nonprion form of Sup35 is a relatively unstable protein and is degraded easily by cellular proteinases in the absence of proteinase inhibitors, even before proteinase K is added. To minimize premature degradation, a modified version of the technique used previously to study the proteinase resistance of mammalian PrP can be applied as described (Y. Chernoff, G. Newnam, I. Derkatch, and S. Liebman, unpublished, 1995).

1. Grow a 10- to 15-ml culture to the middle of log phase (OD$_{600}$ about 1). Precipitate cells by centrifugation, wash them twice with H$_2$O, and resuspend them in 500 μl of lysis buffer P [10 mM Tris–HCl, pH 8.0, 150 mM NaCl, 2 mM MgCl$_2$, 0.5% (w/v) Nonidet P-40 (NP-40), 0.5% (w/v) sodium deoxycholate]. Keep on ice.

2. Lyse cells by vortexing with glass beads (six times for 15 sec each, cooling on ice 1 min in between vortexing cycles). Remove cell debris by centrifugation.

3. Take 50- to 100-μl aliquots. Add the desired amount of the freshly prepared proteinase K in H$_2$O to each sample. Include a control sample without proteinase K. Incubate for various periods of time at 37°. Terminate reactions by adding PMSF to a final concentration of 2 mM. Immediately add SDS sample buffer to 1× and boil for 10 min.

4. Run samples on a 12% SDS–PAGE and detect Sup35 by immunoblotting with a Sup35-specific antibody.

In our hands, incubation with 1 μg/ml of proteinase K for 20–60 min provides the best results when using middle-log cultures of [PSI^+] and [psi^-] isolates of the strain D1142-1A (for strain description, see Refs. 11 and 73). Moreover, [PSI^+] variants of different efficiencies of nonsense suppression exhibit different levels of proteinase K sensitivity (Y. Chernoff and G. Newnam, unpublished, 1996). Thus, concentrations and time periods are specific to the genetic background and the [PSI^+] variant used so both should be titrated carefully.

This approach can be also applied to stationary cultures; however, the proteinase resistance of Sup35 in [PSI^+] or [psi^-] cells from stationary cultures is much higher that that of Sup35 isolated from middle-log cultures. Sup35 from [PSI^+] remains more proteinase K resistant than the nonprion form, but higher concentrations of proteinase K and/or longer incubation periods should be applied. It has been reported that [psi^-] isolates of some strains contain less Sup35 protein than [PSI^+] isolates, especially in the stationary phase.[13] However, we

---

[73] L. P. Wakem and F. Sherman, *Genetics* **124,** 515 (1990).

were unable to observe such a difference in our sets of strains (Y. Chernoff and G. Newnam, unpublished, 1996). It is possible that either this effect is strain specific or this difference is actually due to different levels of degradation of Sup35$^{\text{psi}-}$ and Sup35$^{\text{PSI}+}$ proteins during isolation procedures.

Visual Detection of Prions Tagged with GFP in Living Yeast Cells. Translational fusions of GFP to prion determinants or just their PFDs have provided a model system for monitoring the aggregation state of prion proteins in living cells.[12,15,40,74] In cells without a prion, fluorescence is usually diffuse, apparently reflecting an even distribution of the soluble, normal conformation of the determinant throughout the cytoplasm. In cells with a prion, fluorescence of the fusion protein coalesces into one or more foci, providing an unambiguous difference in the fluorescence pattern between the two conformational states. One caveat to this method is that the prion state can be induced by overexpression of the PFD; therefore, prolonged or high production of such fusions can result in the appearance of fluorescent clumps that will correlate with prion induction *de novo* (e.g., see Refs. 12 and 75).

The appearance of such fusions in prion-containing cells depends on the type of GFP used, on the expression level of the fusion, and how it was made. For example, GFP mutants containing amino acid substitutions (S65T and V164A; superglow GFP, sGFP)[76,77] fluoresce more brightly and are less likely to self-associate in the cell than wild-type GFP.[78,79] Sup35PFD fusions to different forms of GFP in [*PSI*$^+$] are always unevenly distributed in a cell; however, with wild-type GFP fusions, a single large fluorescent clump is usually observed, whereas with sGFP fusions, multiple smaller foci are visible.[74] The number of foci in [*PSI*$^+$] cells expressing sGFP fusions depends on the level of expression. Sup35PFD–sGFP fusions overexpressed under the control of a strong constitutive promoter such as *GPD* or under the control of the *CUP1* promoter in the presence of inducer (50–100 μM CuSO$_4$) usually form a few, relatively large fluorescent clumps in each [*PSI*$^+$] cell (Fig. 5C). In contrast, Sup35PFD–sGFP fusions expressed under the control of the endogenous *SUP35* promoter form many very small clumps in [*PSI*$^+$] cells that are hard to distinguish from the diffuse distribution of Sup35PFD–sGFP observed in [*psi*$^-$] cells[53] (see Fig. 5D). The latter fluorescence pattern more

---

[74] T. R. Serio, A. G. Cashikar, J. J. Moslehi, A. S. Kowal, and S. L. Lindquist, *Methods Enzymol.* **309**, 649 (1999).
[75] P. Zhou, I. L. Derkatch, and S. W. Liebman, *Mol. Microbiol.* **39**, 37 (2001).
[76] R. Heim, A. B. Cubitt, and R. Y. Tsien, *Nature* **373**, 663 (1995).
[77] J. Kahana and P. A. Silver, *in* "Current Protocols in Molecular Biology" (F. M. Ausubel, R. Brent, R. E. Kingston, D. D. Moore, J. G. Seidman, J. A. Smith, and K. Struhl, eds.), p. 9.6.13. Wiley, New York, 1996.
[78] A. B. Cubitt, R. Heim, S. R. Adams, A. E. Boyd, L. A. Gross, and R. Y. Tsien, *Trends Biochem. Sci.* **20**, 448 (1995).
[79] S. R. Kain, M. Adams, A. Kondepudi, T. T. Yang, W. W. Ward, and P. Kitts, *Biotechniques* **19**, 650 (1995).

accurately reflects the actual distribution of the prion form of the Sup35 protein in [PSI^+] cells, as confirmed by direct immunostaining experiments.[64,75] Of course, the appearance of prion fusions to GFP within cells will also depend greatly on the behavior of the prion domain. For example, one large focus per cell is observed when fusions of the PFD of Rnq1 to GFP are expressed in [RNQ^+] cells; however, numerous smaller foci are visible when the same fusion is expressed in cells with a composite prion, termed [RMC^+], composed of the PFD of Rnq1 fused to Sup35MC.[40]

In some cases, how the cells are prepared can affect the appearance of prion–GFP fusions. Treatment of yeast cells with protein synthesis inhibitors (100 μg/ml cycloheximide or 10 mM sodium azide) for 15 min before detection makes Sup35–sGFP clumps more visible (Fig. 5D), thereby increasing the resolution of the procedure and enabling [PSI^+] and [psi^-] cells to be distinguished when Sup35PFD–sGFP is expressed at wild-type levels.[53] Decreased levels or activity of Hsp104 also result in fewer, larger Sup35–sGFP clumps in [PSI^+] cells.[64]

To detect sGFP fusions, collect cells grown to the middle of an exponential phase carrying the appropriate construct and suspend them either in DABCO buffer [2.45% 1,4-diazabicyclooctane, Sigma, in 1× PBS with 7.5% (v/v) glycerol] or in phenylenediamine mounting solution [1 mg/ml p-phenylenediamine, Sigma, in 1× PBS with 90% (v/v) glycerol]. Apply the cells to a polylysine (Sigma)-coated glass slide, cover them with a glass coverslip (Sigma), and seal with nail polish. Fluorescence is examined using a fluorescence microscope at 100–400× magnification with parameters set at excitation 460–490 nm, beamsplitter 505 nm, and emission 515–550 nm. Alternatively, samples can be scanned on a confocal laser-scanning microscope with the excitation wavelength for the argon laser at 488 nm.

Analysis of Prion Aggregation in Vitro

Protein fragments encompassing the PFDs of Sup35[16,80] or Ure2[17–19] can polymerize into amyloid in vitro. Sup35 amyloid formation in vitro has been linked to the propagation of [PSI^+] in vivo. Fragments of Sup35 capable of inducing [PSI^+] in vivo form amyloid in vitro.[9,16] Lysates from [PSI^+] but not [psi^-] strains accelerate the formation of amyloid in vitro,[71b] as do preformed fibers.[16] Deletions within the PFDs of Sup35,[16,81] as well as specific point mutations,[45] slow the process of assembly into amyloid[16,45] and block the induction of new [PSI^+] elements.[9,45] Similarly, the expansion of oligopeptide repeats within the Sup35PFD, which are similar to repeats found in mammalian PrP, accelerates

---

[80] C. Y. King, P. Tittmann, H. Gross, R. Gebert, M. Aebi, and K. Wuthrich, *Proc. Natl. Acad. Sci. U.S.A.* **94**, 6618 (1997).

[81] J. J. Liu and S. Lindquist, *Nature* **400**, 573 (1999).

amyloid formation *in vitro* and increases the efficiency of [PSI^+] induction *in vivo*.[81] Moreover, phenotypic characteristics of hybrid [PSI^+] strains maintained by a chimeric *Candida–Saccharomyces* Sup35PFD protein *in vivo* correlate with physical characteristics of the chimeric Sup35PFD amyloids *in vitro*.[82] Here, we describe the techniques used to generate and characterize Sup35 amyloids *in vitro*. Protocols from Serio *et al.*[74] and Uptain *et al.*[71b] are included for the reader's convenience.

Isolation and Storage of Purified Protein. Sup35 can be divided into three regions based on differences in function and amino acid composition. The C-terminal 432 amino acid region of Sup35 (Sup35C) provides the translational termination function and is conserved evolutionarily among eukaryotes.[41–43] The N-terminal 123 amino acid fragment of Sup35 (Sup35N) is essential for [PSI^+] induction and propagation.[35,44] The 130 amino acid fragment intervening Sup35N and Sup35C is referred to as Sup35M. Sup35N and Sup35NM fragments are frequently (and indeed throughout this chapter) referred to as the prion-forming domain. Although fragments containing only the Sup35N region are capable of ordered amyloid assembly,[16,80] the Sup35NM and complete Sup35NMC fragments most accurately reflect [PSI^+] metabolism *in vivo*. Because Sup35NM and Sup35NMC form amyloid under native conditions,[16] we purify these proteins under denaturing conditions (8 M urea) to maintain the protein in a uniform state that is more amenable to studying the assembly process. The importance of obtaining a denatured, uniform solution of protein prior to initiating a kinetic analysis of amyloid formation cannot be stressed enough. Some authors recommend clearing large protein aggregates by filtration to obtain reproducible results.[45] Methods for purification under denaturing conditions in 8 M urea are discussed next.

EXPRESSION IN *E. coli*. The procedures described here apply to fragments cloned into either pJC45 encoding an amino-terminal 10 residue histidine tag (His$_{10}$) or pJC25, lacking a tag.[83] These plasmids are maintained at a high copy number and have a consensus T7 promoter and the *lacI* operator at the 5' end of the multiple cloning site. Either construct is expressed in *E. coli* strain BL21 [DE3] pAP *lacI*^{q} that contains a [DE3] lysogen for a high-level expression of T7 polymerase following induction with isopropyl-β-D-thiogalactopyranoside (IPTG; Sigma) and expresses a low level of the *lacI* product to repress leaky expression of the polymerase and, therefore, the target protein. To induce expression, we grow transformants in 1 liter of Circle Grow medium (Bio 101, Inc., Carlsbad, CA) containing 50 μg/ml kanamycin and 100 μg/ml ampicillin and incubate at 37° with constant shaking (300 rpm) until an OD$_{600}$ of between 0.4 and 0.8 is reached. Add IPTG to a final concentration of 1 mM and incubate for an additional 2 hr at 37° for the His$_{10}$-tagged version or for 2 to 3 hr at 25° for the nontagged version.

[82] P. Chien and J. S. Weissman, *Nature* **410**, 223 (2001).
[83] J. Clos and S. Brandau, *Protein Expr. Purif.* **5**, 133 (1994).

Collect bacteria by centrifugation (3,000g, 10 min, 4°). The pellet may be stored at −80° or processed immediately.

PURIFICATION OF His-TAGGED Sup35NM AND NONTAGGED Sup35NM. All steps are carried out at 25°. Cells expressing the His-tagged Sup35NM are lysed in buffer containing 20 mM Tris–HCl, pH 8.0, 8 M urea. The lysate is then cleared of debris by centrifugation at 30,000g for 20 min at 10°. Apply the 30K supernatant to a Ni^{2+}-nitrilotriacetic acid agarose column (Ni^{2+}-NTA; Qiagen, Valencia, CA), preequilibrated with lysis buffer. Wash the column with 5 bed volumes of 20 mM Tris–HCl, pH 8.0, 8 M urea, 40 mM imidazole, and elute protein in a single step with 2 bed volumes of 20 mM Tris–HCl, pH 8.0, 8 M urea, 400 mM imidazole. Apply the eluate directly onto a Q Sepharose Fast Flow column (Pharmacia, Piscataway, NJ) preequilibrated with the elution buffer. Wash the resin with 5 bed volumes of 20 mM Tris–HCl, pH 8.0, 8 M urea, 100 mM NaCl and elute the protein in a single step with 2 bed volumes of 20 mM Tris-HCl, pH 8.0, 8 M urea, 300 mM NaCl.

To purify nontagged Sup35NM, apply the 30K supernatant to a Q Sepharose Fast Flow column (Pharmacia) preequilibrated with 10 mM Tris–HCl, pH 7.2, 8 M urea. Wash the column with 5 bed volumes of 10 mM Tris–HCl, pH 7.2, 8 M urea, 85 mM NaCl and elute protein in 3 bed volumes of 10 mM Tris–HCl, pH 7.2, 8 M urea, 150 mM NaCl. Load the eluate from the Q Sepharose directly onto a Macro Prep Ceramic Hydroxyapatite Type I 40-μm column (Bio-Rad) that has been preequilibrated with the same buffer used to elute proteins from the Q Sepharose column. Wash the column with 2 bed volumes of 1 mM potassium phosphate, pH 6.8, 8 M urea, 1 M NaCl and then with 2 bed volumes of 25 mM potassium phosphate, pH 6.8, 8 M urea. Elute the proteins using a linear gradient of potassium phosphate, pH 6.8, from 25 to 125 mM.

The purity of the final product is analyzed by 10% SDS–PAGE followed by staining with Coomassie Brilliant Blue R-250. The predicted molecular mass of Sup35NM is 28.5 kDa; however, due to the presence of the highly charged M region, Sup35NM migrates aberrantly by SDS–PAGE at about 45 kDa. The His_{10} tag causes the fusion protein to migrate slightly higher.

QUANTITATION AND YIELDS. Sup35NM is poorly stained by Coomassie Brilliant Blue G-250, which binds primarily to arginine residues. Protein determination methods based on binding to this dye, such as Bradford, are, therefore, unreliable for the quantitation of protein yields. Sup35NM staining by Coomassie Brilliant Blue R-250, however, is a reliable method for detecting the protein following gel electrophoresis. The concentration of Sup35NM can be determined by the microbicinchoninic acid method (Micro-BCA; Pierce, Rockford, IL) using bovine serum albumin (BSA) as a standard or, alternately, directly from the absorbance at 276 nm in 8 M urea using a molar extinction coefficient (ε) of 29,000 M^{-1} for Sup35NM. Typically, 50 mg of His_{10}-Sup35NM is obtained from a 1-liter culture.

CONCENTRATION AND STORAGE OF PURIFIED PROTEIN. For short-term storage, Sup35NM can be concentrated using Biomax Ultrafree-15 concentrators with a 10-kDa molecular mass cutoff (Millipore). Column fractions containing

Sup35NM are pooled and concentrated at $1500g$ for approximately 2.5 hr at $6°$. The protein may be stored in this state for approximately 1 week at $4°$.

For long-term storage, we dialyze Sup35NM against 8 M urea, 20 mM Tris–HCl, pH 7.5, and then precipitate Sup35NM with methanol to remove any urea. Store the precipitate at $-80°$. Anhydrous methanol (100%) is added to dialyzate containing Sup35NM on ice at a ratio of 5 : 1. The mixture is incubated on ice for 30 min, and the precipitate is collected by centrifugation at $14,000g$ for 30 min at $4°$. The pellet is then washed with 100% methanol (1/2 volume of supernatant) and collected by centrifugation again. The supernatant is removed, and the pellet is stored under 70% methanol (1/2 volume of supernatant) at $-80°$. Prior to use, collect the precipitated protein by centrifugation at $14,000g$ for 30 min at $4°$. The methanol is removed carefully, and the pellet is damp-dried under vacuum without heat for 5 min. The precipitated protein is resuspended in freshly made 20 mM Tris–HCl, pH 7.5, 8 M urea to yield approximately a 30-mg/ml solution. The protein concentration should always be confirmed by one of the just-described methods.

Polymerization Reactions and Propagation of Amyloid in Vitro. The most detailed information regarding the assembly of Sup35 into amyloid has been gleaned from studies of the Sup35NM fragment.[16,84] Full-length Sup35 will form amyloid *in vitro,* but the process is more cumbersome, as quantitative recovery of the protein in amyloid form requires slow dialysis from the denaturant (2 M stepwise decreases in urea until no denaturant remains). In addition, full-length Sup35 in an unpolymerized form binds to the diagnostic amyloid dye, Congo red, eliminating this convenient assay from the repertoire available for monitoring amyloid formation. Therefore, our discussion of amyloid assembly here is restricted to a characterization of Sup35NM, either His-tagged or not.[74] Although for many applications, the His-tagged Sup35NM may be used interchangeably with nontagged protein, the presence of the tag does affect polymerization.[84] Therefore, nontagged Sup35NM should be used in all mechanistic studies of Sup35NM amyloid formation.

Multiple factors influence the efficiency with which Sup35NM will form amyloid *in vitro.* Among these, protein concentration and sufficient dilution from the denaturant are the most crucial. Polymerization reactions, which are in the micromolar range for Sup35NM, form amyloid within a reasonable time frame (30–90 hr).[16,84] We also advise using at least a 100-fold dilution from the denaturant into aqueous buffer, as excess denaturant slows or inhibits polymerization.

Polymerization of Sup35NM occurs over a wide range of buffer, salt, temperature, and detergent conditions.[74,84a] Molar concentrations of monovalent salt and Triton-X 100 up to 10% (w/v) do not to alter the process at all, but even 0.05% (w/v) SDS is sufficient to inhibit polymerization. All of the assays described later were conducted in Congo red-binding buffer (CRBB: 5 mM potassium phosphate, pH 7.4, 150 mM NaCl). For these analyses, Sup35NM is diluted directly

---

[84] T. R. Serio, A. G. Cashikar, A. S. Kowal, G. J. Sawicki, J. J. Moslehi, L. Serpell, M. F. Arnsdorf, and S. L. Lindquist, *Science* **289,** 1317 (2000).

[84a] T. Scheibel and S. Lindquist, *Nat. Struct. Biol.* **8,** 958 (2001).

into CRBB with gentle vortexing to a concentration of $5\mu M$ and is then incubated at 5° without agitation.

Assembly of Sup35NM into amyloid may be accelerated by several conditions. For example, the addition of 1/50 volume of Sup35NM fibers preformed from a 5 μM solution of protein or yeast lysates from [PSI^+] strains will decrease the time of fiber formation to 10–12 hr.[16,71b] Sonication of preformed fibers greatly increases their capacity to seed the assembly of freshly diluted Sup35NM, further decreasing the polymerization to 2 hr. Alternatively, constant gentle agitation on a roller drum (60 rpm) accelerates the assembly of Sup35NM into amyloid to roughly 2 hr at micromolar concentrations.[45,84]

Structural Analysis of Amyloid Assembly. Assembly of Sup35NM into an ordered amyloid can be monitored by spectroscopy, dye binding, sedimentation, and microscopy. These techniques are described next.

BINDING TO 8-ANILINO-1-NAPHTHALENESULFONIC ACID (ANS) (FIG. 6A). ANS (Aldrich, Madison, WI) is a spectroscopic probe that exhibits low fluorescence in aqueous solutions and high fluorescence in hydrophobic environments, with a concomitant blue shift in the wavelength of maximum emission (λ_{max}).[84b] Folding intermediates, such as molten globules, exhibit increased ANS fluorescence relative to either denatured or fully folded proteins. ANS binds preferentially to Sup35NM amyloid. Solutions of 5 μM Sup35NM and 10 μM ANS are excited at 370 nm, and fluorescence emission is monitored between 420 and 570 nm at a 5-nm bandwidth. Structured Sup35NM fibers exhibit a 10-fold increase in ANS fluorescence accompanied by a ~40-nm blue shift in the λ_{max} of emission to 484 nm compared to unpolymerized Sup35NM. This increased fluorescence may indicate the presence of an exposed hydrophobic pocket(s) or groove(s) in mature Sup35NM fibers.

CONGO RED BINDING (FIG. 6B). Similar to many other amyloidogenic proteins, Sup35NM fibers bind to the diagnostic dye, Congo red (Sigma).[85] Monitoring Congo red binding over an assembly time course is a sensitive probe for fiber formation. The absorbance of a solution of 1 μM Sup35NM and 10 μM Congo red in CRBB is monitored between 400 and 600 nm. Sup35NM fibers exhibit a spectral shift in absorbance, with a new peak at 540 nm, in comparison with unpolymerized protein or Congo red alone. The amount of Congo red bound to Sup35NM may be calculated using the following equation:

$$\text{mole Congo red bound/liter solution} = (A_{540}/25{,}295) - (A_{477}/46{,}306)$$

where A_{540} and A_{477} refer to the absorbance at 540 and 477 nm, respectively.[85] Under these conditions, Sup35NM binds roughly 4.4 mol of Congo red per mole of protein with a K_d of 250 nM.[16] In addition, fibers of Sup35N stained with a

---

[84b] L. Stryer, *J. Mol. Biol.* **13,** 482 (1965).
[85] W. E. Klunk, J. W. Pettegrew, and D. J. Abraham, *J. Histochem. Cytochem.* **37,** 1273 (1989).

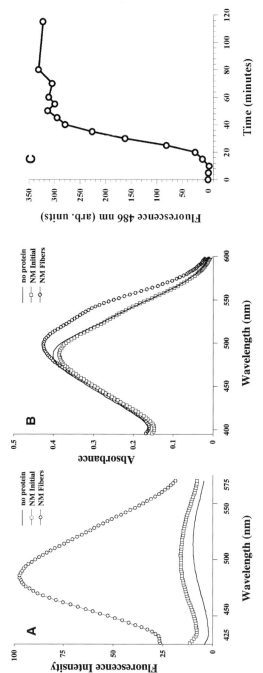

FIG. 6. Spectroscopic detection of the polymerized His-tagged Sup35NM *in vitro*. (A) ANS binding to His-tagged Sup35NM. Fluorescence emission spectrum of ANS alone or bound to unpolymerized His-tagged Sup35NM or to His-tagged Sup35NM assembled into amyloid fibers. (B) Congo red binding to His-tagged Sup35NM. Absorption spectrum of Congo red alone or bound to unpolymerized His-tagged Sup35NM or to His-tagged Sup35NM assembled into amyloid fibers. (C) Thioflavin T binding to Sup35NM during fiber formation. Fluorescence emission of thioflavin T at 486 nm was used to monitor the conversion of soluble Sup35NM into amyloid fibers. (A) and (B) from T. R. Serio, A. G. Cashikar, J. J. Moslehi, A. S. Kowal, and S. L. Lindquist, *Methods Enzymol.* **309,** 649 (1999).

solution of Congo red exhibit apple-green birefringence when viewed by polarized light.[80]

Special consideration should be given to maintaining identical buffer conditions when comparing different samples, as the quantity of Congo red binding to proteins is altered by pH, denaturant, and metals.

BINDING TO THIOFLAVIN T (FIG. 6C). Thioflavin T (ThT) is a yellow dye that has been used as a diagnostic for amyloid and to monitor the kinetics of fiber formation of many amyloids (Sigma).[86,87] We have found that ThT is a sensitive and effective means of monitoring the kinetics of Sup35NM fiber formation *in vitro* (S. M. Uptain and S. Lindquist, unpublished data, 2000). In the absence of amyloid fibers, ThT fluoresces faintly at 438 nm when excited at 350 nm. In the presence of amyloids, including Sup35NM fibers, ThT fluoresces brightly at 481 nm when excited at 450 nm. There are several advantages of using ThT instead of Congo red. Unlike Congo red, ThT does not inhibit Sup35NM fiber formation; thus, fiber formation may be performed in the presence of the dye. Second, ThT is much more sensitive and has a greater dynamic range: we found that the fluorescence change was linear with Sup35NM fiber concentrations from at least 50 nM to 10 μM. Third, fluorescence of ThT is more specific for amyloid than the spectra shift of Congo red.[86] Finally, binding of ThT is very rapid so a prolonged incubation period is not required.

1. We prepare stock solutions of 1 to 5 mM ThT in water freshly for each experiment. The concentration of a ThT stock solution must be determined spectrophotometrically. The molar extinction coefficient at 411 nm is 22,000 M^{-1} in water.

2. We routinely monitor Sup35NM fiber formation of 0.2 μM Sup35NM fibers (molarity based on total monomer) in 1× CRBB, pH 7.4, in the presence of 20 μM ThT. To saturate binding, we recommend maintaining a 20- to 100-fold excess of ThT compared to protein. Although we assay in CRBB at pH 7.4, it has been reported for other amyloids that the sensitivity can be increased further by performing the assays in pH 8.5–9.0.[87]

3. Fluorescence measurements are made with a Fluoromax-3 (Jobin Yvon Inc., Edison, NJ) or any comparable fluorescence spectrophotometer at 25°. The emission wavelength is 450 nm and the excitation wavelength is 481 nm. We use spectral bandwidths of 3 nm for both emission and excitation; however, this will vary depending on experimental conditions and the instrument used.

SDS SOLUBILITY (FIG. 7A). Assembly of Sup35NM into amyloid may also be monitored by the degree of solubilization by 2% (w/v) SDS. Unpolymerized protein remains soluble in 2% (w/v) SDS. In contrast, once amyloid has formed, these structures are largely insoluble in 2% (w/v) SDS at room temperature. This

---

[86] H. LeVine III, *Methods Enzymol.* **309,** 274 (1999).
[87] H. Naiki and F. Gejyo, *Methods Enzymol.* **309,** 305 (1999).

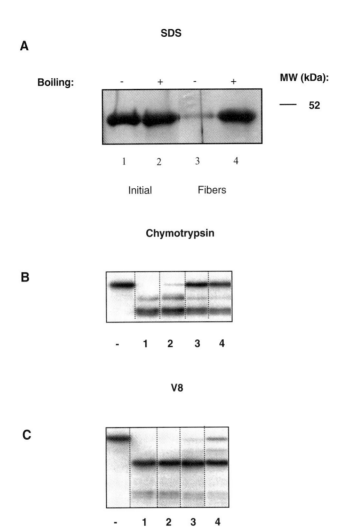

FIG. 7. Biochemical analysis of polymerized Sup35NM *in vitro*. (A) SDS solubility of unpolymerized and polymerized His-tagged Sup35NM. Coomassie Brilliant Blue-stained 10% SDS–polyacrylamide gels of unpolymerized His-tagged Sup35NM (initial: lanes 1 and 2) or amyloid (fibers: lanes 3 and 4) incubated with 2% (w/v) SDS without (−) or with (+) boiling are shown (from Ref. 74). (B and C) Monitoring the progress of fiber formation by limited proteolysis of Sup35NM with chymotrypsin or V8, respectively. Unpolymerized Sup35NM was converted to amyloid fibers within 120 min by incubating at room temperature with rotation on a roller drum at 60 rpm. At 0 (lane1), 30 (lane2), 75 (lane3), or 150 (lane 4) min, samples were digested with chymotrypsin (B) or V8 protease (C) and analyzed by SDS–polyacrylamide gel electrophoresis, electroblotted to Immobilon-P (Millipore), and detected by immunoblotting with an antibody specific for Sup35. (B and C) Reprinted with permission from T. R. Serio, A. G. Cashikar, A. S. Kowal, G. J. Sawicki, J. J. Moslehi, L. Serpell, M. F. Arnsdorf, and S. Lindquist, *Science*. **289,** 1317 (2000). Copyright © 2000 American Association for the Advancement of Science.

difference in SDS solubility, combined with SDS–PAGE, is utilized as a qualitative assay to monitor fiber formation.

Add SDS sample buffer to a final concentration of 1× to two 20-μl aliquots of a 5 μM polymerization reaction. Boil one sample in a water bath for 10 min while incubating the other sample at room temperature. Separate the samples on a 10% SDS–PAGE gel and stain with Coomassie Brilliant Blue R-250. The same amount of unpolymerized Sup35NM enters the gel whether or not the sample has been boiled. In contrast, Sup35NM fibers only enter the gel in boiled samples.

LIMITED PROTEOLYSIS OF Sup35NM (FIGS. 7B AND C). Limited proteolysis of Sup35NM with chymotrypsin and V8 provides sensitive probes for domain-specific structural changes during amyloid assembly. The Sup35N region contains 20 tyrosine residues, which are high-affinity sites for cleavage with the protease chymotrypsin, whereas the Sup35M region contains none. Conversely, the Sup35M region contains 23 glutamic acid residues, high-affinity sites for cleavage with V8 protease, whereas the Sup35N region contains none. Alterations in the digestion pattern reflect either a change in conformation or accessibility for either the Sup35N (chymotrypsin) or the Sup35M (V8) region.

New batches of proteases should be titrated with known samples (both fibers and freshly diluted Sup35NM). Incubate samples (20 μl) of a 5 μM solution of Sup35NM in CRBB with either chymotrypsin (~1/250, w/w) or V8 (~1/25, w/w) at 37° for 15 min. Proteases should be freshly resuspended at a concentration of 1 mg/ml in 1 mM HCl for chymotrypsin (Roche Applied Science, Indianapolis, IN) or 1 mg/ml in water for V8 (endoproteinase Glu-C; Roche Applied Science). Terminate the reaction by adding SDS sample buffer to 1× and boiling immediately for 10 min in a water bath to inactivate proteases. Separate digestion products on 10% SDS–PAGE gels and stain with Coomassie Brilliant Blue R-250. Cleavage of Sup35NM fibers with either protease produces a characteristic digestion pattern that is distinct from that of unpolymerized Sup35NM.[84] Resistance to chymotrypsin cleavage occurs concomitantly with Sup35NM conversion and assembly, whereas a modest resistance to V8 digestion occurs somewhat later.[84]

SEDIMENTATION ANALYSIS. Another convenient assay for monitoring Sup35NM assembly is differential sedimentation. Sup35NM fibers will sediment at high speeds, whereas unpolymerized Sup35NM will not.[16] Centrifuge samples at 100,000g for 10 min at 4° and analyze by 10% SDS–PAGE. Remove supernatant following centrifugation and add SDS sample buffer to 1×. Add an equal volume of 1× SDS sample buffer to the pellet. Boil both samples for 10 min in a water bath, run on a 10% SDS–PAGE, and stain with Coomassie Brilliant Blue or transfer to Immobilon-P (Millipore) for quantitative Western blot analysis using ^{125}I-labeled protein A (Amersham Biosciences, Piscataway, NJ). Partitioning between supernatant and pellet fractions is indicative of the assembly state, unpolymerized or polymerized, respectively.

Alternately, Sup35NM polymer assembly may be monitored using radio-labeled protein. Sup35NM is radiolabeled with [^{35}S]methionine, purified, and added to a polymerization reaction (10,000 cpm/50 μl of reaction, supplemented with unlabeled Sup35NM to 5 μM). Remove samples (50 μl) and separate into supernatant and pellet fractions as described earlier for unlabeled protein. Following centrifugation, measure the soluble counts remaining in the supernatant in a scintillation counter as an indication of the extent of the reaction. The labeling procedure is given here, and the protein is purified by one of the methods described earlier.

1. Grow a single colony of BL21 [DE3] pAP $lacI^q$ harboring the expression plasmid to an OD_{600} of 0.2 at 37° at 300 rpm in 1 liter of Circle Grow medium (Bio 101, Inc.) supplemented with 50 μg/ml kanamycin and 100 μg/ml ampicillin.
2. Collect the cells by centrifugation at 1,500g for 10 min at 4°.
3. Resuspend the pellet in 1 liter of M9 medium (with $MgCl_2$ substituted for $MgSO_4$) supplemented with antibiotics as described earlier. Incubate at 37°, 300 rpm, 1 hr.
4. Collect the cells by centrifugation, as described previously.
5. Resuspend the pellet in 50 ml of M9 (with $MgCl_2$) supplemented with antibiotics as described earlier. Add 3.5 mCi of Tran ^{35}S-label (PerkinElmer Life Sciences, Inc., Boston, MA) and IPTG to 1 mM. Incubate at 37°, 300 rpm for 2 hr.
6. Collect cells by centrifugation as described earlier and store at −80° or proceed with purification. The typical specific activity is $2-3 \times 10^4$ cpm/μg protein, with a total yield of roughly 10 μg.

ELECTRON MICROSCOPY OF Sup35NM FIBERS (FIG. 8). The most important method to date for identifying the presence of amyloid is by microscopy. Transmission electron microscopy (TEM), scanning transmission electron microscopy (STEM), and atomic force microscopy (AFM) were utilized to monitor the assembly of Sup35NM into amyloid. Although each of these techniques provides distinct information about the structure and size of complexes formed by Sup35NM, our discussion here is limited to EM due to the general accessibility of this technique.

Sup35NM fibers are routinely negatively stained[88] for EM analysis. Apply protein (5 μl of a 5 μM solution) to a glow-discharged 400 mesh carbon-coated copper grid (Ted Pella, Inc., Redding, CA). Allow protein to absorb to the grid for 30 sec and then immediately stain with 200 μl of 2% (w/v) aqueous uranyl acetate. Remove excess liquid from the grid with a filter paper wick and then allow to air dry. View samples in a Philips (Eindhoven, The Netherlands) CM 120 transmission electron microscope at an accelerating voltage of 120 kV in low-dose mode, at a magnification of 40,000×. Record images on Kodak (Rochester, NY) SO 163 film.

[88] E. Spiess, H. P. Zimmermann, and H. Lunsdorf, in "Electron Microscopy in Molecular Biology: A Practical Approach" (J. Sommerville and U. Scheer, eds.), p. 147. IRL Press Limited, Oxford, 1987.

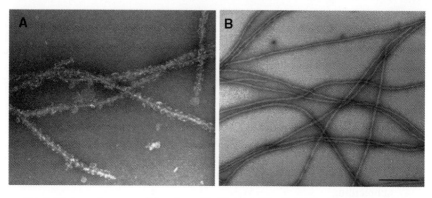

FIG. 8. Electron microscopy of fibers formed by His-tagged Sup35NMC and His-tagged Sup35NM. Protein samples were prepared for visualization by electron microscopy as described. (A) A field of His-tagged Sup35NMC incubated in a high salt buffer consisting of 30 mM Tris–HCl, pH 8.0, 1.2 M NaCl, 10 mM MgCl$_2$ 2 mM GTP, 280 mM imidazole, and 5 mM 2-mercaptoethanol. The base fiber structure has an approximate diameter of 10.6 ± 1.0 nm and displays an amorphous structure along its side, presumably the Sup35C domain. (B) A field of His-tagged Sup35NM dissolved in 20 mM Tris–HCl, pH 8.0, 150 mM NaCl, and 4 M urea and then diluted into 20 mM Tris–HCl, pH 8.0, and 1.2 M NaCl. Fibers from His-tagged Sup35NM exhibit a smooth appearance, whose approximate average diameter is 11.5 ± 1.5 nm. Bar: 200 nm. From T. R. Serio, A. G. Cashikar, J. J. Moslehi, A. S. Kowal, and S. L. Lindquist, *Methods Enzymol.* **309,** 649 (1999).

Fibers formed by Sup35NM have an apparent average diameter of 11.5 ± 1.5 nm.[16] The structure of fibers formed from Sup35NMC is sensitive to buffer conditions. Fibers formed in moderate ionic strength buffer (20 mM potassium phosphate, pH 7.5, 20 mM KCl, 5 mM MgCl$_2$, 2.5 mM 2-mercaptoethanol) are smooth and have an average diameter of 17 ± 2.0 nm.[16] In high ionic strength buffer, fibers of full-length Sup35 are more extended, revealing a interior rod (diameter = 10.6 ± 1.0 nm) and an amorphous outer layer.

Measuring Efficiency of Sup35^{PSI+} Conformational Conversion. Here, we describe a cell-free assay that not only can be used to detect Sup35 in the prion conformation, Sup35^{PSI+}, but to determine the efficiency of conformational conversion. This assay involves fractionating [PSI^+] yeast cellular contents, identifying fractions containing Sup35^{PSI+}, and then measuring how efficiently the prion protein in those fractions stimulates the conversion of soluble recombinant Sup35NM to insoluble amyloid fibers *in vitro*. This approach has been used to show that Sup35^{PSI+} protein from strong [PSI^+] variants converts Sup35NM more efficiently than Sup35^{PSI+} from weak [PSI^+] variants.[71b] Other possible applications of this assay include elucidating the relationship between Sup35NM amyloid formation and Sup35^{PSI+} in prion propagation; investigating the mechanism and intermediates in conversion to the prion form; determining the molecular basis of species barrier; ascertaining the role of molecular chaperones in the conversion process; and identifying modulators of Sup35^{PSI+}.

LYSATE PREPARATION

1. Grow the cells at 25 or 30° to an optical density at 600 nm of 0.4 to 0.6 in the appropriate media.
2. Digest the cell walls enzymatically by treating with 50 μg/ml Zymolyase 100 T (Seikagaku America, Inc.), as described earlier in the section "Analysis of Prion Aggregation *in Vivo*."
3. After washing the cells with 1.2 M sorbitol, incubate the cells in the presence of YPD supplemented with 1.2 M sorbitol and 1 mM NaN$_3$ for 15 min.
4. To initiate lysis, suspend the spheroplasts in buffer containing 25 mM Tris–HCl, pH 7.5, 100 mM NaCl, 5 mM MgCl$_2$, 5% (v/v) glycerol, 4 mM Pefabloc, 5 μg/ml aprotonin, 5 μg/ml leupeptin, and 0.5% (w/v) sodium deoxycholate.
5. Five minutes later, add Brij 58 to 0.5% (w/v). Mix thoroughly.
6. Partially clarify the crude cell lysates by differential sedimentation at 4 to 10,000g for 20 min at 4°.
7. Carefully separate the supernatant fraction (10,000) from the pellet.
8. Determine the concentration of proteins in the supernatant fractions and adjust as needed so that all fractions are the same concentration.
9. Differentially sediment the 10,000 supernatant fractions at 100,000g for 30 min at 4° in a Beckman TLA-100.2 rotor or equivalent.
10. Separate the supernatant (100K) and pellet fractions (100P). Resuspend the 100P pellet fraction in the original volume of lysis buffer without Pefabloc. (We find that Pefabloc inhibits Sup35NM fiber formation).[71b]
11. Repeat the differential sedimentation as in steps 9 and 10.
12. Add 1 volume of 2 M LiCl to the 100,000 pellet fractions and incubate the suspensions for 30 min with gentle rotation at 4°.
13. Layer the 100P pellet suspensions on 30% (w/v) sucrose cushions prepared in lysis buffer without Pefabloc. Differentially sediment the samples at 200,000g for 34 min at 4° in a Beckman SW60 rotor (or equivalent).
14. Fractionate the sucrose cushion into about six fractions of equal volumes. Resuspend the pellet fraction (200,000) in lysis buffer without protease inhibitors.
15. Ascertain the presence of Sup35 in the sucrose cushion pellet fraction by immunoblotting with antisera specific for Sup35.

QUANTITATIVE ASSESSMENT OF EFFICIENCY THAT Sup35^{PSI+} CONVERTS Sup35NM INTO AMYLOID

1. Incubate recombinant 0.3 mg/ml Sup35NM (purified as described in section "Isolation and Storage of Purified Protein") in 1× CRBB, pH 7.4, at 7°, without agitation, in the absence or presence of approximately 5 μg/ml Sup35$^{[PSI+]}$-containing protein fractions (typically a 25-fold dilution). The specific amounts of the pellet fractions needed to detect an effect will vary with the preparation and

prion isolate. To detect nonspecific effects of buffer or pellet components on fiber formation, we strongly recommend including a control with the buffer used to resuspend the sucrose cushion pellet and a control with the 200,000 pellet proteins from a [psi^-] strain.

2. Determine the progress of fiber formation using thioflavin T or Congo red binding assays as described earlier. We advise using at least two unrelated methods to follow Sup35NM fiber formation kinetics; e.g., we routinely use thioflavin T binding and SDS resistance assays. Under these conditions, fiber formation in the presence of just lysis buffer begins after 15 to 20 hr and begins in samples containing Sup35^{PSI+} anytime earlier, depending on the [PSI^+] isolate used and sample preparation.

3. As another control, determine whether any Sup35NM was degraded by contaminating proteases from various sucrose cushion pellet fractions during the incubation period by staining proteins separated by SDS–PAGE with Coomassie Brilliant Blue R-250.

Conclusion: Genetic and Biochemical Criteria of Prions in Yeast

Many approaches are available to identify and study prion proteins in yeast. These include (1) approaches based on phenotypic changes such as nonsense suppression or growth; (2) approaches based on genetic criteria, such as non-Mendelian (cytoplasmic) inheritance; curing by certain stress-inducing agents and by chaperone alterations; and induction *de novo* by overproduction of the protein determinant; and (3) approaches based on biochemical criteria, such as aggregation and proteinase resistance *in vivo* and *in vitro*. Because no one approach is sufficient to establish prion-like behavior, a collection of various approaches should be used. Such a strategy, reviewed here primarily for [PSI^+], is equally applicable to the other yeast prions. It is possible, however, that not all the criteria described here will apply to every new prion candidate, as some differences in the properties of known prions are already apparent. Newly discovered prions may behave in unexpected and dramatically different ways. If so, our hope is that the methods described herein will provide a framework for their discovery and characterization.

Note in proof. Recently, the [PIN$^+$] element was shown to be identical to the [RNQ$^+$] prion in at least some genotypic backgrounds, including 74-D694.[89]

Acknowledgments

We thank T. R. Serio, A. Cashikar, G. Newnam, and R. Wegrzyn for help with preparing the figures. This work was supported by grants from the National Institutes of Health (R01GM58763) and the Amylotrophic Lateral Sclerosis Association to Y.O.C., a Postdoctoral Fellowship (PF-98-135-01-GMC) from the American Cancer Society to S.M.U., a grant from the National Institutes of Health (GM25874) to S.L.L., and the Howard Hughes Medical Institute.

[89] I. L. Derkatch, M. E. Bradley, J. Y. Hong, and S. W. Liebman, *Cell* **106**, 171 (2001).

[31] Assaying Replication Fork Direction and Migration Rates

By ANJA J. VAN BRABANT and M. K. RAGHURAMAN

Introduction

"A picture is worth a thousand words." It was from pictures of replicating yeast DNA, captured by electron microscopy, that the first rough outlines of eukaryotic chromosome replication emerged. Electron microscopy had been used to look at replication in many organisms, but the budding yeast *Saccharomyces cerevisiae* had the unique attribute of being a eukaryote with small chromosomes—small enough to allow isolation and visualization of entire, intact chromosomes. Electron microscopy of such chromosomes demonstrated the presence of replication bubbles and forks and contributed to the notion of fixed origin locations in yeast.[1,2]

However, electron microscopy fell short when it came to detecting some of the dynamic aspects of replication, such as fork migration rates and unidirectional vs bidirectional fork progression from replication origins. For such studies, DNA fiber autoradiography provided a powerful alternative. The most common application of this technique follows a pulse–chase protocol, where cells are incubated with radioactive DNA precursors at one specific activity (the "pulse"), followed by a period during which a radioactive precursor of a different (often higher) specific activity is used (the "chase"). As a consequence, elongating replication forks label the newly synthesized DNA first at one specific activity and then at the other. The labeled DNA can be visualized by spreading the DNA on slides and applying a photographic emulsion layer; the labeled DNA is manifested as tracks of silver grains, and the density of the tracks corresponds to the specific activity of the radioactive label. Thus, fork movement is seen as a transition from a low density to a high density of silver grains along the track (or vice versa, depending on the labeling regimen), and a comparison of track length with the duration of the labeling treatment gives an estimate of fork migration rates. Such studies demonstrated the bidirectionality of replication origins in mammals, as well as in yeast, and led to the first good estimates of fork migration rates.[3,4]

These methods, while powerful, are relatively laborious and time-consuming. A more serious limitation of these methods is that one is reduced to making conclusions about bulk chromosomal DNA by observing anonymous chromosomal DNA

---

[1] C. S. Newlon, T. D. Petes, L. M. Hereford, and W. L. Fangman, *Nature* **247**, 32 (1974).
[2] C. S. Newlon and W. Burke, "ICN-UCLA Symposium on Molecular and Cellular Biology," p. 399. Alan R. Liss, Inc., New York, NY, 1980.
[3] J. A. Huberman and A. Tsai, *J. Mol. Biol.* **75**, 5 (1973).
[4] C. J. Rivin and W. L. Fangman, *J. Cell Biol.* **85**, 108 (1980).

molecules. For a variety of reasons, it has become increasingly important to assay replication forks at specific chromosomal locations. For example, in this age of genomics and so-called transcriptomes, we are still relatively ignorant about if and how the organization of replicons relates to that of transcription units. Such organization may be important for regulating transcription, preventing collisions between RNA and DNA polymerases,[5,6] and genomic stability.[7] Furthermore, checkpoint systems that control S-phase progression have become a hot issue in recent years. In response to DNA damage, cells can delay entry into S phase and slow the rate of S-phase progression.[8-10] Likewise, treatment of cells with hydroxyurea, which inhibits replication fork elongation by blocking ribonucleotide reductase, seems to prevent the initiation of late origins but not of early origins, a response that requires function of the gene *RAD53*.[11,12] How much of the checkpoint control is targeted at origins and how much at elongating forks remains an open question. Last, but not least, there has been a renewed interest in understanding how replication forks are assembled and regulated, especially with the discovery that some of the proteins involved in the control of origin activation may also have roles in fork progression.[13-15] Addressing these questions will require good assays of replication fork movement for specific locations in the genome, particularly when the experiments involve chromosomes or plasmid constructs that were engineered for the purpose.

For the reasons just outlined, techniques such as electron microscopy and fiber autoradiography have largely been superceded by methods, such as two-dimensional (2D) agarose gel electrophoresis,[16-18] that allow examination of specific DNA sequences. This chapter describes two methods that we use routinely to assay replication forks. The first, a modified version of the 2D gel method, allows determination of the direction of fork movement through sequences not containing an origin, as well as determinations of origin location and origin efficiency. The second method is an adaptation of the Meselson–Stahl density transfer method;

[5] B. J. Brewer, *Cell* **53**, 679 (1988).
[6] A. P. Wolffe, *J. Cell Sci.* **99**, 201 (1991).
[7] G. M. Samadashwily, G. Raca, and S. M. Mirkin, *Nature Genet.* **17**, 298 (1997).
[8] A. G. Paulovich and L. H. Hartwell, *Cell* **82**, 841 (1995).
[9] A. G. Paulovich, D. P. Toczyski, and L. H. Hartwell, *Cell* **88**, 315 (1997).
[10] A. G. Paulovich, R. U. Margulies, B. M. Garvik, and L. H. Hartwell, *Genetics* **145**, 45 (1997).
[11] C. Santocanale and J. F. Diffley, *Nature* **395**, 615 (1998).
[12] T. Tanaka and K. Nasmyth, *EMBO J.* **17**, 5182 (1998).
[13] O. M. Aparicio, D. M. Weinstein, and S. P. Bell, *Cell* **91**, 59 (1997).
[14] J. A. Tercero, K. Labib, and J. F. Diffley, *EMBO J.* **19**, 2082 (2000).
[15] K. Labib, J. A. Tercero, and J. F. Diffley, *Science* **288**, 1643 (2000).
[16] B. J. Brewer and W. L. Fangman, *Cell* **51**, 463 (1987).
[17] J. A. Huberman, L. D. Spotila, K. A. Nawotka, S. M. el-Assouli, and L. R. Davis, *Cell* **51**, 473 (1987).
[18] W. L. Fangman and B. J. Brewer, *Annu. Rev. Cell. Biol.* **7**, 375 (1991).

it allows determination of the kinetics of replication for chromosomal restriction fragments of interest. This method, used in concert with the 2D gel method, provides a way of determining the rate of fork progression through specific segments of the chromosome.

Two-Dimensional Agarose Gel Method to Detect Direction of Fork Migration

The use of 2D agarose gel electrophoresis has been instrumental in the identification of replication origins, as well as in the analysis of replication fork movement.[19,20] We focus on the use of a modification of the standard 2D gel method called fork-direction gel analysis to determine the direction of fork movement within a particular DNA sequence. This method is used to determine the relative abundance of forks moving left to right vs right to left through a particular restriction enzyme fragment. We describe how this method can be used to quantify the efficiency of a replication origin or what percentage of cells in a population are initiating replication from a particular replication origin. We also describe how one can discover the location of replication origins themselves within a particular region of interest by performing fork-direction gel analysis at defined intervals within the region.

Principle of 2D Agarose Gel Method

Both standard 2D gel methods and fork-direction gel methods have been described previously.[21,22] As their names imply, the 2D gel techniques rely on the principle of separating DNA restriction fragments in two dimensions. The DNA of interest must contain branched intermediates; we describe a method to isolate DNA from actively growing yeast cells in order to enrich for branched replication intermediates (see later).

The gist of the 2D gel method is as follows. In the first dimension of electrophoresis, restriction fragments are separated based on their mass using conditions in which their shape has minimal effect on migration. In the second dimension, restriction fragments are separated under conditions that emphasize differences in their shape (Fig. 1). The 2D gel is then blotted and probed with the sequence of interest, just as with a standard Southern blot. Consider a particular restriction fragment that does not contain a replication origin. In a population of replicating

---

[19] K. L. Friedman, J. D. Diller, B. M. Ferguson, S. V. M. Nyland, B. J. Brewer, and W. L. Fangman, *Genes Dev.* **10**, 1595 (1996).
[20] A. D. Donaldson, M. K. Raghuraman, K. L. Friedman, F. R. Cross, B. J. Brewer, and W. L. Fangman, *Mol. Cell* **2**, 173 (1998).
[21] K. L. Friedman and B. J. Brewer, *Methods Enzymol.* **262**, 613 (1995).
[22] A. J. van Brabant, S. Y. Hunt, W. L. Fangman, and B. J. Brewer, *Electrophoresis* **19**, 1239 (1998).

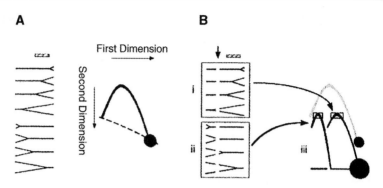

FIG. 1. Diagram of a simple-Y 2D gel and fork direction gel with in-gel digest. (A) Illustration of the 2D gel pattern that would be observed on probing for a fragment that is passively replicated by forks traversing it. The hatched bar represents the probe. The direction of electrophoresis in two dimensions is indicated; the black oval (bottom right) is the "$1n$" spot, representing the unbranched DNA molecules of unit mass. The dashed line represents the arc of unbranched DNA molecules. (B) Illustration of the fork direction gel pattern that would be observed on probing for a fragment that is replicated by leftward (i) and rightward (ii) moving forks. The probe is represented by the hatched bar; the downward arrow marks the location of the in-gel restriction enzyme digest. The gray arc emanating from the upper black oval represents branched molecules that failed to be cut during the in-gel digest. (iii) Boxes indicate the regions of the blot where ^{32}P counts would be measured to quantify the proportions of leftward- and rightward-moving forks.

cells, that fragment will be present as linear molecules (of $1n$ mass corresponding to the filled-in circle or monomer spot in Fig. 1), that did not contain replication intermediates at the time the DNA was collected, and simple Y molecules that comprise a continuum of replication intermediates from $1n$ to $2n$ in mass. Note that the migration of these replicating molecules in the gel will be the same whether the restriction fragment is being replicated from a replication fork moving left to right or moving right to left through the DNA sequence.

In order to distinguish between these two types of replicating molecules, an in-gel digest is performed after the first dimension and before the second dimension gel is run. This in-gel digestion will result in physical separation of rightward-moving forks from leftward-moving forks in the second dimension when probed with the larger of the two fragments generated by the in-gel restriction enzyme digestion. The in-gel digestion will result in displacement of one of the classes of replicating molecules in the gel. For example, for forks moving in one direction, the digestion will remove the small branched, simple Y molecules, leaving only linear, unbranched structures until the replication fork has proceeded beyond the site of the restriction enzyme digestion (Fig. 1B,ii). For forks moving in the other direction, small branched, simple Y molecules will remain detectable by the probe, and these molecules will migrate along the simple Y arc emanating directly from the $1n$ monomer spot resulting from in-gel digestion (Fig. 1B,i).

However, digestion of almost fully replicated molecules within this class will cut off the branched sequences, resulting in only linear fragments being detected by the probe. If the sequence is replicated only by forks moving in one direction, one arc will be detected and the location of the arc relative to the 1n monomer spot will reveal the direction of replication fork movement. However, if a sequence is replicated by forks moving in both directions, two arcs will be detected from the fork-direction gel analysis and the intensity of the two arcs can be compared (Fig. 1B,iii). It is important to note that there is no universal correlation between displaced vs nondisplaced arcs and leftward- vs rightward-moving forks. A determination of which arc represents which direction of fork movement will depend on the restriction site cut during the in-gel digestion and the location of the probe relative to this enzyme site.

Methods

Collection of S-Phase Cells and Isolation of Replication Intermediates. In order to enrich for replication intermediates, yeast cells are synchronized in G_1 phase of the cell cycle and released into S phase. Samples are taken every minute during S phase for a total of 30–40 min and subsequently pooled. Our standard yeast strain, RM14-3a (MATa *cdc7-1 bar1 ura3-52 trp1-289 leu2-3,112 his6*) has mutations in the *CDC7* and *BAR1* genes, making it especially amenable to tight synchronization in G_1. The following method describes the collection of S-phase cells after synchronization with α factor (the *bar1* mutation allows the use of low concentrations of α factor) followed by a high-temperature-induced arrest at the *cdc7* block, which arrests cells just before entry into S phase. Cells will enter S phase immediately after release from the *cdc7* block, which is accomplished by cooling the culture in ice water for a minute or two and returning it to an incubator at room temperature. If the yeast strain does not have the *cdc7* temperature-sensitive allele (*cdc7^{ts}*), cells can be released from α-factor arrest by the addition of Pronase and/or filtering of cells into fresh media without α factor. In this case, cells will enter S phase approximately 30 min after release from the α-factor block at 30° (the actual kinetics of entry into S phase will depend on the temperature of the culture, as well as the strain background). The cells should be monitored for emergence of a bud to be sure that they are in fact releasing from the α-factor block.

Reagents for Cell Collection

α-Factor (custom peptide synthesis from Research Genetics, Huntsville, AL, or alternative vendor): 200 μM solution in water, filter sterilized, and stored at $-70°$ ($= 1000\times$ stock for *bar1* strains) and
3 mM solution ($= 1000\times$ stock for *BAR1* strains)
Pronase (Calbiochem, La Jolla, CA)
0.2 M EDTA

Sodium azide (Sigma, St. Louis, MO): 10% (w/v) solution in water. Use appropriate precautions while handling; store the stock solution in a shatter-proof bottle.

1. In preparation for collecting cells during S phase, freeze 0.1% (w/v) sodium azide, diluted from the 10% stock solution with 0.2 M EDTA, in each of the centrifuge bottles that will be used to collect the cells. Use 40 ml azide–EDTA per 200-ml culture volume. Tilting the bottles in the freezer will increase the frozen surface area, thereby chilling the cells quicker.

2. Grow cells in synthetic media to a density of $\sim 2.6 \times 10^6$ cells/ml (OD_{660} ~ 0.2), with exponential growth for ≥ 4 doublings. We recommend using a culture volume of at least 600 ml in order to get a sufficient amount of DNA for several fork-direction gels. The rated volume of the culture flask should be at least fivefold the volume of the culture (for Erlenmeyer flasks) or twofold the volume of the culture (for Fernbach flasks).

3. Add α factor to a final concentration of 200 nM (*bar1* strains) or 3 μM (*BAR1* strains). Grow the culture for ~ 1.25 doubling times in α factor. Determine the percentage of cells that have arrested by monitoring the percentage of unbudded, "schmooed" cells in the population.

4. This step is only for *cdc7^{ts}* strains; for *CDC7* strains, or if a *cdc7* arrest is not to be used, skip this step and proceed to step 5b. When >90% of the population has arrested, shift the culture to 37°. When the temperature of the culture has reached 37° (check the temperature by dipping an ethanol-wiped thermometer into the culture), add Pronase (10–50 μg/ml of culture) to degrade the α factor. Alternatively, the cells can be filtered out of the α-factor medium, washed two to three times with warm medium, spun down in a warm centrifuge, and resuspended in fresh prewarmed medium at 37°. Check the cell morphology after 90 min at 37°; the *cdc7^{ts}* block will arrest cells as large budded cells.

5a. When >90% of cells in the population have buds, swirl the culture in ice water to quickly cool the cells and return them to a room temperature shaking incubator. The ice–water treatment is to quick chill the culture to 23°. The duration of the ice–water treatment required to bring the culture temperature from 37° to 23° is determined empirically ahead of time for the desired culture volume. Proceed to step 6.

5b. When >90% of the population has arrested, add Pronase (10–50 μg/ml) to degrade the α factor. At 30°, S phase will begin about 30 min after release from the α-factor block; however, the length of the lag will depend on the strain and on the culture temperature and should be ascertained by flow cytometry beforehand.

6. Collect 20-ml samples every minute for 30–40 min beginning with the start of S phase. Pipette the sample into a chilled Corex tube (on ice) containing 0.2 ml 10% sodium azide and then transfer the sample to the centrifuge bottle with the 40 ml frozen azide–EDTA. Cap and shake the centrifuge bottle vigorously

to chill the cells. Repeat with subsequent samples, pooling up to 10 samples per centrifuge bottle. (Because the goal is simply to obtain a collection of cells that represent all stages of S phase, there is no need to keep the individual samples separate.) Be sure to add fresh sodium azide to the Corex tube with each sample and to keep everything on ice. When all the samples have been collected, spin down the cells in a cold (4°) centrifuge at 4000g for 8 min. Resuspend the cells in 50 ml ice-cold water, transfer to a 50-ml conical tube, and spin for 5 min at 4° and 2500g. Discard the supernatant; the cell pellets can be frozen at $-20°$ at this point.

Reagents for Isolation of Replication Intermediates

NIB (nuclei isolation buffer): 17% (v/v) glycerol, 50 mM MOPS, 150 mM potassium acetate, 2 mM magnesium chloride, 500 μM spermidine, 150 μM spermine. Adjust to pH 7.2. Store at 4°.
Acid-washed glass beads (0.45–0.6 mm diameter, Sigma)
TEN (Tris–EDTA–NaCl): 50 mM Tris, 50 mM EDTA, 100 mM NaCl, pH 8.0
25% (w/v) Sodium N-lauroylsarcosine (Sarkosyl)
Proteinase K (Boehringer Mannheim)
Cesium chloride (Gallard-Schlesinger Industries, Garden City, NY, 99.9% pure "Special Biochemical Grade")
5 mg/ml (w/v, dissolved in H_2O) Hoechst 33258 dye (bisbenzimide, Sigma)
"Dummy" CsCl solution: Add 720 μl 25% Sarkosyl to 12 ml TEN. Weigh this solution and bring the volume up to a weight of 16.6 g with TEN. Add 17.2 g CsCl and mix gently to dissolve the CsCl. Add 500 μl Hoechst dye. Seal the cap with Parafilm, wrap in foil, and store at 4°.
5 : 1 (v/v) 2-Propanol : water solution
TE_1: 10 mM Tris, 1 mM EDTA, pH 8.0

General notes. DNA containing replication intermediates is very fragile and susceptible to breakage. Care should be taken to minimize the amount of shearing of the DNA throughout this protocol. After the initial cell lysis step, do not use a vortexer to mix the liquids—just tilt the tube up and down gently to mix. Also, use a pipette with a cutoff tip when resuspending the DNA after ethanol precipitation. The following protocol suggests (in parentheses) volumes to use at each step if the cell pellet was prepared from a 600-ml culture that was synchronized as described previously. Adjust accordingly if the number of cells in the cell pellet differs significantly from this amount.

1. If the pellet was frozen, thaw the pellet on ice. Resuspend the cells at $1.5–2 \times 10^9$ cells/ml (3 ml) in cold NIB.
2. Add an equal volume of acid-washed glass beads. (Having more than an equal volume of beads is preferable to having less.) Keep on ice.

3. Vortex at maximum speed for 30-sec periods with 30 sec on ice between vortexing. Repeat the vortexing 10 to 15 times or until >90% of cells have been broken. Monitor breakage by examining the cell suspension under a phase-contrast microscope, which will distinguish between broken cells or ghosts and intact cells. Be sure to use a good vortexer, which lifts and swirls the contents of the tube.

4. Collect the supernatant with a glass Pasteur pipette or a blue P1000 pipette tip. Stick the tip through the beads to the bottom of the tube and pipette the liquid. You will pick up some beads as well; they will be removed in a later step so do not worry.

5. Rinse the beads at least twice with 1.5 volumes of fresh NIB each time. Pool the supernatants and spin at 20,000g in a refrigerated Sorvall SS-34 rotor or its equivalent for 45 min.

6. Resuspend the pellet (containing nuclei, cell ghosts, unbroken cells, and beads) at a final concentration of 2×10^9 cell equivalents/ml in TEN (3 ml).

7. Add Sarkosyl to achieve a final concentration of 1.5% (180 μl 25% Sarkosyl). Mix gently.

8. Add solid proteinase K to a final approximate concentration of 300 μg/ml (900 μg). Incubate at 37° for 1 hr.

9. Centrifuge at 3000g in a cold SS34 rotor or its equivalent for 5 min to pellet cells and ghosts. The supernatant should have only minor turbidity.

10. Tare a 15-ml Falcon tube on a scale and add the supernatant. Bring the volume up to a weight of 4.15 g with TEN. Add 4.3 g CsCl. Dissolve the CsCl by gentle mixing. Avoid vigorous mixing, which results in foaming.

11. After the CsCl has dissolved, add 125 μl of the Hoechst 33258 solution. Mix gently.

12. Transfer the solution to a 5-ml Quick-Seal ultracentrifuge tube (Beckman Quick-Seal 1/2 × 2 inch tubes) using a Pasteur pipette. Fill the tube to the top with additional "dummy" solution.

13. Spin the CsCl solutions in a Vti65 or VTi65.2 rotor at 55,000 rpm (298,000g) and 20° for 18–24 hr.

14. Visualize the DNA bands under long-wave UV light. Hoechst dye separates DNA based on G+C content; thus, there should be three bands in the gradient, although it might be hard to see all three bands, depending on the amount of DNA. The top diffuse band is mitochondrial DNA, the prominent middle band is chromosomal DNA, and the faint band immediately below the chromosomal DNA is the ribosomal DNA (rDNA) band. Puncture a hole at the top of the gradient with a 22-gauge needle and leave the needle inserted as an air hole. Carefully insert a 16-gauge needle attached to a 3-ml syringe into the side of the tube, just below the rDNA band. Remove rDNA and chromosomal DNA, avoiding the particulate material between nuclear and mitochondrial bands. Try to keep the volume to less than 1 ml. Measure the volume in the syringe. Remove the needle and dispense the material into a sterile 15-ml glass Corex tube.

15. Add an equal volume of 5 : 1 2-propanol:water solution. Swirl the tube well to mix the contents. Do not vortex. After the phases separate, remove the alcohol phase (top) with a Pasteur pipette. We recommend letting the phases settle out in the tip of the pipette and dropping the aqueous phase (bottom) back into the Corex tube until the meniscus is reached; discard the upper alcohol phase. Repeat the extraction two more times.

16. Add 3 volumes of cold 70% ethanol to the DNA slowly, pipetting it down the side of the Corex tube. With the ethanol floating on the CsCl layer, mix the phases using single quick swirling motions. The DNA should come out of solution at the interface as a fibrous network. Continue the single swirls until the phases are mixed and the DNA has fallen out of solution as a "clot."

17. Melt and seal the tip of a Pasteur pipette. Remove the DNA clot using the sealed tip of the Pasteur pipette. The DNA should stick to the glass. Submerge the pipette tip with the clot in 1 ml of fresh 70% cold ethanol to rinse the clot. Tease the DNA from the end of the glass tip by allowing it to stick to the inside of a 1.5-ml microcentrifuge tube. Removing the DNA becomes harder as it begins to dry; if necessary, use a disposable pipette tip to aid transfer of the DNA from the end to the Pasteur pipette to the microcentrifuge tube. Add 100–200 μl TE$_1$ to redissolve the DNA. If the clot is stuck near the top of the microcentrifuge tube, bring it down to the bottom of the tube by pipetting the TE$_1$ onto the clot and letting it roll down with the TE$_1$. If the DNA does not come off the glass tip, break the tip off and keep it in the Eppendorf tube until the DNA has dissolved in the TE$_1$.

18. Spin the ethanol–CsCl solution at 7600g rpm for 20 min in a cold centrifuge. Rinse the pellet with fresh 70% ethanol, recentrifuge briefly, air dry the tube, and resuspend any DNA in 150 μl TE$_1$. Use a cutoff pipette tip when resuspending the DNA to avoid shearing. Pool with the DNA clot and store at 4°. The DNA may take a day or two to go into solution. Encourage dissolving by flicking the tube occasionally. Quantitate the concentration of DNA using a spectrophotometer.

Fork-Direction Gel Analysis

For fork-direction gels, it is important to consider carefully the sizes of the restriction fragments both for the initial restriction fragment from digestion of total DNA and for the subfragment to be probed after the in-gel restriction enzyme digestion. The restriction enzymes chosen for the initial digest should ideally produce a fragment of 3.5–5.0 kb. The in-gel digestion site should be located one-quarter to one-half of the way from one end of this fragment. The probe should hybridize to the larger of the two fragments. (In principle, the smaller fragment could also be probed, but in practice, replication intermediates are harder to detect for fragments smaller than ~2.0 kb.) In addition to the sizes of the restriction fragments, it is also important to choose a "good" enzyme for the in-gel digest: an enzyme that will cleave the agarose-embedded DNA to completion or near completion. Enzymes that work well for in-gel digestion in our hands include AvaI, BglII, ClaI, EcoRV, NcoI, and PstI. As a supplement to the following protocol, we recommend reviewing Friedman and Brewer,[21] as it contains further details of the technique and additional tips for success.

Reagents

Restriction enzymes appropriate for digesting total genomic DNA
Ethanol/potassium acetate: Add dry potassium acetate (0.05 mol) directly to absolute ethanol (100 ml) for a final potassium acetate concentration of 0.5 M. Store the mix at 4°.
70% Ethanol
Restriction enzymes that cut well in agarose for in-gel digestion
Seakem LE agarose (FMC, Rockland, ME)
1× TBE (85 mM Tris base, 89 mM boric acid, 2.5 mM EDTA, pH 8.3)
5× Gel-loading buffer: 30% (w/v) Ficoll, 10 mM EDTA, pH 8.0, 0.1% Sarkosyl, 0.05% (w/v) bromphenol blue, and 0.05% (w/v) xylene cyanol
Costar (Corning Inc., Corning, NY) disposable pipette reagent reservoirs
$TE_{0.1}$: 10 mM Tris, pH 8.0, 0.1 mM EDTA
TE_1: 10 mM Tris, pH 8.0, 1 mM EDTA
~100 ml of 1× restriction enzyme buffer for the in-gel restriction enzyme

1. Digest 5–8 μg of yeast DNA isolated as described earlier in a volume of 300 μl with 5–10 μl enzyme for 5–6 hr at 37°. Because the objective is to obtain complete digestion with the enzyme being used without degradation or star activity, the amount of enzyme required varies depending on the efficiency of the particular restriction enzyme. To minimize the potential for degradation of the DNA and/or collapse of replication intermediates, we recommend against performing overnight digestions.

2. Ethanol precipitate the restriction enzyme digest with 2 volumes of ethanol/potassium acetate. Precipitate the DNA for ~20 min at −70° or at least 1 hr at −20°. Spin for 15 min at full speed in a chilled microfuge. Wash the pellet with 70% (v/v) ethanol and dry.

3. Resuspend the pellet in 15 μl 1× gel-loading buffer (diluted from the 5× stock with 1× TBE). In order to minimize breakage of replication intermediates by multiple pipetting, redissolve the pellet by leaving the tube at 4° for several hours or overnight. Ensure that the pellet is resuspended by gentle pipetting using a cutoff pipette tip before loading the first-dimension gel. Spin the tubes briefly to remove any particulate material.

4. Prepare the first-dimension gel. For fork-direction gels, it is recommended that the first-dimension gel be prepared with Seakem LE agarose, as in-gel digestion

is more efficient in this type of agarose. If the first digestion yields a fragment of interest of 3.5–5 kb, the first-dimension gel should be 0.4% in 1× TBE without ethidium bromide. If the initial restriction fragment is larger or smaller than recommended, the gel percentage may need to be modified. Our first-dimension gels are 20 cm long and ~0.7 cm thick, and we use a comb with narrow wells (0.4 cm wide) in order to obtain narrow gel lanes. We recommend skipping a lane between samples. Run the gel at 1 V/cm for 20–24 hr.

5. Stain the first-dimension gel in 1× TBE with 0.3 μg/ml ethidium bromide. The gel will be very flimsy, slippery, and fragile, so be careful! Photograph the gel under long-wave UV light and avoid overexposure, which will nick the DNA. Determine the position of the unreplicated fragment of interest using the migration of the molecular weight markers. The efficiency of digestion of the DNA can be estimated by analyzing the migration of restriction fragments from multicopy sequences, including the rDNA and 2-μm plasmid, which will be visible as brightly stained bands in the ethidium bromide-stained gel.

6. Carefully cut the lane with a fresh razor blade, using a plastic ruler to maintain a straight cutting edge. Begin cutting 1 cm below the position of the unreplicated fragment of interest and extend 10 cm up the lane. Try to cut the lane as close to the DNA as possible, leaving a minimal amount of agarose on either side of the lane.

7. Carefully transfer the 10-cm agarose slab to a Costar reservoir, labeling which end corresponds to the top and bottom of the first-dimension gel slab. Sliding the gel slab onto a plastic ruler facilitates moving the slab into the reservoir.

8. Fill the reservoir with $TE_{0.1}$ (~50 ml) and rock gently for 30 min at room temperature. Make sure the gel slab is moving in the buffer and is not stuck to the reservoir. Drain the buffer, being careful not to drop the gel slab, add fresh $TE_{0.1}$ and incubate for another 30 min.

9. Prepare ~100 ml of 1× restriction enzyme buffer for the in-gel restriction digest. Drain the $TE_{0.1}$ and add the restriction enzyme buffer. Incubate for 1 hr, rocking gently, at room temperature. Repeat.

10. Drain the buffer. Transfer the gel slab to a strip of Parafilm on a clean glass plate, placing it with one of the "cut" sides of the gel facing up. This orientation allows maximal soaking of the enzyme into the gel. Soak up excess liquid from the gel slab using Kimwipes. Place the glass plate on a sponge foam in a glass Pyrex dish. Pipette 20–30 μl of enzyme along the gel slab, making sure to distribute the enzyme throughout the length of the gel slab. Fill the bottom of the glass Pyrex dish with a small amount of water and seal the dish with Saran wrap in order to create a humidified chamber and prevent evaporation during incubation. Incubate at 37° for 3–4 hr. Pipette an additional 10–20 μl enzyme to the gel slab and continue the incubation at 37° for another 2–3 hr.

11. Return the gel slab to the Costar reservoir and fill the reservoir with TE_1, rocking gently at room temperature for 30 min.

12. Set up the second-dimension gel. For fragments of ~3.5–5 kb prior to the in-gel digest, the second-dimension gel should be 1.1% agarose in 1× TBE with 0.3 μg/ml ethidium bromide. Again, if fragment sizes vary significantly from the recommended size, the concentration of agarose in the second-dimension gel may need to be adjusted. Turn the in-gel-digested first-dimension gel lane 90° from the orientation that it ran in the first dimension. Orient the slab such that the DNA will migrate out of one of the "cut" sides of the gel slab. In order to avoid having the first-dimension gel lane move while the agarose for the second-dimension gel is being poured, pipette a small amount of molten agarose for the second-dimension gel (cooled to ~55°) around the edges of the lane and allow the agarose to harden, thus keeping the gel lane in place. Pour the rest of the agarose in the second-dimension gel mold. Pour the second-dimension gel until it just covers the first-dimension gel lanes. The dimensions of our second-dimension gels are 20 cm wide by 25 cm long. With these dimensions, we can fit two lanes across the top of the gel and have ample length for fragments to separate.

13. Run the second-dimension gel in the cold room at 5–7 V/cm in 1× TBE with 0.3 μg/ml ethidium bromide. Circulate the buffer from anode to cathode to maintain a constant ethidium bromide concentration. Monitor the progress of DNA migration using a hand-held long-wave UV lamp. You should be able to see a prominent arc, consisting predominantly of linear molecules that were not cut by the in-gel digestion, and a haze of DNA fragments below this arc, corresponding to in-gel-digested DNA fragments. Again, the efficiency of in-gel digestion can be estimated by analyzing the migration of multicopy fragments from the rDNA locus or 2 μm plasmid below the arc of linear molecules. Run the gel until the smallest restriction fragments in the arc of linears has migrated 10–12 cm (~3.5 to 6 hr).

14. Photograph the gel on a UV transilluminator.

15. Treat the gel for Southern blotting. We recommend two 15-min washes in 0.25 N HCl followed by a quick rinse in water and two 15-min washes in 0.5 N NaOH, 1 M NaCl to denature the DNA. Neutralize the gels by one 30-min wash in 0.5 M Tris, 3 M NaCl.

16. Transfer the gel to a membrane for hybridization. Continue as you would a standard Southern blot, blotting through the bottom surface of the gel. We label our probe fragments by random priming using fresh [α-^{32}P]dATP at the maximum available specific activity ($\geq$6000 Ci/mmol). Expose the probed and washed membrane to X-ray film. The film may need to be exposed anywhere from several hours to several days. To quantitate the fork-direction gel, an overnight exposure of the membrane on a Packard InstantImager is usually sufficient.

Application of Method

Determining Efficiency of Replication Origin. On a simple scale, replication origins can be categorized as active or inactive. Inactive origins have been identified as ARS that allow autonomous replication of a plasmid, but, when examined by conventional 2D gel techniques at their native chromosomal locations, do not give the characteristic "bubble arc" of active origins; instead, they only have the "simple Y" arc indicative of replication forks moving through the sequence.[23] Within the class of active replication origins, origins can be categorized as very inefficient on the one extreme to very efficient on the other extreme, with all grades in between. This qualitative categorization can be made simply based on the ratio of intensity of the bubble arc compared to the intensity of the complete simple Y arc within the same sequence. In order to obtain a more precise estimate of origin efficiency and to avoid a false estimate because of a possible artifactual increase in intensity of the simple Y arc due to broken bubble molecules, fork-direction gel analysis can be used to analyze sequences flanking a replication origin. In our hands, this type of analysis can be used to quantitate replication origin efficiencies with an accuracy of ±10%.

To minimize error in the quantitation of fork-direction gels, the signal after hybridization should be high and the background should be low. If these conditions are met, the blot can be scanned in an imaging system that detects ^{32}P emissions. We use a Packard InstantImager electronic autoradiography system to quantitate our fork-direction gels, usually after an overnight exposure; other systems (such as the Molecular Dynamics PhosphorImager, Amersham Biosciences, Piscataway, NJ) may be used also. Because X-ray films give a linear response only over a very small dynamic range of ^{32}P intensities, we recommend against quantitation by densitometric scanning of autoradiograms.

After the exposure, the image can be quantitated by outlining rectangular regions that include the tops of each arc and measuring the total counts detected within each region (see Fig. 1B). The height of each rectangle should be kept constant, but the width can be varied such that it contains the top of the entire arc. The rectangles will usually need to be different widths because the displaced arc is more compressed along the horizontal axis than the arc that emanates directly from the monomer spot. A background rectangular box of identical dimensions as the box containing the arc should be determined for each arc. These background boxes should be placed in regions that have similar background levels as the background in the region of the two arcs—ideally, directly above the arcs or on either side of the arcs (but not between the arcs because there will always be higher background in this area). The proportion of forks in each arc is calculated by subtracting the counts in each background box from the counts in the corresponding box containing the arc. Each arc is then expressed as a percentage of the counts in both boxes combined.

---

[23] C. S. Newlon, I. Collins, A. Dershowitz, A. M. Deshpande, S. A. Greenfeder, L. Y. Ong, and J. F. Theis, *Cold Spring Harb. Symp. Quant. B.* **58**, 415 (1993).

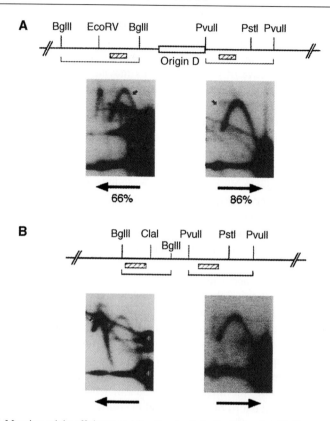

FIG. 2. Mapping origin efficiency and location by fork direction gels. (A) Fragments flanking origin D on a yeast artificial chromosome were examined by fork direction gel analysis. (Left) Fork direction gel of a *Bgl*II fragment cut in-gel with *Eco*RV (probe represented by the left hatched bar). (Right) Fork direction gel of a *Pvu*II fragment cut in-gel with *Pst*I (probe represented by the right hatched bar). Small arrow, arc of leftward moving forks. The percentage forks moving in the direction shown is indicated below each gel image. (B) Example of a fork direction gel where quantitation is not possible. (Left) Fork direction gel of a *Bgl*II fragment cut in-gel with *Cla*I (probe represented by the left hatched bar). Upper asterisk, "1n" spot of linear *Bgl*II fragment (that failed to be cut in-gel by *Cla*I); lower asterisk, "1n" spot of *Cla*I-cut molecules. Small arrow, spot resulting from incomplete digestion of genomic DNA by *Bgl*II. See text for details. (Right) Fork direction gel of a *Pvu*II fragment cut in-gel with *Pst*I (probe represented by the right hatched bar).

In Fig. 2A, in the fork-direction gel of the region to the left of origin D, there are clearly two arcs, indicating forks moving in both directions. The arc resulting from leftward-moving forks (arrow) appears to be more intense, and on quantitation, 66% of forks were found to be in this arc; the remainder, 34%, were moving rightward. In the fork-direction gel to the right of the origin, there is a more distinct difference in intensities of the two arcs. The arc emanating directly from the monomer spot, which corresponds to rightward-moving forks in this case, is clearly more intense than the displaced arc (arrow), representing leftward-moving forks. On quantitation, 86% of forks were found to be moving rightward in this restriction fragment whereas 14% were moving leftward. The fact that the majority of forks are moving outward from the region between the two fragments being examined tells us that there must be an origin within this intervening region. To determine the actual efficiency of this particular origin, we must keep in mind that the outward-moving forks in each fragment consist not only of forks that were initiated by the origin we are examining, but also of forks that were initiated by origins outside the region being examined. In the case of the restriction fragment to the left of the origin, 66% of forks were moving leftward, consistent with forks originating either from origin D or from an origin to the right of origin D. Thirty-four percent of forks were moving rightward and they must be originating from an origin to the left of origin D. Because this 34% of forks will most likely continue to move rightward through origin D and through the region analyzed by the fork-direction gel to the right of origin D, we subtract this 34% from the percentage of forks (86%) that were detected as moving rightward in the region to the right of origin D. Thus, we would conclude that the final efficiency of this origin is (86% − 34%) = 52%. (Calculations that begin with data for the right flank give the same conclusions, of course.) Again, we consider this number to have an error of 10%. We base this error value on the limitations of quantitation, as determined by quantitating a fork-direction gel of a region of a linear plasmid that has a single origin and thus must contain forks moving in only one direction.[22]

Sometimes, the results of a fork-direction gel will be quite clear in broad terms, but will be impossible to quantitate with any accuracy. This difficulty is especially apparent when there is incomplete digestion before the first dimension, during the in-gel digestion, or both. For example, in Fig. 2B (left), in addition to the monomer spot that results from the in-gel digest (lower asterisk), there is still a significant amount of hybridization to the uncut monomer spot (upper asterisk). The gel is further complicated by the fact that the genomic DNA that was run in the first dimension had not been cut to completion, such that in addition to the monomer spot that remains uncut after in-gel digestion, there is another spot of intense hybridization further up along the first-dimension lane (arrow) and a trail of hybridization from this uncut DNA spot. This trail interferes with quantitation of the fork-direction gel arcs, especially with quantitation of the displaced arc whose top lies over this "background" hybridization. Thus, in this case, it is not possible to give a number to the percentage of forks, except for a very rough estimate by eye. Despite not being able to quantify the percentage of forks moving in each direction, quite a bit of information can be gathered from the gel. In particular, it is clear that most forks are moving leftward in this region and together with the right panel, there is also evidence for a replication origin lying between the two restriction fragments.

Identifying Origins of Replication by Fork-Direction Gel Analysis at ~10-kb Intervals. The analysis in Fig. 2 was aimed at determining the efficiency of a known origin. However, it is immediately obvious that this approach can be extended easily to identifying replication origins in previously unmapped regions. One can get clues as to where replication origins exist by virtue of the fact that replication origins are usually bidirectional and forks move in opposite directions in regions flanking an origin. For example, in Fig. 2A, even if we had had no prior knowledge of origin D, the fact that replication forks are moving predominantly away from the center of the region would immediately tell us that at least one origin of replication must be located there. For the purposes of origin mapping, potentially flawed gels as the one in Fig. 2B (left) may serve as well as gels with no background noise—in this example, even with the problems in quantitation, it is obvious that forks are moving away from the center of the region, again indicating that there must be at least one origin located between the two fragments that were examined.

The fragments examined in Fig. 2 were spaced fairly close together. In a mapping strategy that searches for previously unknown origins, fork-direction gel analysis would be performed at 10- to 15-kb intervals. A large region can be scanned with relatively few fork-direction gels as compared to an analysis of overlapping restriction fragments by conventional 2D gel analysis. This approach has been used successfully to map replication origins on native and artificial yeast chromosomes.[24,25] One must of course confirm the existence of a replication origin in the region where fork-direction gel analysis has suggested such an origin by a conventional 2D gel analysis. Restriction fragments can also be subcloned into a plasmid in order to test for autonomous replication by the standard plasmid ARS assay, but one must be cautious in that sequences which provide autonomous replication of a plasmid are not always active as replication origins within their chromosomal context.[23]

In order to locate replication origins using this method of fork-direction gel analysis at spaced intervals, one must have at the very least a good restriction enzyme map of the region of interest, and ideally, the DNA sequence of the entire region. This information allows one to choose the best restriction enzyme sites for fork-direction gel analysis, particularly the enzyme to be used in the in-gel digestion. Additionally, fragments that will be used as probes can be made by polymerase chain reaction (PCR) using the DNA information to design appropriate primers. With some consideration, it is also possible to minimize the number of actual gels that are run by finding restriction enzymes that allow fork-direction gel analysis at multiple locations along a chromosome; i.e., a single fork-direction gel can be run, blotted, and hybridized with different DNA sequences, stripping the blot after each hybridization. If this approach is used, be sure to take into

---

[24] K. L. Friedman, B. J. Brewer, and W. L. Fangman, *Genes Cells* **2**, 667 (1997).
[25] A. J. van Brabant, W. L. Fangman, and B. J. Brewer, *Mol. Cell. Biol.* **19**, 4231 (1999).

account all the corresponding restriction fragments when running and cutting out the first-dimension lane!

Density Transfer Method to Measure Fork Migration Rates

Principle

The rate of fork migration between two locations on a chromosome can be determined by measuring the time of replication for each of those two locations; knowing the difference in replication time and the distance separating the two locations, one can then deduce the rate of fork migration between the two points. Inherent in this approach is the assumption that the region is replicated by a single fork traversing it in one direction. Therefore, this method can only be applied to those regions where the direction of fork movement is already known and where 2D gel analysis has demonstrated the absence of efficient origins within the region being assayed. Sometimes, it may not be possible to avoid the presence of an inefficient origin within the region being analyzed; if so, the analysis will give an approximation rather than an absolute measure of the replication fork rate. Nevertheless, the analysis can still be useful, especially if the goal is to compare two or more strains, or one strain grown under different conditions.

In practice, it is preferable to measure the replication time of several fragments in the region of interest. The inverse slope of the line from a plot of replication time as a function of distance (in kb) then gives the fork migration rate. The deduced fork migration rate is the average rate across the region whether only two fragments or several fragments are assayed. Variation in fork rate within the region can be detected reliably only if multiple fragments are assayed.

Determination of replication time is based on a modification of the Meselson–Stahl density transfer method (Fig. 3[26]). Cells are grown for at least seven population doubling times with [^{13}C]glucose and [^{15}N]ammonia as the sole sources of carbon and nitrogen, respectively, so that the DNA is labeled uniformly with these dense isotopes. If the experiment precludes the use of glucose (e.g., if transcription from the *GAL* promoter is to be induced as part of the experiment), an alternative carbon source such as sodium [^{13}C]acetate may be substituted for [^{13}C]glucose. The cell culture is synchronized at START by treatment with α factor, and the dense labeled medium is exchanged for medium containing normal, light isotopes (^{12}C and ^{14}N). As described in the previous section, if the strain carries a $cdc7^{ts}$ allele, tight synchrony may be achieved by blocking the cells further at the G_1/S-phase boundary after release from the α-factor block. Samples are collected at intervals through S phase after release from the block(s). The DNA is extracted from each sample, cut with a restriction enzyme, and banded by CsCl density gradient centrifugation so that unreplicated, fully dense (heavy–heavy, or HH) DNA is

---

[26] R. M. McCarroll and W. L. Fangman, *Cell* **54,** 505 (1988).

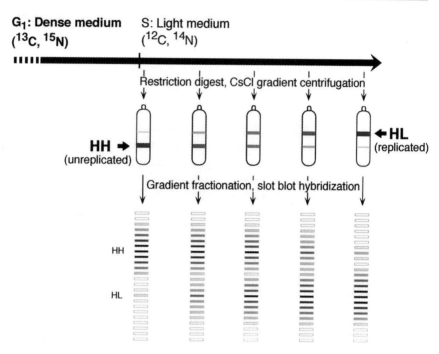

FIG. 3. Diagram illustrating the density transfer method. Cells are grown in isotopically dense (^{13}C, ^{15}N) medium, synchronized just prior to S phase, and released into S phase in the presence of isotopically normal (^{12}C, ^{14}N) medium. Cell samples are collected through S phase, and newly replicated, hybrid density (HL) DNA is separated from unreplicated, fully dense (HH) DNA by centrifugation in CsCl gradients after digestion with a restriction enzyme. The gradients, each of which represents one S-phase sample, are fractionated, slot blotted, and probed with the fragment of interest. The kinetics of replication for that fragment can be deduced from the conversion of the HH form to HL form with increasing times in S phase, seen as a change in the slot-blot hybridization pattern (bottom).

separated from replicated, hybrid density (heavy–light, HL) DNA. Each gradient is fractionated and slot blotted onto a membrane. Hybridization of the slot blot with a probe reveals the relative amounts of sample DNA in the HH vs the HL fractions for the restriction fragment corresponding to the probe, from which the percent replication for that sequence at each S-phase sample time point can be calculated.[26] The time of half-maximal replication for the fragment, the t_{rep}, is used as a measure of the replication time for that fragment. In principle, the average fork migration rate across a region can be deduced from the t_{reps} of two fragments at the boundaries of the region. However, the results are likely to be much more consistent and accurate if several fragments across the region are assayed. Furthermore, it is preferable to measure t_{reps} for fragments spread over a large region rather than a small one because experimental error in measuring replication time will be much less of a factor when comparing widely separated fragments that are replicated by the same fork.

Density Transfer

Reagents and Materials

The following dense isotopes are all obtained from Isotec Inc., Miamisburg, OH (http://www.isotec.com): Sodium acetate-1,2-$^{13}C_2$ 99% atom purity, D-glucose-$^{13}C_6$ 99% atom purity, and ammonium-$^{15}N_2$ sulfate 99% atom purity

-N medium: 1.61 g yeast nitrogen base without amino acids and ammonium sulfate, 11.1 g succinic acid, and 6.67 g NaOH.
Make up to 1 liter with glass distilled water and autoclave.
Dense culture medium: Dissolve the following in -N medium and filter sterilize: 0.1% (w/v) [^{13}C]glucose or sodium acetate and 0.01% (w/v) [^{15}N]ammonium sulfate

Amino acids/supplements (as required, added to the same concentration as in normal synthetic medium)

α Factor (custom peptide synthesis from Research Genetics or alternative vendor): 200 μM solution in water, filter sterilized, and stored at $-70°$ (= 1000× stock for *bar1* strains); 3 mM solution (= 1000× stock for *BAR1* strains)

Preparation of nitrocellulose filters (Pall Gelman Sciences (Ann Arbor, MI) "Supor-200" 47 mm diameter 0.2 μm pore): Nitrocellulose filters are often sold with a coating of glycerol and azide that will have to be washed out before use. Wet several filters by dropping them on glass distilled water in a beaker. Rinse them several times with glass distilled water, submerge them in 50–100 ml glass distilled water, cover the beaker with aluminum foil, and autoclave for 15 min. Prepared filters can be stored in the beaker of water for several months at 4°.

Pronase (Calbiochem)

Sodium azide (Sigma): 10% (w/v) dissolved in water. Use appropriate precautions while handling; store the stock solution in a shatter-proof bottle.

0.2 M EDTA, pH 8.0

1. Grow cells for at least 7 population doubling times in dense culture medium, with exponential growth for ≥ 4 doublings. To reduce the chance of bacterial contamination, antibiotics such as ampicillin (100 μg/ml culture volume) can be added to the culture medium without affecting growth of the yeast cells. The population doubling time of RM14-3a (the strain we have used most extensively for density transfer experiments) is $\sim$2.6 hr at 23° with glucose as the carbon source and $\sim$6.4 hr with sodium acetate as the carbon source; doubling times will be longer with selection for maintenance of a plasmid. We usually aim for about 32×10^6 cells per S-phase sample, which gives enough DNA for at least three slot blots with leeway for errors. Grow the strain in a small volume of the dense medium

(~50 ml) and, while its growth is in logarithmic phase, use it as inoculum for the larger, experimental culture. The small culture may also be used to ascertain the growth rate for the strain in the dense medium. The rated volume of the culture flask should be at least fivefold the volume of the culture (for Erlenmeyer flasks) or twofold the volume of the culture (for Fernbach flasks).

2. When the cell density is 2×10^6 cells/ml (OD_{660} ~0.16), arrest with α factor for 1–1.25 doubling times until the percentage of unbudded cells stops increasing. The concentration of α factor in the culture should be 200 nM for *bar1* strains and 3 μM for *BAR1*$^+$ strains.

3. Collect the cells by filtration through a prewashed, sterile nitrocellulose filter, wash with half of the original culture volume of isotopically normal (^{12}C and ^{14}N, respectively) synthetic medium containing α factor, and resuspend in synthetic medium containing α factor. Note that the final culture volume (in light isotopes) does not need to match the volume of the initial, isotopically dense culture; rather, one should choose a culture volume that is convenient for subsequent manipulation and sample collection, keeping in mind that the cell density should not be higher than ~4×10^6 cells/ml. We typically collect 20 ml of the culture (~32×10^6 cells) per S-phase sample.

4. This step is only for *cdc7*^{ts} strains; for *CDC7* strains, or if a *cdc7* arrest is not to be used, skip this step and proceed to step 5b. Transfer the culture to 37°, wait for the culture temperature to reach 37°, and then add pronase to a final concentration of 0.01–0.05 mg/ml. It is advisable to monitor the temperature of the culture itself using a thermometer that has been wiped with ethanol rather than that of the water in the bath containing the culture—in our experience, the culture temperature is usually about 0.5° lower than that of the water in the bath. Also, when possible, we use glass rather than plastic culture flasks, as heat transfer with plastic flasks is severalfold slower than with glass flasks. Pronase powder may be added directly to the culture, but it is more convenient to dissolve the pronase in 5 ml of synthetic medium first, just to avoid having the dry pronase stick to the sides of the flask.

5a. For *cdc7*^{ts} synchrony and release, hold the culture at 37° for 90–120 min until budded cells accumulate (typically $\geq 90\%$ of the population). Remove one sample ($t = 0$) and then swirl the flask in ice–water and return it to 23° to allow S phase to begin. The ice–water treatment is to quick chill the culture to 23°. The duration of the ice–water treatment required to bring the culture temperature from 37° to 23° is determined empirically ahead of time for the desired culture volume. Continue to collect samples until S phase is complete ($t = $ ~80 min; the length of S phase should be determined beforehand by flow cytometry). We usually collect samples every 5 min until 55 min and then additionally collect samples at 65 and 80 min. See step 6 for details on sample collection.

5b. For release into S phase directly from the α-factor block, remove one sample ($t = 0$) and then add pronase to a final concentration of 10–50 μg/ml to the remainder of the culture. In our standard strain background, S phase begins about 30 min after release from the α-factor block (at 30°). Collect samples every 5 min for ~140–150 min until S phase is complete (approximate S-phase duration should be determined ahead of time for these conditions by flow cytometry). See step 6 for details on sample collection.

6. For each S-phase sample to be collected, freeze 8 ml of 0.1% (w/v) sodium azide (diluted from the 10% stock solution with 0.2 M EDTA) in a 50-ml Oakridge-type plastic, screw-capped centrifuge tube ahead of time. Freeze the tubes slanted to maximize the surface area of frozen azide/EDTA. As each S-phase sample is collected (typically 20 ml), mix the sample with 1/50–1/100 volumes of 10% sodium azide in a 30-ml chilled Corex tube and immediately transfer the mix to a tube of frozen azide/EDTA kept on ice. Alternatively, squirt the 10% azide on the frozen azide/EDTA and immediately add the cell sample. Vortex or shake the tube vigorously to chill the sample and break up the frozen EDTA. Collect cells by centrifugation in a chilled centrifuge (~7000g for 5 min). Discard the supernatant and resuspend each sample in 1 ml of ice-cold sterile, glass-distilled water. Transfer each sample to a 1.5-ml microcentrifuge tube, spin down the cells again by centrifugation for ~15 sec at 14,000g, remove the water by aspiration, and freeze the cell pellets at $-20°$. Cell pellets may be stored at $-20°$ for at least a few weeks.

CsCl Density Gradient Centrifugation and Fractionation

Reagents

$TE_{0.1}$: 10 mM Tris, 0.1 mM EDTA, pH 8.0
RNase A (Sigma), concentrated stock solution: 10 mg/ml RNase A dissolved in 10 mM Tris, pH 7.5, 15 mM NaCl, and boiled 15 min to destroy contaminating DNases. Store in small aliquots at $-20°$.
RNase A, diluted working stock: 5 μg/ml in $TE_{0.1}$ (= 20×), diluted from a concentrated stock solution (see earlier discussion). This diluted stock can be stored on the bench for several weeks.
2× restriction enzyme mix: prepare 50 μl of this solution for each S-phase DNA sample plus 100–150 μl more than the projected need to allow for pipetting error and to have enough left to prepare "dummy" samples: 10 μl 10× restriction enzyme buffer (usually provided by the vendor), 20 units restriction enzyme (but no more than 3 μl), 2.5 μl diluted working stock of RNase A (5 μg/ml), 1 μl of 10 mg/ml bovine serum albumin (often provided by the restriction enzyme vendor), and ice-cold water to bring the volume to 50 μl. Prepare the mix on ice, adding the enzyme last.
TE_{100}: 10 mM Tris, 100 mM EDTA, pH 7.5.

Note. The quality of this TE_{100} solution is critical. In our experience, preparation of TE_{100} by dilution of stock solutions of EDTA gives unsatisfactory results, with the density of the final CsCl solution often being outside the desired range. The following recipe gives consistent results.

Add 18.61 g $Na_2EDTA \cdot 2H_2O$, 1.5 g NaOH pellets, and 5 ml 1 M Tris–HCl, pH 7.5, to 350 ml glass-distilled water. Stir until the solids have dissolved, pH to 7.5 with concentrated NaOH (<0.5 ml of a 12 N stock), and bring the volume up to 500 ml with glass-distilled water. Filter sterilize the solution; do *not* autoclave. The refractive index of the solution should be 1.3393–1.3395. (Calibrate the refractometer to a refractive index of 1.3330 for glass-distilled water.) The TE_{100} solution can be stored on the bench for a few weeks; we take the precaution of lining the threads on the bottle top with Teflon tape before screwing on the cap. The bottle is then further sealed with Parafilm. Always check the refractive index of the solution before use; if the refractive index is not 1.3393–1.3395, discard the solution and prepare a fresh batch.

CsCl Solution: CsCl is dissolved in TE_{100}. Use 1.28 g dry CsCl (Gallard-Schlesinger Industries, 99.9% pure "Special Biochemical Grade") per gram of TE_{100}. The refractive index of this CsCl stock solution should be 1.4052 ± 0.0001.

1. Thaw the cell pellets on ice and extract the DNA from each sample using the standard glass bead/phenol extraction method.[27] Redissolve each nucleic acid pellet in 50 μl $TE_{0.1}$. Add an equal volume of a 2× enzyme mix. (We usually digest the DNA with *Eco*RI.) Digest overnight at 37°. For optimal separation of HH DNA from HL DNA in CsCl gradients, the fragment to be probed should ideally be ≥3 kb. Fragments smaller than ~3 kb tend not to band well, giving poor separation between HH and HL peaks in the CsCl gradient (Fig. 4C).

2. Check for completion of the digest by running 8–10 μl of the reaction mix on a gel. The gel should be blotted and probed for a known fragment. However, if you are familiar with the pattern expected for a particular restriction enzyme (e.g., *Eco*RI), you may be able to decide whether the digest was complete just from rDNA and 2-μm plasmid bands on the ethidium bromide-stained gel.

3. Conditions in steps 3 and 4 are for centrifugation in a Beckman VTi65.2 rotor. Mix each sample (90–92 μl, restriction enzyme included) with 9.141 g (= 5.25 ml) of CsCl stock solution to give a refractive index of 1.4042 ± 0.0001. Load the samples into centrifuge tubes (Beckman Quick-Seal 1/2 inch × 2 inch tubes) using a "dummy" solution (see later) to top off the tubes and eliminate bubbles. Seal the tubes.

Note. We strongly recommend weighing out the CsCl solution (in small, sterile culture tubes, using a good balance) rather than pipetting it out—weighing out the

---

[27] M. D. Rose, F. Winston, and P. Hieter, "Methods in Yeast Genetics: A Laboratory Course Manual." Cold Spring Harbor Laboratory Press, Cold Spring Harbor, NY, 1990.

solution gives more accurate and consistent results. It is worth adding 90 μl of "dummy" restriction digest mix to 9.141 g of the CsCl solution to test for the right refractive index. This dummy CsCl mix then serves to top off the real samples once they are loaded in the centrifuge tubes.

4. Using a dual program, centrifuge at 50,000 rpm (246,000g) for 18 hr, followed by 28,000 rpm (77,000g) for 3.5 hr, with no brake at any step. The temperature is set to 20° throughout.

5. Carefully remove the centrifuge tubes from the rotor. To fractionate a gradient, place the centrifuge tube in the fractionation apparatus (Hoefer Scientific Instruments, Tube Fractionator FS101), cut the top off the Quick-Seal tube (dog toenail clippers, available in any pet store, serve admirably in cutting the tops off), and clamp the tube into place. The gradients are fractionated from the bottom by piercing the bottom of each tube with the hollow fractionator needle and allowing the CsCl solution to drain out through the needle at a rate of ∼1 drop per second.

6. Collect the drops in a 96-well microtiter plate, 10 drops per fraction (i.e., 10 drops per well of the microtiter plate). For convenience in collecting fractions, the plate is moved in opposite directions for alternate rows of wells (i.e., if the first row of fractions is collected left to right, the next row is collected right to left, etc., thereby avoiding the need to move the plate all the way back between fractions). Each gradient typically yields 34–36 fractions of ∼140–150 μl each; two gradients can be dripped into each microtiter plate. Immediately after each plate is filled, it should be sealed with an adhesive Linbro plastic sheet (Hampton Research, Laguna Niguel, CA). Between gradients, clean the fractionator needle by flushing it out with glass-distilled water and forcing air through with a syringe to remove the last traces of water.

7. Read the refractive indices of every third or fourth fraction of each gradient as soon as possible after all the gradients have all been fractionated. Gradient fractions can be stored at 4° indefinitely. Do not freeze them—CsCl tends to crystallize, which makes further handling of the samples difficult.

Note. Other fractionation apparatuses may be used also. However, in our experience, the simplest setups that minimize the path length of sample flow work best. More complex designs that require the sample to flow through lengths of tubing may allow mixing of samples because of turbulent flow within the tubing, thereby blurring the distinction between HH and HL peaks. We strongly recommend practicing with a couple of dummy, CsCl-filled tubes (water does not behave the same as CsCl) before attempting to fractionate a real sample. With the Hoefer apparatus, slight resistance will be felt as the needle pierces the bottom of the Quick-Seal tube. From that point onward, the needle should be advanced very slowly, with a wait of a few seconds between quarter turns of the needle to allow flow to begin. If the needle is advanced too rapidly, the sample may squirt out at a rate that cannot be controlled. Once flow has begun, the flow rate can be controlled by advancing or withdrawing the needle a quarter turn at a time until an appropriate rate is achieved.

Preparation of Slot Blots

Reagents and Apparatus

1 N NaOH
20× SCP: 0.6 M Na_2HPO_4, 2.0 M NaCl, pH 6.8
Slot-blot apparatus: Schleicher and Schuell (Keene, NH) "Minifold II" slot blotting system or its equivalent
Probes for timing markers

We routinely use two or three sequences as our markers for early and late replication. Probe DNAs can be prepared by PCR with yeast genomic DNA (prepared as described in the previous section) as template and the primers listed here.

ARS305 (chromosome III, very early replicating, and may show some "escape" replication at the $cdc7^{ts}$ block, presumably an artifact of premature release from the temperature block.[28]

Primers:

5′-ATTCGCCTTCTGACAGGACG-3′
5′-ATAACGGAGACTGGCGAACC-3′

GAL3 (chromosome IV, *ARS1* adjacent, early/mid-S replication[26])

Primers:

5′-CCTACATGGCCATAGATCCG-3′
5′-AAGGTGGATAGTGCAACCGC-3′

R11 (chromosome V, late replicating[29])

Primers:

5′-GATTGCACGACCAACTTCGG-3′
5′-GTTAGGTAGAGCAGGCAGAC-3′

We routinely use multichannel pipettors to ease the workflow for the following steps, flushing the pipette tips with glass-distilled water between transfers instead of replacing the tips between samples.

1. Instead of using all the fractions to make the slot blots for each gradient, we pick a set of 24 fractions that include the HH and HL peaks (to match the 24 slots

---

[28] A. E. Reynolds, R. M. McCarroll, C. S. Newlon, and W. L. Fangman, *Mol. Cell. Biol.* **9,** 4488 (1989).
[29] B. M. Ferguson, B. J. Brewer, A. E. Reynolds, and W. L. Fangman, *Cell* **65,** 507 (1991).

per lane of the slot blot apparatus). HH DNA peaks at refractive index ≈ 1.4050, and HL at refractive index ≈ 1.4040. Because the density of a DNA fragment depends on its G+C content, different fragments may band at slightly different positions in the gradient, so we use fractions with refractive indices in the range of ~1.4060–1.4030 (Fig. 4). Transfer 42 µl of each of the 24 fractions to a fresh microtiter plate. (The remaining material can be sealed and stored at 4° and used to make additional blots if necessary.)

2. Mix the 42 µl of the CsCl/DNA solution with 28 µl of 1 N NaOH (giving a final concentration of 0.4 N NaOH). Complete mixing may require pipetting the solution up and down a few times.

3. Seal the microtiter plate with an adhesive sheet and incubate at 65° for 1 hr.

4. Cool the plate to room temperature, remove the adhesive sheet, and neutralize the samples with an equal volume (70 µl) of 20× SCP.

5. Prewet the blotting membrane in 10× SCP and set it up in the blotting apparatus along with Whatman (Clifton, NJ) paper backing if recommended by the blotting apparatus manufacturer. Apply vacuum to the blotter, run 140 µl of 10× SCP through the slots, and apply the denatured DNA samples. *Note.* We use an aspirator to create the vacuum, adjusting it to achieve a flow rate of about 30 µl/sec in each well of the slot blotter. To maintain consistency between blots of a series, we do not turn off the aspirator between blots, but keep it running at a constant rate until all the samples have been blotted.

6. Wash out the microtiter wells with 140 µl of 10× SCP each and apply this wash solution to the corresponding slots of the blot.

7. Remove the membrane and UV cross-link it immediately. Cross-linked membranes may be left in 6× SCP, 1% Sarkosyl, 0.1% BSA (diluted from a 30% solution; Calbiochem) until all the blots have been made.

8. Prehybridize and hybridize the membranes using standard conditions. As described earlier, choose two or more fragments in the genomic region of interest to use as probes. We label probes by random priming using ^{32}P as label. It is advisable to probe the blots for the timing markers (*ARS305*, *ARS1*, and R11) to assess the quality of the synchrony of the S phase.

Quantitation of Blots and Data Analysis

The following quantitation method is based on Apple Macintosh applications and file formats; comparable options may be available on the Windows platform as well.

Scanner: Packard InstantImager or Molecular Dynamics PhosphorImager, and the associated software.
Software packages: Microsoft Excel (Microsoft) spreadsheet software or its equivalent, NIH Image (available as a free download at http://rsb.info.nih.

gov/nih-image/) or its equivalent, and KaleidaGraph 3.5 (Synergy Software) graphing software or its equivalent

1. Quantitate the hybridization intensities on the slot blot. We use a Packard InstantImager or a Molecular Dynamics PhosphorImager to detect ^{32}P emissions. In the scanner software, set up a grid so that each cell in the slot-blot corresponds to one cell of the grid and measure the ^{32}P counts for each cell of the grid.

2. For each S-phase sample (i.e., each gradient), plot the hybridization intensity values obtained earlier as a function of the gradient fraction (Fig. 4).

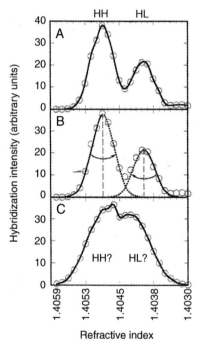

FIG. 4. Samples of slot-blot gradient profiles. Slot blots were scanned for ^{32}P counts on a Packard InstantImager system and the resulting profiles were plotted. Plots show arbitrary hybridization counts for fractions across the gradients. The bottom of the gradients is at the left end. Refractive indices of the gradient fractions are indicated. (A) Typical profile showing good separation of HH and HL peaks. (B) Curves obtained from data in A that would be typically traced to quantitate the areas under the HH and HL peaks using NIH Image software. The outer half of each peak is traced (solid lines), and the inner half (broken lines) is drawn as a mirror image of the outer half. (C) An example of a slot-blot profile with poor separation of HH and HL peaks. Quantitation in such instances is difficult. Poor separation may be the result of jostling of the gradient after centrifugation or of dripping the gradient too rapidly during fractionation. In such cases, the gradient fractions can be pooled and recentrifuged. Poor separation can also occur if the fragment being probed is small or if the initial dense culture medium contained light supplements, such as adenine or uracil.

The plot for each time point is saved as a PICT file. *Note.* If the scanner software allows subtraction of local background, that option should be used; if that option is not available, the value from the cell with the lowest counts in each lane of the slot blot (i.e., each sample) can be used as background to be subtracted from each cell of that lane. The output from the scanner software is usually not in a form that can be used directly by graphing software packages. We transfer data to Microsoft Excel and use a macro to reformat data to a usable form and to perform background subtractions. This automation step is highly recommended.

3. Open the PICT files in NIH Image and measure the area under each peak by tracing the peak with the freehand tool and applying the "Measure" command. Because HH and HL peaks tend to overlap, some judgment will have to be exercised in tracing the overlapping portions of the curves—use the outer half of each peak as a guide and trace a symmetrical line down the overlapping portion (Fig. 4B).

4. The percentage replication at S-phase time t for a given fragment is calculated from the equation

$$\% \text{ replication} = \left(\frac{\frac{1}{2} HL_t}{\frac{1}{2} HL_t + HH_t} \right) 100$$

where HH_t and HL_t represent the areas under the HH and HL peaks, respectively, at time t.

5. Plot the kinetics of replication (percent replication as a function of S-phase time) for each of the probed fragments in the region of interest (Fig. 5B). To facilitate the determination of t_{rep} values, data are fit to a sigmoid curve described by the equation

$$F(a, b, c, d) = d + \frac{at^b}{c^b + t^b}$$

for constants a, b, c, and d, where a is asymptotic maximum (percentage replication), b is slope parameter (positive values give positive slopes for percentage replication), c is the value of percentage replication at inflexion point of the replication curve, d is asymptotic minimum, and t is time in S phase. All plotting and curve fitting can be done in the software package KaleidaGraph or its equivalent.

6. Calculate t_{rep} for each fragment by determining the time of half-maximal replication from the corresponding replication timing curve (Fig. 5B). If the maxima (final percentage replication) for the different fragments are not the same, the mean of the maxima for the different curves can be used to determine the value of half-maximal replication. Alternatively, slot blots can be probed with

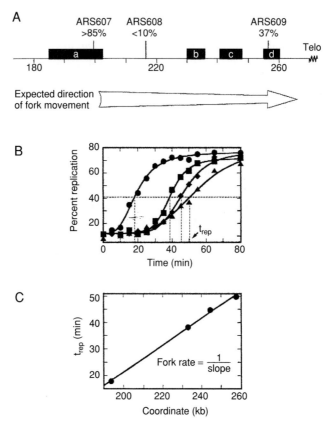

FIG. 5. Deducing the rate of fork migration. (A) Map of the right end of chromosome VI. The locations of three ARSs and their efficiency of activation in the chromosomal context are indicated. The expected direction of fork movement through the region is shown. Tick marks below the horizontal line correspond to 10-kb intervals; numbers are chromosome coordinates (kb) from the left end of the chromosome. Black boxes (a–d), fragments analyzed by hybridization to slot blots. Telo, telomere. (B) Kinetics of replication of *Eco*RI fragments a, b, c, and d (●, ■, ♦, and ▲, respectively; see A for fragment locations). The plot shows percentage replication as a function of S-phase time. t_{rep}, the time of half-maximal replication (shown by the intercept of the broken lines at the X axis). (C) Rate of fork movement deduced by plotting the t_{reps} of fragments a–d as a function of their coordinates. The inverse slope of the line fit to data corresponds to the fork rate.

rDNA-depleted genomic DNA to determine the final percentage replication value for bulk genomic DNA. Genomic DNA depleted of rDNA can be prepared by CsCl banding total genomic DNA with Hoechst dye, as described earlier for isolation of replication intermediates, and excluding as much of the rDNA as possible when extracting the DNA from the ultracentrifuge tube.

7. Plot the t_{rep} of each fragment as a function of the coordinate (in kb) of the fragment center. The inverse slope of the line fit to the points (by linear regression) is the average fork rate across the region (Fig. 5C).

Application of Method

As an example of the application of this method, we show the analysis of fork migration rate for the right end of chromosome VI. *ARS607*, a highly efficient, early activated origin,[24,30] is located on the right arm of chromosome VI, 200 kb from the left end of the chromosome. Although two other potential origins (*ARS608* and *ARS609*) are present in this region, they are inefficient. Thus, the ~64-kb region distal to *ARS607* (Fig. 5A) is replicated predominantly by a rightward-moving fork initiated at *ARS607*,[24,30] which makes it a good candidate for the measurement of fork migration rate. This region has also been examined previously in assessing the role of replication initiation proteins.[14] The left end of chromosome III, which is normally replicated by a fork initiated at *ARS305*, is a good alternative, especially if *ARS305* is mutated such that the entire left end is replicated by a fork from *ARS306*.[14]

Slot blots prepared from a density transfer experiment were probed sequentially with a set of four probes for four different *Eco*RI genomic restriction fragments (labeled a–d in Fig. 5A) at *ARS607* and distal to it. As expected, *ARS607* was found to replicate very early in S phase, with the other three fragments showing progressively later replication with increasing distance from *ARS607* (Fig. 5B). The t_{rep} values for the four fragments were calculated (Fig. 5B), and the chromosome coordinates of the fragment centers were plotted against the corresponding t_{rep} values (Fig. 5C). The inverse slope of the line fit to data gave an average fork rate of 2.0 kb/min, which is consistent with previous estimates of fork rates in isogenic yeast strains.[4,28]

Although the density transfer method is long and requires careful handling of the samples for extended periods of time involving several sequential manipulations, the method is robust overall. Because the percentage replication is calculated separately for each S-phase sample, different S-phase samples do not need to have exactly the same number of cells. However, once the DNA has been centrifuged in the CsCl solution, HH and HL DNA are physically separated. Therefore, all fractions from a given CsCl density gradient must have equivalent volumes of the gradient material, and equal portions must be loaded on the slot blot.

If separation between HH and HL peaks is poor (e.g., Fig. 4C), the experiment may have to be repeated. Poor separation is observed if the DNA fragment being probed is small (<3 kb) or if the dense culture medium was supplemented with

---

[30] M. Yamashita, Y. Hori, T. Shinomiya, C. Obuse, T. Tsurimoto, H. Yoshikawa, and K. Shirahige, *Genes Cells* **2**, 655 (1997).

isotopically light bases (uracil or adenine) to support the growth of auxotrophs. For this reason, we use ADE^+ URA^+ strains whenever possible. Note also that although the relative timing of replication for different fragments is consistent between experiments,[19] the absolute t_{rep} values may differ by up to a few minutes from culture to culture. Therefore, fork rates should be calculated only from t_{rep} values obtained from the same density transfer experiment.

Acknowledgments

We are very grateful to Margaret Hoang for the gift of chromosome VI probes and to Bonny Brewer and Walt Fangman for their comments and suggestions. This work was supported by National Institute of General Medical Sciences Grant 18926 to W. Fangman and B. Brewer and by National Center for Human Genome Research Grant 01298 to B. Brewer and W. Fangman.

[32] Analysis of RNA Export

By CHARLES N. COLE, CATHERINE V. HEATH, CHRISTINE A. HODGE, CHRISTOPHER M. HAMMELL, and DAVID C. AMBERG

Introduction

Nucleocytoplasmic transport plays a critical role in the expression of genetic information in eukaryotic cells. All RNAs except those encoded by the mitochondrial genome are synthesized in the nucleus, but most of these RNAs must be exported to the cytoplasm where they function in protein synthesis. Yeasts have played a critical role in increasing our understanding of this process, with most work focusing on mRNA export and using *Saccharomyces cerevisiae*.

The application of fluorescence *in situ* hybridization (FISH) to detect RNA in yeast has been critical for the analysis of RNA export.[1,2] This technique is the best one available capable of providing information about the subcellular distribution of both total mRNA and individual mRNA species. A satisfactory procedure for the separation of yeast cellular RNAs into cytoplasmic and nuclear fractions has not been developed. The reasons for this are not known but may include the difficulty of removing the yeast cell wall and contamination of the enzymes used for spheroplasting with nucleases. An additional problem may arise from the very small size and volume of the yeast nucleus (~1 μm in diameter), as even a small surrounding sphere of cytoplasmic material will be larger in volume than the nucleus itself. Although not as quantitative as fractionation, FISH analysis

[1] D. A. Amberg, A. L. Goldstein, and C. N. Cole, *Genes Dev.* **6**, 1173 (1992).
[2] T. Kadowaki, Y. Zhao, and A. M. Tartakoff, *Proc. Natl. Acad. Sci. U.S.A.* **89**, 2312 (1992).

also provides information about the distribution of RNA within the cytoplasmic and nuclear compartments. Because different mRNA molecules can have distinct subcellular distributions, it is sometimes useful to detect the location of specific mRNA species and, at the same time, the location of total mRNA. The ability to localize simultaneously both RNA and protein has also increased our understanding of RNA export. Finally, FISH analysis can be harnessed as a screen for mutants defective for RNA export or for distribution of a specific mRNA to its particular subcellular location.

Detection of Poly(A)$^+$ mRNA by *in Situ* Hybridization Using Oligo(dT)$_{50}$ Probes

Preparation of Probes

Materials

DIG oligonucleotide tailing kit (Roche Molecular Biochemicals, Indianapolis, IN)
Digoxigenin-dUTP (Roche)
Terminal deoxynucleotidyl transferase (Roche)
dTTP solution (Roche)
Glycogen (20 mg/ml)
0.2 M EDTA (pH 8.0)
3.0 M sodium acetate
Ethanol, 100%
TE buffer [10 mM Tris–Cl, 1 mM EDTA (pH 8.0)]

In order to detect the intracellular sites at which oligo(dT)$_{50}$ hybridizes, the oligo(dT) must be tagged for subsequent detection. Tagging by direct conjugation of a fluorophore is possible, but we chose to use a digoxigenin tag that could subsequently be detected using an antidigoxigenin antibody. This provides modest amplification of the signal, as antibodies can be conjugated with multiple fluorophore molecules. *In situ* hybridization studies in other organisms have tagged probes with biotin, but *S. cerevisiae* contains a high level of biotin. Therefore, some background problems may occur on detection of sites where biotinylated-oligo(dT) hybridizes using fluorophore-conjugated streptavidin. For these reasons, we have used digoxigenin-tagged oligo(dT)$_{50}$.

We generally use the Roche Molecular Biochemicals DIG oligonucleotide tailing kit. This kit provides several of the reagents essential for making the oligo(dT)$_{50}$ probe. These include the tailing enzyme, terminal deoxynucleotide transferase (25 units/μl; also called terminal transferase), 5× reaction buffer, 25 mM CoCl$_2$ and glycogen. The nucleotide to be added, digoxigenin-11–dUTP (25 nM), is purchased separately. It is possible to order the enzyme and nucleotide separately.

Both $CoCl_2$ and 5× reaction buffer are provided if the enzyme rather than the complete kit is ordered.

The oligonucleotide to be tailed at its 3' end contains 50 Ts and is prepared by standard oligonucleotide synthesis. Using an oligonucleotide of precise length facilitates determining whether the tailing reaction has been successful. Others have successfully used oligo(dT) as short as 25 nucleotides.

Procedure for Probe Preparation

1. Add reagents to a sterile 1.5-ml Eppendorf tube on ice in the following order:

Volume (μl)	Reagent	Final concentration
100	5× reaction buffer	1×
100	$CoCl_2$ (25 mM)	5.0 mM
25	DIG-11-dUTP (25 nmol/μl)	1.25 mM
50	Oligo (dT)$_{50}$ (100 pmol/μl)	0.01 mM
25	dTTP (10 mM)	0.50 mM
50	Terminal transferase (25 U/μl)	2.5 units/μl
150	Sterile doubly distilled H_2O	
Total volume:	500 μl	

The reaction can be performed on a smaller scale (as digoxigenin-11–dUTP is expensive).

2. Incubate at 37° for 15 min.
3. Place tube quickly on ice.
4. Add 25 μl 20 mg/ml glycogen solution and 25 μl 0.2M EDTA (pH 8.0).
5. Mix 0.1 volume 3 M sodium acetate and 2.5–3.0 volumes of chilled 100% ethanol to precipitate the nucleic acids in the reaction mix; add 50 μl.
6. Mix well and incubate at −80° for 30 min.
7. Remove from −80° incubation and thaw briefly at room temperature.
8. Centrifuge at 13,000 rpm for 30 min in a microcentrifuge at room temperature.
9. Decant the ethanol. There should be a white pellet. Carefully wash the pellet with 70% (v/v) ethanol.
10. Centrifuge again at 13,000 rpm for 15 min in a microcentrifuge at room temperature.
11. Decant the ethanol and air dry the pellet. It is best to avoid letting the pellet dry completely, as this will make it difficult to resuspend.
12. Resuspend pellet in 0.5 ml of TE buffer (pH 8.0). If the probe pellet does not go into solution readily, it may be necessary to heat to 37° for 10 min with frequent vortexing. If pellets have been dried completely, this incubation and vortexing will usually be necessary. Aliquot the probe into several tubes and store at −20°. Numerous freeze/thaw cycles result in degradation of the probe, but the

probe can be frozen and thawed a limited number of times. Aliquots put into use generally can be used for 3–6 months.

13. When this assay was developed,[1] we monitored the tailing reaction by electrophoresis to compare the tailed and untailed oligo(dT)$_{50}$. Tailing results in an oligonucleotide migrating more slowly. The shift in migration suggested that there were approximately four to five nucleotides added during tailing. We have had little trouble with this reaction and do not routinely check each probe preparation electrophoretically, but we do test all new probes for their ability to function for *in situ* hybridization.

Growth and Preparation of Cells for in Situ Hybridization

Materials

Slides (12 well, Teflon coated) (Cel-line/Erie Scientific, Portsmouth, NH)
Slide coverslips (Fisher, Pittsburgh, PA, 24 × 60 × 1 mm)
Double-distilled water
Formaldehyde, 37% (w/w) containing 10–15% methanol (Fisher)
Poly(L-lysine), 3% (w/v) in water (Sigma, St. Louis, MO)
Potassium phosphate buffer (0.1 M K$_2$HPO$_4$, pH 6.5)
Potassium phosphate buffer with sorbitol [1.2 M sorbitol, 0.1 M K$_2$HPO$_4$, (pH 6.5)]
Dithiothreitol (DTT) (Angus, Niagara Falls, NY, or GIBCO/BRL, Gaithersburg, MD)
IGEPAL CA-630 (Sigma) (nonionic detergent)
Triethanolamine (Sigma)
Acetic anhydride (Sigma)
Formamide (redistilled) (GIBCO/BRL)
20× SSC (3.0 M NaCl, 0.3 M sodium citrate)
tRNA from baker's yeast (Sigma); 5 mg/ml in doubly distilled H$_2$O
Dextran sulfate (sodium salt, average molecular weight 5000, Sigma)
Salmon sperm DNA, 1 mg/ml in doubly distilled H$_2$O (Sigma)
Denhardt's solution
Bovine serum albumin (BSA) (Sigma)
Fluorescein-conjugated, antidigoxigenin Fab antibody fragments (Boehringer Mannheim, Indianapolis, IN)
p-Phenylenediamine (Sigma)
4,6-Diamidino-2-phenylindole (DAPI) (Sigma)

The source for most of these reagents is not critical, but in many cases, we have only used a reagent obtained from the source(s) listed earlier. In the following procedures, we mention whenever a reagent source has been important. Most reagents and solutions are stable and we have been able to use them repeatedly over periods of months or even years. We indicate those which must be prepared each time these assays are performed.

Procedure

1. In general, cells are grown overnight on a rotator at room temperature in YPD or selective medium. Cells should be grown to early log phase ($\sim$1.0–1.5 $\times$ 10^7/ml; A_{600} = $\sim$0.5–1.0). Best results are obtained if cells are diluted sufficiently the preceding day so that they are at this cell density when they are to be processed for analysis. If cells have grown to a greater density, it is preferable to dilute them and allow them to go through several divisions until the correct cell density is reached rather than use them at a higher than desirable density. It takes some experience and knowledge of the growth characteristics of different strains to ensure that cells will be at the desired density when the assay is started. In addition, preparation of spheroplasts (see later) is affected by density and growth state. Cells can be grown in flasks, although *in situ* hybridization assays require small quantities of cells.

2. In most experiments, cells grown overnight are further manipulated prior to preparation for *in situ* hybridization. Studies of RNA export have relied heavily on temperature-sensitive mutants, which are analyzed at both permissive temperature and following a shift to the nonpermissive temperature. Some studies involve the addition of inhibitors of protein or RNA synthesis. In other studies, the effect on RNA distribution of inducing or repressing expression of a gene of interest involves carbon source alteration (or other manipulation) prior to processing of cells. Studies of the effect of cellular stress on RNA localization have also required shifts in temperature (e.g., to 42°) or the addition of ethanol to growing cells. Cultures should be shifted to the nonpermissive temperature in a shaking water bath. Make certain that the water level in the bath is high enough so that the entire cellular volume is submerged in the water bath. For very brief incubations at another temperature (e.g., heat shock), prewarmed medium can be added to accelerate the temperature shift. This is essential if large volumes of cells are being used; otherwise, an excessively long period would be required for cells to reach the desired temperature after placing them at the higher temperature.

3. Quickly add 1/10 volume of the 37% (w/v) formaldehyde reagent to cells in growth medium and incubate at room temperature on a rotator for 60 min. This dilution results in fixation at a final formaldehyde concentration of 3.4%. We find it necessary to open a new bottle of formaldehyde solution every 4–6 months because its effectiveness decreases over time. If cells that have been temperature shifted are used, the initial 2–5 min of fixation should be performed at the final incubation temperature, and cells then placed in a rotator at room temperature for 60 min to complete fixation.

4. During the fixation period, slides are prepared. We use 12-well/6-mm white Teflon-coated slides. Do not preclean these slides with ethanol, as the polylysine used to hold the cells will not adhere to the slide if they are precleaned. We place

the slides within a covered deep plastic dish (17 1/2 × 14 × 3 cm) in which we have layered wet paper towels on the bottom to retain humidity in the chamber. This prevents solutions on the slides from evaporation, and the volume in each well remains constant. We then place the slides on another piece of plastic to elevate them about 1 cm above the paper towels. To coat wells, we use a multipipetter to add 20 μl of a 0.3% polylysine solution (3% stock solution, filtered with a 0.2-μm syringe filter, and then diluted 1 : 10 with water) to each well and leave for 15–30 min. Wells are then washed twice with sterile twice-distilled H_2O for 5 min each followed by aspiration to remove all traces of water from wells. Allow slides to air dry for several minutes. Once subsequent procedures are started, slides should be kept covered between steps to minimize dust particles settling on the slides. *Note.* Once the cells are on the slides, there are a large number of steps involving sequential addition of different solutions to each well of each slide. Whenever adding or aspirating a solution, we use a house vacuum set up with a trap and a glass Pasteur pipette to which is attached a 200-μl micropipette tip. It is important to avoid touching the wells with the pipette tip, so the tip should be placed at or near the edge of the white Teflon surface.

5. Pellet the fixed cells in a bench-top centrifuge at 3000 rpm for 2 min at room temperature and discard the supernatant.

6. Wash cells twice with 5 ml (or equal volume if cells were grown in a flask) of 0.1 M K_2HPO_4, pH 6.5 (potassium phosphate buffer), and once with 1.2 M sorbitol/0.1 M K_2HPO_4, pH 6.5. Washing cells consists of resuspending them in 5 ml (or equal volume) of wash buffer by gentle vortexing and then centrifugation at 3000 rpm for 2 min. After each centrifugation, carefully discard the supernatant. Cells can be stored overnight at 4° at this point. Storage for longer than 24 hr at 4° generally results in very poor results and should be avoided.

7. Cells should be resuspend so that all strains are at approximately the same cell density. Resuspend cells in 0.1–5 ml of 1.2 M sorbitol/0.1 M K_2HPO_4 (pH 6.5). The volume of buffer to use for resuspension depends on the actual amount of cells to resuspend and is subjective. Even with good planning, different strains will be at different cell densities when the assay is initiated. Poorly growing strains [or use of cells that were initially grown to a lower density (<0.5)] are resuspended in a small volume (e.g., 0.1 ml), whereas wild-type strains or fast-growing mutant strains should be resuspended in a larger volume (1–5 ml). With experience, one can estimate cell quantities through visual examination of the cell pellet. Density of the cells on the well surface definitely influences how well subsequent treatment with Zymolyase works to produce spheroplasts. If cells are too dense, it will be difficult to spheroplast all of the cells. If the density is too low, there may be few cells in any microscope field, and cells may easily become overdigested by Zymolyase. Add ∼20–30 μl of the final cell suspension to each sample well. Allow cells to settle and adhere to wells for at least 30 min in a humid-covered chamber. We usually work up duplicate wells for each strain or condition being analyzed.

8. To 1 ml of 1.2 M sorbitol/0.1 M K_2HPO_4, pH (6.5), add 25 μl of 1 M DTT. Remove nonadhered cells by aspiration and add 20 μl of this solution to each well. Incubate at room temperature for 10 min. We prepare the DTT stock solution in advance and store aliquots at $-20°$. We do not refreeze aliquots.

9. Prepare Zymolyase stock solution [10 mg/ml in 1.2 M sorbitol/0.1 M K_2HPO_4 (pH 6.5)] and centrifuge briefly to clarify. To 1 ml of 1.2 M sorbitol/0.1 M K_2HPO_4 (pH 6.5), add 25 μl of 1 M DTT and 30 μl of the Zymolyase stock solution. Aspirate the previous solution and quickly add 20 μl of the Zymolyase solution. Take care when aspirating so that the wells are never allowed to dry out (until step 8 of the hybridization procedure given later). Incubate at room temperature for 15–40 min. *Note.* Each lot of Zymolyase 100T must be titrated to determine the proper amount to use for spheroplasting. Most lots yield the desired degree of spheroplasting with digestion times of 10–20 minutes. Periodically during Zymolyase treatment, cells can be examined under a microscope until the proper degree of treatment has been achieved. Titrate Zymolyase by treating cells for different periods of time with the same amount of Zymolyase solution and then perform the *in situ* hybridization assay. Some mutant strains may require less or more time for proper Zymolyase digestion than wild-type cells, and this should be tested if visual monitoring of Zymolyase digestion indicates that the strain is behaving differently. Proper Zymolyase treatment is key to a successful analysis. We monitor this step by viewing the slide periodically using differential interference contrast (DIC) microscopy. During Zymolyase treatment, the refractivity of cells changes and cells become considerably darker than prior to treatment. We allow the treatment to continue until $\sim$50% of cells show this change. The Zymolyase treatment is ended by removing the Zymolyase solution and washing cells as described later. With experience, it becomes easy to end the spheroplasting treatment at the proper time. Underdigestion gives weak signals with considerable cell-to-cell variation due to failure to permeabilize all cells, whereas overtreatment damages the cells and leads to final images that show very low contrast and low signal. Overdigested cells appear very flat when viewed by DIC microscopy. The proper treatment duration results in relatively uniform signals for poly(A)$^+$ RNA in all cells.

10. Remove the Zymolyase solution and wash wells twice with 1.2 M sorbitol/ 0.1 M K_2HPO_4 (pH 6.5). The first wash should be rapid, but slides should be incubated at room temperature with the second wash solution for 5 min.

11. Aspirate and wash once for 5 min with 20 μl 0.1 M K_2HPO_4 (pH 6.5) buffer and then wash once for 5 min with detergent solution. This solution contains 0.1 M K_2HPO_4 (pH 6.5) plus 0.1% IGEPAL CA-630. IGEPAL CA-630 is a nonionic detergent substitute for Nonidet P-40 (NP-40), which can also be used. Detergent can be used either directly from the reagent bottle or from a 10% (w/v) solution. To 10 ml of 0.1 M K_2HPO_4 buffer (pH 6.5), add 10 μl of 100% IGEPAL CA-630. Aspirate and wash with 0.1 M K_2HPO_4 (pH 6.5) buffer once for 5 min.

12. Equilibrate cells in *freshly prepared* 0.1 M triethanolamine (TEA) (pH 8.0) for 2 min at room temperature. It is essential to adjust the pH to 8.0. We prepare 5 ml of 0.1 M TEA and adjust the pH with 10–20 drops of 1 N NaOH.

13. Block polar groups by adding 0.1 M TEA (pH 8.0) +2.5% acetic anhydride (1 ml of 0.1 M TEA and 25 μl of acetic anhydride) for 10 min at room temperature. The precise concentration of acetic anhydride is not critical; we have also used 0.25% acetic anhydride with satisfactory results.

Hybridization Procedure

1. Prepare the prehybridization stock solution as follows:

Final concentration	Amount of stock added to prepare 10 ml
50% Deionized formamide	5 ml of 100% Deionized formamide
4× SSC	2 ml of 20× SSC
1× Denhardt's solution	100 μl of 100× Denhardt's solution
125 μg/ml tRNA	250 μl of 5 mg/ml tRNA
10% Dextran sulfate	2 ml of 50% Dextran sulfate
Sterile doubly distilled H_2O	To final volume of 10 ml

The prehybridization stock can be stored at $-20°$ and used repeatedly for at least 6 months. Salmon sperm DNA (obtained from Sigma) comes as a 10% (w/v) solution. To prepare a stock solution, this is diluted to 1 mg/ml with water and stored at $-20°$. It can be refrozen repeatedly. The salmon sperm DNA stock solution is then boiled (incubation in a heat block at 100° for 10 min) and 25 μl is added per 1 ml of prehybridization stock solution.

2. Add 20 μl of the prehybridization solution to each well and incubate in the humid chamber for 1 hr at 37°.

3. To prepare the hybridization solution, add 5–15 μl of the end-labeled oligo(dT)$_{50}$ probe and 7.5 μl of denatured salmon sperm DNA per 300 μl of prehybridization solution. Aspirate the prehybridization medium and add 15–20 μl of hybridization solution to each well and incubate at 37° overnight (12–18 hr) in a humid chamber. We usually add 15 μl of the probe stock per 300 μl of hybridization solution. After the overnight incubation, the volume in each well will have increased to two or three times the original added volume. *Note.* Because of the method for preparing the probe, different probe preparations are likely to have different concentrations. It is necessary to determine empirically the amount of probe required for optimal results (maximum signal, minimum background). We test a range of probe concentrations, usually 5, 10, 15, and 20 μl of probe per 300 μl of prehybridization solution, and then choose the concentration that gives the most consistent and strong overall signal.

4. Remove hybridization solution and wash each well with 20 μl of the following solutions: (a) 1 hr with 2× SSC at room temperature, (b) 1 hr with

1× SSC at room temperature, (c) 30 min with 0.5× SSC at 37°, and (d) 30 min with 0.5× SSC at room temperature.

5. Prior to the addition of antibody to detect probe localization, block the cells by washing sequentially with 20-μl aliquots of the following solutions: (a) 5 min with 1× phosphate-buffered saline (PBS)/0.1% bovine serum albumin (BSA) solution (50 ml of 1× PBS + 50 mg of BSA), (b) 5 min with 1× PBS/0.1% BSA solution + 0.1% IGEPAL CA-630 (5 ml 1× PBS/0.1% BSA + 5 μl of 100% IGEPAL), and (c) 5 min with 1× PBS/0.1% BSA solution.

6. Aspirate the final wash and add 20 μl of a dilution of fluorescein-conjugated, antidigoxigenin Fab antibody fragments in 1× PBS/0.1% BSA solution to each well. Incubate for 2 hr at room temperature. *Note.* Each new lot of the antibody reagent must be titrated to determine the optimal dilution (maximum signal, minimum background). Different lots of the antibody reagent have required dilutions between 1:50 and 1:250. We usually put the covered box containing the slides in the dark during this incubation (e.g., in a drawer).

7. Wash again as in step 5.

8. Aspirate the final wash thoroughly and let the slides air dry completely in a dust-free environment.

9. Prepare the mounting solution, containing 7 ml glycerol, 1 ml 10× PBS, 2 ml doubly distilled H_2O, 0.01 g *p*-phenylendediamine, and 0.5–1.0 μl of the DAPI stock solution (10 mg/ml in sterile doubly distilled H_2O). This solution should be made up fresh each time, ~1 hr before use, and covered with aluminum foil as it is light sensitive. This solution should not be autoclaved. Mix the solution well by rotating on a rotator at room temperature for at least 1 hr. The color should change over time from clear and colorless to light to medium pink. If the solution turns yellow or very dark pink, discard the solution. Eventually, the glycerol will deteriorate and yield a dark pink color in this solution rather than light to medium pink. The glycerol bottle should be discarded and a new bottle opened.

10. To mount slides, add a large drop (~25–50 μl) of mounting solution to every well (including those that were not used for the assay) using a 1-ml pipettor. Center the coverslip over the wells and gently but firmly press down so that the mounting solution spreads evenly over the wells, taking care to prevent air bubbles. Do not remove the coverslip once it is put down. Gently wipe any excess mounting solution with a Kimwipe while holding one-half of the slide down on the bench top. Then repeat on the other half of the slide and seal with nail polish (any brand of clear colorless nail polish should work). Seal all four edges twice and let dry in between. Slides may be viewed immediately and should be viewed within the next 24 hr for the best results. The signal from some fluorophores [e.g., fluorescein isothiocyanate (FITC)] will fade over time. *Note.* Until or between viewings, store the slides at $-20°$ once the nail polish has dried.

11. In initial experiments, we included controls that we rarely perform today. Nevertheless, there may be applications for these techniques, which will require

these controls. The signal generated should be competed by untagged oligo(dT)$_{50}$, but not by unrelated oligonucleotides. To prove that the signal results from hybridization of the tagged oligonucleotide to RNA, we treat cells prior to hybridization, but after placing onto slides, with RNase T2 (100 U/ml), to hydrolyze poly(A). As expected, RNase A is not effective in destroying poly(A) tails, as it is specific for pyrimidines. However, if one wishes to destroy coding regions, other RNases (e.g., RNase A) could be used.

Localization of Individual Species of Yeast mRNA

Introduction

It is possible to detect individual species of mRNA if they are present at a sufficiently high level. We have had excellent success detecting specific mRNAs by *in situ* hybridization if their gene is carried on a high-copy (2 μm) plasmids.[3] Because of its inducibility, low basal level of expression, and thorough characterization, the *SSA4* transcript was chosen first for the detection of individual mRNA species.[4,5] The DNA sequences of the four *S. cerevisiae SSA* genes (encoding hsp70) are closely related, but the 3'-untranslated region of *SSA4* lacks significant identity with the other three *SSA* genes. To obtain a probe specific for *SSA4* mRNA, we employed a polymerase chain reaction (PCR) to generate a 270-bp DNA corresponding to the 3' portion of *SSA4* mRNA.

Except as noted, standard PCR techniques were followed as directed by the Applied Biosystems GeneAmp PCR Core Reagents protocol (Applied Biosystems (Roche), Branchburg, NJ). Digoxigenin-11–dUTP was incorporated directly into probes during the PCR reaction. We routinely include both digoxigenin-11-dUTP and dTTP in the PCR reaction. While using only digoxigenin-11-dUTP maximizes the number of tagged dUTPs per probe molecule, this decreases probe yield, as the polymerase does not use digoxigenin-11-dUTP as readily as TTP. In addition, the T_m of hybrids formed between probe and mRNAs is decreased when probe contains digoxigenin-11-dUTP, with the decrease proportional to the amount of digoxigenin-11-dUTP incorporated. Finally, because an antibody molecule is much larger than a nucleotide pair, the maximum fluorescent signal possible will be achieved before all dTTPs have been replaced by digoxigenin-dUTP. We found that a ratio of 1 : 6 (dTTP : digoxigenin-11–dUTP) provides a relatively high yield of probe, which functions well for *in situ* hybridization, and have generally used 1 mM digoxigenin–dUTP and 160 μM dTTP in the PCR reaction mixtures. To synthesize the *SSA4* probe, a 270-bp fragment was amplified by PCR from a plasmid containing the entire *SSA4* gene, using primers 5'-CAAACCCCATTATGAGTAAA-3'

---

[3] C. Saavedra, K.-S. Tung, D. C. Amberg, A. K. Hopper, and C. N. Cole, *Genes Dev.* **10**, 1608 (1996).
[4] W. R. Boorstein and E. A. Craig, *J. Biol. Chem.* **265**, 18912 (1990).
[5] M. R. Young and E. A. Craig, *Mol. Cell. Biol.* **9**, 5637 (1993).

and 5′-TGGCTTATGACGATGAGAA-3′. Standard primer selection approaches for PCR should be followed in selecting primers.

A stronger signal will result from using probes covering more or all of the mRNA. However, as probe length increases, its efficiency of hybridization decreases, probably due to permeability and accessibility limitations. We avoid use of probes longer than 400 bp and find probes in the 200- to 275-bp range a good compromise between signal intensity and the work involved in preparing a large number of probes against one specific mRNA. Less abundant mRNAs will require more probes of optimal length.

Materials

Applied Biosystems (Roche) Gene Amp PCR Core kit
Digoxigenin-11–dUTP (Roche)
Template for PCR reaction (e.g., *SSA4* ORF in Yeplac111)
Oligonucleotide primers for PCR (for *SSA4*, 5′-CAAACCCCATTATGAGT AAA-3′ and 5′-TGGCTTATGACGATGAGAA-3′)
Agarose for electrophoresis (1.2% agarose gels)

We use the Applied Biosystems (Roche) Gene Amp PCR Core kit. The polymerase in this kit has no 3′→5′-exonuclease activity and provides all of the essentials for making the probe (10× PCR reaction buffer, 25 mM MgCl$_2$, 5 U/μl AmpliTaq DNA polymerase, and 10 mM stock solutions of each deoxynucleotide triphosphate).

Procedure

1. Prepare a master mix in a sterile 1.5-ml Eppendorf tube on ice in the following order:

Volume (μl)	Reagent	Final concentration (after addition of other reagents)
88.75	Sterile doubly distilled H$_2$O	
25	10× Reaction buffer	1×
5	dATP (10 mM)	0.20 mM
5	dCTP (10 mM)	0.20 mM
5	dGTP (10 mM)	0.20 mM
1.25	AmpliTaq DNA polymerase (5 U/μl)	25 Units/ml
25	Oligonucleotide #1 (∼5 pmol/μl)	∼0.5 μM
25	Oligonucleotide #2 (∼5 pmol/μl)	∼0.5 μM
15	MgCl$_2$ (25 mM)	1.5 mM
5	Template (1 μg/ml)	20 ng/ml
250	Total volume of master mix	

2. Aliquot 40 μl of master mix into five separate tubes labeled 1–5 (tube no. 5 is the control tube with no DIG-11-dUTP added).

3. Dilute the 10 mM dTTP stock to 1 mM using sterile doubly distilled H$_2$O.

4. To tubes 1–4 gently add 10 μl of a mixture containing 8 μl of digoxigenin-11-dUTP solution and 32 μl of the 1 mM dTTP solution. To the control tube no. 5, add 8 μl of 1 mM dTTP and 2 μl sterile doubly distilled H$_2$O. The volume in all tubes will be 50 μl.

5. For PCR reactions, we use an Applied Biosystems (Roche) PCR System 2400 machine and prepare our *SSA4* probe using the following PCR conditions: Denature at 94° for 4 min. Cycle at 94° for 1 min/47° for 1 min/72° for 2 min for 30 cycles. Extend at 72° for 7 min. Optimal conditions should be determined empirically for each PCR machine.

6. Pool tubes 1–4 and electrophorese a 10-μl aliquot of the pool and a 10-μl aliquot of control tube 5 using a 1.2% agarose gel. The control PCR product should migrate about 30–50 bp lower on the gel than the pooled tubes containing the digoxigenin-tagged PCR product. The control probe should migrate at 270 bp. The slower migration of the tagged probe reflects the incorporation of digoxigenin-11-dUTP. This probe should then also be aliquoted into three or four tubes because it also will degrade over time and frozen at $-20°$. However, each tube of probe can be used multiple times. The probe can be aliquoted in smaller volumes, depending on how many assays are planned.

7. It is very important that this probe be denatured (temperature block at 100° for 10 min or boiling water bath) before the hybridization solution is added to it. Use of this probe and the washing of slides following reaction with the probe follow the same protocol as is used when the probe is oligo(dT)$_{50}$. We have found that gene-specific probes, stored aliquoted at $-20°$, are stable and produce consistent results for at least 1–3 months.

8. An alternative and more sensitive method for preparing gene-specific probes for mRNA localization by *in situ* hybridization involves the direct incorporation of fluorophore-conjugated deoxynucleotides into the probe.[6,7] No antibodies are used with these probes. It is possible to incorporate sufficient fluorophores into each probe molecule so that the signal generated when one probe molecule hybridizes to an mRNA is considerably stronger than when using digoxigenin-tagged probes. These probes also have better spatial resolution than digoxigenin-tagged probes, but the cost is substantially greater as well.

9. Specificity of the probe can be assessed by comparing results from hybridization of the experimental probe with those obtained using a control probe whose target RNA is either absent or present at so low a level as to be undetectable.

---

[6] E. Bertrand, P. Chartrand, M. Schaefer, S. M. Shenoy, R. H. Singer, and R. M. Long, *Mol. Cell* **2**, 437 (1998).
[7] A. M. Femino, F. S. Fay, K. Fogarty, and R. H. Singer, *Science* **280**, 585 (1998).

10. These probes are stored simply by freezing the PCR reaction at −20° and can be used for up to 3 months. It is best to aliquot the probe to avoid excessive freeze/thaw cycles, but individual aliquots can be thawed and frozen multiple times.

The fluorescence microscope and camera system used to collect image data are also important factors affecting sensitivity of these assays. Femino and colleagues[7] have been able to detect individual mRNA molecules using fluorescent probes and state-of-the-art image analysis. For optimal clarity of the images collected, one can remove out-of-focus light either by using confocal fluorescence microscopy or by using a mathematical algorithm designed to remove out-of-focus light computationally from a series of images collected using a standard fluorescence microscope. This latter approach is called deconvolution and requires that the microscope be equipped with a z-axis motor so that a stack of images can be collected as one focuses through yeast cells. It is also easier to detect mRNAs when they are localized to a small subcellular region than when the same amount of mRNA is located throughout the cell. For example, the same level of mRNA can give a signal just above background when the mRNA is cytoplasmic, and a signal much brighter than background when mRNA export does not occur. In addition to *SSA4* mRNA, we have been able to detect *CYC1*, *CUP1*, and *GAL1* mRNAs expressed from 2-μm plasmids.

A valuable alternative approach to localize specific mRNAs has been developed by Brodsky and Silver.[8] This approach uses a protein construct containing the RNA-binding domain of the U1A 70K protein fused to green fluorescent protein (GFP). Multiple copies of the RNA sequence recognized by the U1A 70K protein are inserted into a gene of interest. The GFP fusion protein binds the message construct *in vivo* and permits direct observation of mRNA localization in living cells. Related approaches using the RNA-binding domain of the phage MS2 coat protein have also been reported.[6]

Although clearly a simpler approach than *in situ* hybridization, this method has some shortcomings. The GFP fusion is present both where the specific mRNA target is localized and in other cellular regions. Thus, one must also examine cells expressing the GFP U1A70K fusion in cells lacking any mRNA construct into which U1A 70K-binding sites have been inserted. It is the difference between this control strain and one with the tagged mRNA that reveals where the RNA is located. This works very well for mRNAs that are directed to discrete cytoplasmic locations (e.g., *ASH1* mRNA[6]) or for detecting nuclear accumulation of the specific mRNA (since the nuclear signal brightens dramatically when a target RNA is present and also accumulates in the nucleus in a mutant defective for mRNA export). By using fluorescent proteins of different colors (e.g., GFP, YFP, CFP; green,

[8] A. Brodsky and P. Silver, *RNA* **6,** 1737 (2000).

yellow, cyan) and both U1A 70K- and MS2 RNA-binding domains, it should be possible to localize simultaneously two different mRNAs within cells. This could be particularly powerful in cases of alternative splicing, where probes specific for different alternatively spliced mRNAs could be labeled with different colored fluorescent proteins and used to localize the two related mRNAs in living cells.

Microscopy

A major factor affecting successful FISH experiments is the microscope. While a state-of-the-art microscope is not required, it is important to choose various microscope parameters in a way to optimize the fluorescent signals detected. The microscope should be equipped for epifluorescence and for either Nomarski or phase-contrast visualization. The choice of filters is particularly important, as this will affect signal strength dramatically and, in two-color fluorescence experiments, will determine how well the signal from one fluorophore can be separated from the signal from a second one. We use filters from Chroma Technology (Brattleboro, VT), but similar filters are also available from many microscope manufacturers. We usually use a 20× objective to search through an entire well on a slide and a 100× oil-immersion objective for image collection. Objectives with a high numerical aperture (NA) enhance sensitivity dramatically.

Initially, we collected images using a 35-mm camera mounted on the fluorescence microscope, but we now collect images exclusively using a cooled charge-coupled device (CCD) camera. These images are digital and analyzed and cropped easily to prepare figures for publication. CCD cameras also boost sensitivity and reduce background signal. If out-of-focus light is to be removed by deconvolution, then a z-axis motor with a small step size (<0.2 μm) is needed to collect a stack of images as the focal plane is shifted through the z axis. Multiple software packages are available for deconvolution, as well as for other aspects of image analysis and presentation. We use Open Lab software (Improvision Inc., Quincy, MA) for our analyses. Images generate large data files and can fill the hard disk drive of a computer rapidly. Images can be archived for long-term storage on either CDs or DVDs. If images are to be compared in a quantitative manner, it is critical to use identical exposure conditions; in any subsequent alteration of the image, exactly the same manipulations should be applied to all images being compared.

Simultaneous Localization of Individual mRNA and Poly(A)$^+$ mRNA

In some of our studies, we wanted to examine the location of *SSA4* mRNA and total poly(A)$^+$ mRNA in the same cells in order to determine the degree to which these were colocalized or localized to different parts of the cell. Therefore, we modified the approaches already described to permit the use of two probes simultaneously. To detect poly(A)$^+$ mRNA in this double-detection assay, we ordered an oligo(dT)$_{50}$ probe directly labeled with Texas Red at its 3′ and 5′ ends (Bio-synthesis, Inc., Lewisville, TX). *SSA4* mRNA was detected as described earlier using a PCR-generated probe containing digoxigenin, followed by the use of FITC-conjugated antidigoxigenin antibodies.

Hybridization was carried out by incubating the fixed permeabilized cells simultaneously with Texas Red-labeled oligo(dT)$_{50}$ and with the digoxigenin-labeled *SSA4* probe. The *SSA4* probe was denatured in a boiling water bath or a heat block at 100° for 10 min and was immediately added to the hybridization solution, which already contained the Texas Red-Labeled oligo(dT)$_{50}$ probe and denatured salmon sperm DNA. Cells were then incubated with probe for 18–24 hr at 37° in a humid chamber.

Cells were washed and processed as described earlier (hybridization procedure, steps 4–10). Although no blocking step would be needed if just the Texas Red-conjugated oligo(dT)$_{50}$ probe were used, the digoxigenin-tagged *SSA4* probe requires antibody detection so the blocking step (hybridization procedure, step 5) is needed. For these studies, it is important to include controls that receive neither probe, each probe separately, and the two probes together. All samples should be viewed by microscopy under both Texas Red and FITC viewing conditions. It is critical that filters be used that eliminate as far as is possible spillover of the Texas Red signal into the FITC channel and spillover of the FITC signal into the Texas Red channel. By viewing samples containing only Texas Red or only FITC under microscopy conditions used for viewing the other fluorophore, there should be very little or no signal distributed in the pattern the signal is distributed when viewed to detect the fluorophore in its own channel. That is, in the FITC channel, there should be no signal when only Texas Red is present and, similarly, in the Texas Red channel, there should be no signal when only FITC is present. One can tolerate limited spillover when the Texas Red and FITC signals are clearly not overlapping. For example, when most mRNA remains nuclear after heat shock and *SSA4* mRNA is induced and exported, the green signal for *SSA4* mRNA is predominantly cytoplasmic, whereas the red signal for poly(A)$^+$ mRNA is primarily nuclear, although there is also a cytoplasmic signal for poly(A)$^+$ RNA, as *SSA4* and other stress–response mRNAs are polyadenylated.

Simultaneous Localization of RNA and a Specific Protein

In some experiments, we have wanted to localize both RNA and a specific protein in the same cells. This involves combining a standard indirect immunofluorescence protocol used for protein antigen detection with the *in situ* hybridization protocol described earlier. This approach is identical to the general method for detecting poly(A)$^+$ mRNA through step 5. Some of the subsequent steps are modified as follows.

6a. In the hybridization procedure step 6 given earlier, the addition of FITC-conjugated antidigoxigenin antibodies is performed as in the protocol given previously, but the solution added to each well should also contain an appropriate (empirically determined) dilution of the antibody being used to localize the protein of interest.

7a. Repeat step 5 washes.

8a. Add the fluorophore-conjugated secondary antibody at an empirically determined dilution in 1× PBS/0.1% BSA and incubate at room temperature for 1 hr in the dark. It is necessary to choose a secondary antibody for protein detection, which is conjugated to Texas Red or another fluorophore other than FITC, whose distribution can be separated readily from that of FITC. Because a secondary antibody is needed to detect the protein of interest, the first antibody (to the protein of interest) must be from a different animal species than used to produce the antidigoxigenin antibody. Otherwise, the secondary antibody would bind both to the antiprotein and to the antidigoxigenin antibodies.

9a. Repeat step 5 washes.

10a. Proceed to aspirate thoroughly, let wells air dry completely in a dust-free environment, and mount as described in the hybridization procedure (steps 9 and 10). *Note.* By substituting a gene-specific probe for oligo(dT)$_{50}$ in this protocol, it is also possible to detect simultaneously the distribution of both the specific mRNA and a protein of interest.

Using FISH Assays to Study mRNA Export

We have employed these techniques often to determine how a mutation of interest affects mRNA export. For this, the mutant strain is analyzed using one of the protocols described earlier and is compared with control strains. Wild-type cells are one control and should not show nuclear accumulation of poly(A)$^+$ mRNA (or specific mRNA species) under any conditions at temperatures between 16° and 37°. We also include a positive control, a mutant strain known to affect mRNA export directly. The *nup159-1*[9] and *dbp5-2*[10,11] strains are used routinely, but strains affecting *PRP20* (*prp20-1*) and *RNA1* (*rna1-1*) have also been satisfactory.[1,12] As a negative control, which should have no signal for poly(A), we use either the *rpb1-1* allele affecting the largest subunit of RNA pol II[13] or the *pap1-1* allele affecting poly(A) polymerase.[14] When specific transcripts are being detected, we

---

[9] L. C. Gorsch, T. C. Dockendorff, and C. N. Cole, *J. Cell Biol.* **129,** 939 (1995).
[10] C. A. Snay-Hodge, H. V. Colot, A. L. Goldstein, and C. N. Cole, *EMBO J.* **17,** 2663 (1998).
[11] C. A. Hodge, H. V. Colot, P. Stafford, and C. N. Cole, *EMBO J.* **18,** 5778 (1999).
[12] D. C. Amberg, M. Fleischmann, I. Stagljar, C. N. Cole, and M. Aebi, *EMBO J.* **12,** 233 (1993).
[13] M. Nonet, C. Scafe, J. Sexton, and R. Young, *Mol. Cell. Biol.* **7,** 1602 (1987).
[14] D. Patel and J. S. Butler, *Mol. Cell. Biol.* **12,** 3297 (1992).

use the *rpb1-1* allele as a negative control or strains lacking the gene of interest if that gene is not essential.

In screening ts mutants for nuclear accumulation of poly(A)$^+$ mRNA, we saw a wide range of phenotypes. In some strains, only a small fraction of the cells show clear nuclear poly(A)$^+$ mRNA. In others, all cells show this phenotypes. It is more likely that a gene is involved directly in mRNA export if ts mutant cells display an mRNA export defect rapidly and in all cells following a shift to the nonpermissive temperature. If a mutant does not show complete penetrance of the phenotype, a likely explanation is that the mutation affects mRNA export indirectly. The largest class of mutants affecting mRNA export indirectly are those affecting chromosome segregation. When a chromosome is lost, cells remain physiologically alive for many hours, but do not divide again. Following chromosome loss, cells become depleted for gene products encoded on the lost chromosome. Genes encoding proteins known to be important for mRNA export are located on many different chromosomes so it is not uncommon for a cell that has lost a chromosome to have lost one containing an mRNA export gene. One of the terminal phenotypes such cells can show is an accumulation of poly(A)$^+$ mRNA in nuclei. In our original screen of 1200 ts mutants, we isolated and further screened more alleles affecting DNA topoisomerase II than any other gene product. In several other mutant strains showing chromosome loss (several *cdc* mutants and a ts mutant affecting *IPL1*), poly(A)$^+$ mRNA accumulated in nuclei of a small fraction of cells (2–20%). In all of these cases, blocking passage through mitosis (e.g., with nocadozole) prevented the mRNA export block from developing. In addition, mutants affecting mRNA export indirectly generally are quite asynchronous in the time after temperature shift when nuclear poly(A)$^+$ RNA accumulation is seen.

A mutant could show low penetrance or asynchronous appearance of the defective phenotype and still affect an important mRNA export gene if, for example, the mutant under study were a weak allele. A related scenario involves mutants on plasmids where different cells contain different plasmid copy numbers. In some mRNA export mutant strains, temperature-sensitive growth is sensitive to copy number with overexpression sometimes permitting growth. Because plasmid number varies, the level of gene products will vary, which may affect the kinetics of appearance of the mutant phenotype. The stability of individual mRNAs obviously also affects the ease with which that RNA can be detected by using these methods.

We have also identified mRNA export mutants using flow cytometry. In this case, we prepared a GFP fusion to Ssa4p (hsp70). *SSA4* mRNA is undetectable in exponentially growing cells, although its expression is induced as cells reach a high density and enter stationary phase. Control experiments indicate that cells unable to export *SSA4–GFP* mRNA (e.g., a *rip1*Δ strain) remain dark, whereas wild-type cells become very bright. A mixture of the two can be separated with 80–90% accuracy. Thus, one can mutagenize cells carrying this construct and then heat shock the cell pool. Any cells defective in the export of *SSA4* mRNA will remain dark, whereas ones retaining mRNA export capability will be bright. Details of this approach will be published elsewhere (C. H. Hammell, C. V. Heath, and C. N. Cole, manuscript in preparation). A secondary screen, involving *in situ* hybridization, is needed to distinguish among strains defective for RNA export, translation, and transcription—the desired strains are those with nuclear *SSA4* mRNA following heat shock.

Studying Export of Other Classes of RNA

Export of other classes of RNA has also been studied in yeast. For tRNA, Sarkar and Hopper[15] took advantage of the presence of introns in some tRNAs to design oligonucleotide probes specific for these introns. They used *in situ* hybridization of digoxigenin-tagged oligonucleotides to examine the distribution of specific tRNAs and their precursors. The approaches they developed were adapted from those described earlier for detecting mRNAs. They were able to detect endogenous levels of specific tRNAs. They also used probes to yeast tRNA introns. Because these introns are removed in the nucleus during the process of tRNA export, there is normally a low nuclear signal for the unspliced tRNA. This signal increases dramatically, without any cytoplasmic signal, when the export of tRNA is blocked.

Assays to examine rRNA export have also been developed. These are complicated by the high level and stability of ribosomes. Moy and Silver[16] took advantage of the fact that the final step in 18S rRNA biogenesis in *S. cerevisiae*, the processing of 20S pre-rRNA into 18S rRNA by removal of the ITS1 sequence, takes place after export, in the cytoplasm.[17] Normally, ITS1 RNA is degraded in the cytoplasm by the Xrn1p exonuclease. Moy and Silver[16] prepared an oligonucleotide specific for ITS1 RNA. In wild-type cells, a signal was detected solely in the nucleolus. In *xrn1*Δ cells, ITS1 RNA accumulated in the cytoplasm. By disrupting *XRN1* in strains of interest, they were able to screen for those that accumulated 20S rRNA (containing ITS1) in the nucleus. Some, but not all, strains defective for mRNA export showed a hybridization signal for ITS1 RNA throughout the nucleus. In others the signal was limited to the nucleolus.

Because the large subunit is processed completely prior to export, an oligonucleotide probe cannot be used to study large subunit export. Two laboratories have used GFP fusions to proteins found in the large ribosomal subunit. Stage-Zimmerman *et al.*[18] used an Rpl11b–GFP fusion, whereas Hurt *et al.*[19] used a

---

[15] S. Sarkar and A. K. Hopper, *Mol. Biol. Cell* **9**, 3041 (1998).
[16] T. I. Moy and P. A. Silver, *Genes Dev.* **13**, 2118 (1999).
[17] S. A. Udem and J. R. Warner, *J. Biol. Chem.* **248**, 1412 (1973).
[18] T. Stage-Zimmermann, U. Schmidt, and P. A. Silver, *Mol. Biol. Cell* **11**, 3777 (2000).
[19] E. Hurt, S. Hannus, B. Schmeizl, D. Lau, D. Tollervey, and G. Simos, *J. Cell Biol.* **133**, 389 (1999).

Rpl25–GFP fusion. When using such a fusion protein, it is critical to determine that the fusion can replace the wild-type protein with little or no negative effect. Because many ribosomal proteins (including Rpl11p) are encoded by two genes, both normal copies must be eliminated in order to study how well the GFP–ribosomal fusion protein functions. Normally, ribosomal proteins not assembled into ribosomes are unstable so these proteins are found predominantly where ribosomes are located or produced. The GFP fusion must behave similarly, with little or no GFP–ribosomal protein fusion present except when part of the 60S ribosomal subunit.

In studying yeast mutants, the wild-type *RPL11A* and *RPL11B* genes were disrupted in mutant strains of interest. The GFP fusions were expressed from plasmids. For these studies, strains of interest were shifted to the nonpermissive temperature for several hours to inactivate the protein defective in the particular mutant strain. Cells were then returned to the permissive temperature, allowing resumption of ribosome biogenesis. It is critical that recovery from a nuclear transport block take more time than resumption of ribosome biogenesis and that the GFP–ribosomal protein reporter be imported into the nucleus under conditions where ribosome subunit export does not take place. For the Rpl25p–GFP fusion, cells were grown to stationary phase on plates (3–4 days), put into liquid media, and incubated at the nonpermissive temperature for several hours (duration depends on the particular mutant strain) before shifting back to the permissive temperature. For Rpl11b–GFP, exponentially growing cells were used, shifted to the nonpermissive temperature, and subsequently returned to permissive growth conditions. Mutants defective for ribosomal subunit export accumulate the GFP fusion in the nucleus; ribosomal proteins are normally found solely in the nucleolus and cytoplasm.

Isolation of RNA Export Mutants

Mutants have played a central role in studies of mRNA export in yeast. We screened 1200 ts strains in pools of 5 and then screened every strain of any pool where more than ~3% of the cells showed nuclear accumulation of poly(A)$^+$ RNA. From this screening, we isolated ~40 strains. We backcrossed these strains to wild-type twice and selected for further study those strains wherein temperature sensitivity and nuclear accumulation of poly(A)$^+$ mRNA cosegregated as a single gene defect. Genes were cloned by complementation using a *CEN* plasmid library. Because not all libraries contain all genes, it was necessary to use two different *CEN* libraries in order to clone each gene of interest. Although we originally performed genetic mapping experiments to confirm that we had cloned the defective gene rather than a suppressor, we now use gap repair to isolate the mutant allele and sequence the DNA to search for mutations. With the mutant gene on a plasmid, it is easy to transform cells that have been disrupted for the gene of interest and are carrying the wild-type version on a *URA3/CEN* plasmid. Standard plasmid shuffling using 5-fluoroorotic acid (5-FOA) yields a strain containing solely the putative mutant allele. It should recapitulate the ts growth and mRNA export defects of the starting strain. We often prepare additional mutant alleles of a gene of interest using error-prone PCR and plasmid shuffling to introduce putative ts mutants into cells. We generally screen ts alleles to find those that have minimal defects in growth and mRNA export at the permissive temperature and a rapid onset of an RNA export block in all cells following a shift to the nonpermissive temperature.

The assays for detection of tRNA, rRNA, or ribosomal protein mislocalization can also be used to screen for mutants affecting tRNA or rRNA/ribosome export, although the assays for rRNA export defects require extended periods at the nonpermissive temperature and therefore do not allow the identification of mutants that affect all cells rapidly. Nevertheless, it is desirable to isolate for further study mutants that do affect all cells in a population.

Acknowledgments

C. M. Hammell was supported by a training grant (CA-09658) from the National Institutes of Health. Research in the authors' laboratory has been supported by grants from the National Institutes of Health (GM33998) and the National Science Foundation (MCB9983378). Many former members of the laboratory contributed to the development of these methods.

[33] Nuclear Protein Transport

By MARC DAMELIN, PAMELA A. SILVER, and ANITA H. CORBETT

Introduction

Work on the mechanism of nuclear transport has led to a fairly detailed understanding of how proteins are targeted for import or export. This work has changed the way that researchers need to think about potential transport mechanisms for their own protein of interest. In addition, it has opened the door to detailed studies of the components that mediate the transport process. The most important realization from these advances is that there are several pathways for both nuclear import and nuclear export. No longer does the identification of a nuclear localization signal consist of recognizing a stretch of basic amino acids by a simple perusal of the protein sequence. Considerable experimental evidence is required to show that any transport substrate enters or exits the nucleus via a specific pathway.

The question of how various substrates are imported into the nucleus has received more attention lately as researchers have realized that dynamic compartmentalization is a regulatory mechanism used extensively by the cell, i.e., many proteins are transported into or out of the nucleus in response to a specific signal. A number of signal transduction pathways specifically stimulate the import of transcription factors into the nucleus to induce gene expression. Well-known examples include both NFAT[1] and NF-κB.[2] Two examples of this regulatory mechanism in *Saccharomyces cerevisiae* are the transcription factor Pho4p, which is imported into the nucleus in low phosphate environments,[3] and the DNA licensing factors Mcm, which are imported into the nucleus prior to Start and then exported after DNA replication commences.[4] This compartmentalization restricts DNA replication to once per cell cycle.[5] With the realization that nuclear transport is a powerful regulatory mechanism for the cell, many researchers have become interested in defining the transport pathway taken by their protein of interest. This chapter provides a detailed experimental approach to address this question. We also describe assays that can be used to assess whether a gene of interest is required for protein import or for protein export. Several assays to determine whether a protein shuttles between the nucleus and the cytoplasm are discussed in [34], this volume. Finally, we describe the use of fluorescence resonance energy transfer (FRET) to study the mechanism of translocation across the nuclear envelope.

Overview of Nuclear Protein Transport

Transport into and out of the nucleus is mediated by nuclear pore complexes (NPCs) embedded in the nuclear envelope.[6] In addition to these transport channels, a number of soluble factors are required to target transport substrates to the nuclear pore and facilitate their delivery.[7] The soluble factors include the small GTPase Ran and the proteins that regulate the Ran GTPase cycle: Rna1p, the GTPase-activating protein (GAP), and Prp20p, the guanine nucleotide exchange factor (GEF). The compartmentalization of these proteins, with the GAP in the cytoplasm[8] and the GEF in the nucleus,[9] establishes a gradient where levels of RanGTP are high in the nucleus and levels of RanGDP are high in the cytoplasm.[7] This compartmentalization facilitates the appropriate formation of transport substrate/receptor complexes in the appropriate compartments.

---

[1] G. R. Crabtree and N. A. Clipstone, *Annu. Rev. Biochem.* **63,** 1045 (1994).
[2] E. N. Hatada, D. Krappmann, and C. Scheidereit, *Curr. Opin. Immunol.* **12,** 52 (2000).
[3] A. Komeili and E. K. O'Shea, *Science* **284,** 977 (1999).
[4] K. Labib, J. F. Diffley, and S. E. Kearsey, *Nature Cell Biol.* **1,** 415 (1999).
[5] B. K. Tye, *Annu. Rev. Biochem.* **68,** 649 (1999).
[6] M. P. Rout and G. Blobel, *J. Cell Biol.* **123,** 771 (1993).
[7] D. Görlich and U. Kutay, *Annu. Rev. Cell Dev. Biol.* **15,** 607 (1999).
[8] A. K. Hopper, H. M. Traglia, and R. W. Dunst, *J. Cell Biol.* **111,** 309 (1990).
[9] M. Ohtsubo, H. Okazaki, and T. Nishimoto, *J. Cell Biol.* **109,** 1389 (1989).

Transport substrates contain targeting signals that direct them to enter the nucleus, exit the nucleus, or even shuttle between the nucleus and the cytoplasm.[10–12] These signals are either recognized directly by soluble RanGTP-binding receptors called importin-βs or karyopherin-βs that target the substrates to the nuclear pore for transport[7] or by adaptor proteins that bind importin/karyopherin-βs. The most extensively studied nuclear targeting signal is the canonical nuclear localization signal (cNLS) exemplified by the SV40 (simian virus 40) large T antigen (PKKKRK).[13] Such signals and their bipartite counterparts, which contain two clusters of basic residues separated by a 10 residue spacer (KRPAATKKAGQAKKKK)[14] are recognized in the cytoplasm by the adaptor protein, importin-α. In this classical nuclear import pathway, a trimeric complex forms between the NLS substrate, importin-α, and importin-β95. The complex is then directed to the nuclear pore through interactions between importin-β95 and nuclear pore proteins.[7] This complex is disassembled in the nucleus when RanGTP binds to importin β95, causing a conformational change that releases the importin-α/substrate complex into the nucleus.[15] Both importin-β95 and importin-α are recycled to the cytoplasm for another round of import.

With the identification of the *S. cerevisiae* importin-β95 protein came the realization that the yeast genome encodes 14 members of this protein family (Table I). These proteins have a conserved N terminus that mediates binding to RanGTP.[16] Subsequent studies have demonstrated that many of these importin-β proteins bind directly to substrate and mediate transport events without an importin-α adaptor protein.[17] In fact, evidence is mounting to suggest that each of these importin-β proteins can mediate distinct import or export events through direct binding to specific substrates.

The most well-studied export receptor is Crm1/Xpo1.[18] This protein recognizes a nuclear export signal (NES) composed of leucines or other hydrophobic amino acids (e.g., LxxLxxLxL). NES-containing proteins form a trimeric export complex with Crm1/Xpo1 and RanGTP. The trimeric complex is then translocated to the cytoplasm where it dissociates when the cytoplasmic RanGAP, Rna1p, induces GTP hydrolysis to yield RanGDP. RanGDP is then reimported into the nucleus by a small protein called Ntf2 to reestablish the RanGTP gradient.[19,20]

[10] L. Gerace, *Cell* **82**, 341 (1995).
[11] C. Dingwall and R. A. Laskey, *Trends Biol. Sci.* **16**, 178 (1991).
[12] W. M. Michael, *Trends Cell Biol.* **10**, 46 (2000).
[13] D. Kalderon, B. L. Roberts, W. D. Richardson, and A. E. Smith, *Cell* **39**, 499 (1984).
[14] C. Dingwall, J. Robbins, S. M. Dilworth, B. Roberts, and W. D. Richardson, *J. Cell Biol.* **107**, 841 (1988).
[15] M. Rexach and G. Blobel, *Cell* **83**, 683 (1995).
[16] D. Görlich, M. Dabrowski, F. R. Bischoff, U. Kutay, P. Bork, E. Hartmann, S. Prehn, and E. Izaurralde, *J. Cell Biol.* **138**, 65 (1997).
[17] R. W. Wozniak, M. P. Rout, and J. D. Aitchison, *Trends Cell Biol.* **8**, 184 (1998).
[18] K. Stade, C. S. Ford, C. Guthrie, and K. Weis, *Cell* **90**, 1041 (1997).
[19] A. Smith, A. Brownawell, and I. G. Macara, *Curr. Biol.* **8**, 1403 (1998).
[20] K. Ribbeck, G. Lippowsky, H. M. Kent, M. Stewart, and D. Görlich, *EMBO J.* **17**, 6587 (1998).

TABLE I
Saccharomyces cerevisiae IMPORTIN-β HOMOLOGUES

Importin-β homologues (alternative names)	Known substrate(s)	Mutant alleles
Importin-β95 (Rsl1, Kap95)	Import of cNLS cargo in conjunction with importin-α[a]	*rsl1-1*[a]
Crm1 (Xpo1, Kap124, exportin 1)	Classical leucine-rich NES export [18]	*xpo1-1*[18]
Cse1 (Kap109)	Importin-α export[b]	*cse1-1* (cs)[c]
Pse1 (Kap121)	Pho4, ribosomal protein import[25, d]	*pse1-1*[e, f]
Kap123 (Yrb4)	Ribosomal protein import[d]	*kap123*Δ[e,f]
Transportin (Kap104)	Nab2, Hrp1 import (hnRNP proteins)[g]	*kap104-16*[g]
Kap120	60S ribosomal subunit[h]	*kap120*Δ[h]
Nmd5 (Kap119)	TFIIS (transcription elongation factor), Hog1p import[i,j]	*nmd5*Δ[j]
Sxm1 (Kap108)	Lha1p (La) import, ribosomal protein import[k]	*sxm1*Δ[k]
Msn5 (Kap142)	Pho4p, Far1p export[l, m]	*msn5*Δ[l]
Kap114 (HRC1004)	TATA-binding protein (TBP) import[n, o]	*kap114*Δ[n, o]
Mtr10 (Kap111)	Npl3p import[p]	*mtr10*Δ[p]
Pdr6 (Kap122)	TFIIA import[q]	*kap122*Δ[q]
Los1	tRNA export[r, s]	*los1-1*[t], *los1*Δ[u]

[a] D. M. Koepp, D. H. Wong, A. H. Corbett, and P. A. Silver, *J. Cell Biol.* **133**, 1163 (1996).
[b] J. K. Hood and P. A. Silver, *J. Biol. Chem.* **273**, 35142 (1998).
[c] Z. Xiao, J. T. McGrew, A. J. Schroeder, and M. Fitzgerald-Hayes, *Mol. Cell Biol.* **13**, 4691 (1993).
[d] M. P. Rout, G. Blobel, and J. D. Aitchison, *Cell* **89**, 715 (1997).
[e] M. Seedorf and P. A. Silver, *Proc. Natl. Acad. Sci. U.S.A.* **94**, 8590 (1997).
[f] Pse1 and Kap123 have overlapping function. A double mutant *pse1-1 kap123*Δ is available.
[g] J. D. Aitchison, G. Blobel, and M. P. Rout, *Science* **274**, 624 (1996).
[h] T. Stage-Zimmermann, U. Schmidt, and P. A. Silver, *Mol. Biol. Cell* **11**, 3777 (2000).
[i] M. Albertini, L. F. Pemberton, J. S. Rosenblum, and G. Blobel, *J. Cell Biol.* **143**, 1447 (1998).
[j] P. Ferrigno, F. Posas, D. Koepp, H. Saito, and P. A. Silver, *EMBO J.* **17**, 5606 (1998).
[k] J. S. Rosenblum, L. F. Pemberton, and G. Blobel, *J. Cell Biol.* **139**, 1655 (1997).
[l] A. Kaffman, N. M. Rank, E. M. O'Neill, L. S. Huang, and E. K. O'Shea, *Nature* **396**, 482 (1998).
[m] M. Blondel, P. M. Alepuz, L. S. Huang, S. Shaham, G. Ammerer, and M. Peter, *Genes Dev.* **13**, 2284 (1999).
[n] L. F. Pemberton, J. S. Rosenblum, and G. Blobel, *J. Cell Biol.* **145**, 1407 (1999).
[o] H. Morehouse, R. M. Buratowski, P. A. Silver, and S. Buratowski, *Proc. Natl. Acad. Sci. U.S.A.* **96**, 12542 (1999).
[p] B. Senger, G. Simos, F. R. Bischoff, A. Podtelejnikov, M. Mann, and E. Hurt, *EMBO J.* **17**, 2196 (1998).
[q] A. A. Titov and G. Blobel, *J. Cell Biol.* **147**, 235 (1999).
[r] K. Hellmuth, D. M. Lau, F. R. Bischoff, M. Kunzler, E. Hurt, and G. Simos, *Mol. Cell Biol.* **18**, 6374 (1998).
[s] S. Sarkar and A. K. Hopper, *Mol. Biol. Cell* **9**, 3041 (1998).
[t] A. K. Hopper, L. D. Schultz, and R. A. Shapiro, *Cell* **19**, 741 (1980).
[u] H. Grosshans, E. Hurt, and G. Simos, *Genes Dev.* **14**, 830 (2000).

Although the mechanism by which substrates are recognized by a receptor and ultimately deposited in the nucleus is now understood in some detail, there is little information about how the receptor/substrate complexes actually move through the NPC. Many components of the *S. cerevisiae* nuclear pore have been identified,[21] but the question of how all these parts are assembled and function in the 60-MDa NPC still remains. We will need a detailed understanding of the three-dimensional structure of the NPC to understand how importin-βs are translocated through the channel. Studying protein interactions *in vivo* presents one way to obtain structural and mechanistic information, despite the complications posed by the large number of proteins involved.

Identification of Functional Nuclear Import/Export Targeting Signal

Overview

Several steps must be taken in order to define a functional targeting signal within the Protein of Interest (POI). The first step is to identify a putative targeting signal within the primary sequence of POI, most often done by identifying different protein motifs by computer analysis. The only signals that can really be identified with any reasonable accuracy through this type of approach are the cNLS, composed of a block of basic residues,[11] and the leucine-rich NES sequence.[10] It is important to remember that sequences identified through simple analysis of the protein sequence are only putative targeting signals. Neither classical NLS nor NES consensus sequences are strict by any stretch of the imagination. For example, NES sequences contain several leucine residues, and leucine is the most common amino acid found in proteins. Thus, identification of a leucine-rich sequence does not necessarily mean that POI is targeted for export via Crm1/Xpo1. Identification of a putative targeting signal is merely a starting point to determine whether the sequence is indeed a functional targeting signal. The second aspect of the analysis is defining the transport pathway *in vivo*. A complete analysis of the nuclear transport of POI involves identification of the functional targeting signal within POI as well as the pathway targeted by that signal.

Detecting Protein of Interest

The analysis described here requires a method to observe the intracellular localization of POI. This can be accomplished in a number of ways. If antibodies are available, the protein can simply be detected through standard indirect immunofluorescence.[22] Alternatively, the protein can be epitope tagged and

---

[21] M. P. Rout, J. D. Aitchison, A. Suprapto, K. Hjertaas, Y. Zhao, and B. T. Chait, *J. Cell Biol.* **148**, 635 (2000).

[22] J. R. Pringle, A. E. Adams, D. G. Drubin, and B. K. Haarer, *Methods Enzymol.* **194**, 565 (1991).

detected with antibodies directed against the epitope tag. The most popular method today is to generate a fusion to the green fluorescent protein (GFP) that can be viewed directly in living cells.[23] A prerequisite for use of any epitope-tagged protein (GFP or other) is that the tagged protein must be functional. Functionality is tested by determining whether the tagged protein complements a mutant or deletion phenotype of the *POI* gene. We have found that the most useful approach to tagging a protein is to express the protein from its own promoter with GFP fused to the C terminus. A number of plasmids are available to create this type of fusion, and technology has also been developed to integrate GFP at the C terminus of any gene.[24] In either case, expression from the endogenous promoter assures that POI is not grossly overexpressed, which could impact the observed intracellular localization.

Criteria for Defining Functional Targeting Signal

Identification of a nuclear-targeting signal within a protein is usually taken to mean that this signal mediates direct binding to a transport receptor. In other words, the implication is that both the targeting signal and the transport pathway have been determined. To define a targeting signal, it is thus essential to demonstrate experimentally (1) that the sequence is *necessary* for transport; (2) that the sequence is *sufficient* to target another protein for transport; (3) that mutations in genes required for that transport pathway inhibit the transport of POI, using *in vivo* analysis; and (4) that there is a *direct interaction* between POI and the putative receptor that is dependent on the identified targeting signal.

Identification of Functional Nuclear Localization Signal

This section outlines the general procedure for identifying a functional nuclear localization signal and, as an example, discusses an analysis of the canonical NLS that mediates import through the importin-α/importin-β95 pathway described earlier. The cNLS is the nuclear-targeting signal that is identified most easily by sequence analysis and thus the most popular to analyze. However, many proteins have noncanonical NLSs and are recognized by importins other than α/β95. Another formal possibility to bear in mind is that POI enters the nucleus in a complex with another protein(s) and thus does not have its own NLS.

We note that analysis of an NES requires that the protein be directed to the nucleus before export can be examined. Protein export can be examined with assays for nucleocytoplasmic shuttling, discussed in a later section.

Necessary. To demonstrate that a putative NLS is necessary for import into the nucleus, the sequence should be deleted or altered by site-directed mutagenesis. If import of the mutant protein is reduced drastically, then the sequence is a functional

---

[23] M. Chalfie, Y. Tu, W. Euskirchen, W. Ward, and D. C. Prasher, *Science* **263**, 802 (1994).

[24] M. S. Longtine, A. McKenzie III, D. J. Demarini, N. G. Shah, A. Wach, A. Brachat, P. Philippsen, and J. R. Pringle, *Yeast* **14**, 953 (1998).

NLS. If no obvious putative NLS is identified, or if multiple putative NLSs exist, systematic deletion analysis may be useful to determine the region of the protein that contains the functional nuclear targeting information. In terms of cNLS, the analysis is generally performed by changing one or more of the basic residues to alanine.

In practice, an NLS is identified by transforming wild-type yeast cells with the wild-type and mutant versions of *POI*. Cells are grown to log phase in selective synthetic medium with 2% glucose. The intracellular localizations of the mutant and wild-type proteins are then compared to determine whether the mutations impact the steady-state localization of the protein. If protein accumulates in the cytoplasm when the putative NLS is mutated, then the sequence is presumed to be necessary for efficient nuclear targeting.

In some cases, it may be necessary to alter the experimental conditions in order to induce protein import. For example, to monitor the import of the transcription factor Pho4, cells must be transferred to low phosphate medium.[25] Nuclear localization of the wild-type protein is thus a critical control for these experiments.

Sufficient. A putative NLS is deemed sufficient to mediate import if it can target an unrelated protein to the nucleus. For this experiment, the most common approach is to fuse the putative NLS to the N terminus of a reporter protein that contains GFP. We have used a modified version of the C-fus plasmid[26] designed to express GFP fusions from the *MET* promoter. Our modified plasmid expresses a chimera of two GFP molecules. The expression of tandem GFPs yields a protein of approximately 60 kDa and minimizes concerns about diffusion into the nucleus.

Wild-type cells are transformed with plasmids encoding the reporter protein with or without the putative NLS (or cNLS) sequence. The cells are grown to log phase in liquid culture, and the steady-state localization of the reporter proteins is examined. If the protein containing the putative NLS is targeted to the nucleus, then the sequence is deemed sufficient for nuclear localization.

In Vivo Analysis. To demonstrate that POI is imported by a certain pathway *in vivo*, it is necessary to demonstrate that import of POI is inhibited when genes in that pathway are mutated. At this point, it is possible only to deduce the import pathway for canonical NLS because noncanonical NLSs have not been matched with specific pathways. However, testing the different pathways is still worthwhile and is feasible by using strains in which one of the importin genes is deleted or mutated (see Table I).

To identify the pathway that recognizes a particular cNLS, one might examine import in cells with mutations in importin-α (*SRP1*) and/or importin-β95 (*KAP95/RSL1*). Both of these genes are essential, but temperature-sensitive mutations can be used in the analysis (e.g., *srp1-31, rsl1-1*; see Table I).

[25] A. Kaffman, N. M. Rank, and E. K. O'Shea, *Genes Dev.* **12,** 2673 (1998).
[26] R. K. Niedenthal, L. Riles, M. Johnston, and J. H. Hegemann, *Yeast* **12,** 773 (1996).

Wild-type and mutant strains are transformed with tagged POI, such as POI–GFP. Proteins that are imported by the pathway being tested, and proteins imported by another pathway, should be used as controls. We have used pGAD–GFP[27] for the classic, importin-α/β95 pathway; Npl3 for Mtr10-mediated import,[28] and Nab2 for Kap104-mediated import.[29] Known substrates for other pathways are listed in Table I.

Transformants are grown to log phase ($\sim 10^7$ cells/ml) at the permissive temperature and then the culture is split, and half is maintained at the permissive temperature and half is shifted to the nonpermissive temperature for 2–3 hr. Times may vary depending on the particular mutant, and for some deletion strains, a temperature shift may not be necessary. Cells are then analyzed to determine the intracellular localization of POI and the control proteins. As discussed earlier, it may be necessary to alter the experimental conditions in order to induce the import of POI.

Direct Interaction. A number of standard biochemical approaches can be used to demonstrate a direct interaction between POI and the putative receptor. The most convincing way to show a direct interaction between POI and a receptor is to use purified recombinant proteins. Full-length or a portion of POI containing the putative NLS can be used. Ideally, wild-type and mutant versions (generated in the "Necessary" portion of the analysis) are compared. A direct interaction dependent on the wild-type NLS sequence demonstrates functional receptor/substrate interaction. This type of analysis could be complemented by a coimmunoprecipitation experiment or by two-hybrid analysis, both of which have been used extensively to analyze substrate/receptor interactions. It should be noted, however, that neither of these techniques demonstrates a direct interaction.

When this analysis is extended to study a putative nuclear export signal, RanGTP must be provided to stabilize the receptor/substrate/RanGTP export complex. This requirement has been somewhat challenging, as most purified Ran is either in the form of RanGDP or has no associated nucleotide. One way to generate RanGTP is to incubate purified recombinant yeast Ran with alkaline phosphatase beads prior to loading with nucleotide (GTPγS or GDPβS).[30]

Analysis of Nuclear Protein Import Process

Overview

Although many aspects of nuclear transport are understood in some detail, there are clearly many more questions to answer. For this reason, researchers may

[27] N. Shulga, P. Roberts, Z. Gu, L. Spitz, M. M. Tabb, M. Nomura, and D. S. Goldfarb, *J. Cell Biol.* **135**, 329 (1996).
[28] L. F. Pemberton, J. S. Rosenblum, and G. Blobel, *J. Cell Biol.* **139**, 1645 (1997).
[29] J. D. Aitchison, G. Blobel, and M. P. Rout, *Science* **274**, 624 (1996).
[30] K. A. Marfatia, M. T. Harreman, P. M. Vertino, and A. H. Corbett, *Gene* **266**, 45 (2001).

have an interest in determining whether a specific gene has an impact on nuclear protein import or nuclear protein export.

Over the years, several approaches have been developed to define genes that play a role in nuclear protein import. The most common approach is to generate or obtain mutant alleles of the gene of interest and then examine the localization of a reporter protein. The most common method of monitoring the localization is to examine the steady-state localization of the reporter protein in cells. Other approaches include an *in vitro* import assay[31] and an assay that examines the kinetics of nuclear protein import.[27]

Choice of Substrate

The diversity of transport pathways presents somewhat of a dilemma for the researcher in that the mutant of interest may disable pathways that do not transport the substrate used in the assay. If the mutation is in a general nuclear transport factor, such as a component of the Ran system or a nuclear pore protein, then theoretically the choice of substrate should not impact the outcome of the experiment. However, if the gene has been implicated in transport via a nonclassical pathway, it may be necessary to choose an appropriate substrate based on experimental data. Several reporter constructs are available for the cNLS pathway. One popular choice is the pGAD–GFP plasmid,[27] which encodes a cNLS-containing fusion protein of approximately 60 kDa, so diffusion is not a concern.

Steady-State Localization

To examine the steady-state localization of a chosen reporter protein or an endogenous substrate, mutant cells and wild-type control cells are grown to log phase at the permissive temperature and then shifted to the nonpermissive temperature. The localization of the substrate is determined either by indirect immunofluorescence or by direct GFP fluorescence.

In Vitro Transport Assay

In vitro import assays in higher eukaryotes have been invaluable in identifying important nuclear transport factors.[32] These assays rely on permeabilized cells that have intact nuclei but lack the soluble cytosolic factors required to target proteins for import.[33] Biochemical purification of these factors using a fluorescently labeled NLS-containing substrate is the way that most classical import factors were identified. A similar assay has been developed in *S. cerevisiae* to combine yeast mutants

---

[31] G. Schlenstedt, E. Hurt, V. Doye, and P. A. Silver, *J. Cell Biol.* **123**, 785 (1993).
[32] S. A. Adam, R. S. Marr, and L. Gerace, *J. Cell Biol.* **111**, 807 (1990).
[33] S. A. Adam, R. Sterne-Marr, and L. Gerace, *Methods Enzymol.* **219**, 97 (1992).

with this powerful biochemical technique.[31] The approach suffers somewhat from the smaller size of the yeast nucleus, which makes the assay less straightforward to perform and the results less amenable to interpretation. For these reasons, this approach has not been employed extensively in yeast.

In brief, the assay involves monitoring the nuclear import of a fluorescently tagged substrate when added to permeabilized cells (nuclei) along with cytosol (soluble factors). Preparation of semi-intact cells and the cytosol from wild-type and mutant strains provides the opportunity to assess the role of certain factors as nuclear or soluble factors in the import process. The experimental details are described elsewhere.[31,34] When evaluating the import at the end of the assay, researchers should consider that (1) scoring must be done rapidly because cell lysis will eventually occur on slides and (2) plus/minus scoring is rather subjective and should be evaluated accordingly.

In Vivo NLS–GFP Protein Import Assay

This *in vivo* assay was developed to quantitate the relative rate of nuclear import of an NLS green fluorescent protein (NLS–GFP) fusion protein.[27] The NLS–GFP fusion protein is small enough to diffuse through the NPC; however, the protein is concentrated in the nucleus in wild-type yeast due to the strong NLS. Treatment of the cells with the metabolic poison sodium azide causes the NLS–GFP to equilibrate across the nuclear envelope. The rate of import is calculated as a function of time after the equilibrated cells are returned to physiological growth medium where the NLS–GFP in the cytoplasm is reimported into the nucleus. Import rates can be compared in wild-type and mutant cells to determine the impact that a specific mutation has on the rate of NLS-mediated import.

Experimental Approach. Wild-type and mutant cells are transformed with pGAD–GFP.[27] Transformed cells are grown to log phase in 10 ml of synthetic medium with 2% glucose and may be shifted to the nonpermissive temperature. Cells are then pelleted, washed once in synthetic medium, resuspended in 1 ml of synthetic medium/2% glucose supplemented with 10 mM sodium azide and 10 mM 2-deoxy-D-glucose, transferred to 1.5-ml microcentrifuge tubes, and incubated with gentle agitation for 45 min. Following incubation, cells are pelleted and washed once with 1 ml sterile ice-cold H_2O to remove sodium azide and 2-deoxy-D-glucose. Cells are pelleted; at this stage, cell pellets can be stored on ice for several hours. At time 0, cells are resuspended in 50 μl of prewarmed synthetic media/2% glucose and incubated in a heat block at the desired temperature for the duration of the assay. At each time point (every 2–3 min) a 2-μl sample is removed and examined under the microscope using standard GFP filters. Individual cells are scored as nuclear or cytoplasmic based on the localization of the NLS–GFP

---

[34] A. H. Corbett, D. M. Koepp, G. Schlenstedt, M. S. Lee, A. K. Hopper, and P. A. Silver, *J. Cell Biol.* **130,** 1017 (1995).

reporter. A cell is generally scored as "nuclear" if the nucleus is sufficiently bright to be clearly distinguished against the cytoplasmic background. At least 50–100 cells should be scored for each time point. Relative import rates can be determined by fitting a linear regression curve through the linear portion of the time course.

Considerations. To examine import rates in temperature-sensitive mutants, the assay temperature used for this procedure can be adjusted. For example, for temperature-sensitive mutants, the assay can be carried out at 37°. Regardless of the assay temperature used, the time course of import in this assay is quite rapid. Thus, samples must be examined under the microscope in rapid succession. We have found that it is best to have one person prepare the slides and another person count the cells.

Analysis of Nuclear Protein Export Process

Overview

The development of assays to examine nuclear protein export has lagged behind the development of import assays because it is easier to measure transport from the site of protein synthesis in the cytoplasm into the nucleus than it is to target a protein into the nucleus and then study its transport back into the cytoplasm. Two assays have been developed in yeast that allow analysis of protein export from the nucleus. The first assay takes advantage of a combination reporter that contains both an NLS and a nuclear export signal (NES). Because this assay uses a leucine-rich NES sequence, it examines export from the nucleus via the Crm1/Xpo1 pathway.[18] The second more general assay, described in the next section on protein shuttling, takes advantage of a specific nuclear pore mutant (*nup49-313*) that blocks protein import but not protein export.[31,35]

The Crm1/Xpo1 pathway is specifically inhibited by the potent drug leptomycin B, which has been used to study nuclear export. Importantly, this pathway is *not* sensitive to leptomycin B in wild-type *S. cerevisiae* cells. Assays using leptomycin B must be performed with yeast (MNY8) harboring a point mutation (T539C) in *CRM1/XPO1* that confers sensitivity to this drug.[36]

Nuclear Export Signal Export Assay

This assay was first developed in the identification of the *S. cerevisiae* NES receptor.[18] The assay takes advantage of the fact that a reporter protein with both an NES and an NLS (NES-NLS-GFP) is dynamic within the cell. For this specific reporter protein, steady-state localization is diffuse throughout the cell. When the reporter is introduced into mutants that have defects in the NES export pathway, the reporter protein accumulates in the nucleus because NES-mediated export is

---

[35] V. Doye, R. Wepf, and E. C. Hurt, *EMBO J.* **13**, 6062 (1994).
[36] M. Neville and M. Rosbash, *EMBO J.* **18**, 3746 (1999).

blocked, whereas NLS-mediated import remains intact. Thus, a wild-type cell has a diffuse localization of the reporter throughout the cell and an export mutant displays a nuclear localization.

Experimental Approach. Wild-type and mutant cells are transformed with the NES-NLS-GFP reporter plasmid. Transformed cells are grown to log phase in synthetic media supplemented with 2% glucose. If temperature-sensitive mutants are used, cells are shifted to the nonpermissive temperature for several hours. Cells are then examined under the microscope to observe localization of the GFP-tagged reporter protein. In each assay wild-type and *crm1/xpo1* mutant cells should be used as controls.

Considerations. At least one yeast protein, Yap1p, has been shown to be exported from the nucleus through direct binding to the NES receptor, Crm1p/Xpo1p.[37] A complementary experiment to the one described earlier with the NES reporter protein is to examine the localization of Yap1p (which has been monitored with a GFP fusion protein[37]) in the mutant being tested. This experiment confirms that export defects observed with a synthetic reporter protein can be replicated with an endogenous yeast NES-containing protein.

Assays for Nucleocytoplasmic Shuttling

Overview

It has become increasingly apparent that proteins are more dynamic than their steady-state localization might indicate. In particular, many proteins constantly shuttle between the nucleus and the cytoplasm. RNA-binding proteins constitute a major class of shuttling proteins. More recently, a protein involved in signal transduction in the pheromone response pathway, Ste5, has been shown to shuttle,[38] indicating that nucleocytoplasmic shuttling is important for a wide range of cellular functions. Detailed methods for the analysis of shuttling proteins are discussed elsewhere.[39]

Protein–Protein Interactions *in Vivo* with Fluorescence Resonance Energy Transfer

Overview

The development of GFP derivatives with shifted excitation and emission spectra has provided an opportunity to study protein–protein interactions in the living yeast cell.[40,41] This method involves the measurement of fluorescence resonance

---

[37] C. Yan, L. H. Lee, and L. I. Davis, *EMBO J.* **17,** 7416 (1998).
[38] S. K. Mahanty, Y. Wang, F. W. Farley, and E. A. Elion, *Cell* **98,** 501 (1999).
[39] E. A. Elion, *Methods Enzymol.* **351,** [34], 2002 (this volume).
[40] A. Miyawaki, J. Llopis, R. Heim, J. M. McCaffery, J. A. Adams, M. Ikura, and R. Y. Tsien, *Nature* **388,** 882 (1997).
[41] R. Heim and R. Y. Tsien, *Curr. Biol.* **6,** 178 (1996).

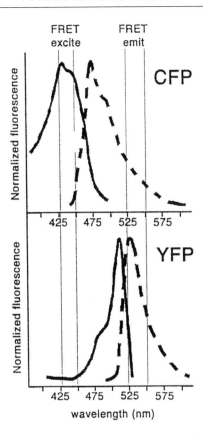

FIG. 1. Excitation (solid) and emission (dashed) spectra for CFP and YFP. The overlap of CFP emission with YFP excitation makes these fluorophores a suitable pair for FRET. Vertical lines represent excitation and emission filters in the FRET filter set (see Table II for exact values). This filter set allows the selective excitation of CFP and the monitoring of YFP emission.

energy transfer between two GFP derivatives. FRET has traditionally been performed *in vitro*, but GFP derivatives have allowed its application in living cells. FRET results in the indirect excitation of an "acceptor" by a directly excited "donor."[42,43] The energy transfer can occur only when the donor and acceptor are very close in space (10–100 Å). Therefore, if two target proteins are tagged with fluorophores, the interaction of those proteins is required for FRET to occur. The excitation and emission spectra of two GFP derivatives, the cyan and yellow fluorescent proteins (CFP, YFP), are shown in Fig. 1; the overlap of CFP emission

---

[42] R. M. Clegg, in "Fluorescence Imaging Spectroscopy and Microscopy" (X. F. Wang and B. Herman, eds.), p. 179. Wiley, New York, 1996.
[43] L. Stryer, *Annu. Rev. Biochem.* **47,** 819 (1978).

wavelengths with YFP excitation wavelengths makes these molecules useful as a donor–acceptor pair in FRET. The CFP–YFP pair has been used to observe FRET in living yeast cells by microscopy[44] and by fluorimetry.[45] We focus on the microscopy-based technique.

FRET has several advantages as a method of studying protein interactions. First, target proteins can be studied under physiological conditions in which they are expressed at their endogenous levels and are localized to the proper cellular compartment. This is generally not the case for the two-hybrid system, a common *in vivo* method for studying protein–protein interactions. Second, positive results in a FRET experiment are more likely to represent a direct interaction than results in two-hybrid or coimmunoprecipitation experiments. The implication of a direct interaction is due to the 100-Å theoretical limit for FRET, which translates to an even smaller value ($\sim$25–35 Å) when accounting for the large size of the GFP molecules and the properties of the CFP–YFP pair.[44] Importantly, the strict requirements for FRET imply that negative results are meaningless: the absence of FRET between two target proteins does not necessarily mean that they do not interact. For instance, if two proteins tagged on their C termini happen to interact via their N termini, CFP and YFP may not come within the required distance for FRET, despite the interaction of the target proteins.

FRET has been useful in studies of nucleocytoplasmic transport because of the advantages discussed previously and because the transport system involves a large number of proteins. The method was first used to map translocation pathways of transport receptors through the nuclear pore complex.[44] The application of FRET was useful because interactions between transport receptors and nearly half of the nucleoporins could be assayed. Additionally, these interactions are presumably transient and thus can be difficult to detect under nonphysiological conditions. FRET has also been used to gain insight into the structure of the NPC.[46] In this case, the advantage of FRET was the ability to dissect a huge macromolecular complex ($\sim$60 MDa), the analysis of which has been severely limited by its sheer size.

Materials

The FRET experiment requires many specific materials: vectors containing the CFP and YFP genes, anti-GFP antibody for confirming expression of the fusion proteins, and an epifluorescence microscope equipped with three specialized filter sets and a digital camera operated by imaging software. CFP and YFP genes are available commercially (e.g., from Clontech, Palo Alto, CA) or can be requested from various laboratories. It may be most convenient to obtain one-step cloning

---

[44] M. Damelin and P. A. Silver, *Mol. Cell* **5**, 133 (2000).
[45] M. C. Overton and K. J. Blumer, *Curr. Biol.* **10**, 341 (2000).
[46] M. Damelin and P. A. Silver, manuscript in preparation (2001).

vectors that contain CFP or YFP along with a promoter or terminator and a yeast selection marker (see later). The anti-GFP antibody, which also recognizes CFP and YFP, is also available commercially (e.g., from Clontech).

We list here the equipment used currently for FRET in this laboratory: a Nikon Diaphot-300 inverted epifluorescence microscope with a 100-W mercury lamp; a liquid-cooled charge-coupled device (CCD) camera (200 series; Photometrics, Tucson, AZ) with a KAF-1400 chip; and a D122 shutter driver (UniBlitz, Rochester, NY). We use a 60× 1.4 NA (not 100×) objective for FRET experiments because brightness can be limiting and is generally related inversely to the magnification level. The MetaMorph Imaging System (Universal Imaging Corp., West Chester, PA) is used for both data acquisition and image analysis. The important features of this software are the ability to (1) interface with the camera in order to acquire precise exposures and (2) calculate the average pixel intensity in a given region of an image.

Three filter sets are necessary for the FRET experiment. Each filter set has three components: an excitation filter, a dichroic mirror, and an emission filter. The "CFP" set includes excitation and emission filters for CFP. Similarly, the "YFP" set includes excitation and emission filters for YFP. The "FRET" set includes an excitation filter for CFP and an emission filter for YFP, and thus is used to monitor the indirect excitation of YFP by CFP. Filters with especially narrow bandpasses have been developed for this application and are available from Chroma Technology Corp. (Brattleboro, VT) or, alternatively, from Omega Optical (Brattleboro, VT). Specifications are shown in Table II. Importantly, the microscope must be able to accommodate two (preferably all three) filter sets at once so that the user can easily change from one to another.

Experimental Design

Accurate interpretation of data requires the comparison of distinct CFP/YFP pairs. Elaborate controls are critical because CFP and YFP are not "ideal" fluorophores for FRET because of their broad excitation and emission spectra. In other words, cells expressing only CFP show significant background with the FRET filter set, as the tail of the CFP emission spectrum extends into the 535-nm range (peak

TABLE II
FILTER SPECIFICATIONS FOR FRET EXPERIMENT

Filter set	Excitation	Dichroic	Emission
CFP	440/20 nm	455 nm LP*	480/30 nm
YFP	500/25 nm	525 nm LP	545/35 nm
FRET	440/20 nm	455 nm LP	535/25 nm

*LP = long pass

YFP emission), as shown in Fig. 1. The YFP background with the FRET filter set is generally much lower than that of CFP and may be significant depending on the expression level of the fusion protein and on the particular microscope being used.

The high background of the CFP fluorophore makes the following strategy necessary; compare Target1-CFP/Target2-YFP with Target1-CFP/Control1- YFP, Target1-CFP/Control2-YFP, etc. In this case, Control1 and Control2 are endogenous proteins comparable to Target2 with respect to *expression level* and *subcellular localization*. Implicit in this approach is the requirement that only strains

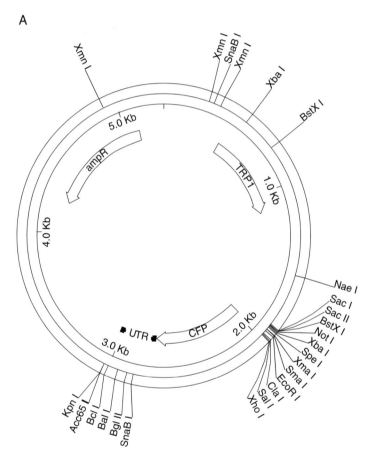

FIG. 2. Plasmid maps for pPS1890 (A) and pPS1891 (B), vectors for integrating CFP and YFP into the yeast genome. Unique restriction sites are shown. A C-terminal fragment of the target gene is cloned into the vector to create an in-frame fusion to CFP or YFP. The resulting vector is linearized within the target gene fragment before transformation. In both vectors, the start codon of CFP and YFP follows the *Xho*I site: CTCGAGCTATG.

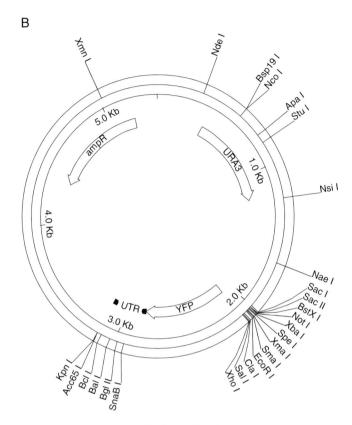

FIG. 2. (continued)

expressing the same CFP-tagged target protein be compared directly. We implemented this strategy in the study of interactions between transport receptors and nucleoporins: FRET was assayed for a transport receptor–CFP fusion when paired with 13 different nucleoporin–YFP fusions, and the signals from the 13 strains were compared to one another.[44]

Three additional controls have been used in FRET experiments in yeast cells: switching the donor and acceptor, performing the experiments in mutants, and performing photobleaching experiments (described under Methods). Switching the donor and acceptor involves constructing an entire new set of strains that includes, following the aforementioned example, Target2-CFP/Target1-YFP, Target2-CFP/Control1-YFP, Target2-CFP/Control2-YFP, etc. This task is straightforward if vectors are used in which CFP and YFP are in the same reading frame, such as pPS1890 and pPS1891 (see later).

The tractable genetics of the yeast system can be used to investigate the effect of *cis* and *trans* mutations on protein–protein interactions. The FRET interaction between two target proteins can be compared in wild-type cells and cells with a mutation in one of the target proteins or in another protein in the pathway. In such experiments, it is important to check that the expression levels of the target proteins are unchanged in the mutant.

Experimental Methods

Strain Construction. FRET strains are constructed by transforming a wild-type *ADE2+* haploid strain such as FY23[47] with different combinations of CFP- and YFP-containing vectors. We recommend using integration vectors, as integration of fusions into the genome provides several advantages. (1) Gene expression is relatively consistent from cell to cell, which facilitates quantitation; (2) protein levels are at endogenous levels; and (3) the fusion protein is the only cellular copy of the target protein and is not competing with endogenous protein for binding sites. An integration vector is usually constructed more easily than a centromeric plasmid because integration requires only a fragment of the target gene. The fragment can be amplified by polymerase chain reaction and inserted into existing vectors to create an in-frame fusion to CFP or YFP. The genomic fragment must contain a restriction site *that will be unique in the final vector,* allowing linearization to stimulate recombination at the target locus. The site should be flanked on either side by ample genomic DNA to allow recombination, preferably >200 bp. It is useful to ensure that the genomic fragment will have a unique site in the CFP and YFP vectors, which have different selection markers and thus some different restriction sites. We routinely use two vectors for C-terminal fusions: pPS1890, based on pRS304,[48] contains CFP and the TRP1 marker (Fig. 2A), and pPS1891, based on pRS306, contains YFP and the URA3 marker (Fig. 2B).[44] Because CFP and YFP are in the same reading frame, generating fusions of both CFP and YFP to a target gene is straightforward with these vectors.

Acceptable results in these experiments depend on the functionality of the fusion proteins. Functionality should be assayed by checking that (1) a full-length fusion protein is expressed, (2) the fusion has the expected subcellular localization, and (3) the fusion protein can replace the endogenous protein in a genetic background in which it is required for viability. The first test is done by preparing whole cell lysates and performing immunoblot analysis with the anti-GFP antibody. Localization is verified by growing cells to midlog phase and examining them under a fluorescence microscope. Finally, and most importantly, expression and localization must be confirmed in a genetic background (or under certain conditions, e.g., 37°, 1 *M* sorbitol) where the target gene is essential. The use of integration

---

[47] F. Winston, C. Dollard, and S. L. Ricupero-Hovasse, *Yeast* **11**, 53 (1995).
[48] R. S. Sikorski and P. Hieter, *Genetics* **122**, 19 (1989).

vectors is very convenient for this purpose. For example, a $nup170\Delta$ strain was used to test the functionality of the Nup188–YFP fusion, as $NUP188$ is required for viability in this background.[44] The $NUP188$–YFP fusion was integrated into the genome of $nup170\Delta$, replacing endogenous $NUP188$. Once the expression and correct localization of Nup188–YFP were verified, the viability of the resulting cells indicated that Nup188–YFP is functional. The same result cannot be achieved with a centromeric plasmid because the cells still have endogenous $NUP188$.

Once the functionality of each fusion has been confirmed, strains can be constructed in which the various combinations of CFP and YFP fusions are expressed, as verified by microscopy and immunoblot analysis. The particular strains needed for the FRET experiments are discussed under "Experimental Design." Cells expressing the individual fusions are also important for the initial analysis.

Data Acquisition. Cells are grown to midlog phase at 25° in synthetic complete medium and then placed directly onto microscope slides without being concentrated. Fresh slides should be prepared every few minutes. For an initial qualitative assessment of the signal, we recommend using the FRET filter set, which allows the excitation of CFP and the monitoring of YFP emission (see Fig. 1). Changes in CFP intensity are nearly impossible to detect by eye against the dark background. Because of the relatively rapid photobleaching of CFP and YFP, any signal will fade substantially within seconds; the user must be examining new fields continually.

FRET is marked by an increase in YFP emission ("sensitized emission") and a coincident decrease in CFP emission ("donor quenching"), as CFP has transferred energy to YFP. In order to account for the changes in YFP and CFP emission as a result of CFP excitation, two images of each cell should be taken: one with the FRET filter set (excite CFP, monitor YFP emission) and one with the CFP filter set (excite CFP, monitor CFP emission). This order of exposures is essential for reducing problems caused by photobleaching. Because the images are quantitated, the exposure settings for all FRET images in an experiment must be identical, and likewise for all CFP images. The exposure time with the CFP set is typically longer than with the FRET set. Additionally, the autoscale functions available on some software applications must be turned off during the experiment.

This type of quantitative two filter set analysis is critical for *in vivo* experiments. Using only the FRET filter set is not sufficient and, in fact, can be misleading, as the concentrations of CFP and YFP are not necessarily constant from strain to strain (a problem not encountered in well-calibrated *in vitro* systems).

An important point that follows from the rapid photobleaching of CFP and YFP is the need to photograph cells before observing the signal. Prior exposure to the 100-W light will influence the signal dramatically. Thus, during data acquisition for any quantitative analysis, locating and focusing on cells should be performed with differential interference contrast (e.g., Nomarski) optics. Finally, we note that taking pictures of individual cells centered in the field both facilitates and improves the analysis.

Image Analysis. We quantitate FRET interactions using a two filter set system in order to reduce systematic errors and cell-to-cell fluctuations in CFP and YFP expression levels. In this analysis, the signal with the FRET filter set is normalized by the signal with the CFP filter set in each cell.

To optimize signal to noise, the average intensity is calculated only in the region of interest, e.g., the nuclear envelope. One relatively simple way of highlighting the region is to use the threshold function on MetaMorph, which selects pixels in a given range of intensities. This technique is generally applicable because the region of interest is the brightest in the cell. The same region (i.e., same number of pixels) should be selected for FRET and CFP images of a given cell. The average intensity in each region, as calculated by the software, is then recorded to a data log. Additionally, for each strain, the average intensity in areas without any cells should be recorded for several FRET and CFP images in order to calculate the background level.

The analysis can be completed on a spreadsheet such as Microsoft Excel. For each strain, the average background signals for the FRET and CFP images are calculated and then the appropriate value is subtracted from each FRET and CFP average intensity. Division of the background-corrected intensities in each cell yields its FRET/CFP ratio. A quick comparison of strains can be achieved by calculating the "mean ratio" of each strain—the average of the ratios for all cells. However, the sets of individual ratios themselves are used in the statistical analysis. The rank sum test can be used to assess the significance level of the potential interaction; this test is preferable in these experiments because the samples are small and the distribution is not always normal.

We have found that imaging 15–20 cells per strain can determine a reproducibly significant interaction using the two filter set method.[44,46] This number may vary depending on the target proteins and the equipment used. Ultimately, the statistics and the reproducibility will determine whether the result is believable.

Photobleaching Experiment. One control for FRET interactions is an experiment that tests the dependence of the signal on YFP emission. The photobleaching experiment discriminates between signals dependent on YFP emission and signals constituted only of CFP background. The FRET assay for a given CFP/YFP pair is compared before and after the YFP has been selectively photobleached. If the signal depends on YFP emission, a decrease will be observed after YFP photobleaching; if the signal depends only on CFP, it will be unchanged. *Note.* In cases where YFP contributes background signal in the FRET filter set, the photobleaching experiment is of little value because both genuine and artificial signals will decrease after photobleaching.

In the experiment, before a given cell is imaged, the YFP is photobleached by excitation with the YFP filter set until YFP emission is no longer detectable under the eyepiece. CFP is not affected by the prolonged exposure with the YFP filter set. Data acquisition and image analysis follow exactly as described earlier, where a given strain is compared before and after YFP photobleaching.

Conclusions

An important realization in recent years is the complexity of the system used by eukaryotic cells to transport macromolecules between the nucleus and the cytoplasm. In our presentation of methods for studying the transport of particular proteins or the nuclear transport process itself, we have tried to emphasize the need to consider the different transport pathways in the cell. Identification of a transport pathway used by a particular protein requires a systematic approach. As we continue to learn more about nuclear transport, new information promises to make this process more straightforward and, at the same time, more elaborate.

Acknowledgments

We are grateful to our colleagues for many reagents and helpful discussions and to Deanna Green for comments on the manuscript. M.D. was supported by NIH Training Grants in Biophysics and in Tumor Biology.

[34] How to Monitor Nuclear Shuttling

By ELAINE A. ELION

Introduction

Many biological processes are controlled by the movement of proteins in and out of the nucleus, often as a means either to turn on or to turn off a signal. For example, nuclear shuttling is used frequently to control transcriptional responses in a variety of systems.[1] Nuclear shuttling is also used to regulate the onset of cell cycle events[2,3] and the access of cytoplasmic proteins to the cell periphery.[4] Some proteins may shuttle continuously in and out of the nucleus, whereas others may undergo regulated entry and exit. The existence of nuclear shuttling as a regulatory mechanism for a particular protein may not always be obvious from the localization pattern of the protein at steady state. This chapter describes a variety

---

[1] A. Kaffman and E. K. O'Shea, *Annu. Rev. Cell Dev. Biol.* **15**, 291 (1999).
[2] R. Visintin and A. Amon, *Curr. Opin. Cell Biol.* **12**, 373 (2000).
[3] C. G. Takizawa and D. O. Morgan, *Curr. Opin. Cell Biol.* **12**, 658 (2000).
[4] E. A. Elion, *Curr. Opin. Microbiol.* **3**, 573 (2000).

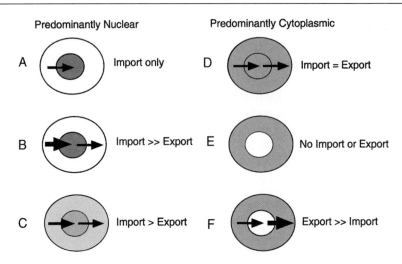

FIG. 1. Possible explanations for nuclear and cytoplasmic protein distribution at steady state. (A) Predominantly nuclear localization as a result of nuclear import with no nuclear export. (B) Predominantly nuclear localization as a result of a much greater rate of nuclear import compared to nuclear export. In this instance, the protein is reimported continuously into the nucleus. (C) Nuclear localization with some cytoplasmic accumulation visible. In this scenario, the protein shuttles continuously, but the rate of nuclear import is somewhat greater than the rate of nuclear export. (D) Only cytoplasmic localization visible. The absence of nuclear exclusion indicates that a pool of protein is in the nucleus, but the rates of nuclear import and export must be similar. (E and F) A predominantly cytoplasmic localization with obvious nuclear exclusion could arise either from a situation in which the protein does not enter the nucleus at all (E) or is exported from the nucleus at a rate that is much greater than its rate of nuclear import (F).

of ways in which one can evaluate whether a protein that appears static at steady state actually shuttles through the nucleus.

Determining Whether the Protein Shuttles through the Nucleus

Clues from Initial Analysis of Steady-State Distribution

A number of issues should be considered when determining whether a protein shuttles in and out of the nucleus. The approaches taken will depend on (1) whether the localization of the protein at steady state is predominantly nuclear or cytoplasmic, (2) whether the protein enters the nucleus by passive diffusion or receptor-mediated events, and (3) whether nuclear shuttling is constitutive or regulated. The steady-state distribution of a protein depends on the relative rates of nuclear import and nuclear export (Fig. 1).[5,6] A protein that appears to be solely

[5] D. Gorlich and U. Kutay, *Annu. Rev. Cell. Dev. Biol.* **15,** 607 (1999).
[6] I. W. Mattaj and L. Englmeier, *Annu. Rev. Biochem.* **67,** 265 (1998).

nuclear at steady state may actually shuttle in and out of the nucleus, but the rate of nuclear import may be greater than the rate of nuclear export. Demonstrating that a nuclear protein shuttles involves finding conditions that reveal its ability to be exported out of the nucleus. A protein that appears to be cytoplasmic at steady state may also shuttle, but the rate of nuclear export may be greater than the rate of nuclear import. Demonstrating that a cytoplasmic protein shuttles involves finding conditions that clearly reveal the existence of a nuclear pool that is exported. Proteins smaller than 50 kDa can enter the nucleus by passive diffusion,[6,7] whereas much larger proteins are most likely to enter via receptor-mediated events. However, small size is not a predictor of whether nuclear entry occurs through passive diffusion because small proteins also undergo receptor-mediated nuclear shuttling events.[8,9] The difficulty in distinguishing between passive diffusion and active shuttling may be circumvented for small proteins by increasing their mass through the addition of single or multiple protein tags such as green fluorescent protein (GFP).

Nuclear shuttling can occur constitutively or be regulated by environmental conditions.[1,10] Evidence of distribution changes from nuclear to cytoplasmic, or vice versa, often provides a first clue that a protein shuttles through the nucleus. If the protein being studied is part of a regulated pathway, then it may be possible that the rates of nuclear import and export will change under varying environmental conditions. Many conditions have been shown to alter the relative steady-state distribution of a protein in the nucleus and cytoplasm, including position in the cell cycle,[11] nutrient changes,[12,13] osmotic stress,[14,15] oxidative stress,[16] and mating pheromone.[17,18] Many of these distribution changes are known to be linked to changes in rates of import and/or export. In addition, many nuclear shuttling events are regulated by phosphorylation[1] and are affected by mutations in key regulatory kinases within a given signal transduction pathway.[11,14,15,19] Thus, initial analysis of protein localization at steady state should involve examining the protein under

---

[7] B. Talcott and M. Shannon Moore, *Trends Cell Biol.* **9**, 312 (1999).
[8] K. Schwamborn, W. Albig, and D. Doenecke, *Exp. Cell Res.* **244**, 206 (1998).
[9] A. Wada, M. Fukada, M. Mishima, and E. Nishida, *EMBO J.* **17**, 1635 (1988).
[10] J. K. Hood and P. A. Silver, *Curr. Opin. Cell Biol.* **11**, 241 (1999).
[11] T. Moll, G. Tebb, U. Surana, H. Robitsch, and K. Nasmyth, *Cell* **66**, 743 (1991).
[12] M. J. DeVit and M. Johnston, *Curr. Biol.* **9**, 1231 (1999).
[13] E. M. O'Neill, A. Kaffman, E. R. Jolly, and E. K. O'Shea, *Science* **271**, 209 (1996).
[14] W. Gorner, E. Durchschlag, M. T. Martinez-Pastor, F. Estruch, G. Ammerer, B. Hamilton, H. Ruis, and C. Schuller, *Genes Dev.* **12**, 586 (1998).
[15] P. Ferrigno, F. Posas, D. Koepp, H. Saito, and P. A. Silver, *EMBO J.* **17**, 5606 (1998).
[16] C. Yan, L. H. Lee, and L. I. Davis, *EMBO J.* **17**, 7416 (1998).
[17] S. K. Mahanty, Y. M. Wang, F. W. Farley, and E. A. Elion, *Cell* **98**, 501 (1999).
[18] M. Blondel, P. M. Alepuz, L. S. Huang, S. Shaham, G. Ammerer, and M. Peter, *Genes Dev.* **13**, 2284 (1999).
[19] A. Kaffman, N. M. Rank, and E. K. O'Shea, *Genes Dev.* **12**, 2673 (1998).

a variety of conditions that are related to its function, as well as cell cycle position in a population of mitotically dividing cells.

The observation that the localization changes of a protein from nuclear to cytoplasmic or vice versa under certain conditions does not prove that a protein shuttles through the nucleus. For example, an apparent increase in nuclear or cytoplasmic distribution could be due to stabilization of the protein in one or both compartments. Once a condition is found to induce a change in protein localization, it is important to determine whether these conditions alter the stability of the protein in question by monitoring the steady-state levels of the protein from cells grown under the same conditions used for localization.

The most convincing evidence that a steady-state distribution change is due to movement of a protein comes from experiments that involve analysis of a previously synthesized pool of protein. This is done by first expressing the protein and then blocking its further synthesis prior to imposing the condition that has been found to induce a change in its localization. An important consideration for this type of experiment is the half-life of the protein and whether it is sufficiently long enough to allow detection of the protein for the duration of the experiment. It is therefore prudent to simultaneously monitor the steady-state level of the protein in whole cell extracts from cultures grown in a mock experiment. The following method provides details of how one should prepare cells to analyze the localization of a previously synthesized pool of a protein of interest.

Method

The two most commonly used approaches to block *de novo* synthesis of a protein are cycloheximide and conditional expression of the protein with a promoter that can be induced and repressed by simple changes in the growth medium. The use of cycloheximide has the advantage that it is simple; cycloheximide is added directly to the medium at a final concentration of 10 μg/ml from a 10-mg/ml stock solution and protein synthesis is blocked almost immediately.[20] The disadvantage of using cycloheximide is that total protein synthesis is blocked, possibly causing indirect effects on the localization event or the protein.

Cycloheximide is therefore more amenable to experiments in which the time frame for the localization studies is minutes rather than hours. Use of a conditional promoter has the obvious advantage that only the protein being studied is varied. The *GAL1* promoter is used frequently for this type of experiment. Cells are pregrown in nonrepressing medium containing 2% (w/v) raffinose and then in medium containing 2% (w/v) galactose to induce expression of the protein. After growth in the galactose medium for a specified length of time (typically 1.5–3 hr), cells are placed in glucose medium to block further expression of the protein for a

---

[20] H. M. Fried and J. R. Warner, *Nucleic Acids Res.* **10,** 3133 (1982).

specified period of time (typically 1–2 hr) before administering the condition that changes the cellular distribution of the protein. An important consideration for this type of experiment is that it involves overexpression of a protein, a condition that can lead to aberrant localization. It is therefore important to choose a length of time of induction that allows for the synthesis of enough protein for detection purposes, without interfering with its normal localization pattern.

Clues from Comparing Different Expression Levels for Cytoplasmic Proteins

In many instances, it may not be possible to find growth conditions that vary the distribution of a protein with respect to the nucleus and cytoplasm, and the protein always appears to be nuclear or cytoplasmic. One clue that a cytoplasmic protein may shuttle through the nucleus is whether it is excluded from nuclei at native or moderate levels of expression (Fig. 1). The absence of nuclear exclusion may indicate that the protein is nuclear as well as cytoplasmic and may shuttle.[17] Further initial evidence of the existence of a nuclear pool can come from increasing the level of expression of a protein, which can lead to the appearance of a more obvious nuclear pool above the background of the cytoplasmic pool in some or all cells of a population.[17,21] However, overexpression of a protein could interfere with detection of a nuclear pool by generating too much of the protein in the cytoplasm. Therefore, at the onset, it is important to compare the localization of the protein at both low and high levels of expression. However, caution must be used in this type of analysis, as overexpression of the protein may affect its localization pattern, as well as the activity of the cellular pathways with which it intersects.

Use of Heterokaryons to Test Shuttling of Nuclear Protein

One way to initially test whether a nuclear protein shuttles is to determine whether it migrates from one nucleus to another in a polykaryon test. This approach is used commonly for assessing protein shuttling in mammalian cells. Cells expressing the protein of interest are fused to cells that do not express the protein and the appearance of the protein in the fused nuclei is monitored.[22] A similar technology has been applied successfully in yeast and takes advantage of zygotic heterokaryons that form in crosses between wild-type and *kar1-1* haploids of the opposite mating type.[23] *KAR1* encodes a protein that is a component of the spindle pole body.[24] *kar1-1* × *KAR1* zygotes are unable to fuse their nuclei as a

[21] P. M. Pryciak and F. A. Huntress, *Genes Dev.* **12,** 2684 (1998).
[22] R. A. Borer, C. F. Lehner, H. M. Eppenberger, and E. A. Nigg, *Cell* **56,** 379 (1989).
[23] J. Flach, M. Bossie, J. Vogel, A. Corbett, T. Jinks, D. A. Willins, and P. A. Silver, *Mol. Cell. Biol.* **14,** 8399 (1994).
[24] L. Marsh and M. D. Rose, *in* "The Molecular and Cellular Biology of the Yeast *Saccharomyces*" (J. R. Pringle, J. R. Broach, and E. W. Jones, eds.), Vol. 3, p. 827. Cold Spring Harbor Laboratory Press, Cold Spring Harbor, NY, 1997.

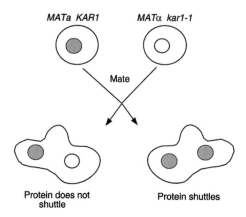

FIG. 2. Use of *kar1* × *KAR1* heterokaryons to determine whether a nuclear protein shuttles. A *MATa KAR1* haploid containing the protein of interest that has been expressed previously from an inducible promoter (shown as gray shading in the nucleus) is mated to a *MATα kar1-1* haploid that does not express the protein. Cells are mated together in equal numbers, the mixture is then fixed and stained for DNA, and the protein is examined for the presence of heterokaryons. Heterokaryons are defined as peanut-shaped cells with two nuclei, with or without an emerging bud. A protein that shuttles will exit the nucleus and be reimported into unmarked nuclei, generating heterokaryons with protein in both nuclei. A protein that does not shuttle will generate heterokaryons that contain the protein in one of the two nuclei.

result of this microtubule defect, which prevents nuclear migration and leads to the formation of heterokaryons with parental nuclei separated by ~3 μm.[24] In this type of experiment (Fig. 2), a tagged version of the protein is conditionally expressed in one of the two parents and then its expression is repressed. These cells are then mated to equal numbers of the mating partner cells under conditions that allow the recovery of significant numbers of heterokaryons (i.e., zygotes with unfused nuclei).[23,25] If the protein shuttles, then it should be possible to detect the tagged protein in both nuclei of the heterokaryons.[23] In contrast, nonshuttling proteins, such as those that contain nuclear targeting sequences, but not nuclear export sequences (NES), will only be detected in one of two nuclei. It is useful to include controls in this experiment, such as a nonshuttling nuclear protein [e.g., a protein that only harbors nuclear import signals, such as β-galactosidase or a GFP–GFP fused to a nuclear localization sequence (NLS)] and a nuclear protein that has been shown to shuttle (e.g., Npl3[23]).

Use of nsp1^{ts} and nup49-313 to Analyze Nuclear Export

More convincing evidence that a protein shuttles comes from showing that mutations that specifically block either the import or the export of the protein

---

[25] M. D. Rose and G. R. Fink, *Cell* **48,** 1047 (1987).

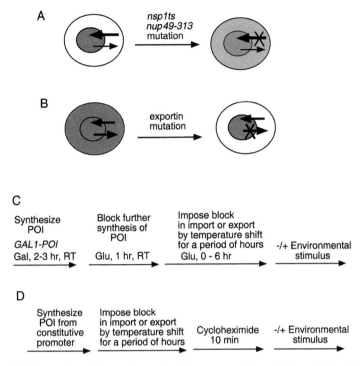

FIG. 3. Using mutations in the nucleocytoplasmic transport machinery to demonstrate nuclear shuttling. (A) *nsp1ts* and *nup49-313* mutations selectively block nuclear import without blocking nuclear export. These mutations can be used to demonstrate nuclear export of a nuclear protein. At the nonpermissive temperature, the protein synthesized previously that has localized to the nucleus is exported and accumulates in the cytoplasm because its reimport into the nucleus is blocked. (B) Exportin mutations can be used to demonstrate nuclear import and reveal the existence of a nuclear pool. When the cognate exportin for the protein in question is mutated, the protein accumulates in the nucleus, leading to a decrease in its cytoplasmic pool. Depending on the exportin, the mutation can either be a temperature-sensitive mutation in an essential exportin [(e.g., *xpo1-1*) K. Stade, C. S. Ford, C. Guthrie, and K. Weis *Cell* **90,** 1041 (1997)] or a null mutation in a nonessential export [(e.g., *msn5Δ*) S. K. Mahanty, Y. M. Wang, F. W. Farley, and E. A. Elion, *Cell* **98,** 501 (1999); A. Kaffman, N. M. Rank, E. M. O'Neill, L. S. Huang, and E. K. O'Shea, *Nature* **396,** 482 (1998)]. (C and D) Two flow sheets for using temperature-sensitive mutations to selectively block nuclear import or export, monitoring *de novo*-synthesized protein using an inducible promoter (C) or cycloheximide (D).

cause a corresponding change in its localization (Fig. 3). This is achieved by using strains that harbor mutations in components of the nucleocytoplasmic transport machinery that regulates nuclear import and export.[26] This machinery includes the nucleoporins that make up the nuclear pore complex, Ran GTPase, Ran GTPase-activating protein (Ran–GAP), Ran guanine exchange factor (Ran–GEF), and the

[26] M. Damelin, P. A. Silver, and A. Corbett, *Methods Enzymol.* **351,** [33], 2002 (this volume).

importin-β family of soluble receptors called importins and exportins that directly mediate the import and export of proteins (see Fig. 4 for a list of the gene names of some of these components). Mutant strains with mutations in many of these components exist and have been characterized.[26] Many of the components of the nucleocytoplasmic transport machinery regulate import and export of both RNA and protein and shuttle through the nucleus themselves. An importin-β family member can regulate both import and export either directly or indirectly by controlling the import or export of other importin-β family members.[5]

One can demonstrate that a nuclear protein is exported by selectively blocking reimport of the protein following its nuclear export, thereby causing the protein to accumulate in the cytoplasm (Fig. 3A). This approach can also be used to establish whether nuclear export occurs constitutively or is regulated, and whether the rate of nuclear export differs under varying conditions. In addition, this approach can be used determine whether a cytoplasmic protein undergoes nuclear export by first converting the protein of interest into a form that accumulates in the nucleus at steady state by adding an NLS to the protein. The most convenient way to analyze export in the absence of import is to use one of two temperature-sensitive nucleoporin mutants, *nup49-313*[27] and *nsp1^{ts}*,[28,29] that preferentially block nuclear import. Both mutants were first used to analyze RNA-mediated nuclear export of the RNA-binding protein Npl3[30] and have been extended to analyze the export of Pho4[31] and Ste5.[17] At nonpermissive temperature, *nup49-313* has been found to block nuclear import of MATα2-lacZ,[27] Npl3,[30] and Pho4,[31] without blocking nuclear export of these proteins or poly(A)$^+$ RNA. This phenotype is allele specific; e.g., the temperature-sensitive *nup49-316* mutation blocks nuclear export of MATα2-lacZ and poly(A)$^+$ RNA in addition to nuclear import.[27] Similarly, at nonpermissive temperature, *nsp1^{ts}* has been shown to block nuclear import of MATα2-lacZ and Pho2-lacZ,[28] Pho4,[31] and Ste5,[17] but does not prevent nuclear export of Pho4[31] or Ste5.[17] The available data indicate that a nuclear protein that lacks nuclear export sequences does not leak out of the nucleus into the cytoplasm in either mutant at the nonpermissive temperature.[27,28] Therefore, accumulation of a given protein in the cytoplasm after temperature shift should reflect its true export from the nucleus. Collectively, these results suggest that both *nup49-316* and *nsp1ts* mutants can be used to assess the export of a wide variety of proteins.

The following considerations should be kept in mind when attempting this type of experiment. First, it is necessary to demonstrate that *nup49-313* and/or *nsp1^{ts}* blocks nuclear import of the protein in question because it is not a given that this will be the case. This can be assessed under conditions in which the protein is expressed

---

[27] V. Doye, R. Wepf, and E. C. Hurt, *EMBO J.* **13,** 6062 (1994).
[28] U. Nehrbass, E. Fabre, S. Dihlmann, W. Herth, and E. C. Hurt, *Eur. J. Cell Biol.* **62,** 1 (1993).
[29] C. Wimmer, V. Doye, P. Grandi, U. Nehrbass, and E. C. Hurt, *EMBO J.* **11,** 5051 (1992).
[30] M. S. Lee, M. Henry, and P. A. Silver, *Genes Dev.* **10,** 1233 (1996).
[31] A. Kaffman, N. M. Rank, E. M. O'Neill, L. S. Huang, and E. K. O'Shea, *Nature* **396,** 482 (1998).

constitutively and should reveal a large increase in the cytoplasmic pool with a corresponding decrease in obvious nuclear accumulation. Second, it is important to use the most optimal conditions for the temperature shift and to monitor the localization of the protein at several time points after the temperature shift. This may vary depending on the protein being studied and the strain background. For example, a shift to 36° for 5 hr was used for *nup49-313* in the analysis of Npl3,[30] whereas shifts to 38.5° for 6 hr and 37° for 3 hr were used for *nsp1^{ts}* in the analysis of Pho4[31] and Ste5.[17] Third, it is important to properly control the expression of the protein in question during the course of the experiment. Ideally, it is best to conditionally express the protein prior to imposing the temperature shift (Fig. 3C) in order to ensure that a cytoplasmic pool does not arise from a *de novo*-synthesized protein that is not imported into the nucleus. This is particularly important if nuclear export occurs constitutively. However, if nuclear export is induced in the presence of an environmental stimulus, it is possible to use a simpler approach of expressing the protein constitutively and using cycloheximide to block protein synthesis after the temperature shift (Fig. 3D). Fourth, it is important to keep in mind that the effect may vary depending on the strain background and the rate of export of the protein being studied. Analysis of the constitutive export of Npl3 in a *nup49-313* strain after 5 hr at 36°[30] showed that only ~30–35% of cells exhibited strong cytoplasmic staining, with an additional ~30–35% of cells showing weak cytoplasmic staining with bright nuclear staining still detected in some cells. Analysis of the constitutive and induced export of SV40-TAgNLS-Ste5 in a *nsp1ts* strain after 3 hr at 37°[17] showed that the majority of cells exhibited fairly weak cytoplasmic staining with obvious nuclear staining in the absence of α-factor mating pheromone, whereas most cells exhibited moderate to strong cytoplasmic staining after brief treatment with α factor. The following experimental approach describes how to analyze a protein of interest (POI) in *nup49-313* or *nsp1^{ts}* cells.

Method

First, create a *POI* gene that is conditionally expressed and yields a protein that accumulates predominantly in the nucleus at steady state and can be detected by an antibody or is fused to green fluorescent protein. If the POI is normally cytoplasmic, it can be engineered to be nuclear through the addition of a heterologous NLS [such as the SV40 (simion virus 40) large T antigen NLS[17]]. Note that the number of NLSs needed to target POI to the nucleus will depend on the relative rate of export.[32] For the purposes of this discussion, make a *GAL1-POI* gene that is tagged with one or more copies of GFP or an epitope (i.e., *GAL1-POI-tagged*). First test that the POI-tagged fusion protein is functional; it is best to evaluate function by expressing the POI-tagged protein at native levels from its own promoter in addition to overexpressing it from the *GAL1* promoter. In

[32] W. Wen, J. L. Meinkroth, R. Y. Tsien, and S. S. Taylor, *Cell* **82**, 463 (1995).

parallel, determine whether *GAL1-POI-tagged* is induced to reasonable levels in 2% galactose medium and fully repressed in 2% dextrose medium by doing an immunoblot analysis of POI-tagged protein in whole cell extracts prepared from cells harboring the *GAL1-POI-tagged* gene.

Second, examine the localization of the *GAL1-POI-tagged* gene in wild type, *nup49-313*, and *nsp1^{ts}* cells to be sure that the POI-tagged protein is (1) readily visualized and localized to the nucleus, (2) requires either Nup49 or Nsp1 for its import into the nucleus, and (3) has a long enough half-life for visualization of nuclear export of a previously synthesized pool during the course of a multihour incubation at nonpermissive temperature, as follows.

Streak the wild-type, *nup49-313*, or *nsp1^{ts}* strains for single colonies on YEPD plates and grow them at room temperature (25°). Inspect the colonies of the mutant strains to see whether they are uniform or not; faster growing segregants are likely to harbor suppressor mutations. Restreak small colonies if necessary and test that they are temperature sensitive. Transform a pure isolate with a plasmid expressing *GAL1-POI-tagged*, taking care to colony purify the transformants at room temperature in SC selective medium containing 2% dextrose (it is preferable to use a centromeric plasmid to maintain reasonably equivalent expression across a population of cells). In general, because of the slow growth of the mutants, it is generally easier to prepare a heavy inoculum from multiple small colonies rather than attempt to grow a culture from a single slow-growing colony. Inoculate transformants into SC selective media containing 2% dextrose and grow overnight to late logarithmic phase (A_{600} of 0.7–1.0) at room temperature. The next evening, adjust the samples so that they are all at A_{600} of 1 and dilute 1 : 100 into SC selective medium containing 2% raffinose and grow overnight at room temperature so that the cells are logarithmically growing the next morning (A_{600} of 0.1–0.7). Note that a higher dilution will be required for the wild-type control. In the morning, pellet the cells and resuspend them in SC selective medium containing 2% galactose at an A_{600} of 0.2. Grow the cells for 2–3 hr at room temperature to induce the expression of POI-tagged.

To determine whether POI-tagged localizes to the nucleus, fix an aliquot for visualization by indirect immunofluorescence (or autofluorescence of a POI–GFP fusion) or look at the cells live if GFP is the tag. Use 4′,6-diamidino-2-phenylindole (DAPI) to show colocalization with nuclear DNA. To determine whether Nup49 and Nsp1 are required for nuclear import of POI-tagged, pellet an aliquot of the galactose-induced cells and resuspend them in prewarmed SC selective medium containing 2% galactose (36° for *nup49-313* and 37° for *nsp1ts*) at an A_{600} of 0.2 and culture the cells at the higher temperature for 5–6 hr. Examine cells at 1-hr intervals and determine whether POI-tagged accumulates in the cytoplasm with increasing time at 37° in the *nup49-313* and *nsp1ts* mutants compared to the wild-type control. Use this experiment to determine an optimal minimal time at nonpermissive temperature that leads to a maximal block in import as assessed by the relative reduction in the nuclear pool and increase in the cytoplasmic pool. Shorter time intervals are better because the cells are slowly dying.

To determine how long one can visualize a previously synthesized pool of POI-tagged, pellet the cells, wash them once in SC medium containing 2% dextrose, and then culture the cells for 5–6 hr in the same medium at room temperature. Follow the localization of POI-tagged at 1-hr time intervals and observe how long it takes for POI-tagged to disappear. A parallel experiment that examines POI-tagged by immunoblot analysis can also be done.

Once initial criteria are satisfied, it is possible to determine whether a shuttling POI-tagged protein accumulates in the cytoplasm when its nuclear import is blocked by *nup49-313* or *nsp1^{ts}* mutations. Grow wild-type and mutant cells as just described to the point where POI-tagged is induced in galactose medium for 2–3 hr at room temperature. Pellet and wash the cells in dextrose medium and then culture them in dextrose medium at room temperature for 1 hr to allow the cells to recover and repress expression of GAL1-POI-tagged. Adjust the culture to an A_{600} of 0.2 with fresh dextrose medium, split the samples in two, and culture one-half at room temperature and the other half at nonpermissive temperature (36° for *nup49-313* and 37° for *nsp1^{ts}*) for a period of 5–6 hr. Remove aliquots at 1-hr intervals and determine whether POI-tagged accumulates in the cytoplasm with increasing time in the mutants at the nonpermissive temperature compared to the wild-type control. Because the amount of cytoplasmic accumulation may vary from cell to cell, it is best to do a careful tally of 200–300 fixed cells. If the protein of interest is thought to have its export induced by certain conditions, it is possible to add the inducing agent (growth factor, nutrient, osmotic shock, etc.) after the cells have been shifted to nonpermissive temperature for an amount of time that has been predetermined to effectively block the nuclear import of POI-tagged. If POI-tagged is suspected to be involved with mRNA metabolism, it may be of interest to test whether its export is dependent on RNA transcription by determining whether addition of the transcription inhibitor thiolutin blocks nuclear export.[26]

Use of Importin-β Family Mutants

One can also analyze the localization of a protein in strains harboring mutations in putative receptors involved in nuclear import and export in order to identify potential importin and exportin receptors. For example, one can demonstrate that a cytoplasmic protein shuttles by inactivating a specific exportin that has been found to be required for nuclear export of the protein.[17,33] This leads to accumulation of the protein in the nucleus with a concomitant decrease in the cytoplasm (Fig. 3B). *Saccharomyces cerevisiae* has 13 proteins with homology to importin-β, the receptor that dimerizes with importin-α and recognizes NLS-containing proteins

---

[33] K. Stade, C. S. Ford, C. Guthrie, and K. Weis, *Cell* **90**, 1041 (1997).

A. Gene names of key regulators of nuclear transport.

		Importin-β Family Members	
Ran:	GSP1, GSP2	RSL1/KAP95	CRM1/XPO1/KAP124
RanGEF:	RNA1	PSE1/KAP121	CSE1/HRC135
RanGAP:	PRP20/MTR1/SRM1	YRB4/KAP123	LOS1
Importin-α:	SRP1/SCM1/NLE1/KAP60	KAP104	MSN5/STE21
Importin-β:	RSL1/KAP95	SXM1/KAP108	LPH2/KAP120 PDR6/KAP122
		HRC1004/TDS2/KAP114	
		NMD5/KAP119	
		MTR10/KAP111	

B. Nuclear Localization Signals

Simple basic NLS   Short cluster of basic amino acids, often preceded by an acidic amino acid or proline residue

　　　　　　　　　　　　SV40 large T antigen[36]   PKKKRKV

Bipartite basic NLS   Two clusters of basic amino acids separated by a flexible spacer of about 10 amino acids, often flanked by neutral and/or acid amino acids

　　　　　　　　　　　　Nucleoplasmin[43]   KRPAATKKAGQAKKKK

Other NLSs   RPL3[44]   MSHRKYEAPRHGHLGFLPRKR

　　　　　　　　　　　　SWI5[11]   EDALVVHRSRMICSGGKKYENV-
　　　　　　　　　　　　　　　　　　VIKRSPRKRGRRKDGTSSVSSSPIKENI

　　　　　　　　　　　　PHO4[19]   SANKVTKNKSNSSPYLNKRKGKPGPDS

Nuclear Export Signals

Simple NES   Short leucine-rich sequence[40] L-X2-3-(F,I,L,V,M)-X2-3-L-X-(L,I)

　　　　　　　　　　　　PKI[32] LALKLAGLDI

　　　　　　　　　　　　Rev[45] LPPLERLTL

FIG. 4. List of gene names of key regulators of nuclear transport and sequences used by proteins to direct nuclear import and nuclear export. This information is compiled from a number of excellent reviews. [A. Kaffman and E. K. O'Shea, *Annu. Rev. Cell. Dev. Biol.* **15**, 291 (1999); D. Görlich and U. Kutay, *Annu. Rev. Cell. Dev. Biol.* **15**, 607 (1999); I. W. Mattaj and L. Englmeier, *Annu. Rev. Biochem.* **67**, 265 (1998); M. Damelin, P. A. Silver, and A. Corbett, *Methods Enzymol.* **351**, [33], 2002 (this volume)]. Descriptions of the different nuclear transport mutants available for experimentation can be found elsewhere in this volume [33].

(Fig. 4).[1,26] Many of these nuclear transport receptors are nonessential, allowing one to analyze strains harboring null alleles in the corresponding genes. In addition, temperature-sensitive alleles exist for several of the essential transport receptors. When embarking on this type of analysis, it is important to keep in mind that the regulation of import and export may involve redundant pathways involving more than one importin and exportin.[17,34,35] Therefore, mutations in candidate exportins

---

[34] N. Mosammaparast, K. R. Jackson, Y. R. Guo, C. J. Brame, J. Shabanowitz, D. F. Hunt, and L. F. Pemberton, *J. Cell Biol.* **153**, 251 (2001).

[35] K. Yoshida and G. Blobel, *J. Cell Biol.* **152**, 729 (2001).

and importins may only cause a partial block in nuclear import or export that has either no effect or only a partial effect on the function of the pathway that utilizes the protein in question. In addition, it is not possible to predict whether a given importin-β family member will regulate import or export of a given protein because the same importin-β family member may regulate both import and export of different classes of proteins. Finally, it is important to use isogenic pairs of wild type and mutant, as the relative amount of nuclear import and export can vary depending on strain background.

Altering the Protein to Study Nuclear Shuttling

Use of SV40 Large TAg NLS Tag and PKI NES Tags to Analyze Nuclear Shuttling of a Protein

The addition of heterologous nuclear localization sequence and nuclear export sequence tags to a protein provides a very useful way to manipulate the localization of a protein for a variety of experiments. The SV40 large T antigen NLS (PKKKRKV[36]) provides a convenient way to import a reporter protein into the nucleus and has been shown to direct nuclear import of a variety of proteins.[17,19,37,38] A single point mutant derivative of the SV40 large T antigen NLS, K128T,[36] generates a partially functional NLS that can be used to shift some but not all of a protein to the nucleus.[17] The NES from the protein kinase A inhibitor PKI (LALKLAGLDI[32]) provides a convenient way to export a reporter protein from the nucleus to the cytoplasm and has been shown to direct nuclear export of a variety of proteins.[32,33,38] Uses of these tags include the following.

1. Monitoring nuclear export of a cytoplasmic protein by using an NLS to shift the steady-state distribution to the nucleus.[17] By shifting the steady-state distribution of a protein from cytoplasmic to nuclear, one can monitor nuclear export under conditions in which nuclear import is blocked (such as in *nsp1^{ts}* and *nup49-313* mutants) or under environmental conditions that increase the rate of export.

2. Using an NLS to suppress a mutational block that prevents nuclear accumulation of a protein. For example, one can test whether an apparant defect in the ability of a protein to accumulate in nuclei is due to a block in nuclear import by determining whether the NLS relocalizes the protein to the nucleus.[19]

3. Using NLS and NES tags to determine the functional consequences of restricting the steady-state distribution of a protein to the cytoplasm or nucleus.[17,38]

---

[36] D. Kalderon, W. D. Richardson, A. F. Markham, and A. E. Smith, *Nature* **311,** 33 (1984).
[37] M. Nelson and P. Silver, *Mol. Cell. Biol.* **9,** 384 (1989).
[38] M. E. Miller and F. R. Cross, *Mol. Cell. Biol.* **20,** 542 (2000).

4. Testing whether nuclear shuttling has functional significance by determining whether coexpression of nuclear and cytoplasmic-restricted forms restores full function.[17] The absence of full complementation under these circumstances may indicate that the nuclear shuttling event itself is important for the ability of the protein to carry out its functions.

Mutating Potential NLS or NES in the Protein

Another way to study a protein that shuttles through the nucleus is to identify *cis* elements that direct its nuclear import and nuclear export, mutate such sequences, observe changes in localization, and analyze these effects on function. Proteins that shuttle into and out of the nucleus often do so through a direct interaction with import or export receptors that bind directly to specific amino acid sequences within the protein, termed nuclear localization sequence[39] and nuclear export sequence.[40] Proteins can contain varying numbers of NLSs or NESs and there is no simple rule as to where these sequences will be located within a given protein.[41] In general, NLSs overlap stretches of basic amino acids, whereas NESs overlap stretches of hydrophobic amino acids, including leucine. NESs may lie in or near regions involved in protein–protein interactions, such as domains involved in oligomerization.[42] Sometimes NLSs and NESs can be identified by homology to canonical NLS or NES recognition sites (Fig. 4).[30,32,36,40,43–45] However, not all proteins will contain obvious NLS and/or NES sequences that match canonical recognition sites. Many examples of NLSs and NESs that do not conform to canonical NLS and NESs exist. In addition, RNA-binding proteins appear to use distinct nucleocytoplasmic shuttling (NS) signals for both import and export, with the sequences required for import and export being partially or directly overlapping.[46] Finally, a protein may not contain a specific targeting sequence at all and may undergo nucleocytoplasmic transport through association with another protein that does contain targeting sequences.[47]

Because it is currently not possible to predict the localization signal of a protein, efforts to identify potential import and export sequences should include nonbiased mutational approaches, in addition to any biased mutation of sequences that may

[39] C. Dingwall and R. A. Laskey, *Trends Biochem. Sci.* **16,** 478 (1991).
[40] L. Gerace, *Cell* **82,** 341 (1985).
[41] S. Nakielny and G. Dreyfuss, *Cell* **99,** 677 (1999).
[42] J. M. Stommel, N. D. Marchenko, G. S. Jimenez, U. M. Moll, T. J. Hope, and G. M. Wahl, *EMBO J.* **18,** 1660 (1999).
[43] J. Robbins, S. M. Dilworth, R. A. Laskey, and C. Dingwall, *Cell* **64,** 615 (1991).
[44] R. B. Moreland, H. G. Nam, L. M. Hereford, and H. M. Fried, *Proc. Natl. Acad. Sci. U.S.A.* **82,** 6561 (1985).
[45] U. Fischer, J. Huber, W. C. Boelens, I. W. Mattaj, and R. Luhrmann, *Cell* **82,** 475 (1995).
[46] W. M. Michael, *Trends Cell Biol.* **10,** 46 (2000).
[47] L. Zhao and R. Padmanabhan, *Cell* **55,** 1005 (1988).

appear to be good bets for targeting signals. Mutations in sequences required for nuclear import should lead to an increase in the cytoplasmic pool of a protein with a commensurate decrease in the nuclear pool. Mutations in sequences required for nuclear export should lead to an increase in the nuclear pool of a protein and decrease the cytoplasmic pool. It is always possible that the mutations will indirectly affect the ability of a protein to be recognized by either import or export machinery, such as by inducing conformational changes in the protein. In determining whether a mutation that affects localization also affects protein function, it is critical to express the protein at native levels. This is because of potential leakiness of mutations affecting protein localization.

Once a region in a protein is defined as being required for import or export, the most compelling proof that this region overlaps either a NLS or NES will come from demonstrating that it is able to target a heterologous protein. A NLS should redistribute a cytosolic protein to the nucleus (e.g., β-galactosidase or GFP_{2-3}) and a NES should redistribute a nuclear heterologous protein to the cytoplasm.[32,33,38]

Optimizing the Study of Nuclear Shuttling

The ability to study nuclear shuttling is dependent on being able to define conditions that differentially affect rates of nuclear import and export. Temperature, mass of the protein, method of protein expression, and method of tagging the protein are four simple parameters that can be varied to optimize the study of nuclear shuttling. First, nuclear import and export are temperature-dependent processes, and modest increases in temperature can increase the rates of nuclear transport.[6,48,49] Therefore, one may be able to selectively enhance the rate of nuclear import or export by changes in temperature. For example, the ability to readily detect a nuclear pool of Ste5 in a logarithmically growing population of wild-type cells increases as the temperature is raised from 24° to 37° in a variety of strain backgrounds (Y. Wang, P. Maslo and E. Elion, unpublished results, 2001). This temperature dependence has been a useful parameter to vary in the analysis of factors that either increase or decrease nuclear import of Ste5. Factors that enhance nuclear import may be discerned more readily at lower temperatures, whereas factors that decrease nuclear import may be discerned more readily at higher temperatures.

Second, one can increase the mass of a particular protein through the addition of multiple protein tags to prevent its passive diffusion into the nucleus. Proteins equal to or less than 50 kDa can diffuse through the nuclear pore complex at rates that are inversely proportional to mass.[6,7] The addition of multiple protein tags (such as 27 kDa GFP) can help focus the analysis on the regulated aspects of

---

[48] N. Shulga, P. Roberts, Z. Y. Gu, L. Spitz, M. M. Tabb, M. Nomura, and D. S. Goldfarb, *J. Cell Biol.* **135,** 329 (1996).
[49] I. Boche and E. Fanning, *J. Cell Biol.* **139,** 313 (1997).

nuclear transport. Increasing the mass of the protein can also slow down its rate of nuclear transport through regulated pathways. This approach was used to be able to distinguish a slower rate of nuclear import for mutant Pho4 proteins.[50]

Third, one can examine the protein shortly after its synthesis from an inducible promoter rather than at steady state after expression from a constitutive promoter. This approach may help reveal differences in rates of nuclear import and be useful when a protein shuttles continuously.[17,50]

Fourth, one can optimize the study of nuclear shuttling greatly by creating derivatives of the protein that are detected more readily in a microscope. For example, it is most advantageous to be able to monitor the localization of a protein when it is expressed at native levels; however, this approach can often be limited by the ability to detect the protein at this level of expression. The most frequently used approach to enhance detection is to fuse multiple copies of an epitope that is recognized by a highly specific monoclonal antibody in fixed cells or to fuse multiple copies of GFP to monitor the protein in either live or fixed cells. These tags are typically added to either the N or the C terminus of a protein in the hopes of causing minimal affects on protein folding.[51] However, it is always important to test whether tagging the protein alters its function or stability when it is expressed at native levels. There is no simple rule for how many copies will help in the detection of a protein. Epitope tags are often inserted in three copies,[52] but can be multimerized to higher numbers to aid in the detection process. For example, both Ash1 and Ste5 were tagged with nine copies of the *myc* epitope.[17,53] The detection of a protein with a GFP tag may be aided by the incorporation of two to three copies of GFP.[33,50] In using GFP, it is important to realize that the signal will reflect the ability of GFP to fluoresce and that this can be affected by a variety of factors,[54] including the ability of the GFP moiety to fold. In this regard, insertion of a flexible spacer sequence between the end of Cdc24 and the end GFP was reported to generate a more readily detected protein.[55]

Acknowledgments

I thank Yunmei Wang, Maosong Qi, Annette Flotho, and Paul Maslo for their helpful comments on the manuscript. This work was supported by a grant from the National Institutes of Health (GM46962).

[50] A. Komeili and E. K. O'Shea, *Science* **284,** 977 (1999).
[51] R. A. Kolodziej and R. A. Young, *Methods Enzymol.* **195,** 508 (1991).
[52] B. L. Schneider, W. Seufert, B. Steiner, Q. H. Yang, and A. B. Futcher, *Yeast* **11,** 1265 (1995).
[53] N. Bobola, R. P. Jansen, T. H. Shin, and K. Nasmyth, *Cell* **84,** 699 (1996).
[54] K. Tatchell, *Methods Enzymol.* **351,** [39], 2002 (this volume).
[55] K. A. Toenjes, M. M. Sawyer, and D. I. Johnson, *Curr. Biol.* **9,** 1183 (1998).

[35] Flow Cytometry/Cell Sorting for Isolating Membrane Trafficking Mutants in Yeast

By THOMAS VIDA and BEVERLY WENDLAND

General Principles of Flow Cell Sorting/Cytometry

Measuring fluorescence within individual cells and isolating cell populations with specific desired properties are the general goals of flow cell sorting. We have chosen to use the more general terms "flow cell sorting" or flow cytometry throughout this chapter. Although fluorescence-activated cell sorting (FACS) is indeed the common principle behind flow cell sorting, the term FACS is trademarked from Becton Dickinson, Inc. and applies to their instruments. Instruments designed to perform flow cell sorting vary in complexity/capability and are far too expensive for most laboratories to own individually. Most research centers have core or fee-for-service facilities that can perform flow cell sorting with excellent technical assistance. For this reason, this chapter does not describe in detail all the various instruments available, but rather gives general guidelines and principles deemed appropriate for isolating novel yeast mutants defective in membrane traffic.

The use of fluorescent cell sorting analysis has been applied extensively to microorganisms and several reviews exist. An entire issue of the *Methods* journal (Volume 21, Issue 3) is devoted to flow cytometry and we encourage the reader to consult this excellent issue on the topic.[1] The majority of applications for yeast flow cytometry involve cell analysis, which measures the numbers of cells with particular properties, as opposed to sorting, in which specific cells with desired properties are collected for further use. For example, analytical flow cytometry can be used to examine genomic ploidy relative to stages in the cell cycle. For this purpose, fluorescent dyes such as 4′,6-diamidino-2-phenylindole dihydrochloride (DAPI) have been used to stain intracellular DNA. The cells are then analyzed via flow cytometry; the amount of fluorescence is proportional to DNA content. In yeast, this is the most common application of analytical fluorescent flow cytometry.[2] Thus far, a limited number of applications in cell sorting have been applied to the isolation of mutants defective in membrane-bound transport or organelle function. The majority of these studies have involved functions of the vacuole or endocytosis (see later).

Different considerations must be taken into account when setting up an experiment involving *analysis* of cells vs *isolation* of mutant cells, although both involve similar principles in distinguishing between populations of cells that vary

---

[1] M. K. Winson and H. M. Davey, *Methods* **21**, 231 (2000).
[2] K. Skarstad, R. Bernander, and E. Boye, *Methods Enzymol.* **262**, 604 (1995).

in fluorescence. When isolating yeast mutants, one must take several measures to select the desired mutants accurately and efficiently. In the simplest terms, fluorescent cells are passed through a very small orifice (as little as 1 μm in the newer instruments) under high pressure to enable very high throughput. Due to the small size of cells, a laser is used to focus intense light on the orifice. Using appropriate filters to select excitation and emission spectra, the spectrum of individual cells flowing through the orifice can be detected and quantified. Depending on the wavelengths that have been set on the instrument, a charge can be imparted onto cells that show the desired spectral properties. These "charge-activated" cells can then be diverted from the flow and collected separately, allowing the desired population to be sorted away from the other cells. Several parameters that are very important in sorting yeast cells are the size of the orifice and the pressure that is used for the flow of cells. Because haploid yeast cells average about 5 μm in diameter, a small orifice is required (much smaller than for mammalian cells). If the orifice is larger, then many false-positive events can take place from multiple cells passing through the orifice together. This will of course decrease the selection properties of cell sorting. The pressure parameter is important for analyzing a large number of cells in a very short time. In at least one of the screens that is described in this chapter, as many as 20,000 events per second were used. This elevated throughput rate may be inappropriate for many of the older cell sorting instruments and result in reduced selectivity of sorting.[3]

Importance of Reference Samples When Setting Up a Screen

To isolate mutants defective in a specific membrane traffic event, fluorescence must be imparted to membranes. The simplest ways to do this are through the use of vital dyes (see later) and membrane protein fusions to the green fluorescent protein (GFP). Several considerations are crucial when setting up a screen for mutants. The most obvious one is that mutant cells must have differences in their fluorescence properties compared to wild-type cells. This difference could involve fluorescence intensity, fluorescence spectral property, or both. Intensity-based screens would involve changes in either the quantity of cell-associated fluor or the emission quantum yield, which might reflect changes in the intracellular environment of candidate mutants. One should also consider the stability of the fluor to both cellular degradation/metabolism and bleaching. Equally important is the availability of treatments or conditions that allow one to mimic the mutant phenotype (phenocopy) for use as reference samples. This is essential for setting up the sort parameters such that these two populations of cells look as different as possible. Another important reference sample is yeast cells that have not been treated with fluorescent dyes or do not express fluorescent fusion proteins. Yeast cells not only give signals due to general light scatter, but they can also have significant autofluorescent properties.

[3] D. J. Recktenwald, *J. Hematother.* **2**, 387 (1993).

In general, the first two samples to be analyzed in any screen should be these unstained cells and cells that exhibit the wild-type fluorescent phenotype. Their patterns can be made as different as possible with appropriate subtraction and gain adjustments on the flow-sorting instrument.

Fluorescent Vital Stains for Use with Yeast

The Molecular Probes catalog (www.probes.com: Eugene, OR) contains a "treasure trove" of reagents for visualizing yeast organelles in living cells. The two vital dyes used most commonly for labeling yeast vacuoles in combination with flow cytometry are the lumenal marker carboxy-2′,7′-dichlorofluorescein diacetate (CDCFDA) or other esterase-dependent fluorescein derivatives and the lipophilic styryl endocytic membrane tracer molecule N-(3-triethylammoniumpropyl)-4-(6-[4(diethylamino)phenyl]hexatrienyl) pyridinium dibromide) (FM 4-64).

CDCFDA is a cell-permeant esterase substrate that not only monitors enzymatic activity, but also reveals the integrity of the cell membrane and, thus, cell viability. These uncharged dyes diffuse into the cell, react with an enzyme, and then become charged and trapped inside an organelle. Because the trapping is dependent on the integrity of the cellular membranes, these reagents exhibit a dual requirement both for production of the fluorescent signal and for its sequestration within the intact cell.

FM 4-64 has several properties that make it particularly useful as a probe for membrane trafficking in yeast. First, it is a styryl dye that is fluorescently invisible in aqueous media, but is highly fluorescent in a lipid environment such as a membrane bilayer. Second, it resembles a phospholipid in that it is amphipathic with one end charged and the other hydrophobic. It is apparently unable to serve as a substrate for flippases and thus remains associated with either the exoplasmic leaflet of the plasma membrane or the lumenal side of internalized membranes. Third, FM 4-64 readily exchanges between the lipid bilayer and the adjacent aqueous medium; thus, labeled uninternalized plasma membrane can be washed to extract the uninternalized dye during the chase period. This allows for pulse/chase style experiments to observe real-time trafficking events similar to those in living mammalian cells.[4]

Using Flow Cytometry to Enrich Trafficking Mutants

At least three different selections for trafficking mutants using FM 4-64 labeling followed by enrichment using flow cytometry have been published.[5-7] The first

---

[4] C. B. Smith and W. J. Betz, *Nature* **380**, 531 (1996).
[5] Y. X. Wang, H. Zhao, T. M. Harding, D. S. Gomes de Mesquita, C. L. Woldringh, D. J. Klionsky, A. L. Munn, and L. S. Weisman, *Mol. Biol. Cell* **7**, 1375 (1996).
[6] B. Wendland, J. M. McCaffery, Q. Xiao, and S. D. Emr, *J. Cell. Biol.* **135**, 1485 (1996).
[7] B. Zheng, J. N. Wu, W. Schober, D. E. Lewis, and T. Vida, *Proc. Natl. Acad. Sci. U.S.A.* **95**, 11721 (1998).

two used FM 4-64 as a tool to search for *vac* mutants that were deficient in vacuole inheritance to the growing bud[5] and for *dim* mutants that were deficient for endocytosis.[6] The rationale for the *vac* mutant screen was to enrich for newly budded daughter cells that lacked a fluorescent signal due to a complete absence of inherited labeled vacuolar membranes. For the *dim* mutant screen, the intent was to enrich for cells that had internalized less dye during the labeling period as a result of reduced internalization of endocytic membranes. A variation on the *dim* mutant screen presently underway is a selection for *brt* (bright) mutants that may represent cells that are deficient in recycling from endosomes or that could correspond to mutations that upregulate or stimulate endocytosis (B. Wendland, unpublished results, 2000). The third screen was done using a double-label technique to enrich mutants that proliferate an endosomal compartment.[7]

Practical Considerations for Setting Up Mutagenesis Followed by Sorting Selection

There are two primary issues to consider in setting up a mutant selection in which mutagenesis is followed by flow cytometry. After mutagenesis treatment, the cells need to recover before enduring the rigors of labeling and sorting. This introduces a problem: in the process of recovering, sibling cells arise in the culture, and ultimately one needs to differentiate between reisolated sibling cells vs cells that represent different mutant alleles of the same gene. There are two solutions that can help to ameliorate this situation. The first is to allow the mutagenized cells to recover for the shortest time possible (e.g., a short overnight incubation of ~10 hr), and preferably the recovery is done at a low temperature, such as at room temperature, to extend the generation time of the cells. Second, it is a good idea to divide the mutagenized cells into pools prior to the recovery period and to keep these pools separate as the cells are labeled and sorted. If two mutants are in the same complementation group but were isolated from different pools, they must represent different alleles and not simply reisolated siblings.

Another factor to consider is the inherent fluorescence of the medium in which the cells are being sorted. Because rich medium (YPD) contains numerous autofluorescent compounds, it is best to use synthetic complete medium (YNB) or buffered solutions.

Generic Protocol for Mutagenesis, Labeling, and Sorting

Ethyl methane sulfonate (EMS) mutagenesis has been used in all previously published studies, but of course other types of mutagenesis, such as UV, hydroxylamine, and transposon mutagenesis, would also be suitable. EMS mutagenesis can be carried out as described previously.[8]

[8] M. D. Rose, F. Winston, and P. Hieter, "Methods in Yeast Genetics: A Laboratory Course Manual," p. 11. Cold Spring Harbor Laboratory Press, Cold Spring Harbor, NY, 1990.

After the mutagenesis, the cells should be separated into pools for recovery. In case the kill rate is greater than or less than expected, it is convenient to set up different dilutions of the mutagenized cells. An example is presented here.

Day One (Preculture). Inoculate a culture of cells to be mutagenized, usually around midday. They should grow to stationary phase, preferably for 20–24 hr, before mutagenesis on Day Two.

Day Two (Mutagenesis). Mutagenize 10 OD_{600} equivalents of wild-type cells ($\sim 2 \times 10^8$ cells). We have used the wild-type strain SEY6210, using 30 μl EMS in 800 μl cells resuspended in 0.1 M potassium phosphate, pH 7.0, at 30° for 1 hr. Set up a mock mutagenesis to calculate the kill rate of the mutagenesis procedure; under these conditions, the mutagenesis should kill $\sim 80\%$ of the cells. After mutagenesis is complete and the mutagen is neutralized or removed, the mutagenized cells should be divided among five flasks for recovery in rich medium overnight at room temperature. For the recovery, the mutagenized cells are resuspended in 1 ml YPD. Then 25, 75, 150, 300 μl, and the remainder are each added to 50 ml YPD in 125-ml Erlenmeyer flasks. Even the most concentrated sample should still be well below stationary phase the following morning. Also remember to inoculate a culture of wild-type cells. These will be used as controls when setting up the parameters on the flow cytometer the next day.

Day Three (Sorting). Harvest 1–3 OD_{600} from each pool of recovered, mutagenized cells and proceed with the desired labeling procedure. With the wild-type control culture, prepare a set of labeled and unlabeled cells. Unlabeled cells will be important for determining the background signal and autofluorescence of the sample. In the event that one might like to recover temperature- or cold-sensitive alleles, the cultures can be preshifted to the restrictive temperature prior to labeling. This pretreatment should not preclude one from isolating mutants that exhibit the desired phenotype constitutively. For a screen for temperature-sensitive endocytosis mutants (e.g., *dim* mutants) using FM 4-64, the culture is preshifted to 37° for 1 hr and maintained at that temperature throughout the 15-min labeling and 45-min chase periods. After the chase period, the cells are harvested, resuspended in synthetic complete medium to a concentration of ~ 0.1–0.5 OD_{600}/ml (~ 1–5×10^6 cells/ml), and held on ice until the sorting begins. The ice treatment is used to cease further trafficking steps; if a particular *ts* allele is reversible, it could begin to internalize dye that is being released from neighboring cells that were killed during the preshift. Dead cells lose plasma membrane integrity and are labeled with the FM 4-64 dye throughout all internal membranes. Thus, they may serve as a continuous "source" of dye in the sample that may not be washed out readily.

Choosing a Sort Window. First, test the wild-type labeled cells and determine the parameters necessary to get the bulk of the cells displayed away from the axes of side scatter and forward scatter. Then display a fluorescence intensity histogram to make sure the appropriate excitation and emission filters are being used. Next, examine the intensity of the unlabeled cells—there will be a peak in the very low intensity region, but there should be little or no overlap between the peaks of unlabeled vs labeled cells. Once the run parameters are set, the sort window is selected from the dot-plot display of fluorescent intensity vs side scatter, where each dot corresponds to a single sort event, and presumably to a single cell (unless the cells are aggregated, or if cell separation or cytokinesis is defective). The sort window should represent a very small fraction of the total population being sorted (i.e., on the order of 0.5% or less). Alternatively, one can set the sort window to be relatively generous and then resort, or culture the cells again and resort the next day on the amplified pool.

After the cell sort is complete, one should have an estimate of the number of "events" that are directed into each tube from the sort window(s). Many of these events represent cell fragments or inviable cells; however, it is safer to assume that the number is accurate. Divide the contents of each tube among the appropriate number of plates to yield ∼500 colonies per 10-cm agar plate. It is best to avoid subjecting the cells to centrifugation in order to concentrate them before plating. Either dilute the contents of the tubes as necessary or use them as is and distribute them onto various plates. Alternatively, sorted cells can be further cultured overnight (with the same considerations as used during the recovery from mutagenesis), relabeled, and resorted the next day to further enrich the population. The strategy used depends on the stringency of the original sort window and on the amount of effort required for the secondary screening to identify the cells among the sorted population that exhibit the desired mutant phenotype(s).

Days Five to Seven (Secondary Screen). After the viable sorted cells have formed colonies, they can be replica plated to identify those with desired secondary growth phenotypes. Alternatively, if a visual screen is being used, they can be examined individually to identify new mutant strains. The success rate (i.e., the percentage of sorted cells that appear to be what was desired) can be quite variable, depending on the parameters used as described earlier, and can range from 1 to 10%.

Dual Vital Dyes for Cell-Sorting Screen

Using two fluorescent probes simultaneously has the advantage of detecting differences in fluorescence other than simply intensity. FM 4-64 and CDCFDA have been used together to isolate mutants that accumulate endosomal intermediates and other nonvacuolar structures.[7] Two vacuolar probes are needed in this screen because cells accumulating endosomal membranes do not have a significant difference in fluorescence compared to cells with vacuolar staining (Fig. 1). The ability to mimic the mutant phenotype comes from the appearance of cells stained with FM 4-64 at 15°. The kinetics of vacuolar delivery are sufficiently decreased ($t_{1/2}$ ∼3 hr at 15° vs ∼15 min at 30° for vacuole staining) to afford an enrichment of intermediate structures. Importantly, CDCFDA labels the vacuole

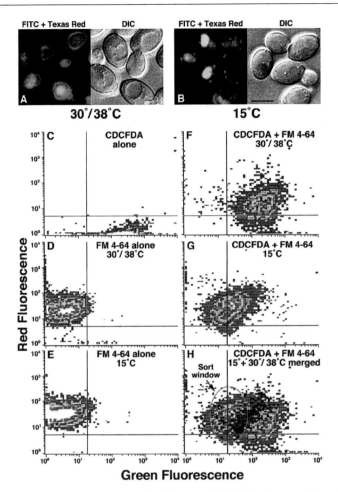

FIG. 1. Flow cytometry can distinguish cells with punctate FM 4-64 staining morphology from cells with vacuolar membrane staining morphology. Diploid yeast cells were stained with FM 4-64 (10 μM) and CDCFDA (10 μM) at the indicated temperatures. At 30°/38° (A), cells were pulsed with FM 4-64 for 30 min and chased for 60 min. After 60 min, CDCFDA was added (at pH 4) for 15 min. At 15° (B), the cells were pulsed simultaneously with both FM 4-64 and CDCFDA (50 μM) for 30 min. For flow cytometry, haploid yeast cells (SEY6210) were stained with CDCFDA alone (C), FM 4-64 alone at 30° (D), and 15° (E) or were stained with both CDCFDA and FM 4-64 at 38° (F) or 15° (G). Fluorescence emission was monitored simultaneously at 525 (green) and 590 (red) nm. After overlaying the patterns from double staining cells at 38 and 15°, a sort window was defined for putative mutants. Each flow cytometry sample represents ∼2–5 × 10^4 cells. White and light gray areas represent the most and least events, respectively. Reprinted with permission from B. Zheng, J. N. Wu, W. Schober, D. E. Lewis, and T. Vida, *Proc. Natl. Acad. Sci. U.S.A.* **95,** 11721 (1998), copyright © 1998 National Academy of Sciences, U.S.A.

lumen efficiently at 15° so that double-labeled cells have "green spheres" surrounded by "red dots."

To stain cells in this manner and to set up reference samples, the following protocol is used. Staining with CDCFDA requires buffering the media to pH 4.0 using 50 mM citric acid (1 M stock solution adjusted with KOH). Typically, CDCFDA is added to 10 μM from a stock solution of 1 or 10 mM in dimethylformamide, and the cells are shaken at 30 or 38° for 15 min. An increased loading concentration of 50 μM and incubation time (30 min) are needed for CDCFDA when staining at 15°. At elevated temperatures (30 and 38°), FM 4–64 is loaded into cells at 10–40 μM for 15 min followed with one wash to remove free dye and a chase period for 45–120 min. At 15°, FM 4–64 is added to 50 μM and incubated for 30 min with *no chase period*. When preparing these reference samples for flow cytometry, all cells are harvested, resuspended in 20 mM HEPES–KOH (pH 7.0), 150 mM potassium acetate, and 5 mM magnesium acetate, and held on ice. Cells should be diluted to ~0.1 OD_{600} unit/ml for flow analysis. With these reference samples, temperature-sensitive mutations are isolated after mutagenesis (as described earlier). A slightly green-shifted sort window is defined for cells double stained at 15° vs 30° (Fig. 1).

Flow cytometry with these samples is performed on a Coulter EPICS Model 753 equipped with a Cyclops high-speed sort system (Cytomation, Ft. Collins, CO). An argon ion laser (Coherent, Palo Alto, CA) tuned to 514 nm and operated at 100 mW travels through a confocal lens block, creating a beam height of 16 mm, and then through a 33 quartz SortSense flow cell tip. Emitted light is collected through a series of optical filters, including a 514–nm long-pass laser-blocking filter and a 550-nm long-pass dichroic filter to reflect FM 4–64 emission to a 590-nm long-pass filter and divert the CDCFDA signal to a 525-nm band-pass emission filter. To compensate for fluorescence overlap between the two fluors, 65% of the CDCFDA signal is subtracted from the FM 4-64 signal and 25% of the FM 4-64 signal is subtracted from the CDCFDA signal. Flow rates are typically 1000–1500 cells/sec. Forward-angle light scatter is collected as a linear signal, and all fluorescence emissions are collected on a four-decade logarithmic scale. Detection with this two-color stain method gives a pseudo fluorescent resonance energy transfer (FRET) by using emission from CDCFDA to excite FM 4-64. This is at least one reason why the two-fluor method is successful in this screen. Two other screen methods are used in this sort (as described earlier): temperature sensitivity and visual inspection under an epifluoresecence microscope.

Conclusions and Future Possibilities

The ability to isolate mutants in any organism represents a powerful method to understand gene function. Having a genetic selection that places pressure of life or death coupled to the function of interest is perhaps the most efficient way to isolate specific mutations. This is not always feasible for many mutations; in these cases, a genetic screen must be performed. Fluorescence cell sorting has great enrichment power for cells with altered properties. For example, the two-color screen described previously is able to select 1 in 10 billion cells. Although many mutants in membrane traffic events exist in yeast, the selections and screens used to obtain them are frequently nonsaturated and could be exploited further. When appropriate fluors are available, flow cell sorting can be an excellent method to isolate additional mutants. With the use of multiple GFP fusions, a two or even three color sort could be performed for pseudo FRET-based screens. For instance, nuclear envelope integrity and endoplasmic reticulum (ER) retention might be explored in a dual-probe screen using variant GFP-tagged ER resident proteins. Mitochondrial function and assembly are also amenable to flow cell sorting, as specifically localized mitochondrial vital stains 4-[4-(dimethylamino)styryl]-N-methylpyridinium iodide[9] and GFP fusions exist.[10] Only our imaginations limit the possible applications of fluorescence-based flow cytometry to the isolation of yeast trafficking mutants and the future development of suitable probes.

Acknowledgments

B.W. thanks Scott Emr, in whose laboratory much of this work was initiated. T.V. thanks Wendy Schober and Dorothy Lewis for contributing greatly to the two-fluor sort method described here. Research in the authors' laboratories is supported by the NIH (GM52092 to T.V. and GM60979 to B.W). B.W. is a March of Dimes Basil O'Connor Scholar and the recipient of a Burroughs Wellcome Fund New Investigator Award in the Pharmacological Sciences.

[9] J. Bereiter-Hahn, K. H. Seipel, M. Voth, and J. S. Ploem, *Cell Biochem. Funct.* **1,** 147 (1983).
[10] B. Westermann and W. Neupert, *Yeast* **16,** 1421 (2000).

[36] Protein Synthesis Assayed by Electroporation of mRNA in *Saccharomyces cerevisiae*

By ANJANETTE M. SEARFOSS, DANIEL C. MASISON, and REED B. WICKNER

Studies of translation rely on a variety of methods to evaluate the role of mRNA sequence and modifications, of protein factors, and ribosomal RNAs and ribosomal proteins in this complex process. The sequence of mRNAs can be varied by altering the structures of shuttle plasmids so that the mRNA expressed *in vivo* is as desired. The 5′ start point and 3′ end point of mRNAs can be largely controlled by varying the position of promoter and termination sequences on the vector, but generally there are multiple start and stop sites for a given plasmid construct. The eukaryotic mRNA is characterized by the presence of a 5' cap structure of the form 7mG5'ppp5'XpY ... and a 3' poly(A) region. The roles of these structures have been studied intensively and are still not completely understood. Producing mRNA lacking a 5' cap or 3' poly(A) from a plasmid *in vivo* is difficult in yeast because no protein-encoding genes are known to produce mRNAs naturally lacking these structures. While RNA polymerase I normally makes rRNA transcripts lacking both cap and poly(A),[1,2] constructs in which RNA polymerase I drives mRNA production are of limited usefulness in assessing translation effects because the cap and poly(A) are probably important for functions other than translation, such as mRNA stability in the nucleus and transport from nucleus to cytoplasm. As a result, RNA polymerase I-promoted genes are very poorly expressed.

Russell *et al.*[3] found that reporter mRNAs could be introduced into spheroplasts of *Saccharomyces cerevisiae* by treatment with polyethylene glycol using conditions similar to those used in some transformation protocols. These mRNAs are expressed and show the usual requirement for the 5' cap structure. The authors investigated the basis of translation of L-A and M_1 viral (+) strands, despite their lack of a 5' cap structure. They found no stimulation of cap-independent translation by including the L-A or M_1 5' noncoding regions or the poliovirus internal ribosome entry site (IRES).[3]

Gallie developed electroporation as a method to directly introduce mRNAs of arbitrary structure into living cells[4] and applied this method to animal cells, plant cells, and yeast.[4,5] He found that both cap and poly(A) structures were critically important to optimal translation rates and that both structures also significantly affected mRNA stability. Effects on translation and mRNA stability were distinguished by kinetic studies, with rates of enzyme synthesis being attributed to translation efficiency and limited duration of expression being accounted for by mRNA turnover.[4,5]

We have used the mRNA electroporation method to examine the effects on translation of chromosomal genes affecting the propagation of the yeast RNA viruses, all of which lack 5' cap and 3' poly(A) structures. We have identified several genes that act to specifically block translation of mRNAs lacking 3' poly(A).[6–9]

---

[1] J. Klootwijk, P. de Jonge, and R. J. Planta, *Nucleic Acids Res.* **8**, 27 (1979).
[2] N. Nikolaev, O. I. Georgiev, P. V. Venkov, and A. A. Hadjiolov, *J. Mol. Biol.* **127**, 297 (1979).
[3] P. J. Russell, S. J. Hambidge, and K. Kirkegaard, *Nucleic Acids Res.* **19**, 4949 (1991).
[4] D. R. Gallie, *Genes Dev.* **5**, 2108 (1991).
[5] J. G. Everett and D. R. Gallie, *Yeast* **8**, 1007 (1992).
[6] D. C. Masison, A. Blanc, J. C. Ribas, K. Carroll, N. Sonenberg, and R. B. Wickner, *Mol. Cell. Biol.* **15**, 2763 (1995).
[7] L. Benard, K. Carroll, R. C. P. Valle, and R. B. Wickner, *J. Virol.* **73**, 2893 (1999).
[8] L. Benard, K. Carroll, R. C. P. Valle, and R. B. Wickner, *Mol. Cell. Biol.* **18**, 2688 (1998).
[9] A. M. Searfoss and R. B. Wickner, *Proc. Natl. Acad. Sci. U.S.A.* **97**, 9133 (2000).

The same method has been used by Preiss and Hentze[10] to study the role of poly(A) in translation initiation. This method will also be of value in determining the effects of RNA sequence and structure and of other cellular components on translation and cytoplasmic mRNA turnover.

Method

Strains

The yeast strains used in these experiments were either generated in our laboratory using standard yeast genetics techniques or generously provided by others. We have not observed variabililty with strain background in suitability for this procedure. As controls for this assay, we have tested mutants in initiation factors that have been shown to have reduced translational efficiency *in vivo*. These included *gcd11-508* (γ subunit of eIF2) and *prt1-1* (eIF3 subunit),[11,12] both of which showed markedly decreased translation efficiency of all mRNAs.[13]

Preparation of Cells

Grow 50 ml of cells to OD_{600} of 0.6. Pellet the cells and resuspend in 4 ml of spheroplast buffer A [50 mM Tris–Cl, pH 7.5, 1 mM $MgCl_2$, 30 mM dithiothreitol (DTT), 15 mM 2-mercaptoethanol, 1 M sorbitol]. Add 200 μl of 5 mg/ml zymolase 20T or 200 μl of 1000 units/ml lyticase (Sigma, St. Louis, MO) in spheroplast buffer A. Incubate at 30° with gentle swirling until the cell wall is enzymatically removed and spheroplasts are formed. The time required for spheroplast formation must be determined empirically for each strain. This can be measured by diluting the cells 1 : 100 in 1 M sorbitol and in H_2O and comparing the OD_{600} for each solution. If the cell wall has been removed, the cells will be lysed rapidly in water as measured by a sharp decrease in the OD_{600} (OD_{600} in H_2O will be $<1/4 \times OD_{600}$ in sorbitol). Generally, incubation of the cells at 30° for 15 min in the presence of zymolase is sufficient.

Once the spheroplasts are formed, wash the cells gently in 5 ml of spheroplast buffer A. The cells must be handled very carefully to avoid lysis. Pellet the cells for 5 min at 1000 rpm in a SS-34 rotor. Resuspend the cells in 1 ml of spheroplast buffer A and add to 9 ml of YPAD–1 M sorbitol. Incubate for 90 min at 30° with gentle swirling to allow the cells to recover from the zymolase treatment.

---

[10] T. Preiss and M. W. Hentze, *Nature* **392**, 516 (1998).
[11] L. H. Hartwell and C. S. McLaughlin, *Proc. Natl. Acad. Sci. U.S.A.* **62**, 468 (1969).
[12] A. G. Hinnebusch, *in* "Translational Control of Gene Expression" (N. Sonenberg, J. W. B. Hershey, and M. B. Mathews, eds.), p. 185. Cold Spring Harbor Laboratory Press, Cold Spring Harbor, NY, 2000.
[13] A. M. Searfoss, T. E. Dever, and R. B. Wickner, *Mol. Cell. Biol.* **21**, 4900 (2001).

Following recovery, gently wash the cells twice using 5 ml of 1 M sorbitol (spinning at 1000 rpm for 5 min in a SS-34 rotor). Resuspend in 1 ml of 1 M sorbitol and proceed with electroporation. The concentration of cells should be around 2×10^8 cells/ml.

Preparation of mRNAs

For the measurement of translation *in vivo*, we use reporter luciferase mRNAs described first by Gallie.[14] Other constructs, including the GUS reporter, can also be used.[14] T7-based luciferase constructs are linearized as follows: LUC A_0 is linearized with *Sma*I and LUC A_{50} is linearized with *Dra*I. LUC A_{50} produces an mRNA with a poly(A)$_{50}$ tail. The capped poly(A)$^+$ (C^+A^+) and capped poly(A)$^-$ (C^+A^-) mRNAs are synthesized with Ambion's mMessage mMachine kit. Uncapped mRNAs (C^-A^+ and C^-A^-) are synthesized with the MegaScript kit (Ambion). These kits allow for the production of large quantities of mRNA to be used in the electroporation assay. Two micrograms of RNA are used per electroporation. Making a large stock is useful for doing multiple experiments with maximum reproducibility.

Electroporation Procedure

Prior to electroporation, keep the electroporation cuvettes (0.2-cm electrode gap, Bio-Rad, Hercules, CA) on ice. Add 2 μg of each reporter mRNA to the cuvettes. Pipette 180 μl of yeast spheroplasts into a cuvette and pulse immediately. The following parameters are optimal to use for introducing RNA into yeast spheroplasts using the Bio-Rad Gene Pulser II: 800 V, 25 Faraday, and 1000 Ω. This should result in a pulse that ranges from 20 to 25 msec in duration. Immediately following the electroporation pulse, add 1.2 ml of ice-cold YPD–1 M sorbitol to the cuvettes. Keep the cuvettes on ice.

Measurement of Reporter Activity

We generally collect samples over a 2-hr time course following electroporation. The spheroplasts are transferred to ice-cold tubes (Falcon 2059) and placed at room temperature or 30° with gentle swirling. Samples are collected at the following time points: 0, 5, 10, 20, 40, 60, and 120 min. One hundred fifty to 200 μl of cells are removed at each time point, the cells are spun down, and the cell pellets are frozen immediately in an ethanol/dry ice bath. The cells can be stored at $-70°$ before assaying for reporter activity. To measure luciferase activity, we use the Luciferase Assay Reporter system (Promega, Madison, WI). The cell pellets are resuspended in 50–100 μl of 1× reporter lysis buffer (Promega) and vortexed vigorously for

---

[14] D. R. Gallie, J. N. Feder, R. T. Schimke, and V. Walbot, *Mol. Gen. Genet.* **228**, 258 (1991).

30 sec to break open the spheroplasts. Twenty microliters of the cell lysate is combined with 200 μl of reconstituted luciferase assay substrate (Promega) and activity is measured immediately in a LKB Wallac 1250 luminometer. The remaining extract is used to measure protein concentration.

Introducing RNAs into the same preparation of spheroplasts results in a 5–15% variability in luciferase activity measured per microgram of protein. It is unknown how much of this variability is due to differences in the amount of RNA introduced into the cells during the electroporation procedure. RNAs that are expressed very efficiently produce a greater range of luciferase activity and thus generate larger errors, particularly in the later time points after the period of linear luciferase synthesis. The incubation temperature during the recovery period, the growth rate of the cells, and their phase of growth also substantially influence reproducibility.

Light output is proportional to luciferase concentration with 1.0 light unit corresponding to 165 fg of luciferase enzyme (as measured in the LKB-Wallac 1250 luminometer). This is comparable to that observed using the Turner 20/20 luminometer (Promega) in which 1.0 light unit corresponds to 200 fg of luciferase enzyme.

Data Analysis

Using the LKB Wallac 1250 luminometer, we generally measure the following activity from wild-type cells after 2 hr.

	Units of luciferase activity/microgram protein
$C^+ A^+$	25 to 30 depending on the strain used
$C^+ A^-$	0.2–0.3
$C^- A^+$	3 to 6 depending on the strain used
$C^- A^-$	0.08–0.1

An example of time course data with this method is shown in Fig. 1. It is very important to normalize the activity measured from each sample to the amount of total protein present (as measured by Bradford assays) in order to accurately compare differences between strains or time points. We generally see about a 4- to 6-fold increase in luciferase activity on addition of a cap to the mRNA and about a 30- to 100-fold increase in the translation of poly(A)$^+$ mRNAs over the same mRNA without poly(A).

One factor that can contribute to changes in luciferase activity measured over time is differences in mRNA stability. Both the cap and the poly(A) tail function to affect mRNA half-lives. The presence of a cap inhibits degradation by the 5′ to 3′ exonuclease Ski1p/Xrn1p, and the poly(A) tail prevents decapping, which itself leads to the 5′ to 3′ Xrn1p degradation. If the Ski1p/Xrn1p system is blocked

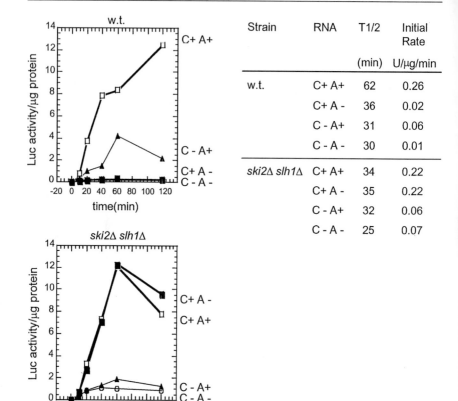

FIG. 1. *ski2Δ slh1Δ* strains translate A⁺ and A⁻ mRNAs alike. Isogenic strains 3221 (wild type) and 4107 (*ski2Δ slh1Δ*) were electroporated with 2 μg of luciferase mRNAs with or without a 5′ cap (C⁺ or C⁻) and with or without 3′ poly(A) (A⁺ or A⁻). Cells were incubated at 25°, and accumulated luciferase activity was assayed at the indicated time points [A. M. Searfoss and R. B. Wickner, *Proc. Natl. Acad. Sci. U.S.A.* **97,** 9133 (2000)]. The functional half-lives of each mRNA were calculated by determining the time required for each to produce 50% of the maximal luciferase activity. The maximum rate of luciferase synthesis was the maximum rate of change of the activity curves (generally during the first 20 to 40 min of the assay).

by mutation or by an mRNA structure, an effect of the 3′ poly(A) in protecting mRNA from degradation by the exosome (a complex of 3′→5′ exonucleases) can be observed (for review, see Schwartz and Parker[15]). One way to measure any differences in mRNA stability is to determine the functional half-life of a given

[15] D. C. Schwartz and R. Parker, *in* "Translational Control of Gene Expression" (N. Sonenberg, J. W. B. Hershey, and M. Mathews, eds.), p. 807. Cold Spring Harbor Laboratory Press, Cold Spring Harbor, NY, 2000.

mRNA. The functional half-life was originally defined by Gallie[4] as the time required, following mRNA delivery by electroporation, for that mRNA to produce 50% of the final amount of protein produced over the time course. This differs from physical half-life in that this is a measure of the time during which an individual mRNA is actively engaged in translation rather then a measure of its longevity in the cell. If the functional half-lives indicate an effect on mRNA stability, the decay profiles of endogenous mRNAs such as *PGK1* and *ACT1* can be determined to distinguish changes in mRNA decay from effects on mRNA translation.[16]

Changes in the rate of luciferase activity produced over time can also reflect changes in the translation rate; generally there is an increase in the translation rate when an mRNA has a poly(A) tail relative to an mRNA that is poly(A)$^-$.[4] The same is true for the presence of the 5' cap.[4] By determining the change in rate, i.e., by measuring the slope of the curve of the graph depicting luciferase activity over time, one can compare changes in the translation rate produced by mRNAs that differ by the presence or absence of a cap or poly(A) tail. Additionally, the slope will indicate a difference in translation rates measured from a given mRNA in two different yeast strains. This could indicate a role for a specific protein in the process of translation initiation. A change in the translation rate in the absence of any difference in functional half-life is indicative of an effect on translation rather than any effect on mRNA stability. This was shown by comparing translation rates measured from A^+ and A^- mRNAs in a wild-type strain and a $ski2\Delta\ slh1\Delta$ mutant (Fig. 1).[9] It was found that the ability of A^- mRNAs in the mutant to be translated, as well as A^+ mRNAs, was due to an increase in their rate of translation rather than due to effects on mRNA stability.

Potential Problems

Cells

Cells must be in early log phase. Cells above OD_{600} of 0.8 may be less active, and cells out of log phase are much less active in translation. Additionally, it is very important that the zymolase treatment of yeast cells is carefully monitored in order to prevent cell lysis. Measuring the OD_{600} in the presence of water or sorbitol as indicated will determine when the cell wall has been removed. Alternatively, microscopic examination of treated cells suspended in water will show lysed cells if spheroplasting is complete. If spheroplasting has proceeded too far and the cells are lysed, there will be no measurable luciferase activity produced following the electroporation. The time required for spheroplast production will vary with strain background and must be determined empirically for each strain.

---

[16] R. Parker, D. Herrick, S. W. Peltz, and A. Jacobson, *in* "Guide to Yeast Genetics and Molecular Biology" (C. Guthrie and G. R. Fink, eds.), p. 415. Academic Press, San Diego, 1991.

Capped mRNA

Only about 80% of the total C^+ mRNA generated by the mMessage mMachine kit (Ambion) is capped. This is generally not a problem when measuring the effect of the cap on translation due to the large effect its presence has on translation (from 5- to 24-fold). However, if a particular mutant increases the translational efficiency of C^- mRNAs, this will be reflected in an increase in the translation of C^+ RNAs. Therefore, any mutation that may preferentially decrease the translation of C^+ mRNAs and increase the translation of C^- mRNAs may not show any observable effect on the translation of C^+ mRNAs.

Additionally, the efficiency of capping during the *in vitro* transcription reaction decreases over time, particularly if the kit has been kept beyond the recommended 6 months (Ambion and unpublished results). This causes a reduction in the translational efficiency of C^+ mRNAs until the efficiency with which they are translated equals that of C^- mRNAs. Therefore, it is important to maintain enzymes and rNTPs at the recommended storage temperatures and to replenish with new components as needed.

Kinetics

Lower incubation temperatures facilitate accurate kinetics compared to incubation at 30°, especially at 37°. This is observed easily when using temperature-sensitive mutants that are incubated at 37°. Their translation rate increases greatly, such that the peak of luciferase activity produced occurs at one-third to one-half the time for the same strains incubated at room temperature.

Temperature-Sensitive Mutants

It is generally sufficient to incubate temperature-sensitive mutants at the nonpermissive temperature following the introduction of mRNAs by electroporation (during the time course). However, depending on the temperature-sensitive mutant that is used, it may be necessary to shift to the nonpermissive temperature during the recovery phase prior to electroporation. This is due to the large increase in luciferase expression during the first 10 min of incubation that is observed with C^+ and A^+ mRNAs. Incubation at the nonpermissive temperature prior to the introduction of mRNA could allow for measurable changes in luciferase expression compared to what is measured in wild-type cells.

Contrast with *in Vitro* Translation Systems

The *in vitro* translation systems developed for yeast are indispensable for studies of translation mechanisms because they offer the possibility of fractionation of the components involved and the combination of components from different mutant strains in the same reaction. However, the RNA electroporation method of

Gallie is particularly useful because it allows study of effects of altering mRNA sequence, structure, and modification under truly *in vivo* conditions. While the *in vitro* translation system has many positive features, it does not necessarily reflect the *in vivo* situation. An analogy can be drawn with early systems of *in vitro* DNA replication in which the synthesis of biologically active DNA was indeed obtained, but there were not the correct requirements for the *dna* gene products or for origins of replication. Because electroporated cells are metabolically active and growing, they have normal amounts of all translation factors and competing cellular mRNAs. They can thus be assumed to give a reasonably accurate reflection of normal translation requirements for cellular components and mRNA structures.

Using the RNA electroporation method, the necessity for both 5' cap structure and 3' poly(A) for translation has been proven.[5] Neither structure alone is sufficient for normal rates of translation. Also using the mRNA electroporation method, it has been found that the requirement for the 3' poly(A) structure is decreased or absent when the nonessential genes *SKI2* and *SLH1* are deleted.[9] The nonessential *SKI3*, *SKI7*, and *SKI8* genes are also involved in this system that represses translation of non-poly(A) mRNAs.[6–8] We expect that the mRNA electroporation method will continue to be useful in studying the mechanisms of translation.

[37] Monitoring Protein Degradation

By DANIEL KORNITZER

Introduction

All cellular proteins have a finite life span and eventually are enzymatically hydrolyzed to their constituent amino acids. Protein degradation processes can be divided in two categories: bulk degradation, such as in the process of autophagy; and selective degradation of specific proteins, which is the subject of this chapter. Individual protein species can differ in stability by several orders of magnitude. Constitutive fast turnover of a protein allows the rapid modulation of its concentration in response to changes in its rate of synthesis.[1] Furthermore, regulated degradation can serve as an additional mechanism to control the concentration of a given protein. Thus, although most cellular proteins are metabolically stable, many proteins that require tight regulation are degraded rapidly. These include key metabolic enzymes and regulatory proteins; cell cycle regulators, in particular, are often subject to regulated degradation.[2] Yeast is especially amenable to the analysis

---
[1] R. T. Schimke, *Adv. Enzymol.* **37**, 135 (1973).
[2] P. Jorgensen and M. Tyers, *Curr. Opin. Microbiol.* **2**, 610 (1999).

of protein degradation *in vivo* because of the wealth of available mutants in specific pathways and the availability of reporter constructs that enable one to perform genetic screens (see later).

Protein Tagging and Protein Fusions

In conjunction with the different protocols used to measure protein stability, immunological methods are typically used in order to detect the protein of interest. Protein instability is generally *cis* dominant, i.e., the fusion of a stable and unstable protein results in an unstable chimeric protein. Therefore, epitope tagging,[3] in which a sequence a few tens of amino acids long is added to the protein of interest, will usually not interfere with its degradation. Nonetheless, care must still be taken not to disrupt the degradation signal; e.g., modification of the N terminus with an epitope tag can inhibit degradation of certain proteins.[4] Because no degradation signals are yet known to be located at the C terminus in eukaryotes, this would be *a priori* the preferred site for epitope tagging.

The *cis* dominance of protein degradation is also useful for designing genetic screens. An unstable protein, or a degradation signal-containing sequence, can be fused to a reporter protein, such as β-galactosidase (for LacZ)[5,6] or orotidine-5'-phosphate decarboxylase (for Ura3).[7] The stability of the construct will then be reflected in the activity of the reporter enzyme, which can be measured with a chromogenic substrate (LacZ) or by uracil prototrophy vs 5-fluoroorotic acid resistance[8] (Ura3). One important caveat, however, is that in some instances the fusion protein will not be degraded to completion, but rather will yield a stable partial degradation product that is still enzymatically active (unpublished observations). This stable product will be detectable in a pulse–chase experiment (see protocol later). In such instances, the activity of the reporter enzyme will not faithfully reflect the degradation of the unstable moiety of the fusion.

Methods for Measuring Protein Degradation

The two common methods for measuring the rate of degradation of a protein are to pulse label the cell with radioactive amino acids and to follow the decay of the labeled protein after chase with an unlabeled precursor or to arrest synthesis and measure the decay of the total protein levels with time.

All the available methods entail a perturbation of cellular metabolism; one aim is to minimize this perturbation, as the degradation of the protein of interest may be

[3] P. A. Kolodziej and R. A. Young, *Methods Enzymol.* **194,** 508 (1991).
[4] K. Breitschopf, E. Bengal, T. Ziv, A. Admon, and A. Ciechanover, *EMBO J.* **17,** 5964 (1998).
[5] M. Hochstrasser and A. Varshavsky, *Cell* **61,** 697 (1990).
[6] S. Irniger, S. Piatti, C. Michaelis, and K. Nasmyth, *Cell* **81,** 269 (1995).
[7] T. Gilon, O. Chomsky, and R. G. Kulka, *EMBO J.* **17,** 2759 (1998).
[8] J. D. Boeke, J. Trueheart, G. Natsoulis, and G. R. Fink, *Methods Enzymol.* **154,** 164 (1987).

responsive to the physiology of the cell. Therefore, inhibition of total cytoplasmic protein synthesis with a translation inhibitor, such as cycloheximide, should be avoided. As an alternative, synthesis of the protein of interest can be specifically arrested using a regulatable promoter. The most widely used promoter for this purpose is the *GAL1,10* promoter, which is active in galactose-containing medium but is repressed up to a thousandfold by the addition of glucose.[9] In a typical protocol, the gene of interest is cloned under the *GAL1,10* promoter, synthesis is induced by shifting the cells to galactose, and, after a few hours of induction, the promoter is shut off by adding glucose to the medium (see later for a detailed protocol). The amount of protein remaining after shutoff of synthesis is followed by Western blot. Although this method entails a milder metabolic perturbation than inhibition of total protein synthesis, shifting carbon sources is not necessarily inconsequential: e.g., glucose was found to affect, via the Ras pathway, the activity of two major ubiquitination complexes.[10] Thus, the possibility that degradation of the protein of interest will be influenced by the carbon source has to be kept in mind. Another complication derives from the fact that although the addition of glucose leads to rapid shutoff of transcription from the *GAL1,10* promoter, shutoff of synthesis of the protein depends on the rate of decay of the transcript. The half-life of natural yeast transcripts can range from 2 to 60 min,[11] and even the hybrid transcripts created for heterologous expression of genes under the *GAL1,10* promoter can be quite stable. For example, we found that mammalian p27^{KIP1} cDNA cloned in the pYES2 vector generated a transcript with a half-life of over 60 min; another hybrid transcript containing a *Candida albicans* gene expressed from the p424*GAL1* plasmid displayed a biphasic decay, with substantial amounts of mRNA persisting even after 2 hr in glucose (T. Shemer, E. Boehm, and D. Kornitzer, unpublished observations, 1999). Thus, stability of the transcript must be taken into account before conclusions can be drawn regarding the stability of a protein. In addition, even if the transcript has a short half-life, the half-life of some proteins can be even shorter; in this case, the decay of the protein band reflects the stability of the transcript rather than that of the protein itself. Therefore, the GAL shutoff method will not yield an accurate measure of protein half-life.

The preferred method for measuring protein degradation accurately is the pulse–chase method. Cells are incubated with labeled precursors for a short time, after which they are washed from the labeled compound and incubated with excess cold precursor. The decay of the amount of protein labeled during the "pulse" period is measured during the "chase" with the cold precursor. For this decay to reflect degradation of the protein accurately, recycling of the labeled precursor during the chase must be minimized and the cytoplasmic pools of precursor must rapidly exchange with the labeled and unlabeled precursors provided in the medium. The

---

[9] M. Johnston and R. W. Davis, *Mol. Cell. Biol.* **4,** 1440 (1984).
[10] S. Irniger, M. Baumer, and G. H. Braus, *Genetics* **154,** 1509 (2000).
[11] A. Jacobson and S. W. Peltz, *Annu. Rev. Biochem.* **65,** 693 (1996).

precursor of choice is [^{35}S]methionine. The cytoplasmic methionine pool equilibrates extremely rapidly (within 1 min)[12] with the methionine in the medium. Furthermore, higher specific activities are achieved with the ^{35}S isotope than with ^{14}C. Using the protocol detailed later, we obtained exponential decay rates as high as $t_{1/2} = 75$ sec with no or very little apparent lag at the beginning of the chase (unpublished observations), also implying that cytoplasmic pools equilibrate with external methionine within the first minute of the chase.

A drawback of the methionine pulse–chase method is the relatively large number of steps required to obtain the labeled protein bands of interest. The experimental error in each of these steps and the potentially variable immunoprecipitation yield cumulatively decrease the accuracy of the determination of labeled protein remaining at each time point. Whereas this is of little significance when, as is often the case, band intensities vary severalfold from one time point to the next, smaller differences will be harder to detect. In principle, an internal control can be used to correct for this type of experimental error, thereby increasing accuracy. The ubiquitin–protein–reference (UPR) technique was developed by Varshavsky and colleagues specifically for this purpose. In this technique, a tripartite fusion protein is used, consisting of (in the amino-to-carboxyl direction) a stable reference protein (e.g., dihydrofolate reductase), ubiquitin, and the protein of interest.[13] The protein of interest is cleaved off of the reference protein–ubiquitin moiety cotranslationally or immediately after translation by the cellular ubiquitin-specific processing proteases. The stable reference protein band can then be used as an internal standard against which the measured amount of radioactivity in the protein band of interest can be normalized.

Protocols

Methionine Pulse–Chase and Immunoprecipitation

1. Overnight cultures are diluted in 10 ml of the appropriate synthetic medium to an OD$_{600}$ of 0.15 and grown for at least two cell doublings.

2. Cells are spun down and washed once with the same medium lacking methionine,* resuspended in 0.3 ml of medium lacking methionine, and transferred to a 1.5-ml tube containing 0.5–1 mCi of [^{35}S]methionine/cysteine labeling mix (e.g., Dupont-NEN "Express" or Amersham "Pro-mix"). To avoid radioactive contamination, we use 1.5-ml tubes with an O ring in the screw cap here as well as for extract preparation (steps 4–6).

---

* We found that methionine incorporation was severalfold more efficient in synthetic medium lacking all but the essential amino acids ("drop-in") than in synthetic complete medium containing all the amino acids except methionine. This is probably due to competition of amino acids with the labeled methionine for the general amino acid permease.

[12] J. R. Warner, S. A. Morgan, and R. W. Shulman, *J. Bacteriol.* **125**, 887 (1976).

[13] F. Levy, N. Johnsson, T. Rumenapf, and A. Varshavsky, *Proc. Natl. Acad. Sci. U.S.A.* **93**, 4907 (1996).

3. The tubes are incubated for 5 min in a 30° water bath and then pelleted for 10 sec in a microcentrifuge at room temperature. The supernatant is discarded, and the cells are resuspended in an Erlenmeyer flask containing 2.5 ml of the original growth medium + 10 mM methionine and 10 mM cysteine (a 200 mM methionine, 200 mM cysteine stock can be kept at $-20°$). The flask is kept shaking in a water bath for the duration of the experiment.

4. One 0.6-ml aliquot is removed immediately and added to a screw-capped 1.5-ml tube on ice containing 102 μl of 1.85 M NaOH, 7.4% 2-mercaptoethanol.[†,14] Three additional 0.6-ml samples are removed at the appropriate times after the chase.

5. After a 10-min incubation on ice, 42 μl of 100% trichloroacetic acid (TCA) is added to each tube. After an additional 10 min on ice, the extracts are centrifuged in the cold for 5 min. The pellets are washed with 1 ml of ice-cold acetone (resuspend the pellets by vortexing and centrifuge in the cold for 5 min) and then dried in a SpeedVac centrifuge.**

6. The pellets are resuspended in 100–200 μl 2.5% SDS, 5 mM EDTA, 1 mM PMSF by vortexing (this can take a while with large pellets). The suspensions are heated to $>90°$ for 1 min and then are cleared by spinning in a microcentrifuge for 5 min at room temperature. Incorporation is measured by counting acid-insoluble radioactivity [add 2 μl of the extract to 20 μl 10 mg/ml of bovine serum albumin (BSA), add 1 ml of 5% TCA, incubate for 15 min on ice, collect the acid-precipitated proteins on a Whatman (Clifton, NJ) GF/C glass fiber filter using a stainless-steel holder-equipped filtering flask, air dry the filter, and measure the radioactivity by scintillation counting]. In practice, if equal amounts of culture are processed, very little variability should occur between different samples of the same culture.

7. Equal amounts of total incorporated radioactivity (in up to 0.1 ml) are added to 1 ml of i.p. buffer[‡,15] + antiproteases + antibody (BSA can be added to 2% to reduce nonspecific binding).

---

†This extraction method was adapted from Yaffe and Schatz.[14] For proteins that precipitate when heated in SDS, such as polytopic membrane proteins, an alternative method is to make a protein lysate by breaking the cells with glass beads in 2.5% sodium dodecyl sulfate (SDS), 5 mM EDTA, 1 mM phenylmethylsulfonyl fluoride (PMSF), and antiproteases from 250× stock (for the pulse–chase protocol) or in 2× protein-loading buffer containing 1 mM PMSF and antiproteases from 250× stock (for the GAL shutoff protocol). Avoid heating the tubes above 37°. Both protocols then proceed from step 7.

**Dry pellets can be kept at $-20°$ overnight before resuspension (longer storage is not recommended because of radiolysis of the proteins in the concentrated pellet).

‡i.p. buffer[15]: 1% Triton X-100, 0.15 M NaCl, 5 mM Na-EDTA, 50 mM Na-HEPES, pH 7.5, 1 mM PMSF, antiproteases from 250× stock. 250× antiprotease stock: 5 mg/ml each chymostatin, pepstatin A, leupeptin, antipain, 1.25 M N-ethylmaleimide in dimethyl sulfoxide. All reagents are from Sigma (St. Louis, MO).

[14] M. P. Yaffe and G. Schatz, *Proc. Natl. Acad. Sci. U.S.A* **81**, 4819 (1984).

[15] B. Bartel, I. Wunning, and A. Varshavsky, *EMBO J.* **9**, 3179 (1990).

8. Tubes are incubated at 4° for 2 hr and then 20 µl of protein A–agarose beads (50% suspension in i.p. buffer) is added to each tube. Tubes are incubated for another 2 hr at 4° while tumbling.

9. Beads are washed three times with cold i.p. buffer containing 0.1% SDS and are then resuspended in 20 µl 2× protein-loading buffer.

10. The tubes are heated to >90° for 5 min, centrifuged for 1 min at room temperature, and the supernatant is electrophoresed on SDS–PAGE.

GAL Shutoff

1. Strains are grown overnight in 20 ml of the appropriate medium with 2% raffinose as the carbon source. To ensure adequate overnight growth, cells should be taken from fresh plates (less than 1 week old) and inoculated in the evening to an OD_{600} of 0.2–0.3.

2. Cultures are diluted in 40 ml of the same medium with 2% galactose as a carbon source to an OD_{600} of 0.3. After 4 hr of induction, glucose is added to 2%.

3. Five-milliliter aliquots of the cultures are taken before the addition of glucose and at each time point afterward, and the cells are pelleted in 15-ml Falcon tubes. The cells are then resuspended in 0.5 ml H_2O, and the suspension is added to an Eppendorf tube on ice containing 85 µl of 1.85 M NaOH, 7.4% 2-mercaptoethanol.[†]

4. Samples are left on ice at least 10 min; 40 µl of 100% TCA is added, and tubes are left on ice for another 10 min.

5. Tubes are centrifuged for 10 min in a refrigerated microcentrifuge; after removal of the supernatant, pellets are washed once in ice-cold acetone.

6. After removal of the acetone, the pellets are dried in a SpeedVac centrifuge and then are resuspended by vortexing in 2× protein loading buffer (this may take a while), 40 µl/OD unit. The suspension is heated to >90° for 10 min.

7. The tubes are centrifuged for 3 min in a microcentrifuge, and 20 µl of the supernatant is loaded per lane.

8. Gel electrophoresis, Western blotting, and detection are performed according to standard protocols.

Derivation of Protein Half-Life

One simplifying assumption in studying protein degradation is that all molecules of a given protein species have an equal chance of being degraded at any given time. As a consequence, the kinetics of protein degradation is expected to be of the first order, and therefore the rate of degradation can be described by a single parameter, the half-life. Plotting the amount of protein remaining against time on a semilogarithmic graph should yield a straight line that crosses the 50% intercept at $t_{1/2}$. An example is shown in Fig. 1: the Gcn4 protein decays exponentially for over two orders of magnitude, with a half-life of 2 min (the apparent slowing of

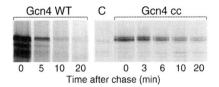
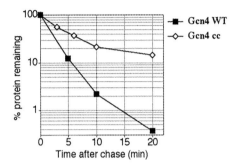

FIG. 1. Degradation of wild-type and mutant Gcn4 proteins. An epitope-tagged version of Gcn4 was used, and degradation was followed in exponentially growing cells using the methionine pulse–chase protocol. Gcn4 WT, the wild-type protein; Gcn4 cc, a mutant protein that is stable during part of the cell cycle.

the decay toward the last time point probably reflects the inaccuracy in measuring the very small amounts of radioactivity left).

In reality, although the observed time course of protein degradation often agrees with first-order kinetics, there are instances where more complex degradation curves are observed. This may indicate a heterogeneity in the population either of protein molecules (e.g., free vs complexed with other proteins) or of cells in the culture (e.g., cells in different cell cycle phases), exhibiting differential degradation. Thus, rather than being discarded as experimental errors, deviations from first-order kinetics of degradation can sometimes give clues regarding the regulation of degradation of a given protein. For example, degradation of the Gcn4 cc protein shown in Fig. 1 is biphasic: the initially rapid degradation slows considerably beyond the 10-min time point. This Gcn4 mutant was subsequently found to be stable during part of the cell cycle (A. Meimoun and D. Kornitzer, unpublished observations, 2001).

Another consideration with regard to the pulse–chase method is that it measures the half-life of newly synthesized proteins. Conceivably, in some instances, a majority of the newly synthesized molecules are degraded rapidly, whereas a small fraction is diverted to a stable subpopulation. Such a stable subpopulation will be detected more readily using a synthesis shutoff protocol, which measures steady-state levels of the protein, rather than with a pulse–chase protocol.

Determination of Proteolytic Pathway for Specific Proteins in Vivo

Two distinct systems for specific protein degradation are known in eukaryotic cells: the ubiquitin/proteasome system and the vacuolar (in yeast) or lysosomal (in animals) system. In the first, target proteins are tagged with covalent chains of the polypeptide ubiquitin and are then degraded by the multicatalytic 26S proteasome.[16] In the second, target proteins are delivered to the vacuole and degraded by vacuolar proteases.[17] Importantly, some overlap between the systems occurs, because tagging of some membrane proteins by ubiquitin can lead to their degradation in the vacuole.[18,19] A number of approaches can be taken, using genetic or pharmacological tools, to determine the degradation pathway of a protein. In practice, although pharmacological tools are preferred in animal cells, the wealth of available yeast mutants and the relative impermeability of yeast to many inhibitors result in the use of mutants being more widespread in yeast. Nonetheless, the specific proteasome inhibitor MG132 has been used successfully to inhibit yeast proteasomal degradation *in vivo,* provided the strain carried an *ise1/erg6* permeability mutation (Ref. 20 and unpublished observations).

Yeast mutants are available in both vacuolar and ubiquitin/proteasome systems. Vacuolar proteinase activity is reduced greatly in the pleiotropic *pep4* mutant.[21] In this mutant, cell growth in standard medium is not affected (but survival under starvation conditions is strongly reduced),[22] making it a convenient strain to assay dependence of a substrate on vacuolar degradation. In contrast, no null mutants exist of either the proteasome or the ubiquitination machinery, which are essential for vegetative growth. Numerous partial loss-of-function mutants of the proteasome are available, but these can be misleading: e.g., Gcn4, a ubiquitin/proteasome substrate,[23] is not stabilized in a *pre1 pre4* proteasome catalytic subunit mutant (unpublished results). Nonetheless, some mutants in the regulatory subcomplex of the proteasome appear to affect degradation of most substrates. Such mutants include *cim5-1*[24] and *rpt2-K229R/S241F*.[25]

The ubiquitin conjugation system is hierarchical, with the ubiquitin-activating enzyme (E1) Uba1 being required—in addition to ubiquitin itself—for all ubiquitination reactions.[16] Although a useful E1 temperature-sensitive mutant has been

[16] A. Hershko and A. Ciechanover, *Annu. Rev. Biochem.* **67,** 425 (1998).
[17] M. Knop, H. H. Schiffer, S. Rupp, and D. H. Wolf, *Curr. Opin. Cell Biol.* **5,** 990 (1993).
[18] J. M. Galan, V. Moreau, B. Andre, C. Volland, and R. Haguenauer-Tsapis, *J. Biol. Chem.* **271,** 10946 (1996).
[19] L. Hicke, *FASEB J.* **11,** 1215 (1997).
[20] D. H. Lee and A. L. Goldberg, *J. Biol. Chem.* **271,** 27280 (1996).
[21] E. W. Jones, *Methods Enzymol.* **194,** 428 (1991).
[22] U. Teichert, B. Mechler, H. Muller, and D. H. Wolf, *J. Biol. Chem.* **264,** 16037 (1989).
[23] D. Kornitzer, B. Raboy, R. G. Kulka, and G. R. Fink, *EMBO J.* **13,** 6021 (1994).
[24] M. Ghislain, A. Udvardy, and C. Mann, *Nature* **366,** 358 (1993).
[25] D. M. Rubin, M. H. Glickman, C. N. Larsen, S. Dhruvakumar, and D. Finley, *EMBO J.* **17,** 4909 (1998).

known for some time in mammalian cells,[26] no analogous yeast mutants were available until recently. A *UBA1* hypomorphic mutant, *uba1-2,* has been described.[27] This mutant reduces ubiquitination and degradation of ubiquitin system substrates, and therefore constitutes a new tool to rapidly confirm the involvement of the ubiquitin system in the degradation of a given protein (a notable peculiarity of the *uba1-2* mutation is that it ameliorates rather than impairs growth at an elevated temperature). Another useful mutant that affects the degradation of many ubiquitin system substrates is the ubiquitin K48R G76A allele.[28] This is a dominant-negative, chain-terminating mutant of ubiquitin that, when overexpressed, was found to inhibit degradation of a variety of substrates, including the synthetic Ub-Pro-LacZ construct,[28] Cln3,[29] Ho,[30] and Gcn4 (Kornitzer and Fink, unpublished observations, 1995). However, monoubiquitination, which suffices in some instances to direct degradation of a protein (e.g., see Ref. 31), is not affected by this mutation.

At the next lower level of the ubiquitin hierarchy are the ubiquitin-conjugating enzymes (Ubcs), which, with ubiquitin ligases, are responsible for the conjugation of ubiquitin to the target proteins. Of the 11 Ubcs known, only Cdc34/Ubc3 is essential, and temperature-sensitive mutants of this gene are available.[32] Reduced degradation of a substrate in a *cdc34-2* mutant or in a null mutant of any of the other 10 Ubcs would constitute strong evidence for degradation via the ubiquitin system, specifically via the subpathway defined by this Ubc. However, because of functional overlap of some of these enzymes (such as Ubc1,4,5),[33] lack of dependence cannot be taken to exclude involvement of this Ubc in the degradation of the substrate. Finally, complete determination of the degradation pathway of a ubiquitin system substrate includes identification of the ubiquitin ligase involved and, ideally, reconstitution of the ubiquitination reaction *in vitro.*[34]

Acknowledgment

Work in the author's laboratory is supported by the Israel Science Foundation.

[26] A. Ciechanover, D. Finley, and A. Varshavsky, *Cell* **37,** 57 (1984).
[27] R. Swanson and M. Hochstrasser, *FEBS Lett.* **477,** 193 (2000).
[28] D. Finley, S. Sadis, B. P. Monia, P. Boucher, D. Ecker, S. T. Crooke, and V. Chau, *Mol. Cell. Biol.* **14,** 5501 (1994).
[29] J. Yaglom, H. K. Linskens, S. Sadis, D. M. Rubin, B. Futcher, and D. Finley, *Mol. Cell. Biol.* **15,** 731 (1995).
[30] L. Kaplun, Y. Ivantsiv, D. Kornitzer, and D. Raveh, *Proc. Natl. Acad. Sci. U.S.A.* **97,** 10077 (2000).
[31] J. Terrell, S. Shih, R. Dunn, and L. Hicke, *Mol. Cell* **1,** 193 (1998).
[32] M. G. Goebl, J. Yochem, S. Jentsch, J. P. McGrath, A. Varshavsky, and B. Byers, *Science* **241,** 1331 (1988).
[33] W. Seufert, J. P. McGrath, and S. Jentsch, *EMBO J.* **9,** 4535 (1990).
[34] J. D. Laxey and M. Hochstrasser, *Methods Enzymol.* **351,** [14], 2002 (this volume).

[38] Analyzing mRNA Decay in *Saccharomyces cerevisiae*

By MICHELLE A. STEIGER and ROY PARKER

Introduction

mRNA turnover is an important process in the regulation of eukaryotic gene expression. The analysis of mRNA turnover often requires knowledge of the rates of mRNA degradation and the pathway by which a particular mRNA is being degraded. This chapter describes experimental procedures that can be used to determine the rates of mRNA turnover in yeast, as well as the specific pathway of degradation for any given transcript. The methods described here are used routinely to examine cytoplasmic mRNA decay events but can also be used to monitor nuclear mRNA turnover. An understanding of the eukaryotic mRNA turnover pathways has facilitated these approaches (Fig. 1).[1] In yeast, mRNA decay occurs predominantly through deadenylation, followed by decapping, and rapid 5' to 3' exonucleolytic decay. Alternatively, a nuclear or cytoplasmic transcript can undergo deadenylation followed by 3' to 5' exonucleolytic decay. Aberrant mRNAs, such as those with premature termination codons or extended 3' UTRs, are degraded via decapping that occurs independent of deadenylation followed by rapid 5' to 3' decay.[2] In more complex eukaryotes, mRNA turnover can also be initiated by specific endonucleolytic cleavage. It is anticipated, but not yet observed, that some yeast mRNAs are targeted by specific endonucleases. Defects in mRNA metabolism may be a secondary result from a defect in other cellular processes (e.g., transcription, translation). This possibility should be considered when analyzing new factors believed to alter mRNA turnover.

Measuring mRNA Half-Life

A mRNA half-life is measured routinely for a mRNA of interest such that the overall stability of that mRNA can be analyzed and quantitated. Several different experimental procedures can be used to measure the decay rates of individual mRNAs in the yeast *Saccharomyces cerevisiae*. Table I gives the methods for determining mRNA half-life and cites the advantages and disadvantages for each method. Transcripts can be radiolabeled *in vivo*, and the rate of mRNA decay can be determined from either the disappearance of specific mRNAs during a chase (pulse–chase) or the kinetics of the initial labeling (approach to steady state).

---

[1] C. A. Beelman and R. Parker, *Cell* **81,** 179 (1995).
[2] P. Hilleren and R. Parker, *RNA* **5,** 711 (1999).

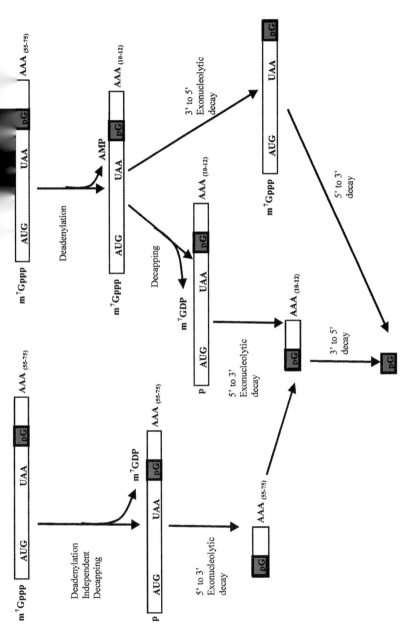

FIG. 1. The three known yeast mRNA decay pathways and decay intermediates. mRNA decay intermediates that accumulate with the placement of a poly(G) tract into the 3' UTR of a transcript are shown. (Left) The nonsense-mediated decay pathway produces a poly(G)→3' end fragment with a poly(A) tail. The predominant decay pathway, 5'→3' decay (shown in the middle), produces a poly(G)→3' end fragment with an oligo(A) tail. (Right) The 3'→5' decay pathway, which produces a 5' end→poly(G) fragment. All intermediates containing the poly(G) tract are degraded further until only the poly(G) tract [or poly(G) "stub"] remains.

TABLE I
METHODS FOR MEASURING mRNA HALF-LIVES IN YEAST

Method	Advantage	Disadvantage
In vivo labeling Approach to steady state	Minimal cell perturbation No need for special strain or construct Can monitor the half-life of many mRNAs simultaneously	Requires large amounts of radioactive material Poor signal-to-noise ratio for minor mRNAs
Transcriptional inhibition drugs (thiolutin, 1,10-phenanthroline)	No need for special strain or construct Can monitor the half-life of many mRNAs simultaneously	May cause loss of labile factors May alter the decay of specific mRNAs May alter other cellular pathways (i.e., transcription or translation)
Temperature-sensitive RNA polymerase II mutant (rpb1-1 mutation)	No need for special construct Can monitor the half-life of many mRNAs simultaneously	May cause loss of labile factors Not useful with other conditional mutants Possible secondary complications caused by heat shock Requires special strain
Regulatable GAL promoter	Minimal cell perturbation Applicable to transcriptional pulse–chase experiments	Requires special construct Allows only mRNAs under control of GAL promoter to be analyzed Changing carbon source may alter mRNA stability

Alternatively, transcription can be inhibited and decay rates derived by determining the relative abundance of individual mRNAs after such inhibition. In vivo labeling methods have several disadvantages, including a requirement for large quantities of radioactively labeled nucleic acid precursors, poor signal-to-noise ratios for low abundance mRNAs, and a failure to provide information on mRNA integrity during the course of an experiment (see Table I). Given these disadvantages, we emphasize protocols that measure mRNA decay rates in yeast subsequent to transcriptional inhibition. Information on in vivo labeling techniques is available in papers published previously.[3]

In the following protocol, mRNA synthesis is inhibited, either in general or for specific genes, and, at various times after such inhibition, the abundance of particular mRNAs is monitored by simple techniques (e.g., Northern blotting or RNase protection). Inhibition of total mRNA synthesis is accomplished by the use of

---

[3] D. Herrick, R. Parker, and A. Jacobson, Mol. Cell. Biol. **10,** 2269 (1990).

either transcriptional inhibitory drugs (such as thiolutin or 1,10-phenanthroline) or using a strain with a temperature-sensitive allele of RNA polymerase II and shifting the culture to the restrictive temperature.[3] These approaches are generally applicable, straightforward, and also provide information on mRNA integrity. The potential disadvantages of these protocols include the nonspecific side effects of drugs used to inhibit transcription and the potential loss of labile turnover factors in the absence of ongoing transcription.

An alternative approach is to inhibit the synthesis of an individual mRNA by placing the gene under control of a regulatable promoter. The *GAL* promoter is often used because this inducible promoter will cause the mRNA to be overproduced in the presence of galactose and transcription will be inhibited rapidly in the presence of glucose. In addition, *GAL*-regulated genes can be used in the transcriptional pulse–chase protocol described in this chapter. The following procedure is outlined for measuring the half-life of a mRNA using either the rpb1-1 mutant or a transcript under control of the *GAL* promoter.

1. Grow 200 ml of cells in the appropriate medium (medium for strains with the *GAL* promoter should contain 2% galactose) until the cells reach midlog phase (OD_{600} 0.5–1.0).

2. Harvest the cells by centrifugation and resuspend in appropriate medium. For rpb1-1 shutoffs, resuspend in 20 ml medium preheated to 37°. For shutoffs of the *GAL* promoter, use medium containing 2% glucose. Immediately remove a 2-ml aliquot of cells as a zero time point. This and all subsequent aliquots are centrifuged for 10 sec in a microfuge, the medium supernatant is removed by suction, and the resulting cell pellet is frozen in dry ice.

3. Continue growing the cells taking subsequent time points as described by the individual protocol. Initially, time points of 5, 10, 20, 30, 40, 50, and 60 min are commonly used.

4. RNA is isolated from these cells.[4] Northern blotting, dot blotting, or RNase protection procedures are useful for quantitating the amount of a particular mRNA at each time point. A semilogarithmic plot of the percentage of mRNA remaining versus time allows for the assessment of the mRNA half-life.[5–7]

Determination of mRNA Decay Pathways

The experimental procedures that were used to define the pathways of mRNA turnover in yeast can now be applied to any transcript to determine the specific

---

[4] G. Caponigro, D. Muhlrad, and R. Parker, *Mol. Cell. Biol.* **13**, 5141 (1993).
[5] C. H. Kim and J. R. Warner, *J. Mol. Biol.* **165**, 79 (1983).
[6] R. Losson and F. Lacroute, *Proc. Natl. Acad. Sci. U.S.A.* **76**, 5134 (1979).
[7] R. Parker, D. Herrick, S. W. Peltz, and A. Jacobson, *Methods Enzymol.* **194**, 415 (1991).

pathway(s) of degradation it undergoes. Three main experimental approaches allow for determining the mRNA decay mechanism. First, inserting a strong secondary structure within the mRNA allows for trapping decay intermediates, which are useful for determining the directionality of decay. Second, putting the gene of interest under a regulatable promoter allows for pulse–chase experiments, thereby allowing the determination of precursor–product relationships during the mRNA decay process. Third, examining mRNA decay in various strains defective in individual steps in mRNA turnover pathways can help define the specific turnover pathway used by the mRNA of interest. Although each experimental approach has its limitations, when used in combination, clear information about the mechanism through which a transcript is degraded can often be obtained.

Trapping mRNA Decay Intermediates

Understanding the mechanism responsible for degrading a specific mRNA can be facilitated through trapping and structurally analyzing mRNA decay intermediates. Normal mRNA transcripts are degraded rapidly and decay intermediates cannot be detected. However, introduction of strong secondary structure into a transcript, which inhibit exonucleases, traps intermediates in mRNA turnover (see Fig. 2 poly(G)→3' end fragment).[8–12]

Introducing a poly(G) tract of at least 18Gs into an mRNA is used commonly as a means for introducing secondary structure.[9–12] The strong secondary structure conferred by the addition of a series of G nucleotides is likely a result of the formation of a G-quartet structure arising from hydrogen bonding of 4 Gs in a planar array.[13] The poly(G) tract serves as a partial block of both 5'→3'- and 3'→5'-exonucleases, thus causing the accumulation of distinct mRNA decay intermediates [see Fig. 2, lanes 6 to 30 (min)].

Determination of Directionality of Decay Pathway. The exact nature of the accumulated mRNA decay intermediate depends on the directionality of mRNA degradation. Determination of the structure of a trapped intermediate, by any appropriate combination of Northern blotting, RNase protections, and primer extensions, can be used to infer the nucleolytic steps that produced the intermediate (Fig. 1). For example, a mRNA undergoing 5'→3' decay will accumulate a mRNA fragment containing the poly(G) tract and extending to 3' end of that RNA (Fig. 2, lanes 6 to 30 min).[9–11] Alternatively, an RNA undergoing 3'→5' decay will accumulate a mRNA fragment from the 5' end to the 3' side of the poly(G) tract.[14]

[8] P. Vreken and H. A. Raue, *Mol. Cell. Biol.* **12,** 2986 (1992).
[9] C. J. Decker and R. Parker, *Genes Dev.* **7,** 1632 (1993).
[10] D. Muhlrad, C. J. Decker, and R. Parker, *Genes Dev.* **8,** 855 (1994).
[11] D. Muhlrad, C. J. Decker, and R. Parker, *Mol. Cell. Biol.* **15,** 2145 (1995).
[12] C. A. Beelman and R. Parker, *J. Biol. Chem.* **269,** 9687 (1994).
[13] J. R. Williamson, M. K. Raghuraman, and T. R. Cech, *Cell* **59,** 871 (1989).
[14] J. S. J. Anderson and R. P. Parker, *EMBO J.* **17,** 1497 (1998).

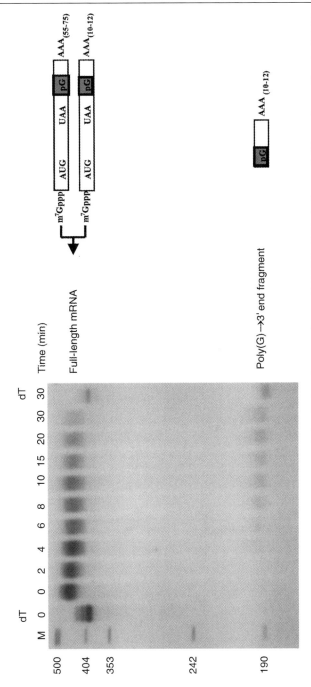

FIG. 2. Polyacrylamide Northern blot of a transcriptional pulse–chase experiment with MFA2 mRNA that has a poly(G) tract inserted into its 3' UTR. The full-length mRNA represents both adenylated and deadenylated capped mRNA. Time refers to the number of minutes after transcription inhibition. Lanes at 0 and 30 min designated dT indicate samples treated with oligo(dT) and RNase H to remove the poly(A) tail. These lanes contain mRNA with no poly(A) tail. The M refers to DNA markers with the nucleotide sizes indicated to the left. Courtesy of Denise Muhlrad, Howard Hughes Medical Institute, University of Arizona.

The placement of a poly(G) tract within a mRNA requires careful consideration. Placing the poly(G) tract in the 5' UTR blocks ribosome scanning and alters translation of the mRNA. Because translation and mRNA degradation are interrelated,[15] an inhibition of translation is likely to alter the decay of that mRNA and is therefore inadequate for turnover analysis. A poly(G) tract placed within the open reading frame of an mRNA may not serve as an efficient block to exonucleases (A. van Hoof and R. Parker, personal communication). Thus, the poly(G) tract is often placed in 3' UTR because this positioning appears not to interfere with mRNA translation or turnover. However, additional analysis is necessary to ensure that insertion of the poly(G) tract does not alter the rate or mechanism of mRNA decay (as discussed later).

A limitation of inferring decay mechanisms from the structures of trapped intermediates is that the initial nucleolytic event cannot be identified. For example, a poly(G)→3' end fragment can be generated by decapping and then 5'→3' decay or possibly by an initial endonucleolytic cleavage followed by 5'→3' degradation. To distinguish these two possibilities, one needs to capture early decay intermediates to examine the initial nucleolytic events. One way to capture early decay intermediates is to use the insertion of multiple poly(G) tracts.[16] When a transcript is degraded first by decapping and followed by 5'→3' degradation, the transcript will degrade to the 5'-poly(G) tract. In contrast, an initial endonucleolytic cleavage between the poly(G) fragments would produce two different fragments. Therefore, these two events can be distinguished by examining the structures of the intermediates derived from a transcript with two poly(G) insertions. Importantly, it should be noted that different mechanisms leading to the same decay intermediates can also be distinguished using specific *trans*-acting mutations (as discussed later).

Utilizing Decay Intermediates to Determine Role of Deadenylation in Decay. Trapped decay intermediates can also be examined directly to determine the role of deadenylation in mRNA decay. The predominant decay pathway accumulates the poly(G)→3' end fragment with an oligo (short) A tail. This is because turnover proceeds through deadenylation, decapping, and rapid 5'→3' decay. However, decapping can occur without poly(A) tail shortening, as has been seen when aberrant mRNAs are degraded through the mRNA surveillance pathway.[17] In this case, because decapping occurs prior to deadenylation, the trapped intermediate contains a longer poly(A) tail, as determined by measuring the poly(A) tail length. Poly(A) tail length is measured by hybridizing a portion of isolated RNA with oligo (dT), followed by RNase H cleavage; this removes the poly(A) tail (see Fig. 2, lanes marked dT). Separation, through a 6% polyacrylamide/8 M urea gel, of isolated RNAs both treated and untreated with oligo (dT)/RNase H allows for direct

---

[15] D. C. Schwartz and R. Parker, *Mol. Cell. Biol.* **20,** 7933 (2000).
[16] D. Muhlrad and R. Parker, *Nature* **370,** 578 (1994).
[17] D. Muhlrad and R. Parker, *RNA* **5,** 1299 (1999).

determination of the poly(A) tail state of the RNA (see Fig. 2, dT lanes).[9] It should be noted that mRNA fragments with a long poly(A) tail can be hard to detect in a steady-state population. After their production, the poly(G)→3' end fragments with a long poly(A) tail undergo both deadenylation and 3'→5' decay.[16] Because the 3'→5' decay rate is usually slower than the deadenylation rate, most fragments originally with a long poly(A) tail will deadenylate rapidly and accumulate at steady state with short or no poly(A) tails. Given this limitation, the best experiment is to use a trancriptional induction or pulse–chase (as discussed later) to generate a pool of newly produced poly(G)→3' end fragments, whose poly(A) tails lengths can then be examined.

Trapped Decay Intermediates as Simple Assay for 3'→5' mRNA Decay. Examining 3'→5' mRNA decay can be examined specifically with the poly(G)→3' end fragment. Investigating 3'→5' decay on the full-length mRNA can be difficult, as 3'→5' decay occurs slower than 5'→3' decay, and very little 5' end→poly(G) fragment is formed.[9,14] Thus, the poly(G)→3' end fragment is the starting mRNA species for monitoring 3'→5' mRNA decay; decay of this RNA will result in a fragment containing only the poly(G) tract.[14] However, determining the rate of 3'→5' decay on the poly(G)→3' end fragment requires that the production of this RNA species be inhibited (as discussed earlier in determining mRNA half-life). Production of the poly(G)→3' end fragment is stopped by inhibiting decapping through the addition of 0.1 mg/ml cycloheximide to cultures. Cycloheximide added to cultures at a concentration of 0.1 mg/ml also inhibits translation elongation; the inhibition of mRNA decapping is likely caused by an indirect effect of cycloheximide addition.[12] Cycloheximide can be added at the same time as mRNA transcription is inhibited and cells can be collected as indicated earlier.[12] Alternatively, a temperature-sensitive allele of the decapping enzyme can be used to block decapping at the nonpermissive temperature.

Limitations in Interpretations of Trapped Decay Intermediates. Trapping and analyzing mRNA decay intermediates can ultimately lead toward an understanding of the mechanism responsible for the mRNA degradation, but this method does have limitations. The trapped intermediates may not reflect the major mRNA decay pathway for a particular mRNA. Specifically, mRNAs that undergo decay through more than one pathway may accumulate intermediates from a minor pathway. There are two main aspects to consider in such as case. First, the relative amounts of an accumulated intermediate should be proportional to the initial amount of full-length mRNA, although this can be misleading if the intermediate itself is very unstable. Second, analyzing the decay rate of a mRNA in cells where one decay mechanism is defective (see earlier) will aid in understanding the predominant decay mechanism. A mutation within the main pathway responsible for degrading a mRNA results in a more stable mRNA than mutations that alter a minor or secondary decay pathway. Another limitation of the methods described earlier is that intermediates trapped in steady state provide limited information about how an individual transcript is degraded. The following experiments overcome this limitation by revealing precursor–product relationships during mRNA decay.

Determination of Precursor–Product Relationships in Transcriptional Pulse–Chase

A powerful method to analyze the pathway of mRNA degradation is a transcriptional pulse–chase experiment. In this experiment, a regulatable promoter is used to produce a homogeneous population of mRNA transcripts, which are then followed during the subsequent steps in mRNA degradation. This reveals the order in which decay events occur and the precursor–product relationships.

In *S. cerevisiae,* the *GAL* promoter has been used to perform transcriptional pulse–chase experiments.[9,11,16] A gene of interest is put under *GAL* control so that a short period of transcription can be accomplished by changing the carbon source. This is performed by first growing cells in medium containing a neutral carbon source, usually raffinose, that does not suppress or induce the *GAL* promoter. Subsequently, the promoter is induced by the addition of galactose followed by suppression by glucose after a short time. After transcriptional repression, a recently synthesized and thus homogeneous population of mRNA is available for study. This method enables the sequence of events in the decay pathway to be examined. For instance, poly(A) shortening and disappearance of full-length mRNA can be monitored to calculate the deadenylation rate and whether mRNA degradation begins before or after deadenylation. When transcripts with a poly(G) tract are used as mRNA decay precursors, the fragments trapped by the poly(G) tract will provide even more information. The level of full-length mRNA and poly(G)→3' end fragments, the appearance and disappearance of poly(G)→3' end fragments, and the poly(A) tail length of the full-length mRNA and fragments can be monitored. These data can provide information about the precursor–product relationships and thus the sequence of events. A detailed experimental procedure is outlined next.

1. Pregrow cells in 5 ml medium containing 2% raffinose. Use this culture to inoculate 200 ml medium containing 2% raffinose until cells reach early log phase (OD_{600}: 0.3–0.4). Adding sucrose to a final concentration of 2% can facilitate cell growth if cells grow poorly in raffinose. Because low pH results in acid hydrolysis of raffinose into galactose and sucrose, the pH of the medium should be adjusted to pH 6.5 with NH_4OH. In some cases, overgrowth of the culture can result in a lowering of the pH and therefore premature induction of transcription.

2. Pellet cells in four 50-ml Falcon tubes by spinning for 2 min at top speed in a table-top centrifuge, resuspend cells in 10 ml medium containing 2% raffinose (or 2% raffinose + 2% sucrose), return to incubator, and shake for 10 min.

3. Transcriptional induction is accomplished by adding 0.5 ml 40% galactose (a final concentration of 2%).

4. Immediately after adding galactose, remove an aliquot of cells (usually 1–2 ml) into a 2-ml Eppendorf tube, briefly centrifuge at top speed for 10 sec at room temperature in a microcentrifuge, and remove the medium supernatant by aspiration. Rapidly freeze cell pellets in crushed dry ice. This is the preinduction sample.

5. After a short time of induction (typically 8 to 10 min), an equal amount of medium containing 4% glucose is added. Immediately remove and quickly harvest an aliquot of cells as just described. This is the t_0 sample. [Transcriptional repression can also be carried out by temperature shift from 24° to 36° using a strain with rpb1-1 mutation (a temperature-sensitive allele of RNA polymerase II, as described earlier).] The addition of glucose to repress *GAL* promoter and temperature shift in rpb1-1 mutants to repress global transcription can be applied at the same time to get a tighter transcriptional repression.[9]

6. Additional cell aliquots are harvested at different time points (as described earlier).

7. mRNA is isolated from the cells. Typically, 10–40 μg of total RNA is separated through a 6% polyacrylamide/8 M urea gel (this gel is 20 cm long and 1 mm thick), typically run at 300 V for 7.5 hr at room temperature. These gels allow direct determination of the poly(A) tail length by comparing the size of the RNA before and after hybridizing to oligo(dT) and RNase H treatment. RNAs too long for the polyacrylamide gel can be cleaved with oligonucleotides specific for the 3' end and RNase H before loading.

Note. RNase H digestion of RNAs requires 10 μg RNA plus 300 ng oligonucleotide dried in a Speed Vac. Resuspend the pellet in 10 μl Hyb mix [25 mM Tris (pH 7.5), 1 mM EDTA, 50 mM NaCl], heat for 10 min at 68°, cool slowly to 30°, and spin down. Mix in 10 μl 2× RNase H buffer [40 mM Tris (pH 7.5), 20 mM MgCl$_2$, 100 mM NaCl, 2 mM dithiothreitol (DTT), 60 μg/ml bovine serum albumin (BSA), and 1 unit RNase H], and incubate at 30° for 1 hr. Stop the reaction by adding 130 μl stop mix (0.04 mg/ml tRNA, 20 mM EDTA and 300 mM sodium acetate). The RNA is prepared for the polyacrylamide gel by phenol/chloroform extracting, chloroform extracting, ethanol precipitating, washing the pellet with 70% (v/v) ethanol, and resuspending in 10 μl formamide gel loading dye (samples are heated to 100° for 3 min prior to loading).

Northern blotting of polyacrylamide gels with an oligonucleotide probe requires transfer of the RNA from the gel to a nitrocellulose membrane. After transfer, the membrane is washed in 0.1× SSC/0.1% SDS for 1 hr at 65°. The blot is incubated for at least 1 hr with prehybridization buffer (10× Denhardt's, 6× SSC, 0.1% SDS) at a temperature 15° below the T_m of the oligonucleotide probe. Then hybridize at the same temperature for at least 6 hr by adding the labeled probe to the blot. Wash the blot three times with 6× SSC, 0.1% SDS for 5 min at room temperature, and once for 20 min at 10° below the T_m of the oligonucleotide probe. Dry the blot and expose.

Analysis of Decay Pathways through Mutations in trans-Acting Factors

Using *S. cerevisiae* as a model organism, the stability of a particular mRNA can be examined in a strain deficient in a specific mechanism of decay. An alteration of an mRNAs half-life caused by a specific mRNA decay defect directly implicates that mRNA turnover pathway in the decay of the mRNA of interest. Analysis of mRNA stability in a strain deficient for a particular mRNA decay pathway should be coupled with direct analysis of the decaying mRNA. The combination of both types of data limits any possible indirect defects on mRNA decay.

Strains Defective in $5' \rightarrow 3'$ Decay. Mutations in several genes has been identified that affect mRNA decapping or $5' \rightarrow 3'$-exonucleolytic decay (see Table II). These include DCP1, which encodes the decapping enzyme,[18] DCP2, which encodes an activator of Dcp1p,[19] PAT1/MRT1 and LSM1, which encode components of a complex that binds mRNA and enhances mRNA decapping rates,[20–22] and XRN1, which encodes the $5' \rightarrow 3'$-exonuclease.[23] Mutations in any of these genes cause a stabilization of transcripts that are degraded by decapping and $5' \rightarrow 3'$-exonuclease digestion. Because Dcp1p and Xrn1p have clear biochemical roles, they are ideal factors to mutate in understanding the role of $5' \rightarrow 3'$ decay in the turnover of a mRNA.

In principle, xrn1Δ and dcp1Δ mutants could be used to distinguish endonucleolytic cleavage followed by $5' \rightarrow 3'$ decay from decapping followed by $5' \rightarrow 3'$ decay. In a case of endonucleolytic cleavage, a dcp1 mutation would be predicted to have no effect and an xrn1 mutant would accumulate the products of the initial cleavage. In contrast, in a case of decapping, a dcp1 mutation would stabilize the transcript of interest and an xrn1 mutant would accumulate full-length decapped RNA.

Strains Specifically Affecting mRNA Surveillance: Nonsense Decay. Three genes are known to be specifically required for the turnover of aberrant mRNAs: UPF1, UPF2 (NMD2), and UPF3.[24–27] Mutations in any of the UPF genes will alter the decay rates of nonsense encoding messages but not the decay rates of normal mRNAs. However, mutations in either DCP1 or XRN1 will alter the decay rates of both normal and aberrant mRNAs.[16,18] Therefore, a transcript that undergoes nonsense-mediated decay is stabilized in upf1, upf2, upf3, dcp1, and xrn1 mutants,

---

[18] C. A. Beelman, A. Stevens, G. Caponigro, T. LaGrandeur, L. Hatfield, D. Fortner, and R. Parker, *Nature* **382**, 642 (1996).
[19] T. Dunckley and R. Parker, *EMBO J.* **18**, 5411 (1999).
[20] L. Hatfield, C. A. Beelman, A. Stevens, and R. Parker, *Mol. Cell. Biol.* **16**, 5830 (1996).
[21] S. Tharun, W. He, A. E. Mayes, P. Lennertz, J. D. Beggs, and R. Parker, *Nature* **404**, 515 (2000).
[22] E. Bouveret, G. Rigaut, A. Shevchenko, M. Wilm, and B. Seraphin, *EMBO J.* **19**, 1661 (2000).
[23] C. L. Hsu and A. Stevens, *Mol. Cell. Biol.* **13**, 4826 (1993).
[24] F. He and A. Jacobson, *Genes Dev.* **9**, 437 (1995).
[25] P. Leeds, S. W. Peltz, A. Jacobson, and M. Culbertson, *Genes Dev.* **5**, 2303 (1991).
[26] P. Leeds, J. M. Wood, B. S. Lee, and M. Culbertson, *Mol. Cell. Biol.* **12**, 2165 (1992).
[27] B. S. Lee and M. R. Culbertson, *Proc. Natl. Acad. Sci. U.S.A.* **92**, 10354 (1995).

TABLE II
GENES THAT AFFECT mRNA DECAY

Pathway	Gene	Function of protein	Accumulating intermediate in loss of function mutant
$5' \rightarrow 3'$ and nonsense-mediated decay	DCP1	Decapping enzyme	Deadenylated capped RNA in $5' \rightarrow 3'$ decay
	DCP2	Activator of decapping enzyme	Deadenylated capped RNA in $5' \rightarrow 3'$ decay
	XRN1	$5' \rightarrow 3'$ exonuclease	Deadenylated decapped RNA in $5' \rightarrow 3'$ decay; Adenylated decapped RNA in NMD
$5' \rightarrow 3'$ decay	PAT1 (MRT1), LSM1-7	Modulate decapping activity	Deadenylated capped RNA
$5' \rightarrow 3'$ and $3' \rightarrow 5'$ decay	CCR4 CAF1 (POP2)	Deadenylation machinery	Adenylated capped RNA
$3' \rightarrow 5'$ decay	SKI6 (RRP41)		
	SKI4 (CSL4)	Component of exosome	
	RRP4		Poly(G)-$3'$ end fragment and a series of $3' \rightarrow 5'$ decay intermediates
	SKI2 SKI3 SKI7 SKI8	Modulate $3' \rightarrow 5'$ exonuclease activity	
Nonsense-mediated decay	UPF1	A putative RNA and ATP-dependent helicase modulates decapping activity	Not relevant—mRNA becomes degraded as normal mRNA
	UPF2 UPF3	Modulates decapping activity	

whereas a transcript degraded by the normal deadenylation dependent decapping and $5' \rightarrow 3'$ decay mechanism is stabilized in dcp1 and xrn1 mutants but not in upf mutants.

Strains Defective in Deadenylation. Specific yeast mutants can also be used to determine if an mRNA requires deadenylation for degradation. This is based on the demonstration that the major cytoplasmic deadenylase requires the products of the *CCR4* and *CAF1* genes.[28] In ccr4Δ or caf1Δ strains, mRNAs that require

---

[28] M. Tucker, M. A. Valencia-Sanchez, R. R. Staples, J. Chen, C. L. Denis, and R. Parker, *Cell* **104**, 377 (2001).

deadenylation show slower mRNA turnover.[28] In contrast, mRNAs that undergo deadenylation-independent decapping show no change in mRNA turnover rates (J. Coller and R. Parker, personal communication). This provides a way to determine if deadenylation is required for the degradation of a given mRNA.

Strains Defective in 3′→5′ Decay. It is also possible to use *trans*-acting mutations to determine if a transcript is degraded primarily in a 3′→5′ direction. Several proteins have been identified that are required for efficient cytoplasmic 3′→5′ decay of mRNA, including Ski2p, Ski3p, Ski4p, Ski6p/Rrp41p, Ski7p, Ski8p, and Rrp4p gene products.[14,29] Rrp4p, Skip6/Rrp41p, and Ski4p/Cs14p proteins are components of a multiprotein complex termed the exosome[30] and are likely to be part of the actual nucleolytic complex that can degrade the mRNA body 3′→5′.[14] Ski2p, Ski3p, Ski7p and Ski8p do not appear to be nucleases and are likely to modulate the activity of the exosome on mRNA substrates. Although all transcripts examined to date are not stabilized significantly in mutants solely defective in 3′→5′ decay, it is possible that there will be specific mRNAs or specific conditions wherein the 3′→5′ decay pathway is predominant. Thus, to determine if a mRNA is being degraded primarily 3′→5′, its decay rate should be examined in some combination of the ski2, ski3, ski4, ski7, ski8, rrp4, and ski6 mutants.

Combination of Approaches

A combination of the aforementioned approaches provides a powerful analysis of the mechanism responsible for the decay of a specific mRNA. A useful first step in combining the experimental approaches described in this chapter is to place the gene of interest under a regulatable promoter. This allows for both easy measurements of the mRNA decay rate and the ability to perform transcriptional pulse–chase experiments. Next, a poly(G) tract can be inserted into the 3′ UTR (or other location within the mRNA that does not perturb mRNA turnover) of the mRNA of interest. Insertion of the poly(G) tract allows for the detection of mRNA decay intermediates, which enables the directionality and role of deadenylation in the decay of a mRNA to be analyzed. Finally, examining the rate and mechanism of mRNA decay in a yeast strain defective for a specific mechanism of mRNA decay can link the turnover of a particular mRNA with a known decay pathway. However, in closing, it should be noted that all the approaches described here are specific for defining the decay of a mRNA of interest within the context of known mRNA decay pathways. The discovery of new mRNA decay pathways will rely on, different experimental approaches, thereby increasing the number of variables to consider.

[29] A. van Hoof, R. R. Staples, R. E. Baker, and R. Parker, *Mol. Cell. Biol.* **20,** 8230 (2000).
[30] P. Mitchell, E. Petfalski, A. Shevchenko, M. Mann, and D. Tollervey, *Cell* **91,** 457 (1997).

[39] Use of Green Fluorescent Protein in Living Yeast Cells

By KELLY TATCHELL and LUCY C. ROBINSON

Introduction

Although protein tags have been used for decades, the introduction of *Aequorea victoria* green fluorescent protein (GFP) into the repertoire of fusion proteins has created a revolution in their use. GFP fusion proteins can be produced and detected like any tagged protein and can be observed directly in live cells. We describe here some considerations when using GFP fusions for localization studies or other assays, discuss available GFP variants and fusion proteins, and briefly review methods of constructing fusion genes for expression in yeast. Reviews have been published that can be very helpful in designing and analyzing GFP fusions.[1,2] As for immunofluorescence, GFP fusion proteins can be visualized by imaging static populations of cells using a conventional epifluorescence microscope. This method has the obvious advantage that the fusion protein can be observed in many cells in a variety of growth states and conditions. However, in many cases, far more information can be gleaned from data if cells are imaged over a period of time. Dynamic changes in the location of a protein can be characterized readily by time-lapse imaging, whereas the same characterization requires considerably more work and some assumptions using static images. We present protocols here for time-lapse imaging starting with little more than an epifluorescence microscope and a sensitive charge-coupled device (CCD) camera. More sophisticated methods are also described for four-dimensional (4D) microscope imaging.

Construction and Use of GFP Fusions

General Considerations

The major concern when using any tag is whether the fusion protein retains function and proper localization. The concern seems greater for GFP than for small peptide tags, given the size of GFP (about 29 kDa) and its propensity to dimerize or aggregate *in vitro*.[3] However, crystals of monomeric GFP have

---

[1] K. F. Sullivan and S. A. Kay, eds., "Green Fluorescent Proteins," Vol. 58. Academic Press, New York, 1999.
[2] J. Thorner, S. D. Emr, and J. N. Abelson, eds., "Applications of Chimeric Genes and Hybrid Proteins," Vol. 327. Academic Press, San Diego, 2000.
[3] R. Y. Tsien, *Annu. Rev. Biochem.* **67,** 509 (1998).

been obtained for both the wild-type protein and an enhanced variant,[3] suggesting that GFP can exist as a monomer even at high concentrations. Also, there are few reports of GFP fusion protein aggregation, although anecdotal evidence shows that fusion proteins can be observed in artifactual aggregates. Examples include GFP–septin proteins, which can be observed in bright bars extending across the cytosol in a small proportion of often older cells (Y. Kweon and L. C. Robinson, unpublished observations, 1998). With GFP-tagged proteins, as with other tagged proteins, a test of function is essential, and function should be assessed as stringently as possible. Biological function provides confirmation that the fusion protein localizes to its normal location(s), although it may be present at additional locations. Thus, complementation of known traits of a null mutant when the fusion protein is expressed at normal levels is essential. It is also helpful to determine growth rates of cells expressing the fusion protein as the only source of that protein and to confirm any known biochemical activity(ies) and protein–protein interactions of the protein. Finally, fractionation studies are useful adjunct methods to confirm the validity of the localization of the GFP fusion.

Empirically, GFP addition has surprisingly little effect on function or on localization of many proteins, although there are reports of proteins for which fusion alters one or both properties.[4] It is likely that additional proteins are affected by GFP addition, but the negative results are not reported. For example, of seven proteins of interest in our laboratories for which GFP fusions have been constructed, two proteins no longer provided biological function even when expressed from a high-copy plasmid, and one of these gave an uninterpretable localization pattern. The lack of function curtailed any further experiments with the two fusion proteins. The other five fusion proteins were functional and were detectable at varying levels when expressed from their native chromosomal locations[5–7] (also L. C. Robinson, unpublished observations, 1997).

As with other observation methods, it is important to keep in mind that no matter how a protein is detected within cells, the major signal observed may not reflect the actual functional location for a given protein. For example, a relatively small amount of protein at its native location may provide function, while the majority of the protein either localizes elsewhere during its lifetime or is mislocalized due to the addition of a tag.

---

[4] A. Brachat, N. Liebundguth, C. Rebischung, S. Lemire, F. Scharer, D. Hoepfner, V. Demchyshyn, I. Howald, A. Dusterhoft, D. Mostl, R. Pohlmann, P. Kotter, M. N. Hall, A. Wach, and P. Philippsen, *Yeast* **16,** 241 (2000).

[5] A. Bloecher and K. Tatchell, *J. Cell Biol.* **149,** 125 (2000).

[6] L. C. Robinson, C. Bradley, J. D. Bryan, A. Jerome, Y. Kweon, and H. R. Panek, *Mol. Biol. Cell* **10,** 1077 (1999).

[7] H. R. Panek, E. Conibear, J. D. Bryan, R. T. Colvin, C. D. Goshorn, and L. C. Robinson, *J. Cell Sci.* **113,** 4545 (2000).

Variants of GFP

The wild-type GFP from *A. victoria* has relatively broad excitation and emission spectra with peaks centered at 396, 475, and 504 nm, respectively. Since the initial report that GFP functions in heterologous cells,[8,9] a plethora of GFP variants have been characterized that fold more efficiently and possess altered excitation and emission spectra, higher extinction coefficients, and higher quantum efficiencies. Different spectra allow two-color imaging with fusions to different proteins, while the other properties together make for a brighter signal. The variants are preferable to the native protein for use in fusion proteins, especially because wild-type GFP folds relatively slowly (hours at 24°), and folding is efficient only at lower temperatures. This property can cause diminution of the fluorescent signal at 30 and 37°, temperatures most often used for yeast growth. Some commonly used GFP variants are listed in Table I. For imaging single GFP fusions, the most useful variants are those based on the S65T variant. This variant is "brighter" than the wild type (two- to sixfold higher extinction coefficient than the wild type), it has a single major excitation peak at 489 nm, and the emission peak is red shifted with respect to the wild type (510 nm). The single 489-nm excitation peak allows S65T and derivatives to be used with standard blue/fluorescin isothiocyanate (FITC) fluorescence filter sets, as well as the common 488-nm line argon laser used for confocal microscopy. The S65T fluorophore folds more rapidly (approximately 30 min) and twice as efficiently as the wild type at 37°. These properties are very important for its use in temperature shift experiments with conditional yeast mutants, especially when observing low abundance fusion proteins for which it is necessary to maximize signal.

A number of GFP variants have been developed that are brighter than S65T and/or fold more efficiently at higher temperatures and have substantially different excitation and emission spectra. A common variant that we use routinely, F64L S65T, or EGFP, has an extinction coefficient slightly higher than that of the S65T GFP, excitation and emission spectra are virtually identical to S65T, and the folding efficiency at 37° is almost twice that for S65T.[3] Other mutants have even higher efficiency folding at 37° as well as higher quantum yield values. Twenty GFP mutants and their spectral characteristics have been tabulated.[3] Even more new variants have been developed, including the citrine YFP variant.[10]

For simultaneous imaging of two different fusion proteins, blue- and red-shifted variants of GFP can be used. With the appropriate filter sets, the two variants can be imaged separately with little overlap. EGFP and YFP (10C; yellow) can be imaged separately in the same cells using custom filter sets with narrow excitation

---

[8] M. Chalfie, Y. Tu, G. Euskirchen, W. W. Ward, and D. C. Prasher, *Science* **263**, 802 (1994).
[9] S. Inouye and F. I. Tsuji, *FEBS Lett.* **341**, 277 (1994).
[10] A. A. Heikal, S. T. Hess, G. S. Baird, R. Y. Tsien, and W. W. Webb, *Proc. Natl. Acad. Sci. U.S.A.* **97**, 11996 (2000).

TABLE I
GFP VARIANTS IN COMMON USE

GFP variant	Name	λ_{ex} (nm)	λ_{em} (nm)	Ref./Source
Wild type[a]	GFP	396, 475	504	Chalfie et al. (1994)[c]
S65T	S65T GFP	489	510	Heim et al. (1995)[d]
F64L, S65T	EGFP	488	508	Cormack et al. (1996)[e]
				Robinson et al. (1999)[f]/Clontech
S65G, S72A[b]	yEGFP	490	510	Cormack et al. (1997)[g]
F64L, S65T, Y66H, Y145F	EBFP	382	445	Heim and Tsien (1996)[h]/Clontech
Y66H, Y145F	P4-3	382	446	Heim and Tsien (1996)[h]
Y66W, N146I, M153T, V163A, N212K	W7	434	476	Heim and Tsien (1996)[h]
K26R, F64L, S65T, Y66W, N146I, M153T, V163A, N164H, N212K	CFP	434	474	Kohlwein (2000)[i]
F64L, S65T, Y66W, N146I, M153T, V163A	ECFP	430–437	475–478	Heim and Tsien (1996)[h]
				Miyawaki et al. (1997)[j]/Clontech
S65G, V68L, S72A, T203Y	YFP/10C/EYFP	514	527	Ormö et al. (1996)[k]/Clontech
S65G, S72A, K79R, T203Y	Topaz	514	527	Cubitt et al. (1999)[l]

[a] Fluorophore formation is temperature dependent.
[b] Yeast (C. albicans)-optimized codons.
[c] M. Chalfie, Y. Tu, G. Euskirchen, W. W. Ward, and D. C. Prasher, *Science* **263,** 802 (1994).
[d] R. Heim, A. B. Cubitt, and R. Y. Tsien, *Nature* **373,** 663 (1995).
[e] B. P. Cormack, R. H. Valdivia, and S. Falkow, *Gene* **173,** 33 (1996).
[f] L. C. Robinson, C. Bradley, J. D. Bryan, A. Jerome, Y. Kweon, and H. R. Panek, *Mol. Biol. Cell* **10,** 1077 (1999).
[g] B. P. Cormack, G. Bertram, M. Egerton, N. A. Gow, S. Falkow, and A. J. Brown, *Microbiology* **143,** 303 (1997).
[h] R. Heim, and R. Y. Tsien, *Curr. Biol.* **6,** 178 (1996).
[i] S. D. Kohlwein, *Microsc. Res. Tech.* **51,** 511 (2000).
[j] A. Miyawaki, J. Llopis, R. Heim, J. McCaffery, J. Adams, M. Ikura, and R. Tsien, *Nature* **388,** 882 (1997).
[k] M. Ormö, A. B. Cubitt, K. Kallio, L. A. Gross, R. Y. Tsien, and S. J. Remington, *Science* **273,** 1392 (1996).
[l] A. B. Cubitt, L. A. Woollenweber, and R. Heim, *Methods Cell Biol.* **58,** 19 (1999).

and emission filters. Although this has been useful,[5,7] the signal levels through these filters are relatively low due to the narrow bandpass filters, and significant bleedthrough can be observed for GFP using the YFP filter set.

Blue-shifted variants (BFPs) with excitation spectra in the UV range and emission in the blue range have also been developed and could, in principle, be used for simultaneous imaging of two fusion proteins. However, the BFPs are not as bright as the less blue-shifted W7 (cyan/CFP) cyan variant and are more subject to photobleaching. Furthermore, the use of BFPs is limited because excitation with UV light can be toxic to cells, especially with a long exposure time(s). Therefore, the cyan and 10C (yellow/YFP) yellow variants are superior partners for double-labeling applications, as they can be imaged with very little bleedthrough. Finally, there are variants selected for their usefulness in flow cytometry and fluorescence-activated cell sorting (FACS) applications (e.g., Ref. 11).

The identification of a new group of fluorescent proteins distantly related to GFP has been reported.[12] This included a protein from a coral genus, *Discosoma* sp., that is now termed DsRed. DsRed has an excitation peak centered around 558 nm and maximal emission in the red at 583 nm. These properties make a standard rhodamine filter set suitable for observation of DsRed. Its extinction coefficient has been reported to be as high or higher than that of *A. victoria* GFP.[12,13] A vector containing the DsRed coding sequence is currently distributed by Clontech, but the folding time of the fluorophore in DsRed is far too long for use in yeast (at least 48 hr for 90% fluorophore maturation[13]). Although Baird *et al.*[13] found no more rapid folding variants on random mutagenesis of DsRed, Bevis and Glick[14] have developed faster folding DsRed variants by a mixture of random and targeted mutagenesis, and some of these have folding halftimes under 60 min. These variants are available for use in yeast as organelle markers, but a further complication for all applications except use as an organelle marker is that the DsRed exists as an obligate tetramer.[10,13] Because all of the Bevis and Glick variants still oligomerize, they may not be useful for protein tagging. Bevis and Glick[14] are currently carrying out further targeted mutagenesis with the goal of identifying monomeric DsRed variants that retain rapid folding times and brightness levels comparable to GFP. A bright monomeric DsRed would make the use of triple-labeling protocols in combination with CFP and YFP possible in yeast, as described by Clontech for mammalian cells (gfp.clontech.com), as well as double-label applications with EGFP.

The ECFP and EYFP variants have been used[15] for the detection of *in vitro* and *in vivo* fluorescence resonance energy transfer (FRET), which reflects protein–protein interaction between two tagged proteins. In the FRET assay, energy transfer can occur from a donor fluorophore to an acceptor fluorophore following illumination with light excitatory to the donor, if the emission spectrum of the donor

---

[11] B. P. Cormack, G. Bertram, M. Egerton, N. A. Gow, S. Falkow, and A. J. Brown, *Microbiology* **143**, 303 (1997).
[12] M. V. Matz, A. F. Fradkov, Y. A. Labas, A. P. Savitsky, A. G. Zaraisky, M. L. Markelov, and S. A. Lukyanov, *Nature Biotechnol.* **17**, 969 (1999).
[13] G. S. Baird, D. A. Zacharias, and R. Y. Tsien, *Proc. Natl. Acad. Sci. U.S.A.* **97**, 11984 (2000).
[14] B. Bevis and B. Glick, personal communication, 2001.
[15] R. Heim and R. Y. Tsien, *Curr. Biol.* **6**, 178 (1996).

overlaps the excitation spectrum of the acceptor fluorophore.[16,17] FRET is monitored by a decrease in emission of the donor and/or an increase in emission of the acceptor on illumination with wavelengths excitatory to the donor fluorophore. For FRET, the two fluorophores must be very close in space (10–100 Å).[17] With CFP to YFP transfer, the FRET distance is approximately 50 Å between fluorophores,[3] corresponding to approximately 30 Å between CFP and YFP molecules. The two fluorophores must also be oriented properly relative to one another for efficient energy transfer. Thus, improper orientation of the two fluorophores in two fusion proteins will abrogate the possibility of using FRET as an assay of interaction. Analysis of FRET requires special equipment, e.g., a scanning fluorometer with a suitable cell for culture analysis or a digital imaging setup that includes a FRET-specific filter set or some combination thereof. This analysis can be a powerful tool for studying protein–protein interactions *in vivo* and has been employed successfully in yeast (see later).

Another variable between commonly used GFP variants is codon usage. Although no problems have been reported for yeast expression of GFP with native *A. victoria* codons or bacterially optimized codons (as opposed to plants[18]), GFP alleles with *Candida*-optimized codon selection (yEGFP) have been developed and are in current use.[11] Commercially available GFP vectors from Clontech, Palo Alto, CA, carry the enhanced alleles of most variants for use in nonmammalian systems. We use alleles optimized for bacterial expression, and the resulting fusion proteins are expressed at levels comparable to wild-type proteins. There are also commercially available GFP fusion vectors (Clontech) that are codon optimized for expression in mammalian cells, the so-called humanized variants. Humanized variants are said to not express well in yeast but, to our knowledge, there are little published data on the subject.

Construction of GFP Fusion Proteins

To be most useful, the GFP fusion protein must retain the properties of the untagged protein when expressed at comparable levels. It is remarkable how many proteins retain biological activity on fusion of the GFP moiety. GFP is extremely stable and folds readily (within approximately 30 min for the S65T variant), even with additional residues appended at either its amino or carboxyl terminus. Addition of GFP to a protein of interest, as for other protein tags, is most commonly at the amino or carboxyl terminus, and the optimal position of addition must be determined empirically for any given protein. Attention must be given to sequences on the protein of interest that could limit the site of addition. For example, signal

[16] L. Stryer, *Annu. Rev. Biochem.* **47,** 819 (1978).
[17] R. M. Clegg, *in* "Fluorescence Imaging Spectroscopy and Microscopy" (X. F. Wang and B. Herman, eds.), p. 179. Wiley, New York, 1996.
[18] J. Haseloff and B. Amos, *Trends Genet.* **11,** 328 (1995).

sequences for ER translocation, organelle targeting or retention, lipid or prenyl group attachment, or other membrane-anchoring sequences prohibit addition at these sites. If the structure–function relationship of the target protein is well defined, it may be possible to add GFP between functional domains of a protein.

Surprisingly, although our laboratories and others have generated a number of functional fusions without any spacer peptide between GFP and the target protein, the presence of such a spacer can affect the stability of a fusion protein *in vivo*. Prescott *et al.*[19] described C-terminal fusions of GFP to four subunits of the mitochondrial ATPase. In each case, a spacer peptide was encoded by vector sequences, and the effects of a short (5–7 amino acids) versus a long (23–28 amino acids) spacer on fusion protein abundance and prevalence of degradation products were examined. The authors reported that for two of the subunits, the short spacer results in high susceptibility to proteolysis, whereas in a third case, the short spacer appears to result in a slightly more stable protein. Therefore, if a fusion protein is not functional or less abundant than the wild-type counterpart, one possible solution, in addition to fusion to the other end of the protein of interest, is to encode a spacer peptide between the protein of interest and GFP.

GFP addition can be accomplished either by introducing GFP into a native or engineered restriction enzyme site within the coding sequence or through the use of a number of GFP fusion vectors. We generally use polymerase chain reaction (PCR) to add convenient restriction sites to the GFP gene and assess function of the cloned product by expression from vector pRSETB (Invitrogen) in *Escherichia coli* strain BL21-DE3 (Invitrogen, Carlsbad, CA) monitoring the green fluorescence of colonies by a handheld UV lamp. GFP fusion vectors can be very useful, especially when constructing large numbers of fusions. Many of these involve PCR-based addition of sequences of interest to various GFP alleles for chromosomal integration, and some include additional tags for protein detection. Kohlwein[20] has collected a list of methods for GFP fusion (as well as other protein tags for fusion) at the N or C terminus of the protein of interest, including both episomal vectors and chromosomal integration methods. One method for chromosomal tagging of a gene of interest with S65T GFP uses a heterologous HIS3 marker or a kanMX marker to select for integrative transformants.[21] Here, vectors carrying GFP-marker modules are used as templates for PCR with primers carrying sequences complementary to the 3′ end of the gene of interest. The PCR product is then integrated at the chromosomal locus. A similar method with the advantage of an excisable integration marker has been described by Kohlwein's group for chromosomal tagging to yield a fusion with GFP at the N terminus.[22] In this method, a kanMX4 cassette

---

[19] M. Prescott, S. Nowakowski, P. Nagley, and R. J. Devenish, *Anal. Biochem.* **273,** 305 (1999).
[20] S. D. Kohlwein, *Microsc. Res. Techn.* **51,** 511 (2000).
[21] A. Wach, A. Brachat, C. Alberti-Segui, C. Rebischung, and P. Philippsen, *Yeast* **13,** 1065 (1997).
[22] B. Prein, K. Natter, and S. D. Kohlwein, *FEBS Lett.* **485,** 29 (2000).

with yEGFP is targeted for integration at the gene of interest by the addition of a chromosomal sequence by PCR. The kanMX cassette is flanked by *loxP* sites, which can be excised on galactose induction of synthesis of Cre recombinase that is expressed from a plasmid.

Once constructed, the GFP fusion can be expressed from a plasmid or from the native chromosomal location under the control of the endogenous or a heterologous promoter. We have had good luck with centromere-based vectors and chromosomal integration by gene replacement for most fusions. For Yck2p and Glc7p, the only obvious difference between chromosomal expression and expression from a centromere-based vector is the uniform expression in the former case. Uniform expression is especially important for time-lapse studies and quantitation of fluorescence levels in a population.

If the protein of interest is expressed at low levels, episomal expression, from standard 2-μm plasmids or plasmids with strong promoters such as *GAL1* or *ADH1*, may be considered. However, as with any overexpression study, it is critical to determine whether overexpression of the fusion protein affects its localization. For example, an epitope or GFP-tagged Yip1 protein expressed from its native chromosomal locus is located predominantly at Golgi membranes, as assayed by cell fractionation[23] or fluorescence microscopy, although the fluorescence signal is very low (L. C. Robinson, unpublished observations, 2000). When overexpressed even from a centromere-based vector, tagged Yip1p becomes detectable more readily but is concentrated at endoplasmic reticulum (ER) membranes. Similarly, overexpression of Yck2p from the *GAL1* promoter not only labels the peripheral plasma membrane, but also internal membranes, including the vacuolar membrane.[7] Such mislocalization may reflect either buildup at normally transient sites or saturation of normal sorting machinery. This can provide useful information about normal sorting of the protein, but is not reliable information about normal location within cells. A possible alternative to multicopy or strong promoter expression in such cases is fusion of two or three tandem copies of GFP to the protein of interest. Such a construct can increase the quantity of signal in a linear fashion.[14,24,25]

Applications for GFP Fusion Proteins

Localization

As for immunofluorescence localization studies, subcellular location of a GFP fusion protein should be assessed using markers for colocalization. Such markers can include proteins for which antibodies suitable for immunofluorescence exist or can include fusions of different GFP variants to proteins of known subcellular

---

[23] X. Yang, H. T. Matern, and D. Gallwitz, *EMBO J.* **17,** 4954 (1998).
[24] I. R. Adams and J. V. Kilmartin, *J. Cell Biol.* **145,** 809 (1999).
[25] O. W. Rossanese, C. A. Reinke, B. J. Bevis, A. T. Hammond, I. B. Sears, J. O'Connor, and B. S. Glick, *J. Cell Biol.* **153,** 47 (2001).

location. In the former case, anti-GFP antibodies can be used in dual-label immunofluorescence. Highly specific monoclonal or polyclonal antibodies can be obtained commercially (Clontech, Molecular Probes, Eugene, OR and others) or from colleagues. These should recognize *A. victoria* variants but not recognize the *Discosoma* DsRed (see Clontech for DsRed antibodies). Also, it is often the case that GFP fluorescence survives mild fixation treatments (10–15 min, sometimes at low temperature), permitting a discernible signal for a fusion protein. Such fixation treatments include 2% (v/v) ethanol, 3–4% (v/v) formaldehyde or paraformaldehyde, and even 0.25% glutaraldehyde.[5,26] For those proteins that survive fixation, their intrinsic GFP fluorescence can be compared with immunodetection of a marker or with the pattern of any compatible fluorescent dye (e.g., Hoechst or 4,6-diamidino-2-phenylindole (DAPI) for chromatin, rhodamine–phalloidin for actin filaments). However, this is not always possible and, in most cases, colocalization with other, spectrally distinct, GFP fusion proteins (e.g., ECFP and EYFP) in living cells is preferable. The double GFP variant labeling method has the distinct advantages that live cells are examined and that the availability of the colocalization marker is unlimited, as opposed to the limited supply of any given polyclonal antibody.

Table II lists representative GFP fusion proteins that decorate most organelles and structures within yeast cells. Table II is not by any means exhaustive, but lists many characterized and reliable fusions. Additional GFP fusions are listed elsewhere.[20] Fusions chosen for Table II specifically label the corresponding organelle or structure. Actin cables are conspicuously absent from this list, for which no reliable GFP fusion marker as yet exists. Table III lists some useful websites for GFP studies.

Fluorescence Resonance Energy Transfer

Test of colocalization is often a prelude to test of interaction. GFP fusions can be ideal for this application also using FRET assay (described earlier). Several examples of this application have been reported for yeast, including an important demonstration that the mating pheromone serpentine receptor Ste2p exists in the membrane as an oligomer,[27] and an elegant study of interactions between nuclear transport receptors and nucleoporins.[28] Both studies utilized CFP as the donor fluorophore with YFP as the acceptor fluorophore. In the former study, a truncated version of the well-characterized Ste2p G-protein-coupled receptor that fails to undergo pheromone-induced endocytosis was fused to CFP or YFP at its C terminus. A similar fusion to GFP was functional by biochemical and genetic

---

[26] O. W. Rossanese, J. Soderholm, B. J. Bevis, I. B. Sears, J. O'Connor, E. K. Williamson, and B. S. Glick, *J. Cell Biol.* **145,** 69 (1999).
[27] M. C. Overton and K. J. Blumer, *Curr. Biol.* **10,** 341 (2000).
[28] M. Damelin and P. A. Silver, *Mol. Cell* **5,** 133 (2000).

TABLE II
GFP FUSIONS DECORATING SPECIFIC INTRACELLULAR COMPARTMENTS AND STRUCTURES

Compartment or structure	Gene/sequence fused to GFP	Ref.
Mitochondria	OSCP/ATP5; N. crassa Fo ATPase subunit 9 (Su9) presequence (vectors with both S65T and P4-3); TRX3	Prescott et al. (1997)[a]; Westermann and Neupert (2000),[b] http://depts.washington.edu/~yeastrc/fm_home5.htm
Nuclear membrane	SEC63 (ER membrane)	Lippincott and Li (2000)[c]
Nuclear pores	NUP1; NUP188; NUP49 NIC96	Damelin and Silver (2000)[d]; Bucci and Wente (1997)[e]; http://depts.washington.edu/~yeastrc/fm_home5.htm
Nucleus chromatin	NLS (nuclear localization signal; both CFP and YFP fusions)	Damelin and Silver (2000)[d]
	HHF2	http://depts.washington.edu/~yeastrc/fm_home5.htm; Hoepfner et al. (2000)[f]
Nucleolus	SSF1	Kim and Hirsch (1998)[g]
Endoplasmic reticulum	ERG4; HDEL (ER retention signal)	Hampton et al. (1996)[h]; Rossanese et al. (2001)[i]
Golgi		
Early	YPT1	Rossanese et al. (2001)[i]
Late	SEC7; RIC1; RGP1	Seron et al. (1998)[j]; Rossanese et al. (2001)[i]; Panek et al. (2000)[k]
Early + late	SEC21	Rossanese et al. (2001)[i]
Late + endosome	KEX2; TLG2	http://depts.washington.edu/~yeastrc/fm_home5.htm; Panek et al. (2000)[k]
COPII vesicles	SEC13, SEC23, SEC24, SEC31	Rossanese et al. (1999)[l]
Vacuolar membrane	VAC8	Pan et al. (2000)[m]
Peroxisomes	SKL (targeting signal)	Gurvitz et al. (1998)[n]
Plasma membrane	HXT2; YCK2; RAS2; STE2	Kruckeberg et al. (1999)[o]; Robinson et al. (1999)[p]; Boyartchuk et al. (1997)[q]; Stefan and Blumer (1999)[r]
Cell wall	CWP1, CWP2	Ram et al. (1998)[s]
Actin cortical patches	CAP2, SAC6	Waddle et al. (1996)[t]
Microtubules	TUB1	http://depts.washington.edu/~yeastrc/fm_home5.htm
Astral microtubules	DYN1	Shaw et al. (1997)[u]
Spindle pole body	SPC42; SPC29	Adams and Kilmartin (1999)[v] http://depts.washington.edu/~yeastrc/fm_home5.htm
Chromosomes—centromeres	lacI-NLS (with array of lacO sites)	Straight et al. (1996)[w]
Septin ring; but neck	CDC3, CDC12; BNI4	Robinson et al. (1999)[p]; http://depts.washington.edu/~yeastrc/fm_home5.htm
Actomyosin ring	MYO1; CYK2	Lippincott and Li (1998, 2000)[c,x]

TABLE II (continued)

Compartment or structure	Gene/sequence fused to GFP	Ref.
Plasma membrane-destined vesicles	SNC1	Lewis et al. (2000)[y]
Polarized growth sites: mitotic pheromone response	CDC24, CDC42, PKC1 STE5	Toenjes et al. (1999)[z]; Andrews and Stark (2000)[aa] Pryciak and Huntress (1998)[bb]

[a] M. Prescott, A. Lourbakos, M. Bateson, G. Boyle, P. Nagley, and R. J. Devenish, *FEBS Lett.* **411**, 97 (1997).
[b] B. Westermann and W. Neupert, *Yeast* **16**, 1421 (2000).
[c] J. Lippincott and R. Li, *Exp. Cell Res.* **260**, 277 (2000).
[d] M. Damelin and P. A. Silver, *Mol. Cell* **5**, 133 (2000).
[e] M. Bucci and S. R. Wente, *J. Cell Biol.* **136**, 1185 (1997).
[f] D. Hoepfner, A. Brachat, and P. Philippsen, *Mol. Biol. Cell* **11**, 1197 (2000).
[g] J. Kim and J. P. Hirsch, *Genetics* **149**, 795 (1998).
[h] R. Y. Hampton, A. Koning, R. Wright, and J. Rine, *Proc. Natl. Acad. Sci. U.S.A.* **93**, 828 (1996).
[i] O. W. Rossanese, C. A. Reinke, B. J. Bevis, A. T. Hammond, I. B. Sears, J. O'Connor, and B. S. Glick, *J. Cell Biol.* **153**, 47 (2001).
[j] K. Seron, V. Tieaho, C. Prescianotto-Baschong, T. Aust, M.-O. Blondel, P. Guillaud, G. Devilliers, O. W. Rossanese, B. S. Glick, H. Riezman, S. Keranen, and R. Haguenauer-Tsapis, *Mol. Biol. Cell* **9**, 2873 (1998).
[k] H. R. Panek, E. Conibear, J. D. Bryan, R. T. Colvin, C. D. Goshorn, and L. C. Robinson, *J. Cell Sci.* **113**, 4545 (2000).
[l] O. W. Rossanese, J. Soderholm, B. J. Bevis, I. B. Sears, J. O'Connor, E. K. Williamson, and B. S. Glick, *J. Cell Biol.* **145**, 69 (1999).
[m] X. Pan, P. Roberts, Y. Chen, E. Kvam, N. Shulga, K. Huang, S. Lemmon, and D. S. Goldfarb, *Mol. Biol. Cell* **11**, 2445 (2000).
[n] A. Gurvitz, H. Rottensteiner, H. Hamilton, H. Ruis, A. Hartig, I.W. Dawes, and M. Binder, *Histochem. Cell Biol.* **110**, 15 (1998).
[o] A. L. Kruckeberg, L. Ye, J. A. Berden, and K. vanDam, *Biochem. J.* **339**, 299 (1999).
[p] L. C. Robinson, C. Bradley, J. D. Bryan, A. Jerome, Y. Kweon, and H. R. Panch, *Mol. Biol. Cell* **10**, 1077 (1999).
[q] V. L. Boyartchuk, M. N. Ashby, and J. Rine, *Science* **275**, 1796 (1997).
[r] C. J. Stefan, and K. J. Blumer, *J. Biol. Chem.* **274**, 1835 (1999).
[s] A. F. Ram, H. Van den Ende, and F. M. Klis, *FEMS Microbiol. Lett.* **162**, 249 (1998).
[t] J. A. Waddle, T. S. Karpova, R. H. Waterston, and J. A. Cooper, *J. Cell Biol.* **132**, 861 (1996).
[u] S. L. Shaw, E. Yeh, K. Bloom, and E. D. Salmon, *Curr. Biol.* **7**, 701 (1997).
[v] I. R. Adams, and J. V. Kilmartin, *J. Cell Biol.* **145**, 809 (1999).
[w] A. F. Straight, A. S. Belmont, C. C. Robinett, and A. W. Murray, *Curr. Biol.* **6**, 599 (1996).
[x] J. Lippincott and R. Li, *J. Cell Biol.* **140**, 355 (1998).
[y] M. J. Lewis, B. J. Nichols, C. Prescianotto-Baschong, H. Riezman, and H. R. Pelham, *Mol. Biol. Cell* **11**, 23 (2000).
[z] K. A. Toenjes, M. M. Sawyer, and D. I. Johnson, *Curr. Biol.* **9**, 1183 (1999).
[aa] P. D. Andrews, and M. J. Stark, *J. Cell Sci.* **113**, 2685 (2000).
[bb] P. M. Pryciak and F. A. Huntress, *Genes Dev.* **12**, 2684 (1998).

TABLE III
SOME USEFUL WEB SITES FOR GFP USERS

http://gfp.clontech.com	Clontech's Living Colors vectors and information, DsRed vectors and information, as well as other GFP information and references
http://www.mips.biochem.mpg.de/proj/yeast/info/tools/hegemann/gfp.html	Yeast GFP fusion vectors
http://depts.washington.edu/~yeastrc	Yeast GFP fusion vectors and strains expressing GFP fusions; microscopy
http://pantheon.cis.yale.edu/~wfm5/grp_gateway.html	GFP applications—links to other pages regarding GFP fusion use and microscopy
http://www.chroma.com/ontercontact.html	Download site for GFP filter set brochure

criteria.[29] Emission spectra were collected by fluorometry from cells expressing either or both fusions (or from plasma membranes from such cells) in response to CFP excitation wavelengths. In the presence of both fusions, a difference spectrum revealed FRET, probably due to receptor oligomerization. The FRET signal decreased when wild-type Ste2p, but not the glucose transporter Hxt2p, was expressed with both fusions. Although the truncated receptors are not internalized in response to pheromone, this defect was corrected by coexpression of wild-type Ste2p, supporting the conclusion from the FRET analysis that the Ste2p molecules self-associate *in vivo*.

The second study[28] identified interactions between nucleoporins and two nuclear transport receptors. ECFP and EYFP fusions were expressed from their respective chromosomal loci as the sole source of corresponding protein and were tested for expression and biological function. FRET was assayed by digital microscopy using a FRET filter set with an ECFP excitation filter and an EYFP emission filter. Fluorescence in the FRET filter set was quantitated across the nuclei of multiple cells to provide values for protein only at the appropriate subcellular compartment. Two nuclear transport proteins and 13 nucleoporins tested for interaction by FRET analysis were not well characterized structurally; however, the FRET signal was observed for a subset of nucleoporins with each transport receptor. Only three nucleoporin fusions failed to show a FRET signal with one or both transport receptors, but it is not clear whether these proteins interact *in vivo*. The biological significance of FRET-indicated interactions was confirmed for a novel interacting pair by coimmune precipitation.

Other Reporter Applications

GFP fusions can be used for many applications in addition to localization studies. Because there are so many fusion proteins available for marking organelles

---

[29] C. J. Stefan and K. J. Blumer, *J. Biol. Chem.* **274**, 1835 (1999).

and processes, one important application is the use of fusion proteins to monitor the integrity of structures and cellular processes as a way of characterizing mutants. Some examples of cellular processes that can be tested simply by observation of specific GFP fusions in living cells are chromosome segregation, nuclear migration and division, spindle pole body separation, organelle inheritance, endocytosis, actomyosin ring contraction, septin ring formation, polarized secretion, vacuolar targeting, GPI anchor addition, actin cortical patch movement, and polarization. Obviously any process can be monitored for which a suitable reporter fusion protein is available.

GFP can be an excellent monitor of gene expression, including kinetic analysis of promoter induction, as it is inherently quite stable in yeast cells, with a half-life of 7 to 70 hr at 30° depending on the variant. The use of GFP to monitor the kinetics and dose–response of galactose induction of the *GAL1* promoter has been reported.[30] However, somewhat surprisingly, GFP fusions can also be used to assess regulated expression. Although GFP itself is highly stable, GFP fusions tend to take on the stability that is characteristic of the fused protein. Chromosome-targeted fusions of yEGFP with C-terminal PEST-rich sequences from the G_1 cyclin Cln2p have been constructed that allow assessment of regulated expression due to the relatively short 30-min half-life of the fusion protein.[31] Expression of these unstable GFPs from any promoter can be analyzed by flow cytometry, as well as by fluorescence microscopy or immunoblot analysis. Another example of an unstable GFP fusion is a functional fusion of the transmembrane phosphate permease Pho84p to S65TGFP. This fusion allowed direct visualization of phosphate-regulated plasma membrane translocation and vacuolar degradation of this enzyme.[32]

As mentioned, GFP is ideal for use with laser excitation so GFP fluorescence can be quantitated for cell populations by flow cytometry analysis and can be used with FACS to obtain populations of cells expressing a desired marker. Fluorescence can also be quantitated by more conventional methods, including analysis of digital images, as well as spectrophotometric methods. An assay for genotoxicity has been reported that uses yEGFP expressed under the control of the *RAD54* and *RNR2* promoters.[33] Here, the fluorescence of GFP expressed in cultures treated with known and potential DNA-damaging agents was quantitated using a plate reader.

---

[30] J. Li, S. Wang, W. J. VanDusen, L. D. Schultz, H. A. George, W. K. Herber, H. J. Chae, W. E. Bentley, and G. Rao, *Biotechnol. Bioeng.* **70,** 187 (2000).
[31] C. Mateus and S. V. Avery, *Yeast* **16,** 1313 (2000).
[32] J. Petersson, J. Pattison, A. L. Kruckeberg, J. A. Berden, and B. L. Persson, *FEBS Lett.* **462,** 37 (1999).
[33] V. Afanassiev, M. Sefton, T. Anantachaiyong, G. Barker, R. Walmsley, and S. Wolfl, *Mutat. Res.* **464,** 297 (2000).

GFP fusions can also be expressed for a variety of reasons in well-characterized mutants affecting specific physiological processes. For example, we have used a number of *sec* mutants in conjunction with inducible expression to assess the trafficking of the GFP–Yck2 fusion protein *in vivo* (J. Bryan and L. C. Robinson, unpublished results, 1999). GFP–Yck2p synthesis was induced at the restrictive temperature for each of the mutants; in each case, internal signal was observed rather than plasma membrane fluorescence. This supported the idea that the polarized plasma membrane distribution of Yck2p,[6] a peripheral membrane protein, reflects delivery via the secretory pathway. We have also used *cdc3* and *cdc12*, mutants in the septin proteins, to assess the requirement for an intact septin ring for assembly of Yck2p[6] and of Glc7p[5] at the bud neck.

Time-Lapse Imaging

One of the most powerful applications for which GFP fusion is ideal is the imaging of protein localization in living cells over time. Static images provide only a limited view, even when imaging an asynchronous population of cells, whereas time lapse can reveal dynamic changes as they occur. Time-lapse studies are especially powerful for relating changes in localization to known cellular events. Furthermore, not only can the dynamic localization of a protein of interest be followed, but also the mechanics of cellular processes, using GFP fusions as markers for those processes. The process of time-lapse imaging can be relatively simple, imaging a single focal plane manually at desired time intervals, or can be complex, imaging multiple focal planes, to cover the entire cell, at desired time intervals. The complexity of a given application depends on a variety of factors, such as the location within the cell of the fusion protein/event, the abundance of the signal, and the time span of the process being followed. The next section describes hardware, software, and protocols used for time-lapse imaging.

Method

Overview

This section presents the components of the microscope system used for localization and time-lapse studies and mentions alternative systems that have been used successfully by others. The importance of each component for two-color imaging and time-lapse studies is discussed. We then describe in detail the protocols used for these studies. Much of the information is also useful for static imaging. A confocal scanning microscope can also be used for GFP imaging and many are equipped to perform time-lapse imaging. However, for the experiments described here, there is no advantage to using a confocal microscope. In most cases, rates of photobleaching are higher with a confocal microscope and the long hours required for imaging can burden a multiuser confocal facility.

Hardware/Software Requirements

Microscope. The microscope system used is diagrammed in Fig. 1. It is based on a standard epifluorescence microscope (Olympus AX70) equipped with differential interference contrast (DIC) optics. For most studies, we use an Olympus UPlan Fluorite 100× (numerical aperture (NA 1.3) objective. Sixty to 63× or 100× objectives with 1.4 NA are usually recommended for imaging, but they

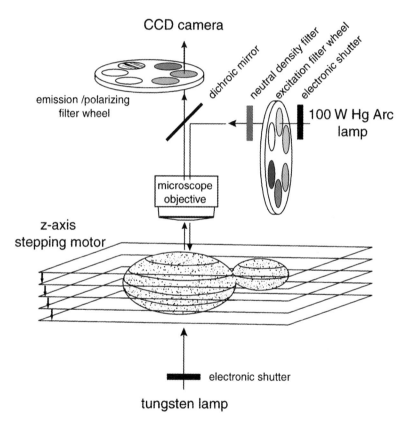

FIG. 1. Schematic diagram of the epifluorescence microscope. Components added to the standard microscope include the following: LEP (Ludl Electronics Products) shutter on the 100-W tungsten lamp; LEP high-speed six-position filter wheel/shutter on the 100-W Hg Arc lamp. This filter contains excitation filters for CFP/YFP imaging; LEP high-speed filter wheel shutter mounted below the camera. This filter wheel contains the polarizing filter for the DIC optics and emission filters for CFP/YFP imaging. A custom mount was required to position the filter set below the camera; An LEP stepping motor attached to the focus controller for computer control of the z-axis position. Filter wheels, shutters, and focus controller are controlled through a 12-inch-wide LEP MAC2000 controller. Images are collected by a MicroMax 1300YHS cooled CCD camera (Roper Scientific). IPLab software (Scanalytics) with custom scripts is used to control the MAC2000 and camera.

are twice as expensive as our UPlan Fluorite objective and our direct comparisons with 1.35 and 1.4 N.A. objectives have not shown much of a difference in terms of brightness. We suggest that users directly compare objectives before buying. The same is true for the microscope brand; Leica, Olympus, Nikon, and Zeiss all have systems with excellent DIC and epifluorescence optics. Our microscope was retrofitted with the following components: an electronic shutter in the light path for the tungsten lamp, a high-speed six-position filter wheel and shutter positioned between the microscope body and the Hg arc lamp that contains excitation filters, another filter wheel and shutter positioned before the camera that holds emission filters and a polarizing filter for DIC optics, and a z-axis stepping motor attached to the fine focus knob to allow computer control of the focus position. These additions were purchased directly from Ludl Electronics Products Ltd (LEP) (200 Brady Ave., Hawthorne, NY 10532) but can also be obtained from companies specializing in image analysis. Appropriate fittings are available for all common microscope brands. The shutters, filter wheels, and z-axis stepping motor are controlled via the LEP MAC 2000 controller, which in turn is controlled via a serial interface to the computer. These components are modular and can be added at different times as funds become available. For example, we first purchased a focus controller and a shutter to control the excitation light, and later added the filter wheels to allow for combination DIC/GFP imaging and two-color GFP imaging. If you choose to add to the system piecemeal, it is important to initially purchase a MAC2000 controller that is large enough to accommodate the additional devices (12-inch-wide controller, not the 6-inch-wide controller).

We use a standard 100-W Hg arc lamp for the excitation light source but a xenon lamp can also be used and has been reported to reduce cell phototoxicity in *Caenorhabditis elegans*.[34] Neutral density filters on two sliders on the microscope are used to reduce the fluorescence excitation light to 1–6% transmission. When imaging a single GFP fusion, we use a Chroma (www.chroma.com) 41001 filter set positioned in the body of the microscope. This filter set is optimized for the S65T variant. Shaw and colleagues[35] and Salmon and co-workers[36] have described a similar microscope system, and those interested in optimum resolution using DIC optics are encouraged to read Salmon *et al.*[36] A complete turnkey imaging system can be purchased from a number of sources, including DeltaVision from Applied Precision (www.api.com) or Everest from Intelligent Imaging Innovations (www.intelligent-imaging.com).

Camera. A sensitive camera is important for most GFP fusion studies and is essential for successful time-lapse imaging. There are many places where one

---

[34] J. Waddle, personal communications, 1999.
[35] S. L. Shaw, E. Yeh, K. Bloom, and E. D. Salmon, *Curr. Biol.* **7,** 701 (1997).
[36] E. D. Salmon, E. Yeh, S. Shaw, B. Skibbens, and K. Bloom, *Methods Enzymol.* **298,** 317 (1998).

can cut costs when assembling an imaging system, but the camera is not one of them. GFP fusions to low abundance proteins may not be visible through the microscope eyepieces, but are detected readily using a sensitive camera. Important parameters to consider when choosing a camera are sensitivity, noise, acquisition rate, resolution, and dynamic range (number of gray levels). Sensitivity and acquisition rate—the time taken for the images to be transferred from the camera to the computer—are important factors in limiting photobleaching and photosensitization. Resolution, determined by the size of individual pixels rather than the number of pixels in the chip, is critical for imaging micrometer-sized yeast cells. High dynamic range is important for quantitation and for visualizing signals over a wide range of brightness levels. Cameras that have proven useful for time-lapse imaging fall into two classes: CCD cameras and silicon-intensified tube (SIT) cameras. In the past, SIT cameras were preferable because of their rapid acquisition rates and high sensitivity. These cameras are still popular, but advances in the acquisition rate and sensitivity of CCD cameras have moved CCD cameras to the forefront. One of the most popular groups of cameras is based on the Sony ICX061 progressive scan CCD, which has an array of 1300 × 1030 very small (6.7 μm) pixels. Cameras that incorporate this chip include the camera we use, the MicroMax 1300YHS, as well as the newer CoolSnap fx, both from Roper Scientific (www.roperscientific.com), the Orca line from Hammamatsu (usa.hammamatsu.com), and the Sensicam SVGA from Cooke Camera (www.cookecorp.com). These cameras have frame transfer rates of greater than three full frame images per second, a quantum efficiency of 50% at 500 nm, and a dynamic range of over 4000 gray levels (for 12-bit images). A new generation of cameras with even greater quantum efficiencies is now available from Hammamatsu (Orca ER), Roper Scientific (CoolSnap HQ), and Cooke Camera (Sensicam QE).

Cameras are available with larger formats, greater sensitivity and dynamic range, and higher transfer rates, but all have limitations for time-lapse imaging. For example, intensified CCD cameras are available that are capable of detecting the fluorescent emission from a single molecule, but their high cost and lower resolution limit their general use. Color CCD cameras are not advisable, as they generally have lower sensitivity than gray-scale cameras. Those who need to acquire color images should investigate a tunable liquid crystal RGB filter system from Cambridge Research and Instrumentation, Inc. (www.cri-inc.com) that allows gray-scale cameras to acquire high-quality color images. This filter system can be added to most microscopes.

Software. The series of events that define time-lapse imaging requires synchronization of shutter opening and closing with image acquisition. This is best achieved with computer software to automate the shutters, filter wheels, the z-axis stepping motor, and the camera. We have used IPLab from Scanalytics Inc (www.scanalytics.com) with an Apple Macintosh G3 computer, but there are a number of other great Mac- and Windows-based software packages. These include (but are not limited to) MetaMorph from Universal Imaging (www.universal-imaging.com), Slidebook from Intelligent Imaging Innovations (www.intelligent-imaging.com), and Openlab from Improvision (www.improvision.com). All of these software packages allow the control of most devices and cameras and also control the new fully automated microscopes; however, these software are expensive for running the 4D imaging system. The public domain software package NIH Image (http://rsb.info.nih.gov/nih-image/) allows both acquisition and analysis of 8-bit images. Of potentially greater utility is the new public domain program Image/J (http://rsb.info.nih.gov/ij/), which will run on virtually any computer running Java 1.1 and will handle up to 32-bit images. Custom acquisition and analysis plugins can be developed to accommodate peripheral devices and cameras, and at the time of this writing a plugin was available to drive Hammamatsu Orca cameras. For the assembly of figures and images for publication, images are converted to 8 bit (256 gray levels) and saved in TIFF format. The resulting files can be opened and manipulated in Adobe Photoshop.

Temperature Control. One advantage of yeast over many other experimental organisms is the range of temperatures at which cells will grow "normally." For many experiments, room temperature is adequate for visualizing cells under physiological conditions. However, for experiments that need to be carried out at temperatures above ambient, visualizing live cells can be more problematic than visualizing fixed cells. This problem can be solved by heating the objective and sample. A slide/sample-warming device fitted to the microscope stage is not usually adequate for this purpose because the high NA objectives used for yeast act as efficient heat sinks. To get around this problem, we have used an objective heater from Bioptechs (www.bioptechs.com) that consists of a small heat tape that surrounds the retractable nosepiece of the objective. A thermal sensor on the nosepiece allows regulation of the temperature to within $0.2°$. We found that the objective heater could not be fitted to our standard Olympus UPlan fluorite $100\times$ (NA 1.3) objective. However, it could be fitted to the much less expensive $100\times$ achromatic (NA 1.25) objective. An achromatic objective is not ideal for full-field observation because the outside portion of the field of view is not in the same focal plane as the center of the field, but we have found that it works surprisingly well for GFP and DIC imaging of cells within the camera field. Figure 2 compares the DIC and fluorescence images of cells taken with a UPlan fluorite objective and a less expensive achromatic objective. We have found that the reduced resolution and reduced fluorescence (less than 20% for GFP) are acceptable for many experiments. Work at temperatures below ambient requires objective and stage/sample cooling devices, which are also available commercially.

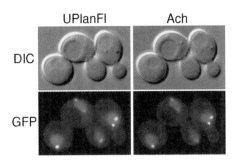

FIG. 2. DIC and fluorescence images of cells containing GFP fusions to spindle pole body protein Spc42[24] and to protein phosphatase I (Glc7) [A. Bloecher and K. Tatchell, *J. Cell Biol.* **149**, 125 (2000)]. GFP-Spc42 decorates the spindle pole body, whereas GFP-Glc7 associates with nucleus and the bud neck. Images on the left and right were taken through a universal plan fluorite (UPlanFl) and a much less expensive achromatic (Ach) objective, respectively. To facilitate a direct comparison, exposure times and normalization values were identical between the two fluorescence and DIC images.

Protocols

Sample Preparation. Cell preparation for microscopy is essentially as described previously.[37] *ade1* and *ade2* auxotrophic strains accumulate a bright fluorescent pigment so it is preferable to use adenine prototrophs for microscopy, although growth of *ade2* strains on high concentrations of adenine (>0.02%) will largely inhibit accumulation of the pigment. Cells are grown to log phase ($A_{600} = 0.3$–0.6) in either YPD (2% glucose, 2% peptone, 1% yeast extract) or synthetic medium (2% glucose, 0.67% yeast nitrogen base, plus necessary amino acids) and are concentrated by centrifugation at 2000 rpm in a microcentrifuge. In some cases, growth in YPD results in higher fluorescence levels. If YPD is the growth medium, the cells should be washed in synthetic medium before mounting because the yeast extract and peptone are highly fluorescent. Without resuspending the pellet, a small number of cells can be aspirated off the surface of the pellet into a micropipette in <1 μl volume and placed on a slab of synthetic complete medium containing 2% agarose. Gelatin at 25%[35] has been used in place of agarose and may give higher quality DIC images. The slab of solid medium is made by placing a drop of molten medium on a glass microscope slide and immediately placing another slide on the top (Fig. 3A). When the agarose has solidified, the top slide is removed carefully by sliding it off to the side, and the exposed agarose slab is trimmed with a razor blade to a square slightly smaller than a standard #1 coverslip. The cell suspension is placed on the slab and covered with a standard coverslip. The edges of the coverslip are then sealed with molten petroleum jelly (Fig. 3B).

[37] J. A. Waddle, T. S. Karpova, R. H. Waterston, and J. A. Cooper, *J. Cell Biol.* **132**, 861 (1996).

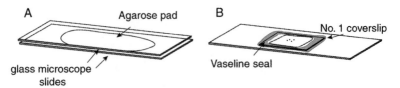

FIG. 3. Sample preparation for time-lapse imaging. (A) A thin pad of molten agarose is formed between two microscope slides. (B) When the agarose has hardened, the top slide is slid off and the pad is trimmed to a square slightly smaller than the coverslip. Cells are placed on the pad and covered with a coverslip, and the coverslip is sealed with molten petroleum jelly.

This is carried out by extruding the petroleum jelly through a syringe fitted with a locking 16- or 18-gauge needle, first passing the needle briefly through the flame of a Bunsen burner to melt the petroleum jelly in the tip. Provided that the cell density is low, the cells will grow for many generations under the coverslip. The petroleum jelly barrier prevents desiccation and can withstand heating the sample to 37°. Valap (a 1 : 1 : 1 petroleum jelly : lanolin : paraffin mixture) has also been used as a sealant.[36]

General Comments. The single most important factor for successful time-lapse imaging is the length of time that cells are exposed to the fluorescent excitation light source. It is critical to minimize the intensity and duration of fluorescence excitation during time-lapse imaging to reduce photobleaching and phototoxicity. Remarkably, exposure to only a few seconds of illumination with blue light at the full intensity of a 100-W Hg arc lamp is sufficient to kill many cells. No obvious change in morphology occurs when this happens, but the cells just stop growing. We reduce the light intensity to between 1 and 6% of maximum transmission using neutral density filters placed in the light path. We use a duration of illumination between 1 and 2 sec per image at intervals between 1 and 5 min. Using this regime, cells can usually be maintained for at least two complete doublings. Although in principle it is possible to illuminate cells manually, acquire an image, and then cut off the light source, it is practical to have an electronic shutter placed in the excitation light path and use computer software to synchronize shutter opening and closing with image acquisition.

Protocol 1. Time Lapse Using a Single Focal Plane. This procedure can be used if the GFP fusion resides at a site that is readily discernible in the DIC image. Examples include bud neck, bud tip, and cell cortex. The shutter on the tungsten lamp is opened and DIC images are captured in a focus mode (DIC images are refreshed at 100-msec intervals on the screen), allowing the sample to be brought into proper focus. Once in focus, an image is saved to disk. The tungsten shutter is closed and the DIC polarizing filter is removed from the light path, as it absorbs at least 50% of the light destined for the camera. The excitation shutter is then opened, with fluorescent light intensity reduced to 1–6% transmission using a neutral density filter(s) to minimize fading and phototoxicity. A 1- to 2-sec fluorescence image is immediately captured from the same focal plane as the DIC image and is saved to the hard disk. The excitation shutter is then closed. This process is repeated at intervals determined by the operator, generally at 2- to 5-min intervals. The major disadvantage to this protocol is that it requires constant attention from the operator. The major advantages are (1) that the fluorescence image is always in the proper focal plane and (2) that a minimum of equipment, mainly the shutter on the Hg arc lamp, is necessary.

Protocol 2. Four-Dimensional Time Lapse with GFP and DIC. Many GFP fusions are mobile within or extend through the cell and a particular structure or organelle may not be entirely in focus in a single image. Examples include microtubules, spindle pole bodies, and cortical actin patches. In such cases, it is necessary to collect images from multiple z-axis positions, which requires automation of the focus controller. The stack of images can be retained as such to give an accurate 3D representation of the GFP fusion[38] or, alternatively, the stack can be flattened into a single image to give a 2D representation.[35,37] The latter is most useful for easily visualizing a long sequence of time-lapse images.

Cells are brought into focus through the objective using only the tungsten light source. A portion of the image is selected for the subsequent time-lapse series. By selecting a subregion of the full frame to save, the time required to transfer the image to the computer is reduced, which in turn reduces the length of time that the cells are exposed to light. We also use 2×2 binning. This process combines the signal from four adjacent pixels, thus increasing the signal fourfold and reducing the necessary exposure time. With the small pixel size of our camera (6.7 μm), resolution is not harmed significantly by this process. A shutter is closed to block the tungsten light and another shutter is opened to expose the cells to light from the epifluorescence source, which has been attenuated with neutral density filters. Images are acquired in different focal planes using the stepping motor attached to the focus control to change the focal plane. We normally collect five to eight images spaced at 0.2- to 0.5-μm increments with a 100- to 200-msec exposure for each image. During this process the cells are exposed to light for 1–2 sec.

The excitation shutter is then closed and images are saved to disk either as a stack of images for 3D reconstruction or as a 2D projection of the set. A polarizing filter below the camera is moved into the light path, the shutter for the tungsten lamp is opened, and one or more DIC images are acquired and saved to disk. This shutter is closed and the polarizing filter is swung out of the light path and the process is repeated after a specified time interval. This interval, ranging from 30 sec to 5 min, is determined empirically by the brightness of the fluorescence, dynamics of the specific GFP fusion, and relative sensitivity of the cells to light exposure. The entire

---

[38] A. F. Straight, W. F. Marchall, J. W. Sedat, and A. W. Murray, *Science* **277**, 574 (1997).

process is controlled by the imaging software, and custom journals or scripts must be written to control all the devices. This can take considerable effort, depending on one's programming skills. Slidebook and Openlab software packages have simplified greatly the process required for setting up time-lapse imaging. Once the time-lapse experiment is complete, additional software manipulation of the images can be used to reconstruct a three-dimensional image of the GFP fusion, to combine DIC and fluorescence images, or to convert the images into a QuickTime movie.

Protocol 3. Imaging Two GFP Variants in the Same Cell. To image CFP and YFP variants in the same cells, we use the 86002 JP4 filter set from Chroma. The excitation and emission filters for the two variants reside in the excitation and emission filter wheels, respectively, while a double bandpass dichroic beam splitter resides in the microscope filter holder. With this filter set, CFP and YFP variants are imaged successively with no bleedthrough. Because the emission and excitation filters reside in the filter wheel, the entire process can be automated to collect images of CFP and YFP in rapid succession.

S65T/EGFP and YFP can also be imaged separately in the same cell,[5,7] but less satisfactorily than with CFP and YFP. This is because the emission and excitation spectra of EGFP and YFP are overlapping and to provide a separate image of each variant, very narrow bandpass filters must be employed, which reduce the level of emitted light. We use Chroma 31039 JP1 and 31040 JP2 filters sets for EGFP and YFP, respectively. These filter sets cannot be automated easily, as each requires a different dichroic beamsplitter and all of the filters are therefore housed within the filter holder in the microscope. We have used a program that prompts the user to switch filter sets for the next exposure, but even under the best of circumstances an elapsed time of at least a second occurs between collection of the EGFP and YFP images. This time is sufficient to allow for noticeable changes in positions of highly dynamic organelles such as Golgi and endosomes.[7] The signals from EGFP and YFP still bleed through the YFP and EGFP filter sets, respectively. The signal from YFP in the EGFP filter set (JP1) is less than 5% of the signal in the YFP filter set (JP2), but the signal from EGFP in the YFP filter set (JP2) is over 30%. In practice, if the YFP fusion is more abundant/brighter, then shorter exposures can be taken with the YFP filter set to minimize the bleedthrough,[5] but a subtraction algorithm may be necessary to visualize the two images independently.

Postacquisition Image Processing. So what do you do with all the data? If only a single image is collected at each time point or if the stack of images from different focal planes is flattened into a 2D representation, data can be visualized simply as a montage of time points or in the form of a movie sequence. The fluorescence image can also be overlaid on the DIC image to simultaneously visualize both the GFP fusion protein and cell morphology.[36] Most software packages can easily create a QuickTime movie from a sequence of images. If images from different focal planes were collected at each time point, this stack of images can be used to create a 3D representation of the GFP fusion. However, because scattered light can greatly obscure the images, a deconvolution algorithm is often used to remove out-of-focus light.[39] A great example of this use can be seen in the work of Straight et al.[38] The process of digital deconvolution, which in its most accurate form uses a point spread function, can be computer intensive and, in the past, required either a mainframe computer or hours of processing time on a PC. However, advances in software and hardware allow digital deconvolution to be carried out with most software packages on a stack of images in minutes or seconds.

Acknowledgments

We thank Jim Waddle for his help and enthusiasm in getting us started with digital imaging and Benjamin Glick, Sepp Kohlwein, and Liz Conibear for helpful and informative discussions. This work was funded by the National Science Foundation (MCB-9974459 to L.C.R.) and the National Institutes of Health (GM-47789 to K.T.).

[39] D. A. Agard, Y. Hiraoka, P. J. Shaw, and J. W. Sedat, *Methods Cell Biol.* **30**, 353 (1989).

Author Index

Numbers in parentheses are footnote reference numbers and indicate that an author's work is referred to although the name is not cited in the text.

A

Abbas-Terki, T., 449, 450(31), 451(31)
Abbey, C. K., 86
Abeijon, C., 334, 363
Abelson, J., 128, 200, 201, 202(15), 207(9), 208(9), 210(12), 214, 215, 215(15; 40; 41), 218(46), 219(46), 661
Abraham, D. J., 530
Abramowitz, M., 22
Achstetter, T., 127, 128, 128(5; 6), 129(5), 130(6), 140(5; 19), 141(19), 147(5; 19)
Adam, G., 444
Adam, R. D., 200
Adam, S. A., 595
Adams, A. E., 50, 53, 68(19), 383, 411, 413(8), 417(8), 418(8), 433, 460, 461, 493, 591
Adams, C., 443
Adams, E., 129
Adams, I. R., 668, 671
Adams, J. A., 19, 664
Adams, M., 525
Adams, S. R., 28, 525
Admon, A., 640
Adrian, G. S., 387(38), 393
Aebersold, R., 131, 281, 301, 312, 313, 320, 387(35), 393
Aebi, M., 352, 526, 527(80), 532(80), 583
Afanassiev, V., 673
Affolter, M., 281
Afzelius, B. A., 121
Agard, D. A., 16, 17(2), 18, 26(2), 27, 27(2), 86, 683
Ahmad, M. F., 229
Aibara, S., 144
Aitchison, J. D., 401, 405, 406(14), 589, 590, 594
Akey, C. W., 82
Akhmaloka, 504
Al, E. J. M., 129

al-Awar, O., 179, 180(18)
Albertini, M., 590
Alberti-Segui, C., 34, 38(3), 39(3), 667
Alberts, B. M., 173, 178, 179, 179(15), 180(17; 18)
Albertus, J., 453
Albig, W., 609
Albright, C., 50, 77(7)
Alepuz, P. M., 609
Alexandrov, M. L., 299
Alexandru, G., 281
Aley, S. B., 200
Allen, J. B., 461
Allen, K. E., 344, 505, 517(33), 519(33)
Altman, R., 173, 174(11), 178(10)
Altmann, M., 222, 236(5), 247(5)
Amati, B. B., 191, 192
Amberg, D. A., 568, 571(1), 583(1)
Amberg, D. C., 172, 568, 577, 583
Amerik, A. Y., 252, 256(10)
Ames, B. N., 334, 363
Amherdt, M., 259, 260, 265, 265(9; 15; 16), 266(9), 268(25), 278(25)
Amit, S., 317
Ammerer, G., 136, 328, 609
Amon, A., 287, 457, 458, 462, 464(9), 607
Amos, B., 666
Anantachaiyong, T., 673
Anders, K., 457, 460(3)
Andersen, J. S., 317
Andersen, S. S., 447
Anderson, J., 222, 227, 236(7), 237(7), 247(7)
Anderson, J. S., 654, 655(14), 660(14)
Anderson, K., 374
Andre, B., 339, 646
Andrews, P. D., 671
Angel, P., 280
Annan, R. S., 279, 281, 282, 283(13; 15), 284(15), 290, 290(13), 292, 292(29), 293(13; 15), 294(13), 307

Anraku, Y., 339, 411, 428
Ansari, A., 206
Antonny, B., 278
Antony, C., 106
Aoi, H., 386
Aoki, T., 447
Aparicio, O. M., 540
Appel, B., 203
Ares, M., 202, 212
Aridor, M., 278
Aris, J. P., 131, 343, 406, 407(18)
Arman, I., 508, 527(41)
Armstrong, W., 55
Arnason, T., 252(20), 255, 256(20)
Arnold, W. N., 55
Arnsdorf, M. F., 529, 530(84), 533, 534(84)
Aroian, R. V., 173
Arvan, P., 342
Asano, K., 221, 222, 227, 236, 236(7), 237(7), 239, 240, 247(7)
Ashby, M. N., 671
Ashford, A., 106
Asthana, S., 31, 32(28)
Astromoff, A., 374
Attfield, P., 444
Auer, M., 343
Aust, T., 671
Austriaco, N. R., Jr., 469, 471(8), 472(8)
Avaro, S., 412
Avery, S. V., 673
Awano, U., 501
Aynardi, M. W., 128, 129(20), 133(20), 149(20)

B

Baba, M., 81, 96, 99, 103(2), 104(7; 8), 115(2), 121(2; 7; 8)
Baba, N., 81, 99, 104(7; 8), 121(7; 8)
Babst, M., 415
Bacallao, R., 4
Bachmair, A., 255
Badaracco, G., 131
Bagnat, M., 344
Bahler, J., 166
Bai, J., 368
Bai, Y., 130
Bailleul, P. A., 517
Bailleul-Winslett, P. A., 513, 525(53), 526(53)
Baird, G. S., 35, 39, 663, 665, 665(10)

Baker, D., 258, 273(6), 353
Baker, R. E., 660
Baker-Brachmann, C., 367(19), 370, 373, 374(19)
Balch, W. E., 278
Bald, R., 203
Baldwin, M. A., 500, 526(19)
Bale, S., 383(13), 384
Ball, C., 443
Ballou, C. E., 473
Banerjee, A. C., 229
Bankaitis, V. A., 352, 421
Bannai, H., 501
Bannykh, S., 278
Banroques, J., 215, 218(47), 219(47)
Banta, L. M., 129
Banuett, F., 339, 340(8)
Bapat, K., 516, 526(64)
Baram, G. I., 299
Barker, G., 673
Barlowe, C., 259, 265(9), 266, 266(9), 358, 363(23)
Barnes, G., 170, 433, 434, 434(5)
Barnett, J., 52
Barrell, B. G., 203
Barrell, G., 297
Bartel, B., 644
Bartosz, G., 473
Baskerville, C., 287
Bassett, D., 368
Basson, M., 102
Bastien, L., 190, 354
Bateson, M., 671
Bauer, M. F., 387(36), 393
Baumann, F., 447, 449
Baumeister, W., 86
Baumer, M., 641
Baxter, B. K., 449
Bays, N. W., 251, 350
Bean, M. F., 290, 292(29)
Becherer, K. A., 346, 356, 364
Beck, K., 154
Beck, T., 342
Becker, J., 447(26), 448, 449(26)
Bednarek, S. Y., 260, 265(15; 16)
Bedwell, D. M., 505
Beelman, C. A., 648, 652, 655(12), 658, 659(18)
Beevers, H., 331
Begueret, J., 516
Beh, C. T., 447, 478, 488(10), 493(10)

Beinert, H., 447
Belenkiy, S. M., 507, 519(37)
Bell, S. P., 185, 540
Belmont, A. S., 671
Belmont, L. D., 434
Benard, L., 632, 639(7; 8)
Bendayan, M., 72
Bengal, E., 640
Benichou, J. C., 319
Benito, B., 340
Ben Neriah, Y., 317
Bentley, W. E., 673
Benz, E. J., Jr., 343
Benzeno, S., 453
Berden, J. A., 671, 673
Bereiter-Hahn, J., 631
Berezney, R., 199
Berger, K. H., 381
Berget, S. M., 200
Berlin, V., 478
Bernander, R., 623
Bernard, F., 339
Berry, G., 37
Berthold, J., 387(36), 393
Bertone, P., 279
Bertram, G., 664, 665, 666(11)
Bertrand, E., 579, 580(6)
Betz, H., 129
Betz, W. J., 625
Bevis, B. J., 58, 665, 668, 668(14), 669, 671
Beyer, A., 374(30; 31), 375(30; 31), 380
Beynon, R., 304
Biddecombe, W. H., 104
Bieber, L. L., 430
Bielinsky, A. K., 184
Bilinski, T., 444
Binder, M., 671
Binkley, G., 443
Black, D. L., 202
Black, M. W., 415
Blair, L., 335
Blake, T., 510
Blanc, A., 632, 639(6)
Bleazard, W., 383(13), 384
Blobel, G., 131, 343, 366, 377(7), 394, 401, 406, 406(3), 407(3; 15; 16; 18), 588, 589, 590, 594, 620
Bloecher, A., 662, 664(5), 669(5), 674(5), 679
Blom, J., 387(37), 393
Blondel, M., 590, 609, 671

Bloom, K., 4, 31, 35, 41(6), 412, 671, 676, 679(35), 680(36), 681(35)
Bloomand, K., 4
Blumer, K. J., 413, 415(17), 600, 669, 671, 672
Boal, T. R., 229
Bobola, N., 622
Bobroff, N., 19
Boche, I., 621
Bochtler, M., 154
Bodnar, A. G., 366
Boeke, J. D., 478, 484, 640
Boelens, W. C., 620
Boguta, M., 514
Bohen, S. P., 451, 452
Bohm, T., 199, 283, 284(18)
Bohni, P. C., 389
Boles, E., 339
Bolliger, L., 446
Bond, J., 304
Bone, Q., 53
Bonetti, B., 505
Bonifacino, J. S., 341
Boone, C., 478
Boorstein, W. R., 444, 577
Boot, R., 370
Borer, R. A., 611
Borkovich, K. A., 448
Boschelli, F., 452
Bossie, M., 611, 612(23)
Botstein, D., 50, 55, 72(11), 77(7), 122, 443, 457, 460(3)
Boucher, P., 252(24), 256, 647
Boucherie, H., 302, 312, 387(41), 393
Bousset, L., 500, 526(18)
Bouveret, E., 658
Bowers, B., 337
Bowers, K., 427
Boyartchuk, V. L., 671
Boyd, A., 28, 353, 358(16), 525
Boye, E., 623
Boyle, G., 671
Boyle, J. A., 203
Boyle, W. J., 280
Boy-Marcotte, E., 129
Bozzola, J. J., 98, 113(3)
Bracegirdle, S. B., 31
Brachat, A., 34, 38(3), 39(3), 166, 418, 662, 667, 671
Bradley, C., 662, 664, 671, 674(6)
Bradley, G., 127, 129(7), 139(7), 140(7), 147(7)

Bradley, M. E., 509, 510, 517, 517(51)
Braguglia, D., 185, 186, 186(12), 187(11; 12), 189, 193(12), 194(13), 195(11), 196(12), 197(12), 198, 199(12)
Brake, A., 132, 335
Brame, C. J., 620
Brandao, I., 56
Brandau, S., 527
Brandt, A., 392
Branlant, C., 211
Branton, D., 351
Braun, B. R., 339, 443
Braunfeld, M. B., 86
Braus, G. H., 641
Breeden, L. L., 280, 457
Breitschopf, K., 640
Brenner, S. L., 513
Bretaudiere, J. P., 84
Bretscher, A., 340, 438
Brewer, B. J., 184, 195, 195(5), 540, 541, 548(21), 553(22), 554, 562, 567(24), 568(19)
Brickner, J. H., 358
Brigance, W. T., 358, 363(23)
Brill, J. A., 478
Bringmann, P., 203
Brini, M., 382
Brizzio, V., 447, 484, 489(21), 491
Brodsky, A., 580
Brodsky, B., 154
Brodsky, J. L., 248, 341, 447
Brody, E., 201, 202(15), 215(15)
Broker, T. R., 200
Bromirski, M., 306
Brower, S. M., 433
Brown, A. J., 664, 665, 666(11)
Brown, C. R., 448
Brown, D. J., 283
Brown, P. O., 129, 457, 460(3)
Brownawell, A., 589
Bruce, A. G., 218
Brunner, M., 387(33; 36), 393
Bryan, J. D., 662, 664, 664(7), 668(7), 671, 674(6)
Bryant, N. J., 343, 350, 353, 358, 358(16), 409, 428
Brzeska, H., 281
Bucci, M., 671
Buchman, A. R., 131
Budd, M., 185

Bukau, B., 443, 446(3), 449(7), 451(7)
Bullitt, E., 82
Buratowski, R. M., 590
Buratowski, S., 590
Burd, C. G., 414
Burke, D., 169, 504
Burke, R. L., 173
Burke, W., 539
Burns, V. W., 4
Bushman, J. L., 229
Bussereau, F., 129
Bussey, H., 203, 297
Butler, D. A., 500, 526(19)
Butler, J. S., 583
Butow, R. A., 383(18), 384, 386(18), 387(31), 392
Butt, T. R., 252(34), 257
Buttle, K., 89, 95
Butty, A.-C., 170, 433, 434(5)
Byers, B., 58, 75(29), 81, 101, 121, 462, 497, 647

C

Cabib, E., 127, 130(13), 133, 337
Callis, J., 252(36), 257
Campbell, C. L., 383(32), 387(32), 393
Campbell, J. L., 131, 132, 185
Campbell, M. J., 457, 460(2), 462(2)
Capaldi, R. A., 430
Capdevielle, J., 294
Caplan, A. J., 450, 452, 453
Caponigro, G., 651, 658, 659(18)
Capucci, L., 131
Caputo, E., 367(19), 370, 373, 374(19)
Cardozo, C., 453
Carlson, M., 129, 341
Carmack, E., 220
Carminati, J. L., 35
Caroff, A., 102, 121(13)
Carr, S. A., 279, 281, 282, 283(13; 15), 284(15), 290, 290(13), 292, 292(29), 293(13; 15), 294(13), 307
Carragher, B., 95
Carroll, C. W., 173, 174(11)
Carroll, K., 632, 639(6–8)
Casamayor, A., 279
Cashikar, A. G., 525, 527(74), 529, 529(74), 530(84), 531, 533, 533(74), 534(84), 536

AUTHOR INDEX

Caspary, F., 219
Causton, H. C., 457
Cech, T. R., 652
Celis, J. E., 508, 527(41)
Cereghino, J. L., 360, 361
Chabot, B., 202
Chae, H. J., 673
Chait, B. T., 313, 320, 405, 406(14)
Chakrabarti, A., 244
Chalfie, M., 16, 592, 663, 664
Chang, A., 339, 342, 343, 344, 344(24; 38), 345(38), 347(38), 349(24), 350(24), 364
Chang, H. C., 450
Chang, L. M. S., 131, 132
Chang, S., 279
Chang, Y.-H., 141, 147(75)
Chant, J., 3, 340, 468
Chao, D. M., 173
Chapman, J. N., 104
Chapman, R. E., 356, 363
Charbonneau, H., 287
Chartrand, P., 579, 580(6)
Chattoo, B. B., 501
Chau, V., 252(15; 24), 253, 255, 256, 647
Chechenova, M. B., 507, 519
Chen, E. J., 325
Chen, H., 86
Chen, J., 659, 660(28)
Chen, J. B., 469, 473
Chen, L., 248
Chen, S., 283
Chen, S. L., 279
Chen, S. Z. W., 287
Chen, X., 434
Chen, Y., 671
Chenevert, J., 478, 484(6)
Cheng, N., 500, 526(17)
Cheng, S. C., 128, 200, 207(9), 208(9), 214, 215(40)
Cheng, Y. C., 447
Cheperegin, S., 508, 527(41)
Chernoff, Y. O., 499, 500, 501, 501(11), 505, 505(2; 25), 507, 507(9), 510, 510(4; 9; 11), 513, 513(9; 11), 514(11; 22), 516, 516(2; 8; 9), 517, 517(8; 9; 25; 33; 51), 519(9; 33; 37), 524(11), 525(53), 526(9; 53; 64)
Chernova, T. A., 507, 519(37)
Cherry, J. M., 443
Chesebro, B., 523
Chi, Y., 257, 279, 285

Chia, W., 468
Chiang, H. L., 448
Chien, P., 507, 527
Cho, J. H., 353, 358(15)
Cho, R. J., 457, 460(2), 462(2)
Choi, S. K., 222, 227(8)
Chomsky, O., 640
Chow, L. T., 200
Chretien, P., 190, 354
Chuang, J. S., 271, 344
Cid, A., 344
Ciechanover, A., 248, 640, 646, 647, 647(16)
Cigan, A. M., 229
Ciosk, R., 31
Clamp, J., 57
Clark, M., 70, 80(32), 122
Clary, D. O., 264
Claus, J., 469, 474(12)
Clay, F. J., 478
Clayton, J., 239, 240
Clegg, R. M., 599, 666
Clipstone, N. A., 588
Clos, J., 527
Cocker, J. H., 199
Cohen, P., 281
Cohen-Fix, O., 457
Cohn, E. J., 162, 163(11)
Cole, C. N., 568, 571(1), 577, 583, 583(1)
Cole, F., 469, 474(12)
Colfen, H., 155(16), 164
Collins, I., 551, 554(23)
Colon, E., 252(36), 257
Colot, H. V., 218, 583
Colvin, R. T., 662, 664(7), 668(7), 671
Company, M., 200, 210(12)
Conde, J., 449, 478, 509
Conibear, E., 351, 408, 409, 421(2), 662, 664(7), 668(7), 671
Conway, A., 457, 460(2), 462(2)
Conway, J., 55
Cooper, A. A., 358
Cooper, J. A., 412, 414(11), 671, 679, 681(37)
Corbett, A. H., 587, 590, 594, 596, 611, 612(23), 613, 617(26), 618, 618(26)
Cordova, L., 89
Cormack, B. P., 664, 665, 666(11)
Correll, C. C., 257
Corsi, A. K., 448
Corthals, G. L., 312
Cosma, M. P., 458, 464(10)

Cost, G. J., 367(19), 370, 373, 374(19)
Costanzo, M. C., 154, 339, 443, 480, 509
Coue, M., 513
Coustou, V., 516
Coverley, D., 184
Cowburn, D., 313
Cowles, C. R., 415, 428(22)
Cox, B. S., 499, 500, 503, 503(5), 508, 508(5), 509, 509(5), 510(10; 28), 511, 527(43; 44)
Crabtree, G. R., 588
Craig, E. A., 442, 443, 444, 446, 447, 447(26), 448, 449, 449(26), 452, 514, 577
Craig, K. L., 253, 257
Craig, S., 105
Crausaz, F., 427
Craven, R. A., 449
Crawford, M. E., 339, 443
Crocker, M. K., 200
Crooke, S. R., 252(24; 34), 256, 257
Crooke, S. T., 647
Cross, F. R., 313, 458, 464(8), 541, 619, 621(35)
Cross, S. L., 49
Crotwell, M., 265, 268(24)
Crowther, R. A., 88
Cubitt, A. B., 25, 28, 525, 664
Cueva, R., 427
Culbertson, M. R., 658
Cullin, C., 500, 526(18)
Culotta, V. C., 342
Cunningham, K. W., 330, 335(13), 392
Cyr, D. M., 447

D

Dagkesamanskaya, A. R., 501, 505(25), 507, 508, 509(35), 517, 517(25), 527(35; 43)
Dahmann, C., 462
Dahmus, M. E., 252(36), 257
Dalbec, J. M., 129, 132(42)
Damelin, M., 37, 49(16), 587, 600, 602(44), 604(44), 605(44), 613, 617(26), 618(26), 669, 671, 672(28)
Danaie, P., 222, 236(5), 247(5)
Daneels, G., 71
Daniels, L. B., 136, 137(59)
Danuser, G., 19
D'Ari, L., 102
Darsow, T., 358

Daum, G., 352, 389
Dave, V., 366
Davenport, J. W., 343
Davey, H. M., 623
Davidson, A. R., 342
Davidson, M. W., 22
Davies, A., 367(19), 370, 373, 374(19)
Davis, C. L., 659, 660(28)
Davis, J. E., 448, 449
Davis, L. I., 406, 407(15–17), 598, 609
Davis, L. R., 184, 540
Davis, M., 317
Davis, M. T., 220
Davis, N. G., 343
Davis, R. W., 131, 132, 203, 297, 457, 460(2), 462(2), 641
Davis, T. N., 34, 131
Dawes, I. W., 671
Dawson, D., 169, 504
Day, R. N., 24, 25, 36, 49
De Antoni, A., 166
Decker, C. J., 565(9), 652, 654(9–11), 655(9), 656(11), 657(9)
De Craene, J.-O., 339
de Duve, C., 377
Deernick, T. J., 95
Defize, L. H., 280
Defossez, P.-A., 469, 472
De Giorgi, F., 382
deJong, C., 430
de Jonge, P., 632
Dekker, P. J., 387(37), 393
de Koning, W., 374
de la Fuente, N., 344
Deleu, C., 516
Demand, J., 448
Demarini, D. J., 418
De May, J., 67, 68(31), 71
Demchyshyn, V., 662
d'Enfert, C., 266
De Nobel, J., 52
DePace, A. H., 509, 526(45), 527(45), 530(45)
DeRisi, J., 51
Derkatch, I. L., 500, 501, 505(25), 507(9), 509, 510, 510(9), 513(9), 516(8; 9), 517, 517(8; 9; 25; 51), 519, 519(9), 525, 526(9; 75)
Dershowitz, A., 551, 554(23)
de Ruijter, W. J., 86
Deschenes, R. J., 169

Deshaies, R. J., 173, 178(8), 257, 279, 281, 282, 283, 283(13; 15), 284(15), 285, 287, 290(13), 293(13; 15), 294(13), 346, 406, 407(19), 413, 415, 452
Devenish, R. J., 667, 671
Dever, T. E., 222, 227(8), 633
Devilliers, G., 671
DeVit, M. J., 609
DeWitt, N. D., 344
Dey, B., 452
Dhruvakumar, S., 352, 647
Didichenko, S. A., 517
Dien, B. S., 387
Dierksen, K., 86
Dietmeier, K., 387(36; 39), 393
Diffley, J. F., 184, 198, 199, 540, 567(14), 588
Dihlmann, S., 614
Diller, J. D., 541, 568(19)
Dilworth, S. M., 589, 620
Ding, R., 81, 99, 105(6)
Dingwall, C., 589, 591(11), 620
DiRisi, J. L., 129
Distel, B., 129, 366, 369(5), 370, 370(5)
Dixon, C. K., 129, 149(29)
Dockendorff, T. C., 583
Doel, S. M., 509, 527(44)
Doenecke, D., 609
Doering, T. L., 342
Dohmen, J., 283
Dolinski, K., 443
Dollard, C., 602
Donahue, T. F., 221, 239, 242
Donaldson, A. D., 31, 102, 121, 541
Donovan, S., 198
Donze, O., 449, 451(32), 453(32)
Dorer, R., 478
Douglas, M. G., 387(38), 393
Douma, A. C., 374
Dowd, G. J., 173, 178(8), 287
Dowzer, C., 462
Doye, V., 595, 596(31), 597, 597(31), 614
Dreyfuss, G., 620
Drubin, D. G., 4, 50, 53, 68(19), 168, 170, 172, 383, 411, 413(8), 417(8), 418(8), 433, 434, 434(1; 4; 5), 435(1), 493, 591
Drugeon, G., 508, 527(41)
Drury, L. S., 198
Duarte, J. A., 504
Dubey, D. D., 199

Dudl, R. J., 130
Duina, A. A., 450, 452, 453(34)
Dujon, B., 203, 297, 319
Duncan, M., 168
Duncker, B. P., 185, 186, 187(11), 189, 194(13), 195(11), 197, 198, 198(24), 199(24)
Dunckley, T., 658
Dunn, K., 17
Dunn, R., 252(23), 255, 256(23), 330, 647
Dunphy, W. G., 264
Dunst, R. W., 588
Duran, A., 337, 341
Durchschlag, E., 609
Dusterhoft, A., 662
Dutcher, S. K., 479, 510
Dutta, A., 185
Dwight, S., 443
Dyer, J., 366

E

Eaglestone, S. S., 511
Ebihara, K., 501
Eccleshall, T. R., 387
Echlin, P., 103, 104(16)
Ecker, D. J., 252(24; 34), 255, 256, 257, 647
Edelmann, L., 98
Edsall, J. T., 162, 163(11)
Edskes, H. K., 500, 507(15), 514, 525(15)
Egelman, E. H., 434
Egerton, M., 449, 664, 665, 666(11)
Egilmez, N. K., 469, 473
Ehmann, C., 127, 128, 128(5; 6), 129(5; 18), 130(6; 10), 139, 140(5; 19), 141(19), 146(18), 147(5; 19)
Einerhard, A. W., 366
Eisen, M. B., 457, 460(3)
el-Assouli, S. M., 184, 540
Elder, H. Y., 104
Elgersma, Y., 374
Elia, L., 478
Elion, E. A., 477, 489, 598, 607, 609, 611(17), 613, 614(17), 615(17), 617(17), 618(17), 619(17), 622(17)
Elledge, S. J., 248, 257, 461
Ellisman, M. H., 84, 95
Ellison, K. S., 252(25), 256
Ellison, M. J., 252(14; 15; 20; 25), 253, 255, 256, 256(20)

Emr, S. D., 129, 330, 335(12; 14), 346, 352, 356, 358, 360, 361, 364, 409, 412, 415, 421, 427(13), 428(22), 491, 625, 626(6), 661
Emter, O., 127, 128(6), 130(6)
Eng, J., 220
England, T. E., 218
Engle, H. M., 343
Englmeier, L., 608, 609(6), 618, 621(6)
Ens, W., 306
Ensink, J. W., 130
Eppenberger, H. M., 611
Epstein, C. B., 366
Erdjument Bromage, H., 307
Erdman, S., 478
Erdmann, R., 365, 366, 369(5), 370, 370(5; 6), 374(6; 30; 31), 375(6; 30; 31), 377(6; 7), 380
Erickson, F. L., 230, 232(17), 247(17)
Erk, I., 102, 121(13)
Escher, C., 395
Espelin, C. W., 19, 29(14)
Espenshade, P., 265, 268(24)
Esser, S., 215
Estruch, F., 443, 444(11), 609
Etcheverry, T., 351
Euskirchen, G., 663, 664
Euskirchen, W., 592
Evans, T., 449
Everett, J. G., 632, 639(5)

F

Fabre, E., 319, 614
Fabrizio, P., 214, 215, 215(41), 218, 218(47), 219(47)
Fahrney, D., 130
Falkow, S., 664, 665, 666(11)
Fang, Y., 452
Fangman, W. L., 184, 195(5), 387, 539, 540, 541, 553(22), 554, 555, 556(26), 562, 567(4; 24; 28), 568(19)
Fanning, E., 621
Farley, F. W., 598, 609, 611(17), 613, 614(17), 615(17), 617(17), 618(17), 619(17), 622(17)
Farquhaaar, M. G., 382
Farr, A., 142
Farr, R., 200
Farrell, A., 452

Farrelly, F. W., 448
Fay, F. S., 579
Feder, J. N., 634
Feldheim, D., 330, 346
Feldman, R. M., 257, 283
Feldmann, H., 203, 297
Femino, A. M., 579
Fenn, J. B., 299
Ferguson, B. M., 541, 562, 568(19)
Fernandez, M., 320
Ferrara, P., 294
Ferrigno, P., 590, 609
Field, C. M., 163, 173, 179, 180(18), 258
Fields, S., 165, 167(18)
Finbow, M. E., 431
Finch, C., 468
Findlay, J. B., 431
Finger, F. P., 340
Fink, G. R., 3, 40, 223, 253, 339, 341, 342, 343, 344(24), 349(24), 350(24), 406, 407(17), 478, 484, 489, 509, 612, 640, 647
Finkelstein, D. B., 448
Finley, D., 252(21; 24), 253, 255, 256, 256(21), 395, 647
Firoozan, M., 504
Firzinger, A., 374(32), 375(32), 380
Fischer, U., 620
Fischere, E. P., 129
Fitch, I., 462
Fitzgerald-Hayes, M., 590
Flach, J., 611, 612(23)
Fleischmann, M., 583
Flessau, A., 374(30), 375(30), 380
Flick, K., 254
Fliss, A. E., 452
Flitney, F., 56
Fluiter, K., 473
Flynn, G. C., 447, 449
Fogarty, K., 579
Fogg, J., 394
Fohlman, J., 299
Foiani, M., 185
Folley, L. S., 480, 509
Ford, C. S., 589, 597(18), 613, 617, 619(32), 621(32), 622(32)
Formosa, T., 173
Forster, J., 449
Fortin, M., 452
Fossa, A., 374(31), 375(31), 380
Fowler, S., 329

AUTHOR INDEX 693

Fox, T. D., 391, 480, 509
Fradkov, A. F., 35, 382, 383, 665
Frank, D., 200
Frank, J., 82, 84, 89
Franse, M., 370
Franza, B. R., 387(35), 393
Franzusoff, A., 50, 222, 258, 394
Frenandez-Bellot, E., 500, 526(18)
Frey, J., 128, 129(17), 140(17), 147(17), 148(17)
Frey, T. G., 84
Fried, H. M., 610, 620
Friedlander, R., 350
Friedman, K. L., 195, 541, 548(21), 554, 567(24), 568(19)
Fries, E., 258
Fritsch, E. F., 175, 193, 213, 218(34)
Frolova, L., 508, 527(41; 42)
Frolova, N. S., 507
Fu, L., 505
Fujii-Nakata, T., 173
Fujiki, Y., 141, 329
Fukada, M., 609
Fuller, M. T., 384, 387(12)
Fuller, R. S., 132, 333, 335, 351, 356, 358, 359, 426
Fung, J. C., 86
Furcinitti, P. S., 83
Futcher, A. B., 460, 622
Futcher, B., 457, 460(3), 462, 647

G

Gaber, R. F., 450, 452, 453(34)
Gabrielian, A. E., 457, 460(2), 462(2)
Galan, J., 255, 256(22), 646
Galibert, F., 203, 297
Galkin, A. P., 507, 519(37)
Gall, L., 299
Gallie, D. R., 632, 634, 637(4), 639(5)
Gallwitz, D., 166, 200, 668
Gambill, B. D., 449
Gammie, A. E., 477, 484, 489(21), 491
Gao, C. Y., 515
Garabedian, M., 452
Garcia, P. D., 394
Gardner, R. G., 251, 257, 350
Garrard, L. J., 374
Garrels, J. I., 339, 443
Garvik, B. M., 220, 540

Gasser, S. M., 184, 185, 186, 186(12), 187(11; 12), 189, 191, 192, 193(12), 194(13), 195(11), 196(12), 197, 197(12), 198, 198(24), 199, 199(12; 24)
Gatto, G. J., 368
Gaume, B., 447
Gaynor, E. C., 352
Gebert, R., 526, 527(80), 532(80)
Geisbrecht, B. V., 374(27), 375(27), 377
Gejyo, F., 532
Gelb, M. H., 313
George, H. A., 673
George-Nacimento, C., 149
Georgiev, O. I., 632
Gerace, L., 589, 591(10), 595, 620
Geraghty, M. T., 368
Gerber, M. R., 452
Gerber, S. A., 313
Gerbi, S. A., 184
Gerrard, S. R., 343
Gerstein, M., 279
Gething, M.-J., 442
Geuze, H. J., 72
Ghislain, M., 647
Ghosh, H. P., 224
Gibson, B. W., 281
Giddings, T. H., 81, 99, 105(5)
Gilbert, M., 185
Gilbert, P. F. C., 88
Gilkey, J. C., 103, 105
Gillin, F. D., 200
Gilon, R., 640
Gimeno, C. J., 341
Gimeno, R. E., 265, 268(23; 24), 325
Girzalsky, W., 366, 369(5), 370(5)
Gite, S., 224
Glauert, A. M., 53, 80(18)
Glick, B. S., 4, 58, 264, 389, 391(26), 392, 446, 665, 668, 668(14), 669, 671
Glickman, M. H., 647
Glover, J. R., 500, 507(12), 514, 519(12), 522(12), 525(12), 526(16), 527(16), 529(16), 530(16), 534(16), 536(16)
Goebl, M. G., 647
Goetsch, L., 58, 75(29), 81, 101, 121, 462, 497
Goffeau, A., 150, 203, 297, 344
Goh, P. Y., 102
Gold, A. M., 130
Gold, R., 17
Goldberg, A. L., 646

Goldfarb, D. S., 621, 671
Goldstein, A. L., 166, 568, 571(1), 583, 583(1)
Gomes de Mesquita, D. S., 625, 626(5)
Gomez, E., 221, 230, 235(19)
Gomez-Ospina, N., 81
Gompertz, B., 468
Gonda, D. K., 255
Gong, X., 343
Goode, B. L., 170, 433, 434, 434(4; 5)
Goodwin, P. C., 17
Gordon, C., 185
Gordon, G. W., 37
Gordon, J. A., 283
Gorenstein, C., 242
Görlich, D., 588, 589, 589(7), 608, 614(5), 618
Gorner, W., 609
Gorsch, L. C., 583
Gorsich, S. W., 384
Goshima, G., 31
Goshorn, C. D., 662, 664(7), 668(7), 671
Gotta, M., 469
Gottschalk, A., 215, 218, 218(47), 219(47)
Goud, B., 70, 351
Gould, K. L., 280
Gould, S. J., 365, 366, 370, 374
Gow, N. A., 664, 665, 666(11)
Grabowski, P. J., 202
Grachev, M. A., 299
Graham, T. R., 330, 335(12; 14), 352, 356, 358, 363(23), 364
Grandi, P., 614
Grant, C. M., 504
Grate, L., 202
Gray, C. C., 104
Gray, V. T., 500, 507(15), 525(15)
Greenberg, J. R., 222, 236(7), 237(7), 247(7)
Greenblatt, J., 173
Greenfeder, S. A., 551, 554(23)
Grewal, A., 307
Griesbeck, O., 38
Griffith, O. W., 160(10), 162, 163(10)
Griffiths, G., 51, 53(13), 56(13), 70(13), 72(13), 121
Gross, D., 443
Gross, H., 526, 527(80), 532(80)
Gross, L. A., 28, 525, 664
Grosshans, H., 590
Gruenberg, J. E., 359
Grunstein, M., 457
Grzelak, A., 473

Gu, Z. Y., 621
Guarente, L., 131, 468, 469, 471(8), 472, 472(8), 474(12; 15)
Guiard, B., 387(34), 393
Guillaud, P., 671
Guillemot, J., 294
Guo, Y. R., 620
Gupta, N. K., 229
Gurvitz, A., 374(32), 375(32), 380, 671
Guthrie, C., 200, 200(14), 201, 202, 214, 589, 597(18), 613, 617, 619(32), 621(32), 622(32)
Gygi, S. P., 301, 312, 313, 387(35), 393

H

Haarer, B. K., 53, 68(19), 383, 411, 413(8), 417(8), 418(8), 493, 591
Haas, A., 252(21), 255, 256(21), 409
Haas, S. M., 430
Haber, J. E., 344, 468
Hach, A., 450, 453(33)
Hadjiolov, A. A., 632
Haenni, A. L., 508, 527(41)
Hagan, K., 448
Hager, K. M., 343
Haguenauer-Tsapis, R., 255, 256(22), 412, 646, 671
Hahn, S., 131
Hailey, D. W., 34
Hale, W. T., 366
Hales, K. G., 384, 387(12)
Hall, M. N., 50, 342, 662
Hallberg, R. L., 392
Hallstrom, T. C., 449
Halvorson, H., 128, 129(16)
Hama, K. O, 95
Hamamoto, S., 259, 260, 265, 265(9; 16), 266(9), 268(23), 278, 334, 352
Hambidge, S. J., 632
Hamilton, B., 374(32), 375(32), 380, 609, 671
Hamilton, T. G., 447, 449
Hammell, C. M., 568
Hammond, A. T., 4, 668, 671
Hampton, R. Y., 251, 257, 350, 671
Hanachi, P., 236
Hanic-Joyce, P. J., 222
Hannavy, K., 446

Hannig, E. M., 221, 230, 232(17), 235(20), 242, 247, 247(17)
Hannus, S., 585
Hansen, R. J., 129
Hansen, W., 394
Harder, W., 374
Hardin, J. A., 203
Harding, S. E., 155(16), 164
Harding, T. M., 427, 625, 626(5)
Hardy, S. F., 200
Harlow, E., 152(3), 153, 167(3), 251
Harper, J. W., 248, 257
Harreman, M. T., 594
Harris, H., 138
Harris, M., 443
Harris, S., 344, 447
Harrison, M. Λ., 431
Hartig, A., 671
Hartl, F., 443
Hartwell, L. H., 478, 479, 540, 633
Harwood, J., 198
Hasan, R., 129
Haseloff, J., 666
Hasilik, A., 136
Hata, T., 130, 144
Hatada, E. N., 588
Hatfield, L., 658
Hatzubai, A. A., 317
Haugland, R. P., 412, 413(16)
Havaux-Copf, B., 515
Haverkamp, J., 312
Hayashi, R., 130, 144
Hayat, M. A., 56, 122
Hayles, J., 194
Hazell, B., 444
He, F., 658
He, S., 391
He, W., 658
He, X., 16, 19, 29(14), 31, 32(28)
Heath, C. V., 568
Hedin, L. O., 480, 509
Heerma, W., 312
Hefner-Gravink, A., 427
Hegemann, J. H., 593
Hegerl, R., 86
Heikal, A. A., 39, 663, 665(10)
Heim, R., 19, 25, 28, 36, 38, 48(11), 382, 525, 598, 664, 665
Heiman, M. G., 102, 106
Heinemeyer, W., 395

Heitman, J., 339
Hellmuth, K., 590
Helmstetter C. E., 473
Hemmings, B. A., 136
Henry, M., 614, 615(30)
Henshaw, E. C., 243, 246
Hentze, M. W., 633
Herber, W. K., 673
Hereford, L., 50, 539, 620
Hereward, F. V., 103, 121(14)
Herman, B., 37
Hermann, G. J., 381, 384, 387(12)
Hernandez, N., 200
Herrick, D., 637, 651
Herring, C. J., 281
Herrmann, J. M., 267, 278, 341, 447, 449
Hershey, J. W. B., 221, 229, 236, 236(16), 247(16)
Hershko, A., 248, 646, 647(16)
Herskowitz, I., 50, 452, 478, 484(6), 489, 500, 526(19)
Herth, W., 614
Hess, S. T., 39, 663, 665(10)
Hessler, D., 84
Hettema, E. H., 366, 369(5), 370, 370(5)
Heun, P., 185, 186, 187(11), 189, 194(13), 195(11), 199
Heuser, J., 119
Heym, G., 144, 145(80)
Hicke, L., 248, 252(23), 255, 256(23), 258, 268, 268(8), 273(6), 330, 341, 352, 353, 427, 646, 647
Hicks, J. B., 40, 253
Hieter, P., 149, 187, 347, 348, 364(29), 380, 480, 481(18), 493(18), 560, 603, 626
Hillenkamp, F., 298
Hiller, M. A., 127, 133(1), 408
Hilleren, P., 648
Hillner, P., 509, 526(45), 527(45), 530(45)
Hiltunen, J. K., 374(31), 375(31), 380
Hinkle, G., 200
Hinnebusch, A. G., 221, 222, 227, 229, 230, 235(20), 236(7), 237(7), 238, 239, 240, 247(7), 281, 633
Hinze, H., 129
Hiraoka, Y., 18, 27, 683
Hirata, D., 447, 449, 452
Hirose, S., 143
Hirsch, J. P., 671
Hirschberg, C. B., 334, 363

Hirschman, J. E., 339, 443
Hjertaas, K., 405, 406(14)
Hoang, C. P., 28
Hobot, J. A., 111, 113(30), 122(30)
Hochstrasser, M., 248, 250, 252, 252(14; 15; 35), 253, 256, 256(10), 257, 640, 647, 648
Hodge, C. A., 568, 583
Hodgins, R. R. W., 252(25), 256
Hoepfner, D., 662, 671
Hoff, K., 447
Hoffman, L., 255
Hoffschulte, H., 130, 132(45)
Hoheisel, J. D., 203, 297
Höhfeld, J., 374, 375(21), 448
Holcomb, C. L., 351, 356, 359
Holder, M. E., 200
Holkeri, H., 342
Holly, J., 458
Holmes, C. F., 281
Holstege, F. C. P., 457
Holthuis, J. C., 352
Holtzman, D. A., 433
Holzer, H., 128, 129, 129(15), 130, 132(45)
Honey, S., 166
Hong, E., 342
Honts, J. E., 433
Hood, J. K., 590, 609
Hope, T. J., 620
Hopkins, K. L., 339, 443
Hopper, A. K., 577, 585, 588, 590
Hori, Y., 567
Horst, M., 446
Horwich, A. L., 443, 446(3)
Hosken, N., 365
Hosobuchi, M., 259, 260, 265(9; 15), 266(9), 406, 407(19)
Hough, L., 57
Howald, I., 662
Howald-Stevenson, I., 411, 426(6)
Howard, R. J., 104
Howell, K. E., 83, 95, 359
Howell, S. H., 374
Howson, R. W., 278, 341
Hsu, C. L., 658
Huang, H., 221, 242
Huang, K., 313, 671
Huang, L. S., 590, 609, 613, 614, 615(31)
Hubbard, A. L., 329
Huber, J., 620
Huberman, J. A., 184, 199, 539, 540

Huddler, D., 447
Huddleston, M. J., 279, 281, 282, 283(13; 15), 284(15), 290, 290(13), 292, 292(29), 293(13; 15), 294(13)
Huls, P. G., 473
Humphries, S., 163, 164(13)
Hunt, D. F., 620
Hunt, R. H., 17
Hunt, S. Y., 541, 553(22)
Hunter, C., 136, 328
Hunter, T., 280
Huntress, F. A., 611, 671
Hurt, E., 81, 319, 585, 590, 595, 596(31), 597, 597(31), 614
Huyer, G., 352
Hwang, E. S., 287
Hyman, A. A., 31, 106, 468

I

Iizuka, N., 222
Ikura, M., 19, 664
Imai, J., 450, 451(35), 453(35)
Inge-Vechtomov, S. G., 500, 501, 501(11), 505(25), 507(9), 508, 510(9; 11), 513(9; 11), 514(11), 516(8; 9), 517, 517(8; 9; 25), 519(9), 524(11), 526(9), 527(42)
Innis, M. A., 136, 137(59)
Inoue, S., 4, 9(11), 12(11)
Inouye, S., 663
Insley, M. Y., 458
Iraqui, I., 339
Irniger, S., 257, 458, 464(9), 640, 641
Ishiguro, J., 19
Ishii, S.-I., 143
Ishino-Arao, Y., 501
Ito, H., 143
Ivantsiv, Y., 647
Iyer, V. R., 129, 457, 460(3)

J

Jackson, K. R., 620
Jacobs, C. W., 461
Jacobsen, K., 449
Jacobson, A., 637, 641, 651, 658
Jacq, C., 203, 297

Jacquet, M., 129
Jain, V. K., 504
James, G. T., 132
James, P., 443, 449
Jan, L., 67, 68(31), 468
Jan, Y. N., 468
Jang, J., 287
Jansen, R. P., 622
Jansson, P. A., 17
Jardine, A. G., 104
Jarosch, E., 350
Jaruga, E., 472
Jaspersen, S. L., 287
Jazwinski, S. M., 469, 470(16), 472, 473
Jeffcoate, S. L., 130
Jenness, D. D., 342
Jennings, E. G., 457
Jenö, P., 302, 387(41), 393
Jensen, O. N., 306, 312, 316
Jensen, R. E., 383, 384
Jentsch, S., 647
Jerome, A., 662, 664, 671, 674(6)
Jiang, J. C., 472
Jigami, Y., 330
Jimenez, G. S., 620
Jin, H., 443
Jinks, T., 611, 612(23)
Joazeiro, C. A., 251, 350
Jochem, J., 647
Johnson, A. D., 173, 174(7), 178(7; 8), 287
Johnson, A. L., 458
Johnson, D. I., 622, 671
Johnson, J. L., 442, 449, 452
Johnson, K. S., 174
Johnson, L. J., 458
Johnson, L. M., 131, 132, 421
Johnson, P. R., 250, 257
Johnson, T., 52, 56
Johnsson, N., 642
Johnston, G. C., 222
Johnston, J. R., 468, 469(5)
Johnston, M., 203, 297, 339, 593, 609, 641
Jolly, E. R., 609
Jones, E. W., 53, 68(19), 127, 128, 129, 129(20), 130(9; 11), 133(1; 2; 11; 20), 134, 136, 136(9; 58), 137(59), 139, 140(73), 145(9), 149, 149(20), 169, 177, 187, 231, 328, 346, 356, 364, 408, 411, 413(8), 417(8), 418(8), 421, 646
Jones, G., 514
Jones, H. D., 4
Jones, J. M., 370
Jones, K. M., 508, 527(43)
Jones, P. C., 431
Jones, W. W., 129, 149(29)
Jong, A. Y. S., 131, 185
Jordan, E. G., 81
Jorgensen, P., 639
Julius, D., 335
Jung, G., 514
Juni, E., 144, 145(80)
Justesen, J., 508, 527(41)

K

Kadowaki, T., 568
Kaffman, A., 590, 593, 607, 609, 609(1), 613, 614, 615(31), 618, 618(1), 619(19)
Kagawa, T., 331
Kahana, J., 525
Kain, S. R., 28, 525
Kaiser, C. A., 50, 77(7), 102, 258, 265, 268(23; 24), 271, 325, 331, 342
Kaiser, P., 254
Kakinuma, Y., 428
Kalderon, D., 589, 619
Kalish, J. E., 367, 368(12)
Kallio, K., 664
Kalton, H. M., 450, 453(34)
Kamminga, A. H., 339
Kanau, W. H., 366, 370(6), 374(6), 375(6), 377(6)
Kanaya, K., 81, 99, 104(7), 121(7)
Kanbe, T., 103
Kane, P. M., 411
Kang, P. J., 449
Kaplan, K. B., 151, 257
Kaplun, L., 647
Karas, M., 298
Karin, M., 280
Karpova, T. S., 412, 414(11), 501, 671, 679, 681(37)
Kastner, B., 215
Katzmann, D. J., 358, 449
Kawano, S., 392
Kay, L. M., 130
Kay, S. A., 661
Kearsey, S. E., 588
Kellenberger, E., 100, 122

Keller, G. A., 365, 367, 368(12), 374
Keller, H. E., 9
Keller, W., 200
Kellogg, D. R., 172, 173, 174(11), 178, 178(10), 179(15)
Kelsay, K., 478
Kennedy, B. K., 469, 471(8), 472(8), 474(12)
Kent, H. M., 589
Kenyon, C., 173
Keranen, S., 344, 671
Khalfan, W., 447
Khan, M. I., 252(34), 257
Khursheed, B., 452
Kielland-Brandt, M. C., 127, 343
Kikuchi, A., 173
Kikuchi, Y., 447
Kilmartin, J. V., 31, 50, 82, 102, 121, 394, 399(2), 460, 493, 668, 671
Kilpelainen, S. H., 374(32), 375(32), 380
Kim, C. H., 651
Kim, J., 478, 671
Kim, S., 447, 469, 470(16)
Kim, U., 200
Kim, Y. I., 431
Kimball, S. R., 230
Kimbrough, T., 478
Kimura, Y., 452
King, C. Y., 526, 527(80), 532(80)
King, E. J., 383(13), 384
Kinney, D., 444
Kipper, J., 394
Kirchman, P. A., 469, 470(16)
Kirkegaard, K., 632
Kirschner, M. W., 457
Kirshnamoorthy, T., 240
Kispal, G., 387(39), 393
Kiss, J. Z., 98, 105
Kisselev, L., 508, 527(41; 42)
Kitts, P., 525
Kjeldsen, T. B., 342
Klar, A., 128, 129(16)
Klasson, H., 339
Klaus, C., 387(36), 393
Kleinschmidt, J. A., 395
Klemic, J. F., 279
Klemic, K. G., 279
Klionsky, D. J., 129, 346, 409, 427, 625, 626(5)
Klis, F. M., 671
Klootwijk, J., 632
Klunk, W. E., 530

Knobel, S. M., 28
Knoblich, J. A., 468
Knop, M., 165, 418, 646
Knorre, V. D., 299
Kobayashi, R., 253
Kochneva-Pervukhova, N. V., 507, 519
Koepp, D. M., 248, 590, 609
Kohlwein, S. D., 4, 34, 35, 38(2), 39(2), 412, 664, 667, 669(20)
Koleske, A. J., 173
Kölling, R., 331, 334(16), 337(16)
Kolodny, M. R., 128, 129(20), 133(20), 149(20)
Kolodziej, P. A., 167, 640
Kolodziej, R. A., 622
Komar, A. A., 500, 526(18)
Komeili, A., 588, 622
Kominami, E., 130, 132(45)
Konarska, M. M., 214
Kondepudi, A., 525
Kondu, P., 339, 443
Koning, A. J., 4, 671
Konings, W. N., 339
Konopka, J., 50, 55, 72(11)
Koonin, E. V., 458, 464(11)
Korn, E. D., 513
Kornberg, R. D., 131
Kornitzer, D., 252, 256(11), 639, 647
Koshland, D., 22, 457
Kosova, B., 81
Koster, A. J., 86
Kotter, P., 662
Kovacech, B., 462
Koval, M., 22
Kowal, A. S., 500, 510, 525, 526(16), 527(16; 74), 529, 529(16; 74), 530(16; 84), 531, 533, 533(74), 534(16; 84), 536, 536(16)
Kozubek, M., 17
Kragh-Hansen, U., 330
Krainer, E., 446
Kranz, J. E., 339, 443
Krappmann, D., 588
Krasnov, N. V., 299
Kraut, R., 468
Krawchuk, M. D., 166
Krawiec, Z., 444
Krebs, D. L., 281
Kremer, J. R., 83
Kress, M., 508, 527(41)
Krieger, M., 130
Krimkevich, Y., 366

AUTHOR INDEX

Krishnamoorthy, T., 221
Kron, S. J., 3, 15, 22, 25(17a), 433, 434(1), 435(1)
Kronidou, N. G., 446
Kruckeberg, A. L., 671, 673
Krutchinsky, A. N., 306
Kryndushkin, D. S., 514
Kubota, Y., 184, 185(4)
Kubrich, M., 446
Kuehn, M. J., 267, 278, 341
Kuhlbrandt, W., 343
Kulka, R. G., 640, 647
Kumar, J., 505, 517(33), 519(33)
Kunau, W.-H., 370, 374, 374(33), 375(21; 33), 380
Kunisawa, R., 335
Kuno, T., 447, 449, 452
Kurihara, L. J., 478, 488(10), 493(10)
Kurihara, T., 265, 268(23; 25), 278(25)
Kuriowa, T., 386
Kurjan, J., 477
Kuroiwa, T., 392
Kushnirov, V. V., 500, 507, 507(9), 508, 509(35), 510(9), 513(9; 13), 514, 516(9), 517, 517(9), 519, 519(9), 523(13), 524(13), 526(9), 527(35; 43)
Kusner, Y. S., 299
Kuster, B., 320
Kutay, U., 588, 589(7), 608, 614(5), 618
Kvam, E., 671
Kweon, Y., 662, 664, 671, 674(6)

L

Labas, Y. A., 382, 383, 665
Labib, K., 540, 567(14), 588
Lacomis, L., 307
Lacroute, F., 499, 508(6), 509(6), 651
Lacy, J. S., 55
Ladinsky, M. S., 83, 95
Ladjadj, M., 84
Lagerwerf, F. M., 312
Lagunas, R., 340
Lai, C. Y., 469, 470(16)
Laloraya, S., 449
Lambertson, D., 248
Lamont, S., 84, 95
Lander, E. S., 457
Landsman, D., 457, 460(2), 462(2)

Lane, D., 152(3), 153, 167(3), 251
Laney, J. D., 248
Langanger, G., 71
Langer, T., 449
Lanker, S., 253
Lappalainen, P., 172, 433, 434(4)
Lark, R., 366
Laroche, T., 199
Larsen, C. N., 647
Laskey, R. A., 184, 589, 591(11), 620
Latterich, M., 259, 264(14), 447, 478, 488(10), 493(10)
Lau, D., 585
Lau, W. T., 278, 341
Lavon, I., 317
Lawrence, C. W., 223
Lawrence, M. C., 86, 88
Laxey, J. D., 648
Lazarow, P. B., 329
Leberer, E., 477
Lechler, T., 433, 434(6), 435(6)
Lee, B. S., 658
Lee, D. H., 646
Lee, J. H., 222, 227(8)
Lee, K. H., 510
Lee, L. H., 598, 609
Lee, M. S., 614, 615(30)
Lee, P., 453
Lee, T. D., 220
Leeds, P., 658
Leff, J., 387
Le Goff, X., 508, 527(41; 42)
Legrain, P., 214
Le Guellec, R., 508, 527(42)
Lehner, C. F., 611
Leith, A., 84
le Maire, M., 330
Lemire, S., 662
Lemmon, S. K., 343, 671
Lengieza, C., 339, 443
Lenney, J. F., 129, 132(40; 42), 141
Lepault, J., 102, 121(13)
Lerner, M. R., 203
Leroux, M., 443
LeSage, A., 15
Lesser, C. F., 214
Lester, R. L., 337
Leunissen, M., 67, 68(31)
Levi, B. P., 427
Levine, B., 37

Le Vine, H. III, 532
Levy, F., 642
Lewis, D. E., 625, 626(7), 628(7)
Lewis, J. B., 200
Lewis, M. J., 358, 671
Lewis, W. H. P., 138
Lewitin, E., 507, 519(37)
Lew-Smith, J. E., 339, 443
Li, J., 173, 367(19), 370, 373, 374(19), 673
Li, L., 507, 508(36)
Li, R., 433, 434(6), 435(6), 671
Li, X., 370
Li, Y., 84, 224, 342
Liang, C., 198
Liang, H., 374
Liang, X. H., 37
Liebman, S. W., 500, 501, 501(11), 507(9), 509, 510, 510(9; 11), 513(9; 11), 514(11), 516, 516(9), 517, 517(9; 51), 519, 519(9), 524(11), 525, 526(9; 75)
Liebundguth, N., 662
Lightbody, J. J., 452
Lill, R., 447
Lin, L., 478
Lin, R. J., 128, 200, 207(9), 208(9)
Lin, S. J., 472
Linder, P., 19
Lindquist, S., 445, 448, 449, 450, 452, 452(40), 453, 499, 500, 501, 501(11), 507, 507(12), 508(36), 510, 510(11), 513(11), 514, 514(11; 22), 516, 516(40), 519, 519(12), 522(12), 524(11), 525, 525(12; 40), 526, 526(16; 40), 527(16; 74; 81), 529, 529(16; 74), 530(16; 84), 531, 533, 533(74), 534(16; 84), 536, 536(16)
Lindsey, S., 84
Ling, R., 252(36), 257
Link, A. J., 220
Linskens, H. K., 647
Lippincott, J., 671
Lippowsky, G., 589
Liu, J. J., 500, 507(12), 519(12), 522(12), 525(12), 526, 526(16), 527(16; 81), 529(16), 530(16), 534(16), 536(16)
Liu, P., 443
Liu, W., 86
Liu, X. D., 452
Liu, X. F., 342
Liu, Y., 370
Ljungdahl, P. O., 278, 339, 341

Llopis, J., 19, 49, 382, 664
Loboda, A. V., 306
Lockhart, D. J., 457, 460(2), 462(2)
London, I. M., 246
Long, R. M., 579, 580(6)
Longtine, M. S., 166, 248, 418, 592
Lopez, N., 514
Lopez, P. J., 202
Lord, J. M., 331
Lord, P. G., 3
Lorenz, M. C., 339
Losko, S., 325, 331, 334(16), 337(16)
Losson, R., 651
Lotscher, H. R., 430
Louis, E. J., 203, 297
Lourbakos, A., 671
Louvion, J. F., 449, 450(31), 451(31), 515
Lowry, O., 142
Lu, Z., 447
Lucchini, G., 185, 458
Luders, J., 448
Lue, N. F., 131
Luger, K., 48
Lührmann, R., 203, 215, 218, 218(47), 219(47), 620
Lui, M., 307
Lukyanov, S. A., 382, 383, 665
Lum, P. Y., 4
Lunsdorf, H., 535
Luo, W.-I., 343, 344(38), 345(38), 347(38)
Luo, Y., 366
Lupashin, V. V., 334, 352
Lustig, A. J., 200
Lusty, C., 444
Luther, P. K., 88

M

Maarse, A., 446
Macara, I. G., 589
MacKay, V. L., 330, 335(14), 356, 364, 458
MacMillan, S. E., 229, 236(16), 247(16)
Madden, D., 278
Maddox, P. S., 31
Mader, A. W., 48
Madhani, H. D., 339
Madura, K., 248
Magrath, I. T., 504

Mahanty, S. K., 598, 609, 611(17), 613, 614(17), 615(17), 617(17), 618(17), 619(17), 622(17)
Maitra, U., 244
Makarow, M., 342
Malczynski, M., 478
Maldonado, A. M., 344
Malhotra, V., 265
Malkus, P., 278, 341
Malnoe, P., 173
Malpartida, F., 329
Mamay, C. L., 81, 99, 105(5)
Mandala, S. M., 343
Mandelkow, E. M., 121
Maniatis, T., 175, 193, 213, 218(34)
Mann, C., 647
Mann, M., 180, 215, 218, 218(47), 219, 219(47), 281, 282, 293, 296, 299, 302, 303, 305, 306, 312, 316, 317, 319, 320, 387(41), 393, 660
Manney, T. R., 129, 458
Manning, A. M., 317
Marchall, W. F., 681, 683(38)
Marchenko, N. D., 620
Marchler, G., 444
Marcusson, E. G., 360, 361
Marek, D., 195
Mares-Guia, M., 130, 132(49)
Marfatia, K. A., 594
Margulies, R. U., 540
Marini, A. M., 339
Markelov, M. L., 382, 383, 665
Markham, A. F., 619
Marko, M., 84, 89
Marks, K. M., 343
Markwell, M. A., 430
Marr, R. S., 595
Marriott, D., 255
Marriott, M., 331
Marsh, B. J., 95
Marsh, J. A., 252(34), 257, 452
Marsh, L., 478, 488(15), 611, 612(24)
Marshak, D. R., 154
Marshall, P., 366
Marshall, T. K., 169
Marshall, W. F., 31
Marszalek, J., 447
Martinez-Pastor, M. T., 609
Martone, M. E., 84, 95
Masison, D. C., 500, 507(14), 508(14), 514, 516(14), 523(14), 631, 632, 639(6)

Masse, S. V., 517
Mastronarde, D. N., 81, 82, 83, 89, 95, 99, 105(5; 9), 113(9)
Mastropaolo, W., 243
Matern, H. T., 668
Matese, J. C., 443
Mateus, C., 673
Mathias, N., 253
Matile, P., 141
Matsuda, Y., 129
Matsumoto, S., 452
Matsuoka, K., 260, 265(16)
Mattaj, I. W., 608, 609(6), 618, 620, 621(6)
Matts, R. L., 246
Matula, P., 17
Matz, M. V., 382, 383, 665
Maunsbach, A. B., 121
Maxfield, F. R., 17
Mayer, M. P., 443, 449(7), 451(7)
Mayes, A. E., 658
McArthur, A. G., 200
McCaffery, J. M., 19, 382, 383, 383(13; 14), 384, 625, 626(6), 664
McCaffrey, G., 478
McCammon, M. T., 387(38), 393
McCann, J. A., 448
McCarroll, R. M., 555, 556(26), 562, 567(28)
McConnell, S. J., 382
McCormack, A. L., 170, 433, 434(5)
McCracken, A. A., 248, 341
McCready, S. J., 509, 527(44)
McCusker, J. H., 166
McDonald, K., 51, 57(12), 81, 82, 84, 85(26), 96, 98, 99, 105, 105(5; 6), 114, 122
McEwen, B. F., 84
McGee, T. P., 352
McGrath, J. P., 647
McGrew, J., 462, 590
McIntosh, J. R., 35, 81, 82, 83, 95, 99, 105(5; 6; 9), 113(9)
McKenzie, A., 418
McKinney, J. D., 458
McLaughlin, C. S., 503, 510(28), 633
McMullin, T. W., 480, 509
McNabb, D. S., 469
McNulty, D. E., 307
McPheeters, D. S., 214, 215(41)
McVey, M., 468
Mechler, B., 136, 139, 646
Mechtler, K., 281

Meeusen, S., 381, 383(21), 386, 387(21), 392(21)
Meijer, M., 387(37), 393, 446
Meinkoth, J. L., 615(36), 619, 621(36)
Melki, R., 500, 526(18)
Mendenhall, M. D., 283, 284(18)
Mercurio, F., 317
Merrick, W. C., 221, 222, 227(8), 236
Mertens, D., 366, 370(6), 374(6), 375(6), 377(6)
Mesiarz, F. R., 131
Mewes, H. W., 203, 297
Meyer, D. I., 331
Meyer, J., 141
Meyers, A., 444
Miao, B., 448, 449
Michael, W. M., 589, 620
Michaelis, C., 31, 257, 640
Michaelis, S., 491
Michell, P., 660
Michimoto, T., 447
Mihalik, S. J., 367, 368(12)
Mikesell, G. E., 280
Milisav, I., 447, 449
Miller, K. G., 173
Miller, M. E., 619, 621(35)
Milligan, R. A., 121
Mills, J. P., 384, 387(12)
Mills, K., 469
Mimura, S., 184, 185(4)
Mini, T., 302, 387(41), 393
Mirkin, S. M., 540
Mishima, M., 609
Misra, L. M., 449
Mitchell, A. P., 127, 130(11), 133(11), 149
Mitchell, D. A., 169
Mitchison, T., 179, 180(17; 18)
Miyakawa, I., 386, 392
Miyakawa, T., 447, 449, 452
Miyawaki, A., 19, 36, 37(14), 38, 38(14), 43(14), 382, 598, 664
Mize, G. J., 220
Moazed, D., 172, 173, 174(7), 178(7; 8), 287
Moczko, M., 387(39), 393
Moehle, C. M., 128, 129, 129(20), 133(20), 136, 149(20; 22; 29)
Moeremans, M., 71
Moll, T., 280, 609
Moll, U. M., 620
Monia, B., 252(24), 256, 647
Monribot, C., 302, 387(41), 393

Montgomery, D. L., 387(38), 393
Monty, K. J., 156, 162(9)
Moon, A., 50
Moon, J., 505
Moor, H., 81, 96, 102(1), 103(1), 105
Moore, C., 200
Moore, M. J., 202, 205
Moore, T. S., 331
Morano, K. A., 443, 452
Moreau, V., 646
Morehouse, H., 590
Moreland, R. B., 620
Moreno, E., 340
Moreno, S., 194
Morgan, D. O., 287, 452, 607
Morgan, G., 81
Morgan, S. A., 642
Moriyama, H., 514
Morrell, J. C., 367, 368, 368(12), 374(27), 375(27), 377
Morris, D. R., 220
Morrison, H. G., 200
Morrissey, J. H., 178
Morsomme, P., 278, 344
Mortensen, P., 312
Mortimer, R. K., 468, 469(5)
Mosammaparast, N., 620
Moslehi, J. J., 525, 527(74), 529, 529(74), 530(84), 531, 533, 533(74), 534(84), 536
Mostl, D., 662
Moy, T. I., 585
Moye-Rowley, W. S., 449
Mozdy, A. D., 383, 383(13; 14), 384
Mueller, S. C., 351, 392
Mühlethaler, K., 81, 96, 102(1), 103(1)
Muhlrad, D., 651, 652, 654, 654(10; 11), 656(11; 16), 659(16)
Mukai, H., 447, 449, 452
Mulero, J. J., 480, 509
Mulholland, J., 50, 55, 72(11), 77(7), 122
Muller, E. G. D., 34
Muller, H., 136, 446, 646
Muller, M., 377
Müller, M., 100, 136
Müller-Reichert, T., 51, 57(12), 84, 85(26), 96, 106
Mundy, C. R., 500, 510(10)
Muñiz, M., 278
Munn, A. L., 625, 626(5)
Munro, S., 356, 363

Murakami, K., 143
Murakami, Y., 203, 297
Murata, S., 143
Murdock, D. G., 127, 133(2)
Murray, A. W., 31, 173, 671, 681, 683(38)
Mursula, A. M., 374(32), 375(32), 380
Myslinski, E., 211

N

Na, S., 344
Nagasu, T., 320
Nagley, P., 667, 671
Naik, R. R., 129, 134
Naiki, H., 532
Najita, L., 222
Nakai, T., 136, 141, 141(64), 147(64)
Nakamura, S., 392
Nakamura, Y., 501
Nakanishi-Shindo, Y., 330
Nakano, A., 339
Nakayama, K. I., 330
Nakayashiki, T., 501
Nakielny, S., 620
Nam, H. G., 620
Naranda, T., 229, 236, 236(16), 247(16)
Nasmyth, K., 31, 199, 257, 280, 281, 283, 284(18), 418, 457, 458, 462, 464(9–11), 540, 609, 622, 640
Nason, T. F., 253
Nasrin, N., 229
Nathan, D. F., 449, 452(40), 453
Natsoulis, G., 640
Natter, K., 34, 38(2), 39(2), 667
Nau, K., 374(27), 375(27), 377
Navarrete, R., 329
Navas, T. A., 461
Nawotka, K. A., 184, 540
Nebes, V. L., 129
Necas, O., 81
Nehrbass, U., 614
Neigeborn, L., 129
Nelson, M., 619
Nelson, R. J., 449
Neubauer, G., 215, 218, 218(47), 219(47), 281, 320
Neumann, K., 98
Neupert, W., 382, 383, 383(5), 387(33; 36), 393, 447, 449, 631, 671

Nevalainen, H., 444
Neville, M., 597
Newlon, C. S., 539, 551, 554(23), 562, 567(28)
Newman, A. J., 128, 200, 207(9), 208(9)
Newman, G. R., 111, 113(30), 122(30)
Newman, S. M., 387(31), 392
Newnam, G. P., 501, 505, 507, 513, 514(22), 516, 517, 517(33; 37), 519(33), 525(53), 526(53; 64)
Ng, R., 200
Nichols, B. J., 352, 358, 671
Nicolas, G., 102, 121(13)
Nicolet, C. M., 444, 449, 452
Niedenthal, R. K., 593
Nielson, K., 239
Nierras, C. R., 508, 509, 527(43; 44)
Nigg, E. A., 611
Nijbroek, G., 491
Nika, J., 221, 230, 235(20), 247
Nikolaev, N., 632
Nikolaev, V. I., 299
Nishida, E., 609
Nishimoto, T., 588
Nishizawa, M., 141
Nixon, J. E., 200
Noble, J. A., 136
Noble, S. M., 200(14), 201
Noda, Y., 353, 358(15)
Nomura, M., 621
Nonet, M., 583
Nothwehr, S. F., 342, 343, 350
Novick, P., 50, 70, 258, 340, 351
Nowakowski, S., 667
Nunnari, J., 381, 383(13; 21), 384, 386, 387(12; 21), 392(21)
Nuoffer, C., 278
Nurse, K. L., 280
Nurse, P., 194
Nyland, S. V. M., 541, 568(19)

O

Obuse, C., 567
O'Connor, J., 58, 668, 669, 671
Oda, Y., 313, 320
O'Dell, J. L., 433
O'Donnell, K. L., 104
Odorizzi, G., 415, 428(22)
Oegema, K., 179

Oehlen, L. J., 458, 464(8)
Oehm, A., 185
O'Farrell, P., 301
Ohashi, A., 136, 141, 141(64), 147(64)
Ohba, M., 447
Ohlson, M., 447
Ohsawa, T., 143
Ohsumi, Y., 81, 99, 104(7; 8), 121(7; 8), 411, 428
Ohtsubo, M., 588
Okamoto, K., 383(18), 384, 386(18)
Okazaki, H., 588
Olesen, J., 131
Oliver, S. G., 203, 297
Olsen, G. E., 200
Olsen, P., 339, 443
O'Neill, E. M., 590, 609, 613, 614, 615(31)
Ong, L. Y., 551, 554(23)
Ono, B., 500, 501, 501(11), 510(11), 513(11), 514(11), 524(11)
Opplinger, W., 446
Orci, L., 259, 260, 265, 265(9; 15; 16), 266(9), 268(25), 278, 278(25)
Orlean, P., 50, 77(7), 334
Orlova, A., 434
Ormö, M., 664
O'Shea, E. K., 278, 341, 588, 590, 593, 607, 609, 609(1), 613, 614, 615(31), 618, 618(1), 619(19), 622
Osherovich, L. Z., 507
Ostermann, J., 449
Osumi, M., 81, 96, 103(2), 115(2), 121(2)
O'Toole, E., 35, 81, 82, 99, 105(5; 9), 113(9)
Overton, M. C., 600, 669
Ozcan, S., 339
Ozkaynak, E., 395

P

Paddock, S., 4
Padgett, R. A., 200
Padmanabhan, R., 620
Pain, V. M., 243
Palmer, E., 501
Palmer, R. E., 22
Palmieri, F., 387(39), 393
Palmisano, A., 387(39), 393
Pan, X., 671
Panaretou, B., 344

Panch, H. R., 671
Pandey, A., 320
Panek, H. R., 343, 662, 664, 664(7), 668(7), 671, 674(6)
Panniers, R., 246
Papa, F., 256
Park, F. J., 128, 129(20), 133(20), 136, 137(59), 149(20)
Park, P. U., 468, 469
Parker, M. L., 458
Parker, R., 136, 636, 637, 648, 651, 652, 654, 654(9–11), 655(9; 12; 14), 656(11; 16), 657(9), 658, 659, 659(16), 660(14; 28), 665(9)
Pasero, P., 184, 185, 186, 186(12), 187(11; 12), 189, 193(12), 194(13), 195(11), 196(12), 197, 197(12), 198, 198(24), 199(12; 24)
Passamaneck, N. Q., 200
Patel, D., 583
Patel, F. I., 427
Patino, M. M., 500, 507(12), 510, 516, 519, 519(12), 522(12), 525(12), 526(16), 527(16), 529(16), 530(16), 534(16), 536(16)
Patterson, B., 200
Patterson, G., 24, 25, 28
Pattison, J., 673
Patton, E. E., 253
Patton, J. L., 337
Pauling, M. H., 202
Paulovich, A. G., 540
Paushkin, S. V., 500, 508, 513(13), 514, 523(13), 524(13), 527(43)
Pavitt, G. D., 221, 230, 235(19; 20), 240
Pavlenko, V. A., 299
Payne, G. S., 330, 335(14), 356, 363, 364, 415, 428(22)
Pearl, L. H., 443, 451(6)
Pearson, C. G., 31
Pelham, H. R., 352, 358, 415, 435, 671
Pellman, D., 150
Peltz, S. W., 448, 637, 641, 651, 658
Pemberton, L. F., 590, 594, 620
Penczek, P., 84, 89
Penman, S., 203
Pennington, R. J., 377
Pereira, G., 418
Periasamy, A., 36
Perkins, G. A., 84
Perkins, S. J., 162, 163(12)

Perlin, D., 344
Perlman, P. S., 383(18), 384, 386(18), 387(31), 392
Perrelet, A., 260, 265, 265(15)
Perrot, M., 302, 387(41), 393
Persson, B. L., 673
Peter, M., 170, 433, 434(5), 609
Peters, J.-M., 457
Peters, R., 193
Peters, T. J., 377
Petersson, J., 673
Petes, T. D., 539
Petfalski, E., 660
Petracek, M. E., 248
Pettegrew, J. W., 530
Pfanner, N., 387(39), 393, 446, 449
Pfeffer, S. R., 264
Pfund, C., 447
Phan, L., 221, 222, 227, 236(7), 237(7), 238, 239, 240, 247(7)
Philippe, M., 508, 527(41; 42)
Philippsen, P., 34, 38(3), 39(3), 203, 297, 418, 662, 667, 671
Philips, J., 478, 489
Phizicky, E. M., 165, 167(18)
Piatti, S., 199, 257, 640
Picard, D., 449, 450(31), 451(31; 32), 452, 453(32), 515
Pickart, C. M., 255
Pikielny, C. W., 214, 215(39)
Pinkham, J. L., 515
Piper, P., 344
Piper, R. C., 428
Piston, D., 24, 25, 28
Planta, R. J., 632
Plevani, P., 131, 185, 458
Ploem, J. S., 631
Plyler, E. K., 17
Podtelejnikov, A. V., 296, 306, 312, 316, 319
Pohlmann, R., 662
Polaina, J., 449
Pollok, B. A., 36
Polozkov, G. V., 517
Pon, L. A., 386, 389, 391(26)
Poolman, B., 339
Popolo, L., 342
Portillo, F., 344
Posas, F., 590, 609
Posewitz, M. C., 281
Potschka, M., 163

Potts, W., 248
Poupart, M.-A., 281, 458, 464(11)
Poznyakovski, A. I., 508, 527(43)
Pozzan, T., 382
Prasher, D. C., 16, 25, 592, 663, 664
Pratt, W. B., 449
Prein, B., 667
Preiss, T., 633
Prescianotto-Baschong, C., 358, 412, 671
Prescott, M., 667, 671
Preston, R. A., 53, 68(19), 411, 413(8), 417(8), 418(8)
Preuss, D., 50, 77(7)
Price, C., 457
Prien, B., 34, 38(2), 39(2)
Pringle, J. R., 3, 50, 53, 68(19), 127, 130(8), 383, 411, 413(8), 417(8), 418, 418(8), 461, 493, 591
Priola, S. A., 523
Prodromou, C., 443, 451(6)
Proud, C. G., 243
Pruliere, G., 173
Prusiner, S. B., 499, 500, 526(19)
Pruyne, D., 340, 438
Pryciak, P. M., 611, 671
Pryer, N. K., 258, 271
Ptyushkina, M. V., 501, 505(25), 517(25)
Puig, O., 166, 218
Pypaert, M., 352, 427

Q

Qin, J., 222, 236(7), 237(7), 247(7), 281
Query, C. C., 202

R

Raboy, B., 647
Raca, G., 540
Rachubinski, R. A., 366
Radermacher, M., 82, 84, 88
Raghuraman, M. K., 199, 539, 541, 652
RajBhandary, U. L., 224
Rakhilina, L., 250, 257
Ram, A. F., 671
Ramaiah, K. V. A., 230
Ramesh, V., 224
Randall, R., 142

Randolph, A., 149
Rank, N. M., 590, 593, 609, 613, 614, 615(31), 619(19)
Rao, G., 673
Rao, J., 452, 453
Rapaport, D., 387(33), 393
Rasmussen, H. H., 508, 527(41)
Rassow, J., 446
Raths, S., 427
Raue, H. A., 652
Ravazzola, M., 259, 260, 265, 265(9; 15), 266(9), 268(25), 278(25)
Raveh, D., 647
Raymond, C. K., 351, 411, 417, 421(24), 426(6), 430(24), 493
Rebischung, C., 34, 38(3), 39(3), 662, 667
Rechsteiner, M., 255
Recktenwald, D. J., 624
Redding, K., 356, 359
Reed, M. A., 279
Reed, S. I., 254, 452
Reedy, M. C., 89
Reedy, M. K., 89
Reggiori, F., 415
Rehling, P., 358
Reich, C. I., 200
Reichert, V., 205
Reinke, C. A., 668, 671
Reizman, H., 427
Remington, S. J., 664
Renkin, C. W., 84
Repetto, B., 387(40), 393
Repnevskaya, M. V., 472
Resnick, D., 363
Reuter, R., 203
Rexach, M. F., 258, 259, 264(14), 265(9), 266(9), 273(6), 277, 353, 589
Reynard, G., 282, 283(15), 284(15), 293(15)
Reynolds, A. E., 562, 567(28)
Reynolds, E., 79, 497
Ribas, J. C., 632, 639(6)
Ribbeck, K., 589
Richardson, W. D., 589, 619
Richmond, R. K., 48
Richmond, T. J., 48
Rickoll, W., 122
Rickwood, D., 334
Ricupero-Hovasse, S. L., 602
Ridge, R. W., 104
Rieder, S. E., 346, 352, 356, 364

Riezman, H., 50, 57(5), 278, 342, 352, 358, 412, 427, 671
Rigaut, G., 166, 180, 219, 282, 658
Riles, L., 593
Rine, J., 50, 57(6), 58(6), 77(6), 102, 257, 671
Rines, D. R., 16, 19, 29(14)
Rinke, J., 203
Rippel, S., 247
Risse, B., 447
Rist, B., 301, 313
Rivin, C. J., 539, 567(4)
Rizzuto, R., 382
Robards, A. W., 103
Robbins, J., 589, 620
Robbins, P. W., 50, 77(7), 334, 363
Roberg, K. J., 265, 268(24), 331, 342
Roberge, M., 191
Roberts, B. L., 589
Roberts, C. J., 351, 417, 421(24), 430(24), 493
Roberts, J. M., 200
Roberts, P., 621, 671
Robertson, L. S., 339, 443
Robinett, C. C., 671
Robins, D. M., 453
Robinson, L. C., 343, 414, 661, 662, 664, 664(7), 668(7), 671, 674(6)
Robitsch, H., 280, 609
Rochon, Y., 312, 387(35), 393
Rodal, A. A., 168, 172, 434
Roepstorff, P., 299, 305
Röhm, K. H., 128, 129(17), 140(17), 147(17), 148(17)
Rohrer, J., 352, 427
Romano, P. R., 281
Roncero, R., 341
Rosbash, M., 213, 214, 215(39), 218, 597
Rose, M. D., 50, 77(7), 187, 348, 447, 449, 477, 478, 480, 481(18), 484, 488(10; 15; 16), 489(21), 491, 493(10; 18), 560, 611, 612, 612(24), 626
Rosebrough, N., 142
Rosenblatt, J., 179, 180(18)
Rosenblum, J. S., 590, 594
Rosenfeld, J., 294
Rossanese, O. W., 58, 414, 416(21), 417(21), 668, 669, 671
Rossi, R., 382
Roth, A. F., 343
Roth, J., 67, 68(30)
Rothblatt, J., 258, 394, 406, 407(19)

Rothe, S., 203
Rothman, J. E., 258, 264, 265
Rothman, J. H., 136, 328
Rothstein, R., 395, 516
Rottensteiner, H., 671
Rout, M. P., 82, 121, 394, 399(2), 401, 405, 406(3; 14), 407(3), 493, 588, 589, 590, 591, 594
Rowlands, A. G., 246
Rowley, N., 331, 342
Rozijn, T. H., 394
Rubenstein, P. A., 434
Rubin, D. M., 647
Ruby, S. W., 214
Ruddock, L. W., 511
Ruis, H., 444, 609, 671
Rumenapf, T., 642
Ruohola, H., 351
Rupp, S., 646
Russell, L. D., 98, 113(3)
Russell, P. J., 632
Rutz, B., 180, 219, 282
Rymond, B. C., 214, 215(39)
Rytka, J., 374(30), 375(30), 380

S

Saari, G. C., 136, 328, 458
Saavedra, C., 577
Sachs, A. B., 222, 227(6)
Sacksteder, K. A., 365, 370
Sadis, S., 252(21; 24), 255, 256, 256(21), 647
Sagliocco, F., 302, 312, 387(41), 393
Saheki, T., 128, 129, 129(15)
Saidowsky, J., 395
Saito, H., 590, 609
Sakakibara, S., 143
Salama, N. R., 259, 265, 265(9), 266(9), 271, 271(20), 272(20)
Salema, R., 56
Salgado, J., 218
Salminen, A., 70, 351
Salmon, E. D., 4, 19, 31, 35, 41(6), 412, 671, 676, 679(35), 680(36), 681(35)
Samadashwily, G. M., 540
Sambrook, J., 175, 193, 213, 218(34)
Samelson, L. E., 281
Sanchez, J. A., 195
Sanders, S. L., 346

Sanderson, C. M., 331
Sando, N., 386, 392
Sandrock, T. S., 433
Santocanale, C., 540
Santoso, A., 507, 509, 526(45), 527(45), 530(45)
Sargent, D. F., 48
Sarkar, S., 585, 590
Sarnow, P., 222
Saupe, S., 516
Savitsky, A. P., 382, 383, 665
Sawicki, G. J., 529, 530(84), 533, 534(84)
Sawyer, M. M., 622, 671
Scafe, C., 583
Scarborough, G. A., 343
Schaefer, M., 579, 580(6)
Scharer, F., 662
Schatz, G., 166, 386, 389, 392, 446, 643
Schauber, C., 248
Scheffer, R. C., 72
Schekman, R., 102, 258, 259, 260, 262, 264(13; 14), 265, 265(9; 15; 16), 266, 266(9), 267, 268, 268(8; 23; 25), 269(29), 271, 271(20), 272(20), 273(6), 277, 278, 278(25), 327, 330, 334, 341, 342, 344, 346, 351, 352, 353, 394, 406, 407(19), 447, 448, 478, 488(10), 493(10)
Schellenberg, M., 141
Schiebel, E., 418
Schieltz, D., 173, 174(11), 220, 283, 383(21), 386, 387(21), 392(21)
Schiffer, H. H., 646
Schilke, B., 447, 449
Schimke, R. T., 634, 639
Schimmoller, F., 342
Schirmer, E. C., 500, 526(16), 527(16), 529(16), 530(16), 534(16), 536(16)
Schlenstedt, G., 447, 595, 596(31), 597(31)
Schleyer, M., 258, 273(6), 353
Schliwa, M., 4
Schlossmann, J., 387(39), 393
Schlumpberger, M., 500, 526(19)
Schmeizl, B., 585
Schmid, R., 359
Schmidt, A., 342
Schmidt, U., 585, 590
Schneider, B. L., 165, 622
Schneider, H. C., 387(36), 393
Schneider, M., 374
Schneider, U., 302, 387(41), 393

Schober, W., 625, 626(7), 628(7)
Schoenfeld, L., 238
Schroder, S., 342
Schroeder, A. J., 590
Schroeter, J. P., 84
Schuller, C., 444, 609
Schultz, L. D., 590, 673
Schulz, K., 374(27), 375(27), 377
Schuster, T., 457, 462
Schuyler, S. C., 150
Schwamborn, K., 609
Schwartz, D. C., 636, 654
Schwarz, E., 447
Schwenke, J., 127, 130(12)
Schwer, B., 206
Schwob, E., 184, 185, 189, 197, 198, 198(24), 199(24), 283, 284(18), 458
Sclafani, R. A., 387
Scott, J. H., 262, 327
Seaman, M. N., 360, 361
Searfoss, A. M., 631, 632, 633, 636, 637(9), 639(9)
Sears, I. B., 58, 668, 669, 671
Sedat, J. W., 16, 17(2), 18, 26(2), 27, 27(2), 31, 86, 681, 683, 683(38)
Seedorf, M., 590
Seeger, M., 330, 335(14), 356, 364
Seelig, L. P., 251, 350
Sefton, M., 673
Ségault, V., 211
Segrev, N., 50
Seiler, S. R., 202
Seipel, K. H., 631
Seki, T., 199
Senger, B., 590
Sentenac, A., 191
Seol, J. H., 287, 413, 415
Serafini, T., 265
Séraphin, B., 180, 202, 214, 218, 219, 282
Serio, T. R., 499, 525, 527(74), 529, 529(74), 530(84), 531, 533, 533(74), 534(84), 536
Seron, K., 671
Serpell, L., 529, 530(84), 533, 534(84)
Serrano, R., 326, 329, 343, 344
Sesaki, H., 383, 384
Seto-Young, D., 344
Seufert, W., 622, 647
Severs, N. J., 81
Sexton, J., 583
Shabanowitz, J., 620

Shah, N. G., 418
Shaham, S., 609
Shalev, A., 239, 240
Shani, N., 374(28; 34), 375(28; 34), 380
Shannon Moore, M., 609, 621(7)
Shapiro, R. A., 590
Sharif, W. D., 28
Sharp, D. J., 122
Sharp, P. A., 200, 202, 214
Sharp, W. J., 449
Shaw, E., 130, 132(49)
Shaw, J. M., 381, 383, 383(13; 14), 384, 387(12)
Shaw, P. J., 683
Shaw, S. L., 4, 35, 41(6), 412, 671, 676, 679(35), 680(36), 681(35)
Shaywitz, D. A., 325
Shen, S. H., 190, 354
Shenoy, S. M., 579, 580(6)
Shepard, C., 130
Shepherd, J. C., 265
Sherlock, G., 443, 457, 460(3)
Sherman, F., 40, 223, 253, 501, 524
Sherwood, P. W., 341
Shevchenko, A., 180, 219, 282, 287, 302, 303, 312, 344, 387(41), 393, 413, 415, 658, 660
Shih, S., 252(23), 255, 256(23), 647
Shilling, J., 449
Shimoni, Y., 258, 265, 268(25), 278(25)
Shin, T. H., 622
Shinoda, S., 501
Shinomiya, T., 567
Shirahige, K., 567
Shkurov, V. A., 299
Shoemaker, D. D., 374
Shore, D., 457
Shou, W., 173, 178(8), 279, 287
Shulga, N., 594, 595(27), 596(27), 621, 671
Shulman, R. W., 642
Shuster, J. R., 149
Sidorova, J. M., 280
Siegel, L. M., 156, 162(9)
Siegers, K., 418
Sikorski, R. S., 347, 364(29), 380, 603
Siliciano, P. G., 200
Silver, P. A., 37, 49(16), 447, 525, 580, 585, 587, 590, 595, 596(31), 597(31), 600, 602(44), 604(44), 605(44), 609, 611, 612(23), 613, 614, 615(30), 617(26), 618, 618(26), 619, 669, 671, 672(28)

AUTHOR INDEX

Simons, K., 344
Simos, G., 585, 590
Sinclair, D., 469, 474(15)
Singer, M. A., 510
Singer, R. A., 222
Singer, R. H., 579, 580(6)
Singer-Kruger, B., 50, 55, 72(11)
Siniossoglou, S., 319
Sipos, G., 333, 351, 426
Sitte, H., 98
Skarstad, K., 623
Skibbens, B., 4, 676, 680(36)
Skierski, J., 473
Skowyra, D., 257
Skrzypek, M. S., 339, 443
Slayman, C. W., 343, 344, 364
Sleytr, U. B., 103
Slilaty, S. N., 190, 354
Slot, J. W., 72
Sluder, G., 4, 9(12)
Small, G. M., 366
Smeal, T., 280, 469, 474(12)
Smirnov, V. N., 500, 507, 509(35), 513(13), 514, 517, 519, 523(13), 524(13), 527(35)
Smith, A. E., 589, 619
Smith, B., 367, 368(12)
Smith, C. B., 625
Smith, D., 174, 279
Smith, J. A., 141, 147(75)
Smith, M. M., 49
Snay-Hodge, C. A., 583
Snyder, M., 131, 132, 279, 478
Soderholm, J., 58, 669, 671
Sogin, M. L., 200
Sogo, L. F., 391
Sollner, T. H., 265
Sommer, T., 341, 350
Sondheimer, N., 507, 514, 516(40), 525(40), 526(40)
Sonenberg, N., 632, 639(6)
Song, J. Y., 84
Sorger, P. K., 16, 19, 29(14), 31, 32(28), 151, 435
Soto, G. E., 95
Soussi-Boudekou, S., 339
South, S. T., 370
Spang, A., 259
Spatrick, P., 342
Spector, I., 513
Speicher, D. W., 343

Spelbrink, R. G., 343
Spellman, P. T., 457, 460(3)
Spence, J., 252(21), 255, 256(21)
Spiess, E., 535
Spitz, L., 621
Spotila, L. D., 184, 540
Spragg, S. P., 163, 164(13)
Sprague, G. F., Jr., 477, 478
Spring, K. R., 4, 9(11), 12(11)
Spudich, J. A., 433, 434(1), 435(1)
Srienc, F., 387
Stade, K., 589, 597(18), 613, 617, 619(32), 621(32), 622(32)
Stadler, R., 191
Staehelin, L. A., 95, 103, 105
Stafford, P., 583
Stage-Zimmermann, T., 585, 590
Stagljar, I., 583
Stahl, D. C., 220
Staley, J. P., 202
Stamnes, M., 265
Standing, K. G., 306
Stansfield, I., 504, 508, 527(43)
Staples, R. R., 659, 660, 660(28)
Stark, M. J., 671
Stearns, T., 35, 53, 68(19), 169, 411, 413(8), 417(8), 418(8), 504
Steen, H., 320
Steenbergen, J. N., 517
Steensgaard, J., 163, 164(13)
Stefan, C. J., 413, 415(17), 671, 672
Steiger, M. A., 648
Steinbrecht, R. A., 100, 103
Steiner, B., 622
Steinmetz, L., 457, 460(2), 462(2)
Steitz, J. A., 202, 203
Stepp, J. D., 343
Sterne-Marr, R., 595
Steven, A. C., 500, 526(17)
Stevens, A., 658, 659(18)
Stevens, S. W., 200, 215, 218, 218(46), 219(46)
Stevens, T. H., 136, 328, 342, 343, 350, 351, 358, 408, 409, 411, 417, 421(2; 24), 426(6), 427, 428, 430(24), 493
Stewart, L. C., 382
Stewart, M., 589
Stillman, B., 198
Stirling, C. J., 406, 407(19), 449
Stommel, J. M., 620
Stotz, A., 19

Straight, A. F., 31, 173, 178(8), 287, 671, 681, 683(38)
Strambio-de-Castillia, C., 394, 406(3), 407(3)
Stroobants, A. K., 370
Stroud, R. M., 130
Stryer, L., 35, 36(10), 599, 666
Studer, D., 105
Styles, C. A., 3, 341
Subramani, S., 365, 366
Sullivan, D. M., 343
Sullivan, K. F., 661
Suprapto, A., 394, 405, 406(14)
Surana, U., 280, 462, 609
Sures, I., 200
Susek, R., 445
Sutterlin, C., 342
Svoboda, A., 81
Swaminathan, R., 28
Swaminathan, S., 252, 256(10)
Swanson, R., 250, 252, 257, 647
Sweder, D., 185
Swedlow, J. R., 16, 17(2), 26(2), 27(2)
Swerdlow, P. S., 253
Swiderek, K. M., 220
Swinkels, B. W., 366
Switzer, R. L., 129
Szaniszlo, P. J., 461
Szczepanowska, J., 281

T

Tabak, F., 129
Tabak, H. F., 366, 370
Tabb, M. M., 621
Tadi, D., 129
Takada, K., 143
Takeshige, K., 99, 104(8), 121(8)
Takisawa, H., 184, 185(4)
Takizawa, C. G., 607
Talcott, B., 609, 621(7)
Talin, A., 382
Tan, P. K., 343
Tanaka, A., 330
Tanaka, C., 447, 449, 452
Tanaka, H., 447, 449, 452
Tanaka, K., 103
Tanaka, M., 501
Tanaka, N., 383(32), 387(32), 393
Tanaka, T., 458, 464(10), 540

Tang, J., 218
Tanner, W., 331
Tartakoff, A. M., 568
Tarun, S. Z., 222, 227(6)
Tarunina, O. V., 501, 505(25), 517(25)
Tatchell, K., 414, 622, 661, 662, 664(5), 669(5), 674(5), 679
Taulien, J., 448
Taylor, A. M., 164
Taylor, D. L., 4, 9(10)
Taylor, K. A., 89
Taylor, K. L., 500, 526(17)
Taylor, S. S., 615(36), 619, 621(36)
Tebb, G., 280, 609
Teem, J. H., 344
Teem, J. L., 213
te Heesen, S., 352
Teichert, U., 646
Teixeira, M. T., 319
Tempst, P., 281, 307
Ter-Avanesyan, M. D., 500, 501, 505(25), 507, 508, 509(35), 513(13), 514, 517, 517(25), 519, 523(13), 524(13), 527(35; 43)
Tercero, J. A., 540, 567(14)
Terrell, J., 252(23), 255, 256(23), 647
Tetrault, J., 172
Tettelin, H., 203, 297
Tharun, S., 658
Thatcher, J. W., 384, 387(12)
Theis, J. F., 551, 554(23)
Thiele, D., 443, 452
Thomas, B., 395
Thomas, D. Y., 477
Thompson, C. M., 173
Thorner, J., 131, 132, 335, 477, 661
Thornton, J., 366
Thorsness, P. E., 383(32), 387(32), 393, 480, 509
Thrower, J. S., 255
Thual, C., 500, 526(18)
Thumm, M., 427
Tieaho, V., 671
Tieu, Q., 383(13; 21), 384, 386, 387(21), 392(21)
Tikhodeev, O. N., 501
Tillberg, M., 339, 443
Tipper, C., 342
Titov, A. A., 590
Tittmann, P., 526, 527(80), 532(80)
Tobias, J. W., 255

Toczyski, D. P., 540
Toda, Y., 330
Toenjes, K. A., 622, 671
Toft, D. O., 449, 450(29), 451(29)
Toh-e, A., 447
Tolbert, N. E., 430
Tollervey, D., 585, 660
Tongaonkar, P., 248
Tonino, G. J. M., 394
Torres, R. J., 449
Tourinho dos Santos, C. F., 344
Trachsel, H., 222, 236(5), 247(5)
Traglia, H. M., 588
Tran, P., 19
Trilla, J. A., 341
Trueheart, J., 478, 484, 489, 640
Trumbly, R., 127, 129(7), 139(7), 140(7), 147(7)
Tsai, A., 539
Tsien, F. Y., 382
Tsien, R. Y., 19, 25, 28, 34, 35, 35(1), 36, 36(1), 37(14), 38, 38(14), 39, 43(14), 382, 525, 598, 615(36), 619, 621(36), 661, 662(3), 663, 663(3), 664, 665, 665(10), 666(3)
Tsuji, F. I., 663
Tsurimoto, T., 567
Tu, Y., 16, 592, 663, 664
Tucker, M., 659, 660(28)
Tuite, M. F., 500, 503, 504, 508, 510(10; 28), 511, 527(43)
Tung, K.-S., 577
Turck, C. W., 173, 178(8), 287
Turecek, F., 313
Tye, B. K., 588
Tyers, M., 253, 257, 284, 639
Typke, D., 86
Tzagoloff, A., 387(40), 393, 444

U

Udem, S. A., 585
Udvardy, A., 647
Ugelstad, J., 359
Uhlenbeck, O. C., 218
Uhlmann, F., 281, 458, 464(11)
Ulane, R. E., 127, 130(13), 133
Uptain, S. M., 499, 519
Urban, J., 350
Urbanowski, J. L., 415

Urdea, M. S., 131
Urrestarazu, A., 339

V

Vai, M., 342
Valasek, L., 238, 239
Valdivia, R. H., 664
Valencia-Sanchez, M. A., 659, 660(28)
Valle, D., 374(28; 34), 375(28; 34), 380
Valle, R. C. P., 632, 639(7; 8)
Valls, L., 136, 328
Valouev, I. A., 519
Valtz, N., 478, 484(6)
Van Arsdell, J. N., 136, 137(59)
van Brabant, A. J., 539, 541, 553(22), 554
vanDam, K., 671
van den Berg, M., 366, 369(5), 370, 370(5), 374
Van den Ende, H., 671
Van den Hazel, H. B., 127
van der Geer, P., 280
van Der Leij, I., 370
van der Ley, P. A., 72
van der Rest, M., 339
van der Sluijs, P., 374
Van Dijck, A., 71
VanDusen, W. J., 673
van Genderen, I. L., 72
van Hoof, A., 660
van Meer, G., 72
van Roermund, C. W. T., 374
van Tuinen, E., 50, 57(5)
Varshavsky, A., 252(15), 253, 255, 257, 395, 640, 642, 644, 647
Vater, C. A., 411, 426(6)
Veenhuis, M., 366, 370(6), 374, 374(6), 375(6; 21), 377(6)
Vega, I., 248
Venkov, P. V., 632
Verdier, J. M., 191
Verkman, A. S., 28
Verma, R., 257, 279, 281, 282, 283, 283(13; 15), 284(15), 285, 290(13), 293(13; 15), 294(13)
Vertino, P. M., 594
Vida, T. A., 352, 412, 427(13), 491, 623, 625, 626(7), 628(7)
Vijayraghavan, U., 200, 210(12)
Visintin, R., 287, 607
Vissers, S., 339

Vogel, J. P., 449, 611, 612(23)
Voisine, C., 447
Volland, C., 646
Voorhout, W. F., 72
Voorn Brouwer, T. W., 366
Voos, W., 387(39), 393, 449
Vorm, O., 303, 305, 312
Vornlocher, H. P., 222, 236, 236(7), 237(7), 247(7)
Vos, A., 374
Vos, M. H., 449
Voth, M., 631
Vowels, J. J., 363
Vreken, P., 652

W

Wach, A., 34, 38(3), 39(3), 166, 418, 662, 667
Wada, A., 609
Wada, Y., 411
Waddle, J. A., 366, 412, 414(11), 671, 676, 679, 681(37)
Wagner, A., 444
Wagner, J. A., 381, 384
Wahl, G. M., 620
Wahls, W. P., 166
Wakem, L. P., 524
Walbot, V., 634
Walczak, C. E., 179
Walmsley, R., 673
Walter, P., 102, 106, 394
Walter, W., 447, 447(26), 448, 449, 449(26)
Walworth, N. C., 70, 351
Wang, C., 450, 453(33)
Wang, Q., 344
Wang, S., 673
Wang, T., 366
Wang, Y., 598
Wang, Y. L., 4, 9(10)
Wang, Y. M., 609, 611(17), 613, 614(17), 615(17), 617(17), 618(17), 619(17), 622(17)
Wang, Y. X., 625, 626(5)
Wange, R. L., 281
Wangh, L. J., 195
Ward, W. W., 16, 525, 592, 663, 664
Warner, J. R., 242, 585, 610, 642, 651
Washburn, M. P., 315

Waterston, R. H., 412, 414(11), 671, 679, 681(37)
Watkins, P. A., 374(28), 375(28), 380
Wattenberg, B. W., 264
Watts, J. D., 281, 320
Webb, G. C., 127, 133(1), 408
Webb, W. W., 39, 663, 665(10)
Weert, M., 312
Wegrzyn, R. D., 501, 513, 514, 514(22), 516, 525(53), 526(53; 64)
Weinberg, R., 203
Weinreich, M., 186, 194(13)
Weinrich, M., 185
Weinstein, D. M., 540
Weis, K., 589, 597(18), 613, 617, 619(32), 621(32), 622(32)
Weiser, U., 129, 138, 146(69)
Weisman, L. S., 4, 625, 626(5)
Weiss, E., 383(21), 386, 387(21), 392(21)
Weissman, A. M., 341
Weissman, J. S., 278, 507, 509, 526(45), 527, 527(45), 530(45)
Welch, S. K., 458
Wen, W., 615(36), 619, 621(36)
Wendland, B., 412, 623, 625, 626(6)
Weng, S., 443
Wente, S. R., 671
Wenzel, B., 374(31), 375(31), 380
Wepf, R., 597, 614
Werner-Washburne, M., 449
Wernic, D., 458, 464(11)
Wertman, K. F., 433
Wesp, A., 50
Westermann, B., 382, 383, 383(5), 387(33), 393, 447, 631, 671
Wheals, A. E., 3
White, C. F., 14
White, J. G., 14, 468
White, J. H., 458
White, K. H., 383(32), 387(32), 393
White, N., 130
Whiteway, M., 477
Whitters, E. A., 352
Wickner, R. B., 499, 500, 500(7), 505(2), 507(14; 15), 508(14), 509(7), 514, 516(2; 7; 14), 523(14), 525(15), 526(17), 631, 632, 633, 636, 637(9), 639(6-9)
Wickner, W., 4, 330, 335(13), 409
Wiebel, F. F., 374(30; 33), 375(30; 33), 380
Wiederkehr, A., 412

Wiemken, A., 141
Wieser, R., 444
Wilkinson, J., 365
Willamson, D. H., 381
Wille, H., 500, 526(19)
Willems, A. R., 253
Williams, C. J., 448
Williams, J. M., 4
Williams, R. H., 130
Williams, R. W., 500, 526(17)
Williamson, D. H., 81
Williamson, E., 58, 669, 671
Williamson, J. R., 652
Willins, D. A., 611, 612(23)
Wilm, M., 180, 219, 282, 293, 299, 303, 312
Wimmer, C., 614
Winey, M., 35, 81, 82, 99, 105(5; 9), 113(9)
Winson, M. K., 623
Winsor, B., 418
Winston, F., 187, 348, 480, 481(18), 493(18), 560, 602, 626
Winter, D., 433, 434(6), 435(6)
Winther, J. R., 127
Winzeler, E. A., 374, 457, 460(2), 462(2)
Wise, J. A., 212
Wittenberg, C., 253, 254
Wittmer, B., 222, 236(5), 247(5)
Wodicka, L., 457, 460(2), 462(2)
Woldringh, C. L., 473, 625, 626(5)
Wolf, D. E., 4, 9(12)
Wolf, D. H., 127, 128, 128(5; 6), 129(5; 18), 130(6; 10), 136, 138, 139, 140(5; 19), 141(19), 146(18; 69), 147(5; 19), 341, 395, 646
Wolffe, A. P., 540
Wolfl, S., 673
Wolfsberg, T. G., 457, 460(2), 462(2)
Wolkwein, C., 350
Wolters, D., 315
Wong, A., 50
Wong, D. H., 590
Wong, E. D., 381, 383(21), 384, 386, 387(21), 392(21)
Wong, J. J., 170, 433, 434(5)
Wong, M. L., 179, 180(17; 18)
Wong, S. T., 243
Wood, J. M., 658
Woodgett, J. R., 280
Woolford, C. A., 134, 136, 137(59)
Woollenweber, L. A., 664

Wozniak, R. W., 589
Wright, R., 4, 50, 57(6), 58(6), 77(6), 102, 671
Wu, J. N., 625, 626(7), 628(7)
Wuestehube, L. J., 258, 259, 264(13)
Wunning, I., 644
Wuthrich, K., 526, 527(80), 532(80)
Wyrick, J. J., 457

X

Xiao, Q., 625, 626(6)
Xiao, Z., 590

Y

Yaffe, M. P., 4, 166, 381, 382, 391, 394, 643
Yaglom, J., 647
Yahara, I., 450, 451(35), 452, 453(35)
Yamamoto, K., 452
Yamashiro, C. T., 351, 417, 421(24), 430(24), 493
Yamashita, M., 299, 567
Yan, C., 598, 609
Yan, W., 443, 447, 447(26), 448, 449(26)
Yanagida, M., 31
Yanagisawa, K., 363
Yang, Q. H., 622
Yang, T. T., 525
Yang, W., 230, 235(20)
Yang, X., 668
Yarar, D., 81
Yaron, A., 317
Yasuhara, T., 136, 141, 141(64), 147(64)
Yates, J. R., 170, 173, 174(11), 220, 283, 315, 383(21), 386, 387(21), 392(21), 433, 434(5)
Yaver, D. S., 427
Ye, L., 671
Yeh, E., 4, 35, 41(6), 412, 671, 676, 679(35), 680(36), 681(35)
Yeung, T., 259, 260, 265, 265(9; 16), 266(9), 268, 269(29), 271(20), 272(20)
Yoda, K., 353, 358(15)
Yokosawa, H., 143
Yoon, H., 242
Yoshida, K., 620
Yoshihisa, T., 258, 259, 265, 268, 268(8; 23), 269(29)
Yoshikawa, H., 567

Young, M. R., 577
Young, R. A., 167, 173, 279, 457, 583, 622, 640
Young, S. J., 84, 95

Z

Zachariae, W., 257, 418
Zacharias, D. A., 35, 665
Zadorsky, S. P., 517
Zanolari, B., 352, 427
Zara, V., 387(39), 393
Zaraisky, A. G., 382, 383, 665
Zavanelli, M. I., 212
Zelenaya-Troitskaya, O., 387(31), 392
Zerial, M., 50, 55, 72(11)
Zhang, B.-Y., 342
Zhang, L., 450, 453(33)
Zhang, M. Q., 457, 460(3)
Zhang, S., 448
Zhang, X., 222, 236(7), 237(7), 247(7), 279, 281
Zhang, Y., 312
Zhao, H., 625, 626(5)

Zhao, L., 620
Zhao, Y., 405, 406(14), 568
Zheng, B., 625, 626(7), 628(7)
Zheng, Y., 179, 180(17)
Zhou, H., 320
Zhou, P., 509, 510, 517(51), 519, 525, 526(75)
Zhou, Z., 461
Zhouravleva, G., 508, 527(42)
Zhu, D., 374(27), 375(27), 377
Zhu, H., 279
Zhu, J., 84
Zhu, X., 344
Ziegelhoffer, T., 449
Zierold, K., 103
Ziman, M., 344
Zimmermann, H. P., 535
Zink, A. D., 505, 516, 517(33), 519(33), 526(64)
Zinser, E., 352
Ziv, T., 640
Zoll, W. L., 222, 227(8)
Zubenko, G. S., 127, 130(11), 133(11), 134, 136, 136(58), 139, 149

Subject Index

A

Actin, purification from yeast
  anion-exchange chromatography, 437
  applications of purified actin, 433–434
  cells
    growth, 434
    high-speed supernatant preparation, 436
    lysis, 434
  cofilin/ADF removal, 437–438
  DNase I affinity chromatography
    column construction, 435–436
    elution, 437
    loading, 437
Actin-associated proteins, purification from yeast
  anion-exchange chromatography, 441
  cell growth and lysis, 439–440
  centrifugation and washing, 440
  overview, 438–439
  salt stripping, 441
  types of proteins, 438
α-factor
  block–release for cell cycle synchronization
    cell growth and treatment, 459
    filtration of cells, 459–460
    materials, 458–459
    monitoring, 460
    overview, 458
  glycosylated prepro-α factor detection, 276–277
7-Amino-4-chloromethylcoumarin, staining of vacuole lumen, 413
Aminopeptidases, yeast
  assays
    fluorescent activity assay, 147–148
    plate assays
      Jones method, 140–141
      Trumbly and Bradley method, 139–140
    cell culture and extraction, 141
8-Anilino-1-naphthalenesulfonic acid, prion binding, 530

ANS, see 8-Anilino-1-naphthalenesulfonic acid
Azocoll, protease B assay, 144–145

B

BrdU, see Bromodeoxyuridine
Bromodeoxyuridine, mitochondrial DNA labeling
  immunofluorescence microscopy, 388
  incorporation, 388
  thymidine kinase requirement, 387
  yeast fixation and processing, 388

C

Carboxy-2′, 7′-dichlorofluorescein diacetate
  flow cytometry sorting of yeast membrane trafficking mutants, 625, 628, 630
  vacuole labeling, 413
  yeast vital staining, 625
Carboxypeptidase S, see also Vacuolar proteases, yeast
  assays
    dipeptide assay
      calculations, 147
      incubation conditions, 147
      principles, 146
      reagents, 146–147
    well test
      incubation conditions, 138
      principles, 137–138
      reagents, 138
      utility, 139
  cell culture and extraction, 141
Carboxypeptidase Y, see also Vacuolar proteases, yeast
  assays
    amidase assay
      calculations, 146
      incubation conditions, 146
      principles, 145
      reagents, 146

715

SUBJECT INDEX

plate assays
  N-acetyl-DL-phenylalanine β-naphthyl
    ester overlay test, 135–137
  well test, 137
  cell culture and extraction, 141
vacuole protein sorting assays
  colony immunoblotting of secreted
    proteins, 421–422
  prevacuolar endosomal compartment
    transport, 427–428
  pulse–chase immunoprecipitation,
    422–425
Cell cycle synchronization, yeast
  α-factor block–release
    cell growth and treatment, 459
    filtration of cells, 459–460
    materials, 458–459
    monitoring, 460
    overview, 458
  cell size correlation, 458
  centrifugal elutriation of G_1 cells
    advantages and limitations, 467
    cell harvesting and loading, 466
    elutriation, 466–467
    materials, 465
    principles, 464–465
  depletion/resynthesis of unstable Cdc proteins
    Gal1-10 promoter, 463–464
    materials, 463
    MET3 promoter, 464
    principles, 463
  DNA replication studies, 187, 189
  experimental design guidelines, 457
  hydroxyurea block–release, 461
  nocodazole block–release, 461
  replication fork analysis
    density transfer assay of rates, 558–559
    S-phase cell collection for fork-direction
      gel analysis, 544
  temperature-sensitive cdc mutants, 462–463
Cesium chloride density gradient centrifugation
  DNA replication intermediates, 546–547
  replication fork migration rate assay
    centrifugation, 560–561
    fractionation apparatus, 561
    materials, 559–560
  yeast DNA replication assays, 194–197
Coimmunoprecipitation
  Hsp70 complexes, 446–447
  Hsp90 complexes, 450–451

spliceosomes, 217–218
yeast protein complexes, 167
Confocal microscopy, three-dimensional
  deconvolution microscopy comparison,
    16–17
Congo red, prion binding, 530, 532
COPII
  components, 265
  Sar1p purification
    affinity chromatography, 267
    glutathione S-transferase fusion protein
      expression in *Escherichia coli*,
      266–267
    materials, 266
    thrombin cleavage of fusion protein, 267
  Sec13/31p purification
    ammonium sulfate precipitation, 272
    anion-exchange chromatography, 273
    cell growth and extraction, 272
    materials, 272
    nickel affinity chromatography, 273
    overview, 271–272
  Sec23p complex purification
    materials, 269
    Sec23/24p, 268–270
    Sec23/Iss1p, 268, 271
    Sec23/Lst1p, 268, 271
  vesicle budding role, 258–259
  vesicle packaging assay
    applications, 277–278
    budding, 276
    detection of glycosylated prepro-α factor,
      276–277
    materials, 275
    principles, 273, 275
    translocation, 275–276
Crm1/Xpo1, *see* Nuclear protein transport, yeast
Cryoelectron microscopy, *see* Electron
  microscopy

D

Differential centrifugation
  membranes from yeast
    applications, 352
    centrifugation, 327
    gel electrophoresis analysis, 327–328
    instrumentation, 327
    marker proteins, 328
    N-linked glycosylation assay, 330–331

SUBJECT INDEX 717

protease inhibition, 328
protein extraction from particulate fraction, 329–330
spheroplast preparation and lysis, 326–328
mitochondria from yeast, 390
peroxisomes from yeast, 376–377
prion factor aggregation
  dot blot analysis, 523
  high-speed centrifugation, 522–523
  low-speed centrifugation, 522
  lysis with glass beads, 521
  overview, 519–521
  spheroplasting, 521–522
DNA replication, yeast
assays for semiconservative DNA replication
  applications, 199
  criteria for physiological replication, 184–185
  media, 187
  nuclei assays
    cesium chloride density gradient centrifugation, 196–197
    chromatin-binding assay, 198–199
    crude nuclei preparation, 192–193
    immunofluorescence microscopy, 197
    replication mixture, 196
    two-dimensional gel electrophoresis, 197
  overview, 184–186
  plasmid assays
    cesium chloride density gradient centrifugation, 194–195
    *Dpn*I resistance analysis, 195
    replication mixture, 194
    two-dimensional gel electrophoresis, 195–196
  S-phase nuclear extract preparation, 190–192
  spheroplasting, 190
  yeast strains, 187
  protein requirements in extracts, 185
  replication fork, *see* Replication fork, yeast
  synchronization of cell cycle, 187, 189
DsRed, fluorescence properties, 665

E

eIF2
  GTPase stimulation by eIF5, assay, 244–245
  guanine-nucleotide exchange by eIF2B, assay, 246–247

purification from yeast
  ammonium sulfate precipitation, 232–233
  materials, 230–231
  nickel affinity chromatography, 231–232
  overexpression in yeast, 230–231
  ternary complex formation assay, 243–244
eIF2B
  eIF2 guanine-nucleotide exchange by eIF2B, assay, 246–247
  purification from yeast
    multiple column method
      extraction, 233–234
      gel filtration, 234–235
      heparin affinity chromatography, 234
      materials, 233
      nickel affinity chromatography, 234
      phosphocellulose chromatography, 234
      time requirements, 235
    single-step nickel affinity chromatography
      extraction, 236
      materials, 235–236
      nickel affinity chromatography, 236)
eIF3
  purification from yeast
    epitope-tagged protein
      chromatography, 239
      extract preparation, 238–239
      materials, 238
    histidine-tagged protein
      chromatography, 237–238
      extract preparation, 237
      materials, 237
    overview, 236
  rescue of Met-tRNA^{Met} and RNA binding to ribosomes in heat-inactivated extracts, 247
eIF5
  eIF2 GTPase stimulation by eIF5, assay, 244–245
  FLAG-tagged protein purification
    overview, 239
    recombinant bacteria, 241
    recombinant yeast, 240
Electron microscopy, *see also* Electron tomography; Immunoelectron microscopy, yeast
  advantages in yeast studies, 96
  cryofixation of yeast
    advantages, 102–103
    double propane jet freezing, 104

fixatives
  materials, 106–107
  Müller–Reichert and Anthony freeze substitution cocktail, 107–108
  preparation, 107
freeze substitution
  dry ice method, 109–111
  Leica AFS device, 112
  low-temperature freezer method, 111–112
high-pressure freezing, 104–105
imaging
  data acquisition, 119–120
  quality evaluation, 120–122
immunogold localization, 119–120, 122
plunge freezing, 104
poststaining
  overview, 117–118
  tannic acid poststaining, 118–119
preparation of cell samples, 105–106
resin infiltration and embedding
  epoxy resins, 113–116
  LR White, 116–117
  resin selection, 113
sectioning, 117
storing samples between freezing and freeze substitution, 108–109
vendors, 122–123
peroxisome immunoelectron microscopy, 372–374
resource requirements
  costs, 97
  equipment, 97
  staff, 97
safety of reagents, 97–98
Sup35 amyloid fibers, 535–536
yeast mating studies
  staining for spindle pole bodies and microtubules, 497–498
  zygote fixation and embedding, 495–497
yeast specimen preparation selection
  conventional method resources, 101–102
  resolution considerations, 98–100
Electron tomography
  instrumentation
    camera, 83
    microscope, 82–83
    software, 83–84, 93
  principles, 82
  yeast

applications, 81–82
data collection, 86–87
large intracellular compartment reconstruction, 91–93, 95
modeling of data, 89, 91
prospects, 95
serial tilt alignment and tomographic reconstruction, 87–89
specimen preparation, 84, 86
Elutriation, G_1 yeast cells
  advantages and limitations, 467
  cell harvesting and loading, 466
  elutriation, 466–467
  materials, 465
  principles, 464–465
Endoplasmic reticulum, yeast
COPII
  components, 265
  role in vesicle budding, 258–259
  Sar1p purification
    affinity chromatography, 267
    glutathione S-transferase fusion protein expression in $Escherichia\ coli$, 266–267
    materials, 266
    thrombin cleavage of fusion protein, 267
  Sec13/31p purification
    ammonium sulfate precipitation, 272
    anion-exchange chromatography, 273
    cell growth and extraction, 272
    materials, 272
    nickel affinity chromatography, 273
    overview, 271–272
  Sec23p complex purification
    materials, 269
    Sec23/24p, 268–270
    Sec23/Iss1p, 268, 271
    Sec23/Lst1p, 268, 271
  vesicle packaging assay
    applications, 277–278
    budding, 276
    glycosylated prepro-α factor detection, 276–277
    materials, 275
    principles, 273, 275
    translocation, 275–276
  cytosol preparation for budding assay, 264–265
  differential centrifugation
    centrifugation, 327

gel electrophoresis analysis, 327–328
  instrumentation, 327
  marker proteins, 328
  N-linked glycosylation assay, 330–331
  protease inhibition, 328
  protein extraction from particulate fraction, 329–330
  spheroplast preparation and lysis, 326–328
donor membrane preparation for budding assay
  materials, 260
  microsome preparation
    centrifugaton, 263
    overview, 262–263
    spheroplast lysis, 263
    storage, 264
  overview, 259–260
  semiintact cell preparation
    radiolabeled membrane preparation, 261–262
    spheroplast perforation, 261
    unlabeled membrane preparation, 262
equilibrium flotation
  cell lysis, 336
  centrifugation, 336
  modification of protocol, 337
  principles, 335–336
  sucrose gradients, 336
flow cytometry sorting of yeast membrane trafficking mutants
  dual vital dyes for screening, 628, 630
  green fluorescent protein fusion constructs, 631
  mutagenesis
    mutagens and mutagenesis, 626–627
    practical considerations, 626
  preculture, 627
  principles, 623–624
  reference samples in screening, 624–625
  secondary screening, 628
  sort window selection, 627–628
  vital stains, 625
protein export machinery, 341
reconstitution of vesicle budding, components, 259
sucrose equilibrium sedimentation
  cell lysis, 331, 333
  centrifugation, 331, 333
  density of membranes, 331, 333
  gel electrophoresis analysis, 333–334
  gradient makers, 334
  weight/weight to weight/volue conversions, 334
temperature-sensitive *sec* mutants, 258
Equilibrium flotation, *see* Sucrose gradient centrifugation
Equilibrium sedimentation, *see* Optiprep gradient centrifugation; Sucrose gradient centrifugation
ER, *see* Endoplasmic reticulum

F

FACS, *see* Flow cytometry
FISH, *see* Fluorescence *in situ* hybridization
Flow cytometry
  yeast membrane trafficking mutant isolation
    dual vital dyes for screening, 628, 630
    green fluorescent protein fusion constructs, 631
    mutagenesis
      mutagens and mutagenesis, 626–627
      practical considerations, 626
    preculture, 627
    principles, 623–624
    reference samples in screening, 624–625
    secondary screening, 628
    sort window selection, 627–628
    vital stains, 625
  yeast mother cell labeling, 473–474
Fluorescence resonance energy transfer
  distance limitations, 35–36, 599
  efficiency, 36
  fusion protein localization in yeast, 34–35
  green fluorescent protein variants in yeast
    applications, 49
    controls, 39–40, 46
    efficiency of transfer, 36
    image acquisition, 44
    imaging parameter optimization, 41–43
    instrumentation, 36–37
    interpretation, 48–49
    presentation of data, 46–48
    principles, 665–666, 669, 672
    ratio quantification, 37–38, 45–46
    sample preparation, 40–41
    spectral overlap, 43–44, 49
    strain construction, 38–40
  nuclear protein transport in yeast using fluorescent protein variants

advantages, 600
controls, 601–602
data acquisition, 605
image analysis, 606
materials, 600–601
nuclear pore complex studies, 600
photobleaching experiment, 606–607
principles, 598–600
strain construction, 602–605
variant pairs, 599–600
Fluorescence *in situ* hybridization
individual messenger RNA localization
fluorescent probe preparation, 579–580
instrumentation, 580
polymerase chain reaction and digoxigenin incorporation in probes, 577–579
SSA4 transcript, 577, 584–585
U1A–green fluorescent protein construct as probe, 580–581
poly(A) RNA detection using oligo(dT) probes
cell growth and preparation, 571–575
controls, 583–584
hybridization conditions, 575–577
probe preparation, 569–571
RNA export studies
advantages, 568–569
microscopy, 581
ribosomal RNA export, 585–586
simultaneous localizations
poly(A) RNA and specific messenger RNA, 581–582
RNA and protein, 582–583
transfer RNA export, 585
FM4-64
flow cytometry sorting of yeast membrane trafficking mutants, 625, 628, 630
vacuole staining, 412
yeast cell wall removal and plasma membrane fusion assay, 491–492
yeast vital staining, 625
Fork-direction gel analysis, *see* Replication fork, yeast
FRET, *see* Fluorescence resonance energy transfer

G

Gcn4, half-life derivation, 645–646
Gel filtration

eIF2B, 234–235
Met-tRNA^{Met}, 226
yeast protein complexes
chromatography conditions, 157
column size, 155
equilibration of column, 156–157
principles, 155
Stokes radius determination, 156
troubleshooting, 158
yeast whole cell extracts, 223
Geldanamycin, Hsp90 interaction inhibition, 451
GFP, *see* Green fluorescent protein
Glutathione *S*-transferase fusion proteins, yeast
advantages of expression system, 168–169
cleavage of tag, 172
induction, 169–170, 172
protease-deficient strains, 169
purification
affinity purification, 170, 172, 174–176
extraction, 170
ion-exchange chromatography, 170
Glycerol gradient centrifugation, spliceosomes, 215–217
Glycosylation, N-linked glycosylation assay, 330–331
Golgi apparatus, yeast
differential centrifugation
centrifugation, 327
gel electrophoresis analysis, 327–328
instrumentation, 327
marker proteins, 328
N-linked glycosylation assay, 330–331
protease inhibition, 328
protein extraction from particulate fraction, 329–330
spheroplast preparation and lysis, 326–328
equilibrium flotation
cell lysis, 336
centrifugation, 336
modification of protocol, 337
principles, 335–336
sucrose gradients, 336
immunoaffinity isolation using magnetic beads
calibration and controls, 360
coated bead preparation, 363
coimmunoisolation of marker proteins and analysis, 359–360, 362–363
GDPase assay, 363–364

SUBJECT INDEX

Kex2p
  antibodies, 359
  assay, 364
  coimmunoisolated markers, 360, 362
  membrane isolation, 363
  principles, 358–360
sucrose equilibrium sedimentation
  cell lysis, 331, 333
  centrifugation, 331, 333
  density of membranes, 331, 333
  GDPase assay of Golgi fractions, 334–335
  gel electrophoresis analysis, 333–334
  gradient makers, 334
  Kex2p assay, 335
  weight/weight to weight/volue conversions, 334
sucrose equilibrium sedimentation
  cell lysis, 331, 333
  centrifugation, 331, 333
  density of membranes, 331, 333
  GDPase assay of Golgi fractions, 334–335
  gel electrophoresis analysis, 333–334
  Golgi separation from endosomal compartments
    high-speed membrane fractions, 354
    marker analysis, 355, 357
    spheroplast preparation, 353–354
    sucrose step gradients, 354–355, 357–358
  gradient makers, 334
  Kex2p assay, 335
  weight/weight to weight/volue conversions, 334
Green fluorescent protein
  flow cytometry sorting of yeast membrane trafficking mutants, 631
  fluorescence resonance energy transfer between variants in yeast
    applications, 49
    controls, 39–40, 46
    efficiency of transfer, 36
    image acquisition, 44
    imaging parameter optimization, 41–43
    instrumentation, 36–37
    interpretation, 48–49
    presentation of data, 46–48
    principles, 665–666, 669, 672
    ratio quantification, 37–38, 45–46

    sample preparation, 40–41
    spectral overlap, 43–44, 49
    strain construction, 38–40
  fusion protein construction in yeast
    aggregation concerns, 661–662
    codon usage, 666
    design of construct, 666–668
    expression, 668
    functional effects, 662
    variant protein selection, 663–666
  fusion protein localization in yeast, 34–35
  mitochondria visualization using targeted fusions, 383–385
  nuclear protein transport in yeast
    fluorescence resonance energy transfer using fluorescent protein variants
      advantages, 600
      controls, 601–602
      data acquisition, 605
      image analysis, 606
      materials, 600–601
      nuclear pore complex studies, 600
      photobleaching experiment, 606–607
      principles, 598–600
      strain construction, 602–605
      variant pairs, 599–600
    nuclear targeting signal fusion constructs for imaging, 592, 596–598, 609, 622
  peroxisome imaging with targeting sequence fusion proteins
    cell growth and mounting, 368–369
    fixed cells, 369
    fluorescence microscopy, 369
    vectors, 367–368
  prion tagging and aggregation analysis, 525–526
  Rp111b fusion protein for ribosome studies, 585–586
  three-dimensional deconvolution microscopy imaging in yeast
    camera settings, 29
    exposure time, 28–29
    growth conditions, 19–20
    instrumentation
      cover glass, 26
      filter sets, 22, 24–25
      Kohler versus critical illumination, 25–26
      temperature controller, 26
    kinetochore analysis, 31–33

mounting
  paraformaldehyde-fixed cells, 21–22
  slide preparation with agar pads, 20–21
  suspended cells, 20–21
oil matching for spherical aberration
  reduction, 26–27
strain construction, 19
viewing and printing of images, 30
U1A–green fluorescent protein construct as
  RNA probe, 580–581
vacuole visualization with targeted
  fusions
  advantages and limitations, 414
  tagging of uncharacterized proteins,
    415–417
  vacuolar protein fusions, 414–415
variants, 19, 663–666
yeast living cell imaging
  advantages, 661
  cellular process monitoring, 672–673
  gene expression analysis, 673
  genotoxicity assay, 673
  image processing, 682–683
  instrumentation
    camera, 676–677
    fluorescence microscopy mode, 674
    microscope, 675–676
    software, 677–678
    temperature control, 678
  intracellular targeting, 669–671
  localization, 668–669
  mutant analysis, 674
  sample preparation, 679–680
  time-lapse imaging
    four-dimensional imaging, 681–682
    overview, 674
    single focal plane, 680–681
    two-variant protein analysis in same cell, 682
Guanidium hydrochloride, prion curing assay, 510–511

H

Heat shock proteins
  classification and resources, 442–443
  Hsp70
    coimmunoprecipitation of complexes,
      446–447
    functions, 446
    genetic interactions and mutant yeast
      strains, 448–449
    Hsp40 interactions, 446
    interaction classification, 448
    subcellular localization, 446
    types, 446
  Hsp90
    binding partners, 449–450
    coimmunoprecipitation of complexes,
      450–451
    functions, 448–449, 453
    genes, 448
    genetic interactions and mutant strains,
      451–452
    pharmacological inhibition of interactions,
      451
    substrate activity assays in mutant yeast
      strains, 452–453
  Hsp104 prion curing assay
    constitutive overproduction, 514–515
    deletion, 516
    inducible overproduction, 515–516
    $[PSI^+]$ interactions, 513–514
  induction assays
    inducers, 443
    Northern blot, 445
    promoter fusions, 444–445
    Western blot, 445
HSPs, see Heat shock proteins
Hydroxyurea, block–release for cell cycle
  synchronization, 461
Immunoaffinity chromatography, yeast protein
  complexes
  antibody preparation, 180–181
  beads
    binding and washing conditions, 182–183
    preparation, 181
  buffers, 181
  controls, 181
  extract preparation, 181–182
  gel electrophoresis analysis, 183
  hemagglutinin tagging, 180
  principles, 179

I

Immunoelectron microscopy, yeast
  aldehyde-fixed cells
    ammonium chloride treatment, 62
    antibodies, 65, 67

SUBJECT INDEX

artifacts and problems, 59, 61
blocking of nonspecific interactions, 69, 81
controls
   importance, 69–70
   positive controls, 71
   primary antibody specificity, 70
   secondary antibody specificity, 70–71
correlative studies, 75
culture density and media, 51–52
dehydration, 58, 82
double labeling
   different surfaces, 72–73, 75
   same surface, 71–72, 78–79
fixation
   buffers, 56
   osmolarity variation studies, 53, 55–56
   principles, 52–53
   small samples, 63–64
   time and temperature, 56–57
gold conjugates, 67–68
immunolabeling, 77–78
lead citrate stain, 80
lectins, 68
metaperiodate treatment, 57–58, 61–62
postlocalization fixation and electron-dense staining, 79–80
reagents, 64–65
resin infiltration and polymerization, 58–59, 62–63
sticky grid solution preparation, 80
ultramicrotomy, 75–77
wash buffer, 81
colocalization of antigens, 51
cryofixation, *see* Electron microscopy
historical perspective, 50
Importins, *see* Nuclear protein transport, yeast
Iss1p/Sec23p complex, *see* Sec23p

K

Kex2p
   assay in Golgi fractions, 335
   Golgi immunoaffinity isolation
     antibodies, 359
     assay, 364
     coimmunoisolated markers, 360, 362
     membrane isolation, 363

Kinetochore, three-dimensional deconvolution microscopy analysis using green fluorescent protein, 31–33

L

Latrunculin A, prion curing assay, 510, 513
Lead citrate, electron microscopy sample staining, 80
Life cycle analysis, yeast
   asymmetric division, 468
   baby machine and daughter cell elution, 473
   bud scar counting, 476
   micromanipulation of mother and daughter cells, 468–472
   mother cell labeling
     flow cytometry, 473–474
     magnetic sorting, 474–477
   mother cell visual identification, 472
   population distribution in culture, 472–473
LR White, cryofixated sample embedding, 116–117
Lst1p/Sec23p complex, *see* Sec23p
Luciferase
   messenger RNA electroporation assay of protein synthesis, 634–635
   peroxisome marker assay *in vivo*
     cell culture, 380
     detection, 380–381
     promoters, 367, 380

M

Macbesin I, Hsp90 interaction inhibition, 451
Mass spectrometry, yeast proteins
   collision-induced dissociation, 298–299
   electrospray ionization principles, 299
   identification of proteins
     database searching, 315–316
     gel analysis
       applications, 320–21
       fixation of gels, 303
       in-gel digestion, 304
       simple protein mixture analysis, 316–317
       staining, 302–303
       two-dimensional gels, 301–302
     liquid chromatography/liquid chromatography–tandem mass spectrometry, 313–315

liquid chromatography–tandem mass
spectrometry, 312–313
mass identification, 301
matrix-assisted laser desorption/ionization
peptide mass mapping
accuracy, 306
calibration, 305–306
data evaluaton, 317–318
orthogonal technique, 306–307
sample preparation, 305
sensitivity, 305–306
overview of techniques, 297–298
peptide mass fingerprinting, 301
posttranslational modification analysis, 317, 319
prospects, 321
sequencing of peptides
mass accuracy, 312
nanoelectrospray sequencing, 307, 310, 312
overview, 301
truncated proteins, 319
trypsin digestion, 304
liquid chromatography coupling, 300–301
matrix-assisted laser desorption and ionization principles, 298–299
peptide fragmentation, 299
phosphorylation site mapping
advantages, 279–281, 319
affinity tagging of proteins, 282
alkaline phosphatase treatment, 319
approaches, 319–320
cell growth and harvesting, 282
immobilized metal affinity chromatography of phosphopeptides, 320
inhibitor cocktails for phosphatase and proteases, 282–283
multidimensional electrospray–mass spectrometry
Net1, 294, 296
precursor ion scanning, 289–290, 293
principles, 290, 292
reversed-phase high-performance liquid chromatography of tryptic digests, 290, 292, 296
Sic1, 292–294
precursor ion scanning, 320
yeast strain design, 281–282
proteomic revolution, 297–297
resolution, 297–298

Mating, yeast
growth phase, density, and timing as critical parameters, 478–479
microscopic analysis
electron microscopy
staining for spindle pole bodies and microtubules, 497–498
zygote fixation and embedding, 495–497
fluorescence microscopy of cell fusion and nuclear fusion defects
cell fusion defect scoring, 488
green fluorescent protein construct analysis, 489, 491
karyogamy mutant scoring, 488
specimen preparation, 487–488
FM4-64 assay for cell wall removal and plasma membrane fusion, 491–492
immunofluorescence of mating mixtures, 492–494
zygote light microscopy, 485
mutant strains, 477–478
plate assays
cyh^r strain isolation, 480
filter matings to assay cytoductants, 479
mating mixture preparation from liquid cultures, 481, 483
micromanipulation of zygotes, 484–485
ρ^0 strain preparation, 480
semiquantitative mating assays, 483–484
MDY-64, vacuole staining, 412
Messenger RNA, see RNA decay, yeast; RNA export, yeast; RNA splicing, yeast; Translation, yeast
Micromanipulation
mother and daughter yeast cells, 468–472
yeast zygotes, 484–485
Microscopy, see Electron tomography; Green fluorescent protein; Immunoelectron microscopy; Three-dimensional deconvolution microscopy; Time-lapse microscopy, yeast growth
Mitochondria, yeast
enrichment
differential centrifugation, 390
equilibrium centrifugation on Optiprep gradients, 390–391
marker protein analysis, 387, 391
spheroplast formation and cell lysis, 389–390

SUBJECT INDEX 725

fusion during mating, vital assay
  bromodeoxyuridine labeling of DNA
    cell fixation and processing, 388
    immunofluorescence microscopy, 388
    incorporation, 388
    thymidine kinase requirement, 387
  DNA fluorescence staining, 386–387
  green fluorescent protein-targeted fusion
    labeling, 384–385
  principles, 384
genome copy number, 381
imaging
  green fluorescent protein-targeted fusions, 383
  immunofluorescence microscopy, 383–384
  vital dye staining
    MitoTracker, 382–383
    selection of dye, 382–383
nucleoid enrichment, 392–393
submitochondrial localization of proteins
  mitoplast preparation by hypoosmotic shock, 391–392
  protease protection assay, 392
mRNA, *see* Messenger RNA
MS, *see* Mass spectrometry

N

Net1
  Cdc14 release and phosphorylation role, 287, 289
  phosphorylation site mapping with multidimensional electrospray–mass spectrometry, 294, 296
  purification
    cell lysis, 287–288
    detergent extraction, 288
    immunoaffinity chromatography, 288–289
  RENT complex, 285
Nocodazole, block–release for cell cycle synchronization, 461
Northern blot
  heat shock protein RNA, 445
  RNA decay in yeast, 657–658
Nuclear envelope, yeast
  isolation
    cell growth
      large-scale preparation, 396–398
      small-scale preparation, 398
    materials, 394–396

modifications of protocol, 408
nuclei isolation
  gradient media, 399–400
  large-scale preparation, 400–402
  small-scale preparation, 402–403
  purity and yield assessment, 405–407
  spheroplast preparation, 398–399
sucrose/Nycodenz gradient centrifugation
  large-scale preparation, 403–404
  small-scale preparation, 404–405
markers, 405–407
Nuclear protein transport, yeast
export
  assays
    green fluorescent protein construct analysis, 597–598
    overview, 597
  Crm1/Xpo1 receptor, 589, 597
  fluorescence resonance energy transfer between green fluorescent protein variants
    advantages, 600
    controls, 601–602
    data acquisition, 605
    image analysis, 606
    materials, 600–601
    nuclear pore complex studies, 600
    photobleaching experiment, 606–607
    principles, 598–600
    strain construction, 602–605
    variant pairs, 599–600
import
  assays
    green fluorescent protein construct analysis, 596–597
    overview, 595–596
    steady-state localization, 595
    substrate selection, 595
  importins, 589–590
  transcription factors, 588
  machinery, 588–589
nuclear pore complex, 588, 591
nucleocytoplasmic shuttling
  export analysis using $nsp1^{ts}$ and $nup49$-313 mutants
    cell culture, 616–617
    constructs, 615–616
    experimental design, 614–617
    fluorescence microscopy, 616–617
    principles, 612–614

green fluorescent protein constructs as substrates, 622
importin-β mutant analysis, 617–619
inducible promoter studies, 622
localization of previously synthesized protein pool
cell growth condition manipulation, 611
cycloheximide inhibition of new protein synthesis, 610–611
overexpression conditions, 611
nuclear targeting signal alteration studies
mutagenesis, 620–621
protein kinase A inhibitor nuclear export signal tagging, 619–620
SV40 large T antigen nuclear localization signal tagging, 619–620
overview, 598, 607–608
passive diffusion avoidance, 621–622
polykaryon test, 611–612
steady-state distribution analysis, 608–610
temperature optimization, 621
targeting signals
identification
criteria, 592–594
difficulty of sequence interpretation, 587, 591
green fluorescent protein construct imaging, 592
nuclear localization signal
canonical signal, 589
identification criteria, 592–594
validation, 587
Nucleus, yeast
fluorescence microscopy of cell fusion and nuclear fusion defects
cell fusion defect scoring, 488
green fluorescent protein construct analysis, 489, 491
karyogamy mutant scoring, 488
specimen preparation, 487–488
protein transport, *see* Nuclear protein transport, yeast
Nycodenz density gradient centrifugation, peroxisome purification from yeast, 377

O

Optiprep gradient centrifugation, mitochondria equilibrium centrifugation, 390–391

P

PCR, *see* Polymerase chain reaction
Peroxisome, yeast
biogenesis, 365–366
green fluorescent protein imaging with targeting sequence fusion proteins
cell growth and mounting, 368–369
fixed cells, 369
fluorescence microscopy, 369
vectors, 367–368
growth conditions and variability in size, abundance, and expression, 366
immunoelectron microscopy, 372–374
immunofluorescence microscopy, 370–372
luciferase marker assay *in vivo*
cell culture, 380
detection, 380–381
promoters, 367, 380
purification
differential centrifugation, 376–377
marker proteins, 374–375
Nycodenz density gradient centrifugation, 377
yield, 374
subperoxisomal distribution of proteins
high-pH extraction of proteins, 379–380
hypotonic lysis, 378–379
salt extraction of membranes, 379
Pheromone receptor, signal transduction in yeast, 340
Phosphoprotein, *see* Mass spectrometry, yeast proteins
Plasma membrane, yeast
differential centrifugation
centrifugation, 327
gel electrophoresis analysis, 327–328
instrumentation, 327
marker proteins, 328
N-linked glycosylation assay, 330–331
protease inhibition, 328
protein extraction from particulate fraction, 329–330
spheroplast preparation and lysis, 326–328
equilibrium flotation
cell lysis, 336
centrifugation, 336
modification of protocol, 337
principles, 335–336
sucrose gradients, 336

SUBJECT INDEX

FM4-64 fusion assay, 491–492
isolation by concanavalin A treatment, 337–338
markers
 overview, 342–343
 Pma1p, see Pma1p
polarization, 340
proteins
 defective protein degradation, 341–342
 endoplasmic reticulum export, 341
 pulse–chase immunofluorescence
  chase, 348
  immunofluorescence staining, 348
  principles, 347
  pulse, 347
  yeast strains, 348–349
 sorting, 342
 stability versus turnover, 340–341
 steady-state levels, 349
 types, 339
 Western blotting, 349–350
Renografin density gradient centrifugation
 cell growth and lysis, 345
 centrifugation, 345
 gel electrophoresis analysis, 345–346
 marker fractionation, 346
sucrose equilibrium sedimentation
 cell lysis, 331, 333
 centrifugation, 331, 333
 density of membranes, 331, 333
 GDPase assay of Golgi fractions, 334–335
 gel electrophoresis analysis, 333–334
 gradient makers, 334
 Kex2p assay, 335
 weight/weight to weight/volue conversions, 334
Pma1p
 abundance, 343
 detection, 343–344
 phosphorylation, 343
 plasma membrane marker, 342
 pulse–chase immunofluorescence
  chase, 348
  immunofluorescence staining, 348
  principles, 347
  pulse, 347
  yeast strains, 348–349
 Renografin density gradient centrifugation, 346

site-directed mutagenesis, 344
steady-state levels, 349
Polymerase chain reaction
 epitope tagging of yeast proteins, 165–166
 probe preparation for fluorescence *in situ* hybridization, 577–580
Prion factors, yeast
 aggregation analysis in cells
  differential centrifugation
   dot blot analysis, 523
   high-speed centrifugation, 522–523
   low-speed centrifugation, 522
   lysis with glass beads, 521
   overview, 519–521
   spheroplasting, 521–522
  green fluorescent protein tagging, 525–526
  proteinase resistance
   cell growth and lysis, 524
   gel electrophoresis analysis, 524–525
   isoform sensitivity, 523
   proteinase K treatment, 524
 amyloid formation *in vitro*
  dye-binding studies
   8-anilino-1-naphthalenesulfonic acid, 530
   Congo red, 530, 532
   thioflavin T, 532
  electron microscopy of fibers, 535–536
  mutant studies, 526–527
  polymerization reactions, 529–530
  protease susceptibility studies, 534
  radioassay of assembly, 534–535
  sedimentation analysis, 534
  sodium dodecyl sulfate solubility studies of structure, 532, 534
 Sup35 isolation
  expression in *Escherichia coli*, 527–528
  fragments of protein, 527
  histidine-tagged protein, 528
  nontagged truncated protein, 528
  quantification, 528
  storage, 528–529
 Sup35^{PSI+} conformational conversion efficiency assay
  lysate preparation, 537
  overview, 536
  quantitative analysis, 537–538
 conformational conversion, 499
 curing assays
  guanidium hydrochloride, 510–511

Hsp104
  constitutive overproduction, 514–515
  deletion, 516
  inducible overproduction, 515–516
  [PSI+] interactions, 513–514
  latrunculin A, 510, 513
  cytoduction testng, 509–510
  diagnostic tests, 499–500
  heredity, 499–500, 508
  induction by Sup35 or Ure2 overexpression, 516–519
  mating assay, 508
  phenotypic detection of [PSI+]
    composite prions, 507–508
    incompatibility with overproduced Sup35, 505
    nonsensense suppression assay
      overview, 501, 503–504
      quantitative analysis, 504–505
    Sup35 inactivation, 501
    tetrad analysis, 509
    [URE3] phenotypic analysis, 505, 507
Protease A, *see also* Vacuolar proteases, yeast
  assays
    fluorescent peptide as substrate, 143–144
    hemoglobin assay, 141–143
    plate assay, 139
  cell culture and extraction, 141
  mutant yeast strains, 136–137
  protease B maturation role, 136
Protease B, *see also* Vacuolar proteases, yeast
  assays
    Azocoll assay
      calculations, 145
      incubation conditions, 145
      principles, 144
      reagents, 145
    hide powder azure overlay assay
      culture, 133–134
      principles, 133
      reagents, 134
      utility, 134
  cell culture and extraction, 141
  deficient mutant strains for artifact prevention
    design, 148–149
    genetic analysis, 132
    precautions, 149–150
    proteolytic artifact causation, 127–128
Protein complexes, yeast, *see also* Fluorescence resonance energy transfer

  genetic analysis, 183
  immunoaffinity chromatography
    antibody preparation, 180–181
    beads
      binding and washing conditions, 182–183
      preparation, 181
    buffers, 181
    controls, 181
    extract preparation, 181–182
    gel electrophoresis analysis, 183
    hemagglutinin tagging, 180
    principles, 179
  protein affinity chromatography
    advantages, 173–174
    column preparation, 176–177
    controls, 174
    extract preparation, 177
    gel electrophoresis analysis, 178–179
    glutathione S-transferase fusion protein preparation, 174–176
    loading and elution conditions, 177–178
    overview, 172–173
    selection od proteins for affinity matrix, 174
  size and shape analysis
    applications, 151
    epitope-tagged proteins
      applications, 164–165
      coimmunoprecipitation, 167
      denaturing lysis, 166–167
      polymerase chain reaction for tagging, 165–166
      Western blot, 167–168
    extract preparation
      glass bead lysis, 152–153
      liquid nitrogen lysis, 151–152
      troubleshooting, 154
    gel filtration
      chromatography conditions, 157
      column size, 155
      equilibration of column, 156–157
      principles, 155
      Stokes radius determination, 156
      troubleshooting, 158
    hydrodynamic calculations, 162–164
    overview, 154–155
    sucrose gradient centrifugation
      data collection, 161
      glycerol gradient comparison, 159
      gradient preparation, 160–161

SUBJECT INDEX

principles, 158–159
standards, 159
troubleshooting, 162
volume load limits, 159–160
Protein degradation, yeast, *see also* Ubiquitination, yeast proteins
assays
  GAL shutoff assay, 641, 644
  half-life derivation, 644–646
  methionine pulse–chase and immunoprecipitation, 641–644
  overview, 640–641
  proteolytic pathway determination, 646–648
  regulators, 639–640
  tagging and fusion of substrates, 640
[PSI^+], *see* Prion factors, yeast

Q

Quinacrine, vacuole staining, 413–414

R

Renografin density gradient centrifugation, plasma membranes from yeast
  cell growth and lysis, 345
  centrifugation, 345
  gel electrophoresis analysis, 345–346
  marker fractionation, 346
Replication fork, yeast
  density transfer assay of rates
    applications, 567–568
    CDC7 strains, 558
    cell culture and harvesting, 557–559
    cell cycle synchronization, 558–559
    cesium chloride density gradient centrifugation
      centrifugation, 560–561
      fractionation apparatus, 561
      materials, 559–560
    materials, 557
    principles, 555–556
    slot blot analysis
      blotting and hybridization, 562–563
      data analysis, 563–567
      materials, 562
      probe preparation, 562
  DNA fiber autoradiography, 539
  electron microscopy, 539
  two-dimensional agarose gel electrophoresis
    analysis of migration direction advantages, 540–541
    applications
      origin efficiency determination, 550–553
      origin mapping, 554–555
    first-dimension electrophoresis and transfer, 548–549
    imaging of gels, 551
    loading of gel, 548
    materials, 548
    principles, 541–543
    replication intermediate isolation
      care in handling, 545
      cell lysis, 545–546
      cesium chloride density gradient centrifugation, 546–547
      materials, 545
      protease digestion, 546
      restriction digestion in gel, 547–548
    S-phase cell collection
      cell cycle synchronization, 544
      harvesting, 544–545
      materials, 543–544
      overview, 543
    second-dimension electrophoresis, 550
    Southern blotting, 550
Ribosome, purification from yeast
  eIF3 rescue of Met-tRNA^{Met} and RNA binding to ribosomes in heat-inactivated extracts, 247
  extraction, 242
  materials, 242
  sucrose gradient centrifugation of subunits, 242–243
RNA decay, yeast
  assay combinations, 660
  half-life determination, 650–651
  intermediate trapping assays
    deadenylation role in decay, 654–655
    directionality of decay pathway, 652, 654
    limitations, 655–656
    pathway determination, 652
    poly(G) tract introduction and secondary structure, 652, 654
    $3' \rightarrow 5'$ decay assay, 655
  mutations in trans-acting factors
    deadenylation-deficient strains, 659–660

5'→3' decay-deficient strains, 658
nonsense decay-deficient strains, 658–659
3'→5' decay-deficient strains, 660
pathways, 648–648
pulse–chase studies
  cell preparation, 656–657
  induction, 657
  Northern blot, 657–658
  principles, 656
  RNA isolation, 657
RNA export, yeast
  challenges in study, 568–569
  fluorescence *in situ* hybridization
    advantages, 568–569
    individual messenger RNA localization
      fluorescent probe preparation, 579–580
      instrumentation, 580
      polymerase chain reaction and digoxigenin incorporation in probes, 577–579
      SSA4 transcript, 577, 584–585
      U1A–green fluorescent protein construct as probe, 580–581
      microscopy, 581
    poly(A) RNA detection using oligo(dT) probes
      probe preparation, 569–571
      cell growth and preparation, 571–575
      hybridization conditions, 575–577
      controls, 583–584
    simultaneous localizations
      poly(A) RNA and specific messenger RNA, 581–582
      RNA and protein, 582–583
    transfer RNA export, 585
    ribosomal RNA export, 585–586
  mutant strain isolation, 586–587
RNA splicing, yeast
  advantages as model system, 203
  assays
    in vitro
      gel electrophoresis, 209
      pre-messenger RNA precursor preparation, 207–208
      splicing reaction, 208–209
      troubleshooting, 209–210
    in vivo
      controls, 212
      gel electrophoresis, 213
      Northern blot, 210

  primer extension analysis, 210–213
  splicing reporter assays, 213–214
  history of study, 200–201
  precursor and product overview, 201–202
  spliceosome
    affinity chromatography, 218–220
    coimmunoprecipitation, 217–218
    glycerol gradient centrifugation, 215–217
    native gel analysis of splicing complexes, 214–215
    small nuclear ribonucleoproteins, 202–203
  splicing extract preparation from *Saccharomyces cerevisiae*
    Dounce extracts, 203–205
    liquid nitrogen extract, 205–207
RNA translation, *see* Translation, yeast

S

Sar1p, purification
  affinity chromatography, 267
  glutathione *S*-transferase fusion protein expression in *Escherichia coli*, 266–267
  materials, 266
  thrombin cleavage of fusion protein, 267
Sec13p, purification of Sec31p complex
  ammonium sulfate precipitation, 272
  anion-exchange chromatography, 273
  cell growth and extraction, 272
  materials, 272
  nickel affinity chromatography, 273
  overview, 271–272
Sec23p, purification of protein complexes
  materials, 269
  Sec23/24p, 268–270
  Sec23/Iss1p, 268, 271
  Sec23/Lst1p, 268, 271
Sic1
  degradation and phosphorylation role, 283–284
  phosphorylation site mapping with multidimensional electrospray–mass spectrometry, 292–294
  purification
    cell growth and lysis, 284–285
    nickel affinity chromatography, 285
    overview, 284
Small nuclear ribonucleoprotein, *see* Spliceosome

Southern blot, fork-direction gel analysis, 550
SPB, *see* Spindle pole body
Spindle pole body, electron tomography, 89
Spliceosome
  affinity chromatography, 218–220
  coimmunoprecipitation, 217–218
  glycerol gradient centrifugation, 215–217
  native gel analysis of splicing complexes, 214–215
  small nuclear ribonucleoproteins, 202–203
Subcellular fractionation, yeast
  differential centrifugation
    applications, 352
    centrifugation, 327
    gel electrophoresis analysis, 327–328
    instrumentation, 327
    marker proteins, 328
    N-linked glycosylation assay, 330–331
    protease inhibition, 328
    protein extraction from particulate fraction, 329–330
    spheroplast preparation and lysis, 326–328
  equilibrium flotation
    cell lysis, 336
    centrifugation, 336
    modification of protocol, 337
    principles, 335–336
    sucrose gradients, 336
  gradient media selection, 352
  marker detection, 326
  plasma membrane isolation by concanavalin A treatment, 337–338
  sucrose equilibrium sedimentation
    cell lysis, 331, 333
    centrifugation, 331, 333
    density of membranes, 331, 333
    GDPase assay of Golgi fractions, 334–335
    gel electrophoresis analysis, 333–334
    Golgi separation from endosomal compartments
      high-speed membrane fractions, 354
      marker analysis, 355, 357
      spheroplast preparation, 353–354
      sucrose step gradients, 354–355, 357–358
    gradient makers, 334
    Kex2p assay, 335
    weight/weight to weight/volue conversions, 334

Sucrose gradient centrifugation
  equilibrium flotation, yeast
    cell lysis, 336
    centrifugation, 336
    Golgi separation from endosomal compartments
      high-speed membrane fractions, 354
      marker analysis, 355, 357
      spheroplast preparation, 353–354
      sucrose step gradients, 354–355, 357–358
    modification of protocol, 337
    principles, 335–336
    sucrose gradients, 336
  equilibrium sedimentation, yeast
    cell lysis, 331, 333
    centrifugation, 331, 333
    density of membranes, 331, 333
    GDPase assay of Golgi fractions, 334–335
    gel electrophoresis analysis, 333–334
    gradient makers, 334
    Kex2p assay, 335
    weight/weight to weight/volue conversions, 334
  ribosomal subunits from yeast, 242–243
  yeast protein complexes
    data collection, 161
    glycerol gradient comparison, 159
    gradient preparation, 160–161
    hydrodynamic calculations, 162–164
    principles, 158–159
    standards, 159
    troubleshooting, 162
    volume load limits, 159–160
Sup35, *see* Prion factors, yeast

T

Tannic acid, poststaining of electron microscopy samples, 118–119
Thioflavin, prion binding T, 532
Three-dimensional deconvolution microscopy
  confocal microscopy comparison, 16–17
  green fluorescent protein imaging in yeast
    camera settings, 29
    exposure time, 28–29
    growth conditions, 19–20
    instrumentation
      cover glass, 26
      filter sets, 22, 24–25

SUBJECT INDEX

Kohler versus critical illumination, 25–26
temperature controller, 26
kinetochore analysis, 31–33
mounting
  paraformaldehyde-fixed cells, 21–22
  slide preparation with agar pads, 20–21
  suspended cells, 20–21
oil matching for spherical aberration reduction, 26–27
strain construction, 19
viewing ad printing of images, 30
image quality factors, 17–19
instrumentation, 17–18
Internet resources, 33
principles, 16–17
prospects, 33
Time-lapse microscopy, yeast growth
bud-site selection patterns, 3–4
culture
  agar layer and sealing, 7, 9
  coverslips, 5
  media, 5, 7
  temperature, 5
differential interference contrast microscopy, 11
digital imaging, 11–13
focus control, 14–15
instrumementation
  camera, 11–13
  light source, 9
  microscope stand, 9
  objective, 10
  software, 13–15
light attenuation, 5
phase contrast microscopy, 10–11
resources, 4
Transfer RNA, *see* Translation, yeast
Translation, yeast
eIF2
  GTPase stimulation by eIF5, assay, 244–245
  guanine-nucleotide exchange by eIF2B, assay, 246–247
  initiation components and overview, 221
  initiation factor purification
    affinity tagging for purification, 230
    eIF2 purification
      ammonium sulfate precipitation, 232–233
      materials, 230–231
      nickel affinity chromatography, 231–232
      overexpression in yeast, 230–231
    eIF2B purification
      multiple column method, 233–235
      single-step nickel affinity chromatography, 235–236
    eIF3 purification
      epitope-tagged protein, 238–239
      histidine-tagged protein, 236–238
    eIF5 FLAG-tagged protein purification
      overview, 239
      recombinant bacteria, 241
      recombinant yeast, 240
messenger RNA electroporation assay
  applications, 633, 639
  cell preparation, 633–644
  comparison with *in vitro* translation systems, 638–639
  electroporation, 634
  luciferase reporter assay and analysis, 634–635
  messenger RNA preparation, 634
  overview, 631–632
  RNA stability considerations, 635–637
  strains, 633
  translation rate measurement, 637
  troubleshooting
    capped RNA, 638
    cells, 637
    kinetics, 638
    temperature-sensitive mutants, 638
Met-tRNA^{Met} preparation for assay
  gel filtration, 226
  materials, 225
  methionyl-tRNA synthetase
    pilot synthesis, 225
    preparative synthesis, 225
    purification from *Escherichia coli*, 224–225
ribosome purification
  extraction, 242
  materials, 242
  sucrose gradient centrifugation of subunits, 242–243
ternary complex formation assay, 243–244
whole cell extract assays
  extract preparation
    cell growth and lysis, 223
    gel filtration, 223
    materials, 222–223

SUBJECT INDEX 733

luciferase messenger RNA translation assay
  luminescence assay, 228
  materials, 227–228
  RNA preparation, 227
  translation reaction, 228
  principles, 222
  ribosome transfer assays
    messenger RNA binding, 228–229
    Met-tRNA^{Met}, 228–229
    overview, 228

U

Ubiquitination, yeast proteins, *see also* Protein degradation, yeast
  deubiquitination in lysates, 249
  enzymes
    conjugating enzymes, 647–648
    E1, 647
  functions, 248
  genetic analysis of machinery, 257
  histidine-tagged ubiquitinated protein affinity chromatography, 254–255
  immunoprecipitation of target protein assays
    immunoblotting from cell lysates
      antibody specificity, 253
      immunoblot sensitivity, 252
      overview, 249, 251
      plasmids for epitope-tagged ubiquitin protein expression, 251–252
    radiolabeled extracts, 253–254
  mutant strains, 646–647
  proteolytic pathway determination, 646–648
  ubiquitin chain conjugation to substrates
    multi-chain conjugation analysis, 256–257
    overview, 255
    ubiquitin mutants for study, 252, 255–256
Ure2, *see* Prion factors, yeast
[URE3], *see* Prion factors, yeast

V

Vacuolar proteases, yeast
  activation, 129–130
  deficient mutant strains for artifact prevention
    design, 148–149
    genetic analysis, 132
    precautions, 149–150
  expression level factors
    growth stage, 128–129
    medium composition, 129
  extract assays
    aminopeptidase fluorescent assay, 147–148
    carboxypeptidase S dipeptide assay
      calculations, 147
      incubation conditions, 147
      principles, 146
      reagents, 146–147
    carboxypeptidase Y amidase assay
      calculations, 146
      incubation conditions, 146
      principles, 145
      reagents, 146
    cell culture and extraction, 141
    protease A
      fluorescent peptide as substrate, 143–144
      hemoglobin assay, 141–143
    protease B Azocoll assay
      calculations, 145
      incubation conditions, 145
      principles, 144
      reagents, 145
  inhibitor cocktails, 130–132
  plate assays
    aminopeptidases
      Jones method, 140–141
      Trumbly and Bradley method, 139–140
    carboxypeptidase S well test
      incubation conditions, 138
      principles, 137–138
      reagents, 138
      utility, 139
    carboxypeptidase Y
      N-acetyl-DL-phenylalanine β-naphthyl ester overlay test, 135–137
      well test, 137
    protease A, 139
    protease B hide powder azure overlay assay
      culture, 133–134
      principles, 133
      reagents, 134
      utility, 134
    requirement of tests, 133
  protease B and proteolytic artifact causation, 127–128
  types and inhibitors, 127–128

734 SUBJECT INDEX

Vacuole, yeast
functons, 408–409
proteases, *see* Vacuolar proteases, yeast
protein sorting
  AP3 pathway, 428
  colony immunoblotting of secreted proteins, 421–422
  cytoplasm-to-vacuole targeting, 427
  prevacuolar endosomal compartment transport, 427–428
  protease-deficient yeast strains, 426
  pulse–chase immunoprecipitation
    membrane-bound proteins, 425–426
    overview, 422
    soluble hydrolases, 423–425
  secretion followed by endocytosis, 427
  transport mutants to dissect trafficking pathways, 426–427
  purification by flotation, 428–430
V-ATPase assays
  coupled spectrophotometric assay, 430–431
  plate reader assay for inorganic phosphate release, 431–432
visualization
  cell mounting, 411–412
  green fluorescent protein target fusions
    advantages and limitations, 414
    tagging of uncharacterized proteins, 415–417
    vacuolar protein fusions, 414–415
  immunofluorescence microscopy
    antibodies, 417–418
    cell preparation, 419–420
    incubations and washes, 420
    materials, 418–419
    mounting, 421
  vital dye staining
    7-amino-4-chloromethylcoumarin staining of lumen, 413
    carboxydichlorofluorescein diacetate labeling, 413
    cell growth, 411
    FM4-64 staining, 412
    MDY-64 staining, 412
    overview, 409, 411
    quinacrine staining, 413–414

W

Western blot
  heat shock proteins, 445
  nuclear envelope markers, 405–407
  plasma membrane proteins in yeast, 349–350
  yeast protein complexes, 167–168

Y

Yeast
  DNA replication, *see* DNA replication, yeast
  Electron microscopy, *see* Electron microscopy; Electron tomography; Immunoelectron microscopy, yeast
  endoplasmic reticulum vesicle budding, *see* Endoplasmic reticulum, yeast
  glutathione S-transferase fusion proteins, *see* Glutathione S-transferase fusion proteins, yeast
  Golgi apparatus, *see* Golgi apparatus, yeast
  green fluorescent protein studies, *see* Green fluorescent protein; Three-dimensional deconvolution microscopy
  growth microscopy, *see* Time-lapse microscopy, yeast growth
  kinase mutants, 279
  life cycle analysis, *see* Life cycle analysis, yeast
  mass spectrometry of proteins, *see* Mass spectrometry, yeast proteins
  mating, *see* Mating, yeast
  mitochondria, *see* Mitochondria, yeast
  mother/daughter cell separation, *see* Cell cycle synchronization, yeast; Life cycle analysis, yeast
  nuclear envelope, *see* Nuclear envelope, yeast
  nuclear protein transport, *see* Nuclear protein transport, yeast
  nucleus, *see* Nucleus, yeast
  peroxisomes, *see* Peroxisome, yeast
  plasma membrane, *see* Plasma membrane, yeast
  prions, *see* Prion factors, yeast
  proteases, *see* Vacuolar proteases, yeast
  protein complexes, *see* Protein complexes, yeast

SUBJECT INDEX

protein degradation, *see* Protein degradation, yeast; Ubiquitination, yeast proteins
RNA, *see* RNA decay, yeast; RNA export, yeast; RNA splicing, yeast; Translation, yeast
secretory porotein abundance, 325
subcellular fractionation, *see* Subcellular fractionation, yeast
synchronization, *see* Cell cycle synchronization, yeast
translation, *see* Translation, yeast
vacuoles, *see* Vacuole, yeast

Z

Zymolase, protease artifacts in yeast, 128

ISBN 0-12-182254-0